Horticulture Today

by
Jodi Songer Riedel
Elizabeth Driscoll

Publisher
The Goodheart-Willcox Company, Inc.
Tinley Park, IL
www.g-w.com

Copyright © 2017

by

The Goodheart-Willcox Company, Inc.

All rights reserved. No part of this work may be reproduced, stored, or transmitted in any form or by any electronic or mechanical means, including information storage and retrieval systems, without the prior written permission of The Goodheart-Willcox Company, Inc.

Manufactured in the United States of America.

ISBN 978-1-63126-245-6

4 5 6 7 8 9 – 17 – 23 22 21

The Goodheart-Willcox Company, Inc. Brand Disclaimer: Brand names, company names, and illustrations for products and services included in this text are provided for educational purposes only and do not represent or imply endorsement or recommendation by the author or the publisher.

The Goodheart-Willcox Company, Inc. Safety Notice: The reader is expressly advised to carefully read, understand, and apply all safety precautions and warnings described in this book or that might also be indicated in undertaking the activities and exercises described herein to minimize risk of personal injury or injury to others. Common sense and good judgment should also be exercised and applied to help avoid all potential hazards. The reader should always refer to the appropriate manufacturer's technical information, directions, and recommendations; then proceed with care to follow specific equipment operating instructions. The reader should understand these notices and cautions are not exhaustive.

The publisher makes no warranty or representation whatsoever, either expressed or implied, including but not limited to equipment, procedures, and applications described or referred to herein, their quality, performance, merchantability, or fitness for a particular purpose. The publisher assumes no responsibility for any changes, errors, or omissions in this book. The publisher specifically disclaims any liability whatsoever, including any direct, indirect, incidental, consequential, special, or exemplary damages resulting, in whole or in part, from the reader's use or reliance upon the information, instructions, procedures, warnings, cautions, applications, or other matter contained in this book. The publisher assumes no responsibility for the activities of the reader.

The Goodheart-Willcox Company, Inc. Internet Disclaimer: The Internet resources and listings in this Goodheart-Willcox Publisher product are provided solely as a convenience to you. These resources and listings were reviewed at the time of publication to provide you with accurate, safe, and appropriate information. Goodheart-Willcox Publisher has no control over the referenced websites and, due to the dynamic nature of the Internet, is not responsible or liable for the content, products, or performance of links to other websites or resources. Goodheart-Willcox Publisher makes no representation, either expressed or implied, regarding the content of these websites, and such references do not constitute an endorsement or recommendation of the information or content presented. It is your responsibility to take all protective measures to guard against inappropriate content, viruses, or other destructive elements.

Cover image: jennyt/Shutterstock.com
Back cover image: Skowronek/Shutterstock.com

Preface

Horticulture Today was designed to meet the need for a comprehensive, thoroughly modern, and visually exciting introduction to the study of horticulture. The authors of *Horticulture Today* set out to provide learners with a breadth of horticultural knowledge while also creating awareness of green industry careers and workplace skills needed to be successful in the twenty-first century. To help readers achieve these goals, the authors have incorporated supplemental activities, such as interactive sidebars, historical and STEM connections, and personalized career connections throughout the text.

The organization and content coverage of *Horticulture Today* was developed with valuable insight from a panel of expert reviewers and with the National Agriculture, Food, and Natural Resources Standards at the forefront. The textbook content also correlates to various state standards and to exam standards for Career Skill Certification™ by Precision Exams. *Horticulture Today* is presented in 33 chapters with six logical sections to help readers develop a fundamental knowledge of plant science and build on this knowledge with ensuing topics.

The initial chapters concentrate on building leadership, communication, and business skills that can be used in both school and work environments. Chapter 1 and chapter 2 are focused on FFA and SAE support fundamentals in leadership skills and experiential learning. Additionally, chapter 3 seeks to help students develop communication and critical thinking skills needed to compete in the twenty-first century. Chapter 5 focuses on horticulture business management and continues to build student professionalism, preparing them for future employment.

The second group of chapters focuses on horticultural biology. Chapter 7 is dedicated to taxonomy, and chapters 8 through 10 cover basic biology as well as plant growth and development. Readers will develop an understanding of basic plant biology as well as how environmental controls can be used to manipulate plant growth. Plant nutrition and soils and media are discussed in chapters 11 and 12. This highly-illustrated section includes extensive coverage of deficiencies, as well as sustainable practices for soil enrichment and preservation. The third group of chapters is dedicated to plant propagation—ranging from traditional to modern micropropagation techniques.

The fourth section encompasses a myriad range of horticultural practices, including greenhouse operations to field cultivation of fruits, vegetables, and ornamental plants. Readers will gain an understanding of practices ranging from traditional methods to modern-day hydroponics. The fifth section covers design practices in the landscaping, floriculture, and turfgrass industries. Hands-on, step-by-step instructions, installation procedures, and management practices are also included. The final section provides detailed coverage of integrated pest management (IPM). Individual chapters cover IPM, insects, disease, weeds, and pesticide safety and management.

We hope you enjoy our presentation of horticulture for our modern society and continue to learn and grow as you pursue a career in the green industry or hone your skills as a home gardener.

About the Authors

Jodi Songer Riedel

Ms. Riedel has more than fourteen years of teaching experience in agricultural education. She has instructed and created curriculum for courses in horticulture, sustainable agriculture, forestry, and environmental and natural resources. Ms. Riedel has also instructed many student teachers and has keenly observed and learned what young teachers, especially those with limited horticultural knowledge, need to successfully instruct agricultural education courses.

In her Wakefield High School classroom, Ms. Riedel employs learning by doing. With Ms. Riedel's instruction, students:

- Have designed and installed prize-winning gardens at the state fair.
- Maintain over eight ornamental gardens and landscapes on school grounds.
- Installed gardens at the YMCA, local day cares, and surrounding schools.
- Present educational workshops for the community.
- Grow and donate vegetables totaling over one ton each year.
- Raise poinsettias and organize fall and spring plant sales.

Ms. Riedel also organizes and supervises urban service learning projects and Supervised Agricultural Experiences (SAEs) in Raleigh, NC. She spends countless hours in and out of the classroom managing the school's 1800-square foot greenhouse and oversees fruit, vegetable, and ornamental gardens within the school grounds. Ms. Riedel funds and sustains her inspirational program and its efforts each year through grants and annual plant sales, cultivated by students. Ms. Riedel's dedication and enthusiasm earned her the 2010 Governor's Environmental Educator of the Year for North Carolina. She has also been awarded a distinguished Kenan Fellowship and assigned to the Governor's Teaching Network.

Elizabeth Driscoll

Ms. Driscoll holds the position of 4-H Subject-Matter Specialist across the Departments of Crop Science, Entomology, Horticultural Science and Soil Science at North Carolina State University with North Carolina Cooperative Extension. She has been working to connect youth and educators to issues in agriculture and natural resources in relevant and meaningful ways for over 15 years. Ms. Driscoll's 4-H "Grow For It" program focuses on developing resources and experiential projects that foster curious and wondering youth, inspire critical thinking and problem solving, build a positive science self-concept, connecting kids to good food and nurturing environmental stewards of the land through gardening and agriculture.

With a BS and an MS in horticultural science from Michigan State University, she actively translates agricultural research from the Land Grant Universities into hands-on, inquiry-based curricular resources to support county-based 4-H youth programs. Serving over 350 Cooperative Extension county field faculty in agriculture, horticulture, 4-H, and Family and Consumer Sciences, Ms. Driscoll uses a strong train-the-trainer approach to broadly and deeply engage audiences in their own innovative delivery of agricultural programs to youth. She leads multiple precollege programs for youth, including the Horticultural Science Summer Institute, Resource Conservation Workshop, Cultivate: Crop Science Teen Weekend, as well as other numerous youth leadership programs. Additionally, Ms. Driscoll cohosts the FoodCorps NC program with The Center for Environmental Farming Systems and many local partners to connect kids to healthy food in schools through school gardens, garden-enhanced food and nutrition education, and access to local food through farm-to-cafeteria projects.

Acknowledgments

The author and publisher would like to thank the following individuals and associations for their valuable input in the development of *Horticulture Today*. In addition, the authors would like to thank their family and friends for their support in this endeavor, as well as their colleagues and key industry contributors for their valuable input in the development of *Horticulture Today*.

Alan Erwin, Panther Creek Nursery
Alberto Pantoja, USDA Agricultural Research Service, Bugwood.org
Alex Ramirez, Design Workshop
Alicia Rittenhouse, AmericanHort
A.M. Leonard Horticultural Tool and Supply Co.
Andy Smith, Eco Turf
Anna Passarelli, Simply Elegant
Baker Creek Heirloom Seed Company/ www.rareseeds.com
Bert Cregg, Michigan State University
Cary Rivard, Kansas State University
Clemson University—USDA Cooperative Extension Slide Series, Bugwood.org
Debbie Roos, North Carolina Cooperative Extension, Chatham County
Denise Etheridge, Homewood Nursery
Dr. Andrea Weeks, George Mason University
Dr. Barbara Fair, North Carolina State University
Dr. Carol Somody, Senior Stewardship Manager, Syngenta
Dr. Matthew Vann, North Carolina State University
Dr. Melodee Fraser, Pure Seed Testing of Oregon
Dr. Rebecca Langer-Curry, Bayer CropScience
Dr. Suzanne O'Connell, University of Georgia
Forest Pathology, USDA Forest Service, Bugwood.org/ Rocky Mountain Research Station
Gary K. England, University of Florida, IFAS Extension
Gavin Sulewski, International Plant Nutrition Institute
Graham Goodenough, Seederman Seeders
H.J. Larsen, Bugwood.org
Home and Garden Information Center, University of Maryland Extension
J. Prabhakaran, International Plant Nutrition Institute
Jennifer Broadwell
Jennifer Nelkin Frymark, Gotham Greens
Jessica Honaker of The Bug Chicks
Joseph O'Brien, USDA Forest Service, Bugwood.org
Joseph Tychonievich
Joshua Bledsoe, National FFA Chief Operating Officer
Kristie Reddick of The Bug Chicks
Kurt Bland, Bland Landscaping
L. Shyamal
Lauren Jolly Photography
Leonora Enking
Leslie Halleck, Halleck Horticultural
Marek Argent
Mark Weathington, J.C. Raulston Arboretum
Marko Penning, Penning Freesia B.V.
Mary Ann Hansen, Virginia Polytechnic Institute and State University, Bugwood.org
Mary Burrows, Montana State University, Bugwood.org
Melanie M. McCaleb, NTU, Inc.
Michael L. Parker, North Carolina State University
National FFA Organization
Neil Devaney, Stuppy Greenhouses
Olivier Pichard
Pam Winegar
Pennsylvania Department of Conservation and Natural Resources - Forestry, Bugwood.org
Randy Beaudry, Michigan State University
Randy Cyr, Greentree, Bugwood.org
Rebecca Nelson, Nelson and Pade Aquaponics
Richard W. Taylor, Taylor Nursery
Riverbend Nursery
Rizaniño Reyes
Robin Hawley, Sokol Blosser
Roger Hanagriff, The AET Record Book

Rosemarie Robson, Robson's Farm LLC
S. Moroni, International Plant Nutrition Institute
Scot Nelson
Scott Bauer, USDA Agriculture Research Service, Bugwood.org
Seny Norishingh
Shep Eubanks, University of Florida, IFAS Extension
Solexx and the Greenhouse Catalog
Spectrum Products Inc.
Stavroula Ventouri-ex www.srgc.net-
T.Yamada, International Plant Nutrition Institute
Tabitha West, Cedar Valley Nursery
Tim Alderton, J.C. Raulston Arboretum
Todd Lawrence
Tony Avent, Plant Delights Nursery, Inc. Apex, NC
Ty Strode, Agri-Starts, Inc.
UC Fruit Report (http://ucanr.edu/sites/fruitreport)
University of Georgia Plant Pathology, University of Georgia, Bugwood.org
University of South Florida Herbarium
University of Arkansas Forest Entomology Lab, University of Arkansas, Bugwood.org
US Department of Agriculture
USDA Agricultural Research Service (ARS)
USDA Forest Service
USDA Natural Resources Conservation Service
V. Arunachalam, International Plant Nutrition Institute
Whitney Cranshaw, Bugwood.org
William Jacobi, Colorado State University, Bugwood.org
William M. Brown, Jr., Bugwood.org
www.ansci.cornell.edu/plants/medicinal/portula.html
YardGreen, Inc.
Yuan-Min Shen, Taichung District Agricultural Research and Extension Station, Bugwood.org
Yuko Frazier, Senior Project Manager at Ambius

Reviewers

The authors and publisher wish to thank the following teaching and industry professionals for their valuable input into the development of *Horticulture Today*.

Julie Anderson
Liberty High School
Frisco, Texas

Crystal Aukema
Oxford Academy and Central School
Oxford, New York

Tim Closs
Naaman Forest High School
Garland, Texas

Becky DeShazo
Bridgeport High School
Bridgeport, Texas

Dave Gossman
Atwater High School
Atwater, California

Christopher Hart
Chatham Central High School
Bear Creek, North Carolina

Rachel Kostman
Oakland High School
Oakland, Oregon

Trisha Lastly
Georgia Agriculture Education
Tifton, Georgia

Kelly Lowery
Penn State Pesticide Education
State College, Pennsylvania

Nancy Meyers
Luella High School
Locust Grove, Georgia

Erol Miller
Northwood High School
Silver Spring, Maryland

Shelley Mitchell, Ph.D.
Oklahoma State University
Stillwater, Oklahoma

JoAnn Pfeiffer
Federal Hocking Secondary Schools
Stewart, Ohio

Lisa Pieper
Caldwell High School
Caldwell, Texas

Melissa Riley
Georgia Agriculture Education
Fort Valley, Georgia

Camber Starling
Heritage High School
Wake Forest, North Carolina

G-W Integrated Learning Solution

Together, We Build Careers

At Goodheart-Willcox, we take our mission seriously. Since 1921, we have been serving the career and technical education (CTE) community. Our employee-owners are driven to deliver exceptional learning solutions to CTE students to help prepare them for careers. Using educators' wisdom and our expertise, we have designed content and tools that will help students achieve success. We begin with theory and applied content based upon a strong foundation of accepted standards and curriculum. To that base, we add student-focused learning features and tools designed to help students make connections between knowledge and skills. We support our instructor with time-saving tools that help them plan, present, assess, and engage students with traditional and digital activities and assets. Because society is in a transition between print to digital, we want to provide the tools and content in any format you need. Our integrated learning solution products come in a variety of digital and online formats and we provide economical bundles so you can select the right product—print, digital, and online mix for your classroom.

Student-focused Curated Content

Goodheart-Willcox believes that student-focused content should be built from standards and/or accepted curriculum coverage. Our authors and Subject Matter Experts (SMEs) are experts in their field and give considerable thought to the best ways to present content: for some it may be a building block approach with attention devoted to a logical teaching progression; for others it may be logical content presentation in small, focused lessons. We call on industry experts and teachers from across the country to review and comment on our content, presentation, and pedagogy. Finally, in our refinement of curated content, our editors are immersed in content checking, securing, and sometimes creating figures that convey key information, and revising language and pedagogy.

Precision Exams Certification

Goodheart-Willcox is pleased to partner with Precision Exams by correlating *Horticulture Today* to their Agriculture Plant and Soil Science Standards. Precision Exams Standards and Career Skill Exams were created in concert with industry and subject matter experts to match real-world job skills and marketplace demands. Students that pass the exam and performance portion of the exam can earn a Career Skills Certification™. Precision Exams provides:

- Access to over 150 Career Skills Exams™ with pre- and post-exams for all 16 Career Clusters.
- Instant reporting suite access to measure student academic growth.
- Easy-to-use, 100% online exam delivery system.

To see how *Horticulture Today* correlates to the Precision Exams Standards, please visit www.g-w.com/horticulture-today-2017 and click on the Correlations tab. For more information on Precision Exams, including a complete listing of their 150+ Career Skills Exams and Certificates, please visit https://www.precisionexams.com.

I earned a CAREER SKILLS™ Certificate in PLANT AND SOIL SCIENCE. You can earn one too!

Ask your instructor how you can earn a CAREER SKILLS™ Certificate for your résumé.

800.470.1215   PRECISION EXAMS   precisionexams.com

©iStock.com/michaeljung

Features of the Textbook

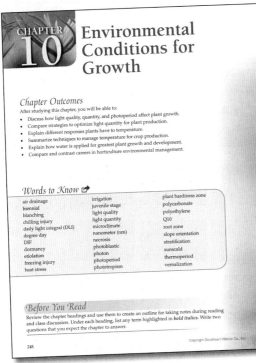

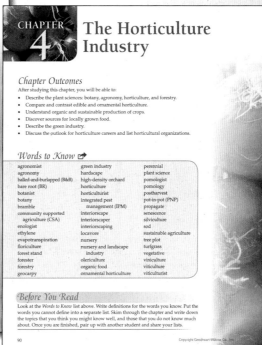

Chapter Outcomes clearly identify the knowledge and skills to be obtained when the chapter is completed.

Words to Know list the key terms to be learned in each chapter.

Before You Read literacy integration activities at the beginning of each chapter encourage development of confidence and skill in literacy and learning.

Thinking Green features highlight key items related to sustainability, energy efficiency, and environmental issues.

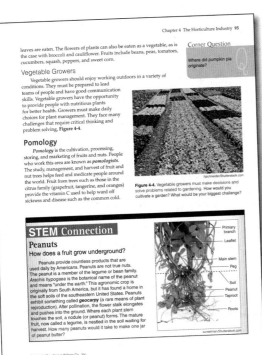

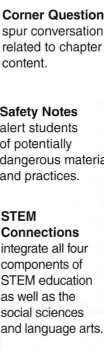

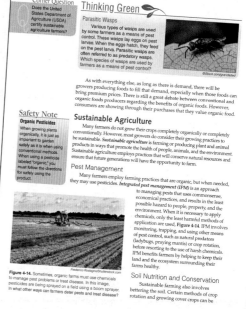

Corner Questions spur conversation related to chapter content.

Safety Notes alert students of potentially dangerous materials and practices.

STEM Connections integrate all four components of STEM education as well as the social sciences and language arts.

Copyright Goodheart-Willcox Co., Inc.

Career Connection features introduce students to people in a variety of careers in the horticulture industry.

AgEd Connection features introduce students to the exciting world of leadership and personal development opportunities including career development events in a variety of areas.

Did You Know? features point out interesting and helpful facts about the agricultural industry.

Illustrated glossaries covering plants, equipment and supplies, pests and disorders, diseases, and weeds.

Illustrations have been designed to clearly and simply communicate the specific topic.

Copyright Goodheart-Willcox Co., Inc.

Chapter Summaries provide an additional review tool for students and reinforces key learning outcomes.

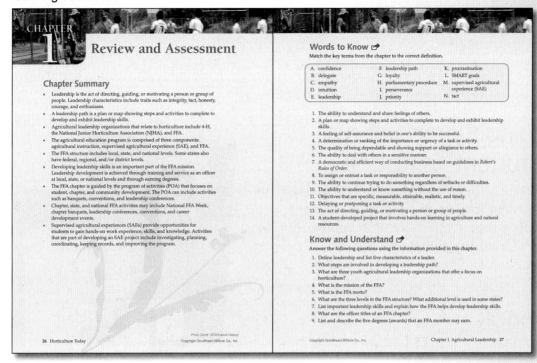

Words to Know matching activities reinforce vocabulary development and retention. All key terms are included in the text glossary and are connected to numerous online review activities.

Know and Understand review questions allow students to demonstrate knowledge, identification, and comprehension of chapter material.

Thinking Critically questions develop higher-order thinking, problem solving, personal, and workplace skills.

Communicating about Horticulture questions and activities help integrate reading, writing, listening, and speaking skills while extending their knowledge on the chapter topics.

STEM and Academic Activities are provided in the areas of science, technology, engineering, math, social science, and language arts.

SAE Opportunities help students make real-life connections to a variety of new and interesting SAE opportunities.

Copyright Goodheart-Willcox Co., Inc.

Student Resources

Textbook

The *Horticulture Today* textbook provides an exciting, full-color, and highly-illustrated learning resource. The textbook is available in a print or online version.

G-W Learning Companion Website/Student Textbook

The G-W Learning companion website is a study reference that contains photo identification and matching activities, animations and videos, review questions, vocabulary exercises, and more! Accessible from any digital device, the G-W Learning companion website complements the textbook and is available to the student at no charge.

Lab Workbook

The student workbook provides hands-on practice with questions and lab activities. Each chapter corresponds to the text chapters and reinforces key concepts and applied knowledge.

Online Learning Suite

Available as a classroom subscription, the Online Learning Suite provides the foundation of instruction and learning for digital and blended classrooms. An easy-to-manage shared classroom subscription makes it a hassle-free solution for both students and instructors. An online student text and workbook, along with rich supplemental content, brings digital learning to the classroom. All instructional materials are found on a convenient online bookshelf and are accessible at home, at school, or on the go.

Online Learning Suite/ Student Textbook Bundle

Looking for a blended solution? Goodheart-Willcox offers the Online Learning Suite bundled with the printed text in one easy-to-access package. Students have the flexibility to use the print version, the Online Learning Suite, or a combination of both components to meet their individual learning style. The convenient packaging makes managing and accessing content easy and efficient.

Instructor's Resources

Instructor's Presentations for PowerPoint®

Help teach and visually reinforce key concepts with prepared lectures. These presentations are designed to allow for customization to meet daily teaching needs. They include objectives, outlines, and images from the textbook.

ExamView® Assessment Suite

Quickly and easily prepare, print, and administer tests with the ExamView® Assessment Suite. With hundreds of questions in the test bank corresponding to each chapter, you can choose which questions to include in each test, create multiple versions of a single test, and automatically generate answer keys. Existing questions may be modified and new questions may be added. You can prepare pretests, formative, and summative tests easily with the ExamView® Assessment Suite.

Instructor's Resource CD

One resource provides instructors with time-saving preparation tools, such as answer key, rubrics, lesson plans, correlation charts to standards, and other teaching aids.

Online Instructor Resources

Online Instructor Resources provide all the support needed to make preparation and classroom instruction easier than ever. Available in one accessible location, support materials include Answer Keys, Lesson Plans, Instructor Presentations for PowerPoint®, ExamView® Assessment Suite, and more! Online Instructor Resources are available as a subscription and can be accessed at school or at home.

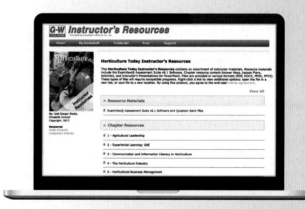

Contents in Brief

1 Agricultural Leadership ... 2
2 Experiential Learning: SAE ... 30
3 Communication and Information Literacy in Horticulture ... 60
4 The Horticulture Industry ... 90
5 Horticultural Business Management ... 120
6 Worker and Tool Safety ... 154
7 Plant Taxonomy ... 178
8 Plant Biology ... 196
9 Plant Growth and Development ... 224
10 Environmental Conditions for Growth ... 248
11 Soils and Media ... 278
12 Plant Nutrition ... 310
13 Seed Propagation ... 338
14 Stem and Leaf Propagation ... 364
15 Layering and Division ... 384
16 Grafting and Budding ... 404
17 Tissue Culture: Micropropagation ... 428
18 Greenhouse Operation and Maintenance ... 448
19 Greenhouse Production ... 478
20 Twenty-First Century Horticulture ... 506
21 Nursery Production ... 536
22 Vegetable Production ... 568
23 Fruit and Nut Production ... 598
24 Landscape Design ... 630
25 Floral Design ... 664
26 Interior Plantscaping ... 690
27 Landscape Installation and Maintenance ... 716
28 Turfgrass Management ... 744
29 Integrated Pest Management ... 778
30 Insects ... 802
31 Disease Management ... 830
32 Weeds ... 858
33 Pesticide Management and Safety ... 876

Contents

Chapter 1
Agricultural Leadership......2
- Leadership Characteristics 4
- Develop a Leadership Path 5
  - Create a Vision and Set Goals 6
  - Become Self-Confident and Healthy.................... 6
  - Reduce Procrastination and Foster Initiative 7
  - Get Organized................. 7
  - Further Your Education 8
  - Leadership Development 8
- Agricultural Leadership Organizations for Youth 9
  - 4-H 9
  - National Junior Horticulture Association (NJHA) 10
- National FFA Organization 11
  - History of the National FFA Organization................ 11
  - FFA Emblem and Motto 12
  - FFA Structure 12
  - Leadership Development in FFA 14
  - Career Connection: Joshua Bledsoe, National FFA Chief Operating Officer 17
  - FFA Program of Activities 18
  - Chapter, State and National FFA Activities................... 20
- Supervised Agricultural Experience................... 23
- Agricultural Leadership Careers...................... 24
  - Agricultural Education Teachers 24
  - Cooperative Extension Service Agents 25
  - Career Connection: Jennifer Broadwell, Agricultural Education Teacher 25
- Review and Assessment......... 26

Chapter 2
Experiential Learning: SAE....................30
- SAE and Agricultural Education.................... 32
  - SAE History 32
  - SAE Purpose 33
- Types of SAEs.................. 34
  - Entrepreneurship SAE 34
  - Placement SAE................. 36
  - Research and Experimentation SAE 38
  - Career Connection: Beekeeping............38
  - Exploratory SAE 40
  - Improvement SAE............... 41
  - Supplemental SAE 42
- The SAE Program Process 42
  - Investigate................... 44
  - Plan....................... 46
  - Coordinate 48
  - Keep Records 49
- SAE Awards and Recognitions.... 50
  - Agricultural Proficiency Awards..... 50
  - Star Awards................... 51

Copyright Goodheart-Willcox Co., Inc.

Careers..................52
    Agricultural Business Manager.....53
    Agricultural Inspector............53
    Career Connection: Alicia Rittenhouse, AmericanHort.....................54
Review and Assessment.........55

Chapter 3
Communication and Information Literacy in Horticulture......60

Written Communication........62
    Purpose......................63
    Audience.....................64
    Presentation..................65
    Topic........................66
    Thesis.......................67
    Project Plan..................67
    Reviewing, Revising, and Proofreading..................68
Critical Thinking and Research...69
    Search......................70
    Preview.....................71
    Find Evidence................71
    Annotate....................72
    Summarize..................72
    Evaluate....................73
Plagiarism and Documentation...74
    Plagiarism...................74
    Documentation...............75
Presentation Methods..........76
    Presentation Types.............78
    Communication with the Audience...79
Information Literacy............80
    Independent Learning...........82
    Social Responsibility...........82
Horticultural Communication Careers....................82
    Horticulture Extension Agent......82
    Garden Writer and Speaker.......83
    Career Connection: Rizaniño Reyes, Garden Blogger.................84
Review and Assessment.........84

Chapter 4
The Horticulture Industry...90

What Is Plant Science?..........92
    Botany......................92
    Agronomy...................93
    Forestry....................93
Edible and Ornamental Horticulture..................93
    Olericulture..................94
    Pomology...................95
    Viticulture...................96
    Floriculture..................97
    Nursery and Landscape Industry...98
    Career Connection: Randy Beaudry, Michigan State University Postharvest Professor.............100
    Interiorscaping................101
    Turfgrass Industry.............102
Organic and Sustainable Production..................105
    Organic Edibles...............105
    Sustainable Agriculture.........106
Eating Local..................107
    Harvesting...................107
    Locavores...................107
    Farmers Markets..............107
    Community Supported Agriculture (CSA)....................108

The Green Industry............109
   Biological and Environmental
      Impacts 110
   Aesthetic Impacts 110
   Economic Impacts 111
   Future of the Green Industry.......111
Careers......................112
   Continuing Education 112
   Horticulture Organizations 112
Review and Assessment........114

Chapter 5
Horticultural Business Management 120
   Small Businesses 122
   Strategic Business Plans........123
      Strategic Environment........... 124
      Internal Resources 126
      Vision and Mission Statement..... 127
   Principal Strategy 128
      Performance Standards 129
   Marketing and Advertising 129
      The Marketing and Advertising
         Process 130
      The Four Ps of Marketing 130
   Professionalism 134
      Professional Traits and Behaviors ... 135
      Unprofessional Traits and
         Behaviors................ 136
   Career Documents.............136
   Job Interviews.................139
      Preparing for a Job Interview 140
      Taking Part in the Job Interview ... 141
      Interview Follow Up............. 143
   School-to-Career Plan..........144
   Horticulture Business Careers ...146
      Horticulture Business Consultant... 147
      Horticultural Sales Representative ... 148
      Career Connection: Leslie Halleck,
         Horticultural Marketing..................147
   Review and Assessment........149

Chapter 6
Worker and Tool Safety....154
   Early American Labor..........156
      Industrial Revolution 156
      Fair Labor Standards Act (FLSA) .. 157
   Safety and Health Agencies......157
      Centers for Disease Control and
         Prevention (CDC)............ 158
      National Institute of Occupational Safety
         and Health (NIOSH).......... 158
      Occupational Safety and Health
         Administration (OSHA)........ 158
      United States Department of Labor
         (DOL)..................... 158
   Safety Hazards 159
      Physical Hazards 160
      Chemical Hazards 160
      Biological Hazards 161
      General Hazards............... 163
      Ergonomic Hazards 163
      Work Organization Hazards 164
   Preventing Accidents 164
      ABCs of Accident Prevention 164
   Workplace Safety Documents...166
      Safety Data Sheets............. 166
      Pesticide Labels 167
   Practicing Safety 167
      SAFE Working Conditions........ 167

Maintaining Tools and
    Equipment.................. 168
    Checking and Maintaining
        Equipment 168
    Maintaining Tools 170
Labor Laws 170
    Minimum Wage and Hours
        Worked.................. 170
    Job Duties for Workers under 18... 171
    Young Worker Responsibility 172
    Taking Action.................. 172
Horticultural Safety Careers 172
    Horticultural Risk Consultant 173
    Horticultural Safety Manager 173
    Career Connection: Kurt Bland,
        Bland Landscaping 173
Review and Assessment........ 174

Chapter 7
Plant Taxonomy.......... 178
History of Plant Taxonomy 180
A System of Botanical
    Classification................. 181
    Domain 182
    Kingdom 183
    Phylum 183
    Class........................ 184
    Order........................ 185
    Family 186
    Genus........................ 186
    Species...................... 187
    Variety....................... 187
    Cultivar 188
Plant Keys 188
Herbaria...................... 188
Careers in Plant Taxonomy 190
    Herbarium Director/Curator....... 190
    Ecologist..................... 191
    Career Connection: Dr. Andrea Weeks,
        George Mason University............... 191
Review and Assessment........ 192

Chapter 8
Plant Biology............. 196
Study of Botany 198
    Botany Disciplines 199
Plant Cells.................... 199
    Cell Membrane 199
    Cell Nucleus 200
    Chloroplasts and Other Plastids... 200
    Mitochondria.................. 201
    Vacuoles..................... 202
    Oil Bodies.................... 202
    Endoplasmic Reticulum 202
    Ribosomes 202
    Golgi Bodies.................. 202
    Lysosomes 203
    Cytoskeleton 203
Plant Tissues 203
    Meristem..................... 203
    Epidermis 204
    Bark (Periderm) 205
    Phloem 205
    Xylem 205
    Parenchyma and Collenchyma 206
    Sclerenchyma 206
Plant Parts and Their Functions... 206
    Roots 206
    Stems 207
    Leaves 210

Flowers 212	Cellular Division 236
Fruits. 214	Sexual Reproduction. 236
Seeds 217	**Plant Breeding Principles. 239**
Careers in Plant Biology. 218	The Plant Breeding Process 240
Botanist. 218	Plant Mutations. 240
Science Teacher 218	**Careers. 241**
Review and Assessment. 219	Plant Physiologist 241
	Plant Breeder 242
	Career Connection: Joseph Tychonievich, Plant Breeder 242
	Review and Assessment. 243

Chapter 9
Plant Growth and Development 224

Photosynthesis 226
  Light-Dependent Reaction 227
  Light-Independent Reactions 228
Respiration 230
  ATP Production 230
Transpiration 231
  Carbon Dioxide Entry 232
  Water Uptake and Nutrient Access 232
  Factors Affecting Transpiration 232
Movement of Solutes 234
  Active Uptake of Inorganic Nutrients 234
  Translocation of Sugars through Phloem. 235
Reproduction 235

Chapter 10
Environmental Conditions for Growth. 248

Light 250
  Light Quality 250
  Light Quantity 252
  Photoperiod 253
Optimizing Light Quantity 255
  Plant Spacing and Orientation 255
  Greenhouse Design 256
  Greenhouse Covers 256
  Supplemental Lighting 257
  Plant Selection 257
Temperature 258
  Q10 Temperature Coefficient. 258
  Plant Responses to Temperature. . . 260
  Thermoperiod. 263
  Degree Days. 263
  Soil Temperatures 263
Managing Temperatures. 264
  Site Selection 264
  Season Extension. 265
  Overhead Irrigation. 267
  Localized Heating 268
  Wind Machines 268
Water 268
  Root Zone. 269
  Consumption Rates 269
  Irrigation 269

Careers in Environmental
    Horticulture................ 271
    Greenhouse Manager........... 271
    Farm Coordinator 271
    Career Connection: Debbie Roos, Sustainable
        Agriculture Extension Agent 272
Review and Assessment........ 273

Chapter 11
Soils and Media.........278
What Is Soil?................. 280
Soil Formation 280
    Soil-Forming Factors........... 280
    Horizons 281
Physical Properties of Soil 282
    Soil Texture 282
    Soil Structure 285
    Soil Water 287
    Soil Color 288
Biological Properties of Soil..... 288
    Microorganisms................ 288
    Soil-Dwelling Insects........... 290
    Burrowing Organisms 290
Chemical Properties of Soil 290
    Soil pH 291
    Soil Testing 292
Soil and Soilless Media......... 293
    Garden or Native Soil 294
    Soilless Media................ 295
Mulch....................... 296
    Inorganic Mulches............. 297
    Organic Mulches.............. 298
Containers................... 300
    Size........................ 300
    Durability................... 300
    Drainage Capability 301
Careers in Soil Science 302
    Soil Scientists 302
    Media Manufacturing 303
    Career Connection: Melanie McCaleb,
        Erosion Control Specialist...............304
Review and Assessment........ 305

Chapter 12
Plant Nutrition 310
Essential Elements............. 312
    Hydrogen (H) 312
    Carbon (C) 313
    Oxygen (O) 313
    Primary Macronutrients......... 313
    Secondary Macronutrients 316
    Micronutrients 317
Mineral Nutrient Uptake 321
    Nutrient Cycle 321
    Nitrogen Cycle 322
    Nutrient Mobility 323
    Soil pH 324
Nutrient Sources 324
    Soil and Tissue Analysis 324
    Organic Materials 325
    Inorganic Fertilizers............ 327
Fertilizer Calculations.......... 328
    Phosphorus (P)................ 329
    Potassium (K) 329
Methods of Fertilizer
    Application 330
    Broadcasting................. 330
    Banding..................... 330
    Side-Dressing................ 330
    Fertigation................... 331
    Slow-Release Fertilizer......... 331

Soil Injection 331
Foliar Application 332
Careers in Plant Nutrition 332
Fertilizer Sales Representative 332
Composting Operator 332
Review and Assessment 333

Chapter 13
Seed Propagation 338
Seed Morphology and Development 340
Stages of Seed Development 340
Apomixis 341
Seed Germination 341
Germination Rates 342
Environmental Conditions for Germination 342
Seed Dormancy 344
Seed Coat Dormancy 344
Chemical Dormancy 345
Morphological Dormancy 345
Physiological Dormancy 345
Seed Propagation Techniques 346
Field Seeding 346
Field Nurseries 348
Greenhouse Production 349
Seed Selection 353
Landraces 353
Wild Populations 354
Hybrids 355
Transgenic Cultivars 355
Seed Production 355
Plant Variety Protection Act 356
Seed Saving 357
Seed Storage 358
Careers in Seed Propagation 358
Seed Sales Representative 358
Nursery Propagator 358
Review and Assessment 359

Chapter 14
Stem and Leaf Propagation . . . 364
Biological Principles of Leaf and Stem Propagation 366
Preformed Roots 367
Wound-Induced Roots 367
Adventitious Shoot and Bud Formation 367
Plant Material Used for Stem and Leaf Cuttings 368
Hardwood Cuttings 369
Semihardwood Cuttings 372
Softwood Cuttings 373
Herbaceous Cuttings 374
Leaf Cuttings 374
Leaf-Bud Cuttings 375
Root Cuttings 375
Hardening Off 376
Rooting Medium 376
Plant Growth Regulators 377
Auxin 377
Cytokinins 378
Careers in Stem and Leaf Propagation 378
Horticulture Illustrator 378
Nursery Inspector 379
Career Connection: Mark Weathington, Arboretum Director 379
Review and Assessment 380

Copyright Goodheart-Willcox Co., Inc.

Chapter 15
Layering and Division 384
- Layering in Propagation 386
- Layering Techniques 388
  - Simple Layering 388
  - Compound Layering 388
  - Air Layering 389
  - Mound Layering 390
  - Trench Layering 391
- Natural Layering 391
  - Tip Layers 391
  - Runners 391
  - Stolons 392
  - Offsets 392
  - Suckers 392
- Crown Division 393
- Division and Separation of Geophytes 393
  - Bulbs 394
  - Corms 396
  - Tubers 396
  - Tuberous Roots 397
  - Rhizome 397
  - Pseudobulb 398
- Careers in Layering and Division 398
  - Fruit Nursery Propagator 398
  - Orchard Manager 398
- Review and Assessment 399

Chapter 16
Grafting and Budding 404
- Benefits of Grafting 406
- Successful Grafting 408
  - Timing Grafting 409
  - Selecting Scion Wood 409
- Types of Grafts 410
  - Whip-and-Tongue Grafting 410
  - Splice Grafting 411
  - Cleft Grafting 412
  - Wedge Grafting 413
  - Saddle Grafting 414
  - Hole Insertion Grafting (HIG) ... 415
  - Bark Grafting 416
  - Bridge Grafting 417
  - Inarch Grafting 418
  - Approach Grafting 418
  - Side-Veneer Grafting 419
- Budding 420
  - Patch Budding 420
  - Chip Budding 421
  - T-Budding 421
  - Successful Budding Tips 422
  - Aftercare 423
- Careers in Grafting and Budding 423
  - Agricultural Communications ... 423
  - Assistant Vineyard Manager 423
- Review and Assessment 424

Chapter 17
Tissue Culture: Micropropagation 428
- History 430
- Advantages and Disadvantages 431
  - Labor and Equipment Costs 432
  - Supply and Demand 432
  - Genetic Diversity 433
- Environmental Requirements ... 433

Growth Media 434
Stages of Micropropagation 434
    Stage 0: Selection and Cultivation of Stock Plants............ 435
    Stage 1: Initiation or Establishment.............. 435
    Stage 2: Multiplication.......... 436
    Stage 3: Rooting 438
    Stage 4: Acclimatization 438
The Future of Tissue Culture and Micropropagation 439
Careers in Micropropagation ... 440
    Micropropagation Lab Technician 441
    Cryopreservation Scientist 441
    Career Connection: Ty Strode, Agri-Starts, Inc. 442
Review and Assessment........ 443

Chapter 18
Greenhouse Operation and Maintenance 448
Greenhouse Planning.......... 450
    Site Location.................. 450
    Crops to Be Cultivated 452
    Climate 452
    Greenhouse Orientation 453
    Operating Costs 453
    Market Opportunities 454
Greenhouse Structures.......... 454
    Even Span 455
    Uneven Span 455
    Lean-To...................... 455
    Quonset 455
    Gothic Arch.................. 455
    Ridge and Furrow 455
    Sawtooth.................... 456
    Other Growing Structures....... 456
Greenhouse Components 457
    Covering Materials 457
    Fans 458
    Louvers, Shutters, and Vents 459

    Evaporative Cooling 460
    Heating Systems................ 461
    Floors 462
    Benches 463
    Irrigation 464
    Additional Greenhouse Equipment... 466
Maintaining Structures and Equipment................. 469
    Structural Maintenance.......... 469
    Greenhouse Equipment Maintenance 470
Greenhouse Structure Careers .. 471
    Greenhouse Engineer........... 471
    Greenhouse Construction Worker 471
    Career Connection: Neil Devaney, Account Executive, Greenhouse Sales.....................472
Review and Assessment........ 473

Chapter 19
Greenhouse Production.... 478
Environmental Requirements... 480
    Light 480
    Air 482
    Nutrients.................... 483
    Temperature 485
    Water 486
    Pest Control 489

Crop Inputs.................489
   Media......................489
   Plant Growth Regulators........490
   Containers, Trays, Tags, and Labels...................490
Plant Materials..............491
   Seeds.....................491
   Unrooted Cuttings............494
   Plugs and Liners.............495
   Bare Root..................495
   Bulbs and Tubers............496
Greenhouse Crops............496
   Container Plants.............496
   Foliage Plants...............497
   Vegetables and Herbs........498
   Cut Flowers................498
   Bedding Plants..............499
   Perennial Plants.............500
Careers in Greenhouse Production..................500
   Plant Tag Technician..........500
   Greenhouse Customer Service..................501
   Career Connection: Denise Etheridge, Homewood Nursery....................501
Review and Assessment........502

Chapter 20
Twenty-First Century Horticulture..............506
Hydroponics.................508
   History of Hydroponics.........508
   Hydroponic Systems..........510
   System Components..........512
   Crops.....................514
Aquaponics..................514
   History of Aquaponics.........515
   System Components..........517
   Growing in an Aquaponic System..................517
   Career Connection: Rebecca Nelson, Nelson and Pade Aquaponics........518

Rooftop Gardening...........519
   Techniques.................519
   Planning a Rooftop Garden......520
Vertical Gardening...........524
   Pockets...................524
   Trays.....................524
   Pot Hangers................525
   Found Objects..............525
   Planters...................525
   Green Walls................526
Raised Bed Gardening........528
   Raised Bed Media...........528
   Intensive Gardening..........529
   Straw Bale Gardening........529
Careers.....................530
   Aquaponic System Manager.....530
   Living Wall Designer..........530
   Career Connection: Jennifer Nelkin Frymark, Gotham Greens..............531
Review and Assessment........532

Chapter 21
Nursery Production.......536
Market Niche................538
   Site Selection...............538
   Crops.....................539
   Types of Nurseries...........539
   Career Connection: Tony Avent, Plant Delights Nursery...................540

- Licensing and Shipping Regulations ... 540
- Market Outlook ... 540
- Production Methods ... 541
  - Container-Grown Production ... 541
  - Field-Grown Production ... 544
  - Pot-in-Pot (PNP) Production ... 547
  - Production Method Advantages and Disadvantages ... 550
- Sustainable Nursery Production ... 552
  - Water Management ... 553
  - Nutrient Management ... 556
  - Substrate Management ... 558
  - Environmental Management ... 559
- Careers in Nursery Production ... 561
  - Equipment Salesperson ... 562
  - Plant Buyer ... 562
  - Career Connection: Alan Erwin, Panther Creek Nursery ... 563
- Review and Assessment ... 564

Chapter 22
Vegetable Production ... 568

- Vegetables for Health ... 570
  - Health Benefits ... 570
  - Nutrients Found in Vegetables ... 570
- Vegetable Markets in the United States ... 571
- Environmental Factors ... 573
  - Growing Site ... 573
  - Water ... 574
  - Soil ... 576
  - Nutrient Management ... 576
  - Temperature ... 577
  - Season Extension ... 579
- Production Methods ... 584
  - Spacing ... 584
  - Crop Rotation ... 584
  - Integrated Pest Management ... 586
  - Good Agricultural Practices ... 586
  - Plant Material ... 587

- Postharvest Handling and Storage ... 589
  - Handling ... 590
  - Storage ... 591
- Careers in Olericulture ... 593
  - Vegetable Grower ... 593
  - Crop Advisor ... 593
- Review and Assessment ... 594

Chapter 23
Fruit and Nut Production ... 598

- Fruits for Health ... 600
  - Health Benefits ... 600
  - Nutrients ... 600
- Fruit Markets in the United States ... 601
- Small Fruits ... 602
  - Site Selection ... 602
  - Site Preparation ... 604
  - Planting ... 605
  - Maintenance ... 607
  - Harvest ... 608
- Tree Fruits and Nuts ... 609
  - Site Selection and Preparation ... 609
  - Planting ... 610
  - Pruning and Training ... 610
  - Nutrient Management ... 614
  - Pest Management ... 615
  - Harvesting and Storage ... 616
- Vine Fruits ... 616
  - Site Preparation and Planting ... 617
  - Planting ... 618
  - Training and Trellising ... 619
  - Nutrient Management ... 622
  - Irrigation ... 623
  - Pest Management ... 623
  - Harvest and Storage ... 623
- Careers in Pomology ... 624
  - Grower ... 624
  - Food Safety Manager ... 624
- Review and Assessment ... 625

Chapter 24
Landscape Design 630
The Design Process 632
- Landscape Design Steps 632
- Drawing Board or Computer-Aided Design 636
- Graphics 637

Elements and Principles of Landscape Design 637
- Line 638
- Form 639
- Texture 640
- Color 641
- Balance 642
- Proportion 642
- Repetition 643
- Emphasis 643
- Unity 643

Tools of Landscape Design 644
- Measurement Tools 644
- Drawing Instruments 644
- Paper 647

Water-Wise Landscape Design . . 648
- Planning and Design 648
- Soil Analysis and Amendments 648
- Plant Selection 648
- Turf Use 649
- Efficient Irrigation 649
- Mulch 651
- Maintenance 651

Water Garden Landscape Design . . 652
- Rain Gardens 652
- Bodies of Water 653
- Container Water Gardens 653
- Water Features 654
- Aquatic Plants 654
- Water Garden Maintenance 656

Careers in Landscape Design . . . 658
- Irrigation Specialist 658
- Aquatic Plant Grower 658
- Career Connection: Alex Ramirez, Design Workshop 659

Review and Assessment 660

Chapter 25
Floral Design 664
Principles and Elements of Design 666
- Principles of Design 666
- Elements of Design 670

Types of Floral Design 673
- Geometric Designs 673
- Line-Mass Continuum Designs 674

Containers, Tools, and Mechanics 676
- Containers 676
- Tools 676
- Ribbon 677
- Bows 678
- Mechanics 678

Flower Arrangements 682
- Corsages and Boutonnieres 682
- Bouquets 682
- Holiday Centerpieces 683
- Wreaths 683

Careers 684
- Wholesale Distributor 684
- Floral Merchandiser 685
- Career Connection: Anna Passarelli, Simply Elegant 685

Review and Assessment 686

Chapter 26
Interior Plantscaping......690
- Interior Plantscaping............692
  - Health Care Facilities.............693
  - Hotels.........................693
  - Shopping Malls.................693
  - Offices........................694
  - Restaurants....................694
  - Homes........................694
  - Botanical Gardens and Zoos.....694
- Design Principles................695
  - Elements of Design.............695
  - Principles of Design............697
  - The Process of Design..........699
- Environmental Requirements...700
  - Temperature...................700
  - Humidity......................700
  - Water.........................701
  - Light..........................703
  - Potting Media..................705
  - Plant Nutrition.................706
- Plant Management..............706
  - Grooming.....................706
  - Training and Pruning...........706
  - Insect Pests...................707
  - Diseases......................708
- Interior Plantscaping Business and Careers....................710
  - Interiorscape Designer..........710
  - Horticultural Technician.........711
  - Career Connection: Yuko Frazier, Project Manager at Ambius.............711
- Review and Assessment........712

Chapter 27
Landscape Installation and Maintenance..............716
- Landscape Design Plans........718
  - Title Block.....................719
  - North Arrow...................719
  - Legend.......................720
  - Specifications..................720
  - Drawing Scale.................720
- Site Preparation.................721
- Hardscape Installation.........721
  - Brick Patios and Pathways.......722
  - Retaining Walls.................723
  - Water Features.................726
- Planting the Design.............727
  - Plant Material..................727
  - Timing of Planting..............728
  - Preparing the Hole and Planting...729
  - Wrapping and Staking..........729
- Landscape Maintenance.......730
  - Watering......................730
  - Fertilizing.....................731
  - Pruning.......................732
  - Edging........................734
  - Mulching.....................735
- Landscaping Business..........736
  - Establishing a Business.........736
  - Product and Services..........736
  - Identifying Customers..........737
  - Assessing Competition.........737
  - Evaluating Risks................737
- Careers in Landscape Installation and Maintenance...738
  - Landscape Contractor...........738
  - Landscape Manager/Groundskeeper.....738
- Review and Assessment........739

Chapter 28
Turfgrass Management 744
- Turfgrass Industry 747
  - Turf Industry 747
  - Benefits 748
  - Turf Sustainability 748
  - Career Connection: Andy Smith, Erosion Control, Eco Turf 749
  - Turf Applications 749
  - Career Connection: Todd Lawrence, Golf Course Superintendent 750
- Turfgrass Morphology and Types 751
  - Turf Morphology 752
  - Types of Turf 754
- Lawn Establishment 757
  - Turf Selection and Timing 757
  - Site Preparation 757
  - Planting Methods 759
  - Irrigation 760
  - Mowing 761
  - Integrated Pest Management for Lawns 762
- Turf Maintenance 763
  - Irrigation 763
  - Fertilization 764
  - Mowing 766
  - Aeration 767
  - Thatch Control 767
  - Integrated Pest Management for Turf 768
- Turf Renovation 769
  - Turf Failure 769
  - When to Renovate 769
  - Pests and Turf Renovation 769
  - Seeding 770
  - Plugging 770
  - Sprigging 770
  - After Care 770
- Careers 772
  - Sod Sales Associate 772
  - Sports Turf Manager 773
  - Career Connection: Dr. Melodee Fraser: Turfgrass Breeder 773
- Review and Assessment 774

Chapter 29
Integrated Pest Management . . 778
- Creating an IPM 780
- Pests 780
  - Weeds 781
  - Insects 781
  - Other Invertebrate Pests 784
  - Vertebrates 784
- Control Measures 785
  - Sanitation 785
  - Habitat Modification 785
  - Plant Material 786
- Inspection and Monitoring 787
  - Scouting 787
  - Pest Identification 790
  - Recordkeeping and Evaluation 790
- Action Thresholds 791
- Corrective Actions 792
  - Physical and Mechanical Controls . . . 792
  - Biological Controls 793
  - Chemical Controls 796
- Careers in Integrated Pest Management 796
  - Research Scientist 797
  - Pesticide Sales 797
  - Manager 797
- Review and Assessment 798

Chapter 30
Insects 802
Anatomy 804
- Exoskeleton 804
- The Head 805
- The Thorax 807
- The Abdomen 810
- Internal Systems 810

Growth and Development 811
- The Egg and the Embryo 811
- Metamorphosis 812

Chemical Signals 813
- Sex Pheromones 815
- Aggregation Pheromones 815
- Alarm Pheromones 815
- Territorial Pheromones 815
- Allomones 815

Taxonomy 816

Agricultural Pests and Beneficials 816
- Feeding Behaviors and Plant Damage 817
- Reproduction Rate 818
- Beneficial Insects 818

Collecting Insects 820
- Bait 821
- Sweep Nets 821
- Pitfall Traps 821
- Light Traps 821
- Killing Jars 821
- Preserving Insects 821

Careers Related to Insects 823
- Agricultural Research Entomologist 823
- Education Entomologist 823
- Career Connection: The Bug Chicks, Entomologists Kristie Reddick and Jessica Honaker 824

Review and Assessment 825

Chapter 31
Disease Management 830
Disease Development 833
- Disease Triangle 833

Organisms That Cause Disease 835
- Viruses 835
- Bacteria 835
- Fungi and Fungal-like Organisms 835
- Organisms Detrimental to Plants 836

Types of Disease 837
- Testing for Disease 837
- Two Types of Plant Disorders 837
- Sick Plants versus Injured Plants 838

Disease Cycle 839
- Inoculation 839
- Entrance 840
- Establishment, Growth, and Reproduction 840
- Dissemination 840

Signs and Symptoms of Disease 841
- Wilting 841
- Abnormal Tissue Color 842
- Defoliation 842
- Abnormal Increase in Tissue Size . 842

Copyright Goodheart-Willcox Co., Inc.

Dwarfing 843
Replacement of Host Plant
  Tissue 843
Necrosis 843
Managing Plant Diseases 843
Cultural Practices 844
Genetically Resistant Plants 844
Beneficial Organisms 844
Quarantines 845
Chemical Applications. 845
Disease Index 846
Anthracnose 846
Blights. 846
Cankers. 847
Club Root 848
Damping Off 848
Galls 849
Leaf Blister and Leaf Curl 849
Leaf Spot 849
Mildew. 849
Mold 850
Root-Knot Nematode 850
Rot 850
Rust 850
Scab 851
Smut 851
Virus 851
Wilts 852
Careers in Disease
  Management. 852
Plant Pathologist. 852
Farm Advisor. 853
Career Connection: Tabitha West,
  Cedar Valley Nursery 853
Review and Assessment. 854

Chapter 32
Weeds 858
Definition of a Weed 860
Impact of Weeds. 860
Costs of Weeds. 861

Weed Characteristics 861
Weed Seeds 861
Environmental Conditions for
  Weeds 862
Benefits of Weeds. 863
Weed Biology 864
Annuals. 864
Biennials 864
Perennials. 864
Parasitic Weeds 865
Weed Identification 865
Grassy Weeds. 866
Sedges 866
Broadleaf Weeds. 866
Weed Management 867
Prevention. 867
Mechanical Control 867
Cultural Control 868
Biological Control 869
Chemical Control 869
Careers in Weed Management. . . 870
Chemical Applicators 871
Crop Consultants 871
Career Connection: Dr. Carol Somody,
  Senior Stewardship Manager,
  Syngenta LLC 871
Review and Assessment. 872

Chapter 33
Pesticide Management and Safety..........876

- Types of Pesticides..........878
  - Insecticides..........878
  - Miticides..........879
  - Herbicides..........879
  - Fungicides..........879
  - Nematicides..........880
  - Molluscicides..........880
  - Biopesticides..........880
  - Rodenticides..........881
  - Algaecides..........881
- Pesticide Formulations..........881
- Pesticide Labels..........883
  - A Legal Document..........884
  - Sections of a Pesticide Label..........884
- Pesticide Application..........886
  - Pesticide Applicator Certification...886
  - Selecting Personal Protective Equipment..........887
  - Determining the Correct Amount to Use..........887
  - Mixing a Pesticide..........888
  - Applying Pesticides..........889
  - Restricted Entry Interval (REI)....889
- Toxicity..........890
  - Types of Toxicity..........890
  - Lethal Dose..........890
  - Lethal Concentration..........891
  - Toxicity Categories..........891
  - Pesticide Poisoning..........891
  - First Aid..........892
- Storage and Disposal..........893
  - Disposal..........893
- Careers..........894
  - Pesticide Chemist..........894
  - Lawyer for an Agricultural Chemical Company..........895
  - Career Connection: Dr. Rebecca Langer-Curry: Bayer Bee Care........895
- Review and Assessment..........896

Illustrated Glossaries
Plant Identification........900
Equipment and Supplies Identification..........913
Pests and Disorders Identification..........924
Disease Identification.....926
Weeds Identification......928

Glossary..........931
Index..........949

Copyright Goodheart-Willcox Co., Inc.

CHAPTER 1

Agricultural Leadership

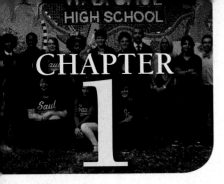

Chapter Outcomes

After studying this chapter, you will be able to:
- Define leadership and identify leadership characteristics.
- Develop leadership skills by planning a leadership path.
- Discuss agricultural youth leadership organizations.
- Describe the National FFA Organization® and its activities.
- Identify the types of supervised agricultural experiences.
- Describe agricultural education teacher and cooperative extension service agent careers.

Words to Know

confidence	leadership path	procrastination
delegate	loyalty	SMART goals
empathy	parliamentary procedure	supervised agricultural
intuition	perseverance	experience (SAE)
leadership	priority	tact

Before You Read

Arrange a study session to read the chapter with a classmate. After you read each section independently, stop and tell each other what you think the main points are in the section. Continue with each section until you finish the chapter.

While studying this chapter, look for the activity icon to:

- **Practice** vocabulary terms with Words to Know activities.
- **Expand** learning with identification activities.
- **Reinforce** what you learn by completing Know and Understand questions.

www.g-wlearning.com/agriculture

USDA/Alyn Kiel; Jodi Riedel; Jodi Riedel; MR. INTERIOR/Shutterstock.com; Iakov Filimonov/Shutterstock.com

Today is an exciting and challenging time in the agricultural industry. Today's agricultural leaders must develop the means to feed a growing population that is expected to reach 9 billion by the year 2050. Within the next 50 years, there must be more food grown and harvested than there has been during the past 10,000 years combined. Achieving this goal will require excellent leadership in governments, health agencies, and especially in agriculture.

Leadership Characteristics

Leadership is the act of directing, guiding, or motivating a person or group of people. Since you were born, there have always been leaders in your life. First your parents and family members were your leaders, then your teachers, maybe members of the clergy, and possibly an employer. Who do you consider a leader? You may immediately think of a president, professional athlete, member of the military, or a teacher, **Figure 1-1**. What is it that makes that person a leader? How can you become a leader? This chapter will help you understand what it takes to become a strong leader and give you guidance into developing your own leadership skills.

When studying successful leaders, you can find many common traits. Listing and defining these characteristics can help you understand where your own strengths and weaknesses lie. Strong leaders share a number of leadership qualities or characteristics, including:

silky/Shutterstock.com

Figure 1-1. Monuments and sculptures are commonly erected to honor great leaders. In what other ways do we honor our leaders?

"Leadership is about others, not ourselves."
—Lee Colan

- *Confidence*—a feeling of self-assurance and belief in one's ability to be successful.
- *Knowledge*—comprehension and awareness of topics important to followers.
- *Positive attitude*—approaching situations or challenges with contagious optimism and confidence.
- *Unselfishness*—the ability to put the needs of others before one's own needs.
- *Ability to inspire*—motivating others with a desire to do something.
- *Commitment*—dedication to a cause or a promise to something.
- *Courage*—the willingness to go forward and show strength when others are weak.
- *Creativity*—using original ideas or unique thoughts to further causes.
- *Enthusiasm*—an eagerness or enjoyment of activities or work.

Corner Question

What industry is the nation's largest employer?

- *Intuition*—the ability to understand or know something without the use of reason.
- *Loyalty*—the quality of being dependable and showing support or allegiance to others.
- Management—the ability to coordinate efforts, people, materials, and goods to reach goals.
- *Perseverance*—the ability to continue trying to do something regardless of setbacks or difficulties.
- Sense of humor—the ability to find amusement in situations or see things as comical.
- *Tact*—the ability to deal with others in a sensitive manner.
- *Empathy*—the ability to understand and share feelings of others.
- Honesty—the quality of being upright, fair, and straightforward.

These are key characteristics of a leader. When identifying these common traits, you may recognize characteristics you already possess. Nurturing these qualities or traits will help you develop the leadership skills necessary for today's competitive workforce.

Develop a Leadership Path

You can begin developing your leadership skills by first planning a leadership path. A *leadership path* is a plan or map showing steps and activities to complete to develop and exhibit leadership skills. Imagine each step of your path as a building block. These steps might include:

- Creating a vision.
- Setting goals.
- Cultivating self-confidence.
- Fostering initiative.
- Getting organized.
- Furthering your education.

Leadership is a skill that is demanded by many, but people often fail to recognize how someone becomes a leader. What does a person have to do to become a recognized leader? There are a series of steps and a pathway that must be followed by the learner. Leadership is not simple, and it is not a skill that is attained in just a short time. Leadership must be built over time and through a series of experiences. Eventually, a true leader will be directing those around him or her, **Figure 1-2**. Leaders must develop a path using leadership building blocks to meet their goals.

"Leadership is about taking responsibility, not making excuses."
—Mitt Romney

Andresr/Shutterstock.com

Figure 1-2. The leadership skills you acquire now will help you succeed in your adult life.

Create a Vision and Set Goals

As a child, you may have pictured yourself as a firefighter or doctor and now envision yourself in a different career. This mental picture is a vision. A series of processes must go into effect to make the vision become a reality. You cannot become a firefighter unless you fulfill the required training time; learn the skills necessary; and have the physical, emotional, and mental capacity to do so. So how do you actually do what you envision? In order to make a vision a reality, one must set goals. One way that people set goals is to make SMART goals. *SMART goals* are objectives that are:

- Specific—stating the who, what, when, where, why, and which of a goal.
- Measurable—able to be evaluated by the numbers, quantities, and data of the goal.
- Attainable—achievable; something that you can make happen.
- Realistic—something that can really happen.
- Timely—completed in an appropriate time frame.

SMART goals can be applied to anything in life. They are extremely applicable to leadership goals as well. In order to become a leader, there a number of SMART goals that you set for yourself, **Figure 1-3**.

Become Self-Confident and Healthy

Most of today's leaders exhibit confidence and lead healthy lifestyles. Some individuals are born with self-confidence, and others need to cultivate it. Males often have more self-confidence in late high school and into their early twenties. They will demonstrate risky behaviors that can sometimes be dangerous because they believe that they can be successful. Although this may not always be a wise choice, this period of self-confidence is believed by many sociologists and psychologists to be a way for males to charter their own paths in life and move away from the comforts of home.

USDA/Bob Nicholas

Figure 1-3. Writing down your SMART goals will help keep you focused. You will need to add new goals periodically as you develop your skills and reach your objectives. Have you ever kept a personal journal? How can keeping such a journal contribute to your success in reaching your objectives?

Self-confidence can be developed in those that are nervous by nature or are not as self-assured as others. Someone who is insecure can expose themselves to risks. Over time, as he or she experiences success after a risk, he or she will gain a feeling of confidence and certainty. For example, if you are afraid to present or speak in front of a class, there is only one way to become sure of yourself—you must practice public speaking. The first time you speak in front of a group could be to give a simple introduction.

Corner Question
What former US president and Nobel Peace Prize winner was also an FFA® member?

Later, you may give a short talk as part of a group. Finally, you may be ready to give an individual presentation. Speaking in public may be uncomfortable in the beginning, but there will be benefits associated with improved self-confidence.

Being healthy can aid in developing self-confidence, **Figure 1-4**. America is in the midst of a health crisis. Issues such as obesity and lack of physical fitness impact mental and emotional health. Many people have developed a sedentary lifestyle. Countless studies show the importance of being active. Getting regular exercise, such as walking or playing sports, promotes mental and physical health. Feeling well can help you be more confident in your abilities. Leaders can model eating healthy and being physically active. What changes could you make today that could improve your health and fitness? Create a SMART goal and see what results you can attain in just a matter of days or weeks.

CLS Design/Shutterstock.com

Figure 1-4. Participating in sports activities in high school will help you stay fit and develop good fitness habits, and becoming the team captain or cocaptain will help you develop your leadership skills.

Reduce Procrastination and Foster Initiative

Procrastination is delaying or postponing a task or activity. Procrastination plagues the young, the old, the strong, the weak, and everyone in between. Chores, tasks, homework, exercise…this endless list of what should or must be done often gets pushed to the side in favor of what people want to do. Some people may only procrastinate once in a while, whereas others may choose to delay important tasks every day. Are you a procrastinator? When procrastination becomes a habit that leads to problems, then you must begin to change your habits, **Figure 1-5**. Determine why you procrastinate, and then promise yourself to change one bad habit at a time.

Demonstrate initiative by creating a SMART goal or plan of action to help you improve work and study habits. Setting a SMART goal aids in getting a goal into action. Initiative is much like a strong work ethic—get up and start doing instead of sitting and dreaming about it.

Get Organized

As you have more commitments and take part in more activities, it is easy to feel confused or become forgetful of deadlines or appointments. Using some simple time-management methods can help you get organized. Here are some suggestions:

- Write down all commitments, appointments, or deadlines.
- Set priorities. A *priority* is a determination or ranking of the importance or urgency of a task or activity.
- Create a plan.

Syda Productions/Shutterstock.com

Figure 1-5. Developing good study habits now will help you do well in high school as well as in college. **How will good study habits carry over into your professional life?**

Using computer programs or apps for mobile devices such as smartphones and tablets can make time management and organization much easier. You can also use a paper calendar to record information. Entering dates into a calendar and creating timely reminders can help you organize tasks and meet deadlines. These tools can only help, however, if you use them properly. Remember that organization is key to achievements in life and in developing leadership skills.

Further Your Education

Leaders appreciate the value of education throughout life. Regardless of what method of education you choose, lifelong learning is essential in leadership. No one would want care from a doctor who stopped studying medicine at the end of medical school. Patients want a doctor who is up-to-date on all technologies and the latest advancements in the field of medicine. This is true of all people, regardless of occupation. In addition to continued formal education, activities such as reading, writing, solving puzzles, and playing chess also help keep the brain active and contribute to one's overall health and success.

> "The quality of a leader is reflected in the standards they set for themselves."
> —Ray Kroc

Leadership Development

People expect leadership to be practiced and developed in the workplace; however, good leadership should be practiced in all areas of life. Whether you are a student in FFA, 4-H, or in the National Junior Horticultural Association (NJHA), you can work toward leadership growth. Follow these tips to promote leadership growth:

- Analyze and solve problems by determining the shortcomings of your organization and coming up with solutions.
- Get experience by taking classes, attending workshops, going to conferences, and getting involved in leadership organizations.
- Learn from the mistakes of others and yourself.
- Create purposeful meetings and get attendees involved.
- Build great teams by identifying strengths of team members and helping to minimize weaknesses.
- Lead and promote teamwork by fostering team building and stronger relationships, **Figure 1-6**.
- Delegate tasks—you cannot do everything alone. *Delegate* means to assign or entrust a task or responsibility to another person. Give team members tasks and coach them through their work if needed. Commend them on their efforts and you will see future rewards.

xuanhuongho/Shutterstock.com

Figure 1-6. Team building exercises build trust and confidence and teach members how to work together to achieve a goal. What would happen if the members of a rowing team did not work together and all row in the same direction or with the same timing?

- Support your team by illustrating your willingness to work with other members.
- Communicate often with your team members and foster a feeling of belonging. Use social media as well as your meeting time to distribute important information (dates, times, fees, etc.) and to compliment and encourage team members.
- Inspire others and motivate them with your enthusiasm and sincerity.

Corner Question

Which long-haired country music legend was also an FFA member?

Agricultural Leadership Organizations for Youth

In the United States, youth leadership organizations are found in nearly every community. These organizations strive to prepare children and youth for their futures. Agricultural leadership organizations, especially those focused on horticulture, hope to impact students while using agriculture as a vehicle for leadership development. Three youth leadership organizations are discussed in this chapter: 4-H, the National Junior Horticulture Association (NJHA), and the National FFA Organization.

4-H

In the early 1900s, several agricultural youth clubs began meeting in Ohio. These organizations focused on canning, raising livestock and poultry, and promoting vocational agriculture in "out-of-school" clubs. Around this same time, canning clubs for girls (an organization founded to help young girls develop the skills to safely preserve food) began to flourish throughout the rural United States. These groups evolved into 4-H, the nation's largest youth development organization.

4-H has simple ideas that worked in 1902 and continue to do so today. The mission to "help young people and their families gain the skills needed to be proactive forces in their communities and develop ideas for a more innovative economy" is successfully reaching more than 6 million 4-H members in urban, suburban, and rural communities. The 4-H organization has led the way for youth to develop leadership skills and also transformed the way youth learned from hands-on experience outside of the classroom, **Figure 1-7A**. The 4-H organization's green four-leaf clover emblem symbolizes the commitment to a person's head, heart, hands, and health, **Figure 1-7B**.

A
oceanfishing/Shutterstock.com

B
blackboard 1965/Shutterstock.com

Figure 1-7. A—Through 4-H programs, students may learn to grow, harvest, and sell the fruits or vegetables they produce. B—The green four-leaf clover emblem symbolizes the organization's commitment to a person's head, heart, hands, and health.

Head

The 4-H promotes the ideas of clear thinking and management. Various activities, competitions, and meetings are used to help develop children's minds. 4-H participants are two times more likely than other youth to participate in STEM (science, technology, engineering, and math) activities outside of school.

Heart

The heart in the 4-H emblem symbolizes concern for the welfare of others. 4-H fosters development of compassion, relationships, and caring in its members. Students who are 4-H members are two times more likely to be civically active. Members interact with volunteers ranging from local business leaders to government employees. More than 540,000 volunteers and 3500 professionals interact with 4-H members to foster relationships that support youth development.

Hands

The hands of 4-H members provide service to their world. Rosalyn Carter, a 4-H alumni and wife of former President Jimmy Carter, formed her zeal for service and giving as a youth. Former First Lady Carter began working with Habitat for Humanity in the early 1980s. Her desire to promote social justice and human rights also prompted her to found the nonprofit Carter Center. With her help, more than 750,000 homes have been built for families in need through Habitat for Humanity. The founding principles of service, developed in 4-H, have contributed to Rosalynn Carter's humanitarian efforts throughout her life.

Health

Healthy lifestyle habits are developed through education and effort. The 4-H organization prides itself on encouraging healthy habits in its members. Members of 4-H are twice as likely to choose healthier options than non-4-H members. Today's childhood obesity epidemic confirms the need for 4-H to continue educating children and helping them to be fit and well.

National Junior Horticulture Association (NJHA)

The National Junior Horticulture Association (NJHA) promotes and sponsors youth horticultural activities and projects. NJHA hopes to increase an understanding of the horticulture industry and support career exploration in this field for children and youth, **Figure 1-8**.

"The ultimate goal of farming is not the growing of crops, but the cultivation and perfection of human beings."
—Masanobu Fukuoka

Srdjan Fot/Shutterstock.com

Figure 1-8. One of the NJHA's goals is to help its young members become good citizens with a basic understanding of nature as well as an acceptance of responsibility for the environment. Volunteering for a public greenhouse or garden will give you hands-on experience, expand your understanding of nature, and allow you to give back to your community.

This youth organization also wishes to develop good citizens with a basic understanding of Earth and its environments. Members of the NJHA participate at a local and national level in leadership and cooperation events.

In 1934, NJHA was the first organization of its kind founded exclusively for youth and the advancement of horticultural science. Individuals who are 22 years of age and younger are welcomed into this organization. Throughout the year, workshops, events, and competitions are hosted at the state and national level.

Corner Question

How many yards of fabric are required for the FFA jackets created for one year?

National FFA Organization

The agricultural education program is comprised of three components: agricultural instruction, supervised agricultural experiences (SAEs), and the National FFA Organization. (A *supervised agricultural experience (SAE)* is a student-developed project that involves hands-on learning in agriculture and natural resources.) FFA's mission is to foster positive developments in its members by advocating premiere leadership, personal growth, and career success through agricultural education.

History of the National FFA Organization

Throughout the United States, visionaries for agricultural education and the FFA collaborated to create a program that would be housed in public education. In 1917, the Smith-Hughes Act (formally named the National Vocational Education Act) was passed by the US Congress and provided the funding needed to create agricultural courses. In 1925, agricultural educators founded the Future Farmers of Virginia (FFV) for boys. This organization served as the model for the FFA. Just three years later, in 1928, the Future Farmers of America® put down its roots in Kansas City. Soon after, national blue and corn gold were adopted as the official FFA colors. The adoption of official dress, which included the signature blue corduroy jackets, was confirmed.

The FFA looks much different now than it did in 1928. The organization includes students from every state and several territories. More than 600,000 members from around the country are members of the FFA. Today's FFA members continue to wear blue jackets and strive to promote agricultural leadership, **Figure 1-9**.

Marie Appert/Shutterstock.com

Figure 1-9. The National FFA Organization offers members opportunities to participate in exciting events such as the Rose Bowl Parade. How are FFA members chosen to decorate and ride on the FFA's Rose Bowl Parade float?

AgEd Connection — FFA Timeline

1917—Smith-Hughes Act is passed, leading to agricultural education in classrooms across the country.
1928—Future Farmers of America is established in Kansas City, Missouri.
1929—National blue and corn gold is adopted as official colors.
1930—Official dress uniform of dark blue shirt, blue or white pants, yellow tie, and blue cap is adopted. Only boys are allowed to be members.
1933—Blue corduroy jacket is adopted as official FFA dress.
1935—New Farmers of America (NFA) is founded at the Tuskegee Institute in Alabama.
1948—First FFA Week is celebrated during the week of George Washington's birthday.
1965—New Farmers of America merges with the Future Farmers of America.
1969—FFA opens membership to girls.
1988—Future Farmers of America changes its name to the National FFA Organization to illustrate the diversity of agriculture. FFA opens to middle school programs.
1991—FFA chapters are founded in Virgin Islands, Guam, and Micronesia.
2007—FFA reaches the half-million members mark and has more than 7000 chapters.

FFA Emblem and Motto

The FFA emblem represents the FFA symbolically. Just as the US flag has a star for each of the 50 states and 13 stripes to represent the original colonies, the FFA emblem has graphical representations of the organization's beliefs and foundations, **Figure 1-10**.

In the early 1900s, an agricultural education teacher coined the motto, "Learning to Do, Doing to Learn, Earning to Live, and Living to Serve." The words mean that students in agricultural education can and should learn by practicing agriculture. In addition, these students must recognize the hard work that is required of them to thrive in the world. Finally, the FFA member should give back to the community and make our nation a better place.

FFA Structure

The FFA structure includes the local, state, and national levels. Some states also have federation, region, and/or district levels. Each part of the organization is equally important to its success.

Local Chapter

The local chapter is housed at a school or a homeschool. Chapters are made up of students and an agricultural educator. The chapter must have members that pay dues, chapter leadership (officers), and a constitution.

> **Corner Question**
> What was the background of the emblem of the New Farmers of America (as opposed to the cross section of corn for the Future Farmers of America)?

Symbols for the FFA Emblem

Ear of Corn	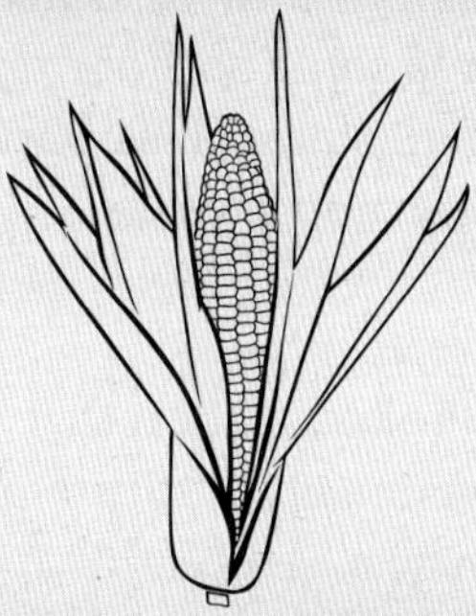 *agrino/Shutterstock.com*	Corn represents unity. This crop is cultivated across the United States and is considered the heart of American agriculture.
Eagle	 *mlorenz/Shutterstock.com*	The eagle represents freedom. Just as this regal bird is a symbol of our nation, the eagle reminds FFA members of the beauty of their freedom.
Owl	 *Eric Isselee/Shutterstock.com*	The owl is a symbol of knowledge and wisdom. The owl reminds FFA members of the power of knowledge in the agriculture industry.
Plow	 *Jose Elias da Silva Neto/Shutterstock.com*	Labor and tillage of the soil signifies hard work and determination. Hard work leads to success in life, school, and work.
Rising Sun	 *america365/Shutterstock.com*	The rising sun symbolizes the promise of a new tomorrow and progress. A new day provides the opportunity for growth.
Agricultural Education and FFA		These two words and those three letters are inscribed to remind us that the National FFA Organization is founded on the principles of agricultural education.

Figure 1-10. The FFA emblem is representative of the history, goals, and future of the organization.

Possible Federation, Region, and/or District

Different states and even regions may operate differently. Some school districts or regions choose to have some sort of organization above the chapter level. This is not true of all states. This part of the organization often facilitates competitions, festivities, and awards.

State

Each state has a department of education that enables agricultural education efforts. State FFA advisors and officers provide leadership for local chapters. The state organization acts as a platform for numerous FFA career development events and award qualifiers for the national level.

National

The National FFA Center is located in Indianapolis, Indiana. Here, a national advisor and national officers provide leadership for all the local FFA chapters and their efforts.

Leadership Development in FFA

The FFA provides avenues for success so that every member can achieve success in the organization. Developing leadership skills is an important part of the FFA mission. Leadership skills include the ability to:

- Direct and guide others.
- Motivate and inspire others.
- Exhibit confidence in yourself and help others build self-confidence.
- Model constructive behaviors.
- Manage and organize tasks and set priorities.
- Delegate tasks and responsibilities.
- Give clear instructions and constructive feedback.

Leadership development is achieved through training and service as an officer at local, state, or national levels and through earning degrees.

Officers

Officers are the heart of the FFA. These individuals guide the organization from the chapter to the national levels. Officers are expected to have a desire to be part of the officer team, possess various leadership skills, and lead by example. In addition, these individuals should work well with their chapter members and be able to lead a meeting using *parliamentary procedures* (a democratic and efficient way of conducting business using *Robert's Rules of Order*). Officers include the president, vice president, secretary, treasurer, reporter, sentinel, historian, parliamentarian, chaplain, and advisor, **Figure 1-11**.

Symbols for the FFA Officers

Officer	Symbol	Description
President	 *america365/Shutterstock.com*	The president presides over meetings according to the rules of parliamentary procedure. The president's symbol is the rising sun, which is a symbol of progress and promise.
Vice President	 *Jose Elias da Silva Neto/Shutterstock.com*	The vice president assumes the duties of the president if he/she is unable. This person coordinates all committee (small, specialized groups of members organized for a specific objective) work. This officer's symbol is the plow, symbolizing hard work and labor.
Secretary	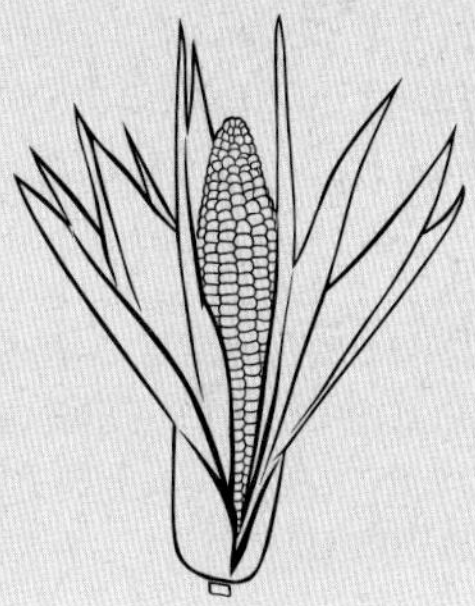 *agrino/Shutterstock.com*	The secretary creates the agenda (list of what will be accomplished) at all meetings and corresponds with all other chapters. This officer is responsible for all issues related to membership. The secretary's symbol is an ear of corn, as this is grown in all 50 states.
Treasurer	 *ledokol.ua/Shutterstock.com*	The treasurer keeps all records of finances, including collecting dues. The symbol of the treasurer is the bust of George Washington. George Washington was a farmer and a financier of his estate.
Reporter	 *nazlisart/Shutterstock.com*	The reporter serves the chapter as a public relations and marketing manager. This officer ensures that all community members are aware of the FFA chapters and all its happenings and accomplishments. The reporter's symbol is a US flag.

(Continued)

Figure 1-11. Each FFA officer is represented by a specific symbol.

Symbols for the FFA Officers (Figure 1-11, continued)

Officer	Symbol	Description
Sentinel	*Hands of friendship (handshake)* — MNSKumar/Shutterstock.com	The sentinel welcomes guests and visitors to all FFA events. This officer assists the president in maintaining order. The symbol of this officer is the hands of friendship.
Historian	*Open book (scrapbook)* — Slobodan Kostic/Shutterstock.com	The historian develops the scrapbook and maintains all memorabilia. The symbol of the historian is the scrapbook.
Parliamentarian	*Gavel* — Leone_V/Shutterstock.com	The parliamentarian is proficient in parliamentary procedure. This officer has the book *Robert's Rules of Order* or the gavel as its symbol. This symbol demonstrates the knowledge of parliamentary law.
Chaplain	*Dove* — Dove sibiranna/Shutterstock.com	The chaplain presents the invocation at banquets and assists the chapter with reflection when called on by its members. A dove, which is a symbol of peace, represents this officer.
Advisor	*Owl* — Taeya18/Shutterstock.com	The advisor supervises the FFA chapter. This individual also offers instruction on leadership and informs the community of the FFA. The owl, which symbolizes wisdom, represents the advisor.

Degrees

The FFA provides an avenue for exploration in agricultural leadership. Active FFA members can be rewarded for certain accomplishments in the FFA. The FFA provides five degrees (awards):

- Discovery FFA Degree (7th and 8th grade award).
- Greenhand FFA Degree.
- Chapter FFA Degree.
- State FFA Degree.
- American FFA Degree.

Career Connection | Joshua Bledsoe

National FFA Chief Operating Officer

National FFA Organization

Joshua Bledsoe has been following the FFA's motto, "Learning to Do, Doing to Learn, Earning to Live, Living to Serve," for most of his life. Josh is the Chief Operating Officer of the National FFA Organization. He is one of the premier leaders of this youth leadership organization, managing seven divisions with 110 FFA team members at the national headquarters in Indianapolis, Indiana. Bledsoe reports to the Chief Executive Officer, Dr. W. Dwight Armstrong.

Bledsoe credits preparation for his current occupation to his roots in his high school FFA chapter. Bledsoe describes his agricultural education teacher as a truly inspirational man who opened Bledsoe's eyes to experiences in agricultural education and FFA. This "eye-opening" was the catalyst for future endeavors that included his work as a college student, teacher, state FFA staff leader, and as a national leader today. These work experiences, along with everyday life experiences, make him who he is today.

Bledsoe claims there is a learning curve for himself working at the National FFA Organization offices. He manages and works with people in every capacity. Josh uses McDonald's founder Ray Kroc's quote, "When you are green you are growing. When you're ripe, you rot," for inspiration. Josh thrives on growing and expects to continue doing so personally, and in the National FFA Organization.

These degrees or awards are much like the Boy Scouts of America's merit badges. The American Degree is the highest honor, much like the Eagle Scout awarded to only a select group of Boy Scouts. These five degrees act as rewards and note superior performance in the FFA. The experiences required to qualify for each degree build on each other, making accurate records for your SAE(s) and FFA participation useful tools in the application process, **Figure 1-12**.

A
Jodi Riedel/Goodheart-Willcox Publisher

B
David Kosling/USDA

Figure 1-12. Your active participation in community projects (A), chapter activities, and other FFA events (B) will help you qualify for FFA degrees.

FFA Program of Activities

To achieve awards and recognitions, an FFA member must participate in a chapter's program of activities (POA). The POA provides opportunities for chapter members to participate in the FFA personally, socially, civically, and competitively. The POA outlines an agenda for the year that active FFA members and officers create before the start of the school year.

The program of activities is organized to develop students, the chapter, and the community. Each month of the calendar includes activities to advance the student, the FFA, and the agricultural education program.

Student Development

Student development is the component of the FFA program of activities that focuses on individual students. It includes activities in areas such as leadership, healthy lifestyle, career success, scholarship, and personal growth:

- Leadership activities promote interpersonal skills and decision making. Activities such as public speaking experiences and leadership competitions fall under this category.
- Healthy lifestyle activities are designed to build personal physical, emotional, and mental fitness. Activities such as kickball tournaments and cooking classes or competitions fall under this category.
- Career success activities may include aspects of SAE development. Recordkeeping workshops and agricultural tours are examples of career success activities.
- Scholarship activities help participants develop positive academic standards. Activities such as filling out scholarship applications and striving to make the chapter honor roll fall under this category.
- Personal growth activities are designed to help students implement career exploration. Activities such as job shadowing and attending agricultural career guest speaker presentations are examples of personal growth activities.

Jodi Riedel/Goodheart-Willcox Publisher

Figure 1-13. North Carolina high school FFA students help elementary school students cultivate a vegetable garden. Together, they grow vegetables for the hungry and homeless in the community. How do the student participants personally benefit from participating in this type of activity?

Chapter Development

Chapter development usually involves committees and groups that work together to reach a goal or solve a problem, **Figure 1-13**.

These committees work to make a difference in the chapter and promote growth of the agricultural education program.

- **Chapter recruitment**—activities that promote and increase enrollment in agricultural education and participation in the FFA. Performing classroom visits and holding National FFA Week activities are examples of chapter recruitment.
- **Financial**—activities that promote sound financial decisions and management. Completing grant applications and holding fund-raising activities are examples of financial activities.
- **Public relations**—activities that promote goodwill and cast a positive light on the FFA and agricultural education. Committees may write and distribute newsletters or submit articles to local media as part of their public relations work.
- **Cooperation**—activities that develop teamwork and collaboration within FFA. Officer training and leadership retreats are examples of cooperation activities.
- **Support group**—improvement of relationships among FFA and parents, community leaders, and the agricultural industry. Committees often work with FFA alumni and the Farm Bureau to foster community relationships.

Community Development

The FFA program of activities encourages members to become involved in their community and fosters FFA service in several ways:

- **Economic development**—improvement of economic health through avenues such as FFA member-owned businesses and job creation.
- **Environmental**—encouraging conservation and responsible citizenship through activities such as community gardening projects; waterway cleanup events; and water, air, and soil quality programs, **Figure 1-14**.
- **Human resources**—events such as after-school programs and mentoring programs for children that provide valuable services to citizens in the community.
- **Citizenship**—activities such as charity work and community service that inspire FFA members to become involved in their community and country.
- **Agricultural promotion**—actions such as Agriculture in the Classroom and National Agriculture Day activities that help the community to become more agriculturally literate.

USDA Natural Resources Conservation Service/Angela Stewart

Figure 1-14. Participation in community development events is an important part of FFA. Members organize and participate in cleanup events to better the community in which they live and to foster relationships between the FFA and the community. *Are there places in your neighborhood that would benefit from this type of event?*

Chapter, State, and National FFA Activities

FFA chapters create the program of activities and include various events to ensure that the needs of the FFA members, chapter, and community are met. To achieve their goals for growth and development, some FFA chapters may include some or all the activities outlined here.

National FFA Week

The National FFA Organization's officer team coordinates an effort to recognize and celebrate FFA for an entire week every February. At the state and chapter levels, additional events are created to observe this week. The focus of the week is to educate communities and America about the prospects available to youth through participation in the FFA.

During FFA week (the week of George Washington's birthday), chapters, state, and the National FFA Organization may choose to complete community service, environmental awareness, education and outreach programs, public relations campaigns, and recruitment activities, **Figure 1-15**. All these activities provide platforms for interacting with the community and spreading the word that the FFA is a national organization that prides itself on agricultural leadership, career exploration, and personal growth.

Chapter Banquet

As the end of the school year approaches, FFA members will find themselves slipping into that blue corduroy once more as they prepare for the chapter banquet. This festivity can take place in school cafeterias, gardens,

Jodi Riedel/Goodheart-Willcox Publisher

Figure 1-15. Community service benefits FFA members and the community. What types of community service activities does your chapter hold?

banquet halls, barns, or restaurants. Regardless of the banquet location, the event is mostly the same from chapter to chapter.

The chapter banquet allows FFA members, parents, agricultural industry members, and the community to gather for a night of celebration. At this event, students may be presented with degree pins and other awards. FFA supporters are recognized with volunteer pins and honorary chapter degrees to illustrate appreciation from the FFA chapter, **Figure 1-16**. Like most FFA events, food plays a central role in the event. The night is capped with a closing ceremony from the FFA officers, and new FFA officers may be installed at this time as well.

Leadership Conferences and Camps

Leadership conferences and camps vary in location and time length. Leadership conferences offer a chance for FFA members to focus on building and fine-tuning leadership skills. Camps are more often offered for recreation. Not all states have camps, but most states have leadership conferences. Activities at the leadership conferences ask students to get out of their comfort zones and try new things.

The Washington Leadership Conference (WLC) takes place at the nation's capital. This weeklong event is the pinnacle of leadership development. This event brings FFA members from all over the country together. FFA members are taught how they can impact their chapter, school, community, and country. Students visit historic sites and often make lifelong friends. FFA members also serve the Washington, DC, community and will often work at area food banks or shelters during their stay.

Jodi Riedel/Goodheart-Willcox Publisher

Figure 1-16. One of the highlights of each school year is the FFA banquet. Students and FFA supporters are recognized for their accomplishments and contributions to the program.

Conventions

Most states have a convention; in addition, the National FFA Convention takes place every October. This event rotates to different cities every 10 years, and is currently scheduled in Indianapolis, Indiana. State conventions and the national convention are well attended. The National FFA Convention is the nation's largest annual student gathering, **Figure 1-17**.

Regardless of location, conventions host students for competition, awards, social events, and service events. There are often workshops, career shows, tours, and professional speakers or entertainment as well. Whether you are attending your state's FFA convention or the National FFA Convention with 50,000 other FFA members, the experience will be unforgettable and it will help you to grow as an agricultural leader.

Career Development Events

Career development events (CDEs) are competitive events put in place to test student knowledge and skills in various agricultural pursuits, **Figure 1-18**. CDEs assess what was learned in agricultural education classes and the FFA. FFA members can participate in team or individual events. Chapter advisors work with the teams. They often pair the team with community members or agricultural experts to practice and fine-tune the skills that will be evaluated in competition. Hard work and practice is rewarded at chapter, state, and national competitions with plaques, ribbons, scholarships, and even cash.

US Department of Agriculture

Figure 1-17. Many universities, colleges, and manufacturers have exhibits and sponsor events at the National FFA Convention. The interactive exhibit illustrated here is a teamwork exercise.

FFA Career Development Events	
Team Events	
Agricultural Communications	Floriculture
Agricultural Issues Forum	Food Science and Technology
Agricultural Mechanics	Forestry
Agricultural Sales	Horse Evaluation
Agriscience Research	Livestock Evaluation
Agronomy	Marketing Plan
Dairy Cattle Evaluation	Meats Evaluation and Technology
Dairy Foods	Nursery/Landscape
Environmental and Natural Resources	Parliamentary Procedure
Farm Business Management	Poultry Evaluation
Individual Events	
Creed Speaking	Prepare Public Speaking
Extemporaneous Public Speaking	Dairy Cattle Handlers' Activity
Job Interview	

Figure 1-18. There are twenty-four career development events and one activity in which FFA members may participate. **Which events interest you most?**

Supervised Agricultural Experience

The supervised agricultural experience (SAE) is one of the three integral parts of the agricultural education program. A supervised agricultural experience (SAE) is a project that involves hands-on learning in agriculture and natural resources that is developed by the student. SAE is about learning by doing or hands-on instruction. The SAE and FFA activities complement one another and enhance classroom and laboratory instructions. The SAE is important to achievement in FFA as students apply skills and knowledge learned in work-based learning to FFA activities. Leadership and career success skills learned in FFA help students achieve success in SAEs.

AgEd Connection — Creed Speaking

Seventh-, eighth-, or ninth-grade students present their interpretation of the FFA creed in this career development event. This five-paragraph work was written by E. M. Tiffany to inspire and remind individuals of the basic and wholesome principles associated with agriculture. It begins: "I believe in a future of agriculture with a faith born not of words but of deeds…" Students recite this work for judges and are asked questions about the meaning of these statements. Students involved in this career development event build public speaking skills that can be carried into future career endeavors.

Students design their own SAE projects and ask their agricultural education teachers only for assistance and guidance. During the SAE, students earn hours that are required by the instructor, and they can also earn money, **Figure 1-19**. Students work to gain skills and knowledge that can be used in their future careers.

Students will develop their own SAE project that should include these activities:

- **Investigation.** Ask yourself what interests you have and what SAE could be a good fit for you. Work with your family, community, local businesses, and your agriculture teacher to identify an appropriate SAE.
- **Planning.** Work with all the stakeholders in your SAE project to develop a timeline and plan your SAE project. What materials will you need? Who will provide these materials? Where will your SAE take place? How will you get to and from your SAE?

cdrin/Shutterstock.com

Figure 1-19. Selling your own products or working for someone else at a farmers market may be a way to earn money and hours toward your SAE. Are there farmers markets near your home or school?

- **Coordination.** Communicate with all the stakeholders in your SAE project and make your SAE become a reality. Learn by doing. This is your chance to do something that you love. If your SAE does not meet your expectations, work with your advisor to make changes so that your project is enjoyable and beneficial.
- **Keeping records.** Record what you have accomplished in the SAE project. Take photographs and catalog all the hours that you spend outside of the classroom in a recordkeeping system such as *The Agricultural Experience Tracker (AET)*™. Ask your teacher about The AET and how it can help make managing your SAE easier. If The AET is not available, use a computer to keep financial and time records.
- **Improving the program.** Take time to reflect and consider your SAE when a portion of your project is completed. What could be done to make your project better? Where are you headed? Where do you want the project to conclude?

Agricultural Leadership Careers

Strong leadership and direction is important for the agricultural industry. Within the agricultural industry, there are leaders at the community, state, national, and international levels. The individuals in these careers have unique and diverse backgrounds and cultures, but they are united in seeking successful solutions to issues and problems facing the agricultural industry.

Agricultural Education Teachers

An agricultural education teacher creates a unique learning environment for his or her students. These teachers are responsible for fostering agricultural literacy in their students through daily instruction. In addition, the teacher

Career Connection

Jennifer Broadwell
Agricultural Education Teacher

Jennifer Broadwell is a high school agricultural education teacher in an urban setting of North Carolina. She has been an instructor for more than seven years. She comes from a rural background, but she has thrived while working with a diverse student population that resides in urban and suburban settings. Jennifer's exemplary FFA group has won several national FFA competitions and has been awarded the National Chapter Award. She teaches students to "do" in the classroom and in the community. She always makes time for her students in and outside the classroom. Her job is definitely not an 8 am to 5 pm kind of job. Every day is unique and alive with opportunity to connect with students and encourage agricultural literacy and development.

Jennifer Broadwell

develops and maintains facilities that can include farms, barns, greenhouses, nurseries, laboratories, and gardens. The agricultural educator advises the FFA and organizes career development event practices and competitions. He or she works with students to craft meaningful SAE projects and strives to submit quality award applications. The agricultural education program is not complete without service learning. The advisor assists in interactions among community members, agricultural stakeholders, the school, and students.

Agricultural education teachers must have a bachelor's degree and often will attain a master's degree by attending graduate school. They should enjoy working with students and community members. Advisors need to be flexible and be prepared to brave harsh environmental conditions. Their work conditions can change by the hour, and their work environment can include being outdoors. Agricultural education teachers strive to make positive differences for their students today so that these students can make positive changes in their world tomorrow.

Cooperative Extension Service Agents

The cooperative extension service educates and serves farmers, agencies, and communities alike in every state of the United States. Each state has at least one college or university dedicated to agriculture. This was established by the Morrill Act of 1862. Within these agricultural institutions are cooperative extension service agents. Agents works as liaisons between the university and both industry and the community.

Extension agents spend a great deal of time learning and then disseminating their knowledge to the public. Within horticulture, extension agents work to help home gardeners and the horticulture industry. Extension agents are problem solvers and are exemplary in seeking solutions and answers to questions. They work during the week and on weekends, hosting workshops and meeting with consumers.

CHAPTER 1

Review and Assessment

Chapter Summary

- Leadership is the act of directing, guiding, or motivating a person or group of people. Leadership characteristics include traits such as integrity, tact, honesty, courage, and enthusiasm.
- A leadership path is a plan or map showing steps and activities to complete to develop and exhibit leadership skills.
- Agricultural leadership organizations that relate to horticulture include 4-H, the National Junior Horticulture Association (NJHA), and FFA.
- The agricultural education program is comprised of three components: agricultural instruction, supervised agricultural experience (SAE), and FFA.
- The FFA structure includes local, state, and national levels. Some states also have federal, regional, and/or district levels.
- Developing leadership skills is an important part of the FFA mission. Leadership development is achieved through training and service as an officer at local, state, or national levels and through earning degrees.
- The FFA chapter is guided by the program of activities (POA) that focuses on student, chapter, and community development. The POA can include activities such as banquets, conventions, and leadership conferences.
- Chapter, state, and national FFA activities may include National FFA Week, chapter banquets, leadership conferences, conventions, and career development events.
- Supervised agricultural experiences (SAEs) provide opportunities for students to gain hands-on work experience, skills, and knowledge. Activities that are part of developing an SAE project include investigating, planning, coordinating, keeping records, and improving the program.

Photo Credit: USDA/Lance Cheung

Words to Know

Match the key terms from the chapter to the correct definition.

- A. confidence
- B. delegate
- C. empathy
- D. intuition
- E. leadership
- F. leadership path
- G. loyalty
- H. parliamentary procedure
- I. perseverance
- J. priority
- K. procrastination
- L. SMART goals
- M. supervised agricultural experience (SAE)
- N. tact

1. The ability to understand and share feelings of others.
2. A plan or map showing steps and activities to complete to develop and exhibit leadership skills.
3. A feeling of self-assurance and belief in one's ability to be successful.
4. A determination or ranking of the importance or urgency of a task or activity.
5. The quality of being dependable and showing support or allegiance to others.
6. The ability to deal with others in a sensitive manner.
7. A democratic and efficient way of conducting business based on guidelines in *Robert's Rules of Order*.
8. To assign or entrust a task or responsibility to another person.
9. The ability to continue trying to do something regardless of setbacks or difficulties.
10. The ability to understand or know something without the use of reason.
11. Objectives that are specific, measurable, attainable, realistic, and timely.
12. Delaying or postponing a task or activity.
13. The act of directing, guiding, or motivating a person or group of people.
14. A student-developed project that involves hands-on learning in agriculture and natural resources.

Know and Understand

Answer the following questions using the information provided in this chapter.

1. Define leadership and list five characteristics of a leader.
2. What steps are involved in developing a leadership path?
3. What are three youth agricultural leadership organizations that offer a focus on horticulture?
4. What is the mission of the FFA?
5. What is the FFA motto?
6. What are the three levels in the FFA structure? What additional level is used in some states?
7. List important leadership skills and explain how the FFA helps develop leadership skills.
8. What are the officer titles of an FFA chapter?
9. List and describe the five degrees (awards) that an FFA member may earn.

10. A chapter's program of activities is organized to promote development of what three groups?
11. What are three examples of chapter development activities that might be part of a chapter's program of activities?
12. What are three examples of community development activities that might be part of a chapter's program of activities?
13. When is National FFA Week and what is its focus?
14. What is the purpose of the FFA chapter banquet?
15. What conventions are hosted by the FFA and what types of activities take place at the conventions?
16. Explain the purpose of a career development event (CDE) and describe a CDE for creed speaking.
17. Explain the importance of the supervised agricultural experience to FFA achievement.
18. What activities are part of developing an SAE project?
19. Describe briefly the duties and activities of an agricultural education teacher.
20. Describe the job of a Cooperative Extension Service agent.

Thinking Critically

1. A member of your FFA chapter has recently posted an inappropriate picture on Instagram. How would you handle this situation?
2. A large, national agricultural chemical supplier is sponsoring a career development event (CDE) team at your school. A member of your CDE team does not want to compete in the event because she does not support this company and believes that it is responsible for polluting the environment. How can you diffuse the situation and ensure that your teammate will compete?
3. Your agriculture teacher was injured recently and will not be back at school for the remainder of the school year. A substitute has been hired and the remainder of your FFA events for the year are up in the air. What could you do to ensure that your chapter is still successful and that everything on your chapter's program of activities is accomplished?

STEM and Academic Activities

1. **Math.** Use your SAE records and analyze your data. Using a program on your computer or the Internet, input all the hours and finance data for your SAE project. Create a statistical analysis to break down how much money was spent and how much money was gained. Analyze the number of hours that was spent helping others through service projects, at school, and in the community. Create a data analysis brochure or presentation to highlight your achievements.
2. **Social Science.** Gather all the agricultural education program's FFA participation information. Enter the FFA members' demographic information and analyze your chapter's population. Compare these demographics to your school's overall population. Do your chapter's demographics mirror the schools? If not, develop a plan to encourage that all students are included in your FFA chapter.

3. **Social Science.** Ask your agricultural education teacher if you can invite a local or state lawmaker to your classroom. Invite the person to visit the program and learn more about what you and your FFA chapter are doing. Write a letter and have your agriculture teacher read through your proposal. Be sure to include your contact information in case the lawmaker wants to arrange to visit your program.
4. **Social Science.** Contact a local 4-H club and ask them to visit your FFA chapter. Host that person or several of the club members at an FFA meeting. Ask them to introduce what they do at their club and what they have in common or how they differ from FFA. Ask the 4-H members if you may go to their club and inform them about your chapter's FFA and the agricultural education program.
5. **Language Arts.** Create a questionnaire for community members to determine their agricultural literacy. Be sure to sample all age groups and demographics of the community's population. With the surveyed responses, determine what topic in agriculture the community knows the least about. On what topic does the community need additional education? Create a program to increase the community's agricultural literacy on this topic.

Communicating about Horticulture

1. **Reading and Listening.** In small groups, discuss the main topics in the chapter. Ask questions of other group members to clarify concepts or terms as needed.
2. **Reading and Speaking.** Select a historical era that interests you. Using at least three resources, research the history of horticulture during that era and write a report. Include any major breakthroughs in horticulture and how they affected the industry. Present your report to the class using visuals, such as PowerPoint®.

SAE Opportunities

1. **Analytical.** Research your local or state government agricultural agencies. Write a speech about the contributions that these individuals make.
2. **Placement.** Intern with your agricultural education teacher.
3. **Analytical.** Research your state's agricultural history and create a book for children to educate them about agriculture.
4. **Research.** Analyze the SAE records of your chapter and create a poster or presentation that highlights your chapter's SAE data. Illustrate how many hours are completed by members in a year for SAE and what your chapter contributes to your local economy and community.
5. **Exploratory.** Job shadow an agricultural leader in your community or state.

USDA/David Kosling

CHAPTER 2
Experiential Learning: SAE

Chapter Outcomes

After studying this chapter, you will be able to:
- Understand the history and purpose of a supervised agricultural experience (SAE).
- Compare and contrast the different types of supervised agricultural experiences.
- Understand the process of investigating, planning, and keeping records for supervised agricultural experiences.
- Describe proficiency awards and recognitions associated with supervised agricultural experiences.
- Describe agricultural business manager and agricultural inspector careers.

Words to Know

agricultural literacy
Agricultural Proficiency Award
agriscience internship
apiculturist
biodiesel
career exploration
entrepreneurship SAE
exploratory SAE
improvement SAE
microgreen
placement SAE
research and experimentation SAE
service learning
Star Award
student interest survey
student resources inventory
supervised agricultural experience (SAE)
supplemental SAE
training agreement
training plan
vermicompost

Before You Read

As you read the chapter, record any questions that come to mind. Indicate where the answer to each question can be found: within the text, by asking your teacher, in another book, on the Internet, or by reflecting on your own knowledge and experiences. Pursue the answers to your questions.

While studying this chapter, look for the activity icon to:
- **Practice** vocabulary terms with Words to Know activities.
- **Expand** learning with identification activities.
- **Reinforce** what you learn by completing Know and Understand questions.

www.g-wlearning.com/agriculture

During your education, you have undoubtedly learned about the relationship of Pilgrims and Native Americans. Did you know that the Pilgrims owed a great deal of their agricultural accomplishments and survival to agricultural education? Squanto, a Native American translator, assisted the Pilgrims after their first winter in the New World. Squanto taught the Pilgrims how to fish and use the fish to fertilize the soil so that the crops would grow. This was most likely the first recorded instance of agricultural education in what is now the United States.

SAE and Agricultural Education

"Don't judge each day by the harvest you reap, but by the seeds that you plant."
—Robert Louis Stevenson

A supervised agricultural experience (SAE) is one of the three primary components of agricultural education, **Figure 2-1**. A *supervised agricultural experience (SAE)* is a student-developed project that involves hands-on learning in agriculture and natural resources. FFA and classroom instruction are the other two components of the agricultural education program.

Supervised agricultural experiences benefit students by allowing them to:
- Gain work experience.
- Explore agricultural industry careers.
- Develop skills in personal and financial recordkeeping.
- Experience differentiated and individualized instruction.
- Take part in community service.

SAEs are one of the most powerful educational tools of the agricultural education program. Students control their personal agricultural education experience and the outcome of their education. Each SAE is tailored to meet the specific objectives of each student.

SAE History

Near the time of the Civil War, Abraham Lincoln and the legislature passed the Morrill Act of 1862. This legislation established land-grant universities in every state of the United States. These institutions were founded to ensure agriculture was researched, studied, taught, and available to all American citizens. The Hatch Act of 1887 built upon the Morrill Act and established agricultural experiment stations. These stations acted as catalysts for agricultural education in public schools.

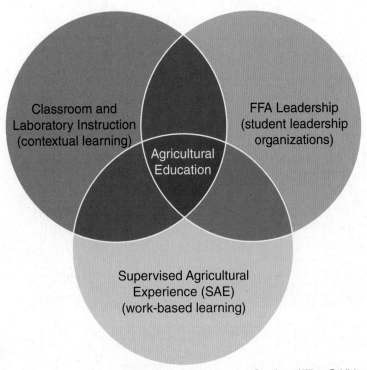

Goodheart-Willcox Publisher

Figure 2-1. The three interlocking circles represent the integral parts of the agricultural education program, which include SAE, FFA, and classroom instruction.

In the early 1900s, boys on farms across America practiced experiential learning. Agricultural students lived on farms and ranches and completed "home projects" that were entrepreneurial in nature and involved crops, livestock, and poultry. Student gardening was also encouraged in schools around the country and teachers used gardening as a medium for education.

Horticultural practices were also linked to consumer and family sciences. Girls' tomato canning clubs began to emerge, **Figure 2-2**. Club members were called "Tomato Club Girls," and ranged in ages from 10 to 20 years old. Each girl cultivated one-tenth of an acre of tomatoes and sold the fruit to community members. The girls also canned any additional fruit for later use. The Tomato Club Girls recorded data about their cultivation and yields into reports. These reports included hand illustrations and photographs, much like the reports that would be used for later supervised agricultural experiences.

USDA/National Archives and Records Administration

Figure 2-2. A canning club from 1925. Young girls were taught how to grow and preserve tomatoes. The club members sold their fresh tomatoes and preserves to earn money. **Have you ever preserved fresh fruits or vegetables?**

In 1917, the Smith-Hughes Act, formally named the National Vocational Education Act, established the nation's first vocational education programs. The Smith-Hughes legislation funded agriculture courses in high schools. Section 10 of the Smith-Hughes Act reads, "...that such schools shall provide for directed or supervised practice in agriculture, either on a farm provided for by the school or other farm, for at least six months per year."

The Smith-Hughes Act determined that supervised agricultural education was an integral part of the agricultural education program. Since 1917, the home-based projects have evolved into today's SAE. SAEs can involve entrepreneurship, analytical or research-based scientific experiments, and agricultural community service. As schools and student populations change, the SAE progresses to meet the needs of its school, students, community, and the agricultural industry.

SAE Purpose

The SAE aids learning outside regularly scheduled classroom sessions. *Supervised* means that the teacher and partners in the SAE should help direct the students. *Agricultural* indicates that the experience involves the food, fiber, and fuel industries. *Experience* is everything that is observed and completed. Together, these make up the SAE program, which includes project plans, activities, experiences, and records.

"The land-grant university system is being built on behalf of the people, who have invested in these public universities their hopes, their support, and their confidence."
—Abraham Lincoln

Corner Question

How many land grant universities are in the United States?

Figure 2-3. An aquaculture operation can make a good SAE project if you have the space for such an installation. Is there space in your school's greenhouse to set up an aquaculture operation?

SAEs also provide an opportunity for communities to contribute to students' educational experiences. Working with parents, teachers, employers, and community members offers students the opportunity to build partnerships. These partnerships may involve a local farmer allotting land for a student experiment, a horticulturist employing a student for seasonal work, a parent purchasing materials to install a pond at home, or a teacher making room in the school's greenhouse for a new aquaculture operation, **Figure 2-3**. All these examples are forms of partnerships between adults and students. In a productive SAE, the project must be driven by the student—not the parent, teacher, or member of the community with whom they have a partnership.

Throughout the course of the SAE process, you will gain valuable life skills and develop good work habits. These life skills will help you succeed throughout your formal education and compete in your future career. Some of the skills and habits you will develop through your SAE projects include:

- Problem solving.
- Analysis.
- Informational literacy.
- Recordkeeping.
- Following instructions.
- Personal habits.
- Work habits.
- Interpersonal relationships.
- Communication.
- Leadership.

Types of SAEs

An important part of your SAE is that it allows you to create your own learning experience. You can decide which activities to include as part of your overall objective. The six types of SAE are entrepreneurship, placement, research and experimentation, exploratory, improvement, and supplemental. You may also include an enterprise, scientific literacy, career discovery, community service, or agricultural skill enhancement in your SAE.

Entrepreneurship SAE

An *entrepreneurship SAE* is a hands-on learning project in which you operate a business and are responsible for all financial risks, **Figure 2-4**. This type of SAE allows you to own, operate, and manage a business in a professional and, hopefully, profitable manner. This type of SAE also gives you the opportunity to earn money while enhancing your agricultural education.

> "When everything seems to be going against you, remember that the airplane takes off against the wind, not with it."
> —Henry Ford

| Entrepreneurship SAEs ||||
Animal	Agribusiness	Crop	Natural Resources
Aquaculture production Bee production Livestock production Poultry production Small animal production	Animal boarding and care Animal feed service Artificial insemination services Custom farmwork Erosion control services Garden service Large engine service Lawn service Small engine service	Agronomic production Christmas tree production Floriculture production Forestry production Fruit production Herb production Mushroom production Nursery production Sod/Turfgrass production Vegetable production	Compost production Soil, water, and air conservation Vermicompost production

Goodheart-Willcox Publisher

Figure 2-4. There are any number of projects that would qualify as an entrepreneurship SAE. If you decide to have an entrepreneurship SAE, you will own and operate the business while assuming all financial risks. How would you determine if there is a market for your products or services?

In an entrepreneurship SAE, you may choose to manage a business in any area of the agriculture and natural resources industry. Examples of business activities are described below.

- **The production of crops for food, fiber, or beauty.** You may choose to become a producer and distributor of *microgreens* (recently germinated plants that are edible and used for food), shear alpaca for their fiber, or develop a cut flower business.
- **Raising livestock or poultry for meat and/or eggs.** You may choose to raise the animals conventionally or organically and develop a niche market product.
- **Converting diesel engines to operate on biodiesel.** *Biodiesel* is a type of fuel made from materials such as vegetable oils or animal fats. You could rebuild old diesel engines and modify them to work on biodiesel and then sell them for a profit.
- **Composting food and landscaping waste to produce soil amendments.** You may choose to start a worm composting venture and gather food scraps from local cafeterias or restaurants to feed the worms. *Vermicompost* is a type of compost in which worms, microbes, and bacteria turn organic matter into fertilizer that can be sold and used as a soil amendment, **Figure 2-5**.

Although most entrepreneurship SAEs involve raising and selling an agricultural commodity, it is not the only way to have an agribusiness venture. If you do not want to raise and sell an agricultural commodity but would like to have an entrepreneurship SAE, you could have an agribusiness in which you act only as the buyer and seller. In this type of SAE, you would purchase agricultural commodities and sell them for a profit. The commodities could include animals, animal products, crops, or services.

wawritto/Shutterstock.com

Figure 2-5. Vermicompost and the worms used to create it can be sold. What types of organic matter should be used to feed the worms? What types should *not* be used?

Thinking Green

Biodiesel

Biodiesel is a fuel manufactured from plants or animal fats that is a sustainable alternative to fossil fuels. Using biodiesel or a blend of biodiesel and conventional diesel fuel has several environmental benefits:

- It reduces greenhouse gas emissions because biodiesel produces less sulfur and carbon dioxide emissions than conventional diesel fuel.
- The emissions from biodiesel may be offset by the carbon dioxide absorbed by the plants grown to produce the fuel.
- Biodiesel produced from waste oils and animal fats prevents these materials from reaching landfills.
- It can be grown, produced, and distributed locally. This reduces transportation costs and the emissions and energy consumption associated with fuel transport.
- Biodiesel has an incredibly low toxicity. According to the EPA, biodiesel is less toxic than table salt and it biodegrades as quickly as sugar.
- As biodiesel production becomes less costly and more readily available, it will help reduce our dependence on fossil fuels.

chiqui/Shutterstock.com

Please note: Biodiesel is available at gas stations around the country and can be used in most vehicles with little or no fuel system alterations. However, it is important to refer to your owner's manual or contact the manufacturer to ensure the use of biodiesel will not damage your vehicle or void its warranty. Also note that oils used to produce biodiesel must be processed before use as a fuel.

Placement SAE

A *placement SAE* is a hands-on learning project in which you have a paid or unpaid internship in the agriculture and natural resources industry. It may also be referred to as an *internship SAE*. A placement SAE provides an opportunity to work for someone and gain experience and knowledge. Recordkeeping for a placement SAE includes tracking the number of hours worked, the amount of income earned, expenses incurred, and the skills acquired. A placement could take place after school, on weekends, or during a school break. Regardless of whether or not you are paid for your placement SAE, the experience of learning from someone else is always valuable.

A placement SAE could take place at a number of facilities such as a(n):

- Farm or ranch.
- Greenhouse or nursery.
- Agricultural company.
- Veterinary hospital or boarding facility.
- Florist.
- Pet store.
- Cooperative extension service, USDA, farm service, or forest service.
- Laboratory.

Corner Question

What is the difference between free-range hens and cage-free hens?

All these locations have professionals who are willing to help you gain skills that can be used for the rest of your life. Identify agricultural businesses in your community and consider contacting them for employment. If you have trouble compiling a list, consult with your teacher or search your state's department of agriculture website for agricultural businesses near you.

Once you have identified a possible placement site, you should let your teacher and parents or guardians know your intentions to pursue this as a site of employment. Your teacher will most likely give you what is called a *training agreement* or *training plan*, **Figure 2-6**. This document is created by your teacher or state FFA to help guide you through the placement

---

SAE Training Plan Placement (Internship)

Student Name: _____
                      Last   First   Middle

Date of Birth: ____/____/____   Phone Number: (____) ____-_____

Student Address: _____
                    Street   City   Zip

Name of Placement (Employer): _____

Name of Supervisor: _____

Employer Address: _____
                   Street   City   Zip

Employer Telephone: _____

Starting Date of Employment: ____/____/____   Ending Date of Employment: ____/____/____

Initial Meetings with Employer, Adviser, and Student

Interview Date: ____/____/____

Wages:

Expectations of Employer (Date: ____/____/____):

Expectations of Student (Date: ____/____/____):

Expectations of Adviser (Date: ____/____/____):
1. Be timely
2. Be courteous
3. Ask questions

Student's Goals for Internship (Date: ____/____/____):
1.
2.
3.

Student's Objectives (Date: ____/____/____):
1.
2.
3.
4.
5.
6.
7.
8.
9.
10.

_____ Date ____/____/____
                Student Signature

_____ Date ____/____/____
                Adviser Signature

_____ Date ____/____/____
                Employer Signature

Goodheart-Willcox Publisher

Figure 2-6. Your training agreement is a useful tool for keeping your SAE on track.

Career Connection

Beekeeping

Steve Oehlenschlager/Shutterstock.com

You can raise bees and rent their hives to landowners for pollination services. **Apiculturists** (people who study and maintain bees) are paid to bring the bees to the site and maintain the hives. Farmers can negotiate for honey, or the honey can be gathered solely by the apiculturist. Bees flourish, crops are pollinated, and honey is produced. Everyone wins with this green venture. It is best to learn from an experienced apiculturist. Find one near you, and you could have a great placement SAE.

Alice Welch/USDA

Figure 2-7. Some schools allow students to work as interns to fulfill their agriscience internship. *What types of intern positions does your school offer?*

"Experience is the past tense of experiment."
—Gregory Alan Elliot

process. The training plan is a signed contract that helps you, your teacher, your parents or guardians, and the employers understand the objectives and goals of the SAE placement. You must discuss the expectations, wages or volunteer status, and a list of learning objectives with all parties involved in order to complete the training agreement. Once this document is filled out, the employer may continue with the interviewing and hiring process.

Your teacher may offer an opportunity for SAE placement. An *agriscience internship* is a job placement working in the school's agricultural education program, **Figure 2-7**. This type of placement SAE exposes you to the professional duties of an agricultural education teacher. In this program, you may help take care of the classroom, laboratory, greenhouse, garden, and animal facilities. Your teacher works with you to help you develop the skills needed to be an exemplary agriculture teacher. You may work before school, during free periods, or after school to complete your tasks. You may also have the opportunity to present a lesson or work with younger students to learn and practice presentation and teaching skills.

Research and Experimentation SAE

Have you ever wondered what type of wood is best for manufacturing a longboard skateboard or what type of turf is best for playing lacrosse? Are you interested in the role moon cycles play in the reproduction rates of insects or the impact of certain chemicals on plant growth? Would you like to determine the best way to market your school's plants or livestock? The answers to these questions can be answered through

Thinking Green

Efficiently Tracking Your SAE Hours

Although modern chargers are designed to draw little or no power when your phone or tablet are not connected and charging, some still use a standby mode that draws a small amount of power. This small amount of power is often referred to as a *parasitic drain*. If you are using a phone or tablet to track your SAE hours, you can conserve energy by unplugging your phone, tablet, or computer cords when your device is not being charged.

Unplugging the cords may save between 10 to 15 cents per cord per month. This may not seem significant, but imagine how much energy could be saved on a daily basis if you and your friends and neighbors all made it a habit to unplug cords when they are not in use. Every little savings counts, and together they can add up to a substantial amount of energy savings.

Take your energy savings a step further and investigate other devices in your home that may be needlessly draining energy. How can you reduce or eliminate parasitic drain in your home?

agricultural research and could all qualify as a research and experimentation SAE. A *research and experimentation SAE* is a hands-on learning project in which you conduct research or use the scientific method to solve a problem related to agriculture, **Figure 2-8**.

Are you a curious person? Do you have questions that need answers? Do you find yourself searching the Internet looking for facts to prove your friends wrong? If you answered yes to one of these questions, then research and experimentation is your SAE answer. Scientific experimentation requires you to pose a question followed by research and/or experiments. You must record data, analyze the information, and illustrate conclusions and recommendations for future experiments similar to what was performed.

A research and experimentation SAE may also be used as the basis for an agriscience fair project. Agriscience fairs are held at local, state, and national levels and qualifying projects compete at the National FFA Convention® each fall. The FFA recognizes students who are studying scientific principles and technologies while conducting research projects. Some student experiments may prove enough validity or stir enough interest that professionals in the agricultural industry or academia may further the study.

A
Kaponia Aliaksei/Shutterstock.com

B
Chris Hill/Shutterstock.com

Figure 2-8. A—Researching and determining what the best type of wood to make a skateboard from qualifies as agriscience research. B—Another project that qualifies as agriscience research is determining what type of grass should be used for a lacrosse field and/or performing experiments to see what improves the field's conditions.

Corner Question

Who was one of the founding scientists of plant genetics?

Agricultural research and experimentation can take place anywhere. Research can be conducted at schools, homes, farms, jobs, or research facilities in industry. This original research can be another way to build partnerships in the community. Scientists at your high school, local businesses, or colleges and universities may be willing to mentor you as you do this research and help you find the equipment and facility you need.

Acquiring research skills is important for many careers. Employers look for individuals who are able to do research, solve problems, and think critically while on the job.

Exploratory SAE

The *exploratory SAE* is a hands-on learning project in which you explore agricultural careers and subjects. You may investigate topics in agriculture by job shadowing and gathering information through reading and viewing experiences. This SAE will help you gain agricultural literacy. *Agricultural literacy* is having the knowledge necessary to synthesize, analyze, and communicate basic information about agriculture.

AgEd Connection: Agriscience Fair Categories

There are six major agriscience fair categories from which to choose:

- **Animal Systems (AS)**—Projects in this category include the study of life processes, health, nutrition, genetics, management, and processing. Subjects may include the study of small animals, livestock, dairy, equines, aquaculture, and poultry.
- **Environmental Services/Natural Resource Systems (ENR)**—Projects in this category include the study of systems, instruments, and technology used in waste management; and the study of the management of soil, water, wildlife, forests, and air as natural resources and their influence on the environment.
- **Food Products and Processing Systems (FPP)**—Projects in this category include the study of product development, quality assurance, food safety, production, sales and service, regulation and compliance, and food service within the food science industry.
- **Plant Systems (PS)**—Projects in this category include the study of plant life cycles, classifications, functions, structures, reproduction, media, and nutrients, as well as growth and cultural practices, through the study of crops, turf grass, trees and shrubs and/or ornamental plants.
- **Power, Structural, and Technical Systems (PST)**—Projects in this category include the study of agricultural equipment, power systems, alternative fuel sources, and precision technology, as well as woodworking, metalworking, welding, and project planning for agricultural structures.
- **Social Systems (SS)**—Projects in this category include the study of human behavior and the interaction of individuals in and to society, including agricultural education, agribusiness economic, agricultural communication, agricultural leadership, and other social science applications in agriculture, food, and natural resources.

Visit the National FFA Organization's website for more information and ideas for your agriscience fair project.

Exploratory SAEs are best for students who wish to gain an overall understanding or appreciation of the agriculture and natural resources industry, **Figure 2-9**. An exploratory SAE will give you a sample of the many aspects of agriculture. This can be accomplished through experiencing an agricultural festival or fair, visiting an agricultural or scientific museum, or traveling to a farm or ranch.

Exploratory SAEs also provide career exploration. *Career exploration* is the investigation of occupations. An exploratory SAE will allow you to job shadow or spend time with professionals while they work. You will see what the professional does and thus better determine whether the career interests you and whether you would like to continue studying the occupation. You may also decide that the job is not one you want to pursue.

Improvement SAE

An *improvement SAE* is a hands-on learning project in which you take something related to agriculture and improve it. Creating a new landscape, installing fencing around a garden, building a trail at a local park, assembling compost bins, or organizing financial records of a local business are all examples of improvement SAEs. An improvement SAE encompasses completing a series of steps to finalize a task. This process can take several hours to several days, weeks, or months to complete. Completing an improvement SAE will help you achieve a sense of accomplishment.

You may have a project in mind but do not have the incentive or know-how to complete it. Maybe there is an old tractor on your property or someone has one nearby, **Figure 2-10**. You could do some research or enlist

USDA

Figure 2-9. With an exploratory SAE, students can explore careers, such as a green roof technician. *What types of plants would thrive in a rooftop garden?*

dvande/Shutterstock.com

Figure 2-10. Fixing up an old tractor is an improvement SAE. You could use the tractor for work, or you could sell it and use the proceeds to fund another SAE.

> "Without continual growth and progress, such words as improvement, achievement, and success have no meaning."
> —Benjamin Franklin

the help of someone knowledgeable to refurbish the tractor. There may be a park in your neighborhood that is full of trash and has equipment that needs repair. You could rally support from the community and renovate the park to make it beautiful and useful for the neighborhood.

An improvement SAE is a wonderful way to incorporate service learning into your experience. *Service learning* is a strategy that integrates community service with instruction and reflection to enrich the learning experience, teach civic responsibility, and strengthen communities. It is also referred to as *community engagement*.

Supplemental SAE

A *supplemental SAE* is a hands-on learning project that enhances agricultural skills and knowledge and takes less than eight hours to complete. This type of SAE (also called supplementary) is normally taught in agricultural education classes and, like other SAEs, involves learning by doing. This project is not related to your main SAE, but is an addition to your overall learning program, **Figure 2-11**.

The content of supplemental SAEs is defined by what is taught in your agricultural education program. Supplemental SAEs could include repairing a fence or gate of a pasture or garden, changing the oil on a piece of equipment, grafting tomatoes, or using a piece of landscaping or farm equipment. It is important to record the hours and understand their impact on the general SAE program. Several supplemental SAEs may be accomplished during the course of an SAE program.

Saklakova/Shutterstock.com

Figure 2-11. Building or staining a fence qualifies as a supplemental SAE. What types of community-based activities would qualify as a supplemental SAE?

The SAE Program Process

When you are first introduced to the SAE program, it can seem overwhelming. Figuring out what an SAE is, understanding the different types of SAEs, trying to determine the top SAE choice for you, taking photos, and understanding finances may seem to be too much for one project. The SAE project may seem more manageable if you look at it as a series of steps. Start by investigating possible SAEs, **Figure 2-12**. Then move to the planning step, thinking about your goals and resources. Consider how you will coordinate your activities with teachers, parents or guardians, and sponsors or employers. Explore methods for keeping records and documenting your experiences. Later on, consider applying for awards. Each of these steps is discussed in more detail in this chapter.

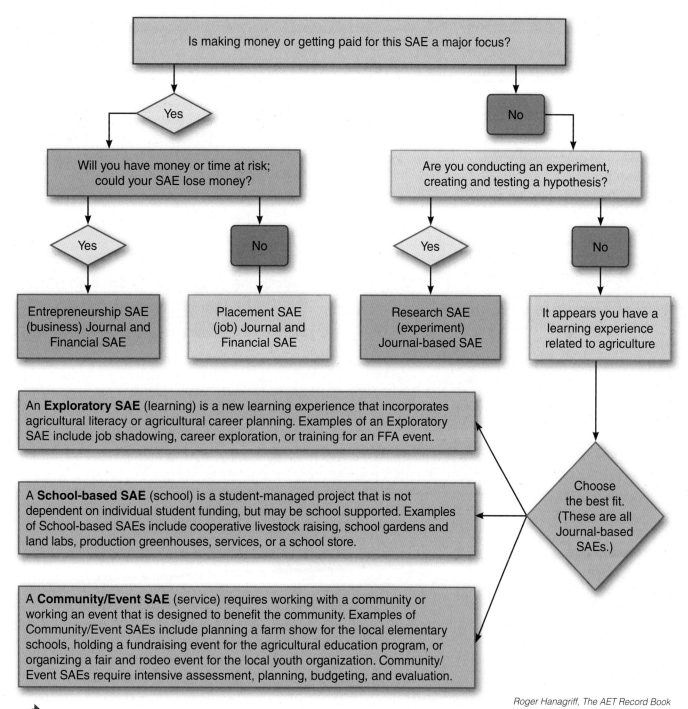

Roger Hanagriff, The AET Record Book

Figure 2-12. If you are having difficulty deciding on an SAE, this decision tree can help you. To use the decision tree, begin answering the questions from the top down and follow the arrows. Complete this exercise for each of your SAE project ideas.

Think about each step separately and plan time to focus on your SAE. You may choose to work on your SAE one day per week or maybe devote a few hours every day to the project. Remember that there is not just one type of project that fits the needs of every student. SAEs can work with everyone's schedule and everyone can have a successful SAE. Make the SAE work for you. Whether you live on a farm or in an apartment, you can develop an SAE to fit your needs.

Corner Question

Can tomatoes and potatoes grow on the same plant?

AgEd Connection: Agricultural Communications

An agricultural communications career development event tests students' skills in all areas of the agricultural communications field. It assesses how well students relate classroom knowledge to relative situations. Participants join a replicated news conference and apply the information collected to complete individual practicums in writing, electronic media, and design. Before the event, students gather media plans related to pioneering agricultural practices, management techniques, and marketing instruments. Each team then produces a 15-minute presentation based on their application. Members also participate in an editing exercise and a general communications quiz.

Bob Nichols/USDA

> "In much of society, research means to investigate something you do not know or understand."
> —Neil Armstrong

Investigate

One of the most crucial steps of the SAE process is the exploration of the project as a whole. What types of SAE projects are out there? What is best for you? Which project will fit your personality and lifestyle?

The first step is to determine your agricultural interests, **Figure 2-13**. No matter who you are, there is something in the agriculture and natural resources industry that can connect to your interests. For your SAE project to be enjoyable and worthy of your time, you must be interested and have a genuine investment in the project. An interest survey can help you to explore your likes and dislikes and what you find appealing.

Complete a *student interest survey*, a questionnaire that will help you identify your interests in agricultural education and the SAE project that would be best suited for you, **Figure 2-14**. Once you have identified your interests, then you can explore the projects that align with your needs as an agricultural education student and as an individual.

USDA

Figure 2-13. Are you interested in forestry? This student is using an increment borer to further investigate a tree for her forestry SAE project. She is considering a career in forestry and is interested in this plant science. This SAE project will help mold her future career aspirations.

Student Interest Survey

General Area	Focus	Level of Interest	Level of Experience
Apiculture (beekeeping)	Honey		
Apiculture (beekeeping)	Hive rental		
Aquaculture (fish)	Hatchery		
Aquaculture (fish)	Harvest		
Aquaculture (plants)	Stock production		
Aquaculture (plants)	Harvest/sell		
Beef cattle	Showing		
Beef cattle	Meat production		
Chickens (broilers)	Meat production		
Chickens (layers)	Egg production		
Companion animals	Breeding		
Companion animals	Service (grooming, walking, daycare)		
Dairy cattle	Showing		
Dairy cattle	Milk and milk by-product production		
Equine	Showing		
Equine	Breeding		
Equine	Training		
Floral design	Arranging/selling		
Flower gardens	Installation and maintenance		
Fruit trees	Planting/grafting		
Fruit trees	Harvesting/selling		
Goats	Showing		
Goats	Meat production		
Goats	Milk and milk by-product production		
Greenhouse production	Produce		
Greenhouse production	Ornamental		
Landscaping	Installation/maintenance		
Lawn care	Maintenance		
Marketing/selling	Products or services		
Mechanics	Refurbishing/maintenance		

(Continued)

Goodheart-Willcox Publisher

Figure 2-14. Your teacher may provide you with a student interest survey or direct you to an online version. Would this type of survey help you decide which college or university is right for you?

Student Interest Survey (Figure 2-14, continued)			
Nontraditional animal production	Meat production		
Nontraditional animal production	By-product production		
Recordkeeping	Financial or other data		
Sheep	Showing/breeding		
Sheep	Meat or milk production		
Small animal production (cavies)	Breeding/selling		
Soils	Testing/improvement		
Swine	Showing/breeding		
Swine	Meat production		
Tree production	Planting/grafting		
Turkeys	Meat/egg production		
Vegetable gardening	Selling/canning		
Welding	Construction/repair service		
Woodworking	Designing/building/repairing		
1. Using the numbers 1–4, rate the areas by your level of interest. (1 for no interest, 4 for highly interested)			
2. Using the numbers 1–4, rate your level of experience in each area. (1 for no experience, 4 for extensive experience)			
Take your top five areas of interest and look into possible SAE projects.			

Did You Know?

According to the National Gardening Association, 43 million Americans garden for food in the United States. Food gardening households spend five hours a week tending to their gardens.

There are eight divisions or systems of the agriculture and natural resources industry that can be explored by the SAE project. As illustrated in the student interest survey, plants, animals, environmental systems, natural resources, agricultural leadership, agribusiness, power systems, and food systems all provide opportunities for further study. Once you have determined which of the eight systems best matches your interests, you can begin to plan your SAE project.

Plan

Once you have identified an SAE area that matches your interests, you may begin to plan for SAE success. One of the first elements to determine is what resources you have available.

Student Resources Inventory

An SAE *student resources inventory* is a questionnaire that will help you identify the tools and supplies that you have or can use for an SAE. The questionnaire includes information about your home, school, work opportunities, and community resources, **Figure 2-15**. Once the SAE student resources inventory is completed, you can see what you have and what you will need to begin and complete your SAE project.

SAE Resources Inventory

Resource	Level of Availability	Location/ownership	Need to Purchase
Academic eligibility			
Arable land			
Family support			
Farm equipment			
Feed (grains, hay)			
Financing			
Garden plot			
Gardening tools			
Hand tools			
Implement trailer			
Livestock facilities			
Livestock trailer			
Mentor			
Positive attitude			
Power tools			
Seedstock			
Time			
Transportation			
Vehicle			
Wood or other construction supplies			
Work ethic			
Workshop			

Questions to Consider

Crop Production. Is land available for you to rent or use for crop production? Who owns the land? How many acres are available? Where is the land located? Will you be able to use the owner's equipment? Do you need insurance?

Livestock Production. Are facilities available for you to rent or use to produce livestock? Who owns the facilities? How much space is available? Where are the facilities located? Do you need insurance?

Gardening. Do you have space for a garden? Who owns the property? Will the land need to be prepared?

Indicate the availability of each resource (home, from a friend or relative, need to purchase). Rate the availability of resources to conduct your SAE program. (1 for available, 2 for limited availability, 3 for not available)

Goodheart-Willcox Publisher

Figure 2-15. Review the SAE resources inventory example illustrated here. Can you think of additional questions or areas of concern?

Thinking Green

Steps to Sustainability

You can make SMART goals to help the environment. For example, your SMART goal could be to decrease your use of plastics. This goal could be accomplished by using a reusable water bottle 100% of the time for this year.

How much of an impact would this simple change have on the environment? Currently, only one of every six water bottles we use makes it to the recycling bin. Where do you think the other five bottles end up?

Vladimir Gjorgiev/Shutterstock.com

Goals

"Whatever words we utter should be chosen with care for people will hear them and be influenced by them for good or ill."
—Buddha

Now that you have identified your resources, you can begin making a list of goals for your SAE project. Creating a goal can be easy, but it must be SMART. As you learned in chapter one, a SMART goal is an objective that is specific, measureable, attainable, realistic, and timely. An example goal would be, "I will provide 20 hours of community service at the homeless shelter by the end of this school year." This goal is definitive, and you can easily determine whether or not you accomplish this goal.

For your SAE project, work with your agricultural education teacher to come up with a set of SMART goals specific to your project. Depending on your project and your agricultural education program at your school, your teacher may ask you to write three to five goals for a semester, year, or the entire course of your project, **Figure 2-16**. Review these goals often to help you stay on track and work toward achieving them.

Coordinate

Although you are the person planning the SAE, setting the goals, and doing the work for the project, several other people may have a role in your SAE. You must coordinate communication among several people for your SAE project. These people may include:

- Teachers.
- Employers or customers.
- Parents or guardians.
- Other individuals who have an interest in your SAE project.

Bob Nichols/USDA

Figure 2-16. A student, such as an FFA officer, may have a goal of giving a speech to an audience. This student, a graduating senior at the Chicago High School for Agricultural Science, gave a speech at her graduation.

Thinking Green

Reusing Calendars

Calendars can be reused. Every so often, calendar dates are replicated. You can save calendars and use them again.

2017: 2006 1995 1989 1978 1967 1961 1950 1939 1933 1922 1911 1905
2018: 2007 2001 1990 1979 1973 1962 1951 1945 1934 1923 1917 1906 1900
2019: 2013 2002 1991 1985 1974 1963 1957 1946 1935 1929 1918 1907 1901

Corner Question

What is the oldest calendar?

You may decide to communicate by making a phone call, sending a text or e-mail message, or talking with others in person. Take time to read and reread texts or e-mail messages before you send them. It is never a good idea to rely on the autocorrect feature and hope that software will catch all your mistakes. You do not want to find yourself in a strange situation with an adult, employer, or customer because of a strange text that you sent. Be aware that not all individuals communicate by texts or e-mails. Ask the adults that you are working with which method of communication they prefer.

Communicating schedules, dates, and times accurately is important, **Figure 2-17**. You need to be aware of deadlines, and members of your SAE team may need to know them as well. Consider using shared calendars on the Internet or sending a copy of your calendar for partners in your SAE project.

tukkata/Shutterstock.com

Figure 2-17. Keeping track of dates is essential for your SAE project. How will you track dates and other important information for your SAE project?

Keep Records

A record is something that gives information or evidence about the past. Records have value for historical, legal, financial, or other reasons. You will need to keep records during your school years and later as an adult. School transcripts, contracts for loans for college, paychecks, bank account statements, tax returns, bills, and receipts are examples of records you may need to keep. As you create and receive more records, you will need to develop a system for managing all this data.

Thinking Green

Saving Power with Computers

Most new computers come with a sleep mode or power management feature. ENERGY STAR®, a government program that rates appliances for energy efficiency, estimates that using these features will keep an extra $30 a year in your wallet. Make sure you have the power-down feature set up on your PC through your operating system software. Power management features are not usually enabled when a computer is purchased and must be enabled by the user.

Did You Know?
The earliest form of human speech took place more than 100,000 years ago.

mallory_mcdevitt1/USDA

Figure 2-18. A student grows organic crops on her one-acre farm in Ohio for her SAE project. She must keep accurate photos and records for her SAE project.

Safety Note

Texting While Driving

When using a mobile device to upload pictures into an SAE software or recordkeeping system, it is important to only do so when appropriate and safe. Never use a mobile device while driving or operating other equipment.

One of your goals for the SAE project should be to develop an effective system for recordkeeping. During your SAE project, your agricultural education teacher may require you to use a specific method of recordkeeping. Teachers often use online systems for records or may have software for records designed for their specific school. Some schools may still use notebooks, pen, and paper for records. No matter what your method, a major part of the SAE project is keeping records. Some of the information that you will record will be:

- Dates and hours for work completed.
- Journal of what was done for the SAE.
- Finance data.
- Inventory data.
- SAE training agreements with partner signatures.
- Goals achieved and lessons learned.
- Photographs, **Figure 2-18**.
- Receipts.

Record your SAE activities immediately after you do something so that you do not forget any piece of information. Many schools are now using The Agricultural Experience Tracker (AET) program. This software allows students to enter data, photographs, and information about FFA activities. The AET is just one example of recordkeeping systems. It can be accessed by computer or mobile device for easy SAE recordkeeping. Students learn a number of skills through work with the SAE that will help them later in life, and recordkeeping is definitely one of them.

SAE Awards and Recognitions

Once your SAE takes flight, then you can begin to apply for awards and recognition.

Agricultural Proficiency Awards

An *Agricultural Proficiency Award* is a prize or recognition given for an exemplary SAE project, **Figure 2-19**. Planning to apply for these awards can aid in defining the focus of SAE projects, establishing goals, and compiling or maintaining records for award applications. These awards can be in the form of plaques, cash, or trips to places around the world. Agricultural Proficiency Awards are available to FFA members in good standing who are enrolled in an agricultural education program and have an SAE project.

There are four categories of SAE programs and nearly 50 proficiency award application areas. The four categories of SAE programs are:

- **Exploratory**—gain agricultural literacy and career knowledge in agriculture and natural resources.
- **Agriscience research and experimentation**—plan and conduct scientific experiments or investigate a question (includes qualitative, quantitative, experimental, descriptive, and quasi-experimental research).

FFA Proficiency Awards	
Agricultural Communications	Equine Science—Entrepreneurship
Agricultural Education	Equine Science—Placement
Agricultural Mechanics Design and Fabrication	Fiber and/or Oil Crop Production
Agricultural Mechanics Energy Systems	Food Science and Technology
Agricultural Mechanics Repair and Maintenance—Entrepreneurship	Forage Production
	Forest Management and Products
Agricultural Mechanics Repair and Maintenance—Placement	Fruit Production
	Goat Production
Agricultural Processing	Grain Production—Entrepreneurship
Agricultural Sales—Entrepreneurship	Grain Production—Placement
Agricultural Sales—Placement	Home and/or Community Development
Agricultural Services	Landscape Management
Agriscience Research—Animal Systems	Nursery Operations
Agriscience Research—Integrated Systems	Outdoor Recreation
Agriscience Research—Plant Systems	Poultry Production
Beef Production—Entrepreneurship	Sheep Production
Beef Production—Placement	Small Animal Production and Care
Dairy Production—Entrepreneurship	Specialty Animal Production
Dairy Production—Placement	Specialty Crop Production
Diversified Agricultural Production	Swine Production—Entrepreneurship
Diversified Crop Production—Entrepreneurship	Swine Production—Placement
Diversified Crop Production—Placement	Turf Grass Management
Diversified Horticulture	Vegetable Production
Diversified Livestock Production	Veterinary Science
Emerging Agricultural Technology	Wildlife Production and Management
Environmental Science and Natural Resources	

Figure 2-19. One of the most common goals for SAE projects is to use them to compete and win a proficiency award. Proficiency awards are given to members that excel in their SAE programs. The list of official proficiency award areas changes on an annual basis, but the list above can be used as a guideline for determining which proficiency category might be right for you.

- **Entrepreneurship**—operate your own enterprise.
- **Placement**—work for an agriculture or natural resource business or individual (paid or unpaid).

When applying for the proficiency award recognition, the focus (categories) will determine the correct proficiency award area in which to apply. Read about the proficiency award areas on the National FFA Organization website and check with your FFA advisor or state FFA to determine when proficiency award applications are due.

Star Awards

Star Awards are prizes and recognitions given to those students who have exemplary SAE projects and are earning an FFA degree. The National FFA Organization® recognizes FFA members who participate in all facets of FFA. Students are recognized as "stars" because of their dedication and exemplary work in all areas of the agricultural education program.

Did You Know?
Nearly 900 million personal computers and 4 million cell phones are in use in the world.

Jodi Riedel/ Goodheart-Willcox Publisher

Figure 2-20. A student who qualifies for the Greenhand degree and has a strong SAE program may also be awarded the Star Greenhand.

Star Awards at the chapter level are distributed at chapter banquets or awards assemblies and are accompanied by a plaque or medal. These awards include:

- **Star Discovery Award.** A student who is in seventh or eighth grade can be awarded the Discovery Degree and the Star Discovery Award. This member demonstrates outstanding leadership.
- **Star Greenhand.** A student who qualifies for the Greenhand degree and has a strong SAE program while also demonstrating superior leadership may be awarded the Star Greenhand, **Figure 2-20**.
- **Chapter Star Farmer.** The student with the best production agriculture SAE and FFA commitment may be awarded the Chapter Star Farmer.
- **Chapter Star in Agricultural Placement.** A student with an outstanding placement SAE may be awarded the Chapter Star in Agricultural Placement.
- **Chapter Star in Agriscience.** A student who has an outstanding SAE in the area of natural resources or agriscience and who is an active member of the FFA may be awarded the Chapter Star in Agriscience.

The State Star Awards are given to FFA members who are earning the State FFA Degree and an Agricultural Proficiency Award. The State Star Awards include:

- Farmer.
- Agribusiness.
- Agricultural Placement.
- Agriscience.

These awards are presented at the state's FFA convention. Members who earn these prestigious awards are also given cash awards.

Each year, sixteen FFA members are recognized for the American Star Awards. The Star Awards that are given to four deserving American Degree award recipients are the highest honor awarded. These awards are given each October at the National FFA Convention. The American Star Farmer, Star Agribusiness, Star Agricultural Placement, and Star Agriscience are distinguished awards that are accompanied by $4000 cash awards. Finalists (three runners-up) are given plaques and $2000 each.

Did You Know?

The American FFA Degree recipients earn and productively invest a total of nearly $103 million through their supervised agricultural experience programs each year.

Careers

The agricultural industry is a driving force behind the American economy. More than 18% of Americans have jobs related to agriculture. With more than 23 million Americans working in the food and fiber industry, there is an occupation that can meet the needs of almost any individual. Two of these job positions are discussed in this chapter: agricultural business manager and agricultural inspector.

Thinking Green

Trophy Recycling Program

Every year, trophies, plaques, and medals are thrown into landfills. However, your FFA chapter can make a difference by purchasing refurbished awards or donating old awards. Trophy shops will simply put on new lettering and make the gently used awards seem like new. This results in keeping awards out of landfills and saves resources needed to manufacture new ones.

focal point/Shutterstock.com

Corner Question

How much do FFA members earn annually from their SAE projects?

Agricultural Business Manager

Agricultural business managers supervise the business operations of a farm, ranch, or other production operation by providing leadership during the production process, **Figure 2-21**. From contracting crop insurance to selecting crops for the planting season to buying greenhouse equipment, it is their responsibility to certify that the production and distribution of produce, grain, or livestock complies with government and environmental regulations while also making a profit.

Agricultural business managers may have a number of duties, including hiring and supervising workers, preparing a budget, organizing regular maintenance, keeping records, and communicating with potential customers. They usually specialize in crops, horticulture, or livestock and may oversee more than one facility.

Lance Cheung/USDA

Figure 2-21. One of a vineyard manager's responsibilities is to put markings on grapevines. What other types of responsibilities does a vineyard manager have?

Agricultural business managers often have farm experience. However, a person in this position may or may not be from a farming background. Agricultural business managers often continue their education and earn a four-year degree in agricultural business, agronomy, or economics. Completing an internship or apprenticeship with a qualified and experienced farmer provides valuable training for an agricultural business manager.

Agricultural Inspector

Agricultural inspectors ensure that agricultural entities obey all government regulations. These workers may review and audit forestry, fishing, or farming operations to ensure that all products and practices conform to safety and health regulations. Agricultural inspectors are accountable for ensuring meat safety and often review meat at

Career Connection

Alicia Rittenhouse
Vice President and Manager for Strategic Engagement at AmericanHort

Alicia Rittenhouse is the vice president and manager for strategic engagement at AmericanHort. She has a horticultural business degree and focuses on branding and marketing of horticultural businesses. She has worked for AmericanHort since college. AmericanHort's focus is on educating, collaborating, researching, and advocating for the horticulture industry. AmericanHort was founded in 2014 upon the merger of the American Nursery and Landscape Association and the Ohio Florist Association (OFA).

Alicia also plays a major role in the HortScholars program. The program educates and exposes horticulture students to facets of the horticulture industry that are not usually experienced during college course work. The program focuses on professional development, which includes networking, education sessions, and working with an industry mentor. The program is open to two- and four-year college undergraduate and graduate students. The HortScholars program takes place at an annual conference in Columbus, Ohio, called Cultivate. More than 10,000 horticulturists gather at this convention setting.

As a horticulture enthusiast in college, Alicia travelled to Chile and studied viticulture and was active in her horticulture club. Today, Alicia enjoys building relationships with young horticulturists and professionals in the industry. She has a passion for advocacy of horticulture and uses her leadership skills to promote AmericanHort to the public and the horticulture industry.

meat-processing facilities, **Figure 2-22**. Agricultural inspectors may also check shipments before they depart or arrive in US ports and cross US borders. Daily duties may include:

- Collecting samples from animals, plants, or products.
- Transporting samples to labs for testing.
- Examining employees working in contact with agricultural products.
- Producing reports on their findings at a particular site.
- Preparing health recommendations to farmers, growers, or authoritative organizations.

There are two ways of preparing to become an agricultural inspector—through education or experience. Agricultural inspectors may have a bachelor's degree in an agricultural science, or they may have relevant work experience combined with some additional course work or training after high school.

Anson Eaglin/USDA

Figure 2-22. A USDA agricultural inspector monitors poultry that is being exported.

CHAPTER 2 Review and Assessment

Chapter Summary

- Experiential (hands-on) learning has been used since the first settlers arrived in what is now the United States. However, it was not until the early 1900s that a formalized standard for agricultural education was put in place.
- The supervised agricultural experience (SAE) program uses hands-on learning in agriculture and natural resources.
- SAEs fall into one of six areas that include entrepreneurship, placement, research and experimentation, exploratory, improvement, or supplemental.
- The SAE process includes investigating SAE opportunities, planning the overall SAE project and experience, coordinating resources, recordkeeping, applying for awards, and sustaining the project through high school or beyond.
- SAE records should be kept throughout the duration of the project and should include all facets of the project, such as money, hours, lessons learned, photographs, and general data related to the project.
- SAE records can be kept through traditional methods or online with computer programs.
- Two types of awards that are available to students for their SAE projects are Agricultural Proficiency Awards and Star Awards. These awards are presented through the chapter, state, and National FFA Organization.

Words to Know

Match the key terms from the chapter to the correct definition.

A. agricultural literacy
B. Agricultural Proficiency Award
C. agriscience internship
D. apiculturist
E. career exploration
F. entrepreneurship SAE
G. exploratory SAE
H. improvement SAE
I. microgreen
J. placement SAE
K. research and experimentation SAE
L. service learning
M. Star Award
N. student interest survey
O. student resources inventory
P. supervised agricultural experience (SAE)
Q. supplemental SAE
R. training agreement
S. vermicompost

1. Having the knowledge necessary to synthesize, analyze, and communicate basic information about agriculture.
2. A job placement working in a school's agricultural education program.
3. A hands-on learning project in which the student makes something related to agriculture better in some manner.
4. A hands-on learning project in which a student has an internship that is paid or unpaid within the agriculture and natural resources industry.
5. A hands-on learning project in which the student operates a business and is responsible for all financial risks.
6. A questionnaire that helps students identify tools, supplies, and other resources they have access to for an SAE project.
7. A signed contract that helps the student, teacher, parents or guardians, and employers understand the objectives and goals of the SAE placement.
8. Prizes and recognitions given to a student who has an exemplary SAE project and is earning an FFA degree.
9. A type of compost in which worms as well as microbes and bacteria are used to turn organic matter into fertilizer.
10. A prize or recognition given by FFA for an exemplary SAE project.
11. The investigation of occupations.
12. A person who studies and maintains bees.
13. A recently germinated plant (sprout) that is edible and used for food.
14. A hands-on learning project in which the student explores agricultural careers and subjects.
15. A strategy that integrates community service with instruction and reflection to enrich the learning experience, teach civic responsibility, and strengthen communities.

16. A hands-on learning project that enhances agricultural skills and knowledge and takes less than eight hours to complete.
17. A hands-on learning project in which the student conducts research or uses the scientific method to solve a problem related to agriculture.
18. A questionnaire that helps a student identify his or her interests in agricultural education and the type of SAE project that would be best suited for that student.
19. A student-developed project that involves hands-on learning in agriculture and natural resources.

Know and Understand

Answer the following questions using the information provided in this chapter.

1. What are the three components of agricultural education?
2. What are five ways in which supervised agricultural experiences provide benefits to students?
3. What three laws discussed in the chapter played a part in establishing agricultural education in today's schools?
4. What is the purpose of a supervised agricultural experience?
5. What are the six types of SAE projects?
6. What is an agriscience internship?
7. Why is career exploration important?
8. What is service learning?
9. What are the steps in the SAE process and how can a student begin an SAE without feeling overwhelmed?
10. What is a SMART goal?
11. What is a record and what are three examples of records most people keep?
12. What are some types of information or records you should keep related to an SAE?
13. Describe how SAE records can be maintained.
14. What are the four categories of SAE programs that can be awarded Agricultural Proficiency Awards?
15. List and describe the Star Awards that can be given at the FFA chapter level.
16. Who may earn a State Star Award and what are the four types of awards at the state level?
17. How much money is given to the National FFA Star Award winners?
18. How much money is awarded to the National Star Award finalists?
19. What education or training does an agricultural business manager need?
20. What are some of the daily duties of an agricultural inspector and what education or training is needed to be an agricultural inspector?

Thinking Critically

1. A local farm manager has more than 50 employees that speak several different languages. Some of the employees have very limited English proficiency. What could this farm manager do to ensure that all his employees are effectively trained and understand safety and protocol on the farm?
2. Your teacher wants to take a field trip to a local farm. The farm has very strict biosecurity measures and will not allow anyone from a farm to enter the facility for fear that they will contaminate the farm with soil, insects, and diseases from other farms. Three students in your class live on farms. What should your teacher do?

STEM and Academic Activities

1. **Engineering.** You or your teacher should identify one structural problem in your classroom, greenhouse, laboratory, barn, farm, or other site around the school. Engineer a solution for this problem. As part of your SAE project, you can design the solution to the problem and then fix the problem. You may be able to do some of the work yourself. Other work may require a trained professional, such as a plumber or electrician.
2. **Math.** Count the number of students in your classroom. Imagine that your teacher is their manager and must create a budget for paying for their work. Each student will get paid $12 an hour for the work in your class. How many hours a day does each of the students in your class work? How many hours in a week? Next, determine how many hours the students will be in class for the school year. How much would your teacher have to pay for the entire class to work for the school year? Do you think receiving money would be better than getting a grade?
3. **Social Science.** Create a video infomercial about your school's SAE and FFA programs. Take video footage of your school's FFA program and interview some of the students. Include coverage of your school's SAE program. The video should be three minutes long. Use video editing software available through your school or online. Share this video with your local FFA and SAE sponsors. Discuss any necessary image release forms with your teacher and obtain signatures as needed.
4. **Social Science.** Develop a form to be used by teachers and students to evaluate the success of an SAE project.
5. **Language Arts.** Write a one-page paper about the land-grant college or university in your state. (You may have more than one.) Research when the school was founded and compare this to the date of the Morrill Act that established land-grant colleges and universities. What role did the Morrill Act play in the foundation of your land-grant college or university?
6. **Language Arts.** Contact a farm manager in your community and arrange to interview this employer. Ask what educational and work history he or she has to prepare for this occupation. Ask the manager questions about what she or he likes or dislikes about the job. Find out more information about how he or she manages people and the farm.

Communicating about Horticulture

1. **Writing and Speaking.** Interview someone local who works in an agricultural field. For example, you could interview a nursery manager/owner, park ranger, or poultry producer. Choose an area you are interested in and/or with which you are not familiar. Ask the person to describe a typical day at work. Prepare a list of questions similar to the following: How long have you been in the _____ industry? Did you go to school? Did you work as an intern? What is the work environment like? What are your job duties? What other types of professionals do you work with? Report your findings to the class, giving reasons why you would or would not want to pursue a career similar to that of the person you interviewed. (Do not forget to send a note thanking the person for their time and help.)

2. **Reading and Writing.** Select an agricultural product, and then determine how you could improve upon it to create a niche market. Write a product description outlining the comparative advantage that your new product would have.

3. **Writing and Speaking.** Using the product you created in question 2, create a 5- to 10-minute presentation to pitch your idea to potential investors, similar to television shows where business hopefuls share their ideas in order to secure money from wealthy investors. Be prepared to share your presentation with the class.

SAE Opportunities

1. **Exploratory.** Job shadow a farm, ranch, or other production facility manager.

2. **Experimental.** Research how to grow microgreens. Grow a few different varieties and perform a taste test with students, staff at your school, family, and community members. Analyze the results and present the information to a local restaurant owner who is using or wants to use microgreens.

michaeljung/Shutterstock.com

3. **Entrepreneurship.** Purchase some red wiggler worms. Start with one bin of worms and feed them your kitchen scraps. Harvest the worm castings and sell them to gardeners. Harvest some of the worms to sell to local fisherman. Add more bins and watch your business grow. Contact the school cafeteria to ask for more food scraps to feed your worms.

4. **Analytical.** Research the Colony Collapse Disorder (CCD) of bees. Create a presentation about your findings and communicate with a local beekeeper or beekeeping organization. Team up with a beekeeper and present your information to local children.

5. **Placement.** Contact your local farm bureau. Set up a paid or unpaid internship. Learn about agricultural communication and the numerous other jobs at the Farm Bureau office.

CHAPTER 3
Communication and Information Literacy in Horticulture

Chapter Outcomes
After studying this chapter, you will be able to:
- Understand the writing process and create effective written documents.
- Demonstrate critical thinking and conduct research.
- Recognize plagiarism and develop methods of documentation.
- Prepare effective presentations and visual aids.
- Describe information literacy.
- List careers related to horticulture and communication.

Words to Know ↪

annotate	demographics	thesis
annotated bibliography	evidence	tone
audience	information literacy	voice
bibliography	paraphrase	works cited
communication	plagiarism	
critical thinking	skim	

Before You Read
Before you read the chapter, interview someone in the workforce (your supervisor, a parent, relative, or friend). Ask the person why it is important to know about communication and information literacy and how this topic affects the workplace. Take notes during the interview. As you read the chapter, highlight the items from your notes that are discussed in the chapter.

While studying this chapter, look for the activity icon **to:**

- **Practice** vocabulary terms with Words to Know activities.
- **Expand** learning with identification activities.
- **Reinforce** what you learn by completing Know and Understand questions.

www.g-wlearning.com/agriculture

Emily J_01/USDA; Bob Nichols/USDA; Wassana Mathipikhai/Shutterstock.com; wavebreakmedia/Shutterstock.com; Bob Nichols/USDA

What separates society today from that of the cavemen? There are many differences, but one of the first that may come to mind is communication. Cavemen had no cellular devices and could not even speak a formal language. Cave dwellers did have drawings and symbols painted on walls of caves as early as 40,000 years ago, **Figure 3-1**. In the twenty-first century, people all over the world are using technology to aid in communication. People are able to connect with one another regardless of time, geographical location, or language.

No matter what the language, to communicate successfully there must be a transfer of sounds, words, and/or images. *Communication* is a process in which a message is sent by one person and received and understood by another person. People communicate using speech or other sounds and using written words and images. People also communicate using nonverbal behavior. Some people are fearless public speakers, others use writing as a means to share ideas, and some create art or music to express their thoughts or feelings to others.

Written Communication

Writing is a part of everyday life. You should leave high school with the communication skills needed to compete in the workplace. Strong writing skills will make future education and training in the horticulture industry easier, **Figure 3-2**. Language skills are also important to developing information literacy. *Information literacy* is the ability to recognize when information is needed and to find, evaluate, use, and communicate information.

> "Either write something worth reading or do something worth writing."
> —Benjamin Franklin

Corner Question
Did you know that cavemen in the Stone Age communicated through symbols just as we do today?

Arpi Miskaryan/Shutterstock.com

Figure 3-1. The cave dwellers communicated with symbols and drawings. Perhaps some of these symbols are still used in today's society. **Do these look like some of the emoticons you use?**

62 Horticulture Today

Copyright Goodheart-Willcox Co., Inc.

Thinking Green

Recycling Paper

Nearly 69 million tons of paper and paperboard are produced in the United States every year according to the Environmental Protection Agency. Approximately 37% of this paper comes from recycled sources. Remember to purchase recycled paper products, and recycle all the paper products that you use, including that used for writing.

Have you participated in paper recycling events at your school or in your community? Does your school have a recycling program?

bikeriderlondon/Shutterstock.com

Many students are insecure about their writing ability and fear critiques of their written work. Other students are comfortable with writing or even enjoy the process. As with other skills, writing skills can be improved with practice and instruction. The writing process involves a series of steps. Some steps, such as evaluating and revising your work, may be repeated during the process until you are satisfied with the finished work. The steps in the writing process include:

1. Determine the purpose of the message.
2. Consider your audience.
3. Select a means of presentation for the message.
4. Write a thesis, or refine the topic and identify the main points of the message.
5. Create a plan for the project and the message. Write the message according to the plan.
6. Review, revise, and proofread the message.

Purpose

The first step in the writing process is to determine the purpose of the message. In general, the purpose of a message is to inform the reader (or listener) about a topic or issue, to persuade the reader to take some action, or to entertain the reader. This chapter discusses the first two purposes, which are typically those used for writing in horticulture. First, identify which of these general goals apply to the message. Next, determine what topic you wish to provide information about or what action you want readers to take. What is the goal of the message? What do you want to accomplish with the message? Identify your reason for writing and you will have a better feeling about the assignment in general. If you have trouble identifying the purpose for writing, talk to the person who assigned you

Bob Nichols/USDA

Figure 3-2. A student is practicing language skills. This contributes to her overall literacy. **How do you practice your language skills on a daily basis?**

the task to gain a better understanding about the project. The purposes of horticultural writing are to:

- Explain a scientific process.
- Educate others about a horticultural topic.
- Report on your findings from a horticultural experiment or research.
- Broadcast information about a topic of interest.
- Persuade others to believe in a horticultural practice or product.

The purpose of your writing directs the way you will deliver your final written message or oral presentation. If the purpose of the message is to inform readers, use a direct approach. Present the main points of the message first and follow with supporting details. If the purpose of the message is to persuade readers to take some action, it may be better to begin by discussing related questions or issues and lead up to the main point of the message.

Audience

After you have identified a goal or purpose for your message, identify the audience for the message. The *audience* consists of the people who will read, hear, or see the message. You need to identify the audience so you can tailor your message to meet the needs or interest of the audience. If you are presenting information about daylilies to a garden club, you can use higher levels of scientific vocabulary than when presenting to a general audience. When presenting that same topic to an elementary school garden club, you will need to use terms that the students will understand. A paper or presentation must be appropriate for the audience receiving the message.

You will need to consider several demographic variables involving your audience. *Demographics* are characteristics or traits of a group, **Figure 3-3**.

"No occupation is so delightful to me as the culture of the earth, and no culture comparable to that of the garden...But though an old man, I am but a young gardener."
—Thomas Jefferson

Bob Nichols/USDA

Figure 3-3. These FFA® officers from all 50 states are part of an audience in Washington, DC. What demographics considerations do you think the speaker made when preparing his speech?

Demographics that you may need to consider include:
- Age and gender.
- Cultural backgrounds.
- Expertise related to the topic.
- Physical and mental capabilities.

You may ask yourself why the physical capabilities of your audience should be considered. When writing horticultural messages, your audience may include those who have limited mobility. Understanding their limits can help you adapt your writing to meet their needs. For example, the instructions for doing gardening tasks may be modified for those with limited mobility.

Corner Question

Who was an author of the US Constitution and a famous garden author?

Presentation

Presentation is the manner in which a message is sent to the audience. What form should the message take? Does the author need to create a printed document, such as a letter, pamphlet, or poster? Should an oral presentation given live or by video be used to deliver the message? What communication media will the author use? In some cases, the author may need to create a formal paper document. In other situations, a blog, website, webinar, or digital presentation created with a software program may be used to communicate words, images, and sounds to the audience, **Figure 3-4**.

Presentation also involves the way authors speak through their writing or speeches. Authors should consider the voice and tone used in their messages. *Voice* is the style of expression or degree of formality used in a message. Does the language need to be formal or informal? To answer this question, consider the purpose of the message and your audience. An article written for a scientific journal should probably use formal language. However, an article about horticulture written for fifth-grade students should probably use informal language. For instance, using pronouns such as *you* and speaking directly to the reader can make the writing informal. In scientific work or formal writing (often academic or industry-related), the author should refrain from using pronouns and write in a more objective manner. *Tone* is the quality of a message that reflects the writer's attitude or mood. Is the message serious or light-hearted, casual or official? Consider the difference in tone between, "A planning meeting will be held to discuss this issue on April 6 at 2 pm" and "Come hang out with us on Friday afternoon to celebrate wrapping up this project." The first example sounds serious and official; the second example uses a light-hearted, casual tone.

Goodheart-Willcox Publisher; monticello/Shutterstock.com

Figure 3-4. A blog is an excellent way to reach an audience interested in a particular topic. How is creating a blog entry different from communicating messages in other ways?

Topic

The topic is the subject of a message. Selecting a topic may not be an issue if the writing assignment is about a specific subject or task. If you are asked to select a topic related to horticulture, you may need to consider several topics and select one. If you are given a broad topic to write about, such as growing vegetables, you may need to select a specific area of the subject. This will allow you to discuss the topic in some detail while keeping the message a reasonable length. In this example, for instance, you might write about growing peas. Activities you can do to help you select a topic include:

- Explore the Internet using a search engine.
- Skim through magazines for images or words to spark an idea.
- Read blog postings on the topic.
- Look at images on image-sharing websites.
- Investigate television channels, especially those that feature gardening or animal shows.
- Ask for suggestions from friends, family, or someone in the horticulture industry.
- Create a concept map related to a topic of interest, **Figure 3-5**.

A concept map is a graphic or diagram used to generate and organize ideas or facts. Start with one idea or word in a circle or "bubble." From that word, draw a line to another bubble and write words or lists related to the main idea. Branch off the main idea and other bubbles, writing freely until you have no more related thoughts. You can create your concept manually using pen and paper. You can also use programs available on the Internet that

Did You Know?
Nobel chemists and physics award winners are also writers. Scientists of all specialties are also writers. Scientists create reports and research papers to convey the findings of their scientific work. These papers help to document their investigations and express to the public what was learned.

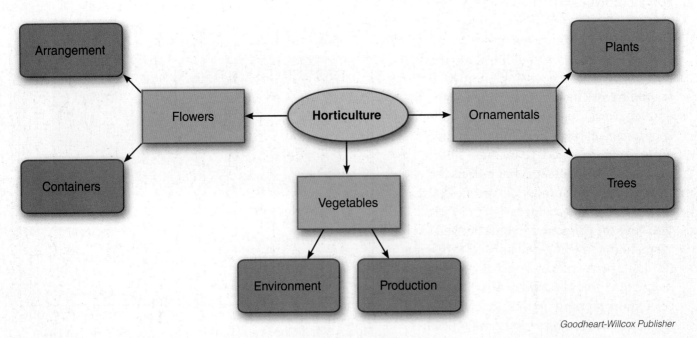

Figure 3-5. If you cannot come up with an idea for writing, try creating a concept map. Which of the activities listed do you think will be most helpful in narrowing down a topic choice?

allow you to create a concept map. This exercise may help you identify or narrow a topic for your message. A concept map can also help you identify the main ideas and supporting details for a message.

Thesis

You may be asked to express the topic of a message as a thesis statement. A *thesis* is a statement or theory that is proposed and then discussed to prove or disprove it. Developing a thesis requires time and critical thinking. The thesis can be a way of letting the reader know the position of the author. Frequently, the thesis responds to a question posed by the author. For example: Are roses the only flower appropriate to give someone you love on Valentine's Day? The response is the following thesis: There is an endless assortment of flowers that can be given to show affection or symbolize romance besides roses.

A thesis should be clear and specific. Many young or novice writers select a topic that is too broad. The resulting thesis statement seems vague and may lead to writing something longer than is appropriate for the assignment. Focus your topic question and the thesis to answer the question. The outcome will be more defined writing that is clear, concise, and specific. Once the thesis is developed, you can begin to research and find evidence to help prove or disprove the statement, **Figure 3-6**. As you begin researching, you may find facts or evidence that may make you rethink your thesis altogether. Be open to the facts or expert opinions found in your research and adjust your message if needed.

> **Corner Question**
>
> How many publications are printed in the United States each year?

Gunnar Pippel/Shutterstock.com

Figure 3-6. Once you have a thesis that is clear and focused, you can begin research to prove or disprove the statement. **What are some methods for researching your thesis?**

Project Plan

Writing a major research project or even a short essay can seem overwhelming to some students. In Chapter 2, you learned how thinking of a supervised agricultural experience (SAE) as a series of steps could make the project seem simpler. The same idea applies to a research paper, essay, oral presentation, or other writing project. A project plan, sometimes called a *plan of attack*, can help you organize and schedule the tasks needed to complete the project. Begin the plan by identifying all the tasks you need to do. For example, you may need to narrow the topic or write a thesis statement, do research, form conclusions, and write the message. You will also need to evaluate and possibly revise the message. You will need to decide the best way to present the message (written paper, oral presentation, or web page) unless that was stated in the assignment. After you have identified all the tasks, create a schedule. Working backward from the date the project must be completed, identify dates for completing each step in the process.

> **Corner Question**
>
> How many roses are produced for Valentine's Day? How many of those are red?

A discovery and research period will follow selecting a topic or writing of the thesis. This process should include a great deal of analysis and critical thinking (discussed later in the chapter) about information and evidence that you find. Upon completion of the research, you should create a plan for your paper. Identify the main points you want to present and arrange them in a logical order. Select supporting details for each main point. You may wish to use a topic outline or an essay map for this purpose. The message should begin by introducing the topic and giving the thesis statement if one is used. Follow the thesis or topic introduction with a brief preview of the main points that will be discussed. The introduction lays out a plan for the body paragraphs. Statements or facts in the introduction set up the skeleton or framework for the body of the document.

Think of the message plan as a list of supporting topics to your thesis or topic statement. Recall the earlier example thesis, "There is an endless assortment of flowers that can be given to show affection or symbolize romance besides roses." The following statements might be, "Instead of roses, consumers can purchase flowers that represent the qualities of love. These qualities can include patience, remembrance, faithfulness, and compassion. Each of these qualities has a specific flower associated with the trait."

In this example, the paragraphs in the body of the message will discuss a certain flower and its association with love. One, two, or three paragraphs may be used to support the thesis for each flower. This should convince the reader that this evidence supports the thesis and that there are indeed valid flower alternatives to roses for Valentine's Day.

The closing of the message should state conclusions or recommendations that can be drawn from considering the body of the message. A few key points from the message body may be repeated in the closing, perhaps using slightly different language. If the purpose of the message is to persuade the audience, a call to take some action should be part of the closing.

Reviewing, Revising, and Proofreading

Authors must pay close attention to every aspect of their writing, **Figure 3-7**. A message that is unclear or that contains errors may not achieve your goals for the message. A poorly written message may also leave the audience with an unfavorable impression of the writer. Once you have written a first draft of the message, set it aside for a day or two (or at least a few hours). Taking a little time away from the project will make reviewing, revising, and proofreading easier. Review your message to see whether you have accomplished the purpose of the project and to see if the message is complete, clear, correct, and concise. Revise (make changes to) the message to correct any problems you find.

Wassana Mathipikhai/Shutterstock.com

Figure 3-7. Always check for spelling errors. Proofreading is essential to professional publications. What other types of errors should you look for when reviewing your writing?

Safety Note

Pesticides

Always follow instructions on the label of a pesticide. Every chemical pesticide sprayed on plants in a greenhouse must have a label. The label is a legal and binding document that must be *read* and *followed* by the user. Be mindful of all safety considerations when using pesticides. What safety precautions do you see in this image?

wellphoto/Shuttestock.com

Corner Question

Can reading aloud to a plant increase its growth? Why or why not?

As your review your work, try to see it from the point of view of the audience. Reading the message aloud may be helpful in finding errors or omissions. Mark changes within the copy (text) and in the margins. You may find it helpful to print or write the message with one or more spaces between the lines or with larger margins on the pages. This will allow more room to write notes.

Do not be afraid to have another person critique your work. Ask someone you trust who is a good writer (a friend, family member, or teacher) to review your work and make suggestions. Remember that someone who reads your work wants you to improve and has your best interests in mind. Take the comments in stride and make changes that will improve the message.

After you are satisfied with the content and organization of the message, proofread to find any errors you may have made in spelling, punctuation, or grammar. In addition to proofreading, use the tools available with your computer software. The spell-check and grammar check features can help you find errors. However, you should not rely solely on the software. After proofreading, make corrections for a final draft of the message. Once you think everything is correct, read carefully one last time before you submit the document.

Critical Thinking and Research

Today's horticulture industry demands professionals who can think critically, **Figure 3-8**. *Critical thinking* is using objective reasoning or consideration before forming a judgment or taking some action. Critically thinking can lead to better decisions and innovative solutions to problems.

wavebreakmedia/Shutterstock.com

Figure 3-8. A horticulturist must use critical thinking when making decisions about how to grow plants. What are some examples of when a horticulturist may need to use objective reasoning?

Thinking Green

Organic Chemical Herbicides

Some herbicides are not made from synthetic chemicals, and they can be used as an organic control to weeds. Citrus oils are one type of chemical weed control that is listed as organic. The oils are toxic and burn the foliage of weeds or unwanted plants.

Antonova Anna/Shutterstock.com

When researching a topic, critical thinkers answer questions using a series of steps. These steps include questioning, discussing or considering, and analyzing information. Horticulturists cannot accept everything presented to them as fact or at face value. Suppose someone told a gullible greenhouse grower that herbicides work to kill insects in the greenhouse and that they should be sprayed all over the plants. Following these directions, the grower would kill all the plants in the greenhouse. (Herbicides kill plants, not insects.) As you have probably learned, just because someone presents information as fact does not mean that it is factual.

To be a successful horticulturist, you must analyze information you research and review the information critically. To successfully research information and form your own educated opinions, you must:

- Search for information.
- Preview search results.
- Find evidence and determine whether sources are credible.
- Annotate documents with questions and comments.
- Summarize the main points of information.
- Evaluate information to determine whether it is credible and factual.

Search

When you have chosen a topic to research or need to find the answers to a question, begin by thinking critically as soon as you begin a search for information. You may search for information using Internet search engines, magazines, newspapers, books, web forums, blogs, and databases. Searching can be a daunting experience. There is so much information available that you may not know where to begin. Start by using search terms or words that are as specific or detailed as possible. Broaden your search terms only if you do not find the resources you were hoping to find. The broader your search parameters, the more time you will need to spend sorting through all the matches that are found.

Preview

Once you have begun to search, the second step is sort and preview search results. When the search engine returns a list of hundreds, thousands, or millions of search matches, you must decide which items from your search results are worth further reading and examination. Try using a table of contents or a summary to help shorten the amount of time you spend determining if this resource needs a more detailed exploration.

Skimming the work can help you determine what the overall text is about and whether to use the source. To *skim* means to read selected parts of a text looking for the main ideas. Skimming allows readers to get the general idea or purpose of a passage quickly without reading every word. A good method for skimming is to read the introductory paragraphs fully, then read any headings and only the first sentence or two of each paragraph after the introduction. Look over the rest of each paragraph to identify key terms, names, or dates. The last one or two paragraphs are usually a summary of the document. Read the entire summary. Skimming works well for nonfictional or informational reading. Skimming helps you save time and determine whether the material is appropriate for further research.

Find Evidence

After you have gathered sources of information, begin finding evidence. *Evidence* is information about a topic indicating whether a belief or position is correct or valid. You will use the evidence to answer a question or to prove or disprove a thesis. Search for graphs, charts, and images that help illustrate information. Search for quotes that relate to the topic or thesis of the message, **Figure 3-9**.

As you examine evidence, you must also determine if the sources you have chosen are credible and reliable. Information can come from a number of sources, some of which are reputable and credible and others that are not. Government documents and educational publications from universities and colleges can usually be considered safe. News agencies, such as the Associated Press or National Public Radio, and the Public Broadcasting Service (television), are also good sources for current events and research.

USDA

Figure 3-9. Including data and graphics helps add credibility to a publication. How might the information in this image look if it were all text copy without graphics?

> **Corner Question**
>
> What movie by famed author and director Orson Wells debuted in 1973 and focused on forgery?

It is important to check and recheck your sources. When evaluating sources, do an Internet search to learn more about the document, organization, or author. Ask questions such as these to help you evaluate a source:

- Is the information relevant to the topic or thesis?
- Is the article or information current? When was it written?
- What is the purpose of the document or article?
- Does the author or organization have a bias or vested interest that may influence the information included in or omitted from the article?
- Are both/all sides of the information presented?
- Does the information include facts or opinions?
- Is the author(s) an expert in the subject area?
- Are conclusions or recommendations supported by credible evidence?
- Are there contradictions in the information?
- Have accepted standards been used in conducting or reporting research?
- Can you find the same information in other credible sources?

There are a number of reputable horticultural and trade journals. These publications include *Greenhouse Grower*, *Grower Talks*, *American Nurseryman*, *HortScience*, *Journal of the Society of American Horticultural Science*, and *HortTechnology*.

Some sources may seem to be legitimate. However, upon further investigation you may find that their "scientific" research does not adhere to many standards. The authors or organizations may want to influence readers, perhaps to buy a product or support a position or candidate. Other sources may not be what they first appear to be. For example, the website *The Onion*® appears to be a news website. However, it is actually a satirical news station that is entirely for entertainment purposes. None of the stories are true. If you did not research this site and took its reports as facts for your research, you could be tricked and your work would contain incorrect information.

Annotate

To *annotate* means to write questions or comments about a document while reading it. When reading information, question what you are reading. Create a list of points, questions, and assumptions as you read. What is the author's viewpoint? What facts or data is presented by the author? Is this really fact or opinion? Answer these questions and you are one step further in the critical-thinking process. You may find it helpful to create notes or annotations in the margins or on a separate page to use when you review the work later. Some computer programs will allow you to add notes or comments to a document.

Summarize

To summarize means to identify the main points of a document or passage. A summary should include the main points made by the author; however, you should state the points in your own words. If the information is very technical, use the same language as in the article. The summary should be concise without minor details. A summary will help you answer questions about a topic or provide evidence to prove or disprove a thesis.

Evaluate

After identifying the main ideas of information, evaluate or analyze the text further. To evaluate means to determine the worth, quality, or significance of something. Just as you evaluate the sources of information, you should also consider questions about the information itself:

- Is the information relevant to the topic or thesis?
- Is the information current?
- What is the purpose of the document or article?
- Is the information biased?
- Are both/all sides of an issue presented?
- Are conclusions or recommendations supported by credible evidence?
- Are there contradictions in the information?
- Does the information agree with other credible sources?

The evaluation of the information found in your research will assist you in proving or disproving your thesis or proving reliable information about a topic. Use information that you find to be credible to support your arguments. Evidence can include images or graphs, charts, and tables, **Figure 3-10**. Often, your audience will be more likely to accept your arguments or conclusions when your writing includes some information from other authors.

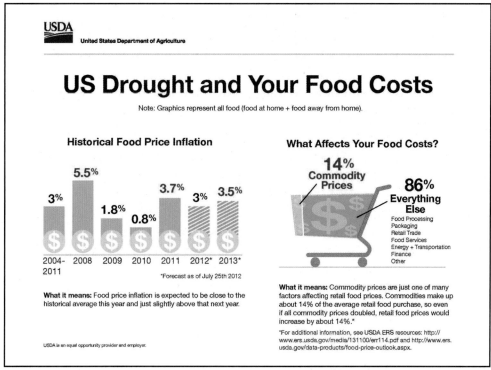

USDA

Figure 3-10. Evidence can include charts and graphs that will enrich your technical writing. **Why does including statistics or graphical data help enhance your arguments or conclusions?**

> **Corner Question**
> What female student was accused of cheating while studying at the Perkins Institute for the Blind?

Plagiarism and Documentation

Writers who are using others' ideas, words, information, and visuals should cite where the information came from and the original creator. As soon as you begin researching information, document the sources you use. Writers, regardless of experience level, must be very careful not to create a document that contains information created by others that has not been cited. Citing a source means telling exactly where the information came from (such as a book, website, or journal article) and who the author is.

Plagiarism

When doing research online, it may be tempting to cut and paste even a small part of another's work. Using work (such as ideas, writing, or images) created by someone else without permission and presenting it as one's own work is *plagiarism*. Plagiarizing is stealing. Whenever you use someone else's work of any kind (words, images, music) without acknowledging their work as the original, this is a form of robbery. Citing who originally made the work, where it is from, and when it originated makes the researcher or current author more credible to the audience.

Follow these guidelines when doing research for information:
- Do not wait until the last minute to start your writing. If you are less panicked, you will not make careless or unintentional plagiarism mistakes.
- Record all sources of information you use.
- Create an *annotated bibliography* (a document that lists citations for sources used and briefly describes each work).
- Never sequence or organize your paper the same way as the author you have researched.
- Do not simply use synonyms of words used in an original work. Your work must be entirely unique or else you must cite the original author to avoid plagiarism.
- Do not cut and paste an author's work into your work. This is dishonest and can lead to disciplinary consequences. (An exception to this guideline is when you will identify the sentence or short paragraph as a direct quote from another source.)

Citing sources gives other authors credit and makes your work appear more credible. Readers will more likely think that your work has depth and that you have taken the time to substantiate your claims if you cite sources carefully. The reader may also appreciate being able to look into the sources that you have cited to further their knowledge.

Plagiarism offenses can include:
- Intentional copying of other work.
- Using passages from other work.
- Copying another work and making a few word changes.
- Using others' ideas.
- Not making clear where an idea from another author ends and your work begins.

- Using images, data tables, graphs, and other visuals without citing the source.
- Not having a work cited.

Plagiarism has real consequences that can include disciplinary action in school or the workplace. As a student, receiving a zero on an assignment or in the course may seem improbable. However, software programs are available that make it easy for instructors or employers to check work for plagiarism. You can search your own work for plagiarism on any level using a tool such as Turnitin, a web-based writing assessment tool. This can help to prevent even an unintentional, but equally punishable, instance of plagiarism. Plagiarism is no different from any other form of cheating, **Figure 3-11**.

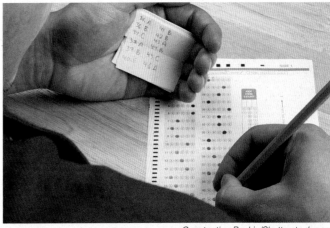

Constantine Pankin/Shutterstock.com

Figure 3-11. Plagiarizing is as serious as any other form of cheating or stealing. How is plagiarizing similar to cheating?

Outside of academics, professionals must be vigilant so as not to plagiarize. Plagiarism of any work can lead to costly lawsuits or loss of employment. Plagiarism is no joke. Stealing is a crime, whether it is words, images, or ideas that are stolen.

Documentation

Citing sources helps to avoid the question of plagiarism. Citing sources gives your reader the essential information about your sources. Who, what, when, and where this information is from is included in every citation. However, there is no need to cite:

- Information that is common knowledge.
- Birth and death dates.
- Events or dates in history.

Keep track of all sources that you use and make a document to help you record your sources and the information that you used. Create a working bibliography with software. This can help you save, organize, label, and comment on excerpts and information from the Internet. Create an annotated bibliography with the same type of software. Simply enter the source information and then include a summary or the essential information that you should remember about that source.

You can *paraphrase* the information for an annotated bibliography. Paraphrasing is using your own words, sentences, and organization to summarize something written (or spoken) by someone else. To paraphrase:

- State the information in your own words and use your own sentence format. (It helps to put the document you are paraphrasing out of your line of vision, forcing you to rephrase.)
- Tell readers what the author presents in a shortened interpretation.
- Cite the author.
- Cite sources and format styles.

bikeriderlondon/Shutterstock.com

Figure 3-12. Some professors are part of the Modern Language Association. What do you think the qualifications are to be a part of the Modern Language Association?

Brief citations should be included within the document. Generally, citations follow the referenced information with the author's name and the page number where it was found (if available). This type of citation follows the guidelines of the Modern Language Association (MLA) format for internal citations, **Figure 3-12**. This is the foremost professional association for researchers in language and literature.

There are countless variables when citing sources. There may be one or more authors. The work may or may not have page numbers. Citation formats vary with the source being cited. When in doubt, consult a reference guide such as the *MLA Handbook for Writers of Research Papers*. This handbook can be found online and in print. This source can help answer your questions regarding citations.

At the end of the paper, a *bibliography*, also called *works cited*, should be used. A bibliography or works cited is a list of all sources used in the work. The full bibliographical reference should include information such as the name of the work, author, and publisher as well as the publication date and page number. Only the documents that are cited within the paper should be included in the works cited. When creating the works cited, remember to:

- List works in alphabetical order by author's last name.
- Use the appropriate information and format for each type of source (book, media, website).
- Consult an authority on MLA citation formats such as the MLA website or use the *MLA Handbook for Writers of Research Papers*.

In addition to MLA, other organizations, such as the American Psychological Association (APA), offer guidelines for writing and formatting references. The APA formatting style is generally used for social science papers. Regardless of the style you use for citations, follow the guidelines carefully for both internal citations and the works cited page.

Presentation Methods

A presentation is a speech or lecture given to an audience, either live or by video. The audience may be a small group or a large one. For some speeches, the presenter may already have a working knowledge of a topic, and there is no need to do research or write a detailed paper before giving a presentation. A simple outline to remind the presenter of the main points of the speech may be all that is needed. For other topics, presenters may need

Corner Question

How many members are in the Modern Language Association?

to do research before giving a presentation. Preparing a speech that requires research and writing is very similar to writing a paper. You should follow the steps of the writing process that you learned earlier:

1. Identify clearly the purpose of the message. Will it be to inform or persuade the listeners?
2. Consider your audience. (Review the information about demographics presented earlier.)
3. Select a means of presentation for the message. Will it be delivered live or via video?
4. Refine the topic, if needed, and identify the main points of the message.
5. Create a plan for the project and the message. Write the message according to the plan.
6. Review, revise, and proofread the message.

In addition to writing the message, you will need to practice delivering the message. You may want to open the message with a question, a quote, or an anecdote to get the attention of the listeners. It is a good idea to memorize this short opening. For the remainder of the presentation, speak from your knowledge of the subject. Study the information you have found in your research or the paper you wrote until you know it well. Then use an outline (on notecards or electronic slides) to list the main points you want to cover. Practice the speech several times until you can talk about each main point, filling in details or giving examples as needed.

You may also want to prepare or organize visual aids to use in the presentation. A visual aid is anything you show the audience to help them understand the message. A garden tool or plant could be a visual aid. Electronic slides that you project onto a screen can be helpful visual aids. You can use text that gives the main point you are discussing or show pictures, charts, or graphs to illustrate a point.

Additional suggestions that you may find helpful for a presentation include:

- Know your audience. Complete an audience analysis.
- Get the audience's attention. Use a "hook." (Hooks can be a shocking statistic, humor, or a quote.) Be sure anecdotes or quotes are appropriate, relevant to the topic, and in good taste.
- Establish a motive. Let the listeners know why it is important for them to hear your presentation. How will they benefit?
- Get to the point. For a message presented in direct order, tell your audience your thesis or topic statement right away. This will help get them interested.
- Summarize. Restate key words or points to emphasize their significance, **Figure 3-13**.

"There are always three speeches for every one you actually gave. The one you practiced, the one you gave, and the one you wish you gave."
—Dale Carnegie

Lance Cheung/USDA

Figure 3-13. Summarizing the key points of your presentation will be helpful for both you and your audience. At what point or points in the presentation would you summarize the key points?

Safety Note

Presentation Safety

- What technological equipment is used for your presentation? Where are the cords? Be sure that cords are taped down or otherwise secured so they do not present a tripping hazard for you or the audience members.
- Are you using a stage? Stay away from the edges so you do not fall. Pay attention to your location on the stage.
- Do not refer to cue cards while you are moving about on the stage. You could fall, trip, or lose your balance. Remain still while you quickly glance at your note cards or electronic slides.

carroteater/Shutterstock.com

Presentation Types

Four basic types of presentations are discussed in this chapter, based on preparation and purpose. The two preparation types of presentations are prepared and extemporaneous. The two purposes for presentations are to inform or to persuade.

Prepared Presentations

With a prepared speech, the presenter knows ahead of time that he or she will be giving a presentation and has time to prepare it. The presenter knows who the audience will be, the date and location of the speech, and the topic of the speech. He or she has a period of time to do research, find answers to questions, write and practice the message, and organize visual aids. The presenter can create a presentation to inform or persuade listeners based on the objectives and the needs or expectations of the audience.

David Foster/USDA

Figure 3-14. The Secretary of Agriculture answers questions after a presentation. When there is little time for preparation, this is known as an extemporaneous speech. What do you think are the most important things to focus on when giving a speech without time for preparation?

Extemporaneous Presentations

An extemporaneous presentation is an impromptu speech that is performed without advance preparation. There are times in industry when a business manager may have little or no time to prepare a statement or speech, **Figure 3-14**. It is common for an employee whose job is public relations to act on the company's behalf with very little preparation. This type of presentation is also often associated with competitions. The FFA and other groups offer an extemporaneous speech event to test the skills and knowledge of FFA members. This is much like being part of a debate team where contestants are posed a question and asked to respond.

Corner Question

What horticultural product contaminated with *Salmonella* led to an outbreak of illness in 2014?

AgEd Connection Extemporaneous Speaking

In this event, a single participant is given only 30 minutes to prepare a speech that is between four and six minutes in length. The purpose of this event is for students to learn how to express themselves verbally and to prepare a speech in very little time. Participants are in official dress and are judged on content presented, tone, eye contact, mannerisms, gestures, poise, articulation, and responses to questions. Contestants can participate at the local, state, and national events.

USDA

Persuasive Presentations

When a landscape design company owner talks to his employees on Monday morning and encourages them to find more clients and reach out to older customers, he is trying to motivate his employees. This is often referred to as a pep talk, but it is literally a mini-motivational speech.

Persuasive speeches are those that try to convince the audience to believe in something or take some action. This is no different from the times in a locker room when a coach tries to convince her team that they are unstoppable and that they can defeat their opponent. A persuasive speech can be formal or informal. One of the most memorable speeches in twentieth century history is that by the late Dr. Martin Luther King, Jr. His "I Have a Dream" speech was persuasive and tried to unify a country ravaged by segregation.

Informational Presentations

Informational speeches present material in a way to increase knowledge of a given topic. When an audience attends a lecture about vegetable gardening, they are expecting to gain knowledge of that horticultural topic. Extension agents and garden center employees often host workshops to help educate the public, **Figure 3-15**. These are considered informative presentations.

Communication with the Audience

The method of delivery and the type of presentation that you choose is based upon the needs of the audience. The topic that is

Bob Nichols/USDA

Figure 3-15. An extension agent presents information to children about apples. This is an example of an informational speech where the audience is very important to the language of the speech.

Copyright Goodheart-Willcox Co., Inc.

Corner Question

What is the ideal speech duration?

chosen must meet the needs of the audience and the format of the presentation must be equally suited for the attendees.

Deciding the best way to create a memorable experience for your audience comes after your audience analysis and topic choice. A presentation is only effective when it impacts the audience in some way. A memorable speech is one that not only is rich with information but also creates a relationship between the audience and the speaker.

A speaker must be believable and relate to his or her audience. There are a number of ways to connect with the audience through stage presence. Stage presence is how a speaker presents herself or himself. To appear more credible and convincing, the presenter should make use of:

- Eye contact. This is easily achieved by scanning the audience and looking at one person, then another, and so on.
- Voice fluctuations. Tones should be changed and volume should vary.
- Mannerisms. Hand movements and overall body movement send a message. Look in a mirror or video yourself to determine your natural movements.
- Articulation. Speak clearly, slowly, and with intention.
- Poise. Stand tall and look confident and professional, **Figure 3-16**.

Lance Cheung/USDA

Figure 3-16. The poise or overall appearance and conduct of a speaker is very important to the audience. What does each person's poise convey to you?

Presenters connect more with their audience when they present their information with some humor and in a manner that appears more like storytelling. A story is even richer when it is accompanied by visuals. It is important to supplement your presentation with images, sounds, and other forms of media. These assist in creating an experience that will leave a lasting impression on the audience.

The presenter should imagine connecting with every person in the audience. If there is no microphone, the person should try to make her or his voice heard by the person furthest away in the audience. Every audience member should be able to hear the presenter clearly. It is okay to check in with the audience and ask if everyone can hear you. This is another opportunity for the speaker to connect with the listeners on a more personal level.

Information Literacy

Today's writers, presenters, and audiences are flooded by various forms of media with entertainment, marketing, and other information. People may feel overwhelmed and simply "tune out" many messages. Presenters must find a way to make an impression on listeners and have their messages heard.

AgEd Connection
What to Know about Presentations
- Know your audience.
- Know your subject.
- Know your goals.
- Know how to organize your material.
- Know how to engage and connect with the audience.
- Know proper stage presence tips and techniques.
- Know that what you learn from this presentation can help you make your next presentation better.

Whether you are a writer, presenter, or audience member, you need to be literate and able to communicate effectively. Literacy was once defined as the ability to read and write. Today, there are a number of forms of literacy that are important for everyone. Information literacy involves recognizing when information is needed and being able to find, evaluate, use, and then communicate the information to others.

An effective communicator must be able to adapt, and must be literate about various information resources. He or she must also be aware of ways to present or communicate information. For example, a presenter does not have to be in front of an audience in person. A presentation can take place using other vehicles of delivery. Virtual presentations can deliver engaging information and motivation for audiences. Presenters can use tools such as:

- Webinars.
- Video software.
- Slideshow software.
- Presentation software.
- Poster software.
- Blogs.
- Websites.

Some tools for virtual presentations provide formats for interacting with audience members. Presenters can create videos or reference videos so that the audience can more or less see a show. Presenters have the ability to include images and audio as well.

A webinar can be a good tool for reaching an audience with horticultural information. Webinars are one of the premiere tools used to educate people. Webinars, also called web conferences, allow audience members to see educational presentations from many locations. This method of delivery offers messaging, video, chatting, and media to be presented interactively. Participants can ask questions and speak just as they would in a face-to-face format. Some webinars require additional software as determined by the vendor of the webinar. Be sure to consider software needs when you are creating or participating in a web conference.

Information literacy is the foundation for lifelong learning regardless of one's occupation or study area. Two additional concepts are often associated with information literacy: independent learning and social responsibility.

> "I have a dream that my four little children will one day live in a nation where they will not be judged by the color of their skin, but by the content of their character."
> —Dr. Martin Luther King, Jr.

Corner Question
What North American culture is known for its rich history in storytelling?

Independent Learning

Independent learning relates to information literacy because a learner wants to gain knowledge based on his or her personal interests or without a formal class or other structured learning program. The learner will still have to follow critical thinking steps to answer a question or find a solution. The person who seeks independent learning will be information literate because of a desire to pursue and gain knowledge through self-directed study or experiences.

Social Responsibility

Social responsibility means acting with consideration for the needs of others or society as a whole. Many students have a good understanding of their social and civic responsibilities. For example, you may feel a responsibility to help protect the environment for the benefit of everyone. You may believe that the right to vote for government officials is both a privilege and a duty. Social responsibility may include actions as simple as teaching younger children how to grow vegetables. Information literacy is needed to be an informed citizen, and it aids in practicing social responsibility, **Figure 3-17**.

Robyn Wardell/USDA

Figure 3-17. A student travels and helps children learn how to garden. Giving back is important for social responsibility. *How do you give back to your community?*

Horticultural Communication Careers

Today's horticultural communicators work around the globe. These individuals are from many cultures, global locations, and horticultural specialties. A horticultural communication specialist may have a specific training (academic or informal) in a horticultural field, but it is a combination of their enthusiasm for plants and their ability to connect with an audience through effective communication that makes these writers, speakers, and conversationalists very powerful in the horticulture industry.

> "The illiterate of the twenty-first century will not be those who cannot read and write but those who cannot learn, unlearn, and relearn."
> —Alvin Toffler

Horticulture Extension Agent

Extension agents are the liaison between a university's horticultural research staff and the industry and the public. A cooperative extension service agent educates industry members and the public about agricultural topics. Horticulture agents use garden cultivation as their medium for connecting with their audiences. Hort agents, as they are often called, have a four-year horticulture science or closely related degree. They often have years of experience in the horticulture industry before becoming an agent.

STEM Connection
Information Literacy

Bikeriderlondon/Shutterstock.com

The following forms of information literacy are needed by people in the horticulture industry and are essential to communicating effectively.

- Tool literacy—the ability to understand and use the practical and conceptual tools of current information technology relevant to education and areas of work and professional life.
- Resource literacy—the ability to understand the form, format, location, and access methods of information resources, especially daily expanding networked information resources.
- Social-structural literacy—the ability to understand how information is socially situated and produced.
- Research literacy—the ability to understand and use the IT-based tools relevant to the work of today's researcher and scholar.
- Publishing literacy—the ability to format and publish research and ideas electronically, in text and multimedia forms. Also, the ability to introduce information into the electronic public realm and the electronic community of scholars.
- Emerging technology literacy—the ability to adapt to, understand, evaluate, and make use of the emerging innovations in information technology and to make intelligent decisions about the adoption of new technology.
- Critical literacy—the ability to evaluate critically the intellectual, human, and social strengths and weaknesses, potentials and limits, and benefits and costs of information technologies.

(Jeremy J. Shapiro and Shelley K. Hughes, "Information Literacy as a Liberal Art." Educom Review, Volume 31, Number 2, March/April 1996.)

Hort agents offer educational seminars and workshops and create publications to disseminate information. They are responsible for administering programs through the agricultural universities of the state in which they work. These agents have flexible hours, but may have to work some weekends. They almost always have to regularly travel in their county, state, or nation. They may even travel internationally for their occupation. An extension agent must enjoy working with people and have excellent written and verbal communication skills. The agent should have a passion for horticulture and for teaching, as this position is a hybrid of both occupations.

Garden Writer and Speaker

A garden writer can be an independent (freelance) author or one that writes for a book, newspaper, magazine, journal, or website. Garden writers must have a wide knowledge of plant material and gardening topics. They must be excellent and fast researchers with a talent for the written word. Today's garden writers can be anywhere in the world since they can submit their work from home or other places to their editor or publisher.

Corner Question
What is the National Association of Landscape Professional's (NALP) Day of Service?

Garden writers must often also be skilled in photography or art, **Figure 3-18**. Some garden writers are also skilled botanical illustrators.

Garden writers may also be lecturers or speakers. Speakers, often called lecturers, must be able to engage their audience and both entertain and educate. Some speakers may work only with small audiences, but others may speak formally in front of thousands at trade shows. Garden speakers may perform on radio, on television, or with live audiences. No matter the format, speakers must leave a lasting impression and communicate professionally.

The Garden Writers Association is an organization specifically for the thousands of garden writers and speakers across the United States. Its website provides information for garden writers, lecturers, and those who seek to employ a garden writer or lecturer.

Ijansempoi/Shutterstock.com

Figure 3-18. This woman is part of the Garden Writers Association. How do you think photography connects to being a garden writer?

Career Connection
Rizaniño "Riz" Reyes
Garden Blogger

Rizaniño "Riz" Reyes grew up on a family fruit plantation in his native Philippines. When his family immigrated to the United States, he was naturally drawn to the produce and floral departments at their local supermarket. Soon he began to research different plants and flowers by viewing old garden catalogs, books, and gardening shows, such as *The Victory Garden*. Riz earned a degree in environmental horticulture and then landed a part-time position through the University of Washington Botanic Gardens. During this time, he ran a small specialty nursery where he grew and marketed his favorite plants and later added garden/floral design, writing, consulting, and maintenance services through his company, RHR Horticulture & Landwave Gardens.

Through his love for collecting and nurturing plants and his ability to share and communicate his growing interests, he has been able to visit numerous gardens and nurseries and meet renowned experts around the world through social media. Today, Riz appears on the Internet as a source for floral inspiration and horticultural opinion. He has several videos on YouTube, a website dedicated to his business ventures, and a blog called *A Next Generation Gardener*, where he chronicles his life experiences with plants and people. He also showcases unique topics for gardeners of all interests and levels of expertise. He was featured in a 2013 article as one of the up-and-coming young gardeners through *Organic Gardening* magazine.

CHAPTER 3

Review and Assessment

Chapter Summary

- Communication is a process in which a message is sent by one person and received and understood by another person(s). People communicate using speech or other sounds and using written words, images, and nonverbal behavior.
- The writing process involves determining the purpose of a message, considering the audience, selecting a way to present the message, refining the topic, identifying the main points, creating a plan for the message, reviewing the message, and revising and proofreading.
- Critical thinking is using objective reasoning or consideration before forming a judgment or taking some action. Critical thinking involves a path of analysis that includes searching for information, previewing results, finding evidence, annotating, summarizing, and evaluating information.
- Plagiarism is using work (such as ideas, writing, or images) created by someone else without permission and presenting it as one's own work. Under no circumstance should you present another's work as your own.
- Authors should follow standard formatting styles, such as MLA or APA, and document or cite all work in any document that is not original. At the end of the paper, a bibliography should be created to list all sources used in the work.
- A presentation is a speech or lecture given to an audience, either live or by video. When you are preparing a speech that requires research and writing, the steps will be similar to those for writing a paper.
- Both prepared and extemporaneous presentations may be given to inform or to persuade listeners.
- Information literacy involves recognizing when information is needed and being able to find, evaluate, use, and communicate the information to others. Independent learning and social responsibility are concepts related to information literacy.
- Several horticultural careers involve effective communication. These can include but are not limited to cooperative extension service agents, garden writers and lecturers, and garden bloggers.

Words to Know

Match the key terms from the chapter to the correct definition.

A. annotate
B. annotated bibliography
C. audience
D. communication
E. critical thinking
F. demographics
G. evidence
H. information literacy
I. paraphrase
J. plagiarism
K. skim
L. thesis
M. tone
N. voice
O. works cited

1. Using objective reasoning or consideration before forming a judgment or taking action.
2. The ability to recognize when information is needed and to find, evaluate, use, and communicate information.
3. The people who will read, hear, or see the message.
4. To write questions or comments on a document as you read it.
5. A summary list of all research and sources that are used to create a text or presentation.
6. A statement or theory that is proposed and then discussed to prove or disprove it.
7. To use your own words, sentences, and organization to summarize something written (or spoken) by someone else.
8. To read selected parts of a text looking for the main ideas.
9. The quality of a message that reflects the writer's attitude or mood.
10. A document that lists citations for sources used and briefly describes each work.
11. A process in which a message is sent by one person and received and understood by another person(s).
12. Using work (such as ideas, writing, or images) created by someone else without permission and presenting it as one's own work.
13. Characteristics or traits of a group.
14. Facts or information about a topic indicating whether a belief or position is correct or valid.
15. The style of expression or degree of formality used in a message.

Know and Understand

Answer the following questions using the information provided in this chapter.

1. What are the steps in the writing process?
2. What are four important characteristics an author should consider about the audience for a message?
3. What are two methods of presentation an author can use to deliver a message?
4. What are four activities you can do to help you select a topic for a written message or oral presentation?

5. Why should writers avoid using a topic that is too broad?
6. How does a writer go about creating a project plan for a writing assignment?
7. How does the introduction of a message lay out a plan for the body of the message?
8. Why is it important to review, revise, and proofread a message?
9. What steps can you take to successfully research information and form your own educated opinions?
10. How is skimming a document different from reading the entire document?
11. What are three types of sources of information that are considered credible or reputable?
12. What are eleven questions you can consider to help you evaluate information from your research?
13. What are three examples of plagiarism?
14. How is paraphrasing different from simply copying a passage from a document?
15. How does a prepared presentation differ from an extemporaneous presentation?
16. What five things can a presenter do to appear more credible and convincing?
17. Name and describe seven forms of information literacy needed by people in the horticulture industry.
18. What is a horticulture extension agent? What education and experience are needed to be a horticulture agent?
19. For what types of employers does a garden writer work? Why is a garden writer often also a garden speaker?

Thinking Critically

1. You are visiting colleges during your senior year. Your parents want you to attend one school, your grandparents have another idea, and your friends would like you to go to yet another school. What methods could you use to determine which school is the best choice for you? Remember, you are the one who is going to college, and the choice needs to be the best for you. List five ways that you will determine your college of choice.
2. A horticulturist has been growing plants organically for the past three years, but she has never been formally certified as an organic grower. There are many expenses associated with becoming a USDA certified organic grower. Some of her customers are now asking that she become a certified organic grower. How can she determine if she really needs to be a certified organic grower, or if it would be best to continue to grow organically without the USDA certification?

STEM and Academic Activities

1. **Technology.** Watch a home gardening show. Find an episode that relates to a scientific process, such as respiration, absorption, photosynthesis, or translocation of plants. Retell the process in your own words. Highlight scientific processes and the inputs and outputs of those processes. Make note of where these processes occur in the plant.

2. **Engineering.** Properly plant a tree or shrub. Research how to properly plant a tree or shrub. Use several resources and determine what are the best suggestions or guides for your area. While planting, engineer a solution to any problem that you encounter (examples: very hard soil, steep grade of soil, pollutants, erosion, or pests).
3. **Math.** Landscape technicians must use math almost every day, specifically geometry. Homeowners will often find themselves lost in a sea of equations when trying to determine simple calculations involving landscape materials. Create a one-page publication for homeowners that explains how to calculate the amount of mulch to purchase in cubic yards for rectangular, circular, and triangular areas. Learners need both text and images to help them through this process.
4. **Language Arts.** Go to a garden or greenhouse and write about one of the flowers that you observe and can identify. Make sure to note as many details as possible. Do not use the name of the plant. When you are finished, have your teacher read this text and see if he or she can determine what plant you are describing. If your teacher can, you have done a good job. If not, your writing needs to include more detail, and you should try again.
5. **Language Arts.** Find a gardening magazine, either online or in print form. Find the section where people write about their problems and someone determines a solution to the problem. Write a letter to this individual about a problem that your school is experiencing in the garden, greenhouse, or classroom related to plants. If you or your teacher cannot find one, interview someone you know (most likely an adult) who has a gardening question. Send this letter to the author of the column and see if you get a response.
6. **Language Arts.** Go to a garden blog and write a comment to the author about one of the posts. Write one to two paragraphs and try to include details about the blogger's post to help jog the author's memory about what he or she wrote.

Communicating about Horticulture

1. **Writing.** Make a two-column chart and list the names and occupation or relationship of twelve people with whom you communicate on a regular basis in the first column. How is the way you communicate with these people influenced by your relationship? Use the second column to identify and write the different ways you speak and behave when in the presence of these individuals.
2. **Listening.** Record a half-hour news broadcast. Before viewing the broadcast, turn on your radio, and have your phone or laptop on. View the broadcast while using your phone or laptop. Once the broadcast is over, write down everything you remember seeing or hearing. Set aside the phone or laptop and turn off the radio. Watch the broadcast again. How much did you remember? How accurate were your observations? How were your observations affected by the distractions? Perform the experiment again (with a different broadcast), but eliminate all distractions. Compare your findings.

SAE Opportunities

1. **Exploratory.** Job shadow a garden writer or lecturer.
2. **Research/Analytical.** Research how to photograph plants or do botanical illustrations. Create a digital presentation or a website illustrating your findings. Create an online slideshow that documents your artwork. For every plant that is included, be sure to include its species so that others can learn about this plant. You could also write a short blurb about each of the plants and where and how they were photographed or illustrated.
3. **Exploratory.** Attend a garden seminar or lecture through your cooperative extension agent. All areas of the country have Master Gardeners. Contact your extension agent or Master Gardener for help in finding a seminar.
4. **Research/Analytical.** Research a topic in horticulture that interests you. Author a paper. Be sure to document your research throughout your text and create a bibliography. When you have completed your work, create a presentation. Present this to an audience at a local nursing home or school for the first round. Fine-tune your work and try participating in the Prepared Public Speaking CDE as well.
5. **Exploratory.** Contact your state's land-grant university and find the horticulture department. Visit the school and meet with one or more of the professors, faculty members, and students. Develop a publication that you can use to educate your school's horticulture program about the opportunities at this university. Highlight those faculty members and students that you spoke to and include quotes and photos of them for the publication. Be sure to get their permission before the interview.

Arthiti Kholoet/Shutterstock.com

CHAPTER 4
The Horticulture Industry

Chapter Outcomes
After studying this chapter, you will be able to:
- Describe the plant sciences: botany, agronomy, horticulture, and forestry.
- Compare and contrast edible and ornamental horticulture.
- Understand organic and sustainable production of crops.
- Discover sources for locally grown food.
- Describe the green industry.
- Discuss the outlook for horticulture careers and list horticultural organizations.

Words to Know

agronomist	green industry	perennial
agronomy	hardscape	plant science
balled-and-burlapped (B&B)	high-density orchard	pomologist
bare root (BR)	horticulture	pomology
botanist	horticulturist	postharvest
botany	integrated pest management (IPM)	pot-in-pot (PNP)
bramble		propagate
community supported agriculture (CSA)	interiorscape	senescence
	interiorscaper	silviculture
enologist	interiorscaping	sod
ethylene	locavore	sustainable agriculture
evapotranspiration	nursery	tree plot
floriculture	nursery and landscape industry	turfgrass
forest stand		vegetative
forester	olericulture	viniculture
forestry	organic food	viticulture
geocarpy	ornamental horticulture	viticulturist

Before You Read
Look at the *Words to Know* list above. Write definitions for the words you know. Put the words you cannot define into a separate list. Skim through the chapter and write down the topics that you think you might know well, and those that you do not know much about. Once you are finished, pair up with another student and share your lists.

While studying this chapter, look for the activity icon to:

- **Practice** vocabulary terms with Words to Know activities.
- **Expand** learning with identification activities.
- **Reinforce** what you learn by completing Know and Understand questions.

www.g-wlearning.com/agriculture

Elizabeth O. Weller/Shutterstock.com; LorenzoT/Shutterstock.com; Juhku/Shutterstock.com; fotocraft/Shutterstock.com; Arina P Habich/Shutterstock.com

When you reach into your backpack to pull out your lunch, you rarely think about much more than the fact that you are hungry or that it is time to eat. Take a moment and think about the fresh fruits and vegetables in your lunch, **Figure 4-1**. Where did they come from? Who is responsible for their journey from the farm to your fork? The person responsible is a horticulturist. *Horticulture* is the science, art, technology, and business of plant cultivation. A *horticulturist* is a person who specializes in this area, and who may raise everything from flowers to potatoes to apples. Horticulturists grow some plants for food and some for landscaping or other aesthetic purposes. Horticulturists nurture plants in homes, businesses, gardens, parks, highways, and even laboratories. Wherever you see a plant that is purposefully grown, you can almost bet it has been touched by a horticulturist's green thumb at some point in its development. This chapter will help you understand what horticulture is, who horticulturists are, and what types of horticulture careers are available.

What Is Plant Science?

Horticulturists usually have a background that includes knowledge in various areas of plant science. *Plant science* is the study of plant growth, reproduction, and adaptation as well as the use of plants for food, fiber, and ornamental purposes. Areas of plant science include the following:

- *Botany* is the scientific study of plants, including their structure, genetics, ecology, classification, and economic importance.
- *Agronomy* is the science and technology of cultivating crops for food, fiber, and fuel.
- *Forestry* is the science or practice of planting, managing, restoring, and caring for forests.
- Horticulture is the science, art, technology, and business of plant cultivation (including fruits, vegetables, flowers, turfgrass, and ornamentals).

Each of the applied plant sciences involves specific education and experiences.

Botany

Botany, or plant biology, is a branch of biology and is known as the foundation of plant science. A *botanist* studies plant sciences and is sometimes called a plant scientist. The science of botany can be traced back to the ancient Greeks and Romans, but it was not until the sixteenth century that it truly began to flourish. This was mainly due to the work of physicians and herbalists who were studying plants to identify those useful in medicine. Training in botany was even part of the medical curriculum.

Rob Hainer/Shutterstock.com

Figure 4-1. The fresh fruits and vegetables we eat everyday are the "fruits of labor" of people working in the horticulture industry. A horticulturist is responsible for the health of plants and trees as well as the processing and shipping of fruits and vegetables he grows. Take a moment and think about the foods you ate for breakfast or lunch today. How many people were involved in their production?

Throughout history, botanical gardens have been maintained around the world, and they continue to be so today. These gardens are intended for the study of plants and educate visitors about the vast world of plants, their usefulness, and their beauty. Cities such as New York, Atlanta, and St. Louis boast about their premiere botanical gardens that are visited by millions annually.

Corner Question

What is a typical forest product that people use when they have a headache or suffer from acne?

Agronomy

As stated earlier, agronomy is the science and technology of cultivating crops for food, fiber, and fuel. In agronomy, crops are grown in rows, harvested, and processed. There are many areas of study in agronomy. An *agronomist*, a person who studies or practices cultivating crops for food, fiber, and fuel, may specialize in areas such as:

- Biotechnology.
- Soil science.
- Weed science.
- Plant breeding.

Crop plants that typically cannot be eaten or used in their raw form (soybeans, cotton, tobacco, field corn, rice, and wheat) are included in agronomy.

Did You Know?

Without the work of agronomists, you may have never had a favorite pair of jeans, peanut butter and jelly sandwiches, or rice in your Chinese take-out food. The next time you enjoy a bowl of cereal or munch on potato chips, remember to thank an agronomist.

Forestry

Forestry, the science or practice of planting, managing, restoring, and caring for forests, is also called *silviculture*. *Foresters* are the dedicated people who plant, manage, and care for forests. Foresters craft and conserve forests to sustain the ecosystems while allowing the nondestructive harvest of lumber and tree products. A *forest stand* consists of trees occupying a specific area that are uniform in species, size, age, arrangement, and condition. Foresters manage forest stands using silvicultural practices to meet the objective of a *tree plot* (a carefully measured area of trees). Whether the forest is grown for beauty or to be selectively cut for specific trees and tree products, forests and their products provide limitless resources.

Edible and Ornamental Horticulture

As with most fields of study, horticulture includes various specialized sciences and practices, and horticulturists can seek both formal and informal career training. Horticulturists can further their studies and focus in specialty areas through community college, university, fieldwork, and global study-abroad programs, **Figure 4-2**.

xuanhuongho/Shutterstock.com

Figure 4-2. Flowers are cultivated around the world for use in medicine, food, and floriculture. If you were a horticulturist based in the United States, how could you benefit from traveling and studying horticulture in places such as this garden in Vietnam?

Corner Question

People have always linked the affinity for growing plants to having a "green thumb." Where did this term originate?

Horticulturists often specialize in either edible or ornamental horticulture. Edible horticulture includes olericulture, pomology, and viticulture, while ornamental horticulture includes floriculture, nursery and landscape production, interiorscaping, and turfgrass management. These areas of horticulture are explained in the following sections.

Olericulture

Olericulture is the science, cultivation, processing, storage, and marketing of herbs and vegetables. People who grow herbs and vegetables are olericulturists.

Herbs

Herbs are the *vegetative* parts of the plant (roots, stems, and leaves) harvested for flavorings, foods, perfumes, or medicines. Botanists and horticulturists may also refer to an herb as an herbaceous plant. In the culinary world, herbs are referred to as spices. Although chefs often use the seeds of these plants and consider them herbs, a true herb is only the root, stem, or leaf of a plant.

Herbs have been cultivated for both medicinal and culinary purposes since the dawn of gardening. They were typically grown outdoors and dried for later use. Today, many people grow herbs in planters and on windowsills as well as in outdoor gardens for easy access to fresh herbs for their tea or coffee and for cooking. Lavender, parsley, chives, chicory, cilantro, lemongrass, oregano, and basil are some of the most popular herbs cultivated, **Figure 4-3**.

thatreec/Shutterstock.com

Figure 4-3. Thai basil is a delightful addition to many dishes. It is grown by olericulturists. How many types of basil can you find at local nurseries or farmers markets?

Vegetables

Consider the age-old debate of the tomato: is it a vegetable or a fruit? The answer, it seems, varies by profession. Chefs traditionally consider fruits and vegetables by the course of the meal in which they are eaten. Fruits are typically sweet and eaten as desserts. Vegetables are typically eaten as appetizers, main courses, and side dishes. Thus, for chefs, the tomato may be considered a vegetable. Horticulturists who grow tomatoes for food may also consider them to be a vegetable. However, botanists consider the tomato to be a fruit because it develops from the fertilized ovary of a flower. Botanists also consider squash, pumpkins, cucumbers, peppers, eggplants, and pea pods as fruits whereas most people call them vegetables. As a tomato grower, you may prefer the Supreme Court's 1893 ruling in favor of vegetable. It all depends on your perspective.

In horticulture, a vegetable can be the roots, stems, leaves, flowers, or fruit of a plant. Therefore, all fruits are vegetables, but not all vegetables are fruits (to further the fruit/vegetable debate). The roots of sweet potatoes, carrots, radishes, and parsnips are harvested for eating raw or for cooking. When you eat celery or asparagus, you are eating the stems of those plants. The leaves of plants such as spinach, lettuce, cabbage, and kale are eaten. With some plants, such as onions, leeks, and shallots, both the stem and leaves are eaten. The flowers of plants can also be eaten as a vegetable, as is the case with broccoli and cauliflower. Fruits include beans, peas, tomatoes, cucumbers, squash, peppers, and sweet corn.

Vegetable Growers

Vegetable growers should enjoy working outdoors in a variety of conditions. They must be prepared to lead teams of people and have good communication skills. Vegetable growers have the opportunity to provide people with nutritious plants for better health. Growers must make daily choices for plant management. They face many challenges that require critical thinking and problem solving, **Figure 4-4**.

Pomology

Pomology is the cultivation, processing, storing, and marketing of fruits and nuts. People who work this area are known as *pomologists*. The study, management, and harvest of fruit and nut trees helps feed and medicate people around the world. Fruit from trees such as those in the citrus family (grapefruit, tangerine, and oranges) provide the vitamin C used to help ward off sickness and disease such as the common cold.

Corner Question

Where did pumpkin pie originate?

hjschneider/Shutterstock.com

Figure 4-4. Vegetable growers must make decisions and solve problems related to gardening. How would you cultivate a garden? What would be your biggest challenge?

STEM Connection

Peanuts
How does a fruit grow underground?

Peanuts provide countless products that are used daily by Americans. Peanuts are not true nuts. The peanut is a member of the legume or bean family. *Arachis hypogaea* is the botanical name of the peanut and means "under the earth." This agronomic crop is originally from South America, but it has found a home in the soft soils of the southeastern United States. Peanuts exhibit something called **geocarpy** (a rare means of plant reproduction). After pollination, the flower stalk elongates and pushes into the ground. Where each plant stem touches the soil, a nodule (or peanut) forms. The mature fruit, now called a legume, is nestled in the soil waiting for harvest. How many peanuts would it take to make one jar of peanut butter?

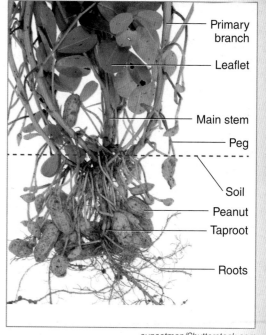

sunsetman/Shutterstock.com

Thinking Green

Espalier Fruit Fencing

A pomologist can create a living fence through the intensive pruning and training techniques known as espalier. This vertical and interlinked growth of fruit trees can provide borders or privacy screens and edible fruit.

Irina Fischer/Shutterstock.com

Almond, pecan, pistachio, and macadamia nut trees produce fruits that provide essential fats and proteins. Pomology also includes the cultivation of fruits that grow on prickly, rambling vines or shrubs called *brambles* (boysenberries, loganberries, raspberries, and blackberries), and smaller, fruit-bearing plants such as blueberries and strawberries. Regardless of whether fruit grows on the ground or up high in a tree, all fruits offer benefits for horticulturists and consumers alike.

Growing Practices

The tree fruit and nut industry is feeding a growing population and meeting this goal through scientific research and application. For instance, apple trees and orchards are managed quite differently today than they were 20 years ago. A *high-density orchard* is a fruit tree orchard with between 150 and 180 trees per acre that bears fruit within 2 to 3 years of planting. Many traditional orchards have been converted to high-density orchards, growing almost four times the number of trees in traditional orchards. Trees may also yield increased harvests with earlier and better quality fruit. In order to feed 9 billion people in 2050, practices similar to high-density orchards must be applied to other crops in pomology and further scientific study must be pursued.

Fruit and Nut Growers

Fruit and nut growers, or pomologists, must be physically fit because the work requires a great deal of physical labor. Head growers manage crops and lead other workers. Pomologists must keep up to date with current research and be excellent recordkeepers. They must decide how best to cultivate their fruits, **Figure 4-5**. Fruits require intense pesticide spray programs or other means of pest management.

Viticulture

Viticulture, sometimes called *viniculture*, is the cultivation of grapes to be eaten fresh and to be used for making juices, raisins, jams, jellies, and wines. Grapes are cultivated by *viticulturists*, then marketed, processed, and stored. Viticulturists must pay close attention to pruning and training grape vines, properly fertilizing soils, and managing pest populations. *Enologists* are people who use grapes to make wine using chemistry and food science knowledge.

A *joloei/Shutterstock.com*

B *Franck Boston/Shutterstock.com*

Figure 4-5. Macadamia nuts (A) are native to Australia but are grown for commercial purposes mainly in Hawaii. Almonds (B) are native to the Middle East and South Asia. *Where else do these trees grow around the world? Are the nuts expensive? Where do other nut trees grow?*

Grape Cultivation

Today, grapes grow all over the world and in nearly every state of the United States. In 2012, more than 7 million tons of grapes were harvested in the United States; nearly 90% were from California alone. Grape cultivation is also strong in Michigan, New York, Virginia, Pennsylvania, and Arkansas. Since people have realized the many health benefits associated with grapes and grape products, the demand for this crop has skyrocketed. Grape cultivation has greater yield and value than many other horticultural crops. The future of grape production promises strong economic returns.

Viticulturists

Grape vines are a complex crop that requires a great deal of study and attention. A viticulturist usually has a college degree or some formal training of grape cultivation. Viticulturists must work outdoors and lead or work with a team of growers, **Figure 4-6**. The work can be strenuous, and critical thinking and problem-solving skills are very important. Viticulturists are rewarded for their efforts with delicious fruit and the opportunity to work in beautiful outdoor settings.

Floriculture

Floriculture is the study, cultivation, and marketing of flowers and ornamental plants. It may also be referred to as *ornamental horticulture* because it includes the creative and decorative aspects of horticulture. Careers that fall under floriculture include buyers, growers, floral designers, florists, interiorscape designers, product developers, wholesalers, brokers, and greenhouse owners and employees, **Figure 4-7**.

Because of the wide range of the floriculture field, people who work in floriculture may or may not be involved with each aspect of the industry. For example, florists who specialize in weddings most likely do not grow the flowers and plants they use in their designs. They probably order from local wholesalers who, in turn, order from local, national, or international growers and brokers. It may seem that a floral designer does not need to know details about growing flowers or ornamental plants. However, many designers do have extensive knowledge about plant morphology, breeding, harvesting, and handling practices. This knowledge helps them make wise choices in designing and purchasing materials for their work.

Did You Know?
Grape seeds are a by-product of grape processing and quite useful. The grape seed oil (extracted by crushing the seeds) is used to create products for cosmetics, medicines, and healthy cooking.

Gyuszko-Photo/Shutterstock.com

Figure 4-6. A viticulturist picks ripe grapes ready for juice making.

Be Good/Shutterstock.com

Figure 4-7. This beautiful wedding bouquet was created by a florist. How much time passes between the harvest of cut flowers and the day of an event?

Plants cultivated for floriculture include bedding plants, houseplants, potted or container plants, and cut flowers and foliage. Many of the plants and flowers are grown to fill seasonal and holiday demands. For example, florists have three mainstay holidays: Valentine's Day, Mother's Day, and Christmas. The primary flowers and plants used for these holidays are:

- Red roses for Valentine's Day.
- Corsages, standard arrangements, dish gardens, container arrangements, and flowering plants for Mother's Day.
- Poinsettias and evergreens for Christmas.

Other popular holiday plants include lilies for Easter and chrysanthemums for both spring and fall decorating. These plants are often sold in pots for the garden or in mixed containers for use on a porch or patio.

Nursery and Landscape Industry

The *nursery and landscape industry* cultivates and arranges outdoor plant materials to create spaces that are inviting, beautiful, and useful to people and the ecosystem. There are countless subdivisions and occupational opportunities within this facet of horticulture. The nursery and landscape industry is often referred to as the *green industry*.

Hands-On Horticulture

Postharvest: Physiology and Technology

Horticulturists are always racing to get harvested products to the consumer because from the moment a crop is harvested, it begins to decompose. People who work and study this area of horticulture are in the field of postharvest. **Postharvest** is the cooling, cleaning, sorting, storing, packing, and shipping of produce, flowers, and other plant materials. Because postharvest principles apply to everything from cut flowers in floriculture to apples in pomology, scientists and technologists studying and developing postharvest techniques may work in any of these areas.

The goals of postharvest include preventing spoilage and prolonging shelf life. The methods used to reach these goals vary by product. For example, all harvested crops have optimum storage conditions that include temperature and humidity ranges. Storing crops at less than optimum temperature or humidity ranges will often hasten ripening or degradation.

Some harvested crops cannot be stored together because one may hasten the ripening or **senescence** (biological aging) of the other. This is usually the result of a natural chemical called **ethylene**. Ethylene is a hormone produced and

©iStock.com/hydrangea100

emitted in varying quantities by different fruits and vegetables. Ethylene is also emitted by decaying plant materials. Some fruits, vegetables, and flowers are more sensitive to ethylene than others, and care should be taken to know which crops should not be stored together. Do you know which fruits and vegetables should not be stored together?

Postharvest goals are also achieved through proper sanitation techniques. These techniques are used to suppress pathogenic exposure and growth and prevent damage that may lead to physical and chemical changes in the product.

Nurseries

A *nursery* is a place where young plants and trees are cultivated for sale and for planting elsewhere. In a nursery, most plants are grown in fields. However, greenhouses and shadehouses are often used when plants and trees are first cultivated from seed or vegetative cuttings (stems, roots, or leaves). Once mature enough, these plants are transplanted in fields and managed to various sizes based upon landscaper and consumer needs.

Plants commonly grown in nurseries include:

- Trees—species that mature to greater than 12′ (3.7 m) tall; most often single-stemmed, woody, and *perennial* (returns yearly) plants.
- Shrubs—species that mature usually to less than 12′ (3.7 m) tall; multi-stemmed, woody and perennial plants.
- Ground cover—plants that create a mat-like growth that spreads to cover the ground; may be woody or herbaceous (includes ivy, ornamental grasses, and vinca).
- Vines—plants that climb other plants, structures, or buildings; may be woody or herbaceous and usually need a support.
- Perennials—flowering or foliage plants of various types (bulbs, tubers, etc.) that have a life span of more than one or two years; often grown in the floriculture industry as well.

Plants are cultivated and harvested differently in nurseries than in the floriculture industry. In the floriculture industry, plants are usually sold in decorative containers. In nurseries, plants are sold as field-grown. This means they are either balled-and-burlapped (B&B), bare root (BR), or pot-in-pot (PNP), **Figure 4-8**. When a plant is harvested using the *balled-and-burlapped (B&B)* method, the root system of the plant or tree is wrapped in burlap

Did You Know?
Harvesting a plant for balled-and-burlap production causes up to 95% of the roots to be lost in the digging process.

A *serato/Shutterstock.com*   B *Swellphotography/Shutterstock.com*   C *USDA*

Figure 4-8. Plants purchased from a nursery may be balled-and-burlapped (A), bare root (B), or pot-in-pot (C). Identify one instance in which each type of "presentation" would be advantageous. What might be some disadvantages to each type?

and tied with twine. This serves to protect the roots during transport. When a plant is harvested using the *bare root (BR)* method, all soil is removed from the plant's root system. The roots may be covered with a plastic bag for shipping purposes. *Pot-in-pot (PNP)* is a method in which a plant is grown in a pot and that pot is placed in another pot that has been sunk into the ground. The system makes watering and moving the plant easier than growing the plant in the ground, and eliminates the problem of the wind knocking over plants in pots sitting on top of the ground.

The method of cultivation depends on what the landscape industry demands and what the public wants. A landscaper may prefer B&B plants because with this method plants or trees of large size can be planted on a landscape site. Bare root plants are easy to ship through the mail to customers across the country. The public may desire plants grown solely in containers because of convenience and transportation issues. The nursery must meet the objectives of all the customers it serves.

Nursery Growers

Nursery growers (often called nurserymen) work in outdoor settings that may include some greenhouse or shade-bearing structures. A nursery grower must understand how to *propagate* (grow plants from seeds or other methods, such as rootings or cuttings), cultivate, and prepare plants for market.

A nursery grower usually manages a team of workers. The job is labor intensive and requires a great deal of lifting and moving of plants. Accurate recordkeeping is critical to a successful nursery. In addition to financial records, growers must keep track of planting, fertilizing, and watering schedules. Growers must be able to reduce or solve problems caused by pests or diseases using proper diagnosis and treatment options.

Career Connection

Randy Beaudry
Michigan State University Postharvest Professor

Dr. Randy Beaudry has worked at Michigan State University assessing and improving the quality of harvested fruits since 1989. He works with Michigan fruit growers who cultivate tomatoes, blueberries, strawberries, and apples. Beaudry and his lab members work to develop technology used to improve fruit quality and prolong shelf life. Dr. Beaudry has also performed packaging studies and developed mathematical models to predict performance of packaged produce.

Professor Beaudry was given Michigan State University's service award for his incredible impact on the apple industry in the state of Michigan. He works with growers to help improve their apple crops and the postharvest technologies associated with this American fruit favorite.

Goodheart-Willcox Publisher

Landscaping

Landscaping challenges horticulturists to produce and maintain visually pleasing and useful environments. To do this, the landscape industry designs, installs, and maintains landscapes for its customers. Career opportunities in landscaping include:

- Landscape architects and designers who create landscape designs to meet the objectives of the customer.
- Landscape managers who facilitate the installation and maintenance of landscape designs. Landscape managers often need to coordinate the work of other contractors and manage all aspects of installation and maintenance contracts.
- Irrigation technicians who design and install irrigation systems to provide water for the plantings.
- Hardscape contractors who design and install *hardscapes* (constructed areas around a building or in a landscape, such as pavers, patios, sidewalks, and retaining walls).
- Water feature designers who design and install water elements (ponds, waterfalls, and pools).
- Snow removal contractors who remove snow from commercial or residential sites using plows and other equipment.
- Landscape lighting contractors who design and install lighting for nightscapes.

These contractors and technicians work together to meet the goals set by the commercial or residential owner of a landscape. They also rely on nurseries to provide quality plant material for establishing successful landscapes.

Landscape work is done outdoors in all types of weather and climates, **Figure 4-9**. Landscape managers and other landscape workers should enjoy working outdoors and with a team of people. Landscape workers usually work long hours during the peak season or during times of a weather-related emergency. One of the best parts of working in the landscape industry is seeing a newly installed landscape and the beauty it provides to a site.

Interiorscaping

Interiorscaping involves the design, installation, and maintenance of plants inside buildings. *Interiorscapes* (indoor landscapes) improve the look and feel of interior spaces. Homes, businesses, classrooms, and hospitals all welcome additions of plants to their spaces.

Joe Dejvice/Shutterstock.com

Figure 4-9. Landscape work is done outdoors in all types of weather. Landscapers often work in teams to accomplish work more efficiently. *What are some advantages and disadvantages to working outdoors?*

Corner Question

Where is the oldest garden center in the United States located?

Studies have shown that interiorscapes provide an abundance of benefits to an interior space:

- Live plants filter the air and produce oxygen.
- Plants are aesthetically pleasing.
- The presence of plants improves employee productivity and discourages absenteeism.

Interiorscaping is a unique division of horticulture. Regardless of what the weather is like, the interior temperature of a building is normally around 70°F (21°C). The plants used in interior designs are usually acclimated to lower light levels and moderate temperatures. They are fed a limited amount of fertilizer to maintain their size and shape. In larger installations, such as shopping malls and large office buildings, the plants are usually changed seasonally and for certain holidays.

Interiorscapers, the people who design and install indoor landscapes, may work with designers and architects to create displays that will work with the architecture and interior design of a building. Interiorscapers usually contract with businesses to maintain and/or replace plant materials throughout the year. Interiorscapers have the unique opportunity of installing landscapes in a controlled environment. They may also encounter unusual challenges, such as installing and maintaining plants at different heights and in unusual spaces.

Turfgrass Industry

Did You Know?
There are more than 40 million acres of tended lawns in the United States. These lawns sequester more than 13.2 million pounds of carbon each year.

Turfgrass is a collection of grass plants that form a ground cover. Many people commonly refer to turfgrass as *sod*. The turfgrass industry includes the cultivation of lawn grasses for homes, commercial sites, athletic fields, and golf courses. Sod farms produce turfgrass that can be grown, harvested, and sold onsite or shipped to retailers. Sod farms may also be associated with university or corporate research. Companies conduct trials and experiment with varieties of turfgrass to improve insect and disease resistance as well as tolerance to salinity, heat, foot traffic, and drought conditions.

Home Lawns

Turfgrass has been cultivated for centuries all over the world. Lawns have been referenced in British history dating to the 1500s. Aristocratic society coveted a lush, green lawn. The same desire for a beautiful lawn is still coveted today in many societies. In the 1940s, after the workweek was decreased to 40 hours, Americans had more free time. Many chose to use this free time to work on their homes and lawns. When people began having more time to establish a lawn, horticultural companies began advertising the therapeutic benefits of working on the home lawn, **Figure 4-10**.

fotocraft/Shutterstock.com

Figure 4-10. The nursery and landscape industry provides products and services that help create beautiful home landscapes.

Thinking Green

Xeriscapes

Landscaping designs that use plants requiring very little water are called xeriscapes. These designs often include succulents, cacti, ornamental grasses, and other drought-tolerant plants. In what areas of the United States would this type of landscaping be most appropriate?

Bruce C. Murray/Shutterstock.com

Corner Question

What is a mole cricket?

Athletic Fields

Like the home lawn, athletic fields may be considered beautiful and are often painstakingly maintained. Many grasses have been bred to withstand heavy traffic and to recover quickly after an afternoon of intense practice or play on the field. Baseball, football, and soccer are just a few of the sports that require green grass for play.

Golf Courses

Golf courses are athletic fields that range from a few acres to 200 or more acres of turf, **Figure 4-11**. Specialists in the turfgrass industry have been experimenting and cultivating different turfgrass species for golf courses for years. Golf course turfgrasses must have adequate insect and disease resistance as well as tolerance to salinity, heat, foot and cart traffic, and drought conditions. Additionally, golf course grasses must meet standards that include everything from their color and texture to the way a golf ball rolls when it lands on the green. Golf course turfgrass needs intensive maintenance. Some parts of the golf course get mowed on a daily basis, regularly removing much of the leaf area used for photosynthesis. Often-mowed grasses need extra fertilizer to make up for the removal of their leaf area.

Safety Note

Walk-Behind Mower

You should mow side-to-side on a hill when using a walk-behind mower. Make sure that the grass is dry before mowing. Wet grass can cause operators to slip and fall into the path of the mower blade, causing serious injury.

cappi thompson/Shutterstock.com

Figure 4-11. Golf courses are maintained by turfgrass workers in the nursery and landscape industry. How many types of turfgrass are used in a single golf course? How and why are different types used?

Sod Growers

Sod growers are responsible for planting and growing turfgrass that will be harvested and sold. Sod growers work outdoors with a team of individuals. Sod cultivation requires the use of heavy machinery. A sod

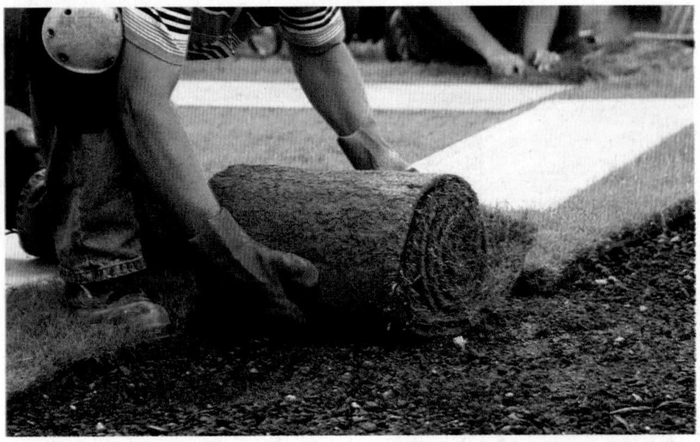

Figure 4-12. Laying sod is labor intensive work, much of which must be done by hand. *What types of machinery are used in the cultivation and harvesting of sod?*

grower must understand how to cultivate and harvest various types of turfgrass. He or she must be able to lift pieces of sod that can weigh more than 50 lb, **Figure 4-12**.

A sod grower may pursue a degree in turfgrass or have some additional formal education. Sod, like almost all crops, must be fertilized and have a pest management plan in place. Additionally, the turfgrass must be cut to a certain height for proper growth and production.

Combining Edible and Ornamental Plants

When designing and planting a landscape, consider using plants that serve two purposes, beauty and nutrition. Vegetable and fruit plants, trees, and shrubs are beautiful just like ornamental trees and shrubs. Edible plants may be placed:

- In mixed containers.
- In planting beds.
- In landscape borders.
- With flowers and ornamental grasses.

As when considering ornamental plants for a design, you should consider the characteristics of fruit or vegetable plants before inserting them into the design. For instance, *Vaccinium angustifolium* 'Tophat' is a blueberry plant that has been bred to grow in a compact form. 'Tophat' blueberry shrubs reach about 24" (61 cm) tall and wide. The plant is blanketed with small blueberries and is perfect for full sun or partial shade. Like most blueberries, it is a perennial in many areas of the country, **Figure 4-13**.

Figure 4-13. This blueberry plant is an excellent edible addition to any garden space. *What types of edibles would you incorporate into a landscape design?*

Using edibles in the landscape adds beauty and can impact your dinner plate and your wallet. Fruits and vegetables grown in the garden may cost a fraction of what they would at a grocery store or farmers market. Fruits and vegetables cultivated at home will also have more nutritional content than those shipped from great distances. Those grown locally or at home are picked when ripe instead of being picked early to allow time for shipping. Eating soon after picking may increase the amount of nutrients available in the food, since there is little time for decomposition. Growing vegetables can help you save money and provide healthy foods.

Corner Question

What is the difference between a reel lawn mower and a rotary lawn mower?

Organic and Sustainable Production

Plants and crops have been grown for many years using conventional methods. These methods may include using practices or harsh chemicals that can harm people, animals, or the environment. Growing foods organically and using sustainable agriculture are two alternatives to using conventional methods.

Organic Edibles

You often hear about eating organic food, but what does that really mean? Most people would agree that *organic foods* are those that are raised using more biological pest control (such as integrated pest management) than chemical control, and without genetic modifications of any kind. The United States Department of Agriculture (USDA) does award a special organic certification to growers after careful evaluation of their farm and growing practices. In part, a farm may be certified if the food's production does not include any synthetic (chemically derived) fertilizers or pesticides. To maintain their organic label after leaving the farm, foods cannot be processed using chemical food additives or industrial solvents. Also, they must not be irradiated to kill bacteria or insects, and they generally cannot be processed using the same machinery as non-organic foods.

There is a general belief among the public that organic foods are safer, and this has fueled the incredible demand for organic food and products. Farmers around the world are working to meet these demands. Unfortunately, organic foods cost the consumer 10% to 40% more than conventionally produced foods because they cost more to produce. Producing organic food is much more labor intensive than conventionally produced food. Organic food production accounts for only 1% to 2% of food grown in the world, but the organic sales market is growing swiftly. Organic food production is the fastest growing division of the food industry in the United States.

Did You Know?
The United States Department of Agriculture (USDA) says that the organic label does not guarantee greater safety, health, quality, or nutritional value for organic foods. The label only ensures that the food was grown organically according to USDA standards.

Safety Note
Food Safety
Just because you are eating something grown locally does not mean you can throw safety aside. Storing foods in the refrigerator is still important. Wash your fruits and vegetables. Cook foods to the appropriate temperatures to kill any pathogens and refrigerate them within two hours after heating.

Thinking Green

Conventional Foods' Carbon Footprint

Conventionally grown food is responsible for 5 to 17 times more carbon dioxide than food that is produced locally or regionally. From where does this additional carbon dioxide come? Make a list of possible sources including those caused by the packaging of seeds to harvesting the produce.

Consider every aspect of crop production when looking for causes of carbon dioxide production. For example, both the combine and the truck used to harvest and transport the rice use fossil fuels and generate carbon dioxide.

©iStock.com/alffoto

Corner Question

Does the United States Department of Agriculture (USDA) certify sustainable agriculture farmers?

Thinking Green

Parasitic Wasps

Various types of wasps are used by some farmers as a means of pest control. These wasps lay eggs on pest larvae. When the eggs hatch, they feed on the pest larva. Parasitic wasps are often referred to as predatory wasps. Which species of wasps are used by farmers as a means of pest control?

©iStock.com/grandaded

As with everything else, as long as there is demand, there will be growers producing foods to fill that demand, especially when those foods can bring premium prices. There is still a great debate between conventional and organic foods producers regarding the benefits of organic foods. However, consumers are showing through their purchases that they value organic food.

Sustainable Agriculture

Many farmers do not grow their crops completely organically or completely conventionally. However, most growers do consider their growing practices to be sustainable. *Sustainable agriculture* is farming or producing plant and animal products in ways that promote the health of people, animals, and the environment. Sustainable agriculture employs practices that will conserve natural resources and ensure that future generations will have the opportunity to farm.

Pest Management

Many farmers employ farming practices that are organic, but when needed, they may use pesticides. *Integrated pest management (IPM)* is an approach to managing pests that uses commonsense, economical practices, and results in the least possible hazard to people, property, and the environment. When it is necessary to apply chemicals, only the least harmful methods of application are used, **Figure 4-14**. IPM involves monitoring, trapping, and using other means of pest control, such as natural predators (ladybugs, praying mantis) or crop rotation, before resorting to the use of harsh chemicals. IPM benefits farmers by helping to keep their land and the ecosystem surrounding their farms healthy.

Soil Nutrition and Conservation

Sustainable farming also involves bettering the soil. Certain methods of crop rotation and growing cover crops can be

Safety Note

Organic Pesticides

When growing plants organically, it is just as important to garden safely as it is when using conventional methods. When using a pesticide labeled "organic," you must follow the directions for safely using the product.

Frederico Rostagno/Shutterstock.com
Figure 4-14. Sometimes, organic farms must use chemicals to manage pest problems or treat disease. In this image, pesticides are being sprayed on a field using a boom sprayer. In what other ways can farmers deter pests and treat disease?

used to improve and preserve the soil's nutrients. Sustainable farmers often practice no-till farming, which also helps to conserve soil.

These methods just briefly outline the mission of sustainable agriculture. Farmers who grow their foods for consumers also grow foods for their families and themselves. Sustainable agriculture practices can ensure that there will be farming tomorrow, while providing food for today.

Eating Local

Have you ever really thought about the freshness of grocery store produce? Traditional paths for fresh fruits and vegetables include a postharvest life of one to two weeks, in addition to a trip of an average of 1500 miles:

- Harvesting, washing, packaging and transportation to a supermarket (five to seven days).
- Storing at the supermarket (one to three days).
- Storing in your refrigerator (for an average of up to seven more days).

By the time you eat the fruits or vegetables, your once nutritious and fresh produce has potentially less nutritional value. To increase the amount of nutrition consumed in fresh fruits and vegetables, buy them locally and eat them within a short time.

Thinking Green

Bacillus thuringiensis (Bt)

Bacillus thuringiensis (Bt) is a soil-borne bacteria that has been used since 1901 and now is important to the organic food industry. This bacteria attacks soft-bodied larva such as caterpillars and helps to save crops from their damage.

Harvesting

How produce is harvested and stored impacts the shelf life and nutritional content significantly. When harvested mechanically, produce can be damaged easily. It will quickly begin to decay and lose its nutritional content at a higher rate. Every part of the harvest and postharvest process plays a role in the nutritional and health benefits of a fruit or vegetable.

Locavores

Many consumers are interested in eating locally grown food because of its wealth of benefits, from less packaging and transportation to supporting local farmers. These individuals are known as *locavores*. Locavores are passionate about buying local ingredients or eating at restaurants that use locally cultivated produce. The common understanding of *local* is within 100 miles.

Farmers Markets

One of the easiest ways to obtain fresh, nutritious, and local produce is to visit your local farmers market, **Figure 4-15**.

Arina P Habich/Shutterstock.com

Figure 4-15. By purchasing locally grown foods, consumers can be assured that less petroleum and fewer carbon emissions are being released due to their choice to buy local. It is a great idea to take your own basket for better transport and less damage to the produce at farmers markets. **Do you and your family use reusable bags or containers when you shop?**

Corner Question

How many farmers markets are in the United States?

Thinking Green

Market Baskets

Taking a basket or reusable bag to the farmers market is a good idea. You will be able to carry produce without damaging or bruising it, and you will not be adding another plastic or paper bag to a landfill. How many plastic bags does your family acquire from a month's grocery shopping?

©iStock.com/ernstboese

Farmers markets were once prevalent across the country, but they lost popularity as convenience, big box, and chain grocers gained acceptance. Farmers markets are finding their place again as a key food supplier in city squares and parking lots across the country.

When visiting a farmers market, you have the opportunity to speak to growers and distributors that know where, how, and when the produce was cultivated and harvested. The seller can provide information about the produce that most grocers cannot.

Purchasing produce from a farmers market gives you the power to better control the nutritional content of the food you consume. Produce purchased from a local grower has not been shipped or been in storage for as long as most of the produce at other supermarkets.

Community Supported Agriculture (CSA)

Did You Know?

CSA farms are usually family-operated, range from 3 to 300 acres, and deliver food for 10 to more than 200 households. CSA farms are highly diversified and usually grow more than 40 different vegetables, herbs, and fruits.

Another venue for purchasing locally grown food is through *community supported agriculture (CSA)*. CSA is a farming practice in which people pay in advance for shares of produce that is delivered at harvest. This arrangement has been a popular way to buy locally grown food for years. A farmer sells "shares" to the public that represent boxes of seasonal vegetables and fruit that are delivered or can be picked up at a location on a schedule. CSA plans are beneficial to both the farmer and the CSA plan members. The grower is given the opportunity to develop relationships with buyers, educate them about the produce and growing methods, and spend less time marketing the food. The members eat fresh and nutritious food, know the cultivation practices used, may get to visit the farm, and are exposed to new types of produce and recipes.

The farmers and the members are both invested in the CSA and the farm and recognize the benefits as well as the risks involved. Members pay for their shares up front. If there is a drought or an infestation of pests and there are slim harvests, the farmer does not reimburse its members for the cost of the shares they purchased. In this way, CSAs are a type of crop insurance for farmers.

The Green Industry

The environmental horticulture industry is also referred to as the green industry. This industry consists of several products and services associated with landscape and nursery production, **Figure 4-16**. Businesses and organizations related to the green industry include:

- Wholesale and retail nurseries.
- Compost, soil, and media producers.
- Sod growers.
- Lawn care services.
- Landscape designers and architects.
- Landscape managers and maintenance firms.
- Greenhouse growers.
- Urban foresters and arborists.
- Masons.
- Public gardens.
- Lawn and garden departments of large stores.
- Landscape and garden suppliers.
- Landscape equipment suppliers.

Members of the green industry specialize and have expertise in growing, maintaining, and managing plants, landscapes, and all components associated with gardening of ornamental plants.

Regardless of global location, the green industry provides biological, environmental, aesthetic, and economic benefits through ornamental plant cultivation.

Thinking Green

The Omnivore's Dilemma

Written by Michael Pollan in 2006, *The Omnivore's Dilemma* was considered a catalyst for consumers to join CSA plans and eat locally grown food. Many people think that this book drove the rise in CSA and farmers markets across the country.

Did You Know?

The state of Virginia has planted more than 500 acres of wildflowers along its highways.

somrak jendee/Shutterstock.com;

Minerva Studio/Shutterstock.com;

Ozgur Coskun; Shutterstock.com;

Jiinna/Shutterstock.com

Figure 4-16. Green industry business activities range from seed production to landscaping. A—Turfgrass worker watering newly installed sod. B—Working in a greenhouse and taking care of geraniums, an annual bedding crop. C—A woman installing a hardscape composed of brick pavers. D—People enjoying the plants at a botanic garden. Can you list five green industry applications you passed on your way to school today?

Did You Know?
Trees can cool the surrounding environment by 10 degrees through shading and evapotranspiration.

Biological and Environmental Impacts

Plants play a major role in any ecosystem. Plainly speaking, without plants, an ecosystem could not exist. The plants that are used in the green industry have biological and environmental impacts as they:

- Act as a food source for other organisms.
- Help to add nutrients to the soil through decomposition.
- Control erosion by holding soil intact with their root systems.
- Add oxygen to the air through the process of photosynthesis.
- Sequester carbon that is emitted into the atmosphere.
- Provide shade.
- Reduce temperatures through the process of evapotranspiration in leaves.

Evapotranspiration is the release of water through plant leaves, which then evaporates and helps cool the air. Plants are essential to the environment and their impacts and benefits are realized by all creatures within that bionetwork.

tommaso79/Shutterstock.com

Figure 4-17. Green spaces do not have to be elaborate to provide aesthetic and health benefits to an urban landscape.

Aesthetic Impacts

Although beauty may be in the eye of the beholder, most people realize the lack of natural beauty in a concrete landscape. Aesthetically, a simple plant on a concrete canvas can make a huge difference in making a space seem more inhabitable and welcoming. As people have recognized both the aesthetic and health benefits of plant materials, there have been many efforts to promote green spaces in urban settings, **Figure 4-17**. Cities often require new construction to include public green spaces as part of a building's design. Many cities and towns are also incorporating green spaces by converting empty lots into public gardens, promoting the installation of rooftop gardens, and by planting trees and plants along public roads.

Thinking Green

Tree Gators

A Tree Gator® is a self-watering tree bag that efficiently waters newly established trees. The bags are made of plastic and have very small holes at their base. The bag wraps around the tree's trunk and slowly releases water over a one- to two-week period, depending on the weather. What do you think inspired someone to invent this product?

Spectrum Products Inc.

Thinking Green

Living Roofs

A green or living roof is one that is partially or completed covered with soil, vegetation, and a waterproof membrane. Aside from being beautiful, living roofs provide insulation, reduce water runoff, and create a habitat. However, roofs must be supported well, as green roofs, especially when wet, are extremely heavy.

Alison Hancock/Shutterstock.com

Corner Question

What automobile manufacturer planted 454,000 square feet atop its roofs?

Studies have shown that human interaction with a beautiful space can improve behavior and mental health. Green spaces can:
- Restore a mind of mental fatigue and help a person focus.
- Help children develop emotional, cognitive, and behavioral connections.
- Encourage creativity and intellectual development.

Economic Impacts

The green industry contributes significantly to the economy of the United States. In recent years, billions of dollars were contributed through annual revenue to the US economy by the green industry. Although many jobs may be seasonal, the landscape service sector of the green industry employs the most workers with more than 1 million employees throughout the United States.

Future of the Green Industry

The green industry continues to be the fastest growing segment of the national agricultural economy. As people express more interest in eco-friendly landscaping and sustainable ways to maintain it, the green industry will continue to grow. Ways in which green industry businesses can capitalize on this opportunity include:
- Cultivating and promoting water-wise plants (plants that use less water).
- Incorporating edibles into landscaping designs.
- Encouraging rainwater harvesting and use.
- Installing sustainable landscapes.
- Using integrated pest management (IPM).
- Offering organic or chemical-free services.
- Promoting new technologies such as permeable pavers.
- Designing affordable, green outdoor living spaces, **Figure 4-18**.

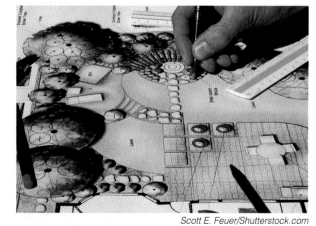
Scott E. Feuer/Shutterstock.com

Figure 4-18. As the number of homes increase, so will the demand for the green industry's services, products, and professionals.

> **Corner Question**
>
> How many states have colleges or universities that offer agricultural or green industry degrees?

Green industry businesses will continue to grow by providing these types of services, designing and marketing products that support these services, and developing even more sustainable agricultural applications.

Careers

In the next decade, the Bureau of Labor Statistics foresees a growth in horticulture industry jobs by 18%. An increase of that size means that there will likely be more jobs than there are qualified employees. There will be many jobs in the horticulture industry for people entering the workforce. Developing unique skills, such as being bilingual or having extensive computer skills, will undoubtedly give prospective employees a competitive advantage in the horticulture industry.

To help you develop your own skills, employment opportunities and skill development are discussed throughout this textbook. Each chapter also features real-life career profiles focusing on people currently involved in different aspects of the horticulture industry.

Continuing Education

Successful people pursue learning opportunities, even after getting a degree and beginning their careers. Learning opportunities include formal and informal classes, professional presentations, industry conferences, and other methods of learning. The main purpose of continuing education is to keep your skills current and to learn about the latest breakthroughs and ideas.

Horticulture Organizations

Professional organizations or societies are groups that seek to promote a particular career area and help or inform people in that career. They can provide friendship, networking opportunities, job leads, learning experiences, professional development, and personal growth. Millions of people are employed in horticulture, and many of them participate in professional organizations. Some of these organizations are described here.

American Horticulture Society

For more than 90 years, the American Horticulture Society (AHS) has catered to nearly 20,000 gardeners and horticultural professionals. The AHS connects people to gardening and the most current research. Through publications and educational workshops here and abroad, the AHS fosters the science and art of horticulture.

American Society of Horticultural Science

The American Society of Horticultural Science (ASHS) is the world's leading horticultural science professional society. There are thousands of members from over 60 countries throughout the world. The organization took root in 1903 and has worked side by side with academic and industry researchers in the field of horticulture. The society wishes to reach academic professionals, industry stakeholders, government employees, and consumers with research, education, and extension activities.

International Society of Horticulture Science

During the nineteenth century, horticulturists around the world recognized the need to exchange information to ensure a global understanding of plants. Thus, they established the first International Horticultural Congress in Belgium in 1854. In 1959, the organization officially became the International Society of Horticulture Science (ISHS).

Today, more than 7500 members around the world participate in ISHS. The organization's objective is to inspire research and education in all branches of horticultural science and to spread knowledge on a global scale through symposiums and publications. Membership is open to all researchers, educators, students, and horticultural industry professionals.

National Association of Landscape Professionals (NALP)

The National Association of Landscape Professionals boasts 100,000 professional landscape and lawn care industry members. The NALP provides industry education, safety education, networking, and business development opportunities. The organization also hosts webinars, conferences, days of service (volunteer opportunities), and industry conventions, **Figure 4-19**.

AmericanHort

In January 2014, AmercianHort was founded by the merger of the American Nursery and Landscape Association and the Association of Horticultural Professionals. This organization represents all horticulture areas and includes students, educators, researchers, breeders, retailers, distributors, greenhouse and nursery growers, interior and exterior landscapers, florists, manufacturers, and all others who are part of the industry. AmericanHort claims to be the leading and the largest association for the horticulture industry. The mission of the organization is to unite members across the country, be influential advocates, support plant businesses, and promote healthy communities.

> **Corner Question**
>
> Which horticulture organization is housed on 25 acres on a historic estate in Virginia?

Marcin Balcerzak/Shutterstock.com

Figure 4-19. Sod workers are installing turfgrass in an NALP project. Are there any NALP projects in your area? Are there other horticultural organizations in your area looking for volunteers?

CHAPTER 4

Review and Assessment

Chapter Summary

- Botany is the study of plants, and there are three botanical sciences: forestry (trees), agronomy (food and fiber crops that are traditionally grown in fields), and horticulture (fresh fruits, vegetables, flowers, grasses, shrubs, landscapes, and the services of this science).
- Horticulture is divided into two divisions, edible horticulture and ornamental horticulture.
- Edible horticulture involves pomology (trees and nuts), olericulture (vegetables and fruits), and viticulture (grapes).
- Ornamental horticulture includes floriculture, nursery and landscape management, interiorscapes, and turfgrass production.
- Eating locally grown produce may increase the amount of nutrition you consume.
- Growers for farmers markets and community supported agriculture cultivate food using organic, sustainable, or conventional practices.
- The green industry consists of several products and services associated with landscape and nursery production and is also known as the ornamental horticulture industry.
- Many career opportunities are available in the horticulture industry and the number of jobs is expected to grow over the next several years.
- Professional organizations or societies are groups that seek to promote a particular career area and help or inform people in that career. There are several professional organizations that work with the horticulture industry.

Words to Know

Match the key terms from the chapter to the correct definition.

A. botanist
B. bramble
C. enologist
D. ethylene
E. evapotranspiration
F. forest stand
G. forester
H. geocarpy
I. green industry
J. hardscape
K. horticulturist
L. interiorscaping
M. locavore
N. nursery
O. olericulture
P. organic food
Q. ornamental horticulture
R. perennial
S. plant science
T. pomology
U. postharvest
V. propagate
W. senescence
X. silviculture
Y. sod
Z. viticulturist

1. A hormone produced and emitted in varying quantities by fruits and vegetables and by decaying plant materials.
2. The portion of the horticulture industry that cultivates and arranges outdoor plant materials to create spaces that are inviting, beautiful, and useful to people and the ecosystem.
3. A consumer who is interested in eating locally grown foods.
4. Produce that has been raised without synthetic chemicals or genetic modifications of any kind.
5. The cooling, cleaning, sorting, storing, packing, and shipping of produce, flowers, and other plant materials.
6. A scientist who studies plants, including their structure, genetics, ecology, classification, and economic importance.
7. The release of water through plant leaves, which then evaporates and helps cool the air.
8. A rare means of plant reproduction in which the flower stalk (after pollination) elongates and pushes into the ground where the fruit matures.
9. A person who specializes in the science, art, technology, and business of plant cultivation.
10. The science, cultivation, processing, storage, and marketing of herbs and vegetables.
11. The study of plant growth, reproduction, and adaptation as well as the use of plants for food, fiber, and ornamental purposes.
12. The study, cultivation, and marketing of flowers and ornamental plants.
13. To grow plants from seeds or other methods, such as rootings or cuttings.
14. The cultivation, processing, storing, and marketing of fruits and nuts.
15. A prickly, rambling vine or shrub.
16. A person who uses grapes to make wine using chemistry and food science knowledge.

17. Trees or other growth occupying a specific area that are uniform in species, size, age, arrangement, and condition.
18. A professional who specializes in the cultivation of grapes to be eaten fresh and to be used for making juices, raisins, jams, jellies, and wines.
19. The ripening or biological aging of harvested crops.
20. A plant that lives longer than one or two years.
21. A collection of grass plants that form a ground cover, often used for sports areas, homes, and industrial sites.
22. The science or practice of planting, managing, restoring, and caring for forests; the cultivation of trees.
23. A place where young plants and trees are cultivated for sale and for planting elsewhere.
24. The constructed areas around a building or in a landscape, such as pavers, patios, sidewalks, and retaining walls.
25. The design, installation, and maintenance of plants inside buildings.
26. A person who plants, manages, and cares for forests.

Know and Understand

Answer the following questions using the information provided in this chapter.

1. Identify and briefly describe the four areas of plant science.
2. Cultivated crops are used for what three main purposes?
3. In what areas might an agronomist specialize?
4. List the three edible horticultural sciences and the four ornamental horticultural sciences.
5. In horticulture, what part(s) of a plant are considered to be a vegetable?
6. Explain how high-density orchards differ from traditional orchards.
7. What is viticulture (also called viniculture)?
8. What is floriculture and what are some careers that fall under this area?
9. How are trees and shrubs different?
10. Describe three methods of cultivating and marketing nursery plants that are field grown.
11. What are four benefits of interiorscapes?
12. Why did growing a home lawn become popular in the 1940s in the United States? How did this affect horticultural companies?
13. Why is growing edibles throughout a garden a good practice?
14. What are some criteria used by the United States Department of Agriculture (USDA) in granting a farm or food organic certification?
15. What is sustainable agriculture?
16. What is integrated pest management (IPM)? What are some strategies associated with IPM and what benefit does it provide?

17. How are farmers markets and CSA plans alike and different?
18. Describe how farms can use community supported agriculture plans to help their businesses be successful.
19. What are three biological and environmental impacts of plants used in the green industry?
20. What is the outlook for jobs in the horticulture industry for the next decade? What learning opportunities exist for continuing education in the horticulture field?

Thinking Critically

1. You purchased avocadoes on Monday to make guacamole. The avocadoes are very firm, and the produce clerk told you they should be fine until Friday, when you plan to make the guacamole. You place them on your counter with all of the produce that does not go in the refrigerator. On Wednesday, you check your avocadoes. They are very soft and the skin is beginning to darken. What could have affected their shelf life?
2. You live on the coast and have had native turfgrass growing on your property. However, you would like to grow a new variety of grass that is softer and better suited for walking on barefoot. The old grass is removed and the new grass, which is sod, is planted in its place. You have been watering regularly, but you notice the grass is looking stressed. When you call the garden center where you purchased the grass, you tell them how much and how often you have been watering. They agree that you have been watering properly. When they ask you where you live, you tell them you live on the beach. They say there is nothing they can do to help. What do you think your location has to do with the suffering of the turfgrass that you purchased and are trying to establish?

STEM and Academic Activities

1. **Science.** Identify an insect in your classroom, garden, greenhouse, or at your home. Determine what the insect is. Will it cause problems or is it beneficial to plants? If it is a pest, find a biological predator that can be used to control it. Compare their life cycles. Why is one the predator or control agent? How did this situation evolve in nature?
2. **Technology.** Find a local garden (this could be at school, on a farm, or in a local neighborhood). Measure the garden site and determine the number of square feet it contains. Record the information and do calculations in a spreadsheet program. Choose a cover crop that could be grown on this site that would help to amend the soil (possibilities are clover, alfalfa, and wheat). Determine how many pounds of seed will be needed for this garden site. Additionally, determine how much this would cost, not only for the seed but for the labor involved.
3. **Engineering.** An urban restaurant wants to plant a garden on the roof of the building. These building structures are called green roofs. When the restaurant owner contacts the company to modify the building to house the green roof, the owner is told that she must first contact an engineer. What are three things an engineer would have to evaluate about the building before a green roof or rooftop garden could be planted?

4. **Math.** A football field measures nearly one acre without the end zones. As a turfgrass consultant, you must determine how much sod must be purchased for this space. An acre measures 43,560 ft². The average size of one piece of sod is 16″ × 24″. (41 cm × 61 cm) How many pieces of sod will you need to purchase if you want to buy 10% extra for error? What is the current price for sod? How much would the sod cost at the current price?

5. **Language Arts.** Search the Internet, horticulture trade magazines, or a local newspaper for three job postings related to the horticulture industry. Choose the position in which you are most interested. Create a résumé with your information for this job posting. Do you have all the qualifications? Make notes of the additional education and experience needed to fulfill the qualifications and how you would go about becoming qualified for the position.

6. **Language Arts.** Use one of the Corner Quotes in this chapter to act as a springboard for a position paper on the importance of horticulture. Create an essay map and graphically lay out a paper. Search the Internet to find the website for ReadWriteThink. Explore this website for tools to help you organize your paper on the horticulture industry.

Communicating about Horticulture

1. **Reading and Speaking.** With a partner, make flash cards of the *Words to Know* listed at the beginning of the chapter. On the front of the card, write the term. On the back, write the phonetic spelling as found in a dictionary. Practice reading the terms aloud, clarifying pronunciations where needed.

2. **Reading and Writing.** Contact one of the horticulture associations via e-mail. Ask for information about a particular topic (related to the organization) in which you are interested. When you receive the information, read it and determine if it answers your questions. If it does, write an e-mail thanking the person and telling what you learned from the information. If you think the information did not answer your questions, write another e-mail requesting additional information. Make sure your queries are clearly stated. Be certain to thank the person for his or her time and attention to your questions.

3. **Speaking.** Debate the topic of food preservatives and synthetic chemicals. Divide into two groups. Each group should gather information in support of either the pro argument (preservatives are necessary for the _____ industry) or the con argument (the chemicals used in preservatives are toxic and can be dangerous for the environment and the workers that handle them/food). Use definitions and descriptions from this chapter, as well as other resources, to support your side of the debate and to clarify word meanings as necessary. Do additional research to find expert opinions, costs associated with horticultural chemicals, and other relevant information.

SAE Opportunities

1. **Exploratory.** Job shadow a florist in your community.
2. **Exploratory.** Contact a local nursery and arrange an interview with the owner or manager. Do some advance research and prepare a list of questions. Contact a big box or chain store with a garden center and arrange an interview with the department manager. Use the same list of questions to interview the department manager. Compare the responses and nurseries and construct a chart or diagram highlighting your findings.
3. **Exploratory.** Visit a grocery store and go to the produce department. Inventory the types of fresh fruits and vegetables. Create a list of what is cultivated by a pomologist and an olericulturist. Determine what plant part is being eaten for each of the produce items.
4. **Experimental.** Harvest or purchase several of the same pieces of produce (examples: heads of lettuce, apples, blueberries, or sweet corn). Develop an experiment with various postharvest treatments for the crop of your choice. Examples of variables could include washing before storage, various temperatures of storage, various humidities, and exposure to lights or gases.
5. **Experimental.** For this activity, you will need several cut flowers from your garden and a separate container for each flower. Although separate, flowers must be placed in the same setting (light, temperature, humidity). Before you harvest the flowers, create a chart to keep track of variables and how each flower's shelf life varied. Variables may include the type of water (municipal, well, with floral preservative, or without floral preservative); the time of harvest (morning, noon, or early evening); and/or the method of harvest (floral knife, steak knife, scissors, or pruning shears). Analyze your results to determine the best methods for harvest and storage.

USDA/David Kosling

CHAPTER 5
Horticultural Business Management

Chapter Outcomes

After studying this chapter, you will be able to:
- Understand the purpose and parts of a strategic business plan.
- Describe marketing and advertising for horticulture businesses.
- Recognize professionalism in employees.
- Explore career documents.
- Prepare for and take part in a job interview.
- Create a school-to-career plan.
- Explore careers related to horticultural business management or marketing.

Words to Know

advertising	letter of application	selective market coverage
brand	marketing	small business
core ideology	mission statement	standard
direct sales	overhead	strategic business plan
entrepreneur	professionalism	value
envisioned future	profit margins	vendor
goal	reseller sales	vision
intensive market coverage	résumé	
job interview	school-to-career plan	

Before You Read

After reading each section (separated by main headings), stop and write a three- to four-sentence summary of what you just read. Be sure to paraphrase and use your own words.

While studying this chapter, look for the activity icon **to:**

- **Practice** vocabulary terms with Words to Know activities.
- **Expand** learning with identification activities.
- **Reinforce** what you learn by completing Know and Understand questions.

www.g-wlearning.com/agriculture

As a child, when you pictured yourself as an adult you most likely imagined yourself as successful and happy. Maybe you thought of yourself as a doctor or lawyer or the owner of a thriving business. Often times, the image may have been you being the boss. Most people do not necessarily want to work for someone else but would rather be in charge and give direction or manage others. After all, that is usually where the financial benefits develop.

Opportunities to join the horticulture industry and own a business increase annually. This multibillion-dollar industry continues to grow along with the global population. There are mouths to feed and houses that need new or renewed landscapes. Opportunities abound for the horticultural entrepreneur looking to enter into the business. But owning a business is not easy. It takes time, hard work, and a great deal of planning.

Small Businesses

Did You Know?

In July 1934, a dozen farmers joined what they called a *village* in Los Angeles on the corner of Fairfax and 3rd Street. This was the first American farmers market.

Every month in the United States 543,000 small businesses open, and during that same month more than that number closes. This statistic is not caused by a single factor. Instead, many factors contribute to the success or failure of any business. A business is a legal entity that produces or buys and sells products and/or services to make a profit. The US Small Business Administration (SBA) defines a *small business* as a company that is independently owned and operated, is organized for profit, and is not dominant in its field.

A small business may have employees or may be identified as a nonemployer. In the United States, there are 28 million small businesses with 500 or fewer employees; 22 million of those are nonemployers. Nonemployers are companies run by someone who is considered self-employed. Other interesting facts about American businesses include:

- Nearly 70% of new firms survive at least 2 years.
- About 50% survive at least 5 years.
- About 30% survive at least 10 years.
- About 25% stay in business 15 years or more.

If you are thinking of going into business for yourself, you should be aware that only one out of four businesses that open their doors today will be around in the years to come. Do not be frightened by these statistics. After all, if all entrepreneurs decided to quit before they even tried, there would not be any farmers markets, garden centers, or landscape services open anywhere in the world, **Figure 5-1.**

sunlover/Shutterstock.com

Figure 5-1. Every year many businesses open their doors for the first time, but not all survive. Farmers markets are thriving in the current business climate, but that trend may not continue. What types of horticulture businesses can survive and even thrive in today's competitive business climate?

Millions of entrepreneurs have their own successful businesses, **Figure 5-2**. An *entrepreneur* is a person who organizes and operates a business. Running a business requires tireless effort that often has numerous challenges. A great deal of effort and time is needed to develop the concept and the framework for the business. Once the business opens its doors, the work has just begun. A business must constantly work to improve and compete with other companies.

Strategic Business Plans

Nearly 80% of businesses fail, and there are many reasons for these failures. Businesses should develop a strategic business plan to help them be successful. A *strategic business plan* is a document that states the mission of the business, examines its current condition, sets goals, and outlines strategies for achieving the goals. A *goal* is an objective to be achieved. A strategic business plan is a road map for the future, and it often helps a business to survive even in an economic downturn. It helps to organize the activities of the business through:

- Establishing company information.
- Creating a mission statement and vision.
- Determining products and services.
- Identifying customers and market.
- Assessing competition.
- Evaluating risks.

A strategic business plan directs a business toward achieving its goals. In addition, the plan also monitors performance and creates solutions for underperformance. For ways in which a traditional business plan differs from a strategic business plan, see **Figure 5-3**.

Monkey Business Images/Shutterstock.com

Figure 5-2. A floral business has a great deal of competition locally and globally. What do you think are the biggest challenges for floral businesses?

"There are no secrets to success. It is the result of preparation, hard work, and learning from failure."
—Colin Powell

Traditional Business Plan	Strategic Business Plan
• Executive summary • Company description • Management and organization • Market and competition evaluation • Product development • Marketing, sales, and service plan • Financial plan	• Assess external environment • Assess value chain and resources • Formulate vision and goals • Develop overarching strategy and implementation plan • Evaluate, monitor performance, adjust, and strategize

Goodheart-Willcox Publisher

Figure 5-3. Although a strategic business plan contains elements of a traditional plan, a strategic plan not only defines company goals but uses those goals to take advantage of available business opportunities.

Corner Question

How many horticultural farms are in business in the United States?

Thinking Green

Start a Business to Save the Planet

Do you have a business idea that could also help the environment? There are "green" businesses opening their doors every month, especially in the horticulture industry. Farmers markets with sustainably or organically grown produce and composting manufacturers are just a few enterprises that are having great success. Read this chapter and start developing a business plan for your own green business.

Did You Know?
Ball Horticultural Company (West Chicago, Illinois) is a world premiere leader in plant development and distribution. The company began as a wholesale cut flower business in 1905.

Strategic Environment

The strategic environment of a business includes the conditions, circumstances, and influences that affect the company. A business must determine what to sell and then gather information about potential customers. To sell a good or a service, there must be a want or a need for the item or service. There must also be a price that both the customer and the business can agree on. Company information, the industry environment, and competitors must also be considered.

Company Information

Company information includes an overview of the company history, structure, management, employees, and other resources. Company information can also include:

- The company's *brand* (a name, label, logo, or image under which a product is sold).
- The company's products or services.
- The company locations and facilities.
- How products will be made.
- How the products or services will be distributed.
- The financial status of the company.

Business Structures

According to the Internal Revenue Service (IRS), several business structures thrive in today's global economy. Those structures are outlined here.

A sole proprietorship is a common business structure in which one individual (or a married couple) is in a business alone. This form of business is simple to operate, flexible, and has fewer taxes and legal controls than other structures. The business owner is solely responsible for all debts incurred by the business. An example of a sole proprietorship is ownership of a lawn care service.

In a general partnership two or more people agree to contribute both money and labor to the business. An example is siblings operating a greenhouse business.

A limited partnership consists of one or more general partners and one or more limited partners who join together in the business. General partners manage the business while limited partners share only in the financial part of the business. An example is a floral design business owned and operated by one person with financial backing from investors.

A limited liability partnership (LLP) business structure is similar to a general partnership, except the business partner does not have financial liability for the negligence of another partner.

A corporation is a complex business structure that has liabilities, rights, and responsibilities beyond that of an individual. A corporation is eligible for many tax benefits but also has increased fees (such as licensing) and decreased personal control.

A nonprofit corporation is an entity established to further an ideal or goal. It is not solely established to create revenue. Nonprofit corporations are often associated with charitable causes like the Leukemia and Lymphoma Society.

A limited liability company (LLC) is formed by one or more people or entities through a written agreement detailing the organization of the LLC. The agreement also lays out the distribution of profits and losses.

A municipality is a public corporation established as a subset of the government. An example is the water municipality in your city.

An association is an organized group of people sharing a common interest, bond, or effort in business. An example of an association is the National FFA Organization.

Industry Information

The industry information that should be considered includes the demographic, sociocultural, and economic factors that impact a business.
- Demographics—the age, income, and gender of your customers. Does your product or service cater to one generation over another?
- Sociocultural—the culture, race, or ethnicity of your customers. The US market is always changing. For example, salsa recently passed ketchup as the highest-selling condiment in the United States.
- Economics—the economic environment. Is there a recession? Is the economy growing?

Competitive Information

The competitive environment is the relationships among businesses in an industry. Other factors that affect how competitive a company can be include technological, political and legal, and environmental factors.

Technological factors include knowing what current technologies could impact your company and what changes in technology are on the horizon. Political and legal factors include knowing your government officials and the political climate and understanding their leadership style. What legal issues are happening that could affect the development of this business today or tomorrow? Environmental factors are concerned with the actual climate or weather for the area. This is a very important variable to consider in horticulture.

> "Plans are nothing, but planning is everything."
> —Dwight D. Eisenhower

Dan Schreiber/Shutterstock.com

Condor 36/Shutterstock.com

Pictureguy/Shutterstock.com

Figure 5-4. Environmental factors are extremely important for horticultural businesses to consider when developing a business plan. *Can a business be insured for damages caused by flooding or drought?*

The US gross domestic product (GDP) can fluctuate yearly due to weather (floods, droughts, excessive heat or cold). Weather and environment can be a liability or an asset to a horticulture business, **Figure 5-4**.

Assessing the competitive environment creates a better business plan and is important for strategic planning.

Internal Resources

When developing a strategic business plan, the founders of the company assess the resources and abilities of the company. Several variables must be considered, including:

- Location—is this a rural or urban location? How will customers get the products or services that are produced? Where are the highways and airports in relation to this company and its products?
- Resources—what natural resources are available? Is there water? What about electrical or other power access for this facility? Is there an affluent population nearby? Where will employees come from? See **Figure 5-5**.
- Production—what are the general products and how are they produced? What services will be offered?
- Value chain—how will the company add value to the products or services? What research and development takes place? What new products or services are in the future? What about marketing and sales? How is their value associated with the processes or the products of the firm?

Milos Muller/Shutterstock.com

Figure 5-5. A horticultural company should locate its business near major roads for easy transport of supplies and products and also be near major utilities such as water and power. *How does a "convenient" location affect the price of real estate or rent?*

Thinking Green

Drip Irrigation

Weather and climate undoubtedly impact horticultural production. There are ways to handle drought issues, including water conservation and drip irrigation. Drip irrigation is the use of very small quantities of water that are applied strategically to plants to ensure adequate growth without excess use of water.

vallefrias/Shutterstock.com

Corner Question

How much of the world's water is consumed by agricultural use?

Vision and Mission Statement

A company's *vision* is a description of its goals for the long term. A company's *mission statement* is a passage that identifies the purpose or the reason for existence of the company. The company's vision and mission statements help managers and employees understand where the company is headed and what it is trying to achieve. Mission statements can also help employees feel they are valuable assets of the company when they are mentioned in the statement, **Figure 5-6**.

A company grows and improves with effective leadership. Company leaders determine the company's basic ideas, standards, and principles, known as the *core ideology*, and what the company plans to achieve, known as the *envisioned future*.

Identifying differences between the company's current position or condition and the goals or vision will help company leaders direct the business. Company leaders also follow a set of principles known as *values*. These values guide company managers as they create goals and make decisions for the company.

"An organization's ability to learn, and translate that learning into action rapidly, is the ultimate competitive advantage."
—Jack Welch

Sun Gro Horticulture Company

Vision
Sun Gro's vision is to be the leading supplier of superior quality plant growing products in North America.

Mission Statement
We will achieve our vision by focusing on customer and employee satisfaction. We will passionately pursue continuous improvements in all aspects of our business through team design. Our responsibility to our shareholders is to deliver sustainable and improving cash flow.

Goodheart-Willcox Publisher

Figure 5-6. The Sun Gro Horticulture Company, located in Agawam, Massachusetts, has an exemplary vision and mission statement.

Corner Question

Who authored *Seven Habits of Highly Effective Teens*?

"Setting goals is the first step in turning the invisible into the visible."
—Tony Robbins

An example of an ambitious goal is from C. Raker and Sons, a horticultural grower: "Our achievements will be the benchmark for which other companies measure themselves." This clear and compelling goal energizes employees and promotes unity in the company. This grand statement means that this company will be so successful that all other companies, not only those in horticulture, will strive to be as good as C. Raker and Sons. This is a huge goal and one that motivates company employees.

Principal Strategy

A horticulture business must develop a strategy and implement a plan of action in business. Three basic business strategies for companies are:

- Be a low cost, big volume distributor. This opportunity is really only for large businesses. This strategy is difficult for the majority of horticulture businesses. It is generally not an option for a new business.
- Differentiate. The business will have specialized products or services. A business can be different from other companies based on quality, price, types of customers, or geographic area.
- Increase value for the customer. The business will acquire products or services that complement or supplement those already offered by the company. Collaborate with other businesses or form cooperatives to offer products or services that one company alone cannot offer.

After a business plan has been created and a strategy selected, the plan must be put into use. Getting the business started may include several steps:

1. Write the strategic business plan.
2. Develop financial security. Acquire loans using lenders and investors until the company is self-sustaining.
3. Secure a location. Lease or purchase property for the business. When determining where to locate a horticulture company, spend a great deal of time researching and analyzing the location. Is the area urban, rural, or suburban? What are the demographics of the population surrounding the business? Can the area support the business and provide workers, **Figure 5-7**?
4. Acquire permits, licenses, and an employer identification number (EIN), which is used for tax-related purposes. Acquire plant certifications, pesticide licenses, and building permits. Verify compliance with handicap laws and worker safety rules.
5. Open a business bank account. Obtain at least two major credit cards.
6. Apply for credit with *vendors* (companies that sell services, goods, or supplies) with whom you will do business.

michaeljung/Shutterstock.com

Figure 5-7. This team of associates must live somewhat locally to the business. The pool of candidates for a horticultural business can be a limiting factor in a company's success.

7. Advertise to the target market. Approximately 6500 residents can marginally support a greenhouse retail business. Use advertising funds wisely by targeting marketing materials to the people or other companies who are most likely to buy your products.

Performance Standards

The last step in the strategic business plan process is developing performance standards. *Standards* are benchmarks or levels for what is acceptable, in this case for company performance. Standards can be different for various types of business and for individual companies. Leaders of a horticulture business must ask, "Does the performance match the standards that were established?" If not, changes should be made.

The horticulture business must first identify key factors for success and growth. These factors include the right team, appropriate skill sets, clear communication methods, and a work and business culture that nurtures success. As time progresses, these factors will change, and it is important to reevaluate the factors periodically. Company leaders must determine which factors are working well and which factors need adjusting or replacing.

A horticulture business must determine what data to measure. The type of data that is analyzed by the business leaders is important. This can include *profit margins* (the amount by which income exceeds costs), company growth, or employee retention rates. Once the data has been identified, standards must be associated with the data. An example standard would be a 5% growth in sales each year. The firm must decide what data to track and set at least one standard associated with the data.

Data and standards lead to performance measurements for a company. Looking at the data, managers can determine if the company's performance matches, surpasses, or does not meet the standards. If the standards are not met, managers need to determine why this is so. Problems resulting in poor performance can be internal or in the market. A corrective action should be put into place. After corrective action, the goal is to meet or surpass standards in the future. Ongoing monitoring must take place along with effective strategies, solutions, and execution of plans for the performance measurements and standards.

Did You Know?
Retail greenhouses must be accessible to all individuals and should have adequate aisle widths to accommodate wheelchairs, as well as ramps for easy entrances and exits.

Marketing and Advertising

Effective marketing and advertising is crucial to making a profit. *Marketing* is the total system of business activities designed to plan, price, promote, and distribute products. The company seeks to satisfy the wants and needs of its potential and present customers while achieving its objectives. *Advertising* is the act of calling public attention to a product or service offered by a company, **Figure 5-8**.

lukeruk/Shutterstock.com

Figure 5-8. A horticultural advertisement should describe what the company is selling and why someone would want to purchase the product. In this advertisement, the consumer can see that the company sells citrus grown on a farm and shipped directly to the customer.

Together, marketing and advertising make up the identity of a business as seen by its customers, the general public, and competitors.

The Marketing and Advertising Process

The marketing and advertising process is made up of several steps that companies follow in their efforts to promote products and services:

1. Target the market selection. Analyze and assess the market. This is the time to develop objectives. The business determines whether to aim at an entire market, several different segments, or one piece of the market. These decisions must be reevaluated periodically.
2. Determine the marketing objectives. Set goals for what the company wants to achieve with the marketing plan. This is steered by profit, sales, volume, pricing, and promotion.
3. Devise an appropriate marketing plan. The firm must decide whether to fill the needs of an existing market, develop or find a new market, develop a new product for a market, or attract new customers with new products and new markets through diversification.
4. Create evaluation and control procedures. Assess data and standards and take prompt corrective actions. Management must determine what is and is not working, update objectives, and change strategies and tactics, if needed. This step then takes the company full circle and back to step one, **Figure 5-9**.

The Four Ps of Marketing

The four Ps of marketing are *product*, *price*, *place*, and *promotion*. These key words indicate factors or activities the company must get right. If just one of these four Ps is not right, there may be limited or no sales of the goods or services. Without enough sales, a company cannot be successful.

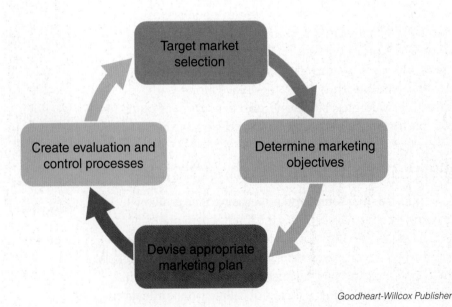

Goodheart-Willcox Publisher

Figure 5-9. The marketing and advertising process is a circular process.

Product

A product is a good (item) or service the business sells. Most businesses start from a simple idea created by an entrepreneur, **Figure 5-10**. The product is usually the result of an idea or an activity that the creator is passionate about. Regardless of the entrepreneur's excitement or belief in the item or service, it must have an appeal to customers or the business will not succeed.

Price

Price is the amount for which a good or service sells. Companies select prices based upon various strategies. A high price is not always best because customers may not be willing or able to pay a high price. A low price is also not always best because it may suggest low quality to some consumers. Company managers consider several factors when setting prices:

- What customers are willing to pay.
- What competing companies charge.
- The profit that is needed from the sale.
- The price needed to cover the cost of buying or making a product and providing some profit.

Sofi Photo/Shutterstock.com

Figure 5-10. A product comes from an idea. Companies launch new products every year based on a simple idea of making life more convenient or a simple need.

Horticulture businesses will often determine the price of an item by adding the cost of materials plus *overhead* (ongoing expenses of operating a business, such as utilities and rent) and labor (wages, employee benefits, and related taxes) to determine the selling price. There are several established pricing strategies that a business can use, as shown in **Figure 5-11**.

Pricing Strategies	
Cost-plus	Price is calculated as a standard percentage above cost.
Value-based	Price is set based on the buyer's perception of the value or worth of the product.
Competitive	Price is set the same or less than the price of another company's product.
Going-rate	Price is based on the common price in the market.
Skimming	Price is set high initially and then later reduced.
Discount	Price is set high but then lowered using coupons or sales.
Loss leader	Price is set at lower than cost in an effort to bring in customers who may also buy other products.
Psychological	Price is set slightly lower than a round number, such as $4.99 instead of $5.00, because the lower number is more appealing to customers.

Goodheart-Willcox Publisher

Figure 5-11. There are many pricing strategies currently used by businesses. Determining the optimum price for a good or service is critical to company success. **Which of these pricing strategies would you adopt in your own business?**

STEM Connection
Japanese Maple Price

What is behind the price of a Japanese maple?
- The cost of plant material is $4.00.
- The cost of the pot, media, and fertilizer is $1.00.
- It takes one hour of labor to maintain the plant until sale, and labor is $10.00 an hour.
- Overhead (rent, taxes, insurance, utilities, and other expenses) is $4.00 per plant.

The cost of the plant is $19.00. The company uses the cost-plus pricing strategy, with the price set at 35% above cost. This makes the selling price $25.65.

ukmooney/Shutterstock.com

Place

Place in marketing is the location at which customers can buy a product or service. The proper placement or distribution of goods or services contributes greatly to sales. Customers, regardless of demographic background, desire convenience. Do customers access your products only through the Internet? Is the product available through department stores? Do you have your own retail store? How can the customer find your product or services, **Figure 5-12**? How will this product get to the customer? Shipping, delivery, and pick-up are also important.

Selling can take place via direct sales or reseller sales. *Direct sales* is a distribution method in which a company sells its products to customers without another party involved. Direct sales can be done through retail stores, door-to-door sellers, mail order catalogs, e-commerce sites, or at other sites. Some companies have representatives that meet customers. This provides an easy and early way to detect market changes. A business can quickly adapt and control how items are sold and the range (geographic location) of sales. In this mode of sale, the business must have effective customer relationships that are considered positive overall.

Reseller sales is a distribution method in which other parties buy a company's products and then sell them to customers. Reseller sales, also called channel sales, can include a wholesaler as well as a retailer. A wholesaler is a company that buys products from manufacturers or distributors and sells them to retail stores. The retail stores sell the products to consumers. The original company has little or no contact with the end customer. In some cases, this can result in a company losing its identity.

InesBasdar/Shutterstock.com

Figure 5-12. A fast and convenient way to search for and purchase horticultural products is on the Internet. This method of interacting with customers became popular in the twenty-first century and has revolutionized all industries.

Corner Question

What company has a brand logo that includes the number 31 along with its name?

For example, a company might have its products resold under another name brand. However, resellers can provide companies with access to customers that they could not normally reach on their own.

The placement or distribution of products and services should be addressed in the strategic business plan. As businesses grow or change over time, however, the company may need to change its distribution plan. Two ways that products reach customers are intensive market coverage and selective market coverage. *Intensive market coverage* is a distribution strategy in which a company attempts to sell its products using all available outlets. This strategy is often used with basic or generic products. An example of this is potting soil that is sold all over the country. *Selective market coverage* is a distribution strategy in which a company sells its products using a limited number of locations. The company narrows it sales sites and establishes relationships with its customers. An example of this is a local landscaping company that offers services based on individual needs, **Figure 5-13**.

Promotion

Promotion primarily includes public relations efforts and advertising. Personal contacts and other communication from a business to its customers can also be related to promotion. Public relations involves activities that create a favorable image of a business in the community, such as donating products to a school or taking part in a community charity fundraising event. These efforts are often recognized via a social or news media outlet. Public relations is a key part of promotion.

As discussed earlier, advertising is calling attention to a company's goods or services. Several methods can be used to advertise products:

- Radio—inexpensive and reaches many potential customers. It is best to advertise during mid- to late-week.
- Television—expensive. Advertisements can be broadcast locally or nationally.
- Print—includes direct mail, printed fliers, magazine advertisements, and coupons.

Alison Hancock/Shutterstock.com

Christina Richards/Shutterstock.com

Rigucci/Shutterstock.com

Figure 5-13. Landscaping companies must determine what their customers need. From debris removal to mulch and evergreen choices, successful landscaping companies provide materials and services that are appropriate for the local climate.

"Many a small thing has been made large by the right kind of advertising."
—Mark Twain

Rob Marmion/Shutterstock.com

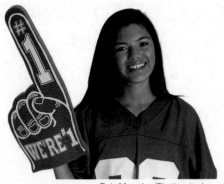

Figure 5-14. A company must decide how it will provide materials to advertise and develop a relationship with their customers. Have you ever received promotional materials? How did that influence your opinion of the company?

- Goods—items that are given free to potential customers. Examples include foam fingers, pens, and T-shirts, **Figure 5-14**. This method can be costly depending on the items used.
- Electronic—websites, social media, and advertisements associated with other products.
- Word of mouth—this method is free. Satisfied customers tell others about the company's goods or services.
- Window displays—can be inexpensive. People passing by a window may notice a seasonal display and come in to shop.
- Generic advertisements—promote goodwill for an entire industry rather than a specific brand. An example is a video, tent, or booth that reaches the public on behalf of citrus growers.

Most companies must advertise and market their products or services. If they choose not to do so, they will most likely be one of the eight out of ten businesses that fail.

Professionalism

What makes someone a professional? Is it getting a job? Is professionalism exhibited only in the workplace? ***Professionalism*** is the exercise of judgment, skill, and polite behavior that is exhibited by someone who is trained to do a job well, **Figure 5-15**. Characteristics of a professional can include a number of the same characteristics of a leader that were discussed in Chapter 1, *Agricultural Leadership.*

Professionalism demonstrated at the workplace can be rewarded with promotions and future opportunities. Employers seek individuals who are problem solvers, courteous, trustworthy, tactful, and committed to their jobs.

Did You Know?

Western Promotions from Medford, Oregon, specializes in fostering relationships with businesses and customers by delivering promotional products, such as tree starters, bags of bulbs, and tins of seeds. The products come labeled with a company's name and logo.

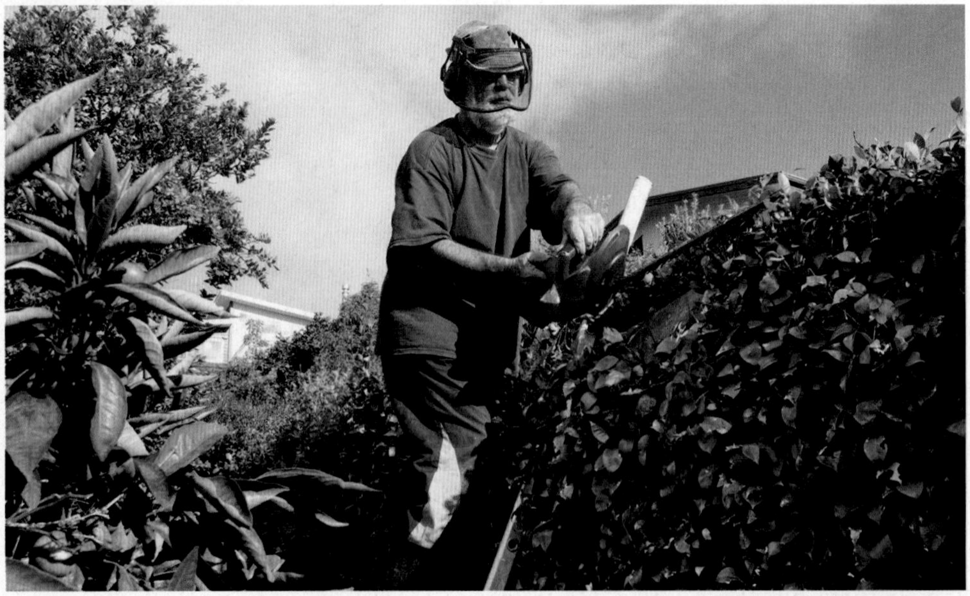

ChameleonsEye/Shutterstock.com

Figure 5-15. Employers seek individuals who are professional, have a good work ethic, and use their training to do the job well.

Corner Question

How many television commercials and advertisements does the average American see by the time she or he is 65 years old?

Career Connection: Professional Certifications in Horticulture

Industry certifications in horticulture demonstrate and verify an excellence and professionalism within the trade that is not exhibited by all professionals. Employers and customers look for these industry credentials and can be satisfied knowing that the individuals with these certifications are some of the best that the horticulture industry has to offer.

Two national horticultural organizations offer various professional certifications. State agencies may also offer additional certifications, so check with your state's extension agency or a member of the horticulture industry to find out more.

- American Society of Horticultural Science. Certified horticulturists are knowledgeable and skilled in all areas of horticulture. Those who are certified have passed a rigorous four-hour exam and must have three years of professional and/or educational experience in horticulture.
- Master Gardener. Each state's cooperative extension agency can offer a master gardener program. Master gardeners are home gardeners who have received a great deal of horticultural training and have volunteered in horticultural efforts in their communities. A master gardener may be more qualified as an employee than someone without any training in horticulture.
- National Association of Landscape Professionals. Some certifications provided by the association include landscape industry certified manager, exterior technician, interior technician, horticultural technician, lawn care manager, and lawn care technician.

Each of these certifications ensures that the person certified has undergone rigorous training and evaluation to demonstrate excellence within their specialty in the green industry.

Professional Traits and Behaviors

Professionals are judged or critiqued every day at their workplace. Whether it is a colleague, a customer, or an employer, there are people always watching and taking note of one's work and contributions to the business. A true professional should employ all of the following actions or traits:

- Adhere to commitments—be timely, dependable, punctual, and dedicated.
- Show respect—be courteous to all people (colleagues, customers, and employers), exhibit kindness, and be polite in speech and body language, **Figure 5-16**.
- Maintain ethical conduct—exhibit honesty and fairness.
- Demonstrate workplace etiquette—dress appropriately, practice good hygiene, turn off or silence phone or headphones, and smile.
- Show leadership skills—take on leadership roles, be organized, express dedication, be competent, admit to mistakes, and be open to giving and receiving apologies.

All of these traits or actions are important and promote excellent performance at a job.

michaeljung/Shutterstock.com

Figure 5-16. This pleasant store representative is working with a customer and answering questions about products.

Unprofessional Traits and Behaviors

Avoiding unprofessional traits and behaviors can help you be a successful employee. Those workers who do not demonstrate professionalism can undermine efforts of others in the workplace. There are several traits and behaviors to avoid that can mark a worker as unprofessional.

Workplace gossip involves talking about others in a less than positive way, which can cause hurt feelings and resentment. Avoid getting involved in workplace problems by refraining from talking about colleagues.

Tardiness means being late to work. Fellow employees may think you are not a team player or are not committed to your work. Employees who are on time may feel resentment toward those who are late to work.

Unprofessional language includes inappropriate body language or verbal language and is offensive to coworkers, customers, and employers. Yawning, gum chewing, and using slang are usually not appropriate on the job. When at work, try to use proper grammar, speak clearly, and avoid showing disrespect toward others.

Avoid unprofessional appearance and poor hygiene. Society is fairly tolerant of long hair, nails, piercings, tattoos, and makeup. At some types of professional jobs, however, it is still important to cover body piercings and tattoos and keep nails trimmed and neat. You want individuals to focus on your job performance rather than your appearance. Strong colognes and perfumes, unkempt hair, body odor, and a generally disorderly appearance may offend coworkers or customers.

> **Did You Know?**
> The Florida Nursery Growers and Landscape Association offers a Certified Horticulture Professional program to recognize authorities in the horticulture industry.

Career Documents

Career documents, such as a letter of application and a résumé, are often required to secure a job. A *letter of application* is a document that requests that the sender be considered for a job opening and introduces the writer's résumé. It is sometimes called a cover letter. A *résumé* is a document that contains a concise summary of a person's education, skills, work experience, and other qualifications for a job. An employer may review résumés from many applicants before selecting candidates to interview. You only have one chance to make a good first impression on the employer. In many cases, the letter of application and the résumé present the first impression the employer will have of you. Both documents should be well written and contain information that will present you in a favorable light.

Letter of Application

A letter of application is a written, formal business document. It should be properly formatted, free of errors, and well organized. The letter should state that the writer wants to be considered as an applicant for a particular job. It may also state how the applicant learned about the job, such as through a newspaper ad or a job posting on the company's website. The letter should highlight key qualifications or experience that qualifies the applicant for the job, mention that a résumé is enclosed, and ask for an interview, **Figure 5-17**.

> Judy Gardener
> 123 Green Thumb Lane
> Plant City, Florida 33564
> 555-867-5309
> jgardener@mail.com
>
> James Smith
> Human Resources Manager
> ABC Horticulture
> 5037 N. Elm Street
> Springfield, IL 62711
>
> Dear Mr. Smith:
>
> I am writing concerning the senior researcher position with your organization. I have a particular interest in working for your company and would appreciate being considered as a candidate for employment.
>
> I have three years experience as a lab assistant in an introductory horticulture class at University of Florida. I am familiar with general lab procedures and protocols. My skill set includes maintaining and cultivating greenhouse plants and cultivating unique plant materials. I have also conducted research involving gibberellic acids on plant reproduction of orchids.
>
> I have enclosed my résumé. I hope it will be helpful in evaluating my qualifications for a position. Please feel free to contact me at your earliest convenience to arrange for an interview. I look forward to meeting with you to discuss this employment opportunity.
>
> Thank you for your time and consideration.
>
> Sincerely,
>
> *Judy Gardener*
>
> Judy Gardener

Goodheart-Willcox Publisher

Figure 5-17. A letter of application introduces you to the potential employer and expresses your interest. Why do you think it is a good idea to make sure this document is professional and well written?

"I think professionalism is important, and professionalism means that you get paid."
—Erica Jong

Résumé

A résumé should be sent with the letter of application. A résumé must be clear, concise, and accurate. It should be written to address a particular job advertisement and show how the candidate is a good choice for the job.

Career Connection: Résumé Tips and Myths

Tips for effective résumés:
- Do not use a premade template. These are often busy and crowded. Creating your own résumé document gives you 100% control over the résumé writing process.
- Ensure that all elements parallel one another and there is uniformity. Be consistent with bold, italics, capitalization, spacing, and margins.
- Avoid noun versions of verbs (words that end in -*tion*.) Example: "Proficiency in creation of floral design." It is better to say, "Created floral designs."
- Remember, creating a résumé is writing, and this is a process. Plan, draft, and get feedback. Revise as needed until you have an effective and error-free document.

Myths about résumés:
- "Since a picture paints a thousand words, I should include one with my résumé." No. Do not include a photograph of yourself.
- "They will never check what I write." Wrong. Some businesses check every detail.
- "Include references on the résumé." No. References use valuable space. Provide them if the employer asks for them.
- "One-page résumés are always best." No. Résumés can be up to three pages. The more experience that a professional acquires the longer the résumé.
- "To make my résumé stand out, I will make it creative or unusual." No. The applicant's qualifications and accomplishments should be the focus of attention. Employers may overlook a great candidate because they think that if the résumé is odd, the applicant is also.
- "Include hobbies and interests." No. While you do want to appear well rounded, the employer only wants to know how you can benefit the company. Mention a hobby or interest only if it relates to the job. For example, if you are currently learning to speak a second language, that interest might make you a more valuable employee.

> "Boxing is the only career where I wouldn't have to start out at the bottom. I had a good résumé."
> —Sugar Ray Leonard

An applicant wants a potential employer to read the résumé and think "I want this person. What a perfect fit." Some guidelines for an effective résumé include:
- Place your contact information at the top.
- Include a work history with names of employers and dates of employment. Mention roles, experiences, and achievements. Include volunteer work if you have no paid work history.
- Include professional qualifications, certifications, education, and memberships to organizations.
- Be consistent in verb tense and use action verbs.
- Make the information concise; it should not be wordy.
- Use simple, standard fonts with 1" margins.
- Left justify the body text.
- Use an appropriate length. Résumés can vary in length (up to three pages), but they are usually just one page for those with little work experience.
- Carefully review the document for mistakes (grammar, spelling, punctuation, format, or style). The document must be error free, **Figure 5-18**.

Judy Gardener

123 Green Thumb Lane
Plant City, Florida 33564
555-867-5309
jgardener@mail.com

OBJECTIVE To obtain a position as a grower in a retail greenhouse.

EDUCATION **University of Florida, College of Agriculture.** Gainesville, FL.
Bachelor of Science, 2014.

EXPERIENCE **Horticulture 100 Lab Assistant, University of Florida.** Gainesville, FL.
May 2011–May 2014
Assisted in horticulture instruction for an introductory course in horticulture. Maintained and cultivated greenhouse plants during fall and spring semesters.

Sales Associate, Nanny's Greenhouse. Plant City, Florida.
June 2009–August 2014
Cultivated unique plant material. Assisted customers with plant selections and horticultural questions.

OTHER EXPERIENCE **Supervised Agricultural Experience—Plant Reproduction.** Plant City, FL.
September 2006–April 2008
Conducted research involving giberellic acids on plant reproduction of orchids. Directed the experiment and shared results with local orchid growers through my high school agricultural education program.

FFA Floriculture Career Development Event Winner. Louisville, KY.
November 2014
Participated in plant identification and floral design contest. Competed against 50 others teams and awarded first place.

SKILLS Proficiency in Spanish.
Microsoft Office, Publisher, Adobe Photoshop.

ACTIVITIES **University of Florida Peer Ambassador.** Mentor for incoming freshmen.
Study Abroad in San Juan, Costa Rica. Participant in fall semester 2013.

Goodheart-Willcox Publisher

Figure 5-18. This résumé is well designed. What type of job do you think this applicant might be interested in?

Job Interviews

If an employer has reviewed your letter of application and résumé and invited you for a job interview, you are ready to move to the next step in the job search process. A *job interview* is a meeting where an employer and a job applicant discuss a job and the applicant's qualifications for the job.

Did You Know?
There are professional résumé writers that you can hire to help you with your résumé.

During this important meeting you will talk with the employer about your suitability for the job. You can also use this meeting to learn more about the employer and the job. Preparing for the job interview and following up after the interview are important parts of the interview process.

Preparing for a Job Interview

Preparing for a job interview requires research and thought by the job applicant. You need to identify information you want to learn about the company and the job. You also need to think about information you may be asked to provide at the interview. To prepare for an interview, job candidates should:

- Determine the employer's needs related to the job.
- Do research to learn about the company and the industry the company is part of.
- Compose answers to possible interview questions.
- Make a list of questions you want to ask the interviewer. Wait for an appropriate time to ask them.
- Rehearse and conduct mock interviews or videotape a practice interview.
- Identify any transferrable skills you have (those that could be used in another position within the company) and may want to point out.

Employers look for a number of skills in job candidates. Some skills are related to a specific job, such as knowing the proper way to prune plants. Other skills, such as the ability to communicate and think critically, are needed in many jobs. Job candidates should be able to respond to interview questions about both job-specific and general skills. It is important for candidates to demonstrate through their answers that they are confident, flexible, self-motivated, leaders, team players, excellent communicators, and proficient in their field.

Dress for Success

The way you look at an interview helps create the employer's first impression of you. Employers may judge a candidate, at least in part, based upon appearance. Dressing appropriately and being clean and well groomed are important for making a good impression. For some jobs, especially those that are almost entirely manual labor, there is no need for men to wear a suit and tie or for women to wear a business suit to an interview. A pressed, button-up shirt or polo shirt with a pair of khaki pants and nice shoes would be appropriate, **Figure 5-19**.

For an office job, however, a suit and tie are preferred for men. Women should dress in a conservative style and wear skirts that are a modest length. For both men and women, hair should be pulled back or away from the face. Employers may consider unusual or overdone makeup, strong colognes or perfumes, piercings, tattoos, facial hair, and very long nails inappropriate for many jobs. Unusual or inappropriate appearance can distract from the candidate's qualifications and skills.

stockyimages/Shutterstock.com

Figure 5-19. Many horticultural jobs require manual labor. For that reason, it is often acceptable to come to an interview dressed in a polo or button-up shirt, khakis, and nice shoes.

Consider your appearance carefully and strive to make a good first impression, **Figure 5-20**.

Arrive on Time

Candidates should not only dress to impress but must also arrive early and be prepared for their interviews. If the candidate is late, he or she almost surely will not be selected for the position. Tardiness is not acceptable for an interview. Arrive no more than 15 minutes early to show that you are prepared and enthusiastic for the interview. Try driving to the location of the interview beforehand to ensure you know where the interview will take place and that there are no surprises on the route. Bring a copy or two of your résumé, a list of references with contact information, and a portfolio of applicable work. Greet the receptionist or others you meet in the office in a friendly and professional manner.

Taking Part in the Job Interview

You have prepared, you arrived a little early, and you are dressed for success. Now it is time to meet with the employer. When the interviewer comes to greet you, stand up and extend your hand for a medium to firm handshake. Smile warmly and make eye contact. Give a brief greeting, such as, "Good morning. I am glad to meet you. Thank you for taking time to discuss the job opening," **Figure 5-21**. Within the first couple of minutes, try to find something in common with the interviewer that you could make small talk about. Scan the office or surrounding environment for ideas. For example, the company may have a display of its products in the reception area. If you have used some of the products, you could mention how effective or useful a product is.

During the interview, answer the questions honestly and to the best of your ability. Give clear and complete answers without talking too much. Try to give answers that will show how you are qualified for the job, but do not misrepresent your skills or experience. Do not fidget, twiddle your thumbs, twist your hair, or yawn during the interview. Do not address the interviewer by first name unless you are asked to do so.

Minerva Studio/Shutterstock.com

Figure 5-20. Dressing for success starts at the interview. Do not wear what you would wear to go out with your friends. Dress professionally to impress your potential employer.

Konstantin Chagin/Shutterstock.com

Figure 5-21. Greeting your interviewer with a firm handshake, a smile, eye contact, and clear speech can set you apart from other candidates. Being pleasant, upbeat, personable, and having good posture are also beneficial. Practice greeting an interviewer before going to your interview.

Career Connection: Job Interview Practice Questions

Consider how you will answer questions such as these as you prepare for a job interview:
- "What are your strengths?" Give a story that illustrates one of your strengths in action.
- "What is one weakness you have?" Either use something insignificant ("I'm color blind." or "I'm a bad dancer.") or tell about a real weakness that you had and how you overcame it. This should not be anything deep into your personal life; it should be related to work. For example, "I used to be disorganized, but then I got a planner and started using apps on my phone to help me get organized. Now I am very organized."
- "Why should our company hire you?" Identify what makes you unique or how you will benefit the company.
- "What is your ideal job?" This should be the job for which you are applying or a job that this job helps you prepare for.
- "How has your education prepared you for this job?" Use concrete examples regarding classes you have taken in agricultural education, math, science, or other areas.
- "What was your last boss like?" Do not say negative things about any previous employers. Keep your comments brief and find something neutral or positive to say. If you did tell them how difficult your previous boss was, they might think that you do not work well with others.
- "Is money important? How much do you expect your salary or pay to be?" You might say that while money is important, so is job satisfaction and learning opportunities. Indicate that you are willing to consider any reasonable wage. Do research to learn what a reasonable wage is for the job before the interview.
- "Do you have any job experience in this field?" Highlight key skills gained from other jobs or other educational experiences that are transferrable.
- "What is your marital status? Do you have a significant other?" These questions are illegal. An interviewer cannot legally ask you questions about your personal life. You could answer by asking, "How does this information apply to this position?"

Remember to be concise, but never respond with just yes or no. Elaborate and let the interviewer understand who you are and why you are the best candidate.

Did You Know?
The average length of an interview is 40 minutes.

Near the end of the job interview, the interviewer may ask if you have any questions. This is a very important point in the interview. You need to ask questions to show your interest in the company and the job. This is your last chance to make a good impression that will stick with the interviewer. Ask the interviewer questions that are positive and cast you as an ideal candidate. Example questions include:
- How would you describe the management style of this company?
- Will any type of training be provided for this job?
- What process is used to evaluate employees?
- When can I expect to hear from you regarding the position?
- Why should someone want to work for your business?
- What are the company's goals and mission?

Do not ask questions similar to these:
- Will I have to work overtime often?
- Do I have every weekend free?
- What else does your company do?
- What are the employee benefits that go with this job?

Career Connection: Job Interview Mistakes

In order to have a positive interview, avoid these mistakes:
- Giving a poor handshake.
- Talking too much.
- Speaking negatively about current or past employers.
- Showing up late or excessively early for interview.
- Being unprepared for an interview.
- Asking about benefits, vacation time, or salary.
- Using verbal ticks (*um*, *uh*, *like*).
- Failing to match the communication style of the interviewer. If the interviewer is all business, do not try to loosen up the interviewer, just be professional. If the interviewer is personable, have a conversation about common interests. If the interviewer asks direct questions, answer with direct information.

The final moments of an interviewer are those that will be the most recent memory in an interviewer's mind. Make these moments count and sway the employer to hire you.

Interview Follow Up

After the interview, write thank-you messages to all individuals who interviewed you. Thank them for the interview and express your continued interest in the job (if you are still interested). Briefly mention skills or experience that qualifies you for the job. You could also mention something beneficial that you will take away from the interview process with that particular company. In your thank-you letter, use a style and format similar to that used for your letter of application. An e-mail thank-you note, rather than a formal letter, is appropriate for some companies. If you do not hear from the company within a reasonable time, it is acceptable to call the interviewer and ask if the job has been filled or when you can expect to hear from the company.

Do not be discouraged if you do not get a job you want. You may have to apply to several companies before you are hired. Think about reasons why you were not the candidate selected for the position. Do you need to acquire more skills, education, or experience? Try to learn from each interview and consider how you can do better in the next interview.

> "Success is not final, failure is not fatal: it is the courage to continue that counts."
> —Winston Churchill

Career Connection: Writing Professional E-mails

1. The subject line should indicate why you are writing the e-mail.
2. In the body of the e-mail, get to the point quickly and clearly.
3. Number or make bullet points in e-mails.
4. Use bold as little as possible.
5. Use proper grammar, punctuation, and spelling. Do not use any slang.
6. Delete every word that is not absolutely necessary.
7. Do not use emoticons or images in the e-mail unless an image is requested by the person you are e-mailing.
8. Avoid using attachments, except when necessary. These often get lost or overlooked by readers.
9. Include your contact information and signature at the bottom of the e-mail.

School-to-Career Plan

If you are in high school, it may seem like a professional career is years and years away. However, you can begin now to prepare yourself for the career and the lifestyle that you want. You may not know what you would like to do as a professional, but usually you have a good idea of the lifestyle you would like to lead. Maybe you see yourself driving a nice car or taking expensive vacations. Whatever you want to do, you will need to have a job that provides income for that lifestyle.

You must consider several factors regarding a career in addition to money it can provide. For example, will the job give you enough free time for family or leisure activities? Will it provide a sense of accomplishment? Do you want to be indoors or outdoors all day for your work? Will there be opportunities for advancement? Answering questions such as these will help you choose a career with which you will be happy. Think about the number of hours you might work over you entire career. A career can span more than 35 years. Working five days a week for 40 hours a week for about 50 weeks a year means that you may work about 70,000 hours in your lifetime. This fact emphasizes the importance of choosing a job that you will enjoy, **Figure 5-22**.

Pressmaster/Shutterstock.com

Figure 5-22. Americans will work about 70,000 hours over the course of their lifetimes. It is extremely important to enjoy doing your work when you spend at least eight hours of each day at your job.

There are a number of factors to consider when choosing a career. It is important to conduct a thorough self-assessment. As you age, your assessment of who you are and what you like will change. In today's work environment, it is uncommon for individuals to work for one company for their entire professional career. People typically change career paths as they evolve.

School-to-Career Plan

Choosing a career path is not always easy. Some people know when they are quite young what career they want to pursue when they are adults. Others still have not found a career they think is right for them after working for many years. Begin carefully exploring career possibilities while you are still in school. If you have no idea where to start, begin with self-study and analysis. Think about who you are, what you like to do, what your interests are. This will help you to begin finding careers that may match your interests.

Corner Question

What famous African-American scientist had a plan for a career despite being a slave?

Then research various careers, job shadow, get an internship and determine if this is a career that would capture your interests and keep you mentally and physically healthy.

Once you have selected a broad career area, you can begin creating a school-to-career plan. A *school-to-career plan* is a document that consists of one or more career goals and the steps or activities needed to achieve those goals. Resources needed to achieve the goals may also be included. For example, your goal may be a career as a high school agriculture teacher. Steps you must complete to achieve the goal might include graduating from high school, getting a college degree in agriculture education, and completing a teaching internship. Resources needed would be funds to pay for college. You can begin your plan with broad goals and steps and then make the plan more specific as you learn more about what is needed.

Your plan will help you map out how to get from where you are now to where you want to be in a career. When beginning a school-to-career plan, follow these simple steps:

1. Consider your interests, abilities, and personal characteristics. It is a good idea to take a personality test (such as the Myers-Briggs personality assessment) to see what kind of personality you have.
2. Narrow the field of jobs to what is suited to your personality.
3. Study the requirements of the job.
4. Prepare a plan for this career.
5. Plan for an alternate career in case the first career does not become a reality.
6. Begin preparing for success in this career by completing educational requirements and developing needed skills.
7. Get work experience (supervised agricultural experiences, job shadowing, internships, and volunteering).

You may find it helpful to look up job listings on career websites of professions that interest you. Determine what you will need to get a particular job by looking at the job requirements posted for the applicants. Create an inventory of the skills and education you have and what you will need to secure a job like this in the future. You can then map out your school-to-career plan from this point, **Figure 5-23**.

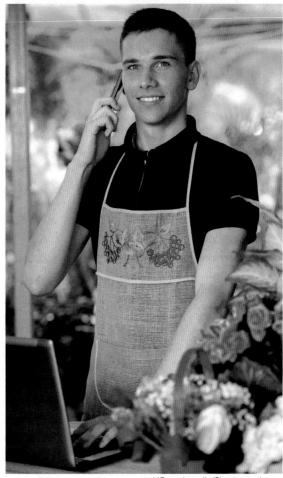

VGstockstudio/Shutterstock.com

Figure 5-23. Take an inventory of the skills that you have to see if you would be a successful candidate for a particular job. If you enjoy talking to others, customer service may be a good job possibility for you.

Career Exploration

Valuable resources exist at your school to help you investigate and find a career that is suited for your personality and interests. Contact school counselors and meet with one of them for career guidance. Schools often have a career development coordinator or counselor who specializes in career counseling. This valuable resource should be used during your high school years.

The National FFA Organization has career resources on its website. There you can find resources that will help you explore careers in agriculture. More than 300 are indexed on this site. Valuable information about each of the careers, including descriptions, educational requirements, skills, qualifications, and salaries, are available, **Figure 5-24**.

Velychko/Shutterstock.com

Figure 5-24. Mapping out a school-to-career plan to become a horticultural employee should include courses in horticulture in high school and college.

No matter what your school-to-career plan may be, it is only effective if you follow it. If your goal is to become a plant pathologist, you cannot simply wish yourself into that occupation. This job requires years of education and job training. Creating a school-to-career map is your first step to acquiring your ideal career.

"Ninety-nine percent of failures come from the people who have a habit of making excuses."
—George Washington Carver

Horticulture Business Careers

Horticulture businesses provide a variety of products and services for consumers and other businesses. Horticulture companies employ various workers and consultants, each with a unique skill set. Workers in some horticulture careers deal with managing the business or marketing. Two such careers are horticulture business consultant and horticulture sales representative.

Corner Question
How many Myers-Briggs assessments, a form of personality tests, are given each year?

AgEd Connection — Marketing Plan CDE

The marketing plan career development event helps students develop skills needed for the marketing industry. Teams of students present a marketing plan for a real business venture to a team of judges. The marketing plan is designed to promote a product, supply, or service. The business venture may be an existing agribusiness or a start-up enterprise.

Horticulture Business Consultant

A horticulture business consultant evaluates horticultural companies with the goal of increasing productivity, efficiency, and profits, **Figure 5-25**. This position requires an understanding of the financial issues associated with a business and a great breadth of horticultural knowledge. The consultant assesses what the company is doing well and what needs to be improved. The consultant will often work with company managers to develop strategies that improve the company and move it toward the company's goals and mission. Horticulture business consultants may assist companies with financial planning,

wavebreakmedia/Shutterstock.com

Figure 5-25. A business consultant for a horticulture company helps to analyze the business and ensure that it is successful.

Career Connection

Leslie Halleck
Horticultural Marketing

When Leslie Halleck began her professional journey, she created a partnership between her love for gardening and her love for teaching others about horticulture. She began her horticultural career in Texas at Northern Texas University and furthered her horticultural studies at Michigan State University. She returned to Texas to work in arboretums and garden centers. There she found her niche for cultivating relationships with people and plants.

Throughout her career Leslie has always focused on helping people nurture their green thumbs. Today, she has her own horticultural marketing business in Texas called Halleck Horticultural. Leslie and her team help people in the green industry (which Leslie believes includes ornamental and edible plant cultivation) to communicate with their customers using creative and innovative mediums.

Halleck Horticultural

Leslie has a wide breadth of horticultural knowledge and an incredible depth of marketing savvy. Her fresh advertising campaigns connect with younger generations of gardeners and help the public to understand why gardening is for everyone. Leslie wants to expose people of all generations, cultures, and backgrounds to gardening.

Halleck assures future horticultural employees that the green industry is one in which anyone can make a comfortable living. "People are deceived into thinking that you can't make money in horticulture," says Halleck. Leslie states that the green industry can provide profits just like other business ventures. Horticulture has the potential of being financially rewarding while allowing workers to provide something the world wants—plants for food or beauty.

tax planning, or quality control. Consultants are often self-employed or employed by a consulting firm. This type of position requires a four-year degree in horticulture or a closely related field.

Horticultural Sales Representative

A horticultural sales representative interacts with customers to provide information about a company's products or services. The sales representative must be knowledgeable about the products or services offered by the company. Sales representatives answer questions for and respond to customer demands. A sales representative should have excellent written and verbal communication skills. Sales representatives are asked to advise customers and make decisions, so they should be good critical thinkers and problem solvers. They often work in retail situations, such as garden centers, florist shops, or landscaping companies; but they may also act as a sales associate for larger companies who are selling to the public, **Figure 5-26**. Experience in sales is always beneficial when applying for these positions. Educational requirements can vary depending on the company's objectives and the customers they target.

"Marketing is a contest for people's attention."
—Seth Godin

wavebreakmedia/Shutterstock.com

Figure 5-26. A sales associate working at a garden center should be friendly, knowledgeable, and enjoy working both indoors and outdoors. *Have you ever encountered a sales associate who was rude? Did your encounter change the way you viewed the business itself?*

CHAPTER 5 Review and Assessment

Chapter Summary

- A small business is a company that is independently owned and operated, organized for profit, and not dominant in its field. In the United States, there are about 28 million small businesses.
- An entrepreneur is a person who organizes and operates a business. Millions of entrepreneurs in the United States run successful businesses.
- A strategic business plan is a document that states the mission of the business, examines its current condition, sets goals, and outlines strategies for achieving the goals.
- In evaluating the strategic environment, managers should consider information about the company, the industry, and competitors. Other areas covered in a business plan include internal resources, the company's vision and mission statement, and performance standards.
- Marketing is the total system of business activities designed to plan, price, promote, and distribute products. A marketing and advertising plan will include the four Ps: product, price, place, and promotion.
- Professionalism is the exercise of judgment, skill, and polite behavior that is exhibited by someone who is trained to do a job well. Characteristics of a professional can include being respectful, ethical, competent, honest, and polite.
- Career documents, such as a letter of application and a résumé, are often required to secure a job. These documents act as an introduction for an applicant and must be free of errors. The purpose of these documents is to secure a job interview.
- A job interview is a meeting where an employer and a job applicant discuss a job and the applicant's qualifications for the job. Job interviews require preparation and planning.
- Job candidates must arrive on time for an interview, dress professionally, and answer questions clearly. They should ask the interviewer appropriate questions as well. After an interview, the applicant should write a thank-you letter to the interviewer.
- A school-to-career plan consists of one or more career goals and the steps or activities that must be completed to achieve those goals. The plan may also include the resources needed to achieve the goals.
- High school counselors are a good source of information for career guidance and exploration. The National FFA Organization has career resources on its website that will help you explore careers in agriculture.

Words to Know

Match the key terms from the chapter to the correct definition.

A. advertising
B. brand
C. core ideology
D. direct sales
E. entrepreneur
F. envisioned future
G. goal
H. intensive market coverage
I. job interview
J. letter of application
K. marketing
L. mission statement
M. overhead
N. professionalism
O. profit margin
P. reseller sales
Q. résumé
R. school-to-career plan
S. selective market coverage
T. small business
U. standard
V. strategic business plan
W. value
X. vendor
Y. vision

1. A distribution method in which a company sells its products to customers without another party involved.
2. A document that requests that the sender be considered for a job opening and introduces the writer's résumé.
3. An objective to be achieved.
4. The exercise of judgment, skill, and polite behavior that is exhibited by someone who is trained to do a job well.
5. A document that lists one or more career goals and the steps or activities and resources needed to achieve those goals.
6. The amount by which income exceeds costs of doing business.
7. A company that sells services, goods, or supplies.
8. The act of calling the attention of the public to a product or service offered by a company.
9. A meeting where an employer and a job applicant discuss a job and the applicant's qualifications for the job.
10. A benchmark or level for what is acceptable.
11. A document that contains a concise summary of a person's education, skills, work experience, and other qualifications for a job.
12. The total system of business activities designed to plan, price, promote, and distribute products.
13. A name, label, logo, or image under which a product is sold.
14. A distribution strategy in which a company attempts to sell its products using all available outlets.
15. A description of an organization's goals for the long term.
16. A distribution strategy in which a company sells its products using a limited number of locations.
17. A distribution method in which other parties buy a company's products and then sell them to customers.

18. A document that states the mission of the business, examines its current condition, sets goals, and outlines strategies for achieving the goals.
19. Basics ideas, standards, and principles.
20. A person who organizes and operates a business.
21. A passage that identifies the purpose or the reason for existence of a company or organization.
22. A company that is independently owned and operated, is organized for profit, and is not dominant in its field.
23. A principle or standard.
24. What a company or other organization plans to achieve.
25. The ongoing expenses of operating a business, such as utilities and rent.

Know and Understand

Answer the following questions using the information provided in this chapter.

1. How does a strategic business plan help to organize the activities of the business?
2. What does the strategic environment for a business include?
3. What are four examples of company information that might be addressed in a strategic business plan?
4. What are three basic business strategies that companies may use?
5. What are some types of data that a company might measure to gauge performance?
6. What are five options for business ownership?
7. What are four steps in the marketing and advertising process?
8. List five examples of ways to advertise a good or service.
9. What are some traits or skills employers seek in employees that mark the employee as a professional?
10. What are some traits or behaviors to avoid that mark a worker as unprofessional?
11. Why is it important that a letter of application and résumé be well written and free of errors?
12. What are two examples of information you should place on a résumé? What are two examples of information you should not place on a résumé?
13. List six activities you should do to prepare for a job interview.
14. Give an example of what would be considered appropriate interview dress for a job that is almost entirely manual labor.
15. What are three examples of questions that would be appropriate to ask the interviewer during an interview?
16. What information should be included in a thank-you letter for an interview?
17. What steps are involved in creating a school-to-career plan?
18. What are two resources you can use for career exploration?
19. What is involved in being a horticulture business consultant and what preparation is needed for the career?
20. Explain briefly what students do in a marketing plan career development event.

Thinking Critically

1. Imagine that you need to advertise for a new horticulture business in your community. The company has a very limited budget but wants to reach as many people as possible to inform them of this new business venture. What method would you use to promote this business? Where would you get the most value for your advertising dollars and why?
2. You have recently started working at a small organic farm as a seasonal worker. You show up on time and act professionally, but your colleague is not doing the same. Your colleague arrives late, claiming he has a hard time getting a ride to work. Today, your colleague texts you. He asks you to clock him in using his time card and says that he will be there shortly. What should you do?

STEM and Academic Activities

1. **Technology.** Develop an online tool to help students create a school-to-career plan.
2. **Math.** Determine how many people live in your town. Knowing that a community of approximately 6500 people can support a retail garden center, how many garden centers could your community support? Now, determine how many retail garden centers are in your community. Do not forget to include big retailers, grocery stores, or hardware stores that sell plants and garden supplies. Do you have more or less stores than could typically be supported?
3. **Math.** Contact a local horticultural company. Ask them how many employees they have. How many are full-time and how many are part-time? Determine how many hours of labor that business supports annually. Now, determine how much revenue must be generated each year to cover the expenses of labor alone. Use minimum wage as a starting point.
4. **Social Science.** Access a personality test online or through your school's guidance or career counselor. Determine what your personality type is according to the results of your test. Write a one-page paper agreeing or disagreeing with the results. Be sure to include several pieces of evidence that justify your response. Include life experiences to better support your case.
5. **Social Science.** Contact a local retail garden shop or florist in your area. Arrange an interview with the owner or manager. Focus your questions on their market. Explore how they determine their market. Ask about the four Ps of their marketing plan.
6. **Language Arts.** Using the knowledge you acquired from this chapter, create an abbreviated strategic business plan for an existing horticulture business of your choice. Evaluate a horticultural product or service that you use. The company most likely already has a vision and mission statement. Access those and include those in your strategic business plan. Contact the company and ask questions about their marketing and advertising. Report your findings to your class.

Communicating about Horticulture

1. **Writing.** Create a poster identifying the various types of businesses that are part of global free enterprise.
2. **Speaking.** Present a mock interview session for your class. Ask your teacher to be the interviewer or ask someone from the local agricultural business to conduct the interviews. Students should dress appropriately for an interview with the company.
3. **Reading and Speaking.** Some organizations or associations provide mentoring services for small businesses as a membership benefit. Form a small group with two or three of your peers and collect informational materials from associations that provide these services. Analyze the data in these materials based on the knowledge gained from this chapter. Make inferences about the services available and recommend the best ones to the class.
4. **Reading and Writing.** Written communication plans are essential for business success. Using an existing marketing plan career development event as a guide, choose an agricultural business in your area and write a business plan outlining how you would help them market their business to their target market.

SAE Opportunities

1. **Exploratory.** Job shadow a sales representative at a horticulture business.
2. **Placement.** Secure a sales job at a horticulture business.
3. **Experimental.** Investigate various advertising techniques. Create several horticultural advertisements using various media formats. Examples include social media, pamphlets, audio files, or videos. Poll a target audience to determine which advertising method they prefer.

Alexander Raths/Shutterstock.com

4. **Exploratory.** Create an educational website for students about professionalism in horticulture. Include pages devoted to career documents, professionalism, and job interview skills.
5. **Entrepreneurship.** Develop a strategic horticulture business plan and create a new company.

CHAPTER 6
Worker and Tool Safety

Chapter Outcomes

After studying this chapter, you will be able to:
- Discuss the history of employee safety and government regulating agencies.
- Describe various types of hazards.
- Recognize strategies for preventing accidents and promoting workplace safety.
- Practice safety in the workplace.
- Identify horticultural tools and equipment and describe their use and maintenance.
- Investigate labor laws related to employee rights and safety.
- Explore careers related to horticultural safety compliance.

Words to Know

- biological hazard
- Centers for Disease Control and Prevention (CDC)
- chemical hazard
- discrimination
- ergonomic hazard
- general safety hazard
- harassment
- material safety data sheet (MSDS)
- migrant worker
- National Institute of Occupational Safety and Health (NIOSH)
- Occupational Safety and Health Administration (OSHA)
- personal protective equipment (PPE)
- physical hazard
- safety data sheet (SDS)
- safety hazard
- strain
- stress

Before You Read

Look at the *Words to Know* list above. Write what you think each term means. Then look up the term in the glossary and compare your definition to the textbook definition.

While studying this chapter, look for the activity icon to:
- **Practice** vocabulary terms with Words to Know activities.
- **Expand** learning with identification activities.
- **Reinforce** what you learn by completing Know and Understand questions.

www.g-wlearning.com/agriculture

In the United Sates, more than 22 million people are employed in agriculture or agriculture-related jobs. And, according to the Food and Agriculture Organization (FAO), more than one billion people are employed in agriculture around the world. These people work in jobs that create the food and fiber that feeds and clothes us, as well as in jobs that collect and process the natural resources that heat our homes and fuel our vehicles. As you can see, agriculture touches nearly every aspect of our lives. Agriculture is also consistently one of the most dangerous industries in which people work. Fortunately, many individuals and organizations continually strive to make agriculture-related work safer.

In this chapter, you will learn about the various organizations responsible for worker safety and how your own responsible behavior will keep you and your fellow students and coworkers safe.

"Before the reward there must be labor. You plant before you harvest. You sow in tears before you reap joy."
—Ralph Ransom

Early American Labor

In the late 1800s and early 1900s, it was not uncommon for working conditions to be hazardous to employees. Workers had little or no say regarding the amount they were paid nor about the length of their workday. Children often worked at an early age as indentured servants or as apprentices, learning a trade. There were no regulations in place regarding age restrictions, working conditions, or fair wages.

Industrial Revolution

As the Industrial Revolution gained momentum in the early 1900s, a larger workforce was needed and child labor was used to fill the void. Children worked long hours on dangerous tasks for little pay, **Figure 6-1**.

Everett Collection/Shutterstock.com

Figure 6-1. A common job in the city for children was selling newspapers. These "newsies" would work long hours (often 12 hours a day) six days a week.

Many children worked on assembly lines operating equipment with very little knowledge about safe ways to do so. Child workers were exposed to harsh chemicals and toxins and often experienced crippling injury and death. The child workers received little or no education. Several attempts were made in the early 1900s to pass legislation regarding labor regulations, but it was not until 1938 that advocates succeeded and Congress passed the Fair Labor Standards Act (FLSA).

Did You Know?
The publication of the book *The Jungle* (1906) by Upton Sinclair was a catalyst for reform in working conditions and production methods.

Fair Labor Standards Act (FLSA)

The Fair Labor Standards Act (FLSA) set rules establishing a minimum wage, granted overtime pay, and prohibited the employment of most minors in industrial jobs. It also set rules on the number of hours an employee can work. The FLSA sets rules to protect employees of all ages and to ensure fair treatment and wages. Since the earlier part of the twentieth century, the FLSA has made a difference in labor conditions and has helped improve the quality of life for many Americans. The FLSA also set the stage for several government organizations created to ensure safe working conditions, provide safety education and training, and promote equal opportunity for employment.

Safety and Health Agencies

As employee safety became a growing concern for employers and the public, the American government responded with legislation, regulation, and the establishment of agencies such as the Centers for Disease Control and Prevention (CDC), the National Institute of Occupational Safety and Health (NIOSH), the Occupational Safety and Health Administration (OSHA), and the US Department of Labor.

"Safety doesn't happen by accident."
—Unknown

History Connection: Cesar Chavez

In the later part of the twentieth century, a man named Cesar Chavez (an American of Mexican ancestry) was a migrant farm worker in California. (A *migrant worker* is a person who moves from place to place to do seasonal work.) He traveled with his family to vineyards and farms across the state to work. Chavez quit school in the seventh grade to focus entirely on farm work to help support his family.

After serving in the US military, Chavez returned to California where he became the voice for migrant workers. He created the United Farm Workers (UFW) in 1962 and helped to stage worker strikes to improve farm labor wages and working conditions. Chavez also successfully demonstrated against the use of certain pesticides in the vineyards. These pesticides were causing health problems for the farm workers and their children.

The UFW still exists following Chavez's death in 1993. The UFW and the state of California made it legal for farm workers to form unions. The farm worker unions continue the struggle to maintain fair wages and ensure safe working conditions.

Centers for Disease Control and Prevention (CDC)

The *Centers for Disease Control and Prevention (CDC)* is a unit of the Department of Health and Human Services. The CDC monitors safety and security risks and health hazards in the United States and abroad. The CDC protects people from health threats, conducts critical research, and delivers health information that protects the United States against expensive and dangerous health hazards. It also responds quickly when the need arises, **Figure 6-2**.

National Institute of Occupational Safety and Health (NIOSH)

In 1970, Congress created the *National Institute of Occupational Safety and Health (NIOSH)*. NIOSH conducts research and makes recommendations dealing with workplace safety, injury, and illness. This federal agency solves health-related safety issues in American workplaces.

NIOSH and the CDC have joined forces and created the CDC-NIOSH Agricultural Safety and Health Centers. These centers are strategically located at eight universities and two additional sites around the country. The safety and health centers were established to conduct research, provide education, and establish prevention projects to address the nation's agricultural health and safety issues.

deepblue/Shutterstock.com

Figure 6-2. When there is cause, the CDC will move in quickly to determine the source of contamination and ensure proper protocol is followed to eliminate the source and prevent additional contamination.

Occupational Safety and Health Administration (OSHA)

The *Occupational Safety and Health Administration (OSHA)* was established in 1970 by the US Congress to ensure safe and healthy working conditions for Americans. This government agency sets and enforces standards for business and industry. It also performs research and provides training and education. OSHA is part of the Department of Labor.

United States Department of Labor (DOL)

The mission of the US Department of Labor (DOL) is to "foster, promote, and develop the welfare of the wage earners, job seekers, and retirees of the United States; improve working conditions; advance opportunities for profitable employment; and assure work-related benefits and rights." The DOL develops or oversees many programs, including:

- Minimum wage.
- Unemployment insurance.
- Work regulations.
- Disability.
- Equal employment opportunities.

The DOL works hand in hand with all the other worker safety agencies that are regulated by the US government.

Corner Question

How many federal OSHA safety inspectors help keep American workers safe?

Safety Hazards

The moment workers enter their job sites they must take time to identify potential safety hazards to prevent injury and illness for themselves and others, **Figure 6-3**. *Safety hazards* are anything on a job site that can cause injury, illness, or death. Common occupational safety hazards include:

- *Physical hazards* are conditions or substances within the work environment that may cause a person harm. Examples include radiation, noise, temperature extremes, falling, electrocution, and receiving cuts or abrasions.
- *Chemical hazards* are toxic substances that can cause a wide range of harmful effects. Examples include cleaning products, fuels, dusts, fertilizers, and pesticides.
- *Biological hazards* are organisms that can cause harm to another living organism. Examples include molds, viruses, bacteria, toxic plants, insects and other pests, animals, and bodily fluids that may contain disease-causing organisms.
- *General safety hazards* are safety hazards that are common among various industries. Examples include slipping on wet floors, falling from heights, injuries from machinery, electrical shocks, and catching a virus from a coworker.
- *Ergonomic hazards* are repetitive movements or positions that may lead to physical stress. *Stress* is a short-term impact of a pressure or tension. Examples include repetitive lifting or keying data.
- Work organization hazards are those caused by emotional or mental stresses.

According to the Department of Labor, these safety hazards cause millions of injuries and illnesses in workers each year. Safe practices are more likely to take place when employees recognize the dangers of their workplace and have proper and up-to-date training in personal protection.

> **Safety Note**
> Get enough rest! Aside from health issues that arise from lack of sleep, excessive sleepiness also contributes to a greater than twofold higher risk of sustaining an occupational injury!

Arthur Eugene Preston/Shutterstock.com

Figure 6-3. To prevent physical injury and possible damage to machinery, landscape and turfgrass workers must follow the manufacturer's safety protocol as well as their employer's safety rules.

> **Corner Question**
> How many American workers are employed full-time in production agriculture?

Did You Know?
Approximately 3 million leaf blowers are sold in the United States annually.

Physical Hazards

All jobs, regardless of location or type of work performed, include physical hazards. Recognizing these hazards is the first step in avoiding injury, illness, or death. Physical hazards are conditions or substances within the work environment that can injure a worker. Examples of physical hazards include:

- Perpetual loud noises.
- Long exposure to sunlight or other sources of radiation.
- Extreme temperatures.
- Wet floors with no tread.

Physical hazards impact many individuals in horticultural occupations. In many horticultural jobs, such as in landscaping and nursery work, individuals are continually exposed to the sun and changes of weather. In addition, landscape and nursery workers use equipment that can be extremely loud and can lead to hearing loss with continued exposure, **Figure 6-4**.

ChameleonsEye/Shutterstock.com

Figure 6-4. A worker using a leaf blower should wear hearing protection. Continual use of loud equipment can create long-term damage to hearing.

Chemical Hazards

Most horticultural and agricultural businesses have chemicals onsite that may be hazardous to employees and the environment. These chemicals are used to promote growth (fertilizers); control or combat diseases, insects, and animal pests (pesticides); and fuel or service equipment (gasoline/diesel fuel). These chemicals may be synthetic or organic and may come in liquid, vapor, or granular form. Exposure to these chemicals may cause a wide range of physical harm.

Safety Note

Leaf Blower Safety

A leaf blower or chain saw produces about 115 decibels of sound and can cause permanent hearing damage after just 30 seconds. It is important to always wear proper ear protection to ensure that you will not suffer temporary or permanent hearing loss. This is a physical hazard to which many landscape and turfgrass professionals are exposed.

©iStock.com/mtnangel

Corner Question
What safety precautions should be taken by horticulture workers who are exposed to the harmful UV rays of the sun?

Safety Note

Vapors and Gases

Vapors and gases can be odorless and colorless but deadly. In 2011, two brothers, ages 16 and 22, died while cleaning an organic waste drainage system at a composting facility in California. High levels of hydrogen sulfide were in the area where the workers were flushing out the drain pipes. The younger brother was overcome by the hydrogen sulfide gas and fell into a shaft. His brother tried to rescue him and also died from the gas. The two young men were unaware of the gas and were not wearing the proper equipment to protect them from this dangerous and ultimately fatal situation.

CandyBox Images/Shutterstock.com

Workers may be exposed to chemical hazards while:

- Applying chemicals to the soil or plants, **Figure 6-5**.
- Cultivating or harvesting plants, vegetables, and fruits.
- Disposing of plant materials.
- Servicing or cleaning machinery.
- Fueling machinery.

When used and stored properly, these chemicals are relatively safe. To avoid misuse or accidental exposure, all chemicals must be identified and stored properly. If chemicals are not in their original containers, they must be properly labeled and the original informative labeling must be readily available. Access and use must also be limited to those authorized and trained to apply or distribute the materials. The proper personal protective equipment must also be worn while working with the chemicals. *Personal protective equipment (PPE)* includes materials and devices worn to provide a shield or defense from dangers. PPE includes earplugs, gloves, respirators, masks, goggles, specific clothing, shoes, and sunblock.

GSPhotography/Shutterstock.com

Figure 6-5. When working with pesticides, workers must follow every detail of instruction on the pesticide label before, during, and after application.

Biological Hazards

Biological hazards are presented by organisms that are or were once living and which can cause harm to another living organism. These hazards are encountered when employees work with animals, plants, bacterial or viral cultures, tissue samples, diseased or infected plant material, compost, manure, and even people who are sick and contagious.

"Safety is something that happens between your ears, not something you hold in your hands."
—Jeff Cooper

Animals

Depending on the area of the country in which they are working, employees may encounter wild or feral animals, ranging from field mice to barn cats to alligators, when they are working outdoors. Some animals pose physical as well as biological hazards. For example, mice may defecate and urinate in and around tools and machinery stored in a barn or shed, as well as bite a hand when startled. Feral cats may carry disease or be infested with ticks and fleas and may also bite if they are startled or feel threatened.

Additional biological hazards from animals include bites or stings from venomous spiders, stinging caterpillars, and venomous snakes, **Figure 6-6**. Spiders, such as the venomous black widow, are often found inside potting sheds and irrigation covers, and venomous snakes may be found between landscaping rocks, in tall grass, and in heavy underbrush. Employees should be instructed on how to handle possible situations involving animals and what steps to take if someone is stung or bitten.

Plants

Workers must also be aware of their environment and pay close attention to where their hands, feet, and body are in relation to plants that may cause allergic reactions. Poison ivy, poison oak, and stinging nettle can cause dermatitis or allergic reactions, **Figure 6-7**. These reactions can be very severe, causing extreme itching and lesions in those who are allergic. These types of plants may be encountered when maintaining existing landscaping or when clearing an area for planting.

Special care should also be taken when handling diseased or infected plant materials. Diseased or infected plant materials must be disposed of properly to prevent spreading of the disease or infection.

A *Peter Waters/Shutterstock.com*

B *neil hardwick/Shutterstock.com*

C *Eric Isselee/Shutterstock.com*

Figure 6-6. Workers need to be aware of natural dangers they may encounter when working outdoors and what actions should be taken if they are bitten by venomous animals. A—Black widow spider. B—Io moth stinging caterpillar. C—Rattlesnake.

A *Stuart Monk/Shutterstock.com*

B *©iStock.com/Derek_Neumann*

C *©iStock.com/toxawww*

Figure 6-7. Plants that can be toxic to human beings vary by region and may be found in both cultivated and uncultivated areas. A—Poison ivy. B—Poison oak. C—Stinging nettle. **Do you know how to treat skin that has come in contact with toxic plants?**

Tools and clothing that come in contact with poisonous and diseased or infected materials should also be thoroughly cleaned and possibly disinfected.

Compost and Manures

In addition to plant material that may be diseased or infected with pests, workers often work with decomposed organic matter, such as compost and manures, which are used to amend soils. These materials may contain harmful bacteria that may cause illness or disease. The best defense against biological hazards is to keep work areas, tools, and equipment as clean as possible, and to regularly wash your hands after handling any biological materials.

Corner Question

How many venomous snake bites occur each year?

General Hazards

General hazards are common dangers that many employees encounter while working at their job site. These hazards can cause illness, injury, or death and are the result of:

- Tripping or slipping on wet surfaces.
- Falling from a height.
- Coming in contact with unguarded machinery.
- Coming in contact with moving machinery parts.
- Receiving shocks or other injuries from electrical cords or improper wiring.
- Becoming trapped or suffocating in a confined space.
- Being run into or bumped by a forklift, tractor, conveyor belt, or vehicle, **Figure 6-8**.

These common hazards can easily be avoided by making the worker's welfare a priority. Always keeping safety in mind will help you avoid illness, injury, or death from workplace accidents.

Ergonomic Hazards

Have you ever done something physical that was repetitive and found that you had muscle cramps or soreness the following day? Maybe you were keying text for an extended period of time and your wrists began to ache, or perhaps you were practicing the guitar and your fingertips became numb. These ergonomic hazards (repetitive motions or positions leading to physical stress) cause temporary and sometimes permanent damage. Examples of ergonomic hazards in the horticulture workplace include:

- Bending over flower or garden beds for long periods of time.
- Kneeling for extended periods of time.
- Hunching over a table while sowing seeds.
- Repeatedly planting or digging.
- Frequently lifting heavy loads.
- Keying or doing computer work.

These tasks, if repetitive and done without following proper safety rules, can lead to acute muscle soreness and fatigue. There can also be lifelong impacts to your physical health.

LUCARELLI TEMISTOCLE/Shutterstock.com

Figure 6-8. When operating large equipment like a tractor, eyes and ears must be open and attentive to the surroundings.

Work Organization Hazards

"The safety of the people shall be the highest law."
—Cicero

Some hazards at a workplace do not always result in physical injuries. Sometimes emotional issues associated with a job lead to stress and *strain* (the long-term impact of a pressure or tension) on one's mental and physical health. These hazards are handled differently by each person. They vary at all jobs just as other physical, chemical, biological, and ergonomic hazards do. Work organization hazards can include:

- Workplace demands.
- Social relationships.
- Lack of respect.
- Emotional stress.
- Violence.
- Pace of work.

Accidents or catastrophic events at a job site can also contribute to stresses or strains of jobs. Exercise and emotional releases can ease the effects of a stressful incident. If strain is occurring, then changes must be made to the workplace environment to improve conditions.

Preventing Accidents

Once safety hazards have been identified, businesses, employers, and employees can begin the process of accident prevention. Preventing and controlling hazards can be accomplished by:

- Maintaining equipment and monitoring and recording maintenance at regular intervals.
- Ensuring that all employees have been trained about safety on the work site.
- Ensuring all personnel are aware of safety procedures associated with all equipment.
- Ensuring all personnel have been trained to operate the equipment, **Figure 6-9**.
- Supplying and training personnel in the use of personal protective equipment (PPE).
- Certifying that employees know how to acquire or implement first-aid treatment for accidents, injuries, and illnesses.

Preventing accidents is much easier than treating injuries or illnesses. Knowing the ABCs of accident prevention can make the workplace a healthier and safer environment for all workers.

ABCs of Accident Prevention

Employers and workers should follow the ABCs of accident prevention to ensure the safety of everyone on a work site. The ABCs stand for *administration*, *barriers*, and *communication*.

A Is for Administration

Follow the rules and procedures established by employers, equipment manufacturers, and government agencies. Wear the correct personal protective gear, regulate workloads, and

Valerie Johnson/Shutterstock.com

Figure 6-9. Pruning, especially at great heights, must take into account the stability and safety of the worker. Do you know how to safely use a ladder?

Thinking Green

Plant a Tree, Relieve Stress, and Save the Planet

Getting outside and planting a tree can alleviate stress. Physical exercise releases serotonin and endorphins that help to make people feel happier. Increased oxygen intake and deeper breathing help to provide a euphoric or happier feeling. Since planting a tree will help reduce the greenhouse effect, filter the air, control erosion, and release oxygen into the air, this is a win-win situation. So breathe it all in. Oxygen can make you feel better, and planting a tree is a great way to ensure we will have cleaner air to breathe.

Make it a family affair and teach your younger siblings the benefits of planting trees. There are often special deals on trees for Arbor Day (the last Friday in April). Many companies or communities sponsor free tree giveaways.

©iStock.com/JackF

rotate jobs. Safety rules were made to prevent injuries and illnesses. Read all warnings and labels before beginning any task.

B Is for Building Barriers

Safety shields should be created to help prevent accidents or harm. These shields include personal protective equipment (PPE) such as hearing protection, eye protection, respiratory protection, specific clothing, gloves, shoes, and sunblock, **Figure 6-10**. Barriers may also be created by locking cabinets or storage facilities that contain chemicals or dangerous materials. Shields and guards should be placed around stationary equipment to prevent injury. Employers and workers should be vigilant and monitor barriers to ensure they are fully functional.

Goodluz/Shutterstock.com

Figure 6-10. Workers use personal protective equipment, such as hard hats, on the job site to prevent injuries.

Safety Note

Machine Safety Devices

A 16-year-old produce market worker in New York City was working alone in a basement using a machine to crush cardboard for disposal. While working, he was caught by the machine's hydraulic arm. Another employee found the worker and called 911; however, the young man could not be revived. He had been crushed by the machine. An inspection of the machine revealed that a safety device had been disarmed to speed up the work process. If the safety device had not been tampered with, the young man would likely still be alive.

It requires a great deal of force to bale cardboard and paper so tightly. Never modify machinery to bypass safety features—it could be a deadly mistake.

©iStock.com/drevalyusha

C Is for Communication

All parties involved must share knowledge about safety and accident prevention. Businesses are required by law to teach their employees to complete all tasks safely, **Figure 6-11**. Employees have a duty to ask questions when they have a concern dealing with safety. Communication about safety also must be in the language that employees understand. If any workers speak another language, safety materials should be provided in a language that the employees will understand. This practice safeguards all individuals and ensures that everyone understands safety procedures at the workplace.

simez 78/Shutterstock.com

Figure 6-11. Your employer may have written guidelines for you to follow.

Workplace Safety Documents

By law, each place of business must have certain safety and employee information posted so that it is visible to all employees. When applicable, these postings may be in multiple languages. The posted documents may state safety or evacuation procedures, explain machine operation, or list who to contact in case of emergency. Other documents, such as safety data sheets, inform people about hazardous substances that are used at the workplace.

Safety Data Sheets

A *safety data sheet (SDS)* is a document that contains information on the potential health impacts of a chemical or other dangerous substance. An SDS outlines the chemistry of the material, identifies potential safety hazards, lists proper handling and storage procedures, explains what to do in case of exposure or ingestion, and explains how to safely handle a spill and cleanup.

Safety Note
Weather Hazards

A 17-year-old Florida landscape worker was unloading rock from a truck bed with his coworkers. There was light rain and some thunder in the distance. Suddenly, there was a lightning strike. The boy was electrocuted, and all the coworkers were injured. The landscape company had not instructed the workers to seek cover when they could hear thunder. If the company had advised the employees of this rule, the teenager's life might have been saved.

Always seek cover during lightning storms. Keep in mind that it does not have to be raining where you are standing for lightning to strike. If you hear thunder or see lightning, find a safe location to wait out the storm.

©iStock.com/Sportactive

Corner Question
What personal protective equipment (PPE) should be worn to mow grass?

All materials that are used on-site must have an SDS available. A safety data sheet was formerly referred to as a *material safety data sheet (MSDS)*.

Pesticide Labels

In addition to the required SDS, all pesticides must have a manufacturer label that is regulated by the Environmental Protection Agency (EPA). The pesticide label serves as a legal agreement for the individual applying the pesticide. When an individual applies a pesticide, he or she is agreeing to read and follow the directions for safe application, storage, and disposal.

Practicing Safety

Greenhouses, nurseries, farms, golf courses, vineyards—each horticultural location will have its own unique safety hazards, **Figure 6-12**. Workplace conditions will vary, but accident prevention and safety practices should always be of the utmost concern for employers and employees.

SAFE Working Conditions

Practicing safety is easy when you remember to apply the acronym SAFE. SAFE stands for *see the safety issue*, *ask for help*, *find a solution*, and *extend optimism*.

CoolR/Shutterstock.com

J. Bicking/Shutterstock.com

Figure 6-12. Each horticultural or agricultural operation will have its own unique hazards. **What types of unique hazards would you find in a vineyard or tree nursery?**

S: See the Safety Issue

Recognize the safety issue. In horticulture, the issue could be the improper storage of leftover pesticides, low tire pressure on a golf cart, or a dull edge on a pruning saw. Whatever the issue may be, the first step is to identify the hazard.

A: Ask for Help

Know who to ask for help. There are people at your workplace who are responsible for safety just as there are at your school. Find out who is responsible for safety issues by speaking with parents, coworkers, school officials, and human resources personnel. If you do not find the answer to your question, contact the Department of Labor, OSHA, CDC, or NIOSH for assistance. A simple search on the Internet may also answer your questions or provide you with a contact number for the government's health and safety agencies.

F: Find a Solution

You may need to approach supervisors, human resources personnel, or employers about safety issues. It is always helpful to approach these individuals with a possible solution to the safety issue if you have one to suggest.

"Safety is a cheap and effective insurance policy."
—Unknown

E: Extend Optimism

Discuss safety situations in a positive manner when talking with an employer or supervisor. For example, you might say, "Should we get a new outlet for that broken electrical switch?" This is a great way to suggest a solution as compared to saying, "You need to buy a new electrical switch or somebody is going to get killed." Your employer will prefer that you ask for help instead of making negative comments. As a result, there will be no animosity, and everyone can benefit from the situation.

Maintaining Tools and Equipment

All tools and equipment can be dangerous if they are not maintained well or are used improperly. Even simple tools, such as screwdrivers, can be dangerous. Many of the tools used by horticulturists are for cutting, digging, planting, and moving; they can be very dangerous. Everyone, employers and employees, should do their part to maintain a safe work environment. Keep the following guidelines in mind to help prevent accidents and injury at school, at work, and at home:

- Read the instruction manual before operating any machine or device.
- Ask for training from your supervisor on how to operate or use machinery or equipment.
- Check all equipment for safety issues before using.
- Maintain and store all equipment and machines properly, **Figure 6-13**.
- Keep records of maintenance on all equipment and machinery.
- Be mindful of all power lines before digging. Call 811 before digging anywhere.
- Be aware of people, vehicles, objects, and road or area conditions around you in all directions while driving any vehicle or machine.
- Do not disarm any safety guards or devices on equipment.
- Wear proper personal protective equipment (PPE).
- Employ SAFE practices as described earlier.

Konstantin Sutyagin/Shutterstock.com

Figure 6-13. Properly storing tools helps keep tools in good condition, makes them easily accessible, and also allows for accurate inventory.

Checking and Maintaining Equipment

Equipment such as lawn mowers, leaf blowers, weed trimmers, chain saws, rototillers, skid steers, and other machines must be cared for to ensure that they can be used safely. Before using any machinery, take steps to determine if it is operable.

- Read the owner's manual. Pay particular attention to anything in the manual associated with the words *hazard*, *caution*, *warning*, or *danger*.
- Wear the proper PPE while servicing any engine.

Safety Note

Call 811 Before You Dig

A federally mandated "call before you dig" number is 811. You can call from anywhere in the United States, and you will be directed to your local 811 center. These individuals will help you make contact with local electrical, gas, and other utility providers. These providers will indicate areas on the site where you should not dig because of underground lines. This free service helps ensure safety for you and avoids interruption of service for others. No matter how big or small the job is, you must call 811 before you dig.

©iStock.com/micke_ovesson

- Disconnect spark plugs and battery cables while servicing engines.
- Check cooling fins and make sure nothing is obstructing movement of the blades.
- Check tire pressure for rototillers, tractors, and other machines with wheels.
- Check the oil and fuel levels. Fill as needed with the appropriate oil and fuel.
- Check belt tensions and chains. Adjust as needed.
- Check and replace air filters after every 25 hours of operation (sooner if in dusty conditions) or according to owner's manual.
- Check for dull, nicked, unbalanced, or broken blades. Sharpen blades (using a grinder), if needed, according to owner's manual or approximately at a 45° angle, **Figure 6-14**.
- Check that the lawn mower blade is balanced.
- Perform all routine maintenance according to the owner's manual suggestions. Record the date of service and what was done.

©iStock.com/craetive

Figure 6-14. Lawn mower blades must be kept sharp to keep the mower efficient and consistent. Although blades may be sharpened with a file, it is more efficient to use a grinder. What PPE should be worn when sharpening blades?

AgEd Connection Equipment and Supplies Identification

Some states offer a career development event (CDE) on tool identification. Although this event does not take place on a national level, many local and state chapters host this contest. Students correctly identify a tool that is commonly used in agricultural practices. Students determine the correct purpose and use of the tool as an additional component of this CDE. Because this CDE is offered as a local option, check with your advisor to determine if this CDE is an opportunity for you and your school's participation. Use the e-flashcards on your textbook's student companion website at www.g-wlearning.com/agriculture to help you study and identify these tools and equipment.

Thinking Green

Electric Lawn Mowers

An electric lawn mower is an excellent way to reduce emissions into the environment. There are battery-operated or electric lawn mowers to choose from. Many of them are not as powerful as a gas-powered lawn mower; however, they can easily cut a maintained lawn. These machines are more suitable for urban and suburban lawns. They reduce both carbon emissions and noise pollution.

Keep in mind that electric mowers are not usually as powerful as gas-powered mowers. If you are using an electric mower, do not allow the lawn to grow too long. Your mower will be much less efficient.

©iStock.com/maXoidos

When the operator has determined that the machine is ready for operation, then she or he is ready to suit up with any needed PPE. The operator must first determine any safety hazards in the area where she or he will be working before work can begin. The operator should start the machine according to directions in the owner's manual. When finished, she or he must properly shut off, clean, maintain, store, and secure the machine to prevent future operation error or injury.

Maintaining Tools

Countless tools hang in the storage areas of horticultural businesses. Pruning saws of every shape and size, grafting tools, shovels, rakes, and cultivators are some of the tools that make a horticulturist's job easier. These tools must be maintained for continued safe use. Follow these guidelines when maintaining tools:

- Wear the appropriate PPE.
- Clean leather parts with glycerin bars or other cleaners made specifically for cleaning leather.
- Clean wood parts, sand if needed, and rub with an oil or sealant to protect them.
- Remove all grease, oil, and rust from metal parts with a solvent or wire brush. Sharpen blades with a file or grinder.
- Coat metal with a medium weight oil and apply a rust-preventive material.

Labor Laws

Workers need to know their rights. Understanding worker rights is just as important as worker safety. Laws and regulations set forth by the US government exist to protect workers. These laws go hand in hand with safety.

Minimum Wage and Hours Worked

The US Department of Labor's Wage and Hour Division administers the federal minimum wage law. This law states the minimum amount that covered employees must be paid per hour. (Some states also have minimum wage laws.) Additionally, this agency also enforces federal laws pertaining to work hours.

These rules are more complicated for youth workers (those under 18 years of age) than for adults.

The hours that a youth worker may work can vary from state to state. In general, youth who are 16 or 17 years of age and are attending school may work as many hours as they choose but only until 11 pm and not before 5 am. Children who are 14 or 15 years old may only work three hours a day during the week between 7 am and 7 pm. They may work eight hours a day on Saturday and Sunday. These students can work no more than 18 hours a week during a school term and no more than 40 hours a week during the summer.

Job Duties for Workers under 18

Youth workers are more limited in job duties than those who are 18 or older. Several laws are in place that prohibit those who are under 18 years old from:

- Operating machinery or power equipment.
- Roofing (any and all things dealing with a roof, both on the ground and on a roof).
- Logging or working at a sawmill.
- Being exposed to radiation.
- Handling, serving, or selling alcohol.
- Driving a forklift.
- Driving as a main part of the job.
- Working in demolition.

Employees at Least 16 Years Old

Employees must be 16 years old to load or unload trucks and work in some construction or manufacturing jobs. In addition, employees must be 16 years old to bake. These jobs are sometimes considered hazardous. A 16-year-old may only be able to work at these sites in limited capacities to comply with OSHA and the Fair Labor and Standards Act provisions dealing with youth employment.

Employees 14 or 15 Years Old

According to the Fair Labor Standards Act, youth must be at least 14 years old to work. Youth workers can do many jobs, but labor laws set some limits. Youth who are 14 or 15 can cook only with electric or gas grills. They are not allowed to:

- Work at dry cleaning or with commercial cleaners.
- Load or unload a truck.
- Work on a ladder or scaffold.
- Build, work construction, or work in manufacturing.

These rules and regulations standardize youth labor to ensure all employees are safe and that hazards are avoided whenever possible. As workers age, more work responsibilities and opportunities are available. These rules are in place because of a history of youth worker accidents. This historical information helps lawmakers determine how to safeguard all workers. It is important that young workers know what types of jobs are legal for them to work, based on their ages.

"Tomorrow is your reward for acting safely today."
—Robert Pelton

Corner Question

How many stitches does the average chain saw injury require?

Young Worker Responsibility

Injury prevention can be difficult, especially for teens. Unsafe work behaviors are often the result of teens trying to work faster and be more productive. Youth will often equate working fast to earning more money. However, working faster or using unsafe methods to increase production may lead to more accidents. Safety should never be disregarded to reach work goals. If a work or production goal conflicts with safety, then there must be a discussion with a supervisor or employer to address the unsafe conditions. Safety laws that are in place to protect workers from illness, injury, or death should always be regarded.

Employees must seek additional information or help when safety rules or precautions are faulty. Workers must use their decision-making skills to avoid work-related accidents. Recognizing risky behaviors and avoiding dangers will lead to better mental and physical health. When supervisors and employers work together to combat safety hazards, everyone at the workplace will benefit.

Taking Action

Workers may be called to take action against hazards at some point in their early careers. These workers must talk with the appropriate people to ensure their safety and that of others. You should always be an advocate for yourself.

Harassment and Discrimination

No employee should ever feel harassed or discriminated against on a job site. *Harassment* is repeated treatment that bothers or annoys another person. *Discrimination* is unfair treatment of another person. It may be based on factors such as age, gender, ethnicity, the presence of a disability, religion, or culture. Harassment and discrimination must never be tolerated. Employees must be their own advocates. They should speak out against individuals, whether coworkers or supervisors, who are bullying, sexually harassing, or otherwise negatively impacting their work environment with emotional, mental, or physical aggression. When in doubt, report the incident. Contact your human resources department or talk to a supervisor or another adult about anything you experience at work that makes you feel uncomfortable.

Employees who feel that they are being discriminated against or note discrimination at their workplace should also contact supervisors or employers about these occurrences. An employee can contact US federal agencies, such as OSHA, NIOSH, or the Department of Labor to report discriminatory practices.

Horticultural Safety Careers

Most employers will agree that safety is a number one priority. A safe workplace ensures fewer accidents, which means more hours working and higher productivity. There are positions within businesses that regulate compliance and train workers to maintain safety standards.

> "I believe discrimination still exists in our society, and we must fight it in every form."
> —Andrew Cuomo

Did You Know?
Employers are not allowed to discriminate against employees who have exercised their right to file a workplace complaint.

Horticultural Risk Consultant

A horticultural risk consultant or manager helps businesses and the industry to understand risk management. A risk consultant explains to employers and employees how to manage and minimize risks at the workplace. These risks may involve worker safety, weather and climate changes, market fluctuations, and financial threats. A risk consultant may also help a business evaluate insurance opportunities. A risk consultant also crafts solutions to all the previously mentioned issues. A risk consultant must be an excellent problem solver and communicator and must have an educational background in business, finance, or an agricultural-related field.

Horticultural Safety Manager

A horticultural safety manager serves as the leadership supervisor for local or regional safety concerns within horticultural companies. This individual focuses on compliance with government agencies, such as OSHA, NIOSH, EPA, CDC, and DOT (Department of Transportation). A safety manager must be proactive, a good educator and communicator, and a motivator. He or she must pay particular attention to preventing safety accidents while promptly handling those that occur in the workplace. A safety manager promotes strategies to decrease workplace illness, injuries, and death. Many companies consider worker safety as the most important variable within the company, making the safety manager an extremely valuable employee.

Career Connection

Kurt Bland
Landscape Company Owner

Kurt Bland, Bland Landscaping

Kurt Bland, the president of Bland Landscaping, is responsible for the safety of his employees. When his employees enter the building, they are greeted with a series of safety reminders before they leave for their job site. In the shop are murals of employees in personal protective equipment with safety standards and expectations. The walls of every room in the building have safety reminders along with OSHA postings. Finally, as the employees exit to their vehicles, they are reminded of vehicle safety and driving rules.

Kurt celebrates with his team of landscape professionals every 100 days of work without any accident. He rewards his employees for their attention to safety. This system promotes a culture of attention to safety. When his employees are safe, with few or no accidents, there are more gains for his company and the employees.

The National Association of Landscape Professionals, or NALP, encourages safety in all facets of the landscape business. This organization and insurance agencies for landscape companies encourage companies such as Bland Landscaping to have a safety committee. A company's safety committee consists of employees from each department. The safety committee helps direct safety decisions and education. Kurt and his employees realize that safety is a priority during the entire day of work. Steering and maintaining this culture of safety is Kurt's responsibility and a duty he gladly accepts.

CHAPTER 6

Review and Assessment

Chapter Summary

- Employee safety has been a major concern of industry, the government, and the public since the early part of the twentieth century. Before this time, child labor was considered a normal practice. After the Fair Labor and Standards Act was passed, several government agencies were established to safeguard workplaces. These agencies are the CDC, OSHA, NIOSH, and the US Department of Labor.
- Types of safety hazards at agricultural workplaces include physical, chemical, biological, general, ergonomic, and work organization hazards.
- Preventing accidents at the workplace is as easy as knowing your ABCs. ABC stands for *administration*, *barriers*, and *communication*.
- Practicing safety can be facilitated by creating a SAFE workplace. SAFE is an acronym that stands for *see the safety issue*, *ask for help*, *find a solution*, and *extend optimism*.
- There are countless tools and equipment that can be used while working in the horticulture industry. Always read safety manuals and employ safe practices before, during, and after operation, use, and maintenance.
- Employee safety laws, especially those impacting youth workers, are important. The US government regulates youth labor. There are strict laws regarding what can and cannot be completed by youth on a work site as well as how many hours can be worked.

Words to Know ↗

Match the key terms from the chapter to the correct definition.

A. biological hazards
B. Centers for Disease Control and Prevention (CDC)
C. chemical hazards
D. discrimination
E. ergonomic hazards
F. general safety hazards
G. harassment
H. migrant worker
I. National Institute of Occupational Safety and Health (NIOSH)
J. Occupational Safety and Health Administration (OSHA)
K. personal protective equipment (PPE)
L. physical hazards
M. safety data sheet (SDS)
N. safety hazards
O. strain
P. stress

1. Repetitious movements or positions that may lead to physical stress.
2. A government agency created to ensure safe and healthy working conditions for Americans.
3. Organisms that can cause harm to another living organism.
4. A government agency that conducts research and makes recommendations dealing with workplace safety, injury, and illness.
5. Anything on a job site that can cause injury, illness, or death.
6. Materials or devices worn to provide a shield or defense from dangers.
7. Short-term impact of a pressure or tension.
8. Toxic substances that can cause a wide range of harmful effects.
9. Repeated treatment that bothers or annoys another person.
10. An operating unit of the Department of Health and Human Services that monitors safety risks and health hazards in the United States and abroad.
11. A document that contains information on the potential health impacts of a chemical or other dangerous substance.
12. Common dangers that most employees encounter at work, such as slipping on wet floors, falling from heights, or electrical shocks.
13. Unfair treatment of another person based on factors such as age or gender.
14. A person who moves from place to place to do seasonal work.
15. Long-term impact of a pressure or tension.
16. Conditions or substances (i.e., radiation, noise) within the work environment that can hurt a person.

Know and Understand ↗

Answer the following questions using the information provided in this chapter.

1. What purpose does the FLSA serve American workers?
2. Who was Cesar Chavez?
3. List the government agencies that are concerned with employee safety.

4. What are six types of common occupational safety hazards?
5. What are three ways to prevent a safety hazard?
6. What are the ABCs of accident prevention?
7. What is an SDS?
8. What does the safety acronym SAFE stand for?
9. What are three actions you can perform to help prevent an accident when using equipment?
10. What is the purpose of the Call 811 service?
11. What are three words you should pay close attention to in any equipment or tool manual?
12. What should be done to maintain the metal parts of machinery and equipment?
13. How does using an electric lawn mower help the environment?
14. How many hours can a student who is 16 or 17 years old work while in school?
15. Can a person who is 16 years old work on a roofing or construction job site?
16. What should you do if you observe or are targeted for harassment or discrimination?
17. What type of work does a risk consultant do for a horticultural business?
18. What are the job duties of a horticultural safety manager?

Thinking Critically

1. Imagine that you and your class have been outside the classroom working on a hot day. One of your classmates collapses. Your teacher is not nearby. What are your first steps in this emergency situation?
2. You are a 15-year-old employee, and your employer has just informed you that one of your coworkers quit without notice. He says that he needs you to work until closing at 10 pm. You walk to work, so transportation is not an issue. Your employer says he will have to pay you *under the table*, or *off the books*, for those extra hours because you were only supposed to work from 4 pm to 7 pm. What should you do in this situation?

STEM and Academic Activities

1. **Science.** Your home may have pesticides and chemicals that sit on a shelf in your garage or other storage area. Catalog those chemicals and determine the proper storage and handling for each one. If your chemicals have not been stored properly, create a system that will ensure safety for these items.
2. **Technology.** Develop an online tool inventory list for your class. Upload an image of each tool, a description, the quantity, and directions of how to use the tools. This will create a helpful database for your school and can be a potential SAE project.
3. **Social Science.** Find an area at home, work, or school that has a potential safety hazard. How can this situation be improved? Describe the solution. Present this information to the person that could make this solution a reality. Ask how you could help to remove this safety hazard using your solution.

4. **Social Science.** Contact a local horticultural business and ask someone to tell you about one recent workplace accident. What could the workers or managers have done to prevent this incident? What steps have been taken to prevent similar issues in the future? Share your information with the class and compare stories of local workplace accidents.
5. **Language Arts.** Recall a time when you had an accident. Write a letter to your younger self explaining what the accident was, the implications, and how to avoid a similar unfortunate event in the future.

Communicating about Horticulture

1. **Writing and Speaking.** Make a series of safety posters for working in various areas on a farm, greenhouse, or nursery. Posters may cover working with machinery, driving a tractor, loading and unloading materials, using an ATV, working with pesticides, or other areas of your choosing. Explain your posters to the class.
2. **Reading and Speaking.** As the foreman in charge of a work crew, it is your responsibility to ensure the safety of your workers in the event of a fire. You must post signs in the work area to educate your workers on the different types of fires and how to extinguish them. Research the types of fires and categorize them as Class A, Class B, or Class C. Create your signs in the form of a presentation. Share the presentation with the class, as though the class were your crew. Ask for and answer any questions your crew may have.

SAE Opportunities

1. **Exploratory.** Talk to your agricultural education teacher about having an OSHA inspector visit your classroom. As an alternative, interview an OSHA inspector. The interview can be in person or by phone or e-mail.
2. **Improvement.** Conduct a safety inspection at your school, classroom, farm, work site, or home. Create a plan to improve safety at the facilities or site and implement the plan.
3. **Analytical/Research.** Investigate a horticultural safety topic. Create a presentation or a video to discuss unsafe methods and demonstrate how to work in a safe manner.
4. **Analytical/Research.** Do research to identify and learn about common pesticides or chemicals used while gardening. Create an informational poster to describe their proper storage and disposal.
5. **Placement.** Find a job (paid or unpaid) in the horticulture industry. Before you begin work, get safety training from your employer.

Monkey Business Images/Shutterstock.com

Chapter 7: Plant Taxonomy

Chapter Outcomes

After studying this chapter, you will be able to:
- Discuss the historical origins of the plant classification system.
- Describe the hierarchical orders of plants.
- Interpret a plant identification key.
- Describe the purpose of herbaria.
- Explore careers related to plant taxonomy.

Words to Know

- angiosperm
- binomial nomenclature
- class
- classification
- common name
- cotyledon
- cultivar
- dichotomous key
- Dicotyledoneae (dicots)
- domain
- ecologist
- family
- genus
- gymnosperm
- herbarium
- International Code of Botanical Nomenclature
- Monocotyledoneae (monocots)
- morphology
- order
- phylum
- scientific name
- species
- specific epithet
- taxonomy
- variety

Before You Read

As you read the chapter, put sticky notes next to the sections where you have questions. Write your questions on the sticky notes. Discuss the questions with your classmates or teacher.

While studying this chapter, look for the activity icon to:

- **Practice** vocabulary terms with Words to Know activities.
- **Expand** learning with identification activities.
- **Reinforce** what you learn by completing Know and Understand questions.

www.g-wlearning.com/agriculture

©iStock.com/vtls; ©iStock.com/bentaboe; ©iStock.com/Vasca; ©iStock.com/Linjerry; ©iStock.com/dkarlsson

Have you ever compared the leaves of hickory, pecan, and walnut trees, **Figure 7-1**? Have you noticed the similarities among the flowers of an apple, pear, and peach? If you observed that these plants resemble each other, it is because they belong to the same plant families. Plants and all living organisms are organized by their relationships to each other. This system of grouping like organisms together is called *classification*. The science of naming and classifying organisms is called *taxonomy*. Taxonomy provides a useful framework for understanding the features of similar plants, including their growth habits, their physical makeup, and their response to environmental conditions. Familiarity with closely related plants helps growers and gardeners know what to expect when growing a new plant for the first time. Taxonomy also allows horticulturists to easily communicate throughout the world by using universal specific names and minimizing confusion that comes from common names.

History of Plant Taxonomy

"What is a weed? A plant whose virtues have never been discovered."
—Ralph Waldo Emerson

The earliest record of categorizing plants into groups belongs to Theophrastus (370–285 BCE), an assistant to Aristotle. Theophrastus divided plants by their growth characteristics into trees, shrubs, half-shrubs, and herbs. He observed a wide range of plant features including the seed structure, germination, and plant habitat range. He described several "plant families" including the parsley (Apiaceae) family, **Figure 7-2**. He noted the umbrella shape of the flower clusters among parsley, fennel, and chervil. His categorization essentially remains the same in today's modern classification.

In the middle of the eighteenth century, a Swedish naturalist, Carl Von Linné (1707–1778), suggested a classification based solely on the flower structures of plants. Better known as Carolus Linnaeus, the Latinized version of his name, he published *Species Plantarum* in 1753 and described more than 1300 plants placed into "classes." The plants were catalogued using the number of stamens, stamen characteristics, and the relationship of stamens to other floral parts. This system forms the basis for plant classification today.

Sergio Schnitzler/Shutterstock.com

Nataly Lukhanina/Shutterstock.com

J. William Calvert/ Hemera/Thinkstock

Figure 7-1. Look closely at the leaves of these trees. A—Pecan. B—Walnut. C—Hickory. **What similarities and differences do you see?**

A Lena Kozlova/iStock/Thinkstock B Ngarto Februana/iStock/Thinkstock C wdhackett/iStock/Thinkstock

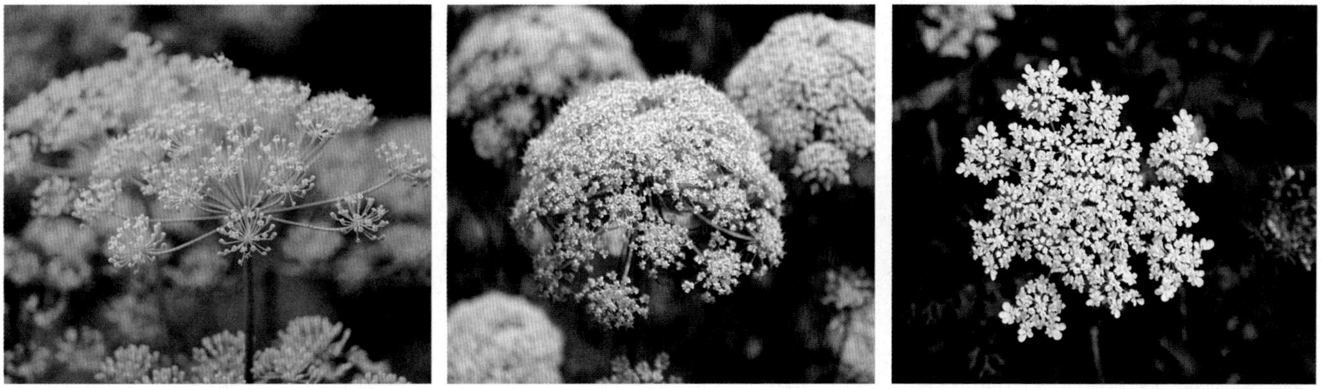

Figure 7-2. Members of the parsley family have similar umbrella-shaped flowers. A—Dill. B—Carrots. C—Queen Anne's lace.

Horticulturists and gardeners generally refer to plants using a ***common name*** (a word or term in everyday language). For example, in the United States, the linden tree is a common landscape feature. In the United Kingdom, the same tree is called a lime tree, **Figure 7-3**. The use of common names can lead to confusion and potential misunderstandings when communicating about living organisms. Carolus Linnaeus pioneered the consistent use of ***binomial nomenclature*** (a two-word naming system) to describe plant species. Prior to this system, different botanists might give a plant species different names. The binomial language developed by Linnaeus provides a methodical and consistent way to describe different plant species. For example, a red maple within this system has the name *Acer rubrum*. Plant names must adhere to specific rules and are governed by the ***International Code of Botanical Nomenclature***. This code is a set of rules that guides the naming or renaming of plant species.

filmfoto/iStock/Thinkstock

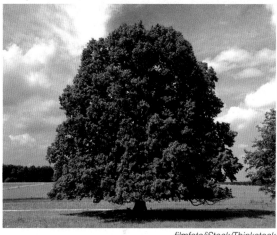

Figure 7-3. Common names make it challenging to communicate correctly about plants. What kind of plant do you envision when someone says "lime tree?" In Great Britain, this name is used for what in the United States is called the linden tree. Can you think of any other plants that go by multiple common names?

A System of Botanical Classification

Considered the father of taxonomy, Linnaeus set the stage for scientists to name plants using a binomial nomenclature and also to organize plants in a hierarchical way that shows the relationships between plant species. While much of taxonomy uses ***morphology*** (the physical form and structure of an organism) to classify plants, many new technologies have emerged in recent years to shift the approach some taxonomists use to organize plant species. Advances in physiological, biochemical, ecological, and molecular techniques make taxonomy a dynamic and changing field.

"A practical botanist will distinguish at the first glance the plants of the different quarters of the globe and yet will be at a loss to tell by what marks he detects them."
—Carolus Linnaeus

Floral morphology of plants has determined how plants are classified since the early days of Theophrastus. It remains an important tool for botanists and horticulturalists to determine plant species. Observations of floral characteristics as well as close examination of leaf shape, margins,

arrangement, and other key features aid in identification. If a new species is found, these characteristics help determine where the plant fits in the current classification scheme. These same features are used to readily identify plants in the horticultural trade.

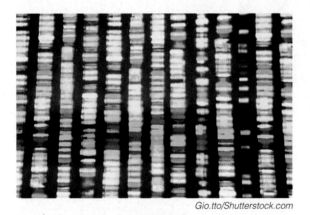

Gio.tto/Shutterstock.com

Figure 7-4. Molecular techniques, such as the DNA barcoding shown in this picture, provide new insights into plant relationships.

Molecular tools developed in the past few years provide easier and quicker means for classifying known and unknown plant specimens. New techniques like DNA analysis help scientists see similarities and differences at a much deeper level than just visually inspecting the plant, **Figure 7-4**. These techniques offer a way to genetically fingerprint a specimen and to examine evolutionary relationships among specimens. These new tools have shifted some taxonomic rankings among horticultural plants. For example, maples used to be in their own family, Aceraceae. Now they belong to a broader family, Sapindaceae, which also includes the horse chestnut and lychee. In plant breeding, these same tools can protect patenting rights for a breeder.

Domain

For years, living organisms were separated into only two broad categories: prokaryotes (single-celled organisms without membrane-bound organelles) and eukaryotes (single-celled and multicellular organisms made of cells with membrane-bound organelles). These two divisions did not accurately reflect significant differences among prokaryotes. Through molecular techniques, scientists have now arranged organisms into three *domains*, which are the highest and most inclusive taxonomic ranking for all living organisms. These domains include Eubacteria (meaning true bacteria), Archaea, and Eukaryotes. **Figure 7-5** shows an example of a specific plant's classifications.

Corner Question

How many plant species have been identified across the world?

Plant Classification Structure	
Classification Level	**Classification Name**
Kingdom	Plantae
Phylum	Tracheophyta
Class	Magnoliopsida
Order	Gentianales
Family	Apocynaceae
Genus	*Asclepias*
Species	*tuberosa*
Common Name	Butterfly milkweed
Scientific Name	*Asclepias tuberosa* L.

Goodheart-Willcox Publisher

Figure 7-5. The hierarchical structure of plant classification goes from most inclusive in domain to least inclusive in species.

Domain is considered the highest rank of classification, with each further subdivision becoming less inclusive and of lower rank order. The ranks include domain, kingdom, phylum, class, order, family, genus, and species. These ranks are discussed in the following sections.

Kingdom

There are six kingdoms, with plants belonging to the Plant kingdom. The Plant kingdom, also called Plantae, includes living organisms that are multicellular, have cell walls, and are autotrophic (able to make their own food supply). The Plant kingdom contains more than 400,000 species. These species include the angiosperms, gymnosperms, ferns, club mosses, hornworts, liverworts, mosses, and green algae. *Angiosperms* are flowering plants that have their seeds enclosed in fruit. *Gymnosperms* are nonflowering plants that produce seeds.

Phylum

After kingdom, plants are further separated into a rank called *phylum*. Plants can be roughly divided into four major groups: nonvascular plants, seedless vascular plants, gymnosperms (nonflowering seed plants), and angiosperms (flowering seed plants), **Figure 7-6**. Within each of these categories, a phylum more specifically defines the group. The rank phylum was called "division," and although this term may still be used today, it is generally agreed by scientists in the field of plant taxonomy to use the term "phylum."

Nonvascular Plants

Mosses are part of a group of nonvascular plants. They lack a vascular system for transporting water and nutrients throughout their structure.

A—Artush/Shutterstock.com B—Jon Bilous/Shutterstock.com C—MIMOHE/iStock/Thinkstock

Figure 7-6. A—The phylum Bryophyta includes mosses, a group of plants that require moist environments for growth and reproduction. B—The phylum Pterophyta is comprised of various fern species, a rich source of shade-loving horticultural plants. C—The gingko tree is an important street and landscape tree and is considered a "living fossil."

They are small in size; lack roots, stems, and leaves; and produce spores rather than seeds. Mosses (phylum Bryophyta), liverworts (phylum Hepatophyta), and hornworts (phylum Anthocerotophyta) make up the group of nonvascular plants. Mosses can be considered a niche market of horticultural plants. They can also be viewed as a weed in the landscape.

Seedless Vascular Plants

D. Kucharski K. Kucharska/Shutterstock.com

Figure 7-7. Horsetail flourishes where it can root in water or clay soil.

Seedless vascular plants are more complex than their nonvascular counterparts. They have a vascular system comprised of xylem and phloem, allowing for movement of water and solutes. They have true roots, leaves, and stems, and their spore-producing structures permit wide dispersal. Like the mosses, hornworts, and liverworts, the seedless vascular plants also require water in or on the soil for sperm to swim to the eggs for fertilization to occur. This group consists of ferns (phylum Pterophyta) and their allies, the club mosses (phylum Lycophyta), horsetails (phylum Sphenophyta), and whisk ferns (phylum Psilotophyta), **Figure 7-7**. The fern (phylum Pterophyta) has the most horticultural value, with many species considered important in the ornamental landscape field.

Gymnosperms

Nonflowering, seed-producing plants are called gymnosperms. The word *gymnosperm* means naked seed. These plants do not produce a seed within a protective structure of a fruit. Gymnosperms are wind pollinated and most (all except one species) have separate male and female reproductive structures called cones. Gymnosperms contain four phyla: the conifers (phylum Coniferophyta), the cycads (phylum Cycadophyta), ginkgoes (phylum Ginkgophyta), and the gnetophytes (phylum Gnetophyta). Conifers contain numerous species of high economic importance in landscape horticulture, including pines, spruces, firs, cedars, junipers, and hemlocks. Ginkgo has only one living species, *Ginkgo biloba* L. It has male and female gametophytes on separate trees that do not produce cones.

Did You Know?

The ginkgo tree, *Ginkgo biloba* L., is called a "living fossil" because nearly identical plants have been found fossilized and dated to nearly 200 million years old. The fossil records suggest that the gingkoes were once a widespread, abundant, and diverse group. The specific epithet, *biloba*, means "two lobes," which describes the leaves well.

Angiosperms

Most seed plants are flowering plants, called angiosperms, and have their seeds enclosed in a fruit. They are vascular plants with complex cellular structures. This group ranges in incredibly diverse sizes, from small perennials to soaring trees. The group hosts flowers of every hue and shape. Angiosperms include flowering annuals and perennials, fruits and vegetables, and woody ornamentals. Angiosperms are in the phylum Anthophyta.

Class

Class is the taxonomic rank that separates or identifies plants within a phylum. Plants in the seeded vascular plant phyla are ranked into two primary classes, Angiospermae (angiosperms) and Gymnospermae (gymnosperms). The class Angiospermae includes all flowering plants. The class Gymnospermae includes all nonflowering, seed-bearing plants (ginkgoes, cycads, gnetophytes, and conifers). Filicinae is another class that includes ferns and fern allies.

Within the Angiospermae class, plants can be placed into subclasses of *Dicotyledoneae (dicots)* or *Monocotyledoneae (monocots)*, **Figure 7-8**.

Figure 7-8. A—Dicots form a subclass with distinct characteristics and include beans, petunias, and geraniums. B—Monocots form their own subclass and include plants such as lilies, grasses, and orchids. Can you see the differences between dicots and monocots?

Dicot members generally have two *cotyledons* (first leaves) in their seeds, net-veined leaves, and flower parts in multiples of fours and fives. Dicots also usually have a vascular cambium and vascular bundles arranged in a ring. Members of this subclass are vast and include beans, roses, magnolias, and geraniums. Monocots have only one cotyledon in the seed, parallel-veined leaves, and floral structures in multiples of three. All roots in monocots are fibrous. They have vascular bundles scattered or in rings of two or more. They usually lack a cambial layer. Monocots include grasses, lilies, orchids, agaves, and palms.

Order

Order is a taxonomic ranking that separates or identifies plants within a class. Most horticulturists and gardeners do not reference the rank order to understand key characteristics that assist in managing plant growth. Rather, order is a category that provides an understanding of evolutionary relationships among plants. Orders end in *ales*. Rosales is an example of an order. The Rosales order is a broad umbrella for plants such as roses (Rosaceae), nettles (Urticaceae), elms (Ulmaceae), and mulberries (Moraceae). **Figure 7-9** shows a phylogenetic tree, which illustrates the genetic relationships of families within an order.

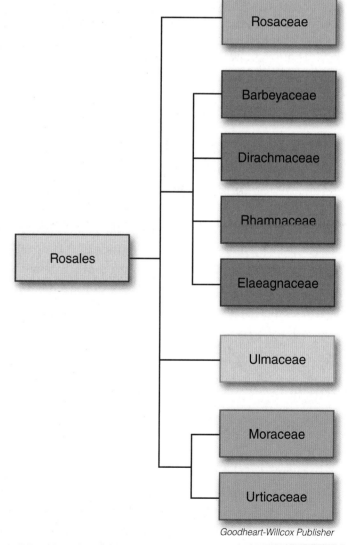

Figure 7-9. A phylogenetic tree shows the evolutionary relationships among plants, helping give order to a huge number of plant specimens.

STEM Connection
Changing Names of Plants

You may have noticed that the scientific names of the plants that you memorized now have different names. Science is a dynamic and fluid field with new information being discovered every day. In plant taxonomy, the relationship between species continues to unfold as more molecular tools evolve to unfold these stories. DNA research has revealed that long-held beliefs about the relationships of one species to another may no longer be true. This means that plants that once belonged in the same genus or even same family may be reclassified. Early taxonomists using keen powers of observation of the physical characteristics of plants were fairly correct in their organization, so many of these plant shifts are not dramatic.

For example, the Japanese maple (*Acer palmatum*) was part of the Aceraceae family until it became part of the broader Sapindaceae family.

Jason Vandehey/Shutterstock.com

This new knowledge is particularly important for taxa that are found to be unique or rare. Resources can be used to protect these species before they are extinct. In other cases, distribution of a new species may be limited and can be a source for further research into understanding the particular local environment that shaped the development of a species.

> "Nature uses only the longest threads to weave her patterns, so that each small piece of her fabric reveals the organization of the entire tapestry."
> —Richard P. Feynman

KPG Payless2/Shutterstock.com

Figure 7-10. Look closely at the flower of this broccoli. If you were to compare this to any other member of the cabbage family, you would find them to be similar.

Family

A *family* is a taxonomic rank that separates or identifies plants in an order. It is one of the most useful rankings in horticulture for plant growth and management. Plants in the same family can be defined as having similar floral structures. The rank of family provides horticulturists a very useful understanding of how to manage these plants, including controlling pests, promoting growth and development, and planning crop rotation. Members of the same family often attract similar insect pests and diseases, need similar nutrient management, and have about the same growth requirements. Family names of plants always end in *aceae*. For example, the cabbage family is Brassicaceae. Most members of this family have similar physical traits, **Figure 7-10**. Cabbage family members also produce similar chemical sulfur compounds and have other genetic similarities. There are more than 600 plant families that include angiosperms, gymnosperms, ferns and their allies, and mosses and liverworts.

Genus

Plants in a *genus* are a subset of organisms within a family that share similar characteristics. The first letter of a genus name is always capitalized and the genus (and specific epithet) of all species is always either italicized or underlined. For example, *Betula* or Betula is the proper way to denote the birch genus. The genus name is used in addition with the specific epithet to comprise a *scientific name*.

Species

A *species* is the lowest and least inclusive ranking of plant classification and is the basic unit of biological classification, **Figure 7-11**. A species has historically been defined as organisms capable of interbreeding and producing fertile offspring. This definition remains a primary tool, although advances in DNA comparisons and other molecular techniques, morphological traits, and ecological niches can also contribute to a definition of a species. The presence of a unique trait or local adaptation may not warrant a new species, but might create a subspecies and possibly a variety. A subspecies is subordinate to species and has enough variation, usually due to geographic isolation, to warrant a taxonomic ranking.

The *specific epithet* is the second half of a binomial name of a plant species. Together, the genus name and the specific epithet form the scientific name of a species. The specific epithet can describe a character trait, identify the person who discovered the plant, or honor a location where it was found. For example, the paper birch species, *Betula papyrifera*, **Figure 7-12**, has a specific epithet, *papyrifera*, that describes the papery texture of its bark. The specific epithet never stands alone because many species may have the same specific epithet. The paper mulberry, for example, also has the specific epithet *papyrifera*, but is defined by its species name, *Broussonetia papyrifera*. The specific epithet is always written in lowercase and either italicized or underlined in the same manner as the genus name.

Variety

A variety is a more specific and distinct subset of a species than a subspecies. A *variety* is a form of a species that is slightly different but not different enough to warrant a new species and hold horticultural value, such as leaf color or pattern or thornlessness. A naturally

Corner Question

What do you think the specific epithet "alba" means?

Figure 7-11. The hollies are part of a large genus. This English holly is its own species, *Ilex aquifolium*.

Figure 7-12. Paper birch has a specific epithet of *papyrifera*, a Latin derivative for paper. Many plants have descriptive specific epithets.

Skorpionik00/Shutterstock.com

Figure 7-13. The honey locust, *Gleditsia triacanthos*, has no thorns.

occurring variation of Japanese maple is *Acer palmatum* var. *atropurpureum*, with *atropurpureum* as the variety. Thornless honey locust, *Gleditsia triacanthos* var. *inermis*, is another example of a variety. The species *Gleditsia triacanthos* has long thorns, but the variety *Gleditsia triacanthos* var. *inermis* has none, **Figure 7-13**.

Cultivar

Many gardeners confuse cultivar with variety. A *cultivar* is a name for a plant that has been bred or selected for horticultural purposes. The word is derived from the words "cultivated" and "variety." The first letter of a cultivar is typically capitalized and the name is enclosed in single quotes. For example, the eggplant cultivar 'Fairy Tale' is written as such.

Plant Keys

You might easily be able to identify a plant as a member of the oak family, Fagaceae. However, it can be much more challenging to identify the plant as an individual species. Plant keys enable horticulturists to identify plants based on specific characteristics through a process of elimination. A *dichotomous key* is a tool that gives users paired choices, called couplets. The user makes a choice, which leads to another set of paired statements. The user continues selecting characteristics that fit the plant being identified. When all choices have been exhausted, the plant species remains. The example in **Figure 7-14** is a sample of a key for identifying maple species. Many online plant keys also exist to help users properly identify their plant species by entering or defining characteristics present in their specimen.

Herbaria

An *herbarium* is a repository for collected plant specimens. Plants are pressed and mounted to archival-quality paper and stored in cabinets that preserve them. Not all plant material lends itself well to being pressed.

STEM Connection | Virtual Herbarium

Many herbaria are now creating digitized collections of their specimens, essentially building an electronic repository to plant researchers across the globe. Specimens are photographed to make high-resolution images that can be made available to wide audiences and used in biodiversity research projects. The digitized images reduce specimen wear and tear and provide a long-term record of the specimen. For each specimen, there is a visual image of the plant or fungi material along with all of the information including the collection information, such as the collector, the location, the date, and botanical nomenclature.

I. Leaves entire
   A. Leaves with obtuse base and three equal triangular lobes—*Acer buergerianum* (trident maple)
   B. Leaves with truncate or cordate base and three or more lobes
      i. Leaves with mostly three lobes
         a. Leaves with silvery underside and red petioles—*Acer rubrum* (red maple)
         b. Leaves with green underside without red petioles—*Acer tataricum* subsp. *Ginnala* (amur maple)
      ii. Leaves with more than three lobes
         a. Fruit mature in late spring and buds red
            1. Leaves with sinuses that are U-shaped and entire—*Acer saccharinum* (silver maple)
            2. Leaves with sinuses that are V-shaped and toothed—*Acer rubrum* (red maple)
         b. Fruit matures in early fall, buds not red
            1. Petiole sap milky, buds green and mostly glabrous—*Acer platanoides* (Norway maple)
            2. Petiole sap not milky
               i. Leaves with five to eleven lobes; double serrate margin—*Acer palmatum* (Japanese maple)
               ii. Leaves with mostly five lobes; coarsely toothed margin
                  a) Leaves pale green, tips horizontal not drooping—*Acer saccharum* (sugar maple)
                  b) Leaves dark green, leaf tips drooping—*Acer saccharum* subsp. *nigrum* (black maple)
II. Leaves compound
   A. Leaves with 3 to 6 leaflets, greater than 8″ long and green stems—*Acer negundo* (box elder)
   B. Leaves with 3 leaflets, less than 8″ long and stems pubescent or flaky—*Acer griseum* (paperbark maple)

Figure 7-14. This is an example of a plant key used to identify maple species.

Photograph by University of South Florida Herbarium

Figure 7-15. This is an herbarium specimen of a pitcher plant, *Sarracenia rubra* C. Walter subsp. *gulfensis* D. E. Schnell and is useful for researchers and in education.

Formaldehyde may be used to preserve bulky items such as fruits or fleshy flowers. All plant material is labeled with essential data, including the plant species name, location, date, and altitude where it was found, and any special environmental conditions, **Figure 7-15**. Specimens are generally organized by their taxonomic relationships, with similar species clustered close to other members of their family.

Herbaria specimens play a central role in documenting plant diversity throughout the world. Herbaria provide a reference collection that enables plant identification, research, and education. Many herbaria may specialize in collecting local flora, creating an inventory of plant species, and showing changes in species over time. Herbaria are similar to libraries and usually allow a loan of specimens for educational or research purposes.

Careers in Plant Taxonomy

New discoveries are being made every day. Because of new techniques in classifying plants, various careers are available in the field of plant taxonomy. To work in the field of taxonomy, you must be able to do extensive research and work well with fellow employees. You may work all over the world to do research, or you may be associated with a specific herbarium. Jobs in this field include herbarium directors or curators and ecologists.

Herbarium Director/Curator

An herbarium director or curator is responsible for the maintenance of the collection of herbaria. Many herbaria are associated with universities, colleges, or botanical gardens. Many directors will remain active in research projects involving local or international flora. The director may mount many of his or her own specimens and supervise staff or students to perform these tasks as well. The tasks require demonstrated abilities to interact and collaborate broadly in research and teaching. Many herbaria provide unique opportunities for outreach education and teaching of undergraduate plant taxonomy courses. For this kind of position, a higher level degree with significant experience is typically recommended and required.

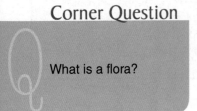

Corner Question

What is a flora?

Ecologist

Ecologists are scientists focused on understanding ecosystems as a whole. This includes the distribution of organisms and the relationships between these organisms and their environment. Many ecologists have a specific focus, such as ecology of desert plants or tropical plants. Ecology involves fieldwork, such as surveying populations and recording data, and policy and management work. The exact purview of an ecologist varies significantly by the employing organization. For example, the national parks system may want an ecologist to assess the environmental impacts of installing a hiking trail through a sensitive area. Many ecologists will use mapping tools including GPS and GIS and write reports that can impact decision making. A degree in ecology, plant biology, or related field is a minimum requirement for many jobs.

Career Connection

Dr. Andrea Weeks, George Mason University
Plant Taxonomist

Dr. Andrea Weeks' earliest memories are rooted in playing outside. Growing up on a farm, she was always fascinated by plants. She would capture twirling maple fruits as they descended from the trees and plant them in rows. In middle school, Andrea started her own business of drying flowers and peddling them to local craft stores. Her interest in the intersection of wild and cultivated plants grew and in her first year of college she found herself enrolled in a plant systematics course that encapsulated everything she loved: plant taxonomy, evolutionary processes, and phylogeny.

From then on, Andrea began to find opportunities and courses in college that let her pursue her interest in understanding plants. From applied internships that put her to work in greenhouses to research positions that led her to the lab to explore plant breeding, tissue culture, and other molecular work, she began to craft a path that would lead her to her career as an associate professor of plant systematics and director of the Ted R. Bradley Herbarium at George Mason University in Virginia.

As a professor, she teaches students and conducts her own research and outreach programs within the field of plant systematics. Her research has focused on the evolutionary biology of plant members within the frankincense and myrrh families and explores how all the species are related and how and when they evolved. As part of her efforts as the director for the herbarium at George Mason, Andrea spearheaded a citizen science project that engaged the public to participate in digitizing old herbarium specimens, making centuries of biological observations of plants widely available and accessible to everyone. Biologists can use this legacy data to begin to understand how plant populations have changed and moved over time and to help answer big questions relating to climate change, among other ideas.

Andrea continues to love her work for the ability to be creative in asking and answering questions in the field of plant biology. She finds it exciting to contribute to the growing body of knowledge about plants in general.

Chapter 7
Review and Assessment

Chapter Summary

- Classification provides an organized framework to understand relationships that plants have to each other. This information is useful for managing plant growth and development.
- The early roots of plant classification began with Aristotle's assistant, Theophrastus, who used visible plant characteristics to cluster plants into families.
- Carolus Linnaeus is considered the father of taxonomy and pioneered the consistent use of a binomial nomenclature for naming plant species.
- Plant taxonomy uses the physical features of a plant as well as molecular, physiological, biochemical, and ecological techniques to classify plants.
- The hierarchical system for organizing plants begins with the least inclusive rank of domain and increasingly becomes less inclusive with kingdom, phylum, class, order, family, genus, and species.
- There are six kingdoms, with plants belonging to the Plant kingdom. The Plant kingdom includes living organisms that are multicellular, have cell walls, and are autotrophic.
- Angiosperms and gymnosperms are important phyla and consist of species that have tremendous horticultural value.
- Class is the taxonomic rank that separates or identifies plants in a phylum. Within the Angiospermae class, plants can be classified as dicots or monocots.
- Order is a taxonomic ranking that separates or identifies plants in a class. Order is a category that provides an understanding of evolutionary relationships among plants.
- The rank of family separates or identifies plants in an order. This rank provides useful information in managing plants, including strategies for controlling pests and diseases, promoting growth and development, and planning crop rotation.
- Plants in a genus are a subset of organisms within a family that share similar characteristics. A species is the lowest and least inclusive ranking of plant classification and is the basic unit of biological classification. A variety is a subset of a species.
- A scientific name for a plant contains both a genus name and a specific epithet. The specific epithet reflects a characteristic of the plant or honors a person or place.
- Cultivars are plants that have been bred or selected for a particular characteristic. A plant species may have several cultivars.
- Plant keys enable users to identify plants based on specific features through a process of elimination. Choices are made based on plant characteristics until all choices are eliminated and the plant species remains.
- Herbaria are repositories for holding collected plant specimens that can be used for research or education.

Words to Know

Match the key terms from the chapter to the correct definition.

- A. angiosperm
- B. binomial nomenclature
- C. class
- D. classification
- E. common name
- F. cotyledon
- G. cultivar
- H. dichotomous key
- I. Dicotyledoneae
- J. domain
- K. ecologist
- L. family
- M. genus
- N. gymnosperm
- O. herbarium
- P. International Code of Botanical Nomenclature
- Q. Monocotyledoneae
- R. morphology
- S. order
- T. phylum
- U. scientific name
- V. species
- W. specific epithet
- X. taxonomy
- Y. variety

1. A form or subclassification of a species that is slightly different but not different enough to warrant a new species.
2. A name for a plant that has been selected or bred for horticultural purposes.
3. The physical form and structure of an organism.
4. A seed producing plant that lacks a protective cover for the seeds.
5. The process of systematically identifying and organizing plant species.
6. A subset of organisms within a family that share similar characteristics.
7. Plants with two cotyledons in their seeds.
8. A word or term for plants that is used by gardeners in everyday language.
9. The first leaf that emerges from a seed.
10. A flowering plant that has seeds enclosed in its fruit.
11. The science of naming and classifying organisms.
12. A two-word naming system, such as that used for plant species.
13. The second half of a scientific name for a plant species, usually descriptive of a plant feature or in honor of someone's name or a place.
14. The taxonomic rank that separates or identifies plants in an order.
15. A taxonomic ranking that separates or identifies plants in a kingdom.
16. The taxonomic rank that separates or identifies plants within a phylum.
17. A repository of collected plant material.
18. The highest and most inclusive taxonomic ranking for all living organisms.
19. A scientist focused on understanding ecosystems as a whole.
20. A set of rules that guides the naming or renaming of plant species.
21. Plants with one cotyledon in their seeds.
22. A taxonomic ranking that separates or identifies plants in a phylum.
23. A tool used to identify plants by pairing choices against each other until all choices have been exhausted and the plant species remains.
24. A two-word name that includes a genus and specific epithet for a plant species.
25. The lowest and least inclusive ranking of plant classification.

Know and Understand

Answer the following questions using the information provided in this chapter.

1. What is plant taxonomy, and why is it used for growers and gardeners?
2. Describe the system for categorizing plants used by Theophrastus.
3. Describe the system for categorizing plants used by Carolus Linnaeus.
4. What problems can result from using common names for plants?
5. Why is the system of binomial nomenclature developed by Linnaeus a better way to name plants than using common names?
6. What are some features of morphology used to identify plants in the horticultural trade?
7. How have molecular techniques changed how plants can be identified?
8. What three categories are used to group all organisms? What are the other ranks or classifications below the domain level?
9. How many kingdoms are used in taxonomy, and what kingdom do plants belong to?
10. What are the four major groups of plants for the phylum rank?
11. What are characteristics of a nonvascular plant such as moss?
12. Seedless vascular plants in what phylum have the most horticultural value?
13. What are some types of conifers that are important in landscape horticulture?
14. What features define the subclasses of Dicotyledoneae (dicots) and Monocotyledoneae (monocots)?
15. List some of the ways that plants within a family are similar.
16. What is a genus? How should a genus name be formatted in writing?
17. What is a species?
18. How does a dichotomous key help correctly identify plants?
19. What role do herbaria play in documenting plants?
20. What are ecologists generally focused on?

Thinking Critically

1. You are hiking in the woods near your home, and you come across a plant that you have never seen before. You think it may be a new plant species. What steps would you take to determine whether this plant is indeed a newly discovered specimen? If it is not a new species, what else might it be?
2. Imagine you are outside on an autumn day and come across two different types of leaves that have fallen. How would you research to find out what kind of plants or trees they came from?

STEM and Academic Activities

1. **Science.** What defines a subspecies? Visit your local herbarium or library and find a plant species that also has a subspecies. What are the characteristics that differentiate the two?
2. **Science.** Research the soil conditions and climate in your area. Find out what flowering and foliage plants grow well in these conditions. Choose another location in the country and do similar research on soil and climactic conditions. Do the same plant species grow in these areas? Why or why not? Are there plants within similar families or different families?

3. **Math.** Find a native plant species in your area. Research the population levels of your plant species. Find out additional population data for your plant species in other areas where it grows. Create a chart that compares plant populations. Do you see differences in their numbers? Why or why not?
4. **Language Arts.** Plants are identified by their unique features. Find a horticultural plant you think is interesting. Write a plant description that would help someone identify it. Use the botanical terms to properly describe the plant characteristics.
5. **Language Arts.** Many poems have been written about the beauty of plants. Plant taxonomy requires the close observation of a plant's features, which can inspire writers to create prose describing their virtues. Write a poem that uses descriptive language about a plant.

Communicating about Horticulture

1. **Speaking.** Working in groups of three, create flash cards for the key terms in this chapter. Each person in the group chooses six terms and makes flash cards for those six terms. On the front of the card, write the term. On the back of the card, write the pronunciation and a brief definition. Use your textbook and a dictionary for guidance. Take turns quizzing one another on the pronunciations and definitions of the key terms.
2. **Reading and Speaking.** Create an informational report on taxonomy as it relates to plants. Explain how the hierarchy system is set up for the Plant kingdom. Explain how the original ranking system was modified due to molecular research. Choose one plant family and list all of its plants and flowers. List the common characteristics that link these plants together. Include drawings or photographs of the most common family members. Present your report to the class.
3. **Listening.** As classmates deliver their presentations, listen carefully to their arguments. Write down any questions that occur to you. Later, ask questions to obtain additional information or clarification from your classmates as necessary.

SAE Opportunities

Teddy Leung

1. **Exploratory.** Visit an herbarium and examine different herbarium specimens.
2. **Exploratory.** Go to your local public garden or arboretum. Find a plant specimen you enjoy and try to identify what it is. Use a dichotomous key to aid in identification.
3. **Exploratory.** Create an herbarium specimen for a plant that you have to know about for a career development event (CDE). Create a classroom collection of specimens to help CDE teams learn about plant material.
4. **Exploratory.** Job shadow a plant taxonomist. Many states have a bio blitz, an event where volunteers can go out and inventory flora in a particular area.
5. **Placement.** Contact your local college or university to see if they have a taxonomist that uses molecular techniques in his or her research. Try to obtain an internship for the summer or for three or four weeks to gain laboratory experience in plant taxonomy.

CHAPTER 8
Plant Biology

Chapter Outcomes
After studying this chapter, you will be able to:
- Describe the study of botany.
- Compare different components of a plant cell.
- Outline the types and functions of a plant tissue.
- Describe plant parts and their functions.
- Compare and contrast careers in plant biology.

Words to Know

achene	cytoskeleton	legume	plastid
adenosine triphosphate (ATP)	dehiscent	leucoplast	pome
adventitious	deoxyribonucleic acid (DNA)	lysosome	ribonucleic acid (RNA)
angiosperm	drupe	meristematic	ribosome
anther	endoplasmic reticulum (ER)	mitochondria	samara
autotroph		node	schizocarp
berry	epidermis	nucleolus	sclerenchyma
capsule	filament	nucleus	sepal
caryopsis	follicle	nut	silique
cell	Golgi bodies	oil bodies	stamen
cellulose	guard cell	parenchyma	stomata
chlorophyll	gymnosperm	periderm	stroma
chloroplast	heterotroph	petal	thylakoid
chromoplast	indehiscent	phloem	tonoplast
collenchyma	inflorescence	pistil	turgor
cortex	internode	pith	vacuole
cytoplasm	leaf mesophyll	plasma membrane	xylem

Before You Read
As you read the chapter, record any questions that come to mind. Indicate where the answer to each question can be found: within the text, by asking your teacher, in another book, online, or by reflecting on your own knowledge and experiences. Pursue the answers to your questions.

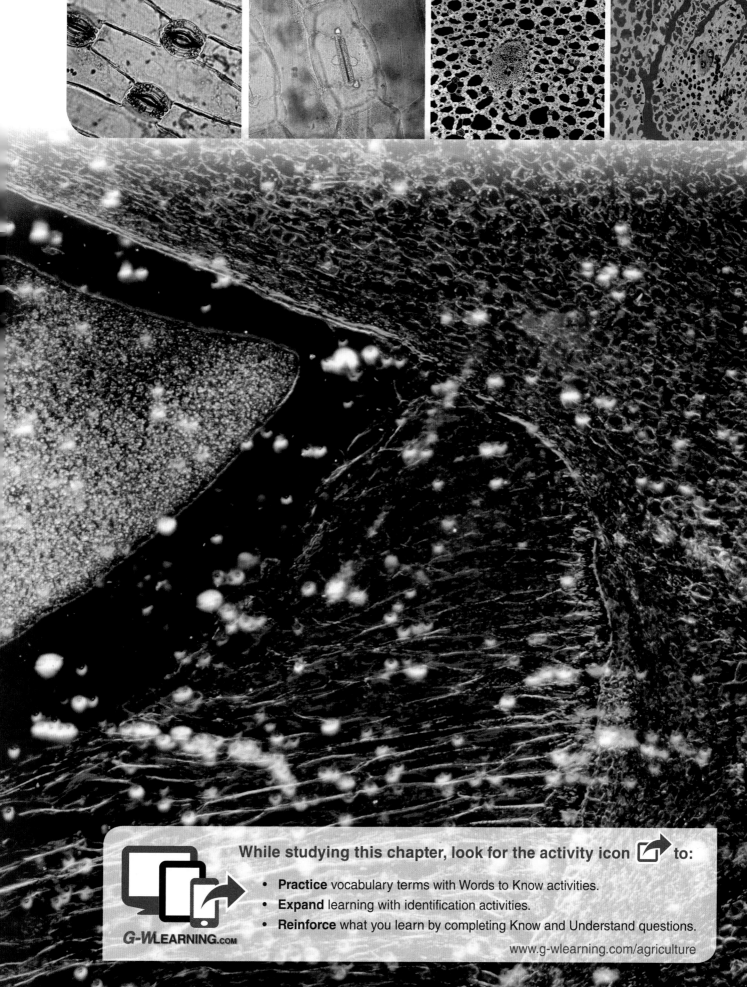

While studying this chapter, look for the activity icon to:
- **Practice** vocabulary terms with Words to Know activities.
- **Expand** learning with identification activities.
- **Reinforce** what you learn by completing Know and Understand questions.

www.g-wlearning.com/agriculture

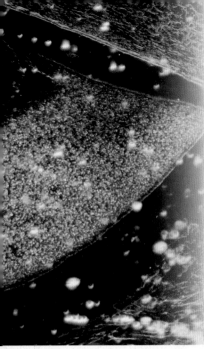

Life on earth depends on plants. Plants nourish our bodies, build houses, provide shade, feed animals, and create clothing. Plants are terrestrial, multicellular organisms that are able to capture light energy from the sun and make their own food (sugar) in a process called photosynthesis (*photo* means light, *synthesis* means putting together). Photosynthesis creates materials that support most other living organisms. Organisms that use the process of photosynthesis to make their own food are called **autotrophs**, or self-feeders. Plants are autotrophs. Some algae and bacteria are also autotrophs. People, animals, fungi, and some bacteria, however, are heterotrophs. **Heterotrophs** are organisms that use external sources of food or energy to fuel their development. These external sources of energy often can be traced back to plants. For instance, when we eat beef, we are sustained by the meat made by the cattle when they grazed on plants.

Study of Botany

The branch of science that deals with the study of plant life is called *botany*. Botany is a rich field of research and learning. Botany has historical roots in medicine, with people pursuing the identification of edible, medicinal, and poisonous plants. Gardens in monasteries provided a collection of medicinal herbs and plants for food, **Figure 8-1**. Horticulture became the practice of transferring scientific knowledge about plants into practical skills for cultivating plants for edible and ornamental purposes.

eFesenko/Shutterstock.com

Figure 8-1. Monks practiced horticulture, tending medicinal plants and cultivating fruits and vegetables on monastery grounds.

Botany Disciplines

Botany has several scientific disciplines. These include:
- Plant physiology—the study of plant growth and development.
- Plant taxonomy and systematics—the study of relationships among plants and their naming and classification.
- Genetics—the study of heredity and variation in organisms.
- Molecular biology—the study of the structure and function of plants at a molecular level.
- Economic botany—the study of past, present, and future uses of plants by people.
- Ecology—the study of the relationships and interactions among plants and organisms in an environment.
- Paleobotany—the study of plant fossils and ancient vegetation.

Botanists have identified a framework of key structures and functions that provide an understanding of how plants grow. From the twisting tendrils of a squash vine to the needle-sharp thorns of a rose, plant parts form the basis for how a plant grows and develops.

Corner Question

What is the world's smallest plant?

Plant Cells

The basic unit of the plant is the *cell*. The cell contains many different organelles (structures) that drive plant processes. The suffix *-elle* means small, so organelles are essentially small organs within the cell, providing similar functions as the organs found in animals. A plant cell is bounded by cell wall made from a material called cellulose. ***Cellulose*** is a polysaccharide (*poly* means many, *saccharide* means sugars) that gives the wall strength and rigidity, Figure 8-2.

Cell Membrane

Inside the cell wall is a permeable *plasma membrane* that controls what substances enter or leave the cell. These include substances, such as the gases carbon dioxide or oxygen, that might passively diffuse across the membrane. Other ions or molecules, such as sugars or salts, require the cell to expend energy in active transport, often with the help of specialized proteins that serve as "gates." The plasma membrane also organizes the production of cellulose that makes up the cell walls and plays a role in cell growth and differentiation.

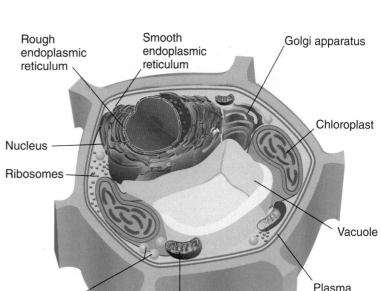

snapgalleria/Shutterstock.com

Figure 8-2. A typical plant cell contains a cell wall, plasma membrane, nucleus, organelle, and other structures.

Cell Nucleus

The *nucleus* is a large organelle found in a plant cell that has two primary functions:

- It controls cellular activities by determining when and which proteins are produced.
- It stores the cell's genetic information and passes this information on to daughter cells through cellular division.

The nucleus is bounded by a pair of membranes and contains the plant's genetic material, *deoxyribonucleic acid (DNA)*. DNA consists of a chain of nucleotides wrapped around a protein. DNA condenses during cellular division into shapes referred to as chromosomes. The order of the nucleotides is important, as that is the genetic code for an individual organism (its genes). Mistakes in the genetic code (caused by chemicals, radiation, or other factors) cause mutations. The nucleus also includes ribonucleic acid and the nucleolus. *Ribonucleic acid (RNA)* manufactures proteins based on the genetic information provided by DNA and plays a significant role in gene expression. The *nucleolus* is responsible for the formation of ribosomes.

The *cytoplasm* surrounds the organelles and nucleus, filling the inside of the cell. Many substances are found in the cytoplasm, such as amino acids, sugars, enzymes, and waste products (on their way to disposal). These substances are used by the cell to produce proteins, maintain the cell metabolism, and keep the cell functioning successfully.

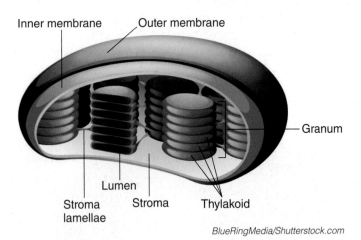

Figure 8-3. Chlorophyll is found in the stacks of thylakoids, surrounded by the stroma.

BlueRingMedia/Shutterstock.com

Chloroplasts and Other Plastids

A *chloroplast* is the most important *plastid* (an organelle that contains food or pigment) because it is the site where photosynthesis occurs. Chloroplasts primarily contain *chlorophyll*, a green pigment that is a receptor of light energy in the red and blue wavelengths. Green light is reflected instead of absorbed, which is why chloroplasts appear green. (*Chloro* means green). Chlorophyll is found in the stacks of thylakoids, surrounded by the stroma, **Figure 8-3**. *Thylakoids* are disc-shaped sacs surrounded by membranes on which the light reactions of photosynthesis take place. The *stroma* is the aqueous space outside the grana (stacks of thylakoids).

The chloroplast also contains carotenoid pigments, which absorb light energy in the blue-green and violet range and reflect the longer yellow, orange, and red wavelengths. A single plant cell may contain as many as 40–50 chloroplasts. Chloroplasts can physically adjust themselves to optimize the amount of light captured or even to deflect damage from high light situations.

History Connection
George Washington Carver

Everett Historical/Shutterstock.com

George Washington Carver (c. 1864–1943) was an agriculturist, inventor, botanist, and chemist. Born into slavery in southwest Missouri, Carver set himself on a path of education that eventually led to his enrollment first at Simpson College and then at Iowa State University. After earning his master's degree at Iowa State, Carver became the first African American faculty member at the university. In 1896, Booker T. Washington invited Carver to be the director of the agriculture department at Tuskegee Institute in Alabama, where he taught for 47 years.

Carver was deeply committed to helping southern farmers succeed through improved agricultural practices. While at the Tuskegee Institute, he developed successful methods of crop rotation, introduced alternative cash crops that also helped improve soil quality, and initiated research into crop products. Carver explored uses for a number of crops, including sweet potatoes and peanuts. He found over 300 ways to transform the peanut into different foods, medicines, plastics, grease, cosmetics, and many other products. This occurred at a time when the boll weevil was devastating cotton crops and the peanut kept farms in business. George Washington Carver is remembered as a humble humanitarian, an advocate for racial equity, and a leading scientist who transformed farming in the United States.

Two other types of plastids are chromoplasts and leucoplasts. *Chromoplasts* are pigmented plastids that contain carotenoids. Chromoplasts are often responsible for the saturated yellow, orange, and red colors found in flowers, aging leaves, fruit, and some roots such as carrots. *Leucoplasts* are colorless plastids that perform functions such as synthesizing starch and forming oils and proteins.

Mitochondria

Commonly considered the powerhouse of the cell, *mitochondria* are the parts of the cell that drive the process of respiration and energy transfer that living organisms need to survive. The glucose produced by plants goes through a series of chemical reactions inside the mitochondria, eventually producing adenosine triphosphate (ATP). *Adenosine triphosphate (ATP)* is a nucleotide found in the mitochondria and the principal source of energy for cellular reactions. ATP provides energy to sustain life. Plants may contain hundreds and even thousands of mitochondria in a single cell. Mitochondria can move within the cell to places where energy is required. Like chloroplasts, mitochondria also have inner membranes where their work is performed. Both mitochondria and chloroplasts also have their own DNA and reproduce on their own, regardless of whether the cell itself is going through cell division at that time.

Vacuoles

A *vacuole* is a large cavity (often taking up to 90% of the space within the cell) found within the plant cell that stores cell sap, waste products, pigments, or other liquids. A permeable membrane called the *tonoplast* surrounds the vacuole and regulates the entrance and exit of these materials. Cell sap, made primarily of water, contains ions (such as calcium, potassium, chlorine, sodium) and solutes (such as sugars and amino acids). This liquid creates pressure, or *turgor*, within the plant cell, and helps provide rigidity and support to plant structures. Turgor within the cell, along with cellulose in the cell wall, provides the stiffness required for leaf and stem support. Without turgor, plants wilt. Turgor also controls the opening and closing of guard cells around the stomata of plants, controlling the exchange of gases and water vapor between the plant and the environment. Vacuoles also may serve as storage units for anthocyanin pigments. These are the blue (*cyan* means blue), violet, purple, dark red, or scarlet colors that give brilliance to cabbage leaves, beet roots, cherry fruit, rose petals, and fall colors of maple leaves, **Figure 8-4**.

HandmadePictures/Shutterstock.com

Figure 8-4. Anthocyanins, the red or blue-violet pigments that are found in many plant parts, including roots, stems, leaves, flowers and fruits, are located within the cell vacuole. What other plants have this coloring?

Oil Bodies

Oil bodies are lipid droplets that are spherical in nature. While found throughout cells in the plant, they are most concentrated in fruits and seeds. Nearly half of the weight of sunflower, peanut, flax, and sesame seeds comes from oil, which provides energy for the developing seedling.

Endoplasmic Reticulum

The *endoplasmic reticulum (ER)* is a complex, folded membrane system that provides a channel for transporting proteins and lipids in the cell. ER can appear rough or smooth. Rough ER has ribosomes attached. Smooth ER is more tubular in form and is involved in lipid synthesis.

Ribosomes

A *ribosome* consists of ribosomal RNA (rRNA) and protein, and is actively involved in the synthesis of proteins. Ribosomes are minute and can be found in large quantities in the cytoplasm and attached to the rough ER. Smaller ribosomes can also be found in the mitochondria and plastids.

Golgi Bodies

Golgi bodies are organelles usually found close to the nucleus that have several functions. Golgi bodies control the flow of molecules in the cell, modifying some before packaging them into vesicles for transport to other parts of the cell or for excretion outside the cell. Golgi bodies serve to collect and bundle proteins, lipids, and other substances for use within the cell. They act like a post office, sorting and delivering vital substances.

Lysosomes

The *lysosome* is the digestive system in the cell, using enzymes to break down large molecules like proteins, polysaccharides, lipids, and nucleic acids. Lysosomes provide a contained place for digestion of large molecules without damaging the molecules of the cell.

Cytoskeleton

The *cytoskeleton* is protein filaments and motor proteins in the cell. It is composed of three major structural fibers:
- Microtubules.
- Microfilaments, also known as actin filaments.
- Intermediate filaments.

These protein filaments form an enormous network that can be cross-linked to other similar filaments and to membranes by means of accessory proteins. The linking of proteins increases cellular rigidity, keep organelles in place and provides a means of transport within the cell. The filaments can easily assemble and disassemble into bundle shapes to assist in the movement of molecules in the cell.

Plant Tissues

Plants develop cells that specialize in specific roles or functions. The *meristematic* cells are responsible for cell formation and growth. Meristematic cells can further develop into specialized cells. The organizations of similarly specialized cells are called plant tissues, which can then collectively be ordered into organs.

Meristem

Meristematic cells divide into new cells, and are most numerous in regions of rapid plant growth. Most growth in plants occurs at the tips of both the main and lateral shoots as well as at the root tips. These growing regions are referred to as the *apical meristems*, **Figure 8-5**. The cells in these regions allow the plant to elongate, or lengthen, in size. Girth, or growth in diameter (think of an expanding tree trunk throughout the years), occurs inside the stem and roots in cells called the vascular cambium, a type of secondary meristem. The exception to these growth patterns can be found in monocots (discussed later in this chapter). Most monocots grow from their leaf bases.

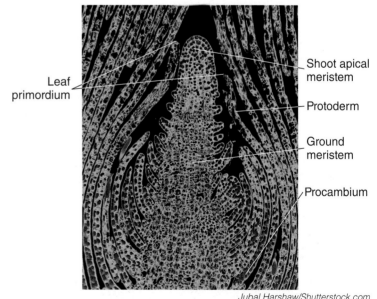

Jubal Harshaw/Shutterstock.com

Figure 8-5. The shoot apical meristem contains stem cells and produces primordial that develop into the aboveground plant organs.

Thinking Green

The Many Uses of Bark

The bark of a tree protects the cambial layers from injury; it also has multiple uses from cork and canoes to baskets and clothing. Across the globe, clothes have been crafted from bark fibers. In parts of eastern Asia and Polynesia, tapa cloth comes from tough interlacing bast fibers extracted from the bark of the paper mulberry, *Broussonetia papyrifera*. The bark strips are soaked, cleaned, and pounded together with mallet. The resulting cloth can be as thin as muslin or as thick as leather.

Bark has also been used for thousands of years to create canoes. Native Americans harvested paper birch (*Betula papyrifera*) in one continuous sheet. The sheet was then used to cover the wooden boat frame. Paper birch is lightweight and durable. Cork has long been harvested from cork oak trees (*Quercus suber*) and used for everything from clothing to floors.

The bark of some trees has potential medicinal properties. The first aspirin came from the bark of a willow tree, which has a chemical compound called salicin. Phenolics in some tree bark may contain anti-inflammatory compounds that can be used to treat arthritis, high blood pressure, asthma, and heart disease.

Epidermis

The *epidermis* covers the outside areas of a plant (stem, leaves, and roots) and protects the plant from environmental stresses, such as heat, insects, and pathogens. It also minimizes water loss and provides a site for gas exchange of carbon dioxide and oxygen. *Guard cells* are specialized epidermal cells surrounding pores called stomata. Guard cells control the opening or closing of the stomata by changing shape and making the opening wider or narrower. The guard cells determine when the stomata should be open or closed. The *stomata* are pores on leaves that open and close to allow for gas exchange and control water loss, **Figure 8-6**. A waxy covering on the epidermis (called the cuticle) also prevents water loss. In dry, hot climates the cuticle on plants is very thick.

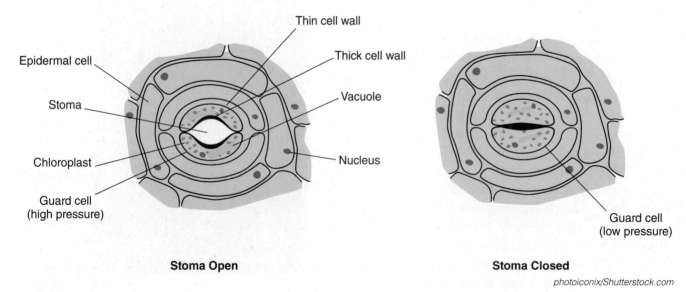

Figure 8-6. Guard cells control the opening or closing of the stomata to allow gas exchange and minimize water loss. They are typically found on the underside of leaves.

Bark (Periderm)

In woody plants, the bark, or *periderm*, replaces the epidermal layer, providing protection and preventing water loss. The cork cambium, found just beneath the bark layer, divides and produces bark tissue. Cork used to make corkboards comes from the bark of the evergreen cork oak (*Quercus suber* L.). It is harvested from the tree every 9–12 years in a way that does not harm the tree, **Figure 8-7**.

Phloem

Phloem is a type of living tissue (called vascular tissue) that carries photosynthetic products synthesized in the chloroplasts throughout the plant. Phloem consists of a series of cells called sieve cells. Collectively, the phloem tissue also provides structural support for the plant. Phloem fibers from flax, hemp, and jute are used in the production of cloth and rope.

LianeM/Shutterstock.com

Figure 8-7. The cork oak produces a spongy periderm that is harvested to produce cork flooring, bulletin boards, and even insulation.

Xylem

Just as phloem tissue conveys sugars and other substances, *xylem* tissue conducts water and nutrients from the roots throughout the plant. Unlike phloem, mature xylem tissue is nonliving. Xylem can also provide structural support to the plant and offer space to store food, **Figure 8-8**. Xylem and phloem are found in the vascular cambium layer of a plant, and are damaged when bark is removed from a tree. If bark is removed in a complete circle around the trunk of a tree (girdling), the tree will die.

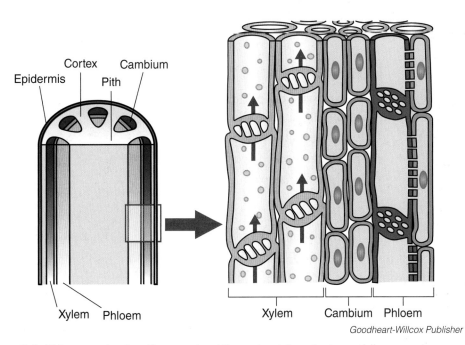

Goodheart-Willcox Publisher

Figure 8-8. Phloem and xylem tissue extend throughout the plant, providing a system that conducts nutrients, water, and food.

Corner Question

What is the world's fastest-growing plant?

Parenchyma and Collenchyma

Did You Know?
When you bite into an apple, the juice is released from damaged parenchyma cells.

Parenchyma are the most common plant tissue. They are made up of the vascular tissue of the *cortex* (outer layers), *pith* (inner layers) of stems and roots, and *leaf mesophyll* (internal layers of leaves). Parenchyma form the greatest portion of the plant. They serve a number of functions including storage, gas exchange, wound healing, secretion and excretion of plant substances, and adventitious growth. *Adventitious* means growing from an unusual place, as in plant roots that grow from stems or leaves.

Collenchyma are closely related to parenchyma cells and specialize in structural support. They have thickened cell walls made of cellulose, but maintain flexibility. They provide support and minimize breakage and can be found near the veins of leaves. When you eat celery, the tough lines of tissue that peel down the outside of the celery are collenchyma cells.

Sclerenchyma

Sclerenchyma plant cells have thick, lignified walls that provide physical strength and support to stems, especially in woody plants. Sclerenchyma can also be found in certain fruits, such as pears, cherries, peaches, and plums. The gritty texture of a pear is due to the presence of sclerenchyma cells.

Plant Parts and Their Functions

Plants may have six different parts: roots, stems, leaves, flowers, fruits, and seeds. Roots, stems, and leaves have specific functions. They have been organized by scientists into three primary organ types of plants. Flowers and fruits are believed to have developed from stem and leaf tissues. However, they have such horticultural importance that they will also be included in this discussion.

Roots

Burrowing deep into the soil, roots provide several essential functions for plants. These functions include:
- Uptake of water and nutrients.
- Anchorage and support.
- Storage of nutrients and food.

Taproots

Root systems are categorized as either a taproot system or a fibrous root system. Taproots originate from the seed root, the radicle. Taproots tend to grow directly downward with lateral roots emerging as branching roots, **Figure 8-9**. Examples of plants that have taproots include alfalfa, oak and maple trees, carrots, and radishes. Other tree species such as spruce, beech, and poplar have much shallower taproots. The length and depth of the taproot depends greatly on soil characteristics, such as

Showcake/Shutterstock.com

Figure 8-9. Taproots are long roots that grow deeply into the soil. What are the advantages of taproots? Are there disadvantages?

texture, structure, and water table depth. Often, lateral roots will reach far greater distances than the spread of the branches of a plant.

Fibrous Roots

Fibrous root systems are shallow and dense. These roots cling firmly to soil, **Figure 8-10**. Fibrous roots emerge as adventitious roots from the stem and provide excellent erosion control. Adventitious roots grow from plant parts other than roots, most commonly from stem or leaf tissues. Monocot plants, such as grasses, are good examples of a fibrous root system. Fibrous roots can also be found in species such as lettuce and petunias.

Adventitious Roots

Adventitious rooting is the primary reason horticultural propagation of numerous plants is possible. Adventitious roots can be found growing out of the stem of corn or tomatoes and may prop or anchor the plant, **Figure 8-11A**. Roots growing out of the nodes of climbing vines, such as ivy, provide support against a surface.

Air Roots

Roots also require oxygen for respiration, thereby limiting many plants from growing in very wet soils. Saturated soils from overwatering or flooding can kill a plant. Specialized roots, called air roots or pneumatophores, provide aeration for mangroves, **Figure 8-11B**.

Stems

Stems are the main axis of the plant that holds the leaves and flowers. The stem extends the vascular system of cells that begins in the root, carrying water and nutrients through xylem and phloem to other parts of the plant. Stems provide support for the plant. They can also store food, usually carbohydrates. A potato (*Solanum tuberosum* L.) is an underground stem (called a tuber) that has been modified to function as storage. Many stems are edible, have aesthetic value (such as the exfoliating bark of many trees), can be used as fuel or building material, or provide a habitat for wildlife.

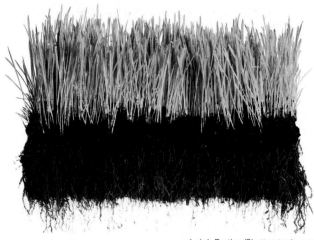

Leigh Prather/Shutterstock.com

Figure 8-10. Many turfgrasses have thick, dense, fibrous root systems, making them ideal for controlling soil erosion. What other benefits do dense turfgrass root systems provide?

A   *Aggie 11/Shutterstock.com*

B   *eye-blink/Shutterstock.com*

Figure 8-11. A—Adventitious roots grow from stem tissue in corn, providing stability for the plant. B—The pneumatophores on these mangroves provide a way for the plant to obtain oxygen needed for respiration. Which other plants have pneumatophores?

Nodes, Internodes, and Buds

A stem has *nodes*, the areas where leaves develop, and *internodes*, or the spaces between nodes, **Figure 8-12**. Buds are a stem's growing point and can be either leaf buds or flower buds. They can be found in the leaf axil, the space between where a leaf and stem intersect. These are called lateral buds. Terminal buds are found at the shoot tip, or apex. Terminal buds exhibit apical dominance. In apical dominance, one stem becomes the leader and lateral growth is suppressed by a hormone called auxin. In horticulture, terminal buds can be pinched off or pruned, encouraging more growth by lateral buds, resulting in bushier growth.

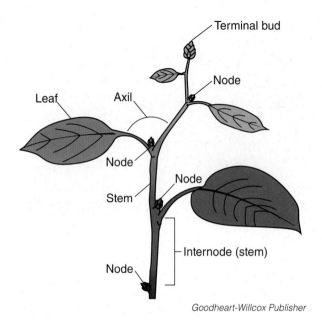

Figure 8-12. Nodes are the intersection on the stem where the leaf is attached. In horticulture, pruning cuts are made immediately above the node. Where on the stem are cuts made for flowers used in floral arrangements?

Bud Arrangement

Bud arrangement on a stem may be alternating (staggered along the stem), opposite (in pairs across from each other), or whorled (in a group of three or more). These can be key identifying feature for many species. Buds can be protected by bud scales, a type of modified leaf. Some buds may undergo dormancy or a rest period during certain seasons, enabling plants to tolerate unfavorable growing conditions.

Specialized Stem Structures

Many stems are modified into specialized structures:
- Corm—a short, solid, fleshy underground stem, **Figure 8-13A**.
- Bulb—a large bud with a short stem fully surrounded by modified leaves called scales, with adventitious roots that emerge from the stem bottom. Tunicate bulbs, such as tulips and onions, are covered with a papery covering. Scaly bulbs, such as lilies, have overlapping leaf bases that do not completely enclose the bulb, **Figure 8-13B**.
- Rhizome—an underground stem that runs horizontally beneath the soil's surface and produces roots and shoots, **Figure 8-13C**.
- Stolon—similar to rhizomes, a stolon is a slender stem that produces roots and shoots at the nodes but runs above ground and emerges from leaf axils. Bermuda grass is an example of a stolon.
- Runner—a specialized stolon that forms new plants at the tip, **Figure 8-13D**. Strawberries are a common example of plants with runners.
- Crown—the area of the stem found at the soil's surface, near the transition of root to shoot. It can be compacted stem tissue. Hostas, rhubarb, and asparagus are considered crowns, **Figures 8-13E** and **13F**.
- Spur—a modified stem on a mature woody plant with short internodes. It often develops into flower and fruiting tissue.
- Sucker—an adventitious shoot that develops from root tissue and can be used to propagate plants such as brambles or figs.

- Watersprout—an adventitious shoot that arises from stem tissue and provides valuable grafting material, particularly for fruit trees.
- Tendril—a modified, flexible shoot that wrap around or hook onto material for support. Members of the squash family, such as cucumbers or pumpkins, use tendrils to climb a trellis, **Figure 8-13G**.
- Thorn, spine, and prickle—plant parts modified for defense. Thorns and prickles are modified stems, and spines are modified leaves. Thorns are sharp, strong, hardened stem tissue. Prickles are small, sharp outgrowths of the plant's outer layer, the epidermis, **Figure 8-13H**. Spines arise from leaf tissue that is modified into a hardened pointed structure.
- Cladophyll—a modified stem that is flattened and leaflike in appearance and is optimized for photosynthesis, **Figure 8-13I**.

Figure 8-13. A—These gladiolus stems are an example of a corm. B—Lily bulbs are nontunicate. C—Ginger is one example of a fleshy rhizome. D—Strawberry plants have runners. E—Hostas are typically propagated by dividing their crown into new pieces for planting. F—Rhubarb plants may also be propagated by dividing the crown. G—Tendrils allow these grapes to climb supports and gain better access to light. H—Roses have prickles, not thorns. I—Asparagus "leaves" are really cladophylls, or modified photosynthetic stems.

Leaves

Figure 8-14. The leaf is the primary organ for photosynthesis and commonly is a flat surface designed to optimize the capturing of light energy. Which plant has the largest leaves?

Leaves are the primary part of the stem where photosynthesis occurs. The leaf is typically comprised of both the blade (flattened portion) and the petiole (leaf stalk), **Figure 8-14**. Sometimes leaves do not have a stalk but instead are attached directly to the plant base, or sessile. Leaves can be arranged a number of different ways on the stem, including alternate, opposite, or whorled, **Figure 8-15**. Alternate leaves are attached singly to the stem at a node. In an opposite arrangement, a pair of leaves is attached to the stem on opposite sides. In whorled arrangements, leaves may be arranged in groups of three or more.

Leaves can be simple or complex. Simple leaves have only one blade. The blade is not divided into smaller leaflets. Maple, oak, and dogwood trees all have simple leaves. Compound leaves have two or more blades known as leaflets. The leaflets look like leaves but do not have a bud at their base. Honey locust, walnuts, and ash have compound leaves, **Figure 8-16**. Compound leaves are in either a pinnate or palmate pattern. Pinnate patterns have leaflets on both sides of the petiole. Palmate patterns have leaflets growing from a single point on the petiole, such as a horse chestnut.

Leaf Venation

Simple leaves can have the following types of venation, or vein pattern:
- Parallel. Veins run in parallel lines throughout the leaf and can be found in all monocots. Grasses, including corn, have parallel venation.
- Palmate. The main veins of the leaf start at a common point. They fan out to the leaf tips and look similar to the palm of a hand. Maples have palmate venation.
- Pinnate. Veins arise in lateral pairs on either side of the main vein. Beech trees have pinnate venation.

Figure 8-15. Leaf arrangement is a common characteristic used in plant identification. A—Alternate. B—Opposite. C—Whorled.

Figure 8-16. Compound leaves can be found on plants like ash (A) and honey locust (B). What other plants can you identify that have compound leaves?

Leaf Anatomy

The anatomy of leaves reflects their function as photosynthetic structures, **Figure 8-17**. A protective upper and lower epidermis surrounds the leaf. These epidermal layers are often covered with a waxy cuticle to prevent water loss. Stomata (singular: *stoma*) most commonly occur on the underside of the leaf and provide openings for gas exchange. The guard cells enclose the stomata and open when full of water (turgid). They collapse and close the opening when the plant is flaccid (drooping due to lack of water).

> "I am constantly roving about, to see what I have never seen before and shall never see again."
> —Thomas Jefferson

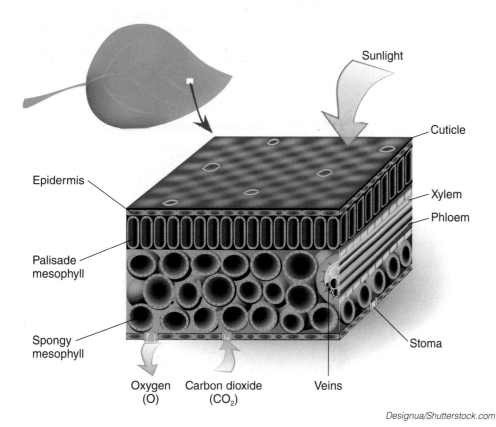

Figure 8-17. The various essential structures of a leaf aid in photosynthesis, prevent water loss, and perform other functions.

"A garden requires patient labor and attention. Plants do not grow merely to satisfy ambitions or to fulfill good intentions. They thrive because someone expended effort on them."
—Liberty Hyde Bailey

Carbon dioxide can be captured through the stomata for photosynthesis and oxygen can be released. The mesophyll is the middle layer of leaf tissue and contains numerous specialized cells including chloroplasts, where photosynthesis occurs.

Leaf Adaptation

Leaves have adapted in a number of ways to optimize capturing light energy or minimizing water loss. Needles found on coniferous plants are short, narrow, and covered in a waxy cuticle to prevent excessive water loss in the winter. Leaves on plants in sunny areas will typically be thicker and smaller, whereas leaves on plants in shady areas will be thinner and larger. Bracts are another leaf modification. They are often immediately below a flower, such as in the poinsettia or dogwood. Some leaves, such as English ivy, change shape as they grow from a juvenile state to adulthood. Other leaves are deciduous and will abscise (fall off) during colder temperatures or due to other environmental factors, such as drought.

Flowers

Flowers are the reproductive organs of a flowering plant, or *angiosperm*. Flowers consist of pistils, stamens, petals, and sepals, **Figure 8-18**. These four floral organs are attached to a flower stalk, a thickened stem called the receptacle.

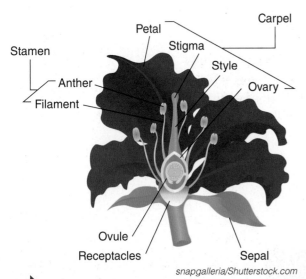

Figure 8-18. Illustration of typical flower parts.

snapgalleria/Shutterstock.com

- The *pistil* is the female organs of the flower and includes the ovary (which contains the ovules), the style (a slender stalk where the pollen tube develops), and the stigma (a sticky appendage on top of the style where pollen is received). The ovary matures into the fruit after fertilization, and the ovules develop into seeds.
- The *stamen* is the male portion of the flower. It contains the *anther* (the part that produces pollen) and a thin stalk called the *filament* that supports the anther.
- A *petal* is a modified leaf that surrounds the female and male reproductive parts of the flower. Petals can be vividly colored or shaped in a way that attracts pollinators. Collectively, the petals are called the corolla. In many flowers, the petals contain nectaries, glands that secrete nectar to attract pollinating insects, birds, or animals.
- The *sepal* is also a modified leaf. It surrounds and protects the flower bud. Sepals are collectively called the calyx.

Imperfect, Staminate, and Pistillate Flowers

Most flowers contain both the pistil and stamens and are considered perfect. If a flower is missing either of these parts, it is called imperfect. A flower with only male reproductive parts is called a staminate flower.

Did You Know?
Pollen comes in many colors, including yellow, white, red, brown, and purple.

Likewise, a flower with only female reproductive parts is called a pistillate flower. If a plant has both male and female flowers on the same plant, it is called *monoecious* (meaning one house). Corn, squash, and begonias are examples of *monoecious* plants, **Figure 8-19**. If a species has separate plants with each plant having only male flowers or female flowers, the plant is considered *dioecious* (meaning two houses). Bittersweet, kiwi, hemp, willow, gingko, and holly are all dioecious species. If fruit is wanted from *dioecious* plants, both a male and female plant are needed for fertilization to occur. If a flower consists of all four floral organs, (pistil, stamen, petals, and sepals), it is termed a complete flower. If any one of the organs is missing, the flower is called incomplete. The numbers of floral parts tends to be consistent within a plant family.

ileana_bt/Shutterstock.com

Figure 8-19. The begonia species is an example of a monoecious plant. It has separate male flowers and female flowers on the same plant. The female flowers are shown here.

Inflorescent Types

Petunias, tulips, and marigolds are single flowers set on a stem and are considered solitary flowers. Clusters of single flowers gathered on a stem are called *inflorescences*. There are a number of different inflorescent types, **Figure 8-20**. Determinate inflorescences have a set number of flower buds with the apex (top) bud opening first, followed by the other flowers on the stem. Indeterminate flowers open on lateral buds first with flowers maturing at the base first then onward to the apex. The floral axis will continue to grow as flowers develop.

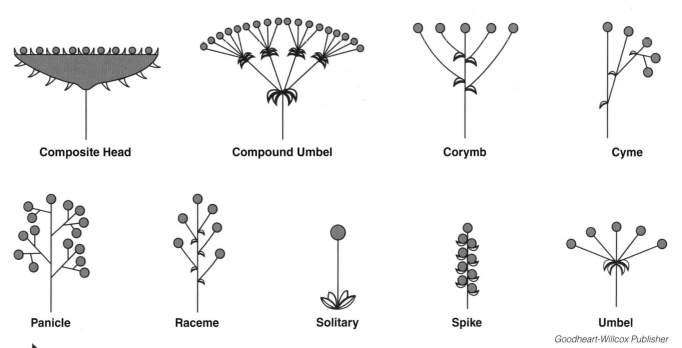

Goodheart-Willcox Publisher

Figure 8-20. Inflorescence types often reflect the relationship with specific pollinators and are important considerations in landscape and floral design.

Corner Question

What is the only fruit to bear seeds on the outside?

Fruits

A fruit is a mature ovary that has formed as a result of the fertilization of the ovules. Fruit tissues surround the seed. The purpose of fruit is to protect the seed inside and to aid in seed dispersal. As a fruit ripens, the fruit tissues are called the pericarp. The pericarp has three parts: the exocarp (the outermost layer), mesocarp (the middle layer), and endocarp (the inner layer), **Figure 8-21**. Different fruits develop these layers in various ways. There are a number of different types of fruits that can be classified as simple, aggregate, or multiple.

- Simple fruits develop from a single pistil or carpel.
- Aggregate fruits develop from flowers that have multiple pistils, such as magnolias and raspberries.
- Multiple fruits result from many flowers fusing together during the maturation process. Pineapple, fig, and mulberry are multiple fruits.
- Parthenocarpic fruits, such as cultivated bananas, are fruits that have developed without seeds.

Simple fruits are further subdivided into different types with many features that have evolved to aid in seed dispersal. Some are fleshy with a soft pericarp; these include most edible horticultural crops. Dry fruits have a hard, dry, nonsucculent pericarp.

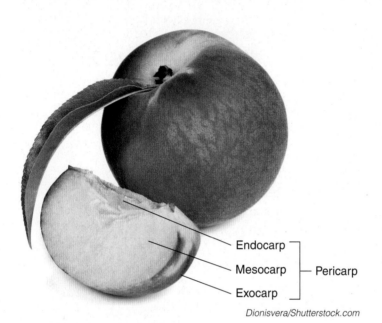

Dionisvera/Shutterstock.com

Figure 8-21. The pericarp is the ripened ovary wall and is comprised of the exocarp, mesocarp and endocarp. Depending on the type of fruit, these layers are differentiated in specific ways.

Fleshy Fruits

Fleshy fruits can have one seed or several seeds. They can be categorized in groups:

- A *berry* is a fruit with fleshy walls (pericarps) and a number of seeds, but without a stone. Common examples of berries are grapes, blueberries, cranberries, oranges, and watermelons, **Figures 8-22A** through **8-22E**.
- A *drupe* is a fruit with a fleshy, soft mesocarp and a seed enclosed by a hard, stony endocarp. Cherries, peaches, plums, and olives are examples of drupes, **Figures 8-22F** through **8-22I**.
- A *pome* is a specialized fruit with a tough endocarp that encloses the seed. Pomes are part of the rose family and include apples, pears, and quince, **Figures 8-22J** through **8-22L**.

Figure 8-22. A–E—Common examples of a berry include grapes, blueberries, cranberries, and watermelons. F–I—Cherries, peaches, plums, and olives are examples of drupes. J–L—Apples, pears, and quince are pomes.

Dry Fruits

Dry fruits can be dehiscent or indehiscent. *Dehiscent* fruits have a seedpod that will split open at maturity and freely release seeds for dispersal. The seeds are released in various ways.

- A *follicle* is a dry dehiscent fruit that splits along the length of the fruit releasing seeds for dispersal, **Figure 8-23A**.
- A *legume* is a dry dehiscent pod that opens along two seams at maturity. It is a member of the bean family (*Fabaceae*), **Figure 8-23B**.
- A *silique* is a dry dehiscent fruit of the mustard family. The seedpod splits along two sides, but the seed often remains attached to part of the fruit structure.
- A *capsule* is a dry dehiscent fruit that splits open to release seeds. It comes from flowers that had many pistils. The seedpods tend to split longitudinally (like okra) or at holes near the top of the capsule (like poppies).

Indehiscent fruits are those that keep seeds within fruit walls after leaving the parent plants.

- An *achene* is a one-seeded dry indehiscent fruit that is held freely within the pericarp, **Figure 8-24**. Examples of achenes include sunflowers, zinnias, and buckwheat.
- A *samara* is a dry indehiscent fruit in which part of the fruit wall is extended to form a wing. Examples include the fruit of maples, elms, and ashes.
- A *caryopsis* is a dry indehiscent fruit in which seeds are firmly attached to the fruit wall, as in grains such as corn or wheat or in other grasses.
- A *nut* is a dry indehiscent fruit that has a hardened pericarp with a loose seed inside. Examples include pecans, walnuts, hickory nuts, and hazelnuts.
- A *schizocarp* is a dry indehiscent fruit that splits at maturity into two or more seeded parts. Schizocarps are found in the carrot family (Apiaceae), and include carrots, dill, and parsley.

A

Jerry Horbert/Shutterstock.com

B

oksana2010/Shutterstock.com

Figure 8-23. A—A common example of a follicle is the milkweed. B—Peas are a common example of a legume.

natuska/Shutterstock.com

Figure 8-24. Zinnias are one example of an achene.

Seeds

A seed is the reproductive unit of plants, both angiosperms and gymnosperms. Angiosperms, as discussed earlier, are a class of flowering plants that develop seeds enclosed within an ovary. *Gymnosperms* are a type of plant that do not have flowers. Instead their developing seeds are borne on the plants without an ovary. Seeds develop as a result of the fertilization process. Coniferous plants and *Ginkgo biloba* L. are gymnosperms.

Monocots and Dicots

Angiosperms have two subclasses: monocots and dicots, **Figure 8-25**. Each class has different seed structures. Developing seeds in monocots have one cotyledon, or seed leaf, when they first emerge. Seeds in dicots have two seed leaves. Both classes have seeds with an embryo protected by a seed coat and endosperm that provides stored energy for developing seeds, **Figure 8-26**.

| Angiosperm Subclasses ||
Monocots	Dicots
Rice	Beans
Wheat	Peas
Barley	Squash
Corn	Sunflowers
Bamboo	Snapdragons
Lilies	Petunias
Tulips	Tomatoes
Garlic	Hydrangeas
Asparagus	Maples

Goodheart-Willcox Publisher

Figure 8-25. Examples of various monocot and dicot species.

Main Parts of a Seed

The seed has three main parts: the embryo, endosperm, and seed coat. The embryo is the immature plant. It draws on food reserves within the endosperm to grow and develop. The seed coat is a protective covering for the developing seed. The functions of seeds are to protect undeveloped or young plants, provide food for early growth, and aid in movement of the undeveloped plant to new locations. Seeds may also prevent growth of the embryo until growing conditions are right.

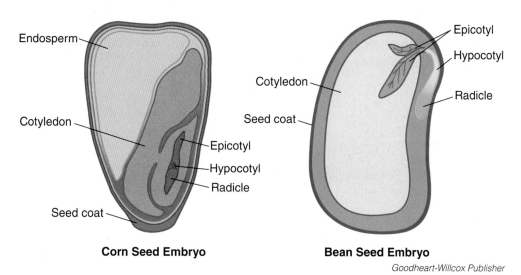

Corn Seed Embryo   **Bean Seed Embryo**

Goodheart-Willcox Publisher

Figure 8-26. A corn seed is a classic example of a monocot seed structure. A bean is a classic example of a dicot seed structure.

In monocots, the seed embryo is comprised of one cotyledon, a radicle, and plumule. The cotyledon is the seed leaf. The plumule develops into the shoot, and the radicle is the seed root. In dicots, the embryo has two cotyledons, plumule, radicle, and a hypocotyl (stem). Seeds are discussed in more detail in Chapter 13, *Seed Propagation*.

Careers in Plant Biology

There are many different specialties and career opportunities in the plant biology area. Careers can vary from ecology to genetics, and can be found in businesses, research centers, industry, and education. A bachelor's degree is required for most careers in plant biology. A graduate degree creates even more career options.

Botanist

Botanists study all forms of plants and their relationship to the environment and other living things. A botanist who enjoys the outdoors and research may search for and identify new plant species, evaluating each plant's parts and uses. Some botanists may do environmental research, studying the effect of climate change on various plant populations. Botanists can also be found working in botanical gardens, arboretums, herbaria, and zoos, studying plants and how they interact in each environment. A botanist has a bachelor's degree in botany.

Science Teacher

Science teachers instruct students in the concepts of science. A degree in plant biology is important for those who wish to teach students about plant science, the environment, natural resources, and earth sciences. The ability to share information about plant growth and development can engage young people in learning. A science teacher with a plant background can create hands-on experiments that help students understand the photosynthetic organisms around them. A bachelor's degree in plant biology, botany, horticulture, or agriculture is required. Additional coursework and experience is required to earn a teaching certificate.

> "I decided that my means were sufficient to enable me to devote myself to botany, a determination which I never, during the long period of my subsequent career, had on any occasion any reason to repent of."
> —George Bentham

Review and Assessment

Chapter Summary

- Organisms on earth are divided into two categories: autotrophs (organisms such as plants that manufacture their own food) or heterotrophs (organisms that rely on external sources of energy).
- Botany is the study of plants.
- The basic unit of a plant is the cell. Cells have different organelles that serve different functions and have different structures depending on their purpose.
- Plant tissues are comprised of specialized cells that provide a framework for plant growth and development, including the transport of water, nutrients, and food.
- Plants have six different parts: roots, stems, leaves, flowers, fruits, and seeds.
- Roots provide plants with anchorage and support, storage of nutrients and food, and uptake of water and nutrients.
- Stems serve as support for the plant and carry water, nutrients, and food throughout. They can be modified for storage of food.
- Leaves serve as the main appendage of the plant where photosynthesis occurs. Leaves have different types of arrangements, vein patterns, and shapes.
- Flowers hold the reproductive portions of plants and consist of four organs: pistils, stamens, petals, and sepals.
- Fruits are mature ovaries that contain seed. Types of fruit include simple, aggregate, and multiple.
- Seeds are the reproductive unit of the plant and consist of an embryo, endosperm, and seed coat.

Words to Know

Match the key terms from the chapter to the correct definition.

A. angiosperm
B. anther
C. autotroph
D. capsule
E. cell
F. dehiscent
G. deoxyribonucleic acid (DNA)
H. drupe
I. endoplasmic reticulum (ER)
J. filament
K. Golgi bodies
L. guard cells
M. heterotroph
N. indehiscent
O. inflorescence
P. mitochondria
Q. nodes
R. nucleus
S. nut
T. pith
U. plasma membrane
V. pome
W. ribonucleic acid (RNA)
X. samara
Y. silique
Z. stomata

1. A fruit with a fleshy, soft mesocarp and a seed enclosed by a hard, stony endocarp.
2. An organism that uses the process of photosynthesis to make its own food.
3. Clusters of single flowers gathered on a stem.
4. A dry dehiscent fruit with a seedpod that splits along two sides while the seed inside remains attached to part of the fruit structure.
5. The inner layer of plant tissue of stems and roots.
6. A fruit with a tough endocarp that encloses the seeds.
7. Cell organelles that control the flow of molecules in the cell, modifying some before packaging them into vesicles for transport to other parts of the cell or for excretion outside the cell.
8. A substance in cells that holds a plant's genetic material and provides genetic information to ribonucleic acid (RNA).
9. A flowering plant.
10. The basic unit of a plant that contains many different organelles that drive plant processes.
11. A complex, folded membrane system that provides a channel for transporting proteins and lipids in the cell.
12. A large organelle found in a plant cell that contains the plant's genetic material.
13. The parts of a cell that drive the process of respiration and energy transfer.
14. Specialized epidermal plant cells with pores that open to allow for gas exchange and control water loss.
15. Pores in epidermal cells that open to allow for gas exchange and control water loss.
16. A substance in cells that plays a role in gene expression.
17. An organism that uses external sources of food or energy.
18. A dry dehiscent fruit that splits open to release seeds and comes from flowers that had many pistils.
19. A thin stalk in the stamen of a flower that supports the anther.

20. A dry indehiscent fruit in which part of the fruit wall is extended to form a wing.
21. Permeable layer inside the cell wall that controls what substances enter or leave the cell.
22. A type of fruit that keeps the seeds within the fruit walls after leaving the parent plant.
23. A dry indehiscent fruit that has a hardened pericarp with a loose seed inside.
24. A type of fruit that will split open at maturity and freely release seeds for dispersal.
25. The part of the stamen of a flower that produces pollen.
26. The places on a plant stem where leaves develop.

Know and Understand

Answer the following questions using the information provided in this chapter.

1. Describe the field of botany and list some specializations that it includes.
2. List the components of a plant cell.
3. Why is a chloroplast the most important plastid? Where is chlorophyll located?
4. What are vacuoles and what is their purpose?
5. In what part of a plant are oil bodies most concentrated?
6. Explain apical meristem.
7. Where does cork come from and how often can it be harvested?
8. Compare and contrast the different types of plant tissues.
9. List the six different plant parts.
10. Summarize the primary functions of the root.
11. Describe two types of root systems.
12. What are the functions of a plant stem?
13. List five types of modified stems and provide an example of each type.
14. What is the primary function of plant leaves?
15. What are two parts of a typical leaf?
16. What is the difference between a simple leaf and a compound leaf?
17. Discuss the three types of venation that can be found on a simple leaf and give examples of plants that exhibit each type.
18. Name the four main floral parts and their functions.
19. Define what comprises a male plant and a female plant.
20. What is the difference between a perfect and an imperfect flower? What is the difference between a staminate flower and a pistillate flower?
21. What is the purpose of fruit?
22. What three parts make up the pericarp?
23. Explain the differences among simple, aggregate, and multiple types of fruits.
24. What are two types or classes of angiosperm seeds? Describe each type.
25. Name and explain the parts of a seed.
26. What are the functions of seeds?

Thinking Critically

1. The leaves of many deciduous plants change color in the fall as the chlorophyll degrades, revealing other pigments present in the leaf. But you notice a maple tree at school is starting to show a color change in July. Why do you think this is happening? Is there anything that can be done to help the tree?
2. You love to garden, so you purchased a hardy kiwi fruit vine to grow on a trellis. Three years have passed. The vine has flowered, but no fruit has ever developed. Can you figure out why? Describe the steps you would take to try to reach a solution.
3. A plant biologist needs to explain the differences between the types of fruit and has asked you to assist. How would you teach your fellow students about the different kinds of fruits? Write a description of the teaching methods you would use.

STEM and Academic Activities

1. **Science.** Onions skin cells provide an easy way to take a look at a real plant cell. Research a protocol to extract and mount the onion cell on a slide. Using a compound microscope, try to identify the different parts of the cell. Can you find the cell wall? What about the cytoplasm and nucleus?
2. **Science.** Gravity helps plant cells orient themselves for growth. This process is called *gravitropism*. Using a potted coleus plant (or plant of your choice), cover the soil with plastic wrap to prevent the soil from falling out. Place your plant upside down and anchor the pot in place. Observe the plant over the next couple of days. What happens? Why? What part of the plant cell is responsible for this reaction? Can the plant return to "normal"? Have you observed this in nature?
3. **Social Science.** This chapter stresses that plant biology is an important science, not only to you, but to society. Do you agree or disagree with this? Use evidence from your life, the lives of others, or from stories you have read to support your position.
4. **Language Arts.** Write notes in a structured journal as you read the chapter. As you take notes, keep track of what you do not completely understand. These might be words or ideas that are unfamiliar or unusual. Also, write what you think are the most important ideas and/or parts of the text. Why do you think so? What evidence is there in the text to support your perspective?

Communicating about Horticulture

1. **Listening and Speaking.** Visit a nearby arboretum, botanical garden, or large flower-growing operation. Ask to interview their botanical expert. Prepare a list of questions before your interview. Here are some questions you might ask: What is your work environment like? What are your job duties? What type of research are you currently doing? What type of facilities do you use for your research? What impact will your research have on the floriculture industry? Ask if you can have a tour of their facilities. Report your findings to the class, giving reasons why you would or would not want to pursue a career similar to that of the person you interviewed.

2. **Writing and Speaking.** Working in groups of three, create flash cards for the Words to Know in this chapter. Each student chooses four terms and makes flash cards for those four terms. Quiz one another on the pronunciations and definitions of the terms. Using your textbook and dictionary for guidance, write the term on the front and the pronunciation and definition on the back.
3. **Reading and Writing.** Draw a perfect flower and label the different functions of each part. Begin by drawing five petals with five sepals attached. Add a pistil and include five stamens. Create another sketch of an imperfect flower, one male flower and one female flower.
4. **Reading and Listening.** In small groups, discuss the main topics in the chapter. Ask questions of other group members to clarify concepts or terms as needed.

SAE Opportunities

1. **Exploratory.** Research local plant conservation societies. Write a speech about the contributions that these organizations make to the community.
2. **Placement.** Intern with a local botanical garden.
3. **Exploratory.** Visit a lab of a plant biologist and write a report about the research being done there.
4. **Experimental.** Create a plant growth study. Make a hypothesis, track your methodologies, observe and record data, and write a report.
5. **Exploratory.** Job shadow a plant biology leader in your community or state.

Lance Cheung/USDA

CHAPTER 9
Plant Growth and Development

Chapter Outcomes
After studying this chapter, you will be able to:
- Identify the stages and products of photosynthesis.
- Discuss the function of respiration and recognize the steps.
- Describe the process of transpiration and the effects of different environmental conditions.
- Recall the movement of solutes through the plant.
- Identify the processes of plant reproduction.
- Describe basic plant breeding principles and reasons plant breeding is important.
- Describe the careers for plant physiologists and plant breeders.

Words to Know

adenosine triphosphate (ATP)	genes	phenotype
alleles	genome	photosynthesis
carbon fixation	genotype	plasmalemma
carotenoid	glycolysis	pyruvate
Casparian strip	grana	recessive
chimera	haploid	relative humidity (RH)
chlorophyll	homozygous	respiration
chromatin	incomplete dominance	sport
chromosome	meiosis	stroma
cohesion	mitosis	thylakoid
cytokinesis	mutation	translocation
diploid	nicotinamide adenine dinucleotide phosphate (NADPH)	transpiration
dominant		turgor
fertilization		zygote

Before You Read
Arrange a study session to read the chapter aloud with a classmate. At the end of each section, discuss any words you do not know. Take notes of words you would like to discuss in class.

While studying this chapter, look for the activity icon to:
- **Practice** vocabulary terms with Words to Know activities.
- **Expand** learning with identification activities.
- **Reinforce** what you learn by completing Know and Understand questions.

www.g-wlearning.com/agriculture

The journey to push roots into the soil and stretch leaves toward the sun begins with a tiny seed. Important plant processes are occurring in each plant cell that drive growth and development, enabling a simple seed to mature into an adult plant. These plant processes are basic biological functions. They are responsible for everything from the growth of a giant sequoia to the production of nectar in honeysuckle. The vital processes for this to occur include photosynthesis, respiration, transpiration, translocation, and reproduction. This chapter explains how plants develop and the ways horticulturists can use this knowledge to influence plant growth.

Photosynthesis

Photosynthesis is a process in which plants capture energy from the sun to convert simple molecules of carbon dioxide (CO_2) and water (H_2O) into carbohydrate molecules that can be used by plants as sources of energy and building blocks for other molecules, **Figure 9-1**. Photosynthesis is one of the most important chemical reactions on earth. Photosynthesis is the pathway by which nearly all of the energy in our lives is ultimately derived and through which oxygen is released. Photosynthesis begins with two reactants:

- Carbon dioxide from the air, which enters through the stomata of the leaves.
- Water, which travels to the leaf through the xylem.

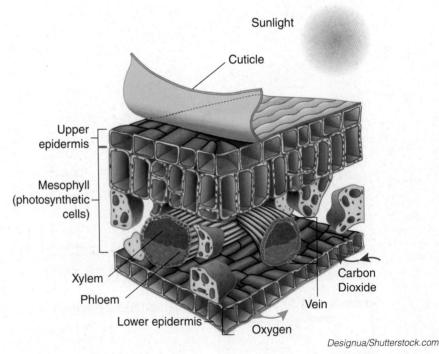

Designua/Shutterstock.com

Figure 9-1. Energy from light drives the process of photosynthesis, converting carbon dioxide and water into sugars used for plant growth and development. Notice that carbon dioxide enters the leaf through the stomata as water leaves the same way.

Oxygen is produced as a by-product of the chemical reaction of photosynthesis. The simplified equation of photosynthesis is:

$$6CO_2 + 6H_2O \xrightarrow{\text{light energy}} C_6H_{12}O_6 + 6O_2$$

Photosynthesis primarily occurs within the chloroplasts inside plant cells. The chloroplasts have double membranes (like mitochondria). Chloroplasts contain *chlorophyll*, a pigment that is important in capturing light energy. Chlorophyll is found inside a *thylakoid*, a disc-shaped sac surrounded by membranes on which the light reactions of photosynthesis take place. Thylakoids tend to be arranged in stacks known as *grana*. The aqueous space outside of the grana is called the *stroma*. Both the grana and stroma provide important spaces for parts of the photosynthetic process to occur.

For light energy to be used by plants, it must first be absorbed by a pigment. Within the chloroplast, these pigments include chlorophyll and carotenoids. Chlorophyll, the principal pigment of the plant that makes leaves appear green, absorbs light primarily in the violet and blue wavelengths and also in the red. Chlorophyll does not absorb green light. Instead, it reflects it and, therefore, makes leaves appear green. This is because there is more chlorophyll in leaves, and it often masks the red-orange *carotenoid* pigments. There are two types of chlorophyll: chlorophyll *a* and chlorophyll *b*. **Figure 9-2** illustrates the part of the light spectrum used by plants for photosynthesis. Note that different chlorophyll and other pigments absorb different wavelengths of light, enabling plants to absorb a greater range of light energy.

Corner Question

What do you call sugar that is produced in the process of photosynthesis?

Light-Dependent Reaction

The process of photosynthesis can be divided into two parts: light-dependent reaction and light-independent reactions. The light-dependent reaction, as the name indicates, occurs in response to light energy being absorbed by chlorophyll. It occurs within the grana, or stacks of thylakoid, within the chloroplasts. Light-independent reaction (discussed in the following section) occurs in the stroma. In the light-dependent process, light energy is converted to chemical energy (sugar).

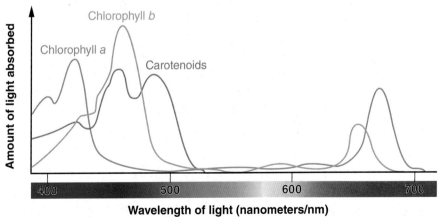

Goodheart-Willcox Publisher

Figure 9-2. Chlorophyll *a* is the principal pigment responsible for capturing light energy. Chlorophyll *b* and accessory carotenoid pigments extend the range for capturing light.

As multiple chlorophyll and other pigment molecules absorb particles of light (photons), the light energy causes electrons to enter an excited state or have a short boost of high energy. These energized electrons move from the chlorophyll molecules to specialized molecules in the thylakoid membrane. In a complex series of reactions, the energy from the excited electrons are passed along an electron transport chain, like a bucket brigade, which is used to create *adenosine triphosphate (ATP)* and *nicotinamide adenine dinucleotide phosphate (NADPH)* molecules. ATP is a nucleotide found in the mitochondria and the principal source of energy for cellular reactions. The NADPH molecule acts as a carrier for electrons in photosynthesis. These molecules provide energy used in light-independent reactions to make sugar.

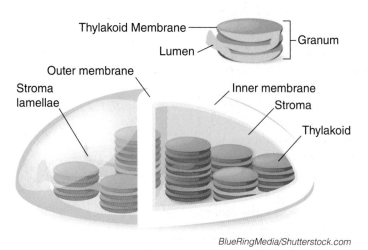

BlueRingMedia/Shutterstock.com

Figure 9-3. The light reaction occurs within the stacks of thylakoid membranes called the grana.

The excited electrons that leave the chlorophyll molecules are replaced by electrons that come from water molecules, which have been split by an enzyme in the thylakoid. When the water molecules are split, chlorophyll molecules take the electron from the hydrogen atoms, leaving H^+ ions. The remaining oxygen atoms from the divided water molecules combine to form oxygen gas (O_2) and are released from the plant. **Figure 9-3** illustrates the process of the light reaction. Note where the processes occur in the cell and the compounds that are used and produced as part of this reaction.

Light-Independent Reactions

In the last stage of photosynthesis, carbon atoms from carbon dioxide are used to make organic compounds in which chemical energy is stored. This process is called *carbon fixation*. Carbon fixation principally occurs through a process in the stroma called the Calvin cycle. The Calvin cycle produces a sugar molecule called glucose through a series of chemical reactions assisted by enzymes. ATP and NADPH provide the energy to drive this process. The Calvin cycle begins with molecules of carbon dioxide obtained from the atmosphere through stomata in the leaves. In a series of phases, six carbon dioxide molecules are enzymatically "fixed" (chemically bonded) to another compound. Through the energy of ATP and NADPH, two three-carbon molecules are created that the plant can use to form sugars, starches, and other compounds, **Figure 9-4**. Most plants species fix carbon this way and are called C3 plants, for the three-carbon molecule that is created.

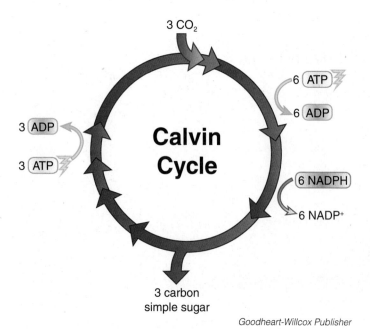

Goodheart-Willcox Publisher

Figure 9-4. Through a series of reactions, the Calvin cycle yields glucose.

STEM Connection: Increase in CO_2 Levels May Increase Plant Production

Researchers estimate that in the year 2050, carbon dioxide (CO_2) levels will be elevated from current levels. The leaves of soybeans grown at these elevated CO_2 levels will photosynthesize and respire more than those grown under current atmospheric conditions. This finding could point to increased crop yields as CO_2 levels rise. The study done by researchers at the University of Illinois and the United States Department of Agriculture found 90 different genes that control respiration were switched on, or expressed, at higher levels in the soybeans grown at high CO_2 levels. This allows the plants to use the increased supply of sugars (from the higher rates of photosynthesis that occur under high CO_2 conditions) to produce energy. The rate of respiration increases significantly (allowing more energy and plant growth) at elevated CO_2 levels.

When stomata remain open to accumulate carbon dioxide, the plant is also losing water through a process called *transpiration*. Under drought conditions, a plant will close its stomata to reduce water loss, resulting in a major decline in photosynthesis. This makes the Calvin cycle an inefficient system because it begins using oxygen in place of carbon dioxide, resulting in no sugars for the plant. Some species of plants that have evolved in drier climates have adapted their carbon-fixation pathways to conserve water without affecting the rate of photosynthesis.

Crassulacean Acid Metabolism (CAM) and C4 Plants

Plants with alternative photosynthetic mechanisms are called C4 plants or crassulacean acid metabolism (CAM) plants, depending upon their carbon-fixation path. The C4 mechanism is found in tropical grasses such as corn, sugarcane, sorghum, or warm-season turfgrasses. C4 plants thrive in high light, high temperatures, and dry conditions. Their leaves concentrate carbon dioxide in particular areas. As a molecule of carbon dioxide enters the leaf, it is temporarily attached to a three-carbon molecule (making a four-carbon molecule, hence the C4 name). The carbon dioxide molecule is shuttled to where it can be used more efficiently, allowing C4 plants to minimize water loss by keeping their stomata closed more often than typical C3 plants.

Many succulents, including cacti and stonecrop, use a crassulacean acid metabolism (CAM) pathway to fix carbon dioxide. (This alternative photosynthetic mechanism was first discovered in *Crassula* species.) Like C4 plants, CAM plants attach a molecule of carbon dioxide to a three-carbon molecule, but they open their stomata at night to minimize water loss. The carbon dioxide molecule is released from the temporary four-carbon molecule during the day so that photosynthesis may resume. Pineapples, Spanish moss, and some orchids are examples of nonsucculent CAM plants, **Figure 9-5**.

In review, photosynthesis is a vital process. It takes carbon dioxide (from the atmosphere) and water and uses energy from light to produce carbohydrates for plant growth. It also processes and releases oxygen as a by-product.

gashgeron/iStock/Thinkstock

Figure 9-5. CAM plants such as cacti can often be found in desert conditions.

Respiration

Take a deep breath and fill your lungs with oxygen. Humans and plants both use stored energy to fuel everyday processes. *Respiration* is a process in which glucose (the product of photosynthesis) combines with oxygen to produce energy in a form that can be used by plants. Plants take in oxygen and use its electrons to power a series of chemical reactions that creates energy for cells to function.

The sugars produced through photosynthesis are used for processes such as plant growth, flower formation, and fruit development. Sugars can be transported and stored throughout the plant, then converted to energy to power the cellular processes that result in new leaves unfolding, roots pushing deeper into the soil, or seeds forming, among other normal functions. A summary equation for respiration is:

$$C_6H_{12}O_6 + 6O_2 \rightarrow 6CO_2 + 6H_2O + energy$$

$$glucose + oxygen \rightarrow carbon\ dioxide + water + energy$$

Respiration happens within the mitochondria of plants, and is basically the reverse of photosynthesis. The respiration process yields energy in the form of ATP. ATP (adenosine *tri*phosphate) provides the energy used for plant growth and development functions when the bond between the second and third phosphate in ATP is broken (converting ATP into adenosine *di*phosphate, or ADP). That release of phosphate provides the energy to keep organisms of all types alive. But plants do not make ATP during photosynthesis; they make glucose. Glucose is needed in order to convert ADP back into ATP so that energy will be available for the plant's use when needed.

ATP Production

To make ATP, a glucose molecule is split into two molecules of a compound called *pyruvate* through a step called *glycolysis* (*glyco* = sugar, *lysis* = splitting). Glycolysis happens in the cell cytoplasm (outside the mitochondria) and produces a small amount of ATP (two ATP molecules). The pyruvate then moves into another series of reactions within the mitochondria called the Kreb's cycle, which in the presence of oxygen produces more ATP. In the third step of respiration, on the inner membrane of the mitochondria, a cycle of reactions called the electron transport system generates more ATP by oxidizing molecules from the earlier reactions in respiration. This is a very broad overview of a very complex process that results in the production of 36 ATP molecules for each glucose molecule, **Figure 9-6**. ATP is sometimes called the currency of the cell, because it provides the energy to drive all cell processes.

Oxygen is a key ingredient in respiration, which can be problematic for plants growing in water-saturated conditions. These conditions can occur through

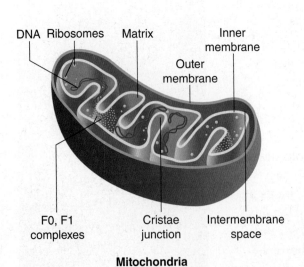

Figure 9-6. Respiration occurs in the mitochondria and produces energy that powers all plant growth and development.

overwatering or in soils with naturally poor drainage. Water replaces air in the soil pore spaces, leaving plant roots with no oxygen to take up and use in the glucose conversion process. As a result, waterlogged soils that are depleted of oxygen can kill root tissues and reduce plant growth or even result in plant death.

Rates of respiration influence plant growth and are linked to temperature. Raising temperatures can increase growth rates and lowering temperatures can reduce growth rates. Growers need to know the effects of temperature on each crop, as the effects vary by plant species. For example, lilies are very responsive to temperature management. However, temperature has little effect on chrysanthemums.

Corner Question

What is anaerobic respiration?

Transpiration

Have you ever wondered how water travels to the top of a tall tree? Recall that leaves have tiny holes called stomata. When the stomata are open, water vapor is released (much like you lose water when you exhale, which is evident on cold winter days). Water is pulled up through the plant as adjoining water molecules exit the stomata of the plant due to evaporation. Roots take in water as a liquid form through the process of diffusion. Water is then drawn upward through xylem cells and released into the air in a gaseous state. This process is called *transpiration*. As much as 90%–99% of water that is taken in by roots is lost through transpiration. Water molecules have a property called *cohesion* in which hydrogen bonding between adjacent water molecules allows the water to be pulled through the plant, **Figure 9-7**.

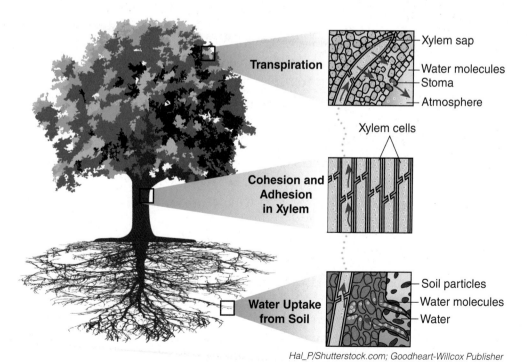

Hal_P/Shutterstock.com; Goodheart-Willcox Publisher

Figure 9-7. The cohesive properties of water allow it to be drawn upward through the plant from the roots to the leaves.

Transpiration provides these key benefits to plants:
- Carbon dioxide entry.
- Water uptake.
- Nutrient access.
- Evaporative cooling.

Carbon Dioxide Entry

Carbon dioxide is needed for photosynthesis. Carbon dioxide molecules enter the plant through the stomata by diffusion. However, carbon dioxide must enter the leaf through a solution in order to pass through the plasma membrane. To do this, carbon dioxide in its gaseous form must come into contact with a moist cell surface. The water from transpiration provides enough moisture for this process to happen. However, any time water encounters conditions where air is unsaturated evaporation occurs, and water is lost from the plant. This is also called *evapotranspiration*. Plant scientists for this reason have called transpiration a "necessary evil."

Miyuki Satake/Shutterstock.com

Figure 9-8. Turgor pressure keeps plant cells rigid. If there is a drop in turgor, wilting occurs.

Water Uptake and Nutrient Access

Transpiration drives the process of water uptake in plants. Water serves numerous functions within the plant, including turgor, or cell rigidity. *Turgor* is the water pressure within a plant cell that helps provide rigidity and support to plant structures. Water fills vacuoles within plant cells and causes the cell membrane to push up to the cell wall, helping the plant tissues stay rigid. If a plant experiences excessive water loss, plant cells lose turgor, become flaccid, and wilt, **Figure 9-8**. This is why grocery stores use misters to spray produce: it keeps the produce from wilting. Water is necessary for biochemical processes and is a medium for dissolved nutrients. Once nutrients enter the transpiration stream, water carries dissolved inorganic nutrients throughout the plant.

Factors Affecting Transpiration

Transpiration rates are affected by various environmental factors:
- Relative humidity.
- Temperature.
- Water.
- Wind.
- Plant structures.

Relative Humidity

Have you ever experienced a summer day where the air is "sticky"? In some areas of the United States, high humidity levels are common. Humidity, or the stickiness in the air, refers to the level of moisture or water vapor present in the air. On rainy days, the humidity levels are near 100%. In plants, relative humidity becomes important when understanding transpiration rates, water loss, and other issues, such as managing diseases and plant propagation.

Relative humidity (RH) is the amount of water vapor in the air compared to the amount of water vapor that air could hold at a given temperature. The higher the temperature, the greater amounts of water vapor the air can hold. The environment inside a leaf often has a relative humidity near 100%. Water exits the leaf through the stomata when the relative humidity in the atmosphere is less than 100%. The different levels of relative humidity (environment versus inside the leaf) create a gradient that allows the diffusion of water from the leaf into the air. Diffusion is movement of a substance from an area of higher concentration to an area of lower concentration. If there is less moisture in the air, the relative humidity is low, resulting in higher rates of transpiration, **Figure 9-9**. Higher levels of relative humidity mean that there is more water vapor in the air and transpiration rates will be lower.

bkkm/iStock/Thinkstock

Figure 9-9. The hairy leaves of lamb's ear create a thick boundary layer around the leaves and minimize water loss.

Temperature

Increases in temperature will increase rates of transpiration. As temperatures rise, air can hold more water, so relative humidity decreases. The lower relative humidity, combined with an increase in temperature, will speed up the amount of water leaving the plant. To see this process in action, try this: On a sunny, warm day, tie a plastic bag tightly around the leaves at the end of a leafy branch or twig on a tree. A few hours later, return and see how much water has accumulated in the plastic bag.

Water

Plants are constantly pulling water from the soil. Water enters the roots of a plant through diffusion. When more water is in the soil than in the plant, water will enter the plant roots. When water needs of the plant exceed the amount of water available in the soil, turgor pressure in the plant will drop and the stomata will close to prevent more water loss. A plant may recover from wilting when the water uptake catches up with transpiration.

Wind

Air movement across the leaf surface reduces a thin layer of water vapor that surrounds the leaf, called the boundary layer. The boundary layer maintains high levels of relative humidity around the leaf to slow water loss to the environment. When wind blows across the leaf, the boundary layer is reduced, creating a situation where transpiration can increase.

Physical Structures

Physical structures of a leaf may also influence transpiration rates. Some plants have thick, succulent leaves, whereas others are covered with "hairs," called *trichomes*. Plants with trichomes will have larger boundary layers. The trichomes provide a structural barrier that decreases the movement of air, allowing for an accumulation of water vapor that will reduce rates of transpiration.

"Gardening requires lots of water—most of it in the form of perspiration."
—Lou Erickson

Figure 9-10. The waxy surface of a kale leaf slows the loss of water. *John E. Manuel/Shutterstock.com*

Leaves with waxy cuticles are hydrophobic, or water repellent, and can slow water movement from inside the leaf to the atmosphere, **Figure 9-10**. Many desert plants have this feature.

Movement of Solutes

Transpiration is the driving force for moving water throughout the plant. The water stream provides a vehicle for transport of inorganic nutrient ions from the soil through the xylem. Sugars produced during photosynthesis move throughout the plant using a different conduit called the phloem.

Active Uptake of Inorganic Nutrients

Nutrients used by plants are the charged forms (ions) of the elements. Ions can have different charges and different forms. Most of the nutrient ions in the soil are surrounded by water, and can travel with water through the epidermis of the plant and through the cortex. Until water and ions get to the endodermis, the innermost level of the cortex, they can travel *between* cells rather than *through* cells. At the edge of the endodermis, however, is a waxy barrier called the *Casparian strip*, **Figure 9-11**. Neither water nor ions can pass through this layer without going through the plasma membrane, or *plasmalemma*. The plasma membrane is permeable to water, but impermeable to ions because of their charges (the interior of the plasma membrane is neutral and repels charges). To get ions into the endodermis and then to the xylem, nutrient ions must be drawn across the plasmalemma using active transport through channels. The plant must use ATP produced during respiration to move the ions into the endodermis, where they can once again travel with the water molecules, this time in the xylem. The transpiration stream moves the water and nutrients throughout the plant.

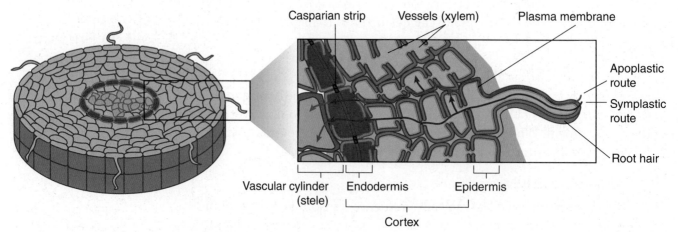

Figure 9-11. Water travels freely into the root until it encounters the waxy Casparian strip. Water and nutrients must be actively transported across a permeable membrane to enter the plant.

Goodheart-Willcox Publisher

Translocation of Sugars through Phloem

Sugars produced through photosynthesis travel through phloem to parts of the plant needing energy, such as growing tips of shoots and leaves and storage organs in roots, leaves, fruits, and seeds. *Translocation* (movement of sugars within the plant) occurs when the *sources* (where sugar is produced or stored in a plant) export sugars to the *sinks* (where sugar is used in a plant). The source can be any photosynthetic tissue, like a leaf, that manufactures sugars, or an organ that has stored sugars, **Figure 9-12**. The sink is any plant tissue that has a need for carbohydrates. Developing fruit is a very competitive sink, as are roots that serve as storage organs, **Figure 9-13**.

Jupiterimages/Thinkstock

Figure 9-12. Leaves and some stems are the principal producers of sugars and are considered sources.

Reproduction

Plant reproduction offers the opportunity to breed better crops that have specific advantages. Plants can be bred or selected to have a number of characteristics that include:

- Higher yields.
- Disease and pest resistance.
- Improved vigor.
- Increased nutritional content.
- Ornamental aesthetics.
- Enhanced stress tolerance.
- Postharvest longevity.

Higher yields mean an increase in harvested products. High yields, together with best production practices, allow for greater profits for growers and more food for the world. Pests and diseases constantly threaten crops. Through plant breeding, cultivars can be produced that allow the plant to use innate resistance to attack, providing a control method for pest management. Plants with strong vigor tend to outcompete other organisms for light, nutrients, and water. An example of vigor is a squash that rapidly leafs out, creating a canopy that shades out weed competition, **Figure 9-14**. Ornamental breeding results in plants with an astonishing variety of leaf colors, shapes, and textures that appeal to

Ryhor Bruyeu/iStock/Thinkstock

Figure 9-13. A sink is a location in a plant that requires sugars. Fruit is a common sink.

blinow61/iStock/Thinkstock

Figure 9-14. Plants that can rapidly leaf out, such as this zucchini plant, have a competitive advantage against weeds. This is a trait that is selected by breeders. **Can you identify any other plants that leaf out rapidly?**

Figure 9-15. Note the color variations on this geranium plant. Ornamental breeding creates new, novel plants for home gardeners.

gardeners, and offers growers a unique commodity, **Figure 9-15**. Increasing the nutritional content in plants results in higher levels of nutrients and minerals available for consumers in the foods they eat. For example, breeders might decide to increase beta-carotene levels in sweet potatoes to impact eye health. Some breeding efforts have improved the shelf life of crops with little loss of quality, increasing postharvest longevity. Understanding plant reproduction and genetics enhances knowledge of how new plants can grow and develop.

Cellular Division

Cells reproduce through a process known as cell division, in which a cell and all of the material contained within it splits into two daughter cells. In multicelled organisms, such as plants and animals, cellular division and enlargement allows for growth, repair, or replacement of wounded or dead cells. Each new daughter cell is exactly the same as its parent cell, inheriting an exact replica of all the genetic information. Vegetative, or asexual, propagation of horticultural crops is possible due to a plant's ability to divide its cells. A vegetative cutting from a plant is therefore genetically identical to the plant it was taken from.

Mitosis and Cytokinesis

Asexual reproduction in plants occurs through a process called *mitosis*. Mitosis is the steps a cell undertakes to duplicate and divide a complete set of chromosomes. Mitosis occurs in four stages: prophase, metaphase, anaphase, and telophase, **Figure 9-16**. In each of these stages, the cell is physically arranging itself to replicate and divide nuclear material. *Cytokinesis*, a process that follows mitosis, is the division of the actual cell into the two new daughter cells. During cytokinesis a cell plate forms between the dividing cells and grows outward to the cellular wall, separating the two new cells. Each new daughter cell has a nucleus with a full set of chromosomes and half of the cytoplasm, containing plant organelles.

Sexual Reproduction

Sexual reproduction in both plants and animals is the union of a female and male gamete that results

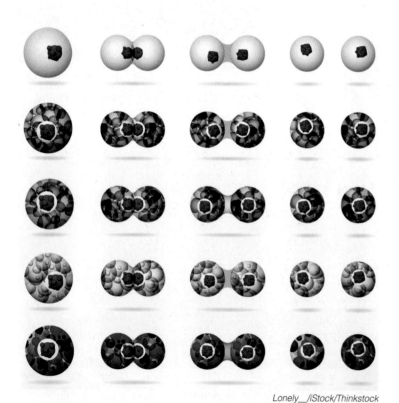

Figure 9-16. Observe how the cell changes through each of the stages of mitosis until two new daughter cells with identical genetic information are formed.

in a new organism that is genetically different than either of its parents. Plant breeding uses the process of sexual reproduction to develop new, novel, and enhanced plant material. Sexual reproduction in plants occurs through meiosis and fertilization. *Meiosis*, like mitosis, involves a nuclear division, but rather than replicate the number of chromosomes, the cell's chromosomes are divided in half. Through *fertilization*, the male gamete combines with the female gamete to create a genetically unique organism.

Chromosomes

Chromosomes are cellular structures that carry the genetic information of a plant. Every species has a specific number of chromosomes, found in the nucleus. For example, humans have 46 chromosomes, dogs have 78 chromosomes, and common wheat has 42 chromosomes, **Figure 9-17**. Chromosomes are found within all the somatic (vegetative) cells of the plant and are considered to be *diploid* (have a full set of chromosomes, one from each parent plant). A plant's entire set of chromosomes is called the *genome*.

Chromosomes are made up of filament-like threads called chromatin. *Chromatin* contains DNA and proteins. *Genes* are specific sequences of nucleotide pairs on a DNA molecule that hold the information to build and maintain cellular processes and pass genetic traits to offspring. Different genes are responsible for any number of physical traits and drive biochemical processes and responses in plants.

In the reproductive cells of the flowers, gametes contain a *haploid* set of chromosomes (a complete set of chromosomes from one parent). During fertilization, male and female haploid nuclei fuse to form a diploid cell called the *zygote*.

Plants		
Organism	Species	Diploid (2N) Chromosome Number
Alfalfa	Medicago sativa	32
Avocado	Persea americana	24
Barley	Hordeum vulgare	14
Bermuda grass	Cynodon dactylon	36
Cashew	Anacardium occidentale	42
Corn (maize)	Zea mays	20
Cotton, upland	Gossypium hirsutum	52
Garden pea	Pisum sativum	14
Grape	Vitis vinifera	38
Mango	Mangifera indica	40
Oats, white	Avena sativa	42
Onion	Allium cepa	16
Papaya	Carica papaya	18
Peanut	Arachis hypogaea	40
Pineapple	Ananas comosus	50
Potato	Solanum tuberosum	48
Rice	Oryza sativa	24
Soybean	Glycine max	40
Squash	Cucurbita pepo	40
Sugarcane	Saccharum officinarum	80
Tomato	Lycopersicon esculentum	24
Wheat, common	Triticum vulgare	42
Animals		
Organism	Species	Diploid (2N) Chromosome Number
Human	Homo sapiens	46
Cat	Felix domesticus	38
Cattle	Bos taurus	60
Chicken	Gallus domesticus	78
Dog	Canis familiaris	78
Fruit fly	Drosophila melanogaster	8
Grasshopper	Melanoplus differentialis	24
Honeybee	Apis mellifera	32
Horse	Equus calibus	64
House fly	Musca domestica	12
Mosquito	Culex pipiens	6

Goodheart-Willcox Publisher

Figure 9-17. Note the vast differences in chromosome numbers among different species.

Meiosis

Prior to the process of meiosis, each chromosome replicates itself. As the cell begins the first step of meiosis I, **Figure 9-18**, pairs of chromosomes line up in the center of the cell where a phenomenon called crossing-over occurs. Meiosis is responsible for genetic diversity. During crossing-over, genes are exchanged between matching chromosomes, so that each chromosome becomes a mosaic of genes from the male and female parent plants. This makes new and unique arrangements that can impact the way a plant grows or looks. This is a critical factor for genetic recombination in organisms. The cell continues to divide in half, but each cell still has a diploid number of chromosomes, with each cell containing a mix of genes from the parent plants. Meiosis II now happens, similar to mitosis, and divides the cells again. The end result of meiosis I and II in one diploid cell is four haploid daughter cells (cells that contains a single set of chromosomes) that will develop into male gametes (sperm) or female gametes (eggs).

Fertilization

Fertilization is the fusion of the male and female gamete haploid nuclei to form the diploid zygote. In plants, the male gamete (sperm) is found protected within pollen grains. Both gymnosperms and angiosperms produce pollen-bearing male gametophytes (the multicellular haploid stage). In gymnosperms, the pollen travels to the female cones, which

"The secret of improved plant breeding, apart from scientific knowledge, is love."
—Luther Burbank

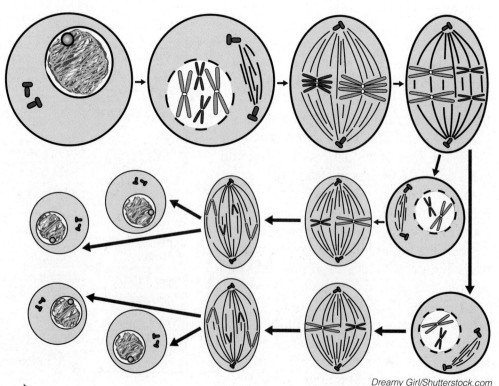

Dreamy Girl/Shutterstock.com

Figure 9-18. The stages of meiosis during sexual reproduction. The chromosome number is halved to create haploid sex cells (sperm and egg cells).

have ovule-bearing scales, **Figure 9-19**. In angiosperms, the pollen travels from the anthers to the stigma. This process is pollination. After pollen lands on the stigma, a pollen tube grows through the style until the sperm can reach the eggs (ovules) contained within the ovary and fertilization occurs.

Plant Breeding Principles

Gregor Mendel's famous cross-pollination of the common garden pea (*Pisum sativum*), **Figure 9-20**, led to the discovery of a few basic principles of gene inheritance. It also played an essential role in establishing early plant breeding methods. Mendel worked with peas that had been inbred, or self-fertilized, for generations and, therefore, were fairly uniform or *homozygous* (having identical pairs of genes for a pair of hereditary characteristics). He took peas with *alleles* (one of a number of variant forms of the same gene) for yellow seeds and cross-pollinated them with peas that had showed the trait for green seeds. The resulting offspring were entirely yellow-seeded peas and were considered hybrids and the first filial, or F1, generation. Because there were no green-seeded peas that resulted from this cross, the yellow-seeded trait is considered *dominant* (the relationship between one allele that is expressed over a second allele). The green-seeded trait is considered *recessive* (the relationship where one allele is only expressed when the second, dominant allele is not present). When an F1 generation is self-fertilized, the progeny are called the F2 generation. The F2 generation of Mendel's crossing experiment produced a 3:1 ratio of yellow-seeded peas to green-seeded peas. A Punnett square, **Figure 9-21**, further illustrates the outcomes of the cross, showing one dominant homozygous yellow seeded progeny (YY), two heterozygous (Yy) offspring, and one recessive (yy) homozygous offspring. Mendel determined that seed color is a trait that is controlled by the genes of both parents.

There are also times when neither gene is dominant, and the results are an intermediate phenotype. For example, in some petunias cross-pollinating a homozygous red petunia with a homozygous white petunia will result in offspring having some pink flowers.

srekap/Shutterstock.com

Figure 9-19. Gymnosperms, unlike angiosperms, do not have flowers. Instead they have male and female cones. This is the female cone of a Douglas fir.

Geo-grafika/iStock/Thinkstock

Figure 9-20. The common garden pea played a key role in understanding basic principles of gene inheritance.

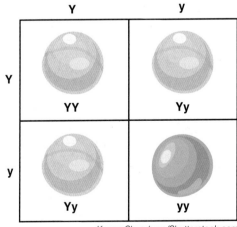
Ksena Shurubura/Shutterstock.com

Figure 9-21. This Punnett square shows the relationship between the cross of a yellow-seeded pea and a green-seeded pea.

Figure 9-22. Incomplete dominance will show traits of both parents. For example, pink petunias may result from cross pollinating red and white petunias.

Figure 9-23. These hybrid peppers have been developed by plant breeders to resist diseases.

This is called *incomplete dominance*. It is a phenomenon in which a plant shows characteristics of both parents, **Figure 9-22**.

The genetic composition of an organism, often in reference to a particular trait, is called the *genotype*. The genotype influences how a plant grows and develops. Environmental conditions such as temperature, water availability, and nutrient availability also influence the way a plant grows and develops. The interaction between a plant's genetic makeup and the environment is called its *phenotype*. The phenotype is a visible expression of an organism's observable characteristics. For example, the yellow and green colorings of Mendel's peas are phenotypic characteristics.

Plant breeders and plant geneticists use this information to determine how to select parents for breeding purposes. Selected parents will have desirable characteristics that, when crossed with other selected parent plants, will result in offspring that have those desirable traits. For example, a pepper breeder may want to decrease the capsaicin (amount of heat) levels in a plant and also have effective disease resistance. To do this, the plant breeder will use parents with these traits and cross-pollinate, **Figure 9-23**.

The Plant Breeding Process

Plant breeders select a parent to be the female, or seed-producing parent, and a parent to be the male. All the male floral parts are removed from the maternal (female) parent. This process is called emasculation and it eliminates the possibility of the flower to self-pollinate. The stamens should be removed before the anthers release the pollen. To cross-pollinate a plant, a breeder removes the stamen from the paternal (male) parent using tweezers and brushes the anther across the stigma of the maternal parent. The seeds are then saved and planted. The offspring are evaluated for the presence of desired characteristics.

Plant Mutations

Mutations are naturally occurring genetic changes that affect the appearance and functions of a plant. Mutations are heritable and can be called a chimera or sport. Mutations can also be induced through radiation or chemicals.

Most mutations are not desirable and have adverse effects on a plant. Examples include a lack of chlorophyll and the inability to photosynthesize. A few mutations can produce desirable traits, such as a unique ornamental or growth habit. *Chimeras* are a type of mutation that allows two genetically distinct tissues to coexist, **Figure 9-24**. *Sports* are tissue mutations that occur in vegetative cells. For example, a branch or stem will show a specific observable trait difference, such as variegated leaves or flower color. Witches' broom is a growth phenomenon that shortens the internode spacing, causing dwarf plants. Although insects, viruses, or diseases can cause witches' broom, it can also be caused by a genetic mutation.

Careers

The study of plant growth and development can lead to careers in the basic and applied sciences. Understanding the plant processes allows researchers to develop best practices for managing and growing plants.

Plant Physiologist

Plant physiologists have a broad systems understanding of whole plant processes, all the way down to the cellular level. They use this knowledge to examine the physical, chemical, and biological functions of living plants. Plant physiologists explore how plants grow and study the components required to drive the development process, from germination to death. Many plant physiologists are researchers in colleges and universities. They perform basic research that can inform applied practices used in horticulture. For example, a plant physiologist might look at the pathways by which a plant can take in toxic pollutants. A horticulturist can use this information to plant species that are most effective in decontaminating polluted soil.

Kristina Gruzdeva/iStock/Thinkstock

Figure 9-24. The variegated snake plant is an example of a chimera. Two genetically different tissues exist within the same plant. Can you think of any other examples of chimeras that you may have seen?

STEM Connection
Induced Mutations in Plant Breeding

The use of radiation, such as X-rays, gamma rays, neutrons, and chemical mutagens to induce variation, has been used to improve both floriculture and agricultural crops. In crops where diversity for a given trait is low or nonexistent, induced mutation provides a way to incorporate novel traits into the traditional plant breeding and selection process. Through mutagenesis (induced mutation), pears have improved disease resistance and spineless pineapples and seedless grapefruit have been developed.

Career Connection

Joseph Tychonievich
Plant Breeder

Joseph Tychonievich is an independent plant breeder and author who was "mesmerized by the magic of plants" as a young child. For the past ten years, he has been breeding nursery plants. Joseph cultivates tree and shrub plants as well as herbaceous perennials and vegetables. He has developed unique alpine plants that are popular with specialty gardeners and also finds old-fashioned annuals, such as heirloom zinnias and hyacinths of the 1820s, really interesting. His work has made him a prominent figure in the horticulture world.

Joseph encourages young gardeners to "Just try stuff. Nursery breeders can find a niche and be very successful. Whether it is growing plants, working in a nursery, blogging, taking photographs, or breeding plants, you have to pursue an idea." As an independent breeder, Joseph has a great deal of flexibility that includes self-employment, independence, traveling, writing, lecturing, and networking with other horticulturists. His blog provides a unique platform to connect with interested gardeners of all ages around the world.

Photograph by Joseph Tychonievich

A bachelor's degree is required for entry-level laboratory positions, both in universities and private industry. Graduate degrees will open doors for high-level research positions at these same institutions, as well as in government agencies such as the United States Department of Agriculture (USDA) or the Environmental Protection Agency (EPA).

Plant Breeder

Plant breeders strive to improve the overall characteristics of various plants for increased production and profitability. Plant breeders work with agronomic, horticultural, and forestry crops. They use traditional breeding methods of cross-pollination as well as newer processes that apply biotechnology and molecular techniques to enhance the breeding process.

Plant breeders generally need a graduate degree (master's or higher) in a related plant science field (horticulture, agronomy, forestry, botany, plant pathology, biology, entomology, soils). Possible careers include working at a public university to create new cultivars that support growers in that state or region. Many private companies hire plant breeders to create desirable hybrids wanted by commercial growers and home gardeners. These types of plants are profitable for the company.

CHAPTER 9
Review and Assessment

Chapter Summary

- The process of photosynthesis enables plants to capture energy from the sun and convert carbon dioxide and water into sugar that can be used by plants and animals as sources of energy. Oxygen is released as a by-product.
- The process of photosynthesis can be divided into two parts: light-dependent reactions and light-independent reactions.
- The sugars produced through photosynthesis are used as energy created by respiration to drive basic plant functions. Respiration converts glucose in the presence of oxygen to a form (ATP) that can be used by plants.
- Transpiration is the process of water being drawn up from the roots through the xylem and released through the stomata. Evaporation powers the process of transpiration through water's cohesive properties.
- The transpiration stream powers the movement of inorganic nutrients throughout the plant. Translocation is the process of moving photosynthetic products (sugars) from sources to sinks.
- Plant reproduction occurs by both mitosis (asexual reproduction) and meiosis (sexual reproduction). Somatic cells divide using mitosis and sexual cells divide through meiosis and allow for genetic variation to occur.
- Plant breeding improves the characteristics of plants by increasing yield, vigor, nutritional content, ornamental value, stress tolerance, and growth habit.
- Plant mutations are naturally occurring genetic changes that affect the appearance and functions in a plant. Mutations can happen at random or can be induced by radiation or chemicals.
- The study of plant growth and development can lead to careers such as plant physiologist and plant breeder.

Words to Know

Match the key terms from the chapter to the correct definition.

- A. allele
- B. carotenoid
- C. Casparian strip
- D. chimera
- E. chlorophyll
- F. chromatin
- G. chromosome
- H. cohesion
- I. cytokinesis
- J. diploid
- K. dominant
- L. fertilization
- M. genotype
- N. grana
- O. haploid
- P. homozygous
- Q. incomplete dominance
- R. phenotype
- S. recessive
- T. relative humidity (RH)
- U. sport
- V. stroma
- W. thylakoid
- X. turgor

1. A filament-like structure that contains DNA and proteins that makes up chromosomes.
2. A cell that contains two complete sets of chromosomes.
3. Cellular structures that carry the genetic information of a plant.
4. A phenomenon where a plant displays characteristics of both parents.
5. The genetic makeup of an organism, often in reference to a particular trait.
6. The aqueous space outside of the grana within the chloroplast.
7. Water pressure within plant cells that create cellular rigidity.
8. The amount of water vapor in the air compared to the amount of water vapor that air can hold at a given temperature.
9. A stack of thylakoid disks found within the chloroplast.
10. Having identical pairs of genes for a pair of hereditary characteristics.
11. The relationship between one allele that is expressed over a second allele.
12. One of a number of variant forms of the same gene.
13. A type of mutation that allows two genetically distinct tissues to coexist.
14. A plant pigment that reflects yellow, orange, and red light and assists in capturing light energy.
15. The process in which the cytoplasm of a single cell is divided to form two daughter cells and the cell plate is formed.
16. The process by which the male gamete combines with the female gamete to create a genetically unique organism.
17. A type of tissue mutation that occurs in somatic (vegetative) cells.
18. A visible expression of an organism's observable characteristics.
19. Folded membrane inside the chloroplasts that serves a key function in the photosynthetic process.
20. The relationship where one allele is only expressed when the second, dominant allele is not present.
21. A cell that contains a single set of chromosomes.

22. A green pigment located in the chloroplasts of plant cells that is a receptor of light energy in the red and blue wavelengths.
23. A waxy barrier that rings the endodermal cells in the roots.
24. A property of water in which hydrogen bonding between adjacent water molecules allows the water to be pulled upward through the plant.

Know and Understand

Answer the following questions using the information provided in this chapter.

1. What is photosynthesis and why is it important to life on earth? What element is created as a by-product of photosynthesis?
2. Where do light-dependent and light-independent reactions occur?
3. Name the three types of light-dependent reactions and plants that use these pathways.
4. Where does respiration occur within the cell?
5. Describe the process of respiration.
6. Why are water-saturated conditions a problem for some plants?
7. How does raising and lowering temperatures affect plant growth?
8. List the benefits of transpiration in plants.
9. What happens if a plant experiences excessive water loss?
10. What is the role of water in plant growth?
11. What factors affect rates of transpiration?
12. Describe the influence of relative humidity on transpiration.
13. How do water and nutrients enter plant roots?
14. Describe the way sugar is translocated throughout the plant.
15. List the characteristics that plant breeders breed for in plants.
16. Summarize the steps of mitosis.
17. Compare and contrast mitosis and meiosis (asexual and sexual reproduction).
18. Identify the role that chromosomes play in a plant.
19. Detail the reasons why meiosis is responsible for genetic diversity.
20. Explain the process of pollination.
21. Explain the process of cross-pollination.
22. What does a plant breeder career involve? What education is needed for this career?

Thinking Critically

1. Bald cypress trees are typically found growing in soggy soil. Water-saturated soils limit oxygen levels needed for respiration. How do you think these trees get enough oxygen to drive the energy-making process of aerobic respiration?
2. Many succulents have a CAM metabolism, opening their stomata at night to minimize water loss. In what kind of environment do you think these plants typically grow? How might they do in another kind of environment?

3. Plant breeders are constantly evaluating and selecting plants for desired characteristics. Describe the ideal plant for your garden and detail the breeding steps you would take to create your plant. What do you need? How long would it take? Is this possible?
4. American chestnut trees have all but disappeared from the Appalachian Mountains due to a fungus from Asia that caused a destructive blight. Can you think of way that scientists might be able to save the chestnut?

STEM and Academic Activities

1. **Science.** Light powers the process of photosynthesis. Plants have multiple pigments that can capture different wavelengths of light. Create an experiment using different types of light and record your observation on the impact light quality has on plant growth.
2. **Science.** Transpiration in plants can be approximately measured by weighing a potted plant before and after a certain period of time. Take a small potted plant and water thoroughly. After the water has drained, weigh the entire potted plant and record the weight. Cover the soil surface with plastic wrap to prevent evaporation from any surface except for the leaves. Observe the plant over time and record the weight each day. How much water is lost after a day? How much is lost after a week? What do you think will happen if you try this in different environmental conditions or with other plants? Experiment and compare and contrast the results.
3. **Science.** Try your skills at plant breeding by cross-pollinating two plants and evaluating the resulting progeny. First find two petunias with different flower colors. Petunias are easy to grow and easy to cross-pollinate. Decide which plant will be the seed parent and which will be the male parent. Emasculate the flowers on the seed parent. Using tweezers, pluck a stamen from the other plant and brush pollen on the sticky stigma of the petunia. Be sure to label your cross with a tag on the pollinated flower. Continue to grow the seed parent until fruit and seeds mature. Plant the seeds and evaluate your new plants for characteristics they might exhibit. How are they the same as the parents? How are they different?
4. **Social Science.** Plant breeding can seem like a science-fiction story, with breeders being able to "design" plants by choosing the specific physical and personality traits desired. Do you think this is a good idea? Why or why not? Use specific reasons and examples to support your position.
5. **Language Arts.** Write a blog post about one of the plant experiments you are conducting from the activities above. Describe the details of your experiment, including the methods, the data, and any conclusions you can draw. Include clear descriptions and explanations that would allow your reader to replicate your work if desired.
6. **Language Arts.** Work with your teacher to set up a mock interview for a job position in plant physiology, plant breeding, or a related field. Prepare yourself to effectively convey the qualities you possess to be a successful applicant. Try to incorporate good communication skills including speaking clearly and enthusiastically about your experiences and skills.

Communicating about Horticulture

1. **Reading and Speaking.** Using your textbook, library resources, and the Internet, research plant breeding of edible horticultural crops. Focus on the history and the way in which plant breeding has changed food security around the world. Create a poster to illustrate your findings and design(s) and present it to your peers.
2. **Listening.** As classmates deliver their presentations on plant breeding, listen carefully to their ideas. Write down any questions that occur to you. Later, ask questions to obtain additional information or clarification from your classmates as necessary.
3. **Reading and Speaking.** Working in groups of three students, create flash cards for the Words to Know in this chapter. On the front of the card, write the term. On the back of the card, write the pronunciation and a brief definition. Use your textbook and a dictionary for guidance. Then take turns quizzing one another on the pronunciations and definitions of various words.

SAE Opportunities

USDA

1. **Exploratory.** Plant physiology is a broad subject that encompasses a number of more narrow disciplines. Molecular biology is a growing field that looks at the molecular functioning of plants. Job shadow a researcher in this industry. What is the purpose of this job position? What are the daily responsibilities? What career path do you need to create to become a molecular biologist?
2. **Exploratory.** Research how to manage relative humidity in a greenhouse. Can it be managed? How does this affect the plants growing in the greenhouse? What are the costs involved? Why might this be important to know as a grower?
3. **Improvement.** Find a location on your school campus that has been neglected. Consider creating a small pocket garden and growing plants that require minimal watering. What kind of photosynthetic pathway do you think these plants have? Why?
4. **Entrepreneurship.** Create a new cultivar to sell at your school plant sale. Find an easy plant to cross-pollinate (make sure they are not patented) and sell the seeds as a niche marketing product.

CHAPTER 10
Environmental Conditions for Growth

Chapter Outcomes
After studying this chapter, you will be able to:
- Discuss how light quality, quantity, and photoperiod affect plant growth.
- Compare strategies to optimize light quantity for plant production.
- Explain different responses plants have to temperature.
- Summarize techniques to manage temperature for crop production.
- Explain how water is applied for greatest plant growth and development.
- Compare and contrast careers in horticulture environmental management.

Words to Know

air drainage	irrigation	plant hardiness zone
biennial	juvenile stage	polycarbonate
blanching	light quality	polyethylene
chilling injury	light quantity	Q_{10}
daily light integral (DLI)	microclimate	root zone
degree day	nanometer (nm)	slope orientation
DIF	necrosis	stratification
dormancy	photoblastic	sunscald
etiolation	photon	thermoperiod
freezing injury	photoperiod	vernalization
heat stress	phototropism	

Before You Read
Review the chapter headings and use them to create an outline for taking notes during reading and class discussion. Under each heading, list any term highlighted in **bold italics**. Write two questions that you expect the chapter to answer.

While studying this chapter, look for the activity icon to:

- **Practice** vocabulary terms with Words to Know activities.
- **Expand** learning with identification activities.
- **Reinforce** what you learn by completing Know and Understand questions.

www.g-wlearning.com/agriculture

©iStock.com/Tadeusz Przbyt; ©iStock.com/fotokostic; ©iStock.com/kot2626; ©iStock.com/Kolidzei; ©iStock.com/NejroN

Have you ever noticed a home gardener putting a cloth sheet over young tomato transplants? Or starting lettuce under a cold frame? These gardeners are attempting to modify the air temperature around the plant in hopes of preventing any injury from low temperatures. Horticulture is a practice of managing environmental conditions to cultivate successful crops. Understanding the growing needs of plants and the ways to control essentials such as light, temperature, and water is key to uncovering the secrets of plant growth and development.

Pat_Hastings/Shutterstock.com

Figure 10-1. Observe how light has shaped the growth habit of this plant, causing it to elongate and become spindly. Would rotating the potted plant each day have helped promote a straighter, stronger stem?

Light

All horticultural plants require light to grow and develop. As discussed in Chapter 9, *Plant Growth and Development*, light is critical for creating chemical energy through photosynthesis. Light is important in other ways, too. It plays a role in pigment formation (chlorophyll, carotenoids, and anthocyanins), some seed germination, growth habit, **Figure 10-1**, shape and size, flowering, fruiting, dormancy, hardiness, plant movement, formation of storage organs, fall color, and leaf drop in temperate climates.

Light varies in its quality, quantity, and photoperiod, and plants respond accordingly. *Light quality* refers to the specific wavelengths of light that a plant receives. *Light quantity* is the amount and duration of light emitted by the light source, whether it is the sun or a lamp. The *photoperiod* is the duration of day length (the amount of time that light is present) and the relationship between the dark and lighted periods.

Light Quality

Imagine a bright yellow flower. As light strikes the flower, the yellow color you perceive is actually the wavelength of light that is reflected by the flower. The only light reflecting from the flower is in the yellow spectrum; the other colors of light are being absorbed by the flower. Plants primarily absorb light in the visible wavelength range of 400 to 700 nanometers, **Figure 10-2**. A *nanometer (nm)* is the unit of measurement used to quantify light wavelengths.

In addition to visible light, the light spectrum includes ultraviolet light (which is not involved in plant processes) as well as far-red and infrared light that play an important role outside of photosynthesis. Although plants absorb across the visible wavelength range of 400 to 700 nanometers, the greatest impact on plant growth is at peaks in red light (650 nm) and blue light (475 nm). Plants respond to the quality of light available. If only blue light is applied to plants through special lighting, plant growth will be shortened, hard, and dark in color. When plants are grown in only red light, the growth becomes elongated and soft. Red and blue light combined promote flowering.

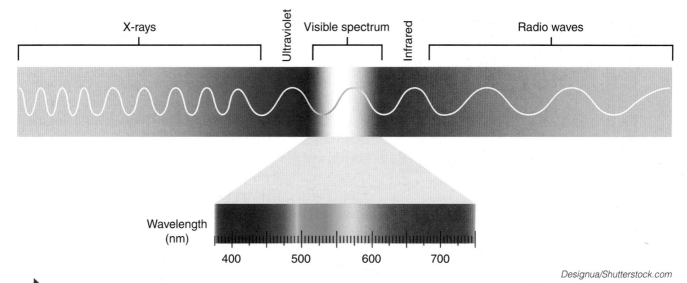

Figure 10-2. This graph illustrates the entire spectrum of light, including both visible light, far-red, infrared, and ultraviolet light. Far-red light is light at the extreme end of the visible spectrum (between red and infrared light). Infrared is invisible radiant energy with longer wavelengths than those of visible light, and ultraviolet light is a form of radiation that is not visible to the human eye.

Light quality management occurs primarily in greenhouses as little can be done to alter ambient light conditions in open fields. The emphasis of light quality management is to provide efficient and sufficient light for photosynthesis. Few practices exist to manipulate light quality in a cost-effective way to influence growth. However, recent research suggests that there may be ways to control growth. For example, it may be possible to constrain the vegetative growth of strawberries to promote flowering and increase fruiting.

Photoblastic Seed Germination

Seeds that are influenced by the presence or absence of light are considered *photoblastic*. Seed responses to light may be positive or negative, meaning that germination can be stimulated or inhibited. Most seeds do not require light to germinate. However, there are a few important exceptions that require light to overcome dormancy. These plants include carrots, lettuce, rubber plant, gloxinia, and zoysia grass. Growers will seed these crops on the soil surface to maximize germination. Unfortunately, many weed species also require light to germinate and the practice of tilling brings these weed seeds to the soil surface. For weed seeds that are positively photoblastic, this increases their germination and potential for competition with horticultural crops, **Figure 10-3**. Seed germination can also be inhibited with exposure to light, as in tomatoes and lilies. These seeds should be planted where no light reaches them.

Figure 10-3. Tillage brings positively photoblastic seeds buried in the soil to the surface where they germinate easily and compete with horticultural crops. Are there methods to prepare the soil that would result in less photoblastic weed seeds brought to the surface?

Figure 10-4. The leaves of this hosta plant are showing necrosis after exposure to high light intensities.

Light Quantity

Have you ever planted a shade-loving plant, such as a hosta, in full sunlight? Does the plant thrive or deteriorate? For many shade-loving plants, too much light can "burn" the plant, causing cellular damage that results in leaf *necrosis* (death), **Figure 10-4**. Light quantity is the number of *photons* (light particles) capable of performing photosynthesis. Light quantity is measured by how much light instantly reaches the plant (light intensity) and how much light a plant has received throughout the day, called the *daily light integral (DLI)*.

Daily Light Integral (DLI)

Increased amounts of light intensity directly increase rates of photosynthesis. Each crop species has an optimal light intensity that maximizes plant growth through photosynthesis. When there is limited light available, plant growth and quality can decline. If there are excessive amounts of light, no real growth gains will be achieved. In some cases, such as the example of the hosta, higher-than-optimal light intensities can actually slow growth of certain plants because damage may occur.

The daily light integral measures the duration of light intensity throughout the day. Light conditions fluctuate during the day with increasing or decreasing cloud cover and sunlight. It is important to know how much light reaches plants during the day. Think about the daily light integral as a gauge that records the amount of light that "falls" in a 24-hour period. The quantity of light a plant receives determines characteristics such as branching, rooting, stem thickness, flower number, and plant height.

Did You Know?
The photic sneeze effect is a genetic tendency to begin sneezing when suddenly exposed to bright light. This reaction is harmless and it is thought to affect about 18% to 35% of the human population!

Etiolation and Blanching

Etiolation is a plant growth response in absence of light. This phenomenon occurs when plants or plant parts are covered to exclude light and plant parts continue to develop. If you have ever removed a tree branch from a lawn and seen long, pale seedlings where the tree branch was, this is an example of etiolation. Etiolation is a common practice when producing certain vegetable crops, such as cauliflower or asparagus. For these items, consumers demand a tender crop with mild flavor. In cauliflowers, growers will tie up leaves around the developing head, **Figure 10-5**. The resulting harvest is a curd (head) that is very white and considered high quality. Keeping plants from light to prevent photosynthesis and to cause white tissue growth is called *blanching*. Cauliflower grown without the blanching process will yield heads with stronger flavors and a slightly yellow color.

Figure 10-5. Growers may tie the leaves of cauliflower heads to promote blanching and mild flavor. How would the labor needed to tie the leaves affect the cost of producing a cauliflower crop?

Photoperiod

Plants have growth responses based on the number of hours of light they receive each day. This phenomenon is called *photoperiodism*. It has horticultural importance primarily for flowering, but also for the development of storage organs, stem elongation, fall color, and leaf fall. Length of light exposure is important, as well as the uninterrupted darkness that follows. Each plant species has its own photoperiod requirements. Horticulturists have determined ways to control lighting to get a desired growth response, such as flowering.

Plants have a *juvenile stage*, which is a period of growth that occurs before they reach adult form. In this stage, flowering will not occur; only vegetative growth happens, regardless of light conditions, **Figure 10-6**. Plants are grouped into three categories that define their light requirements for flower production:

- Day-neutral plants.
- Long-day plants.
- Short-day plants.

Figure 10-6. English ivy has a distinct juvenile leaf compared to its adult leaf.

Day-Neutral Plants

Day-neutral plants can flower under a wide range of day lengths and flower when they reach the adult stage of growth. Examples include cucumber, sweet corn, dandelion, and ever-bearing strawberry, **Figure 10-7A**.

Long-Day Plants

Long-day plants can flower if the light period is as long or longer than the critical day length. Critical day length is the range of light hours per day that plants need to flower; this varies among plant species. For example, many plants, such as spinach, dill, and evening primrose, flower under long days during summer months, **Figure 10-7B**.

Short-Day Plants

Short-day plants can flower when the light period is shorter than the critical day length, which varies by species or even cultivars within a species. Common examples of short-day plants are goldenrod, kalanchoe, poinsettia, and chrysanthemum, **Figure 10-7C**. These plants typically flower under short days, usually in the late summer or fall.

Figure 10-7. A—Ever-bearing strawberries are day-neutral, which allows them to be grown in high tunnels in the winter. B—Evening primrose is a long-day plant and typically flowers during the long days of summer. C—Chrysanthemum is a common short-day plant that requires between 7 to 16 hours for flowering.

Lighting and Forced Flowering

Horticulturists have developed lighting protocols to force flowering for a target date or to encourage increased vegetative growth for larger plants. For example, research at Michigan State University resulted in a growing method to produce flowering summer perennials in the spring, at a time when they would normally only display vegetative growth. Kalanchoe have a critical day length of 14.5 hours and will flower if the length of day is shorter than 14.5 hours. If the plant receives more than 14.5 hours of light a day, it will continue to grow, but it will not flower. Short-day plants can also be kept in a vegetative growth state by interrupting the dark period. This actually requires fewer hours of light (and less energy) than extending the day. An incandescent lamp uses bulbs similar to those traditionally used in homes. These lamps have historically been used to interrupt the dark period. However, LED lighting may soon replace these lights. LED lighting can be used to provide both photosynthesis and photoperiod lighting.

Light Pollution

In many cases, light pollution from various sources (car headlights, streetlamps, athletic field lights, and other sources) will provide unwanted interruption of a dark period. Growers will sometimes pull a shade cloth or black plastic across the plants to prevent unwanted light from reaching the plants, **Figure 10-8**.

Icswart/Shutterstock.com

Figure 10-8. Shade cloth eliminates light disruption during a dark period necessary for flowering (usually for short-day plants). What other methods are used to block light from plants in a greenhouse?

Storage Organ Development

Photoperiod also influences the development of storage organs, such as tubers in potatoes or Jerusalem artichokes. Tuber development in these crops is stimulated by short days, whereas the bulb formation in onions and garlic are a long-day response. The runners of spider plants and strawberries result from long days of 12 to 14 hours or more. In some cases, plants have dual requirements of both photoperiod and temperature for dormancy, fall color, leaf fall, and cold hardiness.

Phototropism

The last growth response related to light is the bending of a plant toward or away from a light source. This is called *phototropism*, photo meaning *light* and tropism meaning *movement*. Recall a time where you may have noticed a houseplant sitting on a window ledge and growing toward the light. This is a positive phototropism, **Figure 10-9**.

Pat_Hastings/Shutterstock.com

Figure 10-9. Plants that grow toward the light are positively phototropic.

Philodendron and monstera are two tropical vines that are often grown as houseplants. These plants exhibit negative phototropism, which is a movement away from a light source. Another common example is the movement of sunflowers in response to the sun. The blooms track the sunlight throughout the day, moving their large flower heads as the sun travels across the sky.

Optimizing Light Quantity

Light quantity can be increased or reduced, depending on the needs of plants. Some light management strategies that growers may use include:

- Plant spacing and orientation.
- Greenhouse design.
- Greenhouse covers.
- Supplemental lighting.
- Plant selection.

Plant Spacing and Orientation

Plant spacing and orientation can maximize a plant's ability to intercept light. In field settings, a grower will use ideal spacing for a particular crop to increase yield as well as to shade weeds. Weeds can significantly reduce light received by crops, resulting in reduced plant quality and yield. If plant spacing is too close, the overcrowding of plants can reduce quality as leaves shade each other and stems elongate. This might happen when transplants are planted too close to one another or seeds are not thinned properly. In a garden situation, planting taller plants at the back of a border with medium and smaller plants toward the front will minimize any potential shading issues, **Figure 10-10**.

Similar spacing needs to occur in a greenhouse as well. Potted plants should be placed close enough together to use space efficiently and to have the proper amount of light reaching all parts of the plants. Most growers consider spacing acceptable when the leaf tips of adjacent plants barely touch. As plants increase in size, containers can be moved to keep proper spacing.

Paul Wishart/Shutterstock.com

Figure 10-10. With the tall plants in the back of the garden and short plants in the front, this garden minimizes any potential shading issues.

Pruning

Pruning can also help increase light interception and flowering in many plants. Thinning branches will allow greater quantities of light to penetrate into the interior areas of a plant. In fruit production, pruning will enhance yield, improve fruit color, increase air circulation, and minimize disease. Poor pruning can inhibit light interception by the plant and leave shaded sections that will have reduced vigor and vegetation. In a hedge, a wider base and narrower top will allow the best light conditions for the plants.

Greenhouse Design

Greenhouse design and orientation can maximize the amount of available light for plant production. Variables such as climate, latitude, and time of year impact levels of light intensity. The primary factor limiting crop production in greenhouses is low light intensity during the winter. Orienting single greenhouses situated above a latitude of 40°N with the ridge running east to west allows for the winter sun to reach into the greenhouse along the sides. Greenhouses located below the 40°N latitude can be oriented north to south because the angle of the sun is higher. Many greenhouses are connected along their length and, regardless of latitude, should be oriented north to south to avoid shadows.

Greenhouses located at high altitudes have an advantage of higher light intensity. Much of the greenhouse production of roses and carnations, both crops that require a high light intensity, are grown in the high mountains of countries such as Columbia and Ecuador, **Figure 10-11**.

Velychko/Shutterstock.com

Figure 10-11. High altitudes provide high light intensity, creating perfect growing conditions for valuable cut flowers such as roses. Does growing plants in high altitudes present additional challenges?

Greenhouse Covers

The material used to cover greenhouses plays a critical role in allowing light transmission. Common materials used to cover a greenhouse include:

- Glass.
- Polycarbonate.
- Polyethylene film.

STEM Connection

Plants Producing Light?

We know that plants need light to drive photosynthesis for normal growth and development, but what about an autotroph that produces its own light? Well, sort of! Clubmoss (Lycopodium) is an herbaceous perennial with "pyrotechnic" spores that can ignite like fireworks. The spores contain volatile oils that are highly flammable and create a brilliant burst of fire when ignited. Where does Lycopodium grow? Is the plant harvested for human use?

Kurkul/Shutterstock.com

Glass allows the highest levels of light to reach the crops. *Polycarbonate* is a type of thermoplastic polymer that can be used as a greenhouse covering. *Polyethylene* is a type of plastic that also can be used as a greenhouse covering. Both polycarbonate and polyethylene provide adequate light for many crops and can be less expensive than glass. In the summer months, light intensity can become too intense and growers may use paint to coat the greenhouse and provide shade. The paint can be removed at the end of the season. Some growers may place a shade cloth over a greenhouse or use it within the greenhouse to minimize light intensity.

Supplemental Lighting

During the dark winter months, greenhouse managers struggle to maintain levels of light intensity needed for most crops. Supplemental lighting helps increase rates of photosynthesis and results in increased plant quality, **Figure 10-12**. Four different types of lamps used in greenhouses include:

- High intensity discharge (HID) lamps include lamps such as high-pressure sodium (most commonly used) or metal-halide, the preferred light sources for plant growth and development in greenhouses. Their light is rich in orange light and deficient in blue and red light.
- Fluorescent lamps offer uniform light intensity with little levels of heat emitted. They are commonly used in seed germination chambers, but rarely used to finish a crop.
- Incandescent lamps are generally not used for supplemental lighting because they give off excessive heat, have poor light quality (red and far-red light), and have low efficiency. They are very useful in managing photoperiodic lighting as discussed earlier.
- LED lamps are a relatively new technology. While much research is still being done, they offer exciting potential for use in both photosynthetic and photoperiodic lighting. They have low energy emissions.

CreativeNature R.Zwerver/Shutterstock.com

Figure 10-12. Supplemental lighting helps growers increase the quality of their plants and allows them to extend growing periods.

Plant Selection

For a home gardener, plants should be chosen to fit a given light situation. If a home has a deeply wooded lot, the plant material that will thrive will be different from plants appropriate for a sunny location.

Figure 10-13. A—Full sun locations are ideal for fruit and vegetable production. B—Partial shade situations create ideal conditions for shade-loving ornamentals.

Gardeners should survey their landscape for lighting conditions and select plants that fit each environment accordingly. Most nurseries label plants according to their lighting needs. This list describes light classifications for a home gardener:

- Full sun areas receive direct sun for at least six hours a day, between 9 am and 4 pm, **Figure 10-13A**.
- Light shade areas receive significant amounts of direct sunlight, with the sun being blocked for two to three hours during the summer months. Light shade areas may also receive constant but filtered light through a fairly open canopy.
- Partial shade areas receive dappled sunlight filtered through trees. The amount of light that reaches plants depends on the density of the overhead canopy, **Figure 10-13B**.
- Full shade areas may receive reflected light, but other plants or structures block direct sunlight. Only shade-tolerant plants will thrive in full shade.
- Dense or deep shade areas receive very little indirect or reflected light. This situation can be found under a dense canopy of mature trees or under an elevated deck.

Temperature

From the icy, arctic winds of the tundra to the hot, arid reaches of the desert, plants have evolved to grow under a variety of temperatures. Most horticultural plants thrive in a temperature range between 50°F to 85°F (10°C to 30°C). Temperature requirements of a crop help determine in which climate they are best suited for growth, the best season in which to produce the plants, and ways to manage temperature. Examine **Figure 10-14** and observe the preferred temperature ranges of certain vegetable crops.

Q10 Temperature Coefficient

As noted in Chapter 9, temperature directly impacts rates of photosynthesis and respiration. For each 18°F (10°C) increase in temperature, the rate of reaction doubles. If a snapdragon is growing at temperatures of 50°F (10°C), it would be growing twice as fast if the temperature was 68°F (20°C). This relationship is known as *Q10*. Growers who want to manage crop timing for a target sale date or a holiday can increase temperatures to promote faster growth.

Corner Question

Can a shade plant grow in the sun?

Vegetable	Germination Temperature		
	Minimum	Optimum	Maximum
Cool-season vegetables prefer temperatures between 50°F to 80°F (10°C to 26.7°C).			
Beets	40°F (4.4°C)	80°F (26.7°C)	90°F (32.2°C)
Broccoli	40°F (4.4°C)	80°F (26.7°C)	90°F (32.2°C)
Cabbage	40°F (4.4°C)	80°F (26.7°C)	90°F (32.2°C)
Carrots	40°F (4.4°C)	80°F (26.7°C)	90°F (32.2°C)
Cauliflower	40°F (4.4°C)	80°F (26.7°C)	90°F (32.2°C)
Kohlrabi	40°F (4.4°C)	80°F (26.7°C)	90°F (32.2°C)
Leeks	40°F (4.4°C)	80°F (26.7°C)	90°F (32.2°C)
Lettuce (leaf types)	35°F (1.7°C)	70°F (21.1°C)	70°F (21.1°C)
Onions, dry (seed)	35°F (1.7°C)	80°F (26.7°C)	90°F (32.2°C)
Onions, green	35°F (1.7°C)	80°F (26.7°C)	90°F (32.2°C)
Parsnips	35°F (1.7°C)	70°F (21.1°C)	90°F (32.2°C)
Peas	40°F (4.4°C)	70°F (21.1°C)	80°F (26.7°C)
Potatoes	45°F (7.2°C)	–	–
Radishes	40°F (4.4°C)	80°F (26.7°C)	90°F (32.2°C)
Spinach	40°F (4.4°C)	70°F (21.1°C)	70°F (21.1°C)
Swiss chard	40°F (4.4°C)	85°F (29.4°C)	95°F (35°C)
Turnips	40°F (4.4°C)	80°F (26.7°C)	100°F (37.8°C)
Warm-season vegetables prefer growing temperatures between 65°F to 90°F (18.3°C to 32.2°C).			
Beans	50°F (10°C)	80°F (26.7°C)	90°F (32.2°C)
Cantaloupe	60°F (15.6°C)	90°F (32.2°C)	100°F (37.8°C)
Corn	50°F (10°C)	80°F (26.7°C)	100°F (37.8°C)
Cucumbers	60°F (15.6°C)	90°F (32.2°C)	100°F (37.8°C)
Eggplant	60°F (15.6°C)	80°F (26.7°C)	90°F (32.2°C)
Peppers	60°F (15.6°C)	80°F (26.7°C)	90°F (32.2°C)
Tomatoes	50°F (10°C)	80°F (26.7°C)	100°F (37.8°C)
Squash, summer	60°F (15.6°C)	90°F (32.2°C)	100°F (37.8°C)
Squash, winter	60°F (15.6°C)	90°F (32.2°C)	100°F (37.8°C)
Watermelons	60°F (15.6°C)	90°F (32.2°C)	110°F (43.3°C)

Goodheart-Willcox Publisher

Figure 10-14. Vegetables are classified as cool-season or warm-season vegetables, depending on their preferred growing temperatures.

Growers can reduce temperatures if a slower growth is preferred. Each species of plant has an optimum temperature for growth and development, along with maximum and minimum temperatures where plant growth stops and permanent injury may occur. Temperatures of either extreme may cause damage, including chilling or freezing injury, thermal-induced cellular damage, or death.

Plant Responses to Temperature

Many plants require specific temperatures to complete their life cycle. Cold temperatures are necessary for some plants to initiate flowers or cold hardiness or even for certain seeds to germinate. Low winter temperatures can be the factor that determines whether certain plant species can survive in a given area. The United States Department of Agriculture (USDA) has determined *plant hardiness zones*. These are areas identified across the United States based on average annual minimum winter temperatures, **Figure 10-15**. Zones are divided into 10° increments. Horticulturists have identified the low temperatures most trees, woody shrubs, and perennials can withstand and have assigned zones to them. A gardener will be able to decide

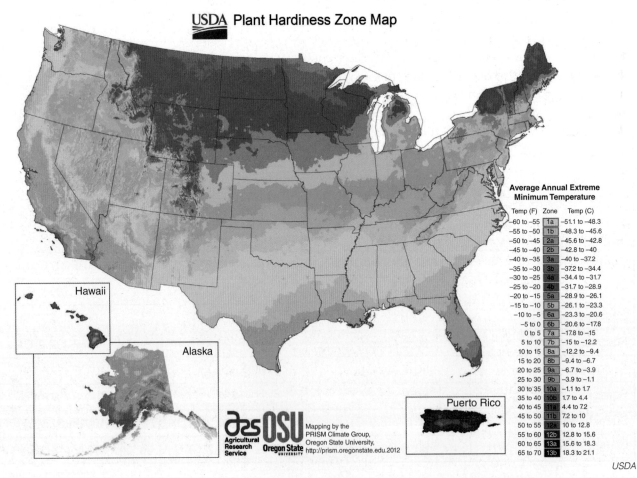

Figure 10-15. The USDA Plant Hardiness Zone map shows the average minimum temperature for each area of the country. Growers can use this information to determine which plants will grow best in a particular area.

whether a plant will be able to survive in the local area using the hardiness information. In certain extreme years, severe damage to plant material can happen if the temperatures are much colder than normal.

Dormancy

Dormancy is a condition in which buds and seeds are inhibited from growing until a certain environmental requirement is met. For many temperate species, the requirement is cold temperatures, **Figure 10-16**. Plants need to be exposed to a certain amount of low temperatures in order to break dormancy. These plants will not grow even if growing conditions such as temperature, water, and oxygen are all available and favorable. This is a survival mechanism to prevent plants from budding and growing at the wrong time, for example in the middle of a warm spell in January.

Montypeter/Shutterstock.com

Figure 10-16. Although buds on woody plants may appear dormant after completing their chilling requirement, they are merely waiting for warmer weather to begin growth.

Bud dormancy on woody, temperate plants occurs as days shorten (photoperiod) and low temperatures are present. The needed number of hours of chilling temperatures depends on the species. For example, apples require between 800 to 1100 chilling hours, whereas a rabbiteye blueberry may require only between 400 and 700 chilling hours. This means that some plants can grow only in certain areas. For example, apples cannot be grown well in many southern states because they cannot receive enough chilling hours.

Damage may also occur to hardy plants when an unusually early frost or freeze finds plants before they are fully dormant or in the spring after they have begun to lose dormancy. Many seeds also have dormancy phases that can be broken through horticultural techniques called *stratification*, a moist, chilling treatment.

Stress

When plants are exposed to temperature extremes, they undergo different stress responses. Stresses of low or high temperatures can cause a number of symptoms and even permanent injury or death in plants. The types of temperature stresses that can occur in plants include chilling injury, freezing injury, sunscald, and heat stress.

Chilling Injury

Chilling injury is a condition in which plants are damaged by low temperatures in the field or in storage. Typical symptoms may include discoloration, death of older leaves, defoliation, wilting, poor growth, lesions, or death. Chilling injury can lead to an increased susceptibility to microorganisms that cause rotting.

Freezing Injury

Freezing injury is a condition in which plants are damaged when low temperatures freeze the water in plant tissue, either in cells or on the surface of plant tissues. Plant death is the most severe freezing injury. The injury may be the death of only certain plant parts from which the plant may be able to overcome.

> "The farmer has to be an optimist or he wouldn't still be a farmer."
> —Will Rogers

> "A good farmer is nothing more nor less than a handy man with a sense of humus."
> —E. B. White

Sunscald

Sunscald is a condition in which plants are damaged due in part to extreme fluctuations and varied combinations of heat, cold, humidity, and sunlight. Sunscald is a fairly common injury where rapid heating and cooling occur on the southwest side of a tree. Damage may be caused to the plant, tree, or fruit. In colder areas, sunscald is a common winter tree bark injury caused by the combination of freezing nighttime temperatures and intense light reflected off the snow or other objects during the day.

Heat Stress

Heat stress is a condition in which plants are damaged due to high temperatures. Prolonged exposure to high temperatures can result in rapid plant death or localized damage to parts of a plant. Many fruits will exhibit damage on the side that is exposed to sunlight.

Hardening Off

Plants can harden, or acclimate themselves, to withstand temperature extremes. This happens naturally in many plants in preparation for winter cold. As fall temperatures decline, light diminishes, and water and nutrients may be limited. These factors can increase a plant's ability to accommodate cold temperatures. Hardening off is a practice used by horticulturists to minimize the potential for plant injury by slowly introducing them to a more extreme temperature environment. For example, tender transplants growing in favorable conditions of a greenhouse may be susceptible to chilling injury if immediately placed in the garden. Hardening a plant means slowly increasing (or decreasing) temperatures and exposure to target temperatures.

Vernalization

Foxglove sends its towering flower stalk forth only after cold temperatures have induced flower formation, **Figure 10-17**. *Vernalization* is exposure of plants to low temperatures in order to stimulate flowering. A *biennial* is a plant with a two-year life cycle, with flowering occurring in the second year followed by plant death. Some biennials, such as celery, onion, and hollyhock, must be exposed to low temperatures for flowering to occur. Horticulturists might force flowering by subjecting a plant to cold temperatures. Then the plant is placed in conditions favorable for growth, and flowering will take place. In some cases, vernalization will result in undesirable flowering. For example, onion sets are often placed at low temperatures for storage. When planted in the field, however, the exposure first to cold temperatures and then to higher temperatures results in flowering rather than bulb formation.

picturepartners/Shutterstock.com

Figure 10-17. Foxglove requires cold temperatures to initiate flower formation, a phenomenon called vernalization.

Thermoperiod

Thermoperiod is the relationship between day and night temperatures and plant growth. When photosynthesis exceeds respiration, growth occurs. If respiration rates are higher than photosynthesis, plant vigor declines and death can occur. Growers have traditionally managed temperatures in greenhouses by providing cooler nights to ensure that respiration rates slow down.

DIF Treatment

The effects of day and night temperatures can influence plant height and flowering times. This relationship has been termed *DIF* and refers to the difference between the night and day temperatures. Simply stated, DIF equals day temperature minus night temperature. If the day temperature is higher than the night temperature, the plants will grow taller through elongation of internodes on the stem. Conversely, if nighttime temperatures exceed daytime temperatures, plant growth can remain shorter and compact. A number of horticultural crops respond to this temperature management, including Asiatic lilies, dianthus, celosia, petunia, poinsettia, rose, salvia, snapdragon, tomato, and watermelon. Some plants do not respond to DIF treatments, such as aster, hyacinth, squash, and tulip. Using DIF methods reduces a need for chemical growth retardants as well. The use of temperature differences can be used to manage flowering time.

> "My dream is to become a farmer. Just a Bohemian guy pulling up his own sweet potatoes for dinner."
> —Lenny Kravitz

Degree Days

Horticultural crops vary widely in growth time from planting to harvest. All organisms, including plants, require a certain amount of heat to develop from one point to another in the life cycle. Each stage of development is calculated in units called degree days, and can be abbreviated as (°D). *Degree days* are the accumulated product of time and temperature for each day. Certain crops can be harvested after reaching a certain number of degree days. For example, vegetable growers can calculate an approximate time of harvest for peas by recording the number of degree days. The approximate number of degree days for peas is between 1000 and 1200. This formula can be used to calculate a degree day:

$$\text{Degree day (°D)} = \frac{\text{daily maximum temperature (°F)} + \text{daily minimum temperature (°F)}}{2} - \text{base temperature}$$

If a grower in Minnesota was managing a calendar for canning peas, he or she would use the base temperature of 40°F in the equation above. Degree days are also used for managing insect pests and can help a grower estimate the arrival of a potential pest.

Soil Temperatures

Soil temperature plays a key role in seed germination, root growth, water uptake, and disease susceptibility. Optimal soil temperatures depend on the crop, but cool-season plants generally prefer a maximum soil temperature of 77°F (25°C). Warm-season crops prefer a minimum soil temperature of at least 10°F (–12.2°C).

If temperatures are too cool for a seed, the seed can have little or reduced germination and may be susceptible to cold temperature diseases. Root growth responds similarly to seeds. If temperatures are too low or too high, growth is diminished. Much of a plant's responses to temperature depend on the plant species.

Managing Temperatures

To a certain extent, growers can prevent damage to outdoor plants caused by temperature. Greenhouses, of course, offer complete protection from damaging cold or hot temperatures through the use of heating and cooling systems. Chapter 18, *Greenhouse Operation and Maintenance*, discusses in greater detail the ways to manage these systems. In a field setting, damage from cold temperatures can be offset in a number of ways:

- Site selection.
- Season extension.
- Overhead irrigation.
- Localized heating.
- Wind machines.

Sedlacek/Shutterstock.com

Figure 10-18. The wine grape vines were planted in this vineyard to encourage air drainage, moving potentially damaging cold air out of the field.

Site Selection

Factors such as air drainage, orientation of slope, altitude, and location of large bodies of water influence temperature. These factors should be considered when selecting a location to raise crops. Cold air is heavier than warm air and will travel downhill. Warm air is less dense and rises uphill through convection. *Air drainage* is the process by which cold air sinks, flowing downhill to the lowest available point where it accumulates until dispersed by heat or wind. Orienting crops to provide proper air drainage minimizes potential cold damage. For example, in a vineyard, growers can take advantage of geography by planting on higher points of elevation and hills with a slight to moderate slope. This encourages cold air to move out of a field and to settle in an area of lower elevation, **Figure 10-18**. These low-lying areas (sometimes called frost pockets) accumulate cold air and can cause freezing or chilling injury to crops.

Microclimates

Localized areas with temperature differences are called *microclimates*. Microclimates can be helpful or harmful, depending on the crop being grown. A cold-tender fig tree may survive if grown against a brick wall that radiates heat and raises the air temperature a few degrees. Conversely, imagine the strip of land that often straddles the space between a sidewalk and street. The heat that may radiate from these concrete surfaces can make the area a very challenging environment in which to grow plants, **Figure 10-19**.

Slope Orientation

Growers with hilly land must consider slope orientation for the greatest benefit of their intended crop. *Slope orientation* refers to the direction the slope faces (north, south, east, or west). In cooler climates, growers plant on slopes facing south to maximize light exposure and heat accumulation. Northern, eastern, and northeastern slopes are preferred in climates with warm or hot summers and cold winters. These locations can reduce damage caused by rapid heating and cooling of stems that may cause bark splitting or trunk injury.

Joe Klune/Shutterstock.com

Figure 10-19. Planting between the sidewalk and street creates a microclimate that is hot, dry, and challenging. Are there microclimates around your home or school that present planting and growing challenges?

Bodies of Water

Large bodies of water can provide a more moderate temperature for the surrounding land. Water has a high heat capacity and offers slower temperature fluctuations, creating favorable conditions for many crops. In the United States, many fruit growers locate along large bodies of water, such as the Great Lakes, including along the eastern shores of Lake Michigan, Lake Ontario, and Lake Erie.

Season Extension

Many growers have learned that covering their crops with a protective barrier, such as fabric, plastic, or glass, can extend the season by increasing temperatures earlier or later into the season. At its simplest, growers might cover their crops with a special fabric that traps heat from the sun and raises the temperature a few degrees, **Figure 10-20**. It allows light and water to penetrate and can be easily applied and removed each day. It also has been used as an effective measure against pests.

Low and High Tunnels

Low tunnels are slightly more sophisticated structures that allow for winter production of hardy cool-season vegetable crops. A low tunnel is created by inserting plastic or metal conduits into the soil on either side of the bed. A row cover is placed on the crops, and plastic is pulled over the metal or plastic hoops.

Zeljko Radojko/Shutterstock.com

Figure 10-20. Row covers trap heat and provide frost protection to tender plants.

STEM Connection
Growing Strawberries in the Winter

If you live in California or Florida, the availability of locally grown strawberries in the winter is not unusual. But in other parts of the southeast and southwest, strawberry growers are beginning to produce this much loved and scrumptious crop during the dark days of the coldest months. Using a high tunnel to modify winter temperatures, strawberries can be planted inside and grown nearly eight to ten months out of the year. Challenges to develop a strawberry that is best suited for short days and fluctuating temperatures and to develop best management practices for winter production are still being ironed out. In the meantime, a taste of spring and sunshine can now be found locally throughout the south as cold wind blusters and even snow may sprinkle the ground.

Lowe Llaguno/Shutterstock.com

Crops can be maintained through the cool weather and be ready for an early spring harvest. High tunnels or hoop houses are large greenhouse-like structures that are often unheated. They provide a longer fall growing season and an earlier spring/summer season by providing warmer air and soil temperatures for crops to grow, **Figure 10-21**.

A Sever180/Shutterstock.com   B bibiphoto/Shutterstock.com

Figure 10-21. A—Low tunnels provide a way to keep plants from freezing during cold temperatures. B—High tunnels are unheated structures that extend the growing season by creating a warmer production environment during late fall, winter, and early spring seasons.

Gardeners have long used homemade systems to protect against cold temperatures. Many gardeners use cold frames, **Figure 10-22**, that warm the soil and permit seeds for plants such as spinach to germinate and begin to grow. Milk jugs with the bottom cut off or glass cloches (small glass domes) may also be used to shelter tender transplants. Both trap radiant heat from the soil and warm the air inside.

Overhead Irrigation

Irrigation systems, especially overhead sprinklers, may provide some protection against frost in early spring or fall. As water falls on the plants in cold temperatures, heat energy is released as it changes to ice. As long as liquid water is constantly freezing on the plant, the surface temperature will remain at or near 32°F (0°C). With variables like wind or inadequate irrigation rates, damage may be more severe than if no irrigation was provided. Using overhead irrigation as a means of protecting against early frost is a common practice in the southeastern United States, particularly to shield tender flowers of strawberries, **Figure 10-23**.

jeff gynane/Shutterstock.com

Figure 10-22. Cold frames are perfect spaces for starting cool-season vegetables a few weeks before your neighbor!

Ulrich Mueller/Shutterstock.com

Figure 10-23. Using sprinkler irrigation can prevent damage to strawberry crops that have already set flowers.

Localized Heating

Heaters deliver localized heat to raise air temperatures and protect crops. Heaters are commonly used for fruits in the citrus family as well as in avocado groves, **Figure 10-24**. Fuels such as propane, liquid petroleum, and natural gas are placed in small heaters spread throughout a grove. Heating fuel may also be distributed via a pipeline system. Heaters tend to be used only in situations where the costs of crop loss outweigh the costs of fuel and heating.

Wind Machines

Wind machines are large fans that mix the air within an orchard to raise the air temperature near the ground. This can fluctuate the temperature by two or three degrees, just enough to reduce injury from cold temperatures. Wind machines can be more economical than heaters. They may also be used with heaters to reduce the number of heaters needed. Wind machines are used primarily in the tree fruit industry.

Water

Did You Know?
Horticulture has been practiced since humankind's early days. A small clay tablet dated to 3500 years old was found in Iraq and gives early instructions on irrigation.

Did you know that 95% of a raw tomato is water? Water makes up 80% to 90% of herbaceous plants. Nearly half of woody plants are comprised of water. Water is essential for functions that drive plant growth and development. In Chapter 9, you learned that the water stream provides a vehicle for transport of inorganic nutrient ions from the soil through the xylem. Water enters plants through the roots. Once water is inside a plant, transpiration channels water and its solutes to where it is needed. Water is a key ingredient in photosynthesis as it combines with carbon dioxide to create simple sugars. Water also provides the rigidity needed for plant form and structure, called turgor pressure.

Linda Armstrong/Shutterstock.com

Randy Miramontez/Shutterstock.com

Figure 10-24. Using heaters in fruit orchards and vineyards can be an effective means to prevent freezing injury.

Root Zone

The soil surrounding the plant's roots, called the *root zone*, serves as a natural reservoir from which the plant draws moisture and nutrients. For vegetables, this area immediately surrounds the plant. For most trees and shrubs, the leaf canopy sheds rain and creates an area called a dripline. This roughly circular area is much like the drip area of an open umbrella. For these plants, the most active water absorption is in the dripline area and beyond. This area is where watering should occur. Most of the roots spread two to four times as wide as the plant's canopy.

Corner Question

How much land in the United States is irrigated for agricultural production?

Consumption Rates

Water consumption rates vary greatly among plant species. These rates are influenced by soil type, temperature, light intensity, rainfall, humidity, and wind speed. A good rule of thumb is to water deeply and less frequently (rather than watering lightly and often). Watering deeply will encourage deeper rooting of plants. For vegetables, this means about 1″ to 2″ (2.54 cm to 5.08 cm) of water per week. A cool-season turf may need about 2.25″ (5.7 cm) a week. A warm-season and drought-tolerant turf, such as buffalograss, may not need to be watered for a couple of weeks. Plants use three to five times as much water during the hot summer months as they do during the winter. By adjusting a watering schedule with the season and as significant changes in the weather happen, growers and gardeners can conserve water resources and maintain optimal plant growth.

Irrigation

Horticulturists have a long history of delivering water to plants through a practice called irrigation. *Irrigation* is applying water to land or soil to assist in growing plants. From the flooding of fields in ancient China to aquifers in Rome, water has long been understood as a vital need for optimal plant growth and development. The main types of irrigation are surface, sprinkler, and drip irrigation.

Surface Irrigation

Surface irrigation includes both furrow and flood irrigation. Furrows are small ditches between planted rows in fields. Water flows from a central supply source and floods the ditches, **Figure 10-25**. Crops are planted in raised beds to provide good aeration and drainage.

ksb/Shutterstock.com

Figure 10-25. Furrows create a pathway for water to flood a field and irrigate crops.

Flood irrigation is an overall flooding of an entire area. Water is released from a central basin or reservoir and allowed to flood the desired areas. This is the least efficient use of water, but it requires less labor than furrow irrigation. Flooding of greenhouse floors has proven to be a very efficient method of watering pots. Water flows in for a set amount of time and then is recaptured and reused.

Sprinkler Irrigation

Sprinkler irrigation uses overhead watering of crops. This method applies water to both the plant foliage and the soil. A number of different mechanisms distribute water to plants. These devices may include simple sprinkler attachments to a garden hose or much more complex systems with underground piping and pop-up spigots for lawns. For valuable crops, many fields have permanent piping for irrigation. A pump will pull water from a reservoir and provide water to the main line. Lateral lines run off the main line to reach the crops. Risers are smaller pipes that come off the lateral lines and hold a single or double nozzle that sprays the water. If you have ever traveled through the Midwest, you may have seen many large agricultural fields that use center pivot irrigation, **Figure 10-26**. The equipment has wheels that slowly rotate from a center point in a circle to provide water to a field.

Sprinkler irrigation can increase the chances for disease because it wets the foliage and can splash soil, and any spores or microbes it contains, onto the foliage. Black spot on rose plants is a fungal disease spread by spores, often splashed onto foliage through sprinkler irrigation.

Drip Irrigation

Low-volume drip irrigation systems provide water to plants in a targeted, efficient way. Water is slowly, but frequently, applied through irrigation tubes

> "Heliotrope. to be sowed in the spring. a delicious flower, but I suspect it must be planted in boxes & kept in the house in the winter. the smell rewards the care."
> —Thomas Jefferson

Jim Parkin/Shutterstock.com

Figure 10-26. Center-pivot irrigation mechanisms provide overhead watering to crops.

under low pressure, **Figure 10-27**. Less water is used, reducing costs and conserving water supplies. Fewer diseases occur because the foliage is not being wetted. Weed seeds that are outside the irrigation zone will not get watered, so weed growth is greatly reduced in areas watered through drip irrigation. Drip irrigation is used both in greenhouses and field settings. In the field, drip tubes are often laid under plastic mulch.

Careers in Environmental Horticulture

Whether your educational focus or practical experience is in science, business, conservation, or design, you can find a rewarding professional career in environmental horticulture. Two of these jobs, greenhouse manager and farm coordinator, are described in the following section.

Max Lindenthaler/Shutterstock.com

Figure 10-27. Drip irrigation uses less water than other irrigation methods, conserves water supplies, and reduces disease incidence.

Greenhouse Manager

A greenhouse manager is responsible for the daily operation of the greenhouses associated with a nursery or vegetable production operation. Daily operation tasks include watering, planting, fertilization, space allocation, growth media and soil mix preparation, insect and disease control, and establishing and monitoring environmental conditions. A greenhouse manager may also be responsible for developing and monitoring pest management programs as well as training the staff who will implement the program.

Depending on the operation, a greenhouse manager's responsibilities may also include financial and staff management. These responsibilities may include record maintenance, budget management, billing, hiring and supervising staff, safety training, establishing work schedules, and performance evaluations. An associate's or bachelor's degree in horticulture, with a minimum of two to three years of greenhouse experience and experience supervising other people, is often required for this job.

Farm Coordinator

A farm coordinator serves as a liaison between farm production and sales. As part of a farm that does direct marketing through CSAs, farmers markets, produce boxes, restaurant sales, and so on, the farm coordinator plans and manages its product procurement and sales, as well as the distribution of produce. Farm coordinators may also be responsible for bookkeeping, payroll, and some administrative tasks for the farm. This role is the farm's primary communicator and outreach with shareholders, members, clients, and the general community. Experience or a related degree in business management, agriculture economics, or horticulture marketing is recommended.

Career Connection
Debbie Roos
Sustainable Agriculture Extension Agent

On any given day, Debbie Roos can be found with a camera around her neck capturing the mysteries of the natural and agricultural worlds. She takes vivid images that reveal her fierce fascination with the environment, from nests of native digger bees to the recently unfurled wings of a monarch late in the season. Her agricultural documentation highlights the tastes of the season—spring onions, summer heirloom tomatoes, or a heap of greens ready for the stewpot. Posting these wonders to social media, Debbie strives to both spark curiosity and to inform her friends and followers about the intersection of stewarding our natural resources and supporting local farms in Chatham County where she works and lives. Holding the enviable position of Sustainable and Organic Agriculture Agent with North Carolina's Cooperative Extension Service, this is her mission: to engage the citizens of her area in understanding the importance of eating and buying good food as well as supporting the farmers that produce it in a way that builds environmental health.

With early roots in the Peace Corps, Debbie found she loved agriculture and she loved working with people. Seeking out experiences on organic farms, internships, formal training

Courtesy of Debbie Roos

in both anthropology and later in horticulture, she began to craft a skill set that found a home with Extension. Translating research from the land grants of NC State and NC A&T State into practical solutions for her local growers, she is constantly giving lectures, workshops, farm tours, and even camps for kids about sustainable agriculture. She is always busy: connecting with the community in her pollinator paradise demonstration garden, talking with a grower about a new crop or innovative practice, or simply thinking about new ideas. Debbie claims that "she would wilt on the vine," if she was stuck in an office. Over the years, her determined efforts have woven together a vibrant community of farmers, ranchers, restaurateurs, and consumers that enthusiastically support their local food economy and maintain the vitality of their natural resources.

CHAPTER 10 Review and Assessment

Chapter Summary

- Understanding the growing needs of plants and the ways to control essentials, such as light, temperature, and water, is key to uncovering the secrets of plant growth and development.
- Light is critical to plants for photosynthesis, some seed germination, growth habit, flowering, fruiting, dormancy, hardiness, plant movement, formation of storage organs, fall color, and leaf drop.
- Light quality influences photosynthesis as well as the type of growth a plant will exhibit. In some seeds, certain wavelengths of light are necessary for germination to occur.
- Light quantity measures the number of light particles that reach a plant. Each plant species has an optimal light quantity that maximizes photosynthesis for plant growth.
- Plants have growth responses based on the number of hours of light they receive each day. Each plant species has its own photoperiod requirements. Photoperiod is related to flowering, development of storage organs, stem elongation, fall color, and leaf fall.
- Horticulturists have developed strategies that optimize a crop's opportunity to intercept light. These strategies involve plant spacing and orientation, greenhouse design and covers, supplemental lighting, and plant selection.
- Warmer temperatures will increase rates of plant growth through increased photosynthesis and respiration.
- Plants have specific responses to temperature, including dormancy, stress to temperature extremes, and vernalization.
- Horticulturists manage temperature using different tools and techniques to create a desired growth response. In the field, this includes site selection, season extension, irrigation, heating, and air mixing. In the greenhouse, managing temperature includes heating and cooling.
- For plants, water is key for photosynthesis, nutrient transport, and support for cellular form and structure.
- Horticulturists apply water through various irrigation methods to deliver water efficiently to plants and to conserve water resources. The main types of irrigation are surface, sprinkler, and drip irrigation.

Words to Know

Match the key terms from the chapter to the correct definition.

A. air drainage
B. biennial
C. blanching
D. chilling injury
E. daily light integral (DLI)
F. degree day
G. dormancy
H. etiolation
I. freezing injury
J. heat stress
K. irrigation
L. juvenile stage
M. light quality
N. light quantity
O. microclimate
P. nanometer
Q. necrosis
R. photoblastic
S. photon
T. photoperiod
U. phototropism
V. plant hardiness zone
W. stratification
X. sunscald
Y. thermoperiod
Z. vernalization

1. A condition in which plants are damaged by low temperatures in the field or in storage.
2. The wavelengths of light that a plant receives.
3. A plant growth response in absence of light.
4. A small area with different environmental conditions than the surrounding area.
5. The duration of day length (the amount that light is present) and the relationship between the dark and lighted periods.
6. Death of plant tissue, usually resulting in dark brown or black coloration.
7. Exposure of plants to low temperatures in order to stimulate flowering.
8. A condition in which plants are damaged due to heat, cold, humidity, or intense sunlight.
9. A condition in which plants are damaged due to high temperatures.
10. A plant with a two-year life cycle, with flowering occurring in the second year followed by plant death.
11. The required number of heat units that a plant needs to have to reach a certain point of development, usually flowering or harvest.
12. The process by which cold air sinks, flowing downhill to the lowest available point where it accumulates until dispersed by heat or wind.
13. A condition in which buds and seeds are inhibited from growing until a certain environmental requirement is met.
14. A period of growth that occurs before plants reach adult form.
15. The unit of measurement used to quantify light wavelengths.
16. One of several areas identified across the United States based on average annual minimum winter temperatures.
17. The relationship between day and night temperatures and plant growth.
18. The physical movement of a plant or its parts toward or away from a light source.

19. A moist, chilling treatment used to break dormancy in seeds of certain plant species.
20. A characteristic of seeds that have a germination response to the presence or absence of light.
21. The practice of applying water to land or soil to assist in growing plants.
22. A condition in which plants are damaged when low temperatures freeze the water in plant tissue.
23. Keeping plants from light to prevent photosynthesis and to cause white tissue growth.
24. The amount of light intensity a plant receives throughout a given day.
25. The amount and duration of light emitted by the light source.
26. A light particle and a measure of light quantity.

Know and Understand

Answer the following questions using the information provided in this chapter.

1. In what plant processes does light play a role?
2. What is light quality and what wavelengths of light do plants absorb?
3. How does light affect seed germination?
4. What are two components of light quantity?
5. Why is the daily light integral important?
6. How do growers promote etiolation in plants such as cauliflower?
7. What is photoperiodism and how is it important for plant growth?
8. What is critical day length and what are two examples of plants that require a short-day photoperiod to flower?
9. What are some light management strategies that growers may use?
10. How does plant spacing impact light interception by plants?
11. What different types of supplemental lighting are used in greenhouses?
12. How does an increase in temperature impact rates of plant growth?
13. How is dormancy broken in plants?
14. What types of stresses can occur when plants are exposed to temperature extremes?
15. What does it mean for a plant to harden?
16. How do growers use the concept of DIF to manage growth?
17. How does tracking degree days help growers?
18. What are five ways that damage from cold temperatures can be offset?
19. Describe the functions of water in plant growth.
20. Briefly describe the three types of irrigation practices.

Thinking Critically

1. Why is it important to understand how environmental conditions affect plant growth and development?

2. You are visiting a botanic garden after recent freezing temperatures and notice that some bananas that had been growing are blackened and show signs of rot. A gardener is working nearby and cuts some other bananas down to the ground. Why did she not remove the bananas? What do you think happened to the banana physiologically and what management practices could the gardeners use to prevent this damage?

STEM and Academic Activities

1. **Science.** Research the soil conditions in your area. What kind of soil do you have? How might this influence your watering practices? What properties of soil can impact how water reaches the plant roots?
2. **Technology.** LED lighting is a promising area for providing light needed for photosynthesis and reducing energy costs. Create an experiment by growing plants under different types of light. What light quality and quantity do they provide? How will this impact the growth of your plants?
3. **Math.** A grower has asked you to help her be more efficient in using supplemental lighting. Visit at least three greenhouse supply companies online that sell these items. Compare their prices, including any discounts for the number of items needed. Make a chart listing the name of each company and its price for each item. Identify the lowest price for each item and calculate the total cost for the items.
4. **Language Arts.** Assume that you are a vegetable grower, and you keep a daily journal to help you remember information that you might need later. Today, you held consultations at two local restaurants about what crops they would be interested in purchasing. Write a journal entry about these consultations that includes all the information you might need in the future.
5. **Language Arts.** Plants have different growing requirements. Select a plant that you enjoy cultivating in your garden. Research the environmental conditions it needs to grow. Create a descriptive label that defines its growing needs so any gardener could grow it.

Communicating about Horticulture

1. **Reading and Writing.** Research a position description for a farm manager. Consider the skills and experiences that you would need to have to be a highly qualified candidate for the position. Create a résumé and cover letter that highlight these skills and experiences.
2. **Writing and Speaking.** Use line drawings on poster board to explain the skeleton and basic shape of at least three types of season extensions covered in this chapter. Indicate which types of materials are used to establish the growing conditions needed for your crops. Display the drawings as reference tools for the class and be prepared to explain to the class how each season extension technique works.
3. **Reading and Listening.** Divide into groups of three to four students. Have each person choose a concept in the chapter. Each person should report to the group about the main ideas of their concept. The other people should then share one idea they heard back to the group. Repeat until each person has a chance to report.

SAE Opportunities

1. **Exploratory.** Visit a vegetable grower and observe the ways the field is planted. Take note of the row orientation, plant spacing, and other methods that take advantage of managing environmental conditions.

2. **Experimental.** Create a plant growth experiment with different types of light quantity. What plant will you use? How will you create a spectrum of light quantity? What growth responses do your plants exhibit? Write a report on your findings and submit it to a national youth horticulture organization.

tony4urban/Shutterstock.com

3. **Experimental.** Conduct a postharvest experiment using bananas. Determine five different temperatures that you will place your bananas in. What do you expect to find? How will you collect data? What symptoms, if any, did you see the bananas show in response to the different temperatures?

4. **Exploratory.** Job shadow a horticulture production researcher. Ask questions about the research projects and the ways he or she is trying to improve production practices that bring down grower costs and increase yields.

5. **Placement.** Visit a local greenhouse and apply for a position or volunteer. Learn the different strategies for optimizing light, water, and temperature used at the greenhouse.

CHAPTER 11 Soils and Media

Chapter Outcomes

After studying this chapter, you will be able to:
- Discuss the importance of soils and media in horticulture.
- Describe the factors that shape the formation of soil.
- Describe the physical properties of soil.
- Describe the biological properties of soil.
- Identify the chemical properties of soil.
- Compare and contrast soil and soilless media.
- Describe inorganic and organic mulches.
- Discuss containers used in horticulture production.
- Discuss careers related to soil science.

Words to Know

adhesion	clayey soil	macropore	rhizobia
adsorbed	coconut coir	micropore	sandy soil
aeration	cohesion	mulching	saturation
A horizon	compost	mulchmat	slow-release fertilizer
anion	cover crop	mycorrhiza	soil auger
bark	desorbed	organic matter	soilless media
B horizon	electrical conductivity (EC) meter	parent material	soil pH
biochar		pedologist	soil pore space
bioplastic	field capacity	pedology	soil probe
brown waste	geotextile	peds	soil structure
bulk density	gravitational water	perlite	soil survey
capillary water	green waste	pH paper	soil texture
cation	horizon	plasticulture	surface horizon
cation exchange capacity (CEC)	hydrometer	porosity	topsoil
	infiltration	pyrolysis	vermiculite
chlorosis	ion	relief	weathering
C horizon	loamy soil	respire	

Before You Read

Write down everything you know about soil and planting media. It is okay if you do not know much about the topic; just write down what you are certain you know. Look at the Words to Know list above. Write down definitions of the words you know. Put the words you cannot define into a separate list. Finally, skim through the chapter and write down the topics that you think you might know well and those that you do not know much about. Once you are finished, work with another student and share your lists.

While studying this chapter, look for the activity icon to:

- **Practice** vocabulary terms with Words to Know activities.
- **Expand** learning with identification activities.
- **Reinforce** what you learn by completing Know and Understand questions.

www.g-wlearning.com/agriculture

When you look around your indoor environment, many things you see are related to the soil. Some of the clothes you are wearing come from cotton grown in a southern soil. The paper you write on comes from a tree, which once had its roots deep in the soil. Even the computer you use contains silicon, obtained from sand. Soil is the foundation of life because everything grows in or feeds off what grows in the soil.

Soil is an integral part of human existence and is important for both agricultural and nonagricultural uses:

- Soil allows us to grow food, such as delicious, juicy strawberries.
- Soil provides a home to billions of earth-dwelling organisms.
- Soil serves as a filter of natural and man-made wastes.
- Soil helps regulate water supplies.
- Soil provides a foundation for roads, buildings, and other structures.

As you can see, soil is an important part of everyday life. So, what exactly is this substance, and how is it formed?

What Is Soil?

Soil is one of the world's most important natural resources. Soil, air, and water are the foundation of life on planet Earth. Soil consists of four main components: air, water, minerals, and *organic matter* (living and dead organisms). The minerals and organic matter make up the solid parts of soil. Air and water are in the *soil pore spaces*, or the gaps among solid soil components. A typical soil will be 50% pore space, 45%–50% minerals, and 0%–5% organic matter. Variations of these components give soil its physical, chemical, and biological properties, **Figure 11-1**.

Soil Formation

Soil formation is a long, slow process. For example, some of the soil in your garden was beginning to form more than 500 years ago, right around the time Leonardo da Vinci was painting the *Mona Lisa* and Michelangelo was sculpting the famous *David*.

Soil-Forming Factors

Five soil-forming factors shape the development of soil. These factors are the reasons soil differs from the top to the bottom of a hill or why soil found on a beach has different characteristics from the soil found in a wetland or farm field. Soil formation begins with *weathering*, which is the physical,

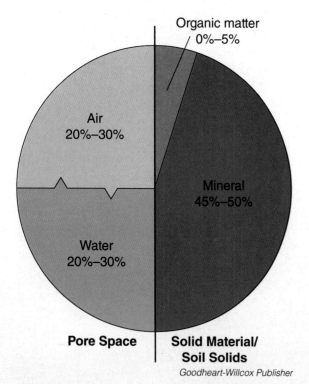

Figure 11-1. This pie chart shows percentages of soil components.

chemical, and biological process that creates soil. The rate of weathering is determined by five factors:

- Climate—weather conditions, including both temperature and rainfall.
- Organisms—microorganisms that break down organic matter.
- *Relief*—an environment where the rocks and soil sit on the landscape.
- *Parent material*—the substances (bedrock, sediment, or organic material) that are weathered to form soil.
- Time—the length of time during which soil is formed, impacting soil composition.

Corner Question

How long does it take to form an inch of soil?

Climate

In addition to temperature fluctuations and variations in rainfall, the climate includes factors such as sun and wind exposure. Consistent exposure to the heat of the sun and high winds will hasten the breakdown of parent material.

Organisms

Organisms, such as plants, animals, and microorganisms, all play a role in soil formation. Microorganisms break down organic matter and animals and plants help mix the soil. The combination of these actions forms or changes the soil.

Relief

Relief, or topography, contributes to soil formation by influencing soil wetness, soil temperature, and the erosion rate. Parent material on the ridge of a mountain may break down quickly due to its complete exposure to the elements.

Parent Material

Soil formation begins with the weathering of rocks, which are considered a parent material. The parent material comes from bedrock or transported sediments. Parent material can also be organic material (such as in deep peat bog soils). Parent material greatly influences a soil's physical and chemical properties.

Time

The length of time a soil has to form will impact its composition. Some elements may be washed or blown away before contributing to soil formation in a particular area.

Horizons

As soils develop, layers begin to form. Have you ever dug a really deep hole or observed a cut in the landscape along a road that exposes a profile of soil? Look closely, and you will see layers of soil called *horizons*, **Figure 11-2**. A horizon is a layer of soil distinguished by properties and characteristics developed through the five factors of soil formation.

Antonov Roman/Shutterstock.com

Figure 11-2. The horizons or layers of a soil profile can vary within inches of one another. The top layers of the horizons have the most organic matter. Topsoil is an extremely valuable natural resource.

AgEd Connection: Land Judging CDE

Land Judging is a national career development event. Local teams judge soils by identifying soil properties, such as texture, slope, erosion class, permeability, and surface runoff, and make a determination of land capability class. Teams use this information and decide on recommended land treatments that include vegetative, mechanical, fertilizer, and soil amendments. The USDA Natural Resources Conservation Service has a rangeland resource program that provides soil education for students. This NRCS soil scientist is giving student participants a practice day in preparation for a land-judging competition.

USDA Natural Resources Conservation Service

Soil horizons have crucial implications for plant growth. The fertile, upper, outermost layer of soil is called the *topsoil*, *surface horizon*, or *A horizon*. This layer is the most important for plant growth. Below the surface horizon is the *B horizon*, or subsoil layer, which contains much of the original parent material that has been weathered. The final horizon is the *C horizon*, the subsoil layer that contains much of the original parent material that has not been weathered.

The factors that affect soil formation also influence physical properties, such as soil texture, structure, and color, as well as its biological properties and chemical properties, such as pH.

Physical Properties of Soil

Soil physical properties include texture, structure, soil water, and color. They provide key information about a given soil and ways in which the soil might be used or managed.

Soil Texture

Soil texture refers to the proportion of different sizes of mineral particles present in a soil. Soil texture is the most important physical property to consider when managing soils for cultivation. The three soil particles are sand, silt, and clay. Have you ever built a sandcastle? Remember the gritty feel of the sand? The coarse feel of sand comes from its large particle size. Have you ever stepped in mud and had the mud cling to your boots? Likely, the mud contained a certain amount of clay. Clay has the smallest particle size, but it has a physical structure that makes it sticky. Sand has the largest soil particles at 2 mm–0.05 mm, followed by silt at 0.05 mm–0.002 mm, and then clay at less than 0.002 mm. Most soils are combinations of these particles. The particles of various sizes and their proportions have different physical characteristics that significantly influence plant growth potential. These characteristics include:

- Water-holding capacity—how much water the soil can hold at any given time so it will be available for uptake by plants.

- *Aeration*—the process by which air is circulated, in this case to provide oxygen for root respiration.
- Drainage rate—the rate at which water passes through the soil.
- Compaction potential—how easily the soil compacts or compresses.
- Erosion potential—how easily topsoil is lost through erosion or tillage after rainfall.
- Nutrient retention—the ability to store plant nutrients and resist pH changes.

Textural Classes

Soil textures can be divided into three broad groups: sandy soils, clayey soils, and loamy soils. These groups provide a framework for horticulturists to understand the general nature of these soils and their requirements for management.

Sandy soils are dominated by a significant proportion of sand. They have large-sized particles (2.0 mm–0.05 mm) and pore spaces that allow good aeration. Sandy soils allow water to infiltrate and permeate the soil. Sandy soils till easily and warm up quickly in the spring. Certain horticultural crops, such as sweet potatoes, blueberries, and sugar beets thrive in sandy soils, **Figure 11-3**. Sandy soils, however, usually have low organic matter, low native fertility, and low capacity for holding both water and nutrients.

Clayey soils have the finest particle size (less than 0.002 mm), which contributes to slow drainage and poor aeration. Clayey soils are also easily compacted. However, clay has a high water-holding capacity and an excellent ability to hold plant nutrients. It does not erode easily if it is well-aggregated. Plants that thrive in clay soils include hollies and pines.

Loamy soils have a fairly even mixture of sand, silt, and clay particles. A loamy soil has moderate water-holding capacity and a strong ability to store plant nutrients. Loamy soils present the most ideal medium for horticultural production. Loamy soils support the broadest range of plant possibilities, including most fruit and vegetable crops. Loamy soils do not aggregate well, making them the most susceptible to erosion by water or wind.

Figure 11-3. Tap-rooted vegetables, such as sugar beets, perform well in sandy soils.

Determining Soil Texture

Suppose you decide you would like to grow sweet potatoes, and you know they thrive best in a fine, sandy loam. How would you figure out what kind of soil texture you have? Soil texture can be analyzed through laboratory methods using a hydrometer. A *hydrometer* is an instrument used to measure the percentages of sand, silt, and clay in a sample to determine the soil textural class, **Figure 11-4**. You could also use a field technique called *soil texture by feel*, **Figure 11-5**.

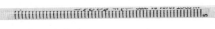

Figure 11-4. A hydrometer is used to measure particle sizes of soil: sand, silt, and clay.

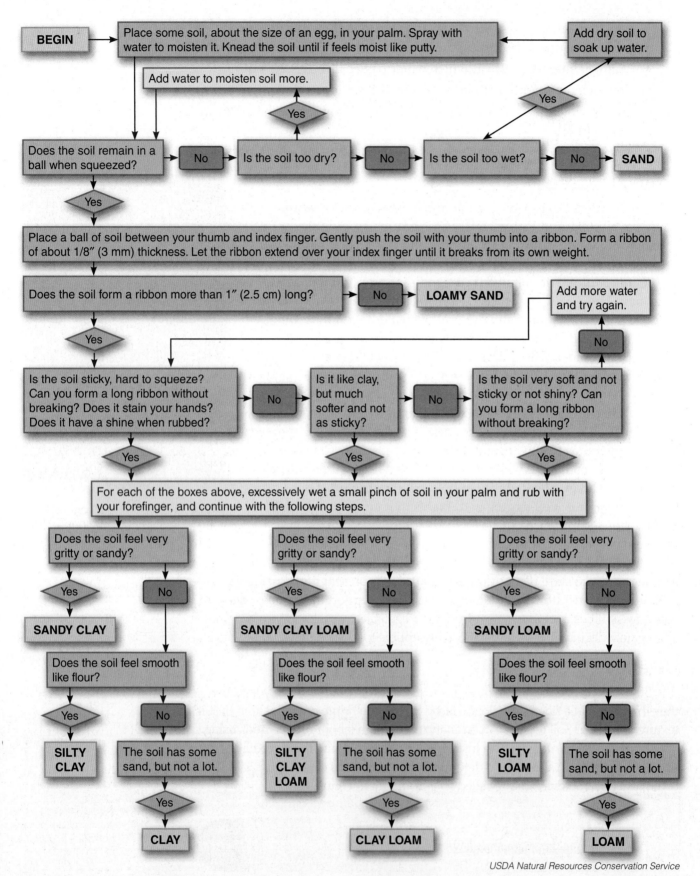

Figure 11-5. This diagram will help you learn to identify your soil texture using the *soil texture by feel* technique. What type of soil texture is in your home or school garden?

Scientists who study the soil, or *pedologists*, have years of experience and practice in the field classifying soils for *soil surveys* (comprehensive studies of the soil of an area), on-site waste applications (septic systems), or land classification. They can tell you almost the exact percentage of sand, silt, and clay in a sample. To determine soil type, start by calibrating your figures with known soil textures.

It is important to understand that the textural properties of soil do not influence soil management decisions. Rather, management decisions are focused on the improvement of soil through soil amendment. As a horticulturist, it is important to understand how to use a soil's texture to optimize crop production, **Figure 11-6**.

Soil Structure

Soil structure refers to binding together of sand, silt, and clay particles into aggregates called *peds*. *Soil texture* refers to the amount of sand, silt, and clay present in a given soil; whereas *soil structure* refers to the aggregation of these soil particles.

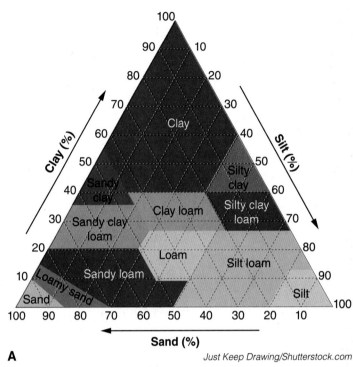

A

Just Keep Drawing/Shutterstock.com

Influence of Soil Texture on Properties and Behavior in Soil			
Property/Behavior	**Sand**	**Silt**	**Clay**
Water-holding capacity	Low	Medium to high	High
Plant available water	Low	High	Medium
Aeration	Good	Medium	Poor
Organic matter decomposition	Fast	Medium	Slow
Organic matter content	Low	Medium to high	High to medium
Water erosion	Low	High	Low
Ability to store plant nutrients	Poor	Medium to high	High
Drainage rate	High	Medium to slow	Slow to very slow
Porosity	Low	Medium	High
Pollutant leaching	High	Medium	Low
Bulk density	High	Medium	Low
Compactibility	Low	Medium	High
Warming in spring	Fast	Moderate	Low
Infiltration	Fast	Medium	Low

B

Goodheart-Willcox Publisher

Figure 11-6. A—Properties of sand, silt, and clay are shown here. B—Soil texture influences soil behavior.

Corner Question

Where does the term *heavy soil* come from?

Soil structure affects water movement, including *infiltration* (how the water moves through the soil) and surface runoff. Soil structure plays a role in aeration, heat transfer, and *porosity* (the state of having space or gaps). Consider a soil structure that limits rooting depth of a landscape tree. What will happen to the tree? With a diminished depth for roots to grow, shallow root development will occur. This increases the likelihood that the tree may fall due to wind. Rainfall will not easily infiltrate the soil surface, increasing potential for runoff and erosion.

A well-structured soil has ample pore space, which allows water and air to move easily into and through the soil. This enhances a plant's uptake of water and nutrients. Good structure increases water infiltration through the soil's surface and minimizes erosive potential.

Managing Soil Structure

Unlike soil texture, structure can be strengthened through good conservation practices or destroyed by mismanagement. Organic matter can be added to soil by using cover crops or applying compost. A *cover crop* is a crop planted to add nutrients to the soil and manage erosion. *Compost* is decomposed organic matter, such as animal manure or food wastes, used to add nutrients to soil. This material acts as a binding agent, clumping soil particles together and improving the structure. Agricultural practices such as conservation tillage, where a farmer limits tillage by planting directly into the soil with a seed drill and reduces the passes of a tractor across a field, also give soil the opportunity to build structure. Mismanagement of soils can result in compacted soil that has little pore space for air and water. In addition, poor management practices contribute to weak soil structure that falls apart and causes limited infiltration, ponding, or standing water. Poor structure may also be caused by heavy construction or agricultural machinery and rainfall on bare soil. Wet soil is more susceptible to compaction than dry soils.

Soil Bulk Density

Bulk density is the mass of a given volume of dry soil that takes into consideration the solid and pore spaces of a soil. This is helpful for evaluating soil for attributes, such as total porosity, aeration, and water content for plant uptake. An increase in bulk density through soil compaction adversely affects plant production. Soil compaction diminishes the capacity of soil to take in water. This decreases the water available to the plants and also limits the potential for root growth, **Figure 11-7**. Long-term, intense tillage may increase bulk density by destroying soil structure and pore space and by depleting organic matter that helps bind soil particles together.

Svend77/Shutterstock.com

Figure 11-7. Long-term, intense tillage of soil can negatively impact the health of the soil. Today's farmers use a number of tillage practices and production methods to promote the health of the soil.

> ### Hands-On Horticulture
> **Water Movement through Soil**
>
> Take two empty tennis ball containers, and drill five holes in the bottom of each one. Fill one tennis ball container with a sandy soil and the other with a clayey soil. Make sure the soil particles are completely dry. Place each soil-filled container on top of a clear cup. Make a hypothesis about which container water will flow through the fastest. Why?
> Pour equal amounts of water into each container and observe the movement of water as it infiltrates the soil surface and as it moves downward through the soil profile. What is happening? Was your hypothesis correct? Why or why not?

Good soil structure may be promoted and maintained through soil management techniques such as:
- No-till practices.
- Minimizing traffic and confining it to dry soil.
- *Mulching* (adding a layer of material over soil surfaces).
- Adding crop residues, compost, or manures to stabilize soil aggregates.
- Using cover crops to promote good soil structure and lower bulk densities.

Soil Water

Plants rely on the availability of water in the soil to maintain vital processes for growth and development. A soil's physical properties, including texture, structure, and the nature of the pores (bulk density), greatly influence the movement of water and its availability for plant uptake. There are two categories of water in the soil: gravitational and capillary. *Gravitational water* is free water that moves through the soil by the force of gravity. Gravitational water infiltrates the soil or runs off the soil surface and is generally not available to plants. Gravitational water occupies the *macropores*, or large pore spaces in the soil. *Capillary water* is held in the soil against gravitational pull, and most of this water is available to plants. Capillary water is found in the *micropores*, or small pore spaces. It moves as a thin film around soil particles through the attractive forces of *adhesion* (water molecules binding to soil particles) and *cohesion* (water molecules binding to themselves). The plant roots use this water until the attraction between the water molecules and soil particles is stronger than the pull of the roots. *Field capacity* is the maximum amount of water that a soil can hold against the pull of gravity. A soil's water-holding capacity is the amount of water that a given soil can hold for crop use. For example, a sandy textured soil has a low water-holding capacity and would be affected more by drought conditions. A clayey soil has a high water-holding capacity.

Soil Color

Soil characteristics are important when considering how to use and manage land for horticultural production. Consider a handful of soil. If you observe it closely, the first thing you might notice would be the color. Is it brown? Red? Yellow? Yellowish-brown? Soil color ranges from green-blues to browns and blacks, to reds, yellows, and oranges. Although the color itself will not determine how you may use the soil, what it reveals about the soil will.

Soil colors give clues about the soil's composition and site characteristics, such as drainage and water content, mineral weathering, and biological activity. For example, well-drained soils have bright colors. Soils with a high water content and poor drainage have very gray colors. Soils with a deep, rich, black color usually have a healthy presence of organic matter that has coated the soil particles, rendering them a dark, black-brown color, **Figure 11-8**.

So how does a soil's color affect the grower's decision on what to plant? If the soil cannot be amended and changed easily or quickly enough to accommodate a desired crop, the grower may need to plant something else. For example, when growing tomatoes, you will want the soil to have a deep, rich, black color that indicates the presence of organic matter, good drainage, and high nutrition content. If you plant tomatoes in a gray soil with a high water content, there will be poor growth or crop failure.

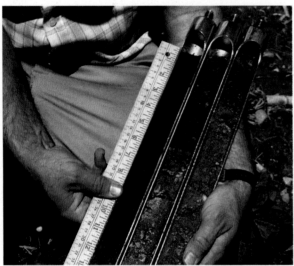

Photo by Lynn Betts, USDA Natural Resources Conservation Service

Figure 11-8. Here are three soil samples taken from different topsoils. Besides color, what do you think would be some of the differences between the soils?

Biological Properties of Soil

The living organisms that dwell in the soil influence soil properties. They are responsible for the biological activity within the soil that provides services such as:

- Decomposition.
- Carbon recycling.
- Nutrient production.
- Pore formation.

The multitude of organisms that call the soil their home include microorganisms, soil-dwelling insects, and burrowing animals, **Figure 11-9**. Their individual behaviors and interactions keep soil healthy and help plants grow.

Microorganisms

Microorganisms abound in soil. In fact, there are more microorganisms living in a teaspoon of soil than there are people living on Earth. These microorganisms include bacteria, actinomycetes, fungi, algae, protozoa, and nematodes. Many microorganisms produce gums and gels that help bind soil particles together and improve soil structure. Many break down not only plant and animal residue, but also chemical structures such as pesticides.

Corner Question

What does soil color tell you?

Bacteria, particularly *rhizobia*, play a vital role in horticultural production. This bacteria converts elements such as nitrogen into a usable form that can be taken up by a plant. Fungi spread throughout the soil and actively recycle nutrients by breaking down organic matter. *Mycorrhiza* is a fungus that grows in association with the roots of a plant in a symbiotic or mildly pathogenic relationship. Such fungi form a relationship with plants by providing plants with nutrients, such as nitrogen and phosphorus, in return for direct access to carbohydrates. Scientists continue to uncover the roles of living organisms in the soil and the impact they have on soil health as well as its potential for plant production.

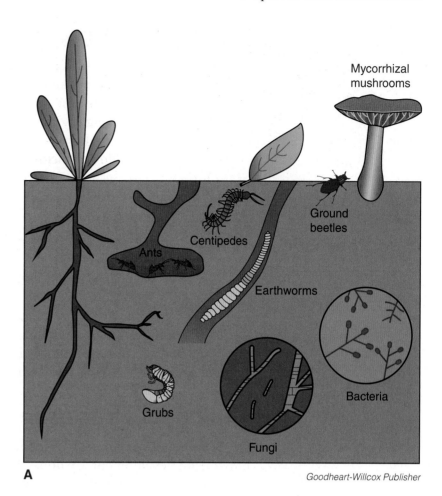

Goodheart-Willcox Publisher

Soil Organisms Concentrations in Soil per Square Meter		
Soil Organisms	**Examples**	**Quantity (per square meter)**
Vertebrates	Gophers, mice, moles	1
Arthropods	Ants, adult beetles and beetle larva, centipedes, grubs, maggots, millipedes, spiders, termites, woodlice	5000
Annelids	Earthworms	3000
Mollusks	Snails, slugs	100
Nematodes	Nematodes	5,000,000
Rotifers	Rotifers	10,000
Protozoa	Amoeba, ciliates, flagellates	10,000,000,000
Algae	Greens, yellow-greens, diatoms	1,000,000,000
Fungi	Yeasts, mildews, molds, rusts, mushrooms	100,000,000,000
Actinomycetes	Many kinds of actinomycetes	1,000,000,000,000
Bacteria	Aerobes, anaeraobes	10,000,000,000,000

Goodheart-Willcox Publisher

Figure 11-9. A—Various organisms living in soil. B—Population counts of soil organisms.

Soil-Dwelling Insects

Soil-dwelling insects, such as termites and ants, move soil around and aid in soil mixing, which can lead to improved soil aeration and water infiltration. Other soil-dwelling insects, dung beetles for example, will take animal manure from the soil surface and bury it deep into the soil to provide nutrients for their developing brood, **Figure 11-10**. Plants are able to draw nutrients from the manure as it decomposes and is processed by the young beetles.

Burrowing Organisms

The tunneling of burrowing rodents, such as moles, gophers, and prairie dogs, can provide spaces for gas exchange and for water flow. Their excrement and bodies also contribute to soil health as they decompose below the surface horizon.

In addition to moving soil around and aiding in soil mixing, earthworms ingest soil particles and organic residues. They enhance the stability of soil aggregates and increase the availability of plant nutrients.

Alta Oosthuizen/Shutterstock.com

Figure 11-10. Dung beetles will take animal manure and roll it to a desired location before burying it to provide nutrients to their young.

Chemical Properties of Soil

Soil provides a platform for plants to obtain essential nutrients for growth and development. By understanding the cycles of nutrients in the soil, one can better understand his or her own role in managing this system and optimizing it for plant production. Nutrients are added to the soil through biological processes as soil-dwelling organisms break down organic matter or transform minerals to forms plants can use. Nutrients are also added to soil with synthetic fertilizers.

Nutrient uptake is a function performed by a plant's roots. The roots absorb cations and anions that have dissolved in the soil water. *Ions* are atoms or molecules that have lost one or more valence electrons and become positively charged *cations* or negatively charged *anions*.

An *electrical conductivity (EC) meter* is a tool that measures the amount of soluble salts in a sample of soil or media. The salts are dissolved nutrient ions. This information can be used in management decisions that impact plant growth.

Mobile anions, such as forms of nitrogen, sulfur, and chlorine, are not strongly attracted to soil particles, but they are very soluble in water. Immobile nutrients do not readily move through the root zone to root surfaces. As a root pulls nutrients from a solution, nutrients are *adsorbed* (taken up and held by a soil particle) and *desorbed* (released) to the soil water solution for availability by plants to uptake. This process is the key chemical reaction for making nutrients available to plants and is known as the *cation exchange capacity (CEC)*, or the amount of cations that a soil can hold. The CEC is tempered by the ability or capacity of a soil particle's charge to hold cations. Soils with clay particles and organic matter have higher CECs.

Corner Question

What do you think a soil called *fairy dust* looks like? Where do you think this type of soil could be found?

Soil pH

The nutrient ions previously discussed have positive and negative charges that influence the acidity or alkalinity of a soil, otherwise known as the *soil pH*. Soil pH may be determined through the use of *pH paper*. Once exposed to the soil sample, the paper test strip reacts and changes color to indicate the pH level. In soil solutions where cations (hydrogen ions, H+) predominate, the soil will be acidic or have a pH lower than 7. In alkaline soils, or soils with a pH higher than 7, anions (hydroxyl ions, OH−) are more numerous. If an equal proportion of both ions is present, the soil is considered to be neutral.

Soil pH influences plant growth by indirectly affecting the availability of plant nutrients for uptake. It also influences the activity of microorganisms in the soil. Some nutrients are available to plants over a broad range of pH, whereas other nutrients are more sensitive to fluctuations in pH, **Figure 11-11**. For example, nutrients such as iron, zinc, manganese, and copper are very soluble at low pH and readily available to plants. However, aluminum, an essential plant nutrient, becomes toxic to plants at a low pH as it becomes more soluble. Most horticultural plants can thrive in a pH range of 5.5 to 7.0, **Figure 11-12**.

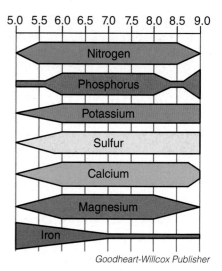

Figure 11-11. Availability of nutrients as mediated by pH. What types of organic materials can you add to soil to increase the pH? What types would you use to lower the pH?

Optimal pH Ranges for Plant Growth			
Herbaceous Plants			
Asparagus	6.0–7.5	Muskmelon	6.0–6.8
Beans	6.0–7.0	Oat	6.0–6.8
Beet	6.5–7.5	Okra	6.0–6.8
Bluegrass, annual	6.5–7.5	Onion	6.0–7.0
Broccoli	6.0–7.0	Pea	6.0–7.5
Cabbage	6.5–8.0	Pea, sweet	5.5–7.5
Cantaloupe	6.0–7.5	Peanut	6.0–6.8
Carrot	5.5–7.0	Pepper	5.5–7.0
Cauliflower	7.0–8.0	Potato	4.8–6.5
Celery	5.8–7.0	Pumpkin	5.5–7.5
Corn	5.5–7.5	Radish	6.0–7.0
Cucumber	5.5–7.5	Rhubarb	5.5–7.0
Eggplant	5.5–6.0	Spinach	6.0–7.0
Fescue	6.0–7.0	Squash	6.0–7.0
Garlic	5.5–8.0	Strawberry	5.5–6.5
Kale	6.0–6.8	Sunflower	6.0–7.5
Kohlrabi	6.0–6.5	Sweet potato	5.2–6.0
Lettuce	6.0–7.0	Tomato	5.5–7.5
pH ranges and level of acidity: 4.0–5.0 strongly acidic; 5.0–6.0 moderately acidic; 6.0–8.0 slightly acidic and slightly alkaline			

Goodheart-Willcox Publisher (Continued)

Figure 11-12. Optimal pH ranges for common garden plants. Can you estimate the pH of soil by the types of natural flora growing in the area?

Optimal pH Ranges for Plant Growth (Figure 11-12, continued)

Trees and Shrubs

Plant	pH Range	Plant	pH Range
American holly	4.5–5.5	Lilac	6.0–7.5
Apple	5.5–6.5	Magnolia	5.0–6.0
Arborvitae	6.0–8.0	Maple, sugar	6.0–7.5
Ash	6.0–7.5	Oak, black	6.0–7.0
Azalea	4.5–5.5	Oak, pin	4.5–5.5
Beech, American	5.0–6.5	Oak, red	4.5–5.5
Birch	5.0–6.0	Peach	6.0–6.8
Blueberry	4.5–5.5	Pear	6.0–6.8
Cherry, sweet	6.5–8.0	Pine, red	5.0–6.0
Clematis	5.5–7.0	Pine, white	4.5–6.0
Crabapple	5.5–7.0	Raspberry, black	5.5–6.5
Dogwood	6.0–7.0	Raspberry, red	6.0–7.5
Douglas fir	6.0–7.0	Rhododendron	4.5–7.0
Hemlock	5.0–6.0	Sumac	5.0–6.0
Honey locust	6.0–8.0	Walnut, black	6.0–8.0
Hydrangea, blue-flowered	4.0–5.0	White spruce	5.0–6.0
Hydrangea, pink-flowered	6.0–7.0	Willow	6.0–8.0
Juniper	5.0–6.0	Yew	6.0–7.0

pH ranges and level of acidity: 4.0–5.0 strongly acidic; 5.0–6.0 moderately acidic; 6.0–8.0 slightly acidic and slightly alkaline

Soil pH is dynamic and can be altered with the addition of acid or base materials. In certain areas, the climate naturally creates acidic soils that need to be managed with the addition of a base such as lime. In other parts of the world, alkaline soil or water raise the pH slowly over time. Through incorporation of an acid such as sulfur (also a plant nutrient), optimal plant growth can be achieved. Horticulturally important crops that prefer strongly acidic soil environments include members of the heath family (*Ericaceae*), such as blueberries, cranberries, lingonberries, azaleas, and rhododendrons.

Soil Testing

Knowing the nutrient levels and pH of a soil can help in attempting to manage its fertility. Using a *soil probe* (a tool used to pull a soil sample from the earth), horticulturists extract soil samples and send them to a lab for analysis. The analysis will reveal nutrient deficiencies or toxicity and provide suggestions for management. Nutrient deficiencies are more common than toxicities.

Nutrient deficiencies limit plant development and growth. Plants experiencing nutrient deficiencies show symptoms that range from stunted growth to *chlorosis* (a yellowing of the leaves) or necrosis (deadening of plant tissue). The results from a soil test will indicate the lack or abundance of different nutrients. This information will help determine the nutrients and concentrations that need to be applied for an optimal balance for the intended horticultural use (landscape, turf, vegetables, or flowers).

Horticulturists may also use a soil auger to pull a soil sample. A *soil auger* is similar to a probe, but can pull samples from as deep as 10′ within the soil profile. A soil auger enables soil scientists to examine soil color, texture, and structure.

Soil and Soilless Media

In agriculture, soil and specially formulated *soilless media* (also known as potting soil) are critical components for plant growth. Soilless media is a sterile mix of natural ingredients used to raise plants in greenhouses, cutting beds, and containers. It is lighter than soil with more pore spaces. The type of soil or media you use and how it is managed greatly influences how well a plant will grow.

STEM Connection
Taking a Soil Test and Reading a Soil Report

Most inexpensive soil test kits from local garden centers are not reliable and do not accurately measure pH or give recommendations for soil amendments. Soil texture, organic matter content, plants to be grown, target pH, soil acidity level, CEC, clay content, and current pH are factors to consider in determining the amount of lime needed to raise the soil pH. Consistently reliable results can be obtained only by submitting samples to a soil-testing laboratory. Test your soil several months before planting or fertilizing so that you can implement recommendations before the growing season. For example, if you would like to grow a spring vegetable garden, submit a soil sample in the fall or winter. As a general rule, test sandy soils every two to three years and clayey soils every three to four years.

To take a soil sample, begin by moving leaves, mulch, and other debris from the soil surface. Use a trowel or shovel to dig a hole:
- 6″–8″ (15 cm–20 cm) deep for gardens.
- 2″–4″ (5 cm–10 cm) for lawns.
- 6″–10″ (15 cm–25 cm) for trees and shrubs.

Repeat this procedure in six to eight areas (subsamples) to obtain a more representative sample for testing. Avoid sampling in areas that are obviously different, including wets spots, compost piles, under eaves, or under brush piles. Mix the subsamples together thoroughly and remove any large roots, stems, leaves, rocks, or other materials. Fill the soil sample box to the fill line (approximately a pint) to ensure there is enough soil for the necessary tests. Include information on what you intend to grow for appropriate recommendations.

Photo by Tim McCabe, USDA Natural Resources Conservation Service

As part of a cooperative soil survey program by the state of Iowa, Iowa State University, and the USDA Natural Resources Conservation Service, all the soils in Iowa have been surveyed and catalogued. A technician is testing the pH of this soil sample.

The soil test report will give information on your soil's nutrient levels and pH. Recommendations will be given on how much fertilizer to apply based on what you are growing. If the soil needs to be amended to adjust the pH, liming calculations per square foot will be given. Adding an acidic material to lower the pH may be suggested as well. Contact your local cooperative extension service office for additional resources and support on nutrient management calculations and pH adjustments.

Across farms, in gardens, and in greenhouses, soil and soilless media serve as the foundation for growing fruits, vegetables, landscape ornamentals, cut flowers, and bedding plants. Soil anchors a plant's root system, providing structural support so the plant will not fall over, **Figure 11-13**. Soil also has a network of air spaces, which allow for plant roots to *respire* (give off carbon dioxide and take in oxygen). These same soil pore spaces also provide a place for water to be held and used by plant roots.

Soil and soilless media can moderate temperature changes, shielding plant roots from extreme temperature changes. Plants require nutrients to complete essential functions for growth and development. Soil and various media formulas provide mineral nutrients, such as nitrogen, phosphorous, potassium, calcium, and iron. (Further discussion of plant essential nutrients will be covered later in the text.)

Garden or Native Soil

One of the most common gardening mistakes made by beginners is trying to grow plants in a container filled with garden soil. (In this text, *garden soil* refers to the soil found naturally in a landscape without significant amendments.) Why is this a bad practice? If plants can thrive in a garden full of garden soil, why is it a bad idea to use it in containers? Garden soil is a heavy soil that holds too much water to be confined in small structures. No matter how durable the container, or how healthy the plant is to begin with, if you use garden soil, your plant will have health problems.

Using garden soil in a container undoubtedly causes water-related problems. When garden soil is used in containers, it may never have the chance to dry out. Therefore, the plant roots never get the chance to breathe (respire). If roots cannot respire, they cannot perform other life functions and will ultimately die. Remember, plants are like people and need water, air, nutrients, light, and space to grow. Additionally, once garden soil does dry out, it may become hard to rehydrate. It can become as hard as a brick and will barely hold water again.

Aside from water-holding and drainage issues, garden soil confined to a container can become a thriving habitat for pathogens and insects, **Figure 11-14**.

MyImages - Micha/Shutterstock.com

Figure 11-13. Soils anchor the roots of plants. Poor soil structure and other contributing factors can lead to the capsizing of trees, which can be damaging or deadly to people.

Margaret M Stewart/Shutterstock.com

Figure 11-14. Various beneficial and problematic insects and microorganisms inhabit garden soil. Here, white grubs that will become beetles as adults rest in this handful of soil. What other types of living organisms do you think are in this handful of soil?

Pathogens are disease-causing organisms such as bacteria, viruses, and fungi. Garden soil is also home to insects of all shapes and sizes, some of which may be beneficial whereas others are simply pests. Garden soils will also have unique chemistries with variations in pH that can lead to nutritional issues. As discussed earlier, if the pH is too acidic or alkaline, plants will not have access to nutrients in the soil and can then become starved of the essentials they need to thrive, **Figure 11-15**.

Soilless Media

Garden soil should never be used in a planting container, but what should be used instead? Instead of garden soil, horticulturists use specially formulated potting soil, or soilless media, in containers. Soilless media can have various ingredients, may or may not be lightweight, is full of nutrients, is able to hold water while allowing for proper drainage, and provides suitable space for root growth. Examples of naturally occurring soilless media include water, gravel, and sand, **Figure 11-16**. Scientists also formulate soilless media in laboratories for specific horticultural objectives.

The combination of ingredients in soilless media behaves similarly to garden soil without the negative impacts garden soil may have on plant growth. Typical potting soils contain the following ingredients in various ratios:

- *Vermiculite*—a lightweight material made of small pieces of mica that readily absorbs water.
- *Perlite*—a very lightweight, pea-sized (or smaller) rock that is white and comes from volcanoes. This material is used to help increase drainage.
- Peat moss—a brown, water-holding material harvested from peat bogs that provides space for root growth. Peat moss is made of aged, decomposed plant and animal matter and forms the foundation of most potting soil mixes.
- Sphagnum moss—a once-living moss that comes from wet bogs. The moss is harvested, dehydrated, and often compressed. The quality of the sphagnum moss is dependent on the length of the strands. This product is used to hold moisture and provides space for root growth.
- *Coconut coir*—a brown fibrous material that is the result of shredding coconut husks. This is a renewable resource that acts like peat moss but is more sustainable. Coconut coir also holds water and provides space for root growth.

Alena Brozova/Shutterstock.com

Figure 11-15. On the left, a strawberry plant lacks iron. The interveinal chlorosis on the younger leaves is an indicator of this nutrient deficiency. What do you think will happen to a plant that lacks chlorophyll and is yellowing?

Yurchyks/Shutterstock.com

Figure 11-16. A plant growing in a water solution shows the capacity of roots to inhabit a space, searching for water and nutrients.

- *Bark*—pieces from the outermost layer of trees that help add drainage to soilless media mixes. The bark particles range in size depending on the type of material that will be grown in the mix.
- *Slow-release fertilizer*—small, pellet-sized material added to soil to increase fertility. It ranges in color and composition. The mostly synthetic fertilizer pellets are coated in a polymer that releases a little of the fertilizer every time the soil is watered. These polymers usually last for two to four months.

Today's horticultural manufacturers are always reworking their soilless media formulas to meet the needs of growers and gardeners. Whether you grow orchids, succulents, or fairy gardens, there are potting soil mixes specially formulated for optimal plant performance.

Mulch

Mulching is adding a layer of material over the garden soil. This material can be organic or inorganic, **Figure 11-17**. Mulch is usually spread three to four inches deep over the soil surface. Mulch can include a number of materials, such as bark, straw, leaves, pine needles, compost, gravel, rocks, or even shredded tires. Mulch provides more than just aesthetic value. Mulch can suppress weeds, moderate soil temperature, help the soil retain water, add nutrients, and provide a platform for traffic throughout a garden.

Organic and Inorganic Mulches		
Mulch Type	**Advantages**	**Disadvantages**
Organic	• Readily available • Can be made at home • Increases microbial activity • Adds nutrients to the soil • Suppresses weeds • Retains moisture in soil • Insulates plants to lessen temperature fluctuations	• When purchased, it is expensive • Not all organic material adds the same value to the soil • Will need to be replaced
Inorganic	• Does not decompose • There is no need to reapply (unless using plasticulture) • New purpose for some items that would otherwise be thrown away	• Expensive initial cost • Weeds will grow through rocks or on top of geotextiles • Most always purchased (not available at garden site) • Geotextiles and plasticulture are not recyclable

Goodheart-Willcox Publisher

Figure 11-17. Advantages and disadvantages of organic and inorganic mulches. Can you think of other waste materials that might be suitable for mulch?

Thinking Green

Coconut Coir vs. Peat Moss

Coconut coir (pronounced core) is a coarse fiber created when the coconut husk is removed from the shell and shredded. Coconut coir is a sustainable resource and a renewable way of amending soils. It may also serve as a soilless media foundation. Coconut coir fibers have advanced structural stability that allows for good drainage and water absorption. The coconut fibers do not decompose as quickly as peat moss in soil.

Peat moss comes from peat bogs, mostly in Ireland and Canada. These bogs store over 562 billion tons of carbon (more than all of the trees in the world). When peat is mined, the bogs must be drained of the water, and carbon is released into the atmosphere. Harvesting this resource is often considered contradictory to sustainable practices. When available, more renewable sources for media and containers should be used.

xuanhuongho/Shutterstock.com

Gabriela Insuratelu/Shutterstock.com

Inorganic Mulches

Inorganic mulches can include a number of synthetic materials that are used in the landscape. Geotextiles, plasticulture, and tires are inorganic mulch materials commonly used by landscapers and growers. *Geotextiles* are products such as permeable, weed control fabric made from a plastic material, **Figure 11-18**. They may be applied over a soil surface. Plants can be planted in holes made in the fabric, or another mulching material can be placed over the fabric. The permeability of geotextiles allows water to readily infiltrate the soil.

Plasticulture is an impermeable, plastic, inorganic mulch material commonly used in vegetable production. Sheets of black, white, or red plastic are laid over the soil surface after drip irrigation lines are installed. Plants are then placed into holes made through the plastic. The irrigation pipes water the plant, and there is no need to weed that space. The plastic mulch has to be replaced with each new crop.

Eugene Sergeev/Shutterstock.com

Figure 11-18. A geotextile product can be used as a mulching material or to prevent erosion. It is still permeable and allows water to seep into the ground, but weeds and other materials cannot enter.

You may have observed shredded tires used as a mulch on playgrounds or areas of traffic in parks. All these inorganic mulches have a purpose in gardening, **Figure 11-19**.

Organic Mulches

Organic mulches, such as bark and leaves, provide a bounty of benefits to gardeners by improving soil health. These mulches help to increase the amount of microbial activity in a soil (mainly bacteria and fungus) that converts organic matter into nutrients for plant root uptake. The mulched soil has increased worm activity, which leads to more pore spaces where water can be held and travel and for roots to grow into.

The addition of organic materials, such as tree bark, leaves, and straw, all help improve the soil's structure. These materials aid in aggregation of soil particles, which increases the surface area available to hold water and nutrients. The soil's structure improves greatly with each addition of organic mulch.

Photo by Lynda Richardson, USDA Natural Resources Conservation Service

Figure 11-19. This layer of black plastic is used to cover the soil. Here, tomatoes sit on top of the plastic, while beneath the plastic are drip irrigation lines. The plastic helps to mulch the soil, detour weeds, and insulate the soil from temperature fluctuations and water loss.

STEM Connection
Mulch and Soil Nutrition

A wealth of benefits occurs when organic mulch is applied to the soil; however, not all organic mulches are created equal. Bacteria and fungi break down the carbon and help to produce a useable form of nitrogen for the plants. High amounts of carbon and low amounts of nitrogen force microbes to break down nitrogen reserves in existing soils, robbing plants of their nutrient source. A carbon to nitrogen ratio exists in all organic matter. Materials with ratios greater than 30:1 (such as wood chips) do not have enough nitrogen to support microbial growth. Gardeners could add fertilizer to increase nitrogen levels and decrease competition among plants and microbes. A gardener, however, can also use mulches with a low carbon-to-nitrogen ratio on nutrient-poor soils, new landscapes, or spaces where there is a need for rich plant growth. Keep mulch products such as wood chips on paths or areas with established trees or shrubs. Place mulches with low carbon-to-nitrogen ratio values where you desire lush growth.

Hannamariah/Shutterstock.com

Thinking Green

Cover Crops

Cover crops, also called green manures, are crops that are planted in gardens and crop fields between commercial plantings to keep the soil covered year-round. Cover crops may be cut just before flowering and then worked into the soil. The practice of using cover crops increases soil nutrients, improves soil quality, decreases soil erosion, improves water-holding capacity, helps with compaction problems, and suppresses weeds.

Grigorev Mikhail/Shutterstock.com

Compost is decomposed organic matter that may include animal manures, food wastes, and vermicompost (worm castings), **Figure 11-20**. Compost can be used as a mulch. Compost is rich with nutrients and can be made by recycling organic materials that may otherwise be thrown into a landfill. You can purchase compost or make it with very little effort.

To make compost, mix two to three parts of *green waste* (nitrogen-rich materials such as leaves, fresh manures, grass clippings, and coffee grounds) to one part *brown waste* (carbon-based materials such as sawdust and shredded cardboard). Place the materials in a simple compost bin where they can be easily turned and mixed. Depending on rainfall, you may need to add water on a regular basis. Allow naturally occurring microorganisms and time to decompose the organic matter.

Evan Lorne/Shutterstock.com

Figure 11-20. A compost bin can be fabricated with sides whereas a pile lies without boundaries. Layers of green and brown organic matter decompose and make a ready amendment to soil. What other types of compost bins have you seen? Does one type work more efficiently than another?

Safety Note

Compost Material

Do not add pet feces or animal waste such as bones or meat to your compost pile. Pet feces and animal waste may contain pathogens that can cause illness and death in humans. These pathogens could infect the person handling the compost or someone ingesting food that has come in contact with the compost. You should also avoid adding diseased plant materials and invasive weeds to prevent spreading diseases and weeds when distributing compost.

Thinking Green

Mulchmat

Mulchmat is a nonwoven wool or cotton matting material referred to as an agri-textile. This matting may be used in horticulture, landscaping, and re-vegetation of soil. Mulchmat is available in different colors, shapes, and thicknesses and may also be cut to size. This is a fully biodegradable product that suppresses weeds, retains water, insulates plants, adds nutrients to the soil, and is sustainable.

Compost is ready to be used when it looks and smells like earth. Compost may be used just as you would any other mulch or it may be incorporated by hand or machine into the top 6″ of soil.

It is important to realize the objective of the gardener before selecting and applying mulch in a garden. Using inorganic materials may seem to require less maintenance or cost than organic materials. However, the plants may require periodic weed control and the application of fertilizer.

Moolkum/Shutterstock.com

Figure 11-21. Plants can grow in containers of all sizes, shapes, colors, and materials. The type of container used depends on the objective of the grower and the type of media that is used.

Containers

The majority of growing containers used in plant production are made of plastic and are available in a range of sizes, shapes, and colors, **Figure 11-21**. For both small- and large-scale production, growing containers should be durable and lightweight. They should also provide adequate drainage. Growers should consider those three factors when selecting a container for a plant. Having a container that is aesthetically pleasing or that advertises the grower or product is an added benefit.

Size

In general, the container should only be slightly larger than the plant's root ball. By using a pot that is only slightly larger than the root ball, the roots will be able to adjust to the new mass of the soil. If the pot is too large, the excess soil will retain more moisture than the roots can absorb or use, and the plant may die from being overwatered.

Durability

The container must be able to withstand continuous water movement. Choose plastic, metal, clay, and ceramic containers when possible. If you want to minimize watering, use plastic, metal, or ceramic. Clay pots are porous, and water continuously evaporates through them.

Thinking Green

Biochar

K.salo.85

Biochar is a type of charcoal used for agricultural purposes that has high nutritional content. Biochar is created through **pyrolysis** (using heat to decompose organic material in the absence of oxygen). This agricultural practice was used by pre-Columbian Amazonians. The natives would simply cover their smoldering agricultural waste in pits to create biochar that they would then use to amend their soil. Since those times, people around the world have seen the positive impacts that biochar has in soil. Today, methods that are more environmentally sound are used to recycle agricultural waste and amend soil.

There are a wide variety of hot-pressed fiber containers made of rice hulls, wheat, peat, wood pulp, spruce fibers, coir, and bamboo. They can be biodegradable and are often made with natural or synthetic binding materials. Fiber containers are semi-porous and increase water and air circulation. These containers vary in rigidity and resiliency.

Drainage Capability

The container should have drainage holes or allow for the addition of drainage holes through drilling. If it is not possible to add drainage holes, you may want to consider another container. If the container limits drainage, excess water can cause serious root health problems. Plants must also have access to air. If the excess water cannot drain, *saturation* will occur (water will fill pore spaces in the soilless media). If a soilless media is continually saturated, there is no room for air in the pores that is necessary for respiration. The roots will begin to rot and will not be able to pull up the water or nutrients needed for optimal growth, **Figure 11-22**. If you still love the container and there are no drainage holes, do your best to monitor how much water is applied and make sure the plant is allowed to dry out between watering.

In addition to the typical plastic pots and flats for plant production, horticulturists should be aware that almost anything can be used as a container for growing plants. As long as the container can hold potting media and water and allow proper drainage, it can be used as a growing container.

ronstik/Shutterstock.com

Figure 11-22. This planter does not provide enough drainage for this Cyclamen plant. As a result, the plant's soil is continuously saturated, and there is no air in the soil for the plant's roots. Eventually, the plant cannot respire, and the plant's health suffers.

Maria Dryfhout/Shutterstock.com

Figure 11-23. Plants can be placed just about anywhere. An old pair of boots has been reclaimed as a home for these succulent plants. What other unusual planters have you seen? Were they modified to accommodate growing needs?

Photo by John Kelley, USDA Natural Resources Conservation Service

Figure 11-24. Soil scientists, also known as pedologists, study the soil. A soil scientist working for the NRCS is conducting a field test to determine the fluidity class of the soil sample.

Items ranging from an old pair of shoes to old bathtubs to buckets may be used to grow and display plants, **Figure 11-23**.

Careers in Soil Science

Pedology, or the study of soils in their natural environment and soil conservation, offers endless career opportunities. There are currently not enough qualified personnel to meet the demand for employees in this industry. Universities and government agencies employ soil scientists to study and regulate soil standards and research conservation practices, **Figure 11-24**. Many employees work with private industry to help prevent erosion or contamination at construction and agricultural sites. Another division of this industry involves providing soil and garden educational resources to garden enthusiasts.

Soil Scientists

After a large region of the Midwestern United States experienced a severe drought in the 1930s, often referred to as the Dust Bowl, the USDA Natural Resources Conservation Service (NRCS) was established. Its purpose was, in part, to control and restore natural resources in the United States. Soil scientists, also called pedologists, are the foundation of the NRCS. Soil scientists employed by the NRCS travel around the country. They work with both public and private landowners to map the soil and complete soil surveys. Modern technology allows the scientists to precisely map the soil with advanced instruments. Scientists are able to catalog detailed information and develop accurate soil maps for the country. The information is made available to the public through government websites. Landowners can access the information and use it to help them to make decisions about the land. For example, landowners can determine whether or not the site under analysis should be used for agriculture based on soil characteristics. Soil scientists are often employed by the NRCS. Soil scientists, however, can also be employed by colleges, universities, and in private industry. They may work in a laboratory or in consulting jobs that will include on-site work.

Thinking Green

Bioplastic Sleeves

Bioplastics are a type of biodegradable plastic made from organic components. Bioplastic sleeves are biodegradable plastic tube-like plant holders. They are not true containers because they must be kept in a tray until the plant's roots hold the substrate together. They are plantable and fully degrade within a year.

Bioplastics appear like traditional plastics but are made of biopolymers or a blend of bio- and petrochemical-based polymers. The blend may include palm fiber, beet, potato, corn, cassava, sugar cane, proteins from soy, and keratin from waste poultry feathers. Bioplastics may also use lipids from plant oils and animal fats. Advantages of bioplastics include durability, weight, rigidity, and resistance to immediate decay.

zirconicusso/Shutterstock.com

Media Manufacturing

There are numerous companies throughout the United States and the world that manufacture media and mulches, **Figure 11-25**. These types of companies manufacture professional growing mixes for greenhouse, nursery, interiorscape, and retail markets. Facilities may be equipped to include the compost production, packaging, and storage operations. These facilities can be ideal for media and compost production. Areas of employment at this type of facility include sales and marketing, distribution, soil scientist, chemist, or manufacturer.

Rikard Stadler/Shutterstock.com

Figure 11-25. Companies can manufacture their own potting soil or media using organic components such as peanut shells, coconut coir, and other sustainable materials. What types of organic materials should not be used when manufacturing potting soil or other growing media?

Career Connection
Melanie McCaleb
Erosion Control Specialist

Melanie M. McCaleb

On any given day, you can find Melanie McCaleb knee-deep in soil. As a soil scientist that specializes in erosion, sediment, and turbidity control, Melanie plays a key role stewarding the environment by protecting water quality through minimizing pollution from sediments. She designs and implements measures that keep soils in place to be productive for plant growth. A typical week finds her in the field consulting with construction firms to ensure they have proper best practices in place that minimize storm water runoff and encourage rainfall infiltration into the soil surface. She might also work with a homeowner to design a rain garden that holds storm water that is laden with soil particles. Melanie also works within her office to draw and design soil erosion and sediment control plans that help businesses and homeowners manage and conserve their soil resources.

Melanie graduated from North Carolina State University with a bachelor's and a master's degree in soil science. She applies that research-based knowledge to serve the citizens of North Carolina.

Melanie provides solutions for erosion and sediment control problems as the owner of her own small business, NTU, Inc., which provides innovative solutions for superior water quality. This job requires her to have a fluid understanding of federal, state, and local regulations that allow her to make practical, safe, and economically responsible decisions. She successfully collaborates with many government, state, and local agencies throughout the southeast region of the United States. She serves as the president for the Southeast Chapter of the International Erosion Control Association (IECA). "I love soil because it's a good excuse to get dirty! And the exploration and discovery is unlimited," proclaims Melanie.

CHAPTER 11 Review and Assessment

Chapter Summary

- Soils are living and dynamic media that are the foundation of life.
- Soil consists of four main components: air, water, minerals, and organic matter.
- Five soil-forming factors shape the development of soil: climate, organisms, relief, parent material, and time. As soils develop, layers called horizons are formed.
- Soil can be described by its properties: texture, structure, soil water, and color. Soil texture is the most important physical property to consider when managing soils for cultivation.
- There are three basic soil textures: sand, silt, and clay. Each has unique characteristics that contribute to plant growth. Field tests or laboratory tests can be used to determine soil texture.
- Soil structure refers to the aggregation of soil particles. Soil structure can be managed to improve water-holding capacity, porosity, infiltration, and resistance to erosion.
- Plants rely on the availability of water in the soil to maintain vital processes for growth and development. There are two categories of water in the soil: gravitational and capillary.
- Soil colors give clues about the soil's composition and site characteristics, such as drainage and water content, mineral weathering, and biological activity.
- The living organisms that dwell in Earth's soil influence soil properties. They are responsible for activities such as decomposition, carbon recycling, nutrient production, and pore formation.
- Soil provides essential chemical nutrients for plant growth and development. The optimum pH to sustain plant growth is between 5.5 and 7.0. Soil probes or augers can be used to take soil samples for testing.
- Soil and soilless media serve as the foundation for growing fruits, vegetables, landscape ornamentals, cut flowers, and bedding plants. Soil and soilless media can moderate temperature changes and provide required nutrients needed for growth and development.
- While soil is used for growing plants in the field, soilless media is used to grow plants in a container. It is lightweight, full of nutrients, and able to hold water while allowing for proper drainage.
- Mulches can be made from inorganic or organic materials, including compost. They are used to suppress weeds, retain moisture, moderate soil temperature, and amend the existing soil.
- Several types of containers can be used to grow plants, but lightweight containers that are durable and have drainage holes are preferred.
- Careers related to soil science include soil scientist, erosion control specialist, and horticultural media manufacturing.

Words to Know

Match the key terms from the chapter to the correct definition.

A. adhesion
B. aeration
C. biochar
D. brown waste
E. capillary water
F. compost
G. field capacity
H. geotextile
I. gravitational water
J. green waste
K. hydrometer
L. infiltration
M. mulching
N. organic matter
O. parent material
P. pedology
Q. peds
R. perlite
S. porosity
T. respire
U. saturation
V. slow-release fertilizer
W. soil pH
X. topsoil
Y. vermiculite
Z. weathering

1. A permeable, inorganic mulch material made from a plastic.
2. A type of charcoal used for agricultural purposes with high nutritional content.
3. The physical, chemical, and biological process that creates soil.
4. The process in which a plant gives off carbon dioxide and takes in oxygen.
5. Aggregates produced through the binding of sand, silt, and clay particles.
6. Small, pellet-sized material of various compositions that is added to soil to increase fertility.
7. An instrument used to measure the percentages of sand, silt, and clay in a sample to determine the soil textural class.
8. Living and dead organisms.
9. Water that is held in the soil against gravitational pull and is typically available to plants.
10. The binding of water molecules to soil particles.
11. Decomposed organic matter that may include animal manures, food wastes, and vermicompost.
12. The process by which air is circulated, for example in soil to provide oxygen for root respiration.
13. The maximum amount of water that a soil can hold against the pull of gravity.
14. The material (bedrock, sediment, or organic material) that is weathered to form soil.
15. The study of soils in their natural environment.
16. The fertile, upper, outermost layer of soil.
17. A very lightweight, pea-sized (or smaller) rock that is white and comes from volcanoes.
18. The acidity or alkalinity of a soil.
19. Adding a layer of material (organic or inorganic) over the garden soil for aesthetic value, weed suppression, soil temperature moderation, water retention, or as a platform for traffic.
20. Free water that moves through the soil by the force of gravity.

21. Carbon-based materials, such as sawdust and cardboard, that are a common ingredient used to create compost.
22. The manner in which water moves through the soil.
23. Organic materials rich in nitrogen, such as leaves, fresh manures, grass clippings, and coffee grounds, used to create compost.
24. A situation in which water fills pore spaces in soil or soilless media to the point that there is no room for air in the pores.
25. The state of having space or gaps.
26. A lightweight material made of small pieces of mica that readily absorbs water.

Know and Understand

Answer the following questions using the information provided in this chapter.

1. Why is soil the foundation of life?
2. What are some agricultural and nonagricultural uses of soil?
3. What are the four main components of soil?
4. What five factors shape the development of soil?
5. List four physical properties of soil and briefly describe each one.
6. Compare the three textures of soil (sandy, clayey, loamy).
7. Describe a land judging career development event.
8. How can using cover crops help improve soil structure? What other methods can be used to improve soil structure?
9. Describe two types of water movement through soil.
10. How do soil colors in wet conditions differ from those in well-drained conditions?
11. What are three types of living organisms that are active in soil? What are four types of activities they help bring about?
12. What is the key chemical reaction for making nutrients available to plants?
13. How does soil pH influence plant growth? In what pH range can most horticultural plants thrive?
14. Why is it important to have soils tested?
15. Describe the procedure for taking a soil sample.
16. What are three items found on a soil test report?
17. How is soilless media different from garden soil?
18. What are three examples of inorganic mulches and three examples of organic mulches? What benefits can mulch provide?
19. Why is it important for plant containers to have drainage holes?
20. Describe the activities of a soil scientist employed by the USDA Natural Resources Conservation Service.
21. What areas of employment are offered at a media and mulch company?

Thinking Critically

1. You find a soil that smells rotten and is gray in color. Hypothesize whether this soil would be appropriate for growing blueberry bushes. Justify your reasoning.
2. The florist's hydrangea (*Hydrangea macrophylla*) is often grown in a greenhouse in parts of the United States and South America. It is cultivated to be a sharp blue or pink color. Sometimes, flowers are cultivated and end up a color known as "blurple" (undesirable shades of mauve). The color of the hydrangea is dependent on aluminum. The absence of aluminum ensures pink color, and high levels of aluminum lead to a blue color. Growers know to maintain a pH of 6.0 for pink hydrangeas and a pH of 5.2 to 5.5 for blue hydrangeas. What role does pH play in the availability of aluminum to hydrangeas?
3. A grower would like to produce a crop of cacti. What ingredients should he or she include in the media mix and why?

STEM and Academic Activities

1. **Science.** Experiment with the effect of a soilless media component (such as vermiculite, perlite, bark, coconut coir, or peat moss) on germination rates of plants. Create mixes with unique ratios or use various blends of commercially made mixes to determine which is best as a germination media.
2. **Science.** Go to the NRCS website. Find the soil survey for the site where your school is located. Begin discovering vital information about the soil at your site. What crops could be grown on the soil? What makes the soil a good or bad choice for having a school housed there?
3. **Science.** Choose a site at the school where you and other students would like to plant a cover crop. Determine which plant should be planted at the site based on the time of year and site location. As a class or individually, take a soil sample and send it to the area cooperative extension service agent. Once the soil report is returned, determine what amendments need to be made to the soil for the crop that will be planted. Make the amendments to one half of the site and leave the other half untouched. Plant the cover crop and analyze growth of both sides of the site.
4. **Social Science.** Find a public service announcement about soil, most likely through the NRCS website. Read or listen to this announcement, and then create your own public service announcement related to soil. Write an outline of the announcement. Then make a video, create an audio file, or make a unique presentation such as a comic strip. Upload the file to a video sharing site so others can learn from the announcement.
5. **Language Arts.** Use one of the Corner Quotes in this chapter to act as a springboard for a position paper on the importance of soil as a natural resource. Create an essay map and graphically organize the information. Search the Internet for the Read-Write-Think website for a tool that will help you organize a paper about soil.

6. **Language Arts.** Contact local growers at a greenhouse, nursery, or landscape company. Survey these people about their use of bioplastics, pressed fiber containers, biosleeves, and other containers. Create a questionnaire and ask them about their experiences with these container products. Ask the growers about their customers and their experiences with these products. Summarize the information and create a data table or chart to express the information in a graphical manner.

Communicating about Horticulture

1. **Writing and Speaking.** Interview a local soil scientist. Ask the person to describe a typical day at work. Prepare a list of questions similar to the following: How long have you been in the soil science industry? Did you go to school or did you learn as an intern? What is the work environment like? What are your job duties? What other types of professionals do you work with? Report your findings to the class, giving reasons why you would or would not want to pursue a career similar to that of the person you interviewed.
2. **Reading and Speaking.** After reading this chapter, you should have a good understanding of the different types of soil and how soil impacts plant growth. Determine the type of soil that you have at school through an NRCS website called Web Soil Survey. Create a report in which you describe how the specific type of soil can be used—for agriculture, septic systems, houses, and so on. Share your findings with your peers.

SAE Opportunities

1. **Improvement.** Develop a soil conservation plan for your school.
2. **Exploratory.** Job shadow a district soil conservationist with the NRCS.
3. **Exploratory.** Visit a mulch distributor. Ask about the types of inorganic and organic mulch that are available from this source. Which type of mulch does the distributor sell most and to what types of customers?

Chris Byrne/Shutterstock.com

4. **Exploratory.** Take a soil sample at your home and determine the pH. Create a list of plants that would be well suited for your home. Explain how you could change the pH to accommodate most plants.
5. **Experimental.** Create a test plot and sow various cover crops to improve the soils. Incorporate the plants into the soil. Then plant the same crop to see which cover crop improved the soil and its nutrition the best.

CHAPTER 12 Plant Nutrition

Chapter Outcomes

After studying this chapter, you will be able to:
- Describe essential elements used by plants.
- Summarize the mineral nutrient uptake pathway into plants.
- List organic and inorganic sources of nutrients.
- Solve fertilizer calculations.
- Describe methods of fertilizer application.
- List careers related to nursery production.

Words to Know

ammonification	fertigation	phytotoxicity
assimilation	fertilizer grade	ratio
banding	foliar application	side-dressing
broadcasting	interveinal area	soluble
chelate	macronutrient	sorption
complete fertilizer	micronutrient	superphosphate
cytochrome	necrotic lesion	tilth
deficiency	nitrification	
denitrification	phytoremediation	

Before You Read

Take two-column notes as you read the chapter. Fold a piece of notebook paper in half lengthwise. On the left side of the column, write main ideas. On the right side, write subtopics and detailed information. After reading the chapter, use the notes as a study guide. Fold the paper in half so you see only the main ideas. Quiz yourself on the details and subtopics.

While studying this chapter, look for the activity icon to:

- **Practice** vocabulary terms with Words to Know activities.
- **Expand** learning with identification activities.
- **Reinforce** what you learn by completing Know and Understand questions.

www.g-wlearning.com/agriculture

What is your favorite meal? Pizza? Burger and fries? Roast turkey? Kale chips? The food we eat nourishes our body and gives us the vitamins and minerals we need to stay healthy. Most of our vitamins and minerals come from plants (or animals that ate plants), which is why a diet rich in grains, fruits, and vegetables is important for normal growth and development. Plants also require minerals for normal development. In this chapter, you will learn which elements are essential for plant growth, how plants take in these elements, what occurs when elements are lacking, and how these elements are added to growing media.

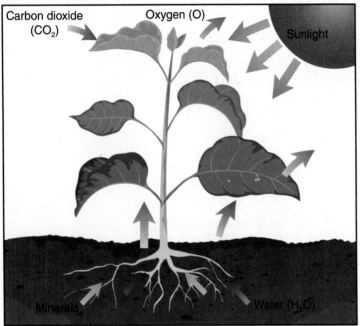

Figure 12-1. Plants obtain most of the carbon, hydrogen, and oxygen used during photosynthesis from water and carbon dioxide in the atmosphere. Most of the other needed elements are found in the soil.

Essential Elements

There are 17 elements essential for plant growth: carbon (C), hydrogen (H), oxygen (O), nitrogen (N), potassium (K), phosphorus (P), calcium (Ca), magnesium (Mg), sulfur (S), boron (B), chlorine (Cl), copper (Cu), iron (Fe), manganese (Mn), molybdenum (Mo), nickel (Ni), and zinc (Zn). Most of these elements are found in the soil. However, most of the carbon, hydrogen, and oxygen that plants use during photosynthesis comes from water and carbon dioxide in the atmosphere, **Figure 12-1**.

Nitrogen, phosphorus, and potassium are considered *macronutrients* because they are needed in large quantities by plants for normal growth and development. Macronutrients may also be referred to as primary nutrients. Calcium, magnesium, and sulfur are also macronutrients, but they are termed secondary nutrients as they are taken up in fewer quantities than nitrogen, phosphorus, and potassium. The remaining nutrients—boron, chlorine, copper, iron, manganese, molybdenum, nickel, and zinc—are termed *micronutrients*. These nutrients are essential for plant growth, but they are needed in small or minute amounts.

Hydrogen (H)

As discussed in Chapter 9, *Plant Growth and Development*, plants create their own food in the form of sugars through photosynthesis. Carbon, hydrogen, and oxygen are used by the plant to manufacture sugar ($C_6H_{12}O_6$) and are considered the building blocks for plant growth. Because plants get their hydrogen from water (H_2O), hydrogen is never a limiting element as it is constantly available through the uptake of water.

312 Horticulture Today

Carbon (C)

Almost half of a plant's dry matter consists of carbon. Plants take carbon in the form of carbon dioxide (CO_2) from the air. They convert it to simple sugars that are used to build carbohydrates, starches, cellulose, lignin, and proteins. With widespread availability in the air, carbon is rarely limited in field settings. However, if plants are spaced too closely together, there can be a CO_2 shortage. In greenhouses, high rates of photosynthesis by plants can swiftly lower CO_2 levels. Some growers practice CO_2 enrichment. They may burn natural gas; release CO_2 from dry ice; or use compost, manure, or organic mulches in the greenhouse. Plants respond favorably to such treatments. However, this practice is not widespread because of cost.

Oxygen (O)

Plants obtain oxygen (O) by breaking down carbon dioxide during photosynthesis and saving a small portion for driving cellular respiration. Most oxygen remains unused by plants and is released as a by-product. The oxygen that is taken in as carbon dioxide is only available to leaf and stem tissues, requiring root tissue to acquire oxygen from the soil environment. Plants absorb oxygen in the chemical forms O_2 (oxygen) and H_2O (water). (Oxygen is written as O_2 because oxygen is highly reactive and disconnected oxygen atoms have a tendency to form bonds.)

Primary Macronutrients

Macronutrients are elements that plants need in significant amounts for healthy growth. A shortage of these elements can lead to stunted growth, discoloration, poor yields, or death of plant parts.

Nitrogen (N)

Nitrogen (N) is commonly considered the most important mineral nutrient in plants. It is used in the highest quantities and is a component of amino acids (used to build proteins), chlorophyll, nucleotides in DNA, coenzymes, and some lipids. Nitrogen *deficiency* (lack or shortage) results in severe stunting of growth and chlorosis, or yellowing of the leaves. Because nitrogen is a component of chlorophyll production, a lack of nitrogen inhibits the plant's ability to form green chlorophyll molecules. Nitrogen is mobile in the plant, meaning that it can break down in its current location and move to where it is needed. Often, chlorosis may be observed in older leaves because nitrogen has been taken to produce younger, actively growing leaves, **Figure 12-2**.

Nitrogen is cycled among living and nonliving things and exists in different chemical forms. Nitrogen gas (N_2) makes up 78% of the atmosphere, but, despite its abundance, few living things can use this form of nitrogen.

IPNI Image/J.Prabhakaran

Figure 12-2. Symptoms of nitrogen deficiency of this cocoa plant include yellowing of older leaves.

STEM Connection

Carnivorous Plants

Have you ever wondered why some plants devour insects? Carnivorous plants literally digest prey from insects to small mammals (depending on the species) as a way to supplement their nutritional needs. Most carnivorous plants inhabit nutrient-poor environments, such as bogs, where the availability of nitrogen and phosphorus is limited. The plants have specialized cells that secrete digestive enzymes and allow these fascinating plants to thrive. How many types of carnivorous plants exist? What is the largest species and where does it grow?

Matthijs Wetterauw/Shutterstock.com

Plants normally absorb nitrogen as nitrate (NO_3^-); some plants can also use ammonium (NH_4^+) or amino ions (NH_2^+). Nitrogen fixation is the process of converting atmospheric nitrogen gas (N_2) to ammonium (NH_4^+) by certain bacteria (for example, *Rhizobium*) that are either free-living (nonsymbiotic) or that live in a symbiotic relationship with leguminous plants such as beans, alfalfa, clover, and peanuts.

Nitrogen Application

Nitrogen can be applied in fields and greenhouse settings in various ways. Chemical fertilizers commonly contain nitrogen, phosphorus, and potassium. Many different concentrations are available and are tailored to crop needs. For example, a fertilizer for the lawn will meet the growing requirements for turf, whereas a fruit tree fertilizer would be specific for developing fruit. Nitrogen is highly *soluble* (dissolves in water) and readily leaches through the soil. Therefore, proper timing and application rates are required to minimize negative effects to the environment. Organic growers will often use leguminous cover crops, such as clover, vetch, or cowpea, to fix nitrogen and then turn it back into the soil as a green manure, **Figure 12-3**. Raw or composted manure is another source of nitrogen. Manure and cover crop applications provide a slow release of nutrients and can be more difficult to manage than commercial fertilizers.

Digoarpi/Shutterstock.com

Figure 12-3. Crimson clover is used as cover crop to fix nitrogen and increase nutrient levels in the soil. Are other types of clover used as cover crops?

Potassium (K)

Potassium (K) is another macronutrient that plants need in large amounts. Potassium is a catalyst for enzyme reactions, and is important for protein synthesis and sugar and starch formation. It influences osmosis through the permeability of cellular membranes. (Osmosis is the movement of molecules in a solution through a semipermeable membrane into a region of higher solute concentration, in the direction that equalizes the solute concentrations on the two sides.) Potassium plays a critical function in photosynthesis by playing a key role

Corner Question

What are good sources of potassium for human health?

in the opening and closing of stomata. Potassium helps maintain turgidity, reducing water loss and wilting. Plants deficient in potassium show chlorotic or mottled leaves, usually in older tissue. The leaf tips and margins may have *necrotic lesions* (spots of tissue death). Growth is usually stunted, and yields are reduced, **Figure 12-4**.

Potassium is present in soils as a natural component of weathered soils. Potassium is more mobile than phosphorus but less mobile than nitrogen. Because parent material and the level of weathering vary dramatically, the amount of naturally occurring and available potassium also varies. Plants commonly use the potassium ion (K^+) form. Growers and gardeners apply it as part of a fertilizer or manure after reviewing soil test results. Potassium can be broadcast or applied in bands as well as part of a fertigation program. *Fertigation* is the process of adding fertilizer to irrigation water.

IPNI Image/S.Moroni

Figure 12-4. These eggplant leaves show potassium deficiency symptoms of chlorotic and mottled leaves with necrotic lesions.

Phosphorus (P)

Phosphorus plays a crucial role in photosynthesis and respiration as a component of adenosine triphosphate (ATP), which provides energy for the chemical reactions necessary to support plant life. Phosphorus is an element in DNA and a part of coenzymes and phospholipids. It is needed in smaller amounts than nitrogen, but it is still extremely important for the functioning of the plant. Phosphorus is important for root growth, seed development, and stem strength. If phosphorus is unavailable to a plant, leaf tips can appear burnt and older leaves will turn dark green and accumulate anthocyanins, which are pigments that change leaf tones to reddish-purple, **Figure 12-5**. Phosphorus is highly mobile in plants and, if deficiencies occur, it may move from old plant tissue to young, actively growing areas.

Photography by Home and Garden Information Center, University of Maryland Extension

Figure 12-5. Phosphorus is highly mobile in plants, which means deficiency symptoms tend to appear in older tissue.

The decomposition of organic matter in the soil serves as natural reservoir of phosphorus, but much of it is unavailable to plants. Plants take up phosphorus mainly as orthophosphate ions ($H_2PO_4^-$), and sometimes as monohydrogen phosphate (HPO_4^{2-}). The activity of microorganisms in breaking down phosphate is influenced greatly by soil temperature and moisture as well as soil pH. The amount of phosphorus in the soil that is available to plants is very small compared to the total amount present in the soil. Most growers will provide supplemental phosphorus through the addition of inorganic fertilizers or manures. Phosphorus generally has a low solubility, and it is limited in its ability to leach through the soil and pollute groundwater. However, because it is adsorbed onto soil particles, it can be washed to waterways through soil erosion. Excess phosphorus in water will increase algal growth and eventually cause reduced oxygen levels, which can lead to eutrophication.

> **Did You Know?**
> Once water is removed, one third of the mass of the human body comes from calcium.

Secondary Macronutrients

As stated earlier, calcium, magnesium, and sulfur are also macronutrients but are taken up in fewer quantities than nitrogen, phosphorus, and potassium. Calcium and magnesium are rarely added to soils as fertilizers and may be supplied by mineral weathering or through the addition of limestone powder used to correct soil acidity. Sulfur may be added to soil through the atmosphere or as impurities in fertilizers.

Calcium (Ca)

Calcium (Ca) strengthens cell walls, providing structure and strength to the plant. It is involved in permeability of cellular membranes, viscosity of cytoplasm, and aids in the communication between cells. It is also part of enzyme cofactors. Abnormal terminal bud growth and poor root growth are common symptoms of a calcium deficiency. In severe circumstances, the growing point dies and roots blacken and rot. Calcium does not translocate (move to other places) in the plant, causing new tissue, such as young leaves, to develop symptoms first. Deficient plants might shed blossoms and buds prematurely. A common example of calcium deficiency is blossom end rot in tomatoes, **Figure 12-6**.

alybaba/Shutterstock.com

Figure 12-6. Blossom end rot is a typical result of calcium deficiency. Many home gardeners may need to add lime or calcium carbonate to their soil to adjust pH and and calcium.

When calcium levels are low, soils tend to be acidic. Calcium ions (Ca^{+2}) are commonly added to these soils in the form of limestone. They replace hydrogen ions (H^+), lowering their concentration and raising the pH level.

Magnesium (Mg)

Secondary macronutrients are needed in significant quantities, but not to the same extent as nitrogen, phosphorus, and potassium. Magnesium (Mg) is a central component of the chlorophyll molecule. A magnesium deficiency results in a shortage of chlorophyll and yields poor and stunted plant growth. The magnesium salt of ATP is the active form of ATP that helps build, modify, or break down compounds as part of a plant's normal metabolism. Magnesium is highly mobile between older and younger plant tissues. Deficiency symptoms include chlorotic or mottled leaves, leaf tips and margin turned upward, and color loss between the veins, leaving a striped appearance, **Figure 12-7**.

Photograph from UC Fruit Report (http://ucanr.edu/sites/fruitreport)

Figure 12-7. A lack of magnesium can result in a shortage of chlorophyll, which displays as chlorosis and poor and stunted plant growth.

Magnesium is abundant in many soils and becomes available to plants as minerals weather and break down. Magnesium is held on the surface of clay and organic matter particles and does not readily leach.

Plants can use magnesium in the form of the magnesium ion (Mg^{2+}). Acidic soils decrease the availability to plants, but the addition of dolomitic limestone, which contains magnesium, supplies sufficient magnesium for crop growth. Magnesium influences the uptake of potassium when large quantities of magnesium are present.

Sulfur (S)

Sulfur (S) is a part of a few amino acids and is important to protein synthesis. Sulfur is also a component of many brassicas (mustards) and alliums (onions and garlic), giving them their characteristic odor and flavor. Sulfur aids in seed production and plays a role in winter hardiness. Sulfur does not readily translocate in the plant. Deficiencies show up first in young leaves that display light green veins and *interveinal areas* (space between veins), **Figure 12-8**.

Sulfur is supplied to plants from the soil by organic matter and minerals, often through rainwater. It is readily soluble and can be leached with heavy irrigation or rainfall. It can also move upward to the soil surface through evaporation. In many parts of the country, sufficient rainfall provides enough sulfur needs. However, sulfur can be inconsistently available at crucial times for optimal plant growth. Therefore, it is often supplemented through the addition of a fertilizer. Sulfur is also applied to lower soil pH. Plants absorb sulfur in the form of sulfate (SO_4^{2-}).

Photographed by Richard W. Taylor

Figure 12-8. Sulfur-deficient plants may show light green veins and chlorotic interveinal areas.

Did You Know?
Sulfates are salts of sulfuric acid and many are prepared from that acid.

Micronutrients

Boron, chlorine, manganese, iron, nickel, copper, zinc, and molybdenum are termed micronutrients. Although elements that are considered micronutrients may be needed in only small amounts by plants, they are important for healthy plant growth. Stunting, dieback, and spotting on leaves are some problems that lack of micronutrients may cause.

STEM Connection

A Helpful Fungus among Us

Mycorrhizae are fungi that form mutualistic relationships with plants. Benefitting both organisms, the fungi gain carbohydrates necessary for growth and development and the plant host has increased access to the rhizosphere for nutrient and water uptake. The hyphae of the fungi colonize the roots of the plants and then extend through the soil, functionally amplifying the plant's reach through the soil. Mycorrhizae have other reported benefits to plants such as increasing drought tolerance and pathogen resistance, enhancing vigor and transplant establishment, and the potential for improved phytoremediation. **Phytoremediation** is the use of living plants to remove organic and inorganic contaminants from soil.

Figure 12-9. Boron deficiencies arise in new tissue, usually at the growing tips, and can result in dieback on terminal buds.

Figure 12-10. Copper is the least mobile nutrient in plants, and deficiency symptoms appear in the young tissues as twisted, misshapen leaves, and yellowing leaves that wither easily.

Boron (B)

Boron (B) is a crucial component of cell walls and is involved in DNA and RNA synthesis. Boron deficiencies are prevalent throughout North America. Because of its immobility, symptoms are first observed in new growth. Stunting occurs due to the termination of root and stem growth, and often dieback can be seen at terminal buds. Other deficiency symptoms may include diminished flowering; thickened, curled or twisted chlorotic leaves; and soft or necrotic spots in fruits and tubers, **Figure 12-9**.

Organic matter in the soil holds boron, providing a reservoir for plant use. Plants uptake boron in the form of boric acid (H_3BO_3), which is available through a broad pH range of 5.0 to 7.5. In more alkaline soils, boron needs to be added to a fertilizer to ensure that it is available for plants. Boron becomes readily soluble in acidic pH levels. Leaching may transpire in sandy soils, especially with heavy irrigation or rainfall. Many crop species have sensitivities to boron toxicities. Careful selection of crops for rotation in fields after boron applications should be considered.

Chlorine (Cl)

Chlorine's role in the plant is not fully understood. Chlorine (Cl) appears to support root development and to be essential in the electron transport chain of photosynthesis. It has also been found to help with stomata regulation, acclimation to water availability, and transport of other nutrients in the plant, such as calcium, magnesium, and potassium. Deficiencies are challenging to diagnose because of the widespread availability of chlorine in the soil. Yellowing and necrotic spotting on the leaves can occur, and leaves can become bronze in appearance.

Precipitation naturally adds chlorine to the soil, and it is commonly a component of most fertilizers. Potassium chloride (KCl) is the most practical source, and the form that plants can use is chloride (Cl^-). Chloride is a highly mobile ion in the soil, and management practices to minimize leaching should occur.

Copper (Cu)

Copper (Cu) is a catalyzer and component of many plant enzymes. It is one of the electron carriers in photosynthesis and necessary for chlorophyll and lignin synthesis. Copper is the most immobile of the micronutrients. Deficiency symptoms appear on young plants or young tissue. Plants may exhibit twisted or misshapen leaves, leaf yellowing, pale green leaves that wither easily, necrotic spotting, stunted growth, and stem and twig dieback, **Figure 12-10**. Copper deficiencies tend to be rare, but they appear most commonly in plants grown on organic or sandy soils.

Usable forms of copper are the cuprous ion (Cu^+) or the cupric ion (Cu^{2+}), with copper uptake decreasing as soil pH levels rise. Copper uptake may decrease as available levels of phosphorus and iron increase. Copper is generally applied as copper sulfate ($CuSO_4$) or as finely ground copper oxide (CuO). Build-up of copper in the soil can cause toxicities.

Iron (Fe)

Iron plays an active and critical role in normal plant growth and development. It is necessary for chlorophyll synthesis and the electron transport chain of respiration. It is a component of *cytochromes* (proteins that carry electrons during respiration). Iron is one of the most commonly observed micronutrient deficiencies. It primarily causes yellowed leaves due to low concentrations of chlorophyll, **Figure 12-11**. Interveinal chlorosis also appears in young leaves, and stems can be short and slender.

Iron becomes unavailable at high pH levels. Although it is abundant in most soils, it can be limited in sandy soils or tightly bound in organic soils. Cool, wet weather and poorly aerated or compacted soils limit iron uptake. Iron is available to plants in the form of ferrous ions (Fe^{2+}) and ferric ions (Fe^{3+}). Iron can be applied as foliar spray or in the form of a chelate. A *chelate* is a large organic molecule that can hold metallic cations (positively charged ions) and prevent them from reacting with anions (negatively charged ions). By keeping them from reacting, chelates allow metallic cations to remain soluble and therefore available for plant uptake. In horticulture, iron is often chelated to make the iron soluble in water and accessible for plant uptake.

Photograph from UC Fruit Report (http://ucanr.edu/sites/fruitreport)

Figure 12-11. Iron is a key component of chlorophyll synthesis; a deficiency results in yellowed leaves as visible on this plum tree's leaves.

Manganese (Mn)

Manganese (Mn) is responsible for activating enzymatic processes and is a part of some enzymes. It aids in photosynthesis and upholds chloroplast membrane integrity. Manganese has been shown to accelerate germination and maturity and increase the availability of phosphate and calcium. Manganese is immobile in the plant. Signs of deficiencies first appear in new tissue with yellowing between the veins, **Figure 12-12**. Necrotic spots may also show up in the leaves.

Manganese is available to plants in the form of manganous ions (Mn^{2+}). If it needs to be applied to the soil, it is usually banded as magnesium sulfate ($MnSO_4$). Deficiencies tend to occur in high pH soils. They can be corrected by keeping the pH below 6.5.

Photograph by Shep Eubanks

Figure 12-12. This peanut plant shows manganese deficiency as interveinal chlorosis.

Soil moisture may affect manganese availability. The most severe deficiency symptoms materialize in cool temperatures on soils with high organic matter and waterlogged soils. As temperatures warm and soil moisture decreases, symptoms will often disappear.

Molybdenum (Mo)

Molybdenum (Mo) is involved in phosphorus metabolism and nitrogen fixing with symbiotic and free-living, nitrogen-fixing soil bacteria. Minute amounts of molybdenum are needed, and it is usually available in most mineral soils. Deficiencies arise when soil pH is too low, the opposite for most micronutrients. Symptoms appear as general yellowing or interveinal chlorosis and stunting of the plant. Necrosis along leaf margins and cupping or distorting of the leaves may also be present, **Figure 12-13**. Plants absorb molybdenum in the form of simple molybdate ions (MoO_4^{2-})

IPNI Image/V.Arunachalam

Figure 12-13. Soils deficient in molybdenum may be corrected with applications of as little as 35 to 70 grams (1.0 ounces per acre).

Nickel (Ni)

The enzyme urease in legumes contains nickel, which is necessary for the conversion of urea to ammonia (NH_3) and part of the nitrogen metabolic pathway. Nickel deficiency symptoms appear as necrosis or necrotic spotting in leaf tips. In nursery crops, mouse ear (small, curled leaves) and stunted growth arise due to nickel deficiencies, **Figure 12-14**.

Similar to many of the micronutrients, nickel is unavailable at high pH levels. In these soils nickel fertilization may be necessary for good quality crop growth and yield. Plants use nickel in the form of nickelous ions (Ni^{2+}). This provides a challenge to growers as nickel readily oxidizes to unavailable forms in the soil. The simple and most effective practice to correct nickel deficiency is to spray leaves with a solution of nickel sulfate ($NiSO_4$) or other water-soluble nickel fertilizer.

Photograph Gary K England

Figure 12-14. Mouse ears or a curling of the leaves is a symptom of nickel deficiency in nursery crops.

Did You Know?
The United States Mint has produced the nickel coin since 1866 with a composition of 75% copper and 25% nickel. Valued at five cents, the nickel actually cost nine cents to mint in 2013.

Zinc (Zn)

Zinc is a critical component of many enzymes. It plays a role in the synthesis of the amino acid tryptophan, which is a precursor to the plant growth hormone auxin. Zinc is also important for fruit, seed, and root system development; photosynthesis; and crop stress protection.

Zinc deficiencies reveal the impacts of low levels of auxin by showing stunted growth and shortened internodes, **Figure 12-15**. Many plants develop a rosette stem habit. Leaf sizes can be small, margins may be distorted, and interveinal chlorosis on older leaves could appear. Plants absorb zinc in the form of zinc ions (Zn^{2+}).

Zinc tends to be widely available in mineral soils. Similar to iron and manganese, it becomes fairly insoluble at high pH levels. Some sandy soils low in organic matter and organic soils in cold, wet weather will have zinc deficiencies.

Photograph from UC Fruit Report (http://ucanr.edu/sites/fruitreport)

Figure 12-15. Plants deficient in zinc, such as this peach, show interveinal chlorosis.

Mineral Nutrient Uptake

Plants have relatively simple needs for growth and development. Through photosynthesis, water and carbon dioxide are transformed into organic compounds for energy. The uptake of essential elements from the environment provides the building blocks for essential biochemical reactions, the distribution of these substances throughout the plant, and the use of these substances in metabolic pathways that drive growth. Most of the inorganic nutrients are absorbed through the roots as ions from the soil solution, as discussed in Chapter 9.

Did You Know?
Sphalerite (zinc sulfide) is the primary ore mineral from which most of the world's zinc is produced.

Nutrient Cycle

All nutrients are cycled through the environment. By understanding how nutrients move through the soil, water, plants, and atmosphere, growers can use effective and sustainable nutrient management practices. Generally nutrients follow a basic cycle with these key activities:

- Roots uptake water containing dissolved nutrient ions from the soil.
- Plants absorb nutrients, grow, develop, and eventually die, returning nutrients to the soil through plant residue.
- As nutrients are removed from the soil solution by the plants, several reactions occur to buffer or resupply nutrients to the solution.
- Ions adsorbed to the surface of the soil desorb from the soil to resupply the soil solution. This is a critical step to the availability of nutrients to plants.
- Nutrients added through fertilizer, organic materials (manure/compost), or other substances increase the ion concentration in the soil solution.
- Some ions do not stay in solution; rather they are adsorbed to soil particles or precipitated as solid minerals. Often this happens at a pH not favorable to the nutrient.
- Minerals can also dissolve to resupply the soil solution. For example, a change of pH will allow manganese to dissolve.
- Soil organisms decompose plant residues and can absorb nutrients from the soil solution (and sometimes compete with plants).

- As plants, microorganisms, and other organisms die and decompose, nutrients are released back into the soil solution.
- Nutrients can be lost through leaching by rainfall or heavy irrigation.
- Nutrients can be added through rainfall.
- Nutrients can be lost in gaseous form to the atmosphere.

Nitrogen Cycle

All nutrients interact with the environment in specific ways. Nitrogen, one of the most important macronutrients for plant growth, has a unique cycle. This cycle illustrates how nutrients change forms with interaction with the soil, microbes, atmosphere, and plants, **Figure 12-16**.

The nitrogen cycle has three principle stages:

1. *Ammonification* or nitrogen mineralization is the process by which ammonium ions (NH_4^+) are made by saprophytic bacteria and fungi by incorporating nitrogenous compounds into amino acids and proteins, with excess nitrogen (in the form of ammonium ions) being released as a by-product of their metabolism.

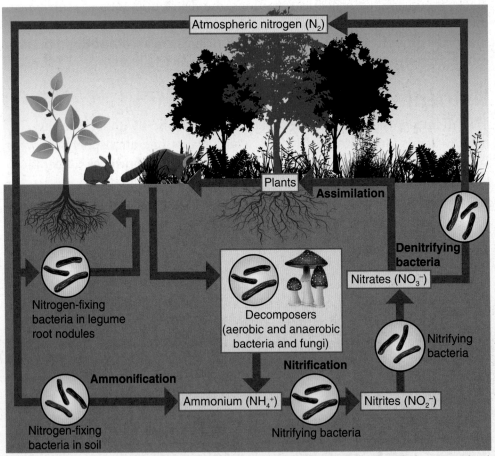

Nikitina Olga/Shutterstock.com (plant); Ihnatovich Maryia/Shutterstock.com (background plants); Glasscage/Shutterstock.com (roots); iHonn/Shutterstock.com (bacteria); Macrovector/Shutterstock.com (raccoon/rabbit); La Gorda/Shutterstock.com (mushrooms)

Figure 12-16. The nitrogen cycle is dynamic, with forms of nitrogen changing through the metabolic processes of soil microbes and their interaction with the plant and its environment.

2. *Nitrification* also involves bacteria. Several species of soil-dwelling bacteria will oxidize ammonium ions into nitrite ions (NO_2^-) in a process called nitrification. Nitrification is an energy-yielding process for these bacteria. They use oxygen to reduce carbon dioxide for energy. This is similar to the process in which plants use light energy to reduce carbon dioxide.

    The nitrifying bacterium oxidize ammonium to nitrite ions (NO_2^-).

    $$2NH_4^+ + 3O_2 \rightarrow 2NO_2^- + 4H^+ + 2H_2O$$

    Nitrite is a toxic form of nitrogen for plants, but it is rarely present in amounts to impact growth. *Nitrobacter*, another soil-dwelling bacteria, oxidizes the nitrite to form nitrate ions (NO_3^-), also with a release of energy.

    $$2NO_2^- + O_2 \rightarrow 2NO_3^-$$

3. *Assimilation* is the formation of organic nitrogen compounds from inorganic nitrogen compounds present in the environment. Nitrate is the form by which almost all nitrogen is absorbed by most crop plants. In fact, most chemical fertilizers contain ammonium ions (urea), which need to be converted to nitrate ions (NO_3^-) by nitrification.

Denitrification

Denitrifying bacteria reduce nitrate to a volatile form of nitrogen, such as nitrogen gas (N_2) and nitrous oxide (N_2O), which then return to the atmosphere. This process, called **denitrification**, tends to occur in anaerobic conditions such as swamps, but it is not limited to such conditions.

Industrial Nitrogen Fixation

Industrial nitrogen fixation is a process in which nitrogen gas (N_2) reacts with hydrogen gas (H_2) at high temperature and pressure to form ammonia. This process is energy expensive and requires the use of fossil fuels to make chemical fertilizers for agricultural and garden applications.

Nitrogen Loss

Nitrogen loss from the cycle happens when plants are harvested. Ultimately, the nitrogen is returned to the cycle. It also happens if nitrogen leaches from the soil where it can enter groundwater or runs off into the waterways, requiring a need for thoughtful fertilizer applications.

Nutrient Mobility

How readily nutrients move through the soil solutions impacts the supply available for uptake by the plant. Nutrient ions have a positive (cation) or negative (anion) charge associated with them. Mobile ions such as nitrate (NO_3^-), sulfate (SO_4^{2-}), and chloride (Cl^-) are not attracted to soil particles (called exchange sites); are soluble; and easily move through the soil solution for plant uptake. These ions have a larger sorption zone within the volume of soil that roots occupy, making access fairly straightforward.

"Life on earth is such a good story; you cannot afford to miss the beginning... Beneath our superficial differences, we are all of us walking communities of bacteria. The world shimmers, a pointillist landscape made of tiny living beings."
—Lynn Margulis

Sorption is a term used to describe both adsorption of ions to soil particles and absorption of nutrient ions into a plant. A sorption zone describes both of these processes that occur simultaneously. For example, if a gardener is fertilizing her garden, she can broadcast a nitrogen fertilizer. The mobility of nitrogen will allow it to reach the plant roots.

Immobile ions have a much more limited movement within the soil. Plant roots get these nutrients from a much smaller sorption zone that surrounds individual roots. In practical applications, fertilizer containing the phosphate ion ($H_2PO_4^-$) should be placed near the active root zone of the plant. Other immobile ions include ammonium (NH_4^+), potassium (K^+), calcium (Ca^{2+}), magnesium (Mg^{2+}), and molybdate (MoO_4^{2-}).

Soil pH

As discussed in Chapter 11, *Soils and Media*, nutrient ions have positive and negative charges that influence the acidity or alkalinity of a soil, known as the soil pH. Soil pH influences plant growth by affecting the availability of plant nutrients for uptake as well as the activity of microorganisms in the soil. Some nutrients are available to plants over a broad range of pH, whereas other nutrients are more sensitive to fluctuations in pH. Plant nutrients such as iron, zinc, manganese, and copper are very soluble at low pH and readily available to plants. However, most other plant nutrients are available at a slightly higher pH. Growers frequently test soil pH to identify nutrient deficiencies or toxicities and manage their soil and fertilizer applications accordingly.

Nutrient Sources

Proper nutrient management for plants demands an understanding of the nutrient needs by crop, the specific nutrient, the soil, and local environmental conditions. Soil and tissue analysis is important to gain information for managing nutrients to promote healthy plants. Nutrients may be supplied using organic materials or inorganic fertilizers.

Soil and Tissue Analysis

The soil provides a source for many of the nutrients required by plants. However, because the agricultural and garden ecosystem is open (as crops are harvested or as leaching or runoff may occur), nutrients need to be replaced by adding fertilizer and organic materials. Soil and plant tissue testing provide information to help growers make decisions about nutrients.

A soil test analyzes soil for nutrient quantity (generally nitrogen, phosphorus, and potassium), organic matter, soil texture, and pH. Some tests will provide deeper data, such as levels of additional macronutrients, calcium, and magnesium. Information about micronutrients, such as iron, zinc, and other elements, may be included. Soil tests provide recommendations for growers on fertilizer application rates for specific crops and adjustments for pH if needed.

Plant tissue analysis offers another method for testing nutrient levels in a plant, giving data for fertilization decisions. Plant tissue analysis uses leaves or parts of leaves to determine the amount of nutrients in a plant.

Samples can vary depending on the age of the tissue (older leaf versus younger leaf); where the tissue is located on the plant; and the type of organ (leaf or stem) as well as the environmental conditions. Despite these variables, tissue analysis can be critical for crops such as those grown in the floriculture industry.

Organic Materials

Organic materials, such as plant residues, animal manures, and compost, can be a source of nutrients when added to the soil. As these materials break down, they provide a slow release of nutrients and also improve soil structure.

Plant Residues

After harvest, crop residue, such as stalks, leaves, vines, and other parts, can be left behind in the field, **Figure 12-17**. Microbes, such as bacteria, fungi, and other decomposers, break down this plant material. When these microorganisms die, they return the nutrients to the soil for plant uptake. A *ratio* is the relative value of one thing compared to another. If plant material has a high carbon to nitrogen (C:N) ratio, the microorganisms in the soil will consume the soil's nitrogen for their growth. This can potentially cause a nitrogen deficiency for plants.

Alex Ionas/Shutterstock.com

Figure 12-17. Corn stover (non-grain part of harvested corn) has a high carbon to nitrogen ratio that can impact the availability of nitrogen to the plants. However, it minimizes soil erosion and builds soil structure.

Straw and wood chips have a high C:N ratio (50:1). Although they will eventually decompose and add nutrients to the soil, the decomposition process will create a temporary nitrogen deficiency. Gardeners should include an application of nitrogen fertilizer to ensure both crop growth and microbial decomposition of the straw or wood chips. Plant wastes, such as cottonseed meal, sawdust, or bark, may also be added to amend the soil.

Green Manures

Cover crops, such as alfalfa, clover, vetch, and buckwheat, are planted in fields to improve soil structure, minimize soil erosion, and add nutrients, **Figure 12-18**. Leguminous cover crops can add significant amounts of nitrogen to the soil through their ability to fix nitrogen. These plant materials are termed green manures. They are tilled into the soil to increase the availability of nutrients for plant uptake.

HildaWeges Photography/Shutterstock.com

Figure 12-18. Buckwheat is an ideal cover crop for warm, summer conditions.

"I'm queen of my own compost heap, and I'm getting used to the smell."
—Ani DiFranco

Compost

The process of composting uses naturally occurring microorganisms to break down organic materials into a useful soil amendment. The resulting compost can be added to the farm field, nursery container, or home garden. Adding composted material, rather than raw plant residues (such as straw), reduces the competition for nitrogen and rapidly improves the physical structure of soil. This improved structure allows better drainage, increased water-holding capacity, and improved infiltration. Successful composting requires a few key components, including:

- Plant material such as leaves, stems, crop waste, lawn clippings, or straw.
- Soil or compost in small amounts that contain starter microorganisms for decomposition.
- Aeration to provide the oxygen required for microbes to decompose plant material efficiently. This is generally accomplished through turning the pile.
- Moisture provided through rainfall or irrigation to maintain the decomposition process.
- Fertilizer—depending on the C:N ratio of plant material present, the addition of nitrogen may be necessary for optimal decomposition.

These materials can be added to a compost bin, **Figure 12-19**, or in large-scale operations, piled in long windrows. The bacteria and fungi will begin to actively break down materials and give off heat. The heat generated creates internal pile temperatures in the range of 130°F to 150°F (55°C to 65°C). This heat kills off undesirable pest materials, such as weed seeds, insects, and plant pathogens. The composting process creates a soil material that efficiently recycles nutrients and enhances the pool of nutrients available for uptake.

Arina P Habich/Shutterstock.com

Figure 12-19. Compost bins provide a space to decompose plant material into a rich soil amendment for the garden. What other types of compost bins are used by home gardeners? Which designs are most efficient?

Animal Wastes

Animal manure is usually a combination of both animal excrement (feces and urine) and bedding material, such as straw or wood shavings. Manure nutrients can help build and maintain soil fertility. They can improve soil *tilth* (quality for crop growth), increase water-holding capacity, diminish risk of wind and water erosion, improve aeration, and promote beneficial organisms. There are two principal objectives in applying animal manure to land:

- Maximize use of the manure nutrients by crops.
- Minimize water pollution potential.

Animal manures vary greatly depending on the species of livestock, their diet, the proportion of bedding material included, and the storage and age of the manure. Properly composted manure provides nitrogen, phosphorus, potassium, and a number of micronutrients. Much of the nitrogen present in animal waste comes through the urine and can be lost by volatilization, runoff, and possibly leaching if the manure is stored without protection from the weather, **Figure 12-20**. Fresh manure is often applied to a commercial field to minimize nitrogen losses, and it is applied before crops are planted. Raw manure can burn plants with salts and ammonia, so usually only well-rotted or composted manure is added to a home garden. In this case little nutrient value remains in the manure, but it is highly valued as a soil amendment.

Animal manures do bring a significant source of phosphorus to plants. However, the amount of phosphorus in manure is lower when compared to the phosphorus content of inorganic fertilizers. Phosphorus fertilization with animal manure is most efficient in cooperation or as part of a livestock operation.

Industrial and Municipal Wastes

Wastewater treatment facilities and other commercial industries create waste every day, some of which has potential as fertilizers. Many municipal sewage plants have created systems that use microbes to recycle the solid waste materials (human waste) into fertilizers that can be used in a variety of settings from forestry applications and lawns to home gardens. The process removes human pathogens and captures the mineral nutrients for plants, **Figure 12-21**.

Svend77/Shutterstock.com

Figure 12-20. Organic materials vary widely in the amount of nitrogen available for plant uptake. The nitrogen in animal waste may be lost through leaching if the manure is stored without protection from the weather.

Perfect Gui/Shutterstock.com

Figure 12-21. Some municipal districts have systems designed to recycle wastewater and waste into a fertilizer product.

Inorganic Fertilizers

Fertilizers that are developed from materials other than plants and animals are termed inorganic. Many of the organic fertilizers release nutrients slowly over time, as microbes change ions into a form that plants can use. Most inorganic fertilizers are made of mineral nutrients in a form that is readily available for plant use. Also termed synthetic fertilizers, most nutrients in these fertilizers come from raw mineral sources and are refined in ways that make them consistent and inexpensive.

Corner Question

How much nitrogen (N), phosphate (P_2O_5), and potash (K_2O) would a bag of 18-6-12 fertilizer that weighs 50 lb contain?

Nitrogen Sources

Nitrogen is the mineral nutrient required in the highest quantity by plants to drive growth and optimize yields. Blood meal, cottonseed meal, composted manures, and other organic sources of nitrogen can be expensive to produce and are not available in large quantities. The most economical and widely used source of nitrogen comes from ammonia, a gas produced when nitrogen gas is combined with hydrogen gas pulled from natural gas. This method is called the Haber-Bosch process and follows the equation $N_2 + 3 H_2 \rightarrow 2 NH_3$. Ammonia is the base product for many other fertilizers, including anhydrous ammonia (NH_3), ammonium nitrate (NH_4NO_3), urea [$(NH_4)_2CO_3$], ammonium hydroxide (NH_4OH), nitric acid (HNO_3), and ammonium phosphates. Nitrogen may be combined with other elements such as sodium nitrate ($NaNO_3$).

Phosphorus Sources

Phosphorus in fertilizer is primarily delivered in a form called *superphosphate*. The raw material rock phosphate is treated with phosphoric acid to create superphosphate. This product delivers a high and predictable phosphorus percentage to plants, ranging from 20% for superphosphate to 45% for concentrated superphosphate.

Potassium Sources

Potassium comes from salt mining operations that extract crude (which is subsequently refined) or relatively pure salts. The two principal potassium salts are muriate of potash (KCl) and potassium sulfate (K_2SO_4). Both are quite soluble and tend to be mixed with other elements in mixed fertilizers.

Mixed Fertilizers

In many of the chemical processes to synthesize nitrogen, phosphorus, and potassium fertilizer, secondary macronutrients, such as calcium or sulfur, are part of the chemistry to fix the nutrients into a stable form that plants can use. Many micronutrients can also be present in small quantities as impurities.

Mixed fertilizers containing multiple nutrients are commonly used for field, nursery, and greenhouse crop production. A *complete fertilizer* is one in which nitrogen, phosphorus, and potassium are all present in a mixture. The guaranteed analysis (or *fertilizer grade*) is a listing of nutrients contained in the bag by weight. The first number of the analysis lists the percentage by weight of nitrogen (N). The second number represents the percentage by weight of phosphorus (P_2O_5 or phosphate). The third number is the percentage by weight of potassium (K_2O or potash). Knowing the fertilizer grade is important in determining how much fertilizer to apply to your lawn, garden, nursery crop, or vegetable field.

Fertilizer Calculations

The ratio of a fertilizer is the proportion of nitrogen (N) to phosphate (P_2O_5) to potash (K_2O) content. For example, a 5-10-5 fertilizer has a ratio of 1-2-1, and an 18-6-6 fertilizer has a ratio of 3-1-1. Fertilizer recommendations from a soil test are given in ratios. For example, if a soil test recommended a 3-1-2 ratio for a lawn, the ideal fertilizer would be something such as 18-6-12. A fertilizer with a 3-1-2 ratio contains 1.5 times as much nitrogen as potash and three times more nitrogen than phosphate.

The fertilizer ratio does not usually reflect nitrogen, phosphorus, and potassium content. Rather, nitrogen is expressed on an actual elemental basis and phosphorus and potassium are expressed as phosphate and potash compounds. Thus, an 18-6-12 analysis fertilizer contains 18% nitrogen (N), 6% phosphate (P_2O_5), and 12% potash (K_2O). A bag of 18-6-12 weighing 100 lb would contain 18 lb of nitrogen, 6 lb of phosphate, and 12 lb of potash.

Phosphorus (P)

Phosphate (P_2O_5) contains 44% actual phosphorus. To determine how much phosphorus (P) is in 100 lb of 18-6-12, multiply 6 lb P_2O_5 by 0.44 (44% of P_2O_5 is actual P):

$$(P_2O_5\ 6\ lb) \times (0.44) = 2.6\ lb\ actual\ P$$

Potassium (K)

Potash (K_2O) contains 83% actual potassium. To find out how much potassium (K) is in 100 lb of 18-6-12, multiply 12 lb K_2O by 0.83 (83% of K_2O is actual K):

$$(K_2O\ 12\ lb) \times (0.83) = 10\ lb\ actual\ K$$

Therefore, 100 lb of 18-6-12 contains these actual nutrients:
- 18 lb of nitrogen (N).
- 2.6 lb of phosphorus (P).
- 10 lb of potassium (K).

One can calculate how much of any fertilizer to purchase to supply any nutrient with the following information:
- Area of land on which fertilizer will be applied (in square feet).
- Recommended application rate.
- The analysis of the fertilizer.

Example: Lush Lawns Fertilizer Company has been selected by a homeowner to provide 1 lb of total nitrogen per 1000 ft² of area as recommended by the soil test results. The homeowner's lawn area is 10,000 ft². How much 18-6-12 fertilizer should the fertilizer company apply to the area?

1. Area to be fertilized × recommended rate of nitrogen = total pounds of nitrogen needed
   10,000 ft² × 1 lb of N per 1000 ft² = 10 lb of N needed
2. Pounds of nutrient needed ÷ percentage of nutrient in 1 lb of fertilizer = pounds of actual fertilizer needed
   10 lb of N needed ÷ .18 lb N/lb of fertilizer = 55.6 lb of fertilizer needed
   About 55.6 lb of 18-6-12 fertilizer should be applied to supply 1 lb of N per 1000 ft² to a 10,000 ft² lawn.

Methods of Fertilizer Application

Fertilizers can be applied as solids, liquids, or even gasses. The methods of application used vary depending on the equipment available, the environmental conditions, and the crop requirements. Methods include broadcasting, banding, side-dressing, fertigation, slow-release fertilizers, soil injection, and foliar fertilization.

Broadcasting

Spreading fertilizer uniformly over a field is method called *broadcasting*. Broadcasting is a common practice with agronomic row crops. Fertilizer is cast across a field and tilled or incorporated into the soil prior to planting. Fertilizers may be in liquid or dry forms, such as a manure slurry or granular fertilizer. With crops such as turfgrass, a granular or powdered fertilizer can be simply top-dressed or spread right on top of the grass. Irrigation usually follows this practice to dissolve the fertilizer into the soil and minimize risk of salt damage to the plants. **Figure 12-22** shows a broadcast spreader that is used by a farmer to fertilize the fields prior to planting. Broadcasting can be fast and economical, but there may be high nutrient losses and diminished nutrient efficiency uptake, especially with phosphorus.

Leonid Ikan/Shutterstock.com

Figure 12-22. A broadcast spreader is ideal for nutrients that have high mobility in the soil solution.

Banding

In *banding* applications, the fertilizer is spread in bands or lines where developing roots will easily reach it; either to the side and below the seed rows, slightly below the seeds, or in between rows. A common practice is to band fertilizer 2″ to the side and 2″ deeper than the seeds or plants, although specific crops have optimal distance and position preferences. Banding provides the plants with a concentrated zone of nutrients and can improve nutrient use efficiency. For phosphorus, banding increases phosphorus uptake by plants, but the process can be more costly than broadcasting and takes more time.

Side-Dressing

In the *side-dressing* method, fertilizer is applied between rows of young plants. This method can promote growth and nutrient uptake, especially with crops that have a heavy nutrient demand, such as corn and potatoes. Although side-dressing can promote high nutrient use, it can be costly. Using it effectively requires good judgment and knowledge of the crop for proper timing.

Fertigation

Fertigation is used in greenhouse operations to distribute water-soluble fertilizers through an irrigation system, **Figure 12-23**. In greenhouse settings, fertigation can be a very economical and efficient method for nutrient delivery and uptake. Many nursery practices also use fertigation in combination with a slow-release fertilizer in the soilless potting media. Research has suggested that as much as 70% of the water and nutrients applied to a container crop by overhead fertigation can be lost to runoff and leaching.

Mediagram/Shutterstock.com

Figure 12-23. Fertigation is a common practice in greenhouses. Fertilizer mixes with the irrigation water and is distributed in the watering systems. *Can fertigation be used by home gardeners?*

Slow-Release Fertilizer

Plants require nutrients throughout their growth and development. Slow-release fertilizers release nutrients over an extended period and, theoretically, should meet all the nutrient demands of a crop during active growth, **Figure 12-24**. Slow-release fertilizers also limit the amount of nutrients lost to leaching or runoff. No special equipment is needed for application. Many organic sources (manures and compost) of fertilizers already slowly release nutrients due to the mineralization process of soil microbes. Fertilizer producers have also manufactured slow- or controlled-release fertilizers that bind nutrients in some sort of coating that slowly decomposes with the help of microbes and soluble materials. Many slow-release granules are incorporated into the soilless media of nursery containers before planting. This allows growers to do a single application of fertilizer.

Soil Injection

Anhydrous ammonia in a liquid form can be injected into the soil near the plant roots to precisely apply nutrients. The ammonia combines with soil moisture to form ammonium (NH_4^+), a plant-available form of nitrogen. This practice is not common, but is used in the landscape industry for tree and shrub nutrient applications.

Photograph by Matthew Vann

Figure 12-24. These tobacco float trays sit in a nutrient solution that provides constant moisture to germinating seeds.

Foliar Application

Foliar application is a method of providing micronutrients to plants by spraying fertilizer directly onto the leaves. This technique is generally used only to correct a nutrient deficiency. It is costly and has a high risk of phytotoxicity. *Phytotoxicity* is a poisonous effect by a substance on plant growth. For example, too many nutrients can cause abnormal growth or death in plants.

Careers in Plant Nutrition

The horticultural industry provides a variety of positions for those interested in plant nutrition, depending on your interests and level of education. For example, you may choose to research and develop organic or inorganic fertilizers, design methods of fertilizer application, or sell and distribute fertilizers for agronomic or floriculture operations. Two career positions you might consider in the field of plant nutrition are working as a fertilizer sales representative or a composting operator.

Fertilizer Sales Representative

Fertilizer sales representatives develop and implement fertilizer and related product sales to nursery, greenhouse, vegetable, grain, and fruit market segments. As a sales representative, you must have a fluid knowledge of your company's products, their use, and applicability in a given situation. It is also helpful to have hands-on experience or education in one or more of the market segments to which you will be selling your products.

A range of companies employ sales representatives in the agricultural market, including chemical companies supplying fertilizers or pesticides or agricultural supply companies. An associate's degree or bachelor's degree in horticulture or a related discipline is preferred. Five or more years of experience as a grower/manager may sometimes be substituted for formal education.

Composting Operator

Many landscaping companies, public gardens, municipalities, or independent composting businesses hire operators to manage the production of high-quality compost. Compost operators are responsible for daily operations of the production of compost, including the retrieval or receiving of materials such as soil waste, leaves, feed stocks, food waste, and organic debris. The storage and delivery of compost may also be a responsibility. Operators ensure proper production and creation of high-quality soil mixes through monitoring and recordkeeping.

Operators must be able to demonstrate capacity in the safe operation of loaders, tractors, plows, trucks, forklifts, manure spreaders, screeners, and other composting equipment. A high school diploma and a minimum of three years of experience working at a farm-scale composting facility and an understanding of soil science are preferred for this position. Applicants must also have or be willing to obtain a CDL Class A driver's license and a forklift operator's license.

> "Don't judge each day by the harvest you reap, but by the seeds that you plant."
> —Robert Louis Stevenson

CHAPTER 12 Review and Assessment

Chapter Summary

- Plants require minerals for normal development, to drive cellular and organ functions, and for various plant processes. There are 17 elements essential for plant growth.
- Nitrogen, phosphorus, potassium, magnesium, calcium, and sulfur are considered macronutrients because they are needed in large quantities by plants for normal growth and development.
- The nutrients boron, chlorine, manganese, iron, nickel, copper, zinc, and molybdenum are termed micronutrients. They are essential for plant growth, but they are needed in small or minute amounts.
- Carbon, hydrogen, and oxygen are used by the plant to manufacture food and are considered the building blocks for plant growth.
- Plants can only use certain forms of a nutrient ion, and most inorganic nutrients are absorbed through the roots as ions in the soil solution.
- Nutrients are constantly cycled through the soil, water, plants, and atmosphere. An example of a nutrient cycle is the nitrogen cycle. Through soil dwelling bacteria and other microbes, nitrogen changes form in order to be available to plants.
- Nutrient ions have positive and negative charges, which can affect their mobility and availability for plant uptake.
- A soil test analyzes a field soil for nutrient quantity, organic matter, soil texture, and pH. Soil tests provide recommendations for growers on fertilizer application rates for specific crops and adjustments for pH if needed.
- Plant nutrients can come from organic materials or inorganic sources (raw minerals). Organic materials include plant residues, green manures, compost, animal wastes, and municipal wastes.
- Fertilizers that are developed from materials other than plants and animals are termed inorganic. Most inorganic fertilizers are made of mineral nutrients in a form that is readily available for plants to use.
- Mixed fertilizers containing multiple nutrients are commonly used for field, nursery, and greenhouse crop production. A complete fertilizer is one in which nitrogen, phosphorus, and potassium are all present in a mixture.
- Fertilizer calculations consider the recommended rate of application per given square footage and the fertilizer grade to determine how much fertilizer should be applied.
- Fertilizer can be applied using multiple methods including broadcasting, banding, side-dressing, fertigation, slow-release fertilizers, soil injection, and foliar fertilization.

Words to Know

Match the key terms from the chapter to the correct definition.

A. ammonification
B. assimilation
C. banding
D. broadcasting
E. chelate
F. complete fertilizer
G. cytochrome
H. deficiency
I. denitrification
J. fertigation
K. fertilizer grade
L. foliar application
M. interveinal area
N. macronutrient
O. micronutrient
P. necrotic lesion
Q. nitrification
R. phytoremediation
S. phytotoxicity
T. ratio
U. side-dressing
V. soluble
W. sorption
X. superphosphate
Y. tilth

1. A fertilizer containing all three primary macronutrients: nitrogen, phosphorus, and potassium.
2. The process by which ammonium ions (NH_4^+) are made by saprophytic bacteria and fungi by incorporating nitrogenous compounds into amino acids and proteins with excess nitrogen (in the form of ammonium ions) being released as by-product of their metabolism.
3. Tissue death that often occurs in plant leaves as spots along the margins or in entire leaves as a result of a nutrient deficiency or plant pathogen.
4. A listing of nutrients contained in a fertilizer by weight.
5. A poisonous effect by a substance on plant growth.
6. A general term to describe good soil quality for crop growth, including texture, structure, and pore space.
7. Able to be dissolved in water.
8. Rock phosphate that has been treated with phosphoric acid to increase the amount of plant-available phosphorus.
9. The process by which several species of soil-dwelling bacteria will oxidize ammonium ions into nitrite ions.
10. The formation of organic nitrogen compounds from inorganic nitrogen compounds present in the environment.
11. A process in which bacteria reduce nitrate to a volatile form of nitrogen, such as nitrogen gas (N_2) or nitrous oxide (N_2O), which then return to the atmosphere.
12. A nutrient that plants need in high quantities for normal growth and development.
13. The process of adding fertilizer to irrigation water.
14. A chemical compound composed of a metal ion and a substance whose molecules can form several bonds to a single metal ion.
15. A lack or shortage of something, such as plant essential nutrients.

16. A term used to describe both adsorption of ions to soil particles and absorption of nutrient ions into a plant.
17. The relative value of one thing compared to another.
18. A nutrient that is essential for plant growth but is needed in small or minute amounts.
19. The space between veins in a leaf.
20. A protein that carries electrons during respiration.
21. A fertilizer application method in which the fertilizer is spread in lines either to the side and below the seed rows, slightly below the seeds, or between rows.
22. A fertilizer application method in which fertilizer is spread uniformly over a field and tilled or incorporated into the soil prior to planting.
23. The use of living plants to remove organic and inorganic contaminants from soil.
24. A method of providing micronutrients to plants by spraying fertilizer directly onto the leaves.
25. A fertilizer application method in which fertilizer is applied between rows of young plants.

Know and Understand

Answer the following questions using the information provided in this chapter.

1. What 17 elements are essential for plant growth?
2. Which elements are considered macronutrients and why?
3. Where do plants obtain carbon and is this ever a limiting element for plant growth and development?
4. Roots require oxygen for normal cellular respiration. Where do roots obtain oxygen?
5. Why is nitrogen such an important element in the plant?
6. Describe the symptoms of phosphorus deficiency.
7. Where would symptoms of calcium deficiency occur in a plant and why?
8. Which elements are considered micronutrients and why?
9. How is soil pH related to manganese deficiencies? How can manganese deficiencies be managed?
10. Summarize the key activities of a mineral nutrient uptake cycle.
11. Detail the ways that bacteria play a role in the nitrogen cycle.
12. Explain the sorption zone in the soil.
13. How does soil pH affect the nutrient balance of the soil and affect plant growth?
14. What are the reasons for analyzing soil samples and plant tissues?
15. Why does the carbon to nitrogen ratio matter in plant nutrition?
16. What are the key elements for successful composting?
17. What are the benefits of using an animal manure or waste in crop production?
18. Inorganic fertilizers are made from what type of materials?
19. In a complete fertilizer, what do each of the three numbers in the analysis mean?
20. What methods are used in applying fertilizers?

Thinking Critically

1. Your friend has started a small farm and is growing muscadine grapes. She noticed a yellowing in the older leaves on the vine. She called you to help her determine what is going on and what can she do to save her crop. Write a short summary on what you think is happening and why.
2. You have a backyard vegetable garden that you fertilize regularly with the recommended grade, rate, and timing of fertilizer application. You observe yellowed leaves and interveinal chlorosis in young leaves and stunted and slender stems. What is occurring and what can be done?

STEM and Academic Activities

1. **Science.** Most plant species have different nutrition requirements. Research a plant that you find interesting and write a report that summarizes your findings.
2. **Science.** Phytoremediation is the use of living plants to remove organic and inorganic contaminants from soil. Research the types of plants used in phytoremediation and the types of contaminants that can be removed. Write a report that summarizes your findings.
3. **Engineering.** Design a three-dimensional model that represents the relationships of nutrients to the soil solution and plant roots.
4. **Math.** As a backyard rose gardener, you know that roses are "heavy feeders." It is important to provide proper nutrition for optimal growth and flowering. Your Cooperative Extension Service agent office recommends that you use a complete fertilizer of 10-10-10 at a rate of 3 lb of actual nitrogen per 1000 ft² three times a year. How much fertilizer should you apply?
5. **Language Arts.** The ability to read and interpret information is an important workplace skill. Organic certified growers can only use specific products to provide for plant nutritional needs. Evaluate and interpret research to determine what these products are and the philosophy of some organic growers. Locate three reliable resources for the most current information and read and interpret the information. Write a report summarizing your findings in an organized manner.
6. **Language Arts.** Make a collage. Using pictures from magazines or free online resources, create a collage that helps you remember the role of each essential nutrient and its deficiency symptoms. Show and discuss your collage in a group of four to five classmates. Are the other members of your group able to determine the nutrient and functions that you tried to represent?

Communicating about Horticulture

1. **Reading and Speaking.** Create an informational pamphlet on how to apply for a job in the floriculture industry. Research résumé strategies and portfolio organization and download a sample job application. Present your pamphlet to the class. After your project has been graded and returned to you, review the instructor's comments. List the type of changes you could make to improve your project.
2. **Listening and Speaking.** Make a collage. Using pictures from magazines or free online resources, create a collage that helps you remember the function of each nutrient within a plant. Show and discuss your collage in a group of four to five classmates. Are the other members of your group able to determine the system and functions that you tried to represent?

SAE Opportunities

1. **Exploratory.** Visit a local vegetable grower and scout the fields for signs of nutrient deficiencies. Learn how the grower manages plant nutrition. Ask what techniques are used to optimize growth and yield and to conserve natural resources.

Lance Cheung/USDA

2. **Exploratory.** Create a plant nutrition calendar for your garden plants. What nutrient needs do they have? Are they similar or different? How will you supply the nutrients the plants need? What sources of nutrients will you use? Why?
3. **Experimental.** Conduct a plant nutrition experiment using your favorite tree or shrub. How will you set up your experiment? What variables are you testing? What methods will you use to gather data? Share your findings with your class or enter your experiment in a science fair.
4. **Exploratory.** Job shadow a fertilizer sales representative. What is his or her role in supporting the plant nutrition needs of growers? How does he or she determine what product might be best for a given situation?
5. **Placement.** Identify a horticultural researcher in your area who studies plant nutrition. Where does this person work? What does he or she do? How is he or she improving understanding of the role of nutrients in horticultural growing systems?

CHAPTER 13
Seed Propagation

Chapter Outcomes
After studying this chapter, you will be able to:
- Describe seed morphology and development.
- Understand the environmental conditions needed for optimal seed germination.
- Compare strategies to break dormancy responses in different seeds.
- Explain different seed propagation techniques.
- Summarize categories used for selecting seeds.
- Discuss production of high-quality seeds and seed saving and storage.
- Identify careers related to seed propagation operations and maintenance.

Words to Know

abscisic acid	lag phase	radicle
apomixis	landrace	radicle protrusion
cell expansion	maturation drying	scarification
desiccation	phenol	seedbed
genetically modified organism (GMO)	photodormancy	seedlot
	plug	thermodormancy
germplasm	plumule	transgenic
histodifferentiation	priming	viable
imbibition	quiescent	vivipary

Before You Read
Read the chapter title and tell a classmate what you have experienced or already know about the topic. Write a paragraph describing what you would like to learn about the topic. After reading the chapter, share two things you have learned with the classmate.

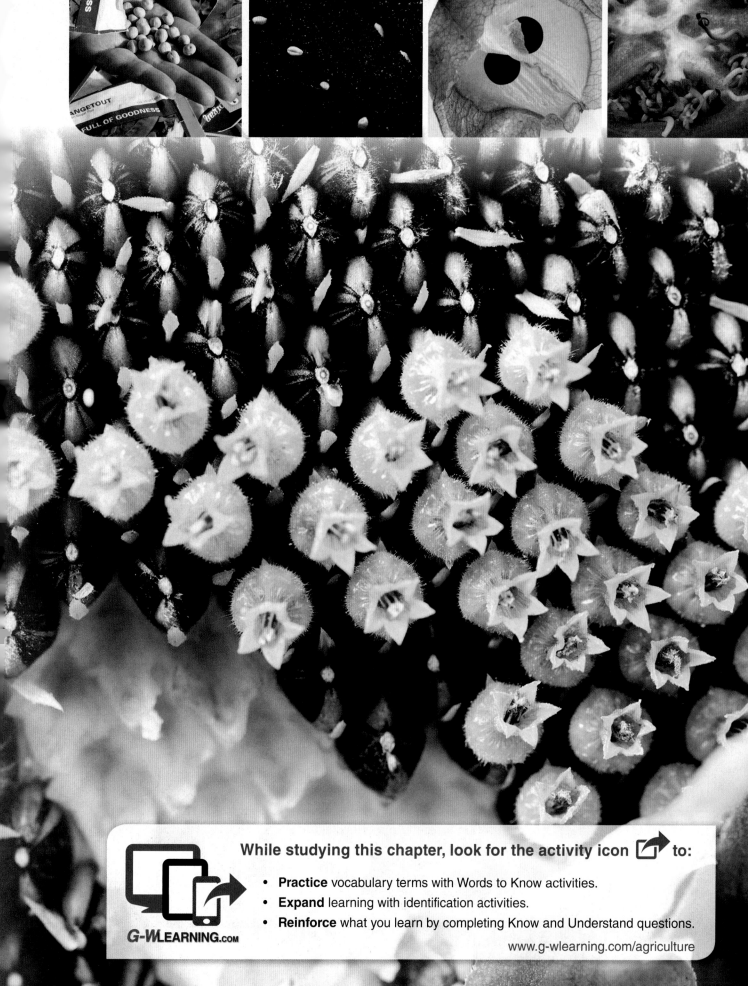

While studying this chapter, look for the activity icon to:

- **Practice** vocabulary terms with Words to Know activities.
- **Expand** learning with identification activities.
- **Reinforce** what you learn by completing Know and Understand questions.

www.g-wlearning.com/agriculture

The magic of gardening begins with sowing a simple seed and watching it burst from the soil, sprouting into something wonderful. Horticulture is the pursuit of growing plants and it all starts with seeds. Consider the zinnia, an ornamental annual. Zinnias are a much-loved garden flower and a prolific bloomer. They produce many seeds that can be saved and planted next year, **Figure 13-1**.

Where do you start when trying to grow a seed? What do seeds need in order to grow? What if your seed does not grow? Seed propagation is the science and art of understanding a seed's biology and being able to coax it to germinate and flourish.

Amawasri Pakdara/Shutterstock.com

Figure 13-1. Zinnias can be sown directly in the garden and seeds from the flowers can be saved for the following year.

Seed Morphology and Development

A seed is the mature reproductive unit of gymnosperms and angiosperms. Seeds are formed as a result of the fertilization process. In addition to the formation of a seed, fertilization is also responsible for the genetic variation of a species. All seeds have an embryo, a protective outer covering, and storage tissue.

The protective covering of a seed may be its seed coat or the fruit (pericarp) that surrounds the seed. The storage tissue in dicots is contained in the cotyledons. For monocots, energy is stored as starchy tissue known as the endosperm. The storage tissue provides the initial energy needed to drive germination. Seeds and fruits come in a wide array of shapes, sizes, and appearances.

Stages of Seed Development

The three stages of seed development are histodifferentiation, cell expansion, and maturation drying.

Histodifferentiation

Histodifferentiation is a stage of seed development when the embryo and endosperm develop distinct characteristics. Seed size increases rapidly due to the process of cellular division. In dicots, this is the stage where cotyledons are formed. In monocots, more specialized structures are formed to aid in germination.

Cell Expansion

During *cell expansion*, seeds undergo a phase of swift cell enlargement due to the accumulation of food reserves in the form of carbohydrates, fats, oils, and proteins.

Corner Question

What two plant families are responsible for almost all the food and feed crops that comprise the world's diet?

Maturation Drying

Maturation drying is the last stage of seed development. It is the point at which seeds have reached physiological maturity. Seeds experience rapid water loss through the vascular separation from the mother plant. The *desiccation*, or drying of seeds, is a preparation for seeds to be *quiescent* or dormant. Quiescent seeds are in a state of inactivity, but will readily germinate if given water. Dormant seeds will fail to germinate under even favorable conditions, and require some sort of treatment to grow. *Vivipary* is a unique, natural phenomenon in which seeds germinate inside their fruit without maturation drying. Have you ever seen a seed within the fleshy fruit of a tomato begin to germinate, **Figure 13-2**? This is vivipary. In most plants this is an undesired genetic mutation. In others, such as the mangrove, vivipary allows a plant to compete for survival.

Kathy Clark/Shutterstock.com

Figure 13-2. This tomato is displaying vivipary, a condition in which seeds germinate before they are fully mature.

Apomixis

Some plants are able to spontaneously produce seeds without going through the fertilization process. The process, called *apomixis*, occurs asexually and creates genetic clones of the parent. Apomixis might happen when an embryo develops from an unfertilized egg nucleus that has not undergone meiosis. Many plants from the genus *Citrus* form embryos adventitiously. They can sometimes produce as many as six seedlings from a single embryo. It is nearly impossible to distinguish a clonally produced seedling from one that results from sexual fertilization.

Seed Germination

Germination is the process by which seeds begin to grow. Three conditions must be met in order for germination to occur:
- The seed must be *viable* (the embryo must be alive).
- Environmental conditions must be favorable. Water, proper temperature, oxygen, and sometimes light must be present.
- Dormancy conditions must be overcome.

If these conditions have been fulfilled, then the early phases of germination can begin. In early germination, three phases describe the process of a seed beginning to increase its water uptake:
- *Imbibition* is a physical process where water is rapidly taken up by the seed, hydrating the inner tissues.
- The *lag phase* is a period with little or no water uptake, but with many cellular activities that prepare the seed to grow.
- *Radicle protrusion* is the last period of early germination characterized by the emergence of the seed root, or *radicle*, **Figure 13-3**.

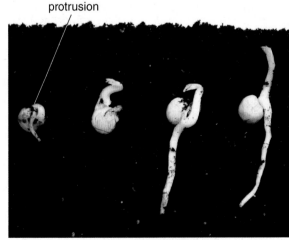

Bogdan Wankowicz/Shutterstock.com

Figure 13-3. The seed root, or radicle, emerges during the last step of early germination.

The new seedling experiences root and shoot elongation. The growing point for the root is the radicle and the growing point for the developing shoot is the *plumule*. As the seed embryo begins to grow, it draws upon its reserves of carbohydrates (starch), proteins, and lipids (oils) until photosynthesis starts to occur.

"The love of gardening is a seed once sown that never dies."
—Gertrude Jekyll

Germination Rates

Many seeds have variable germination rates and can be measured by three indicators: percentage, rate, and uniformity. A *seedlot* is made up of seeds of a particular crop gathered at one time, with similar germination rates and other characteristics. The number of seeds that germinate in a given seedlot defines the germination percentage. For example, if there are 100 seeds and 70 germinated, the germination percentage would be 70%. The germination rate measures the speed at which seeds will grow. Germination uniformity identifies the time frame in which the seeds emerge. In some seed species, these parameters are highly variable and also differ from season to season. Many plant breeders have breeding objectives that include better percentage, rate, and germination uniformity.

Environmental Conditions for Germination

Certain conditions must be present for seeds to germinate. Temperature, water, oxygen, and sometimes light can affect seed germination.

Temperature

Temperature is the most critical factor that regulates the timing of germination. Temperature influences seed dormancy control, germination percentage, and rate. Seed germination temperatures have been identified for most horticultural species and include a minimum, maximum, and optimum range. Minimum temperature is the lowest temperature at which germination will occur; maximum, the highest. Optimal temperature is where the largest percentage of seeds germinates at the fastest rate. Most seeds can be broadly grouped based on their temperature requirements.

STEM Connection

Love in a Puff

Love in a Puff (*Cardiospermum halicacabum*) is a delicate vine that trails up trellises using tendrils. After the vine bursts with dainty white flowers, they slowly form into an interesting fruit. Inflated with air like a beach ball, the "puff" is unique. Inside the puff is the fruit, little black seeds that have a white heart on them.

wasanajai/Shutterstock.com

Cool Temperature

Seeds in the cool temperature group prefer cooler soil temperatures for germination to occur. They can be further subdivided into *cool temperature tolerant* and *cool temperature requiring*. Cool temperature tolerant seeds germinate from 86°F (30°C) to about 104°F (40°C), with an optimum temperature of 77°F–86°F (24°C–30°C). Crops in this class include broccoli, cabbage, alyssum, and carrots. Cool temperature requiring crops will not germinate at temperatures higher than 77°F (25°C) and include many winter annuals such as celery, lettuce, onion, and primrose, **Figure 13-4**.

Figure 13-4. Primrose requires cool temperatures and will not germinate at temperatures above 77°F (25°C).

Warm Temperature

Plants in this category will not germinate at temperatures below 50°F (10°C) (asparagus, sweet corn, and tomato) or 60°F (15°C) (beans, eggplant, peppers, and cucumbers). Many of these plants will show symptoms of chilling injury if exposed to low temperatures.

Water

Water starts the germination process. The rate of water movement into the seed depends on the germination media's texture, media compaction, and proximity of soil-to-seed contact. Seeds have negative water potential, creating a gradient for the water in the soil (which has a high water potential) to move into the seed. Most seed germination media have a fine texture that constantly provides water to the seed as it is taken up.

Seed *priming* is the process of hydrating and then drying out a seed for greater germination rate and uniformity. Seeds are allowed to imbibe water and continue into the lag phase of early germination but halted before radicle emergence. Seeds are then dried to nearly their original water content. Primed seed storage may be shorter, and the benefits of priming may be lost if seeds are stored for too long. Seed priming is commercially used for crop production of bedding plant plugs where uniform germination is critical to profits. *Plugs* are seeds that are grown in small containers to transplantable size, **Figure 13-5**.

Figure 13-5. The seeds of bedding plant plugs are often primed and planted with machinery to ensure uniformity.

Oxygen

Oxygen is required for respiration of germinating seeds. Gas exchange between the germination medium and seed embryo are necessary for rapid and uniform germination. Seed beds can become saturated due to heavy rains or irrigation. These beds will have limited oxygen available to seeds and result in poor or reduced germination rates.

Figure 13-6. The lodgepole pine requires the high temperatures of a fire to melt away the resin that covers the cone in order to release seeds.

Figure 13-7. The seed coat of the Kentucky coffeetree can be as thick as two millimeters.

Light

Light quality and photoperiod play a role in seed germination, particularly as it relates to dormancy. Light-sensitive species have a dormancy that requires light in order to germinate. Many of these plant species evolved to survive in shallow soils. Examples include begonia, *Calceolaria*, coleus, *Kalanchoe*, primrose, and African violets. Germination is inhibited by light in some species, including *Allium*, amaranth, and phlox.

After seedlings have emerged, high levels of light are important for sturdy and vigorous transplant growth. Low light levels produce reduced photosynthesis rates and poor seedling growth. Supplementary lighting often fulfills these lighting needs for seedling growth.

Seed Dormancy

Once harvested, most seeds only require favorable temperatures and water for germination to occur. These seeds are considered quiescent. Other seeds have a physiological adaptation that allows seeds to delay germination, called *seed dormancy*. Dormancy allows a seed to germinate only when environmental conditions favor seedling survival. For temperate species, this often requires a moist, chilling period (winter conditions) before germination in the spring. Some species require extremely high temperatures before germination, **Figure 13-6**. Desert species can survive long droughts, and then germinate after a rain. This kind of dormancy is called *primary dormancy*, and it occurs as the seed is shed from the plant.

When quiescent seeds or seeds that have emerged from primary dormancy are exposed to an extreme stress of drought, high temperatures, or low oxygen levels, they can become dormant. This is called *secondary dormancy* or *induced dormancy*. It provides a survival mechanism for seedlings if adverse environmental conditions are present. Some species even have a double dormancy and require multiple treatments to germinate.

Seed Coat Dormancy

The seed of the Kentucky coffeetree has a hard, bony seed coat that resists any efforts by water to penetrate, **Figure 13-7**. In nature, these seed coats break down slowly over time through exposure to weather, microorganisms, or other substances. In horticulture, the process of physically removing part of the seed coat to allow imbibition is called **scarification**.

Many horticultural species require scarification. Scarification methods include:
- Mechanical scarification. Seed coats are chipped or weakened (but not damaged) by rubbing with sandpaper, scraping with a file, or cracking with a vise or hammer. Large-scale commercial operations may have a drum lined with sandpaper that removes the seed coats of small seeded legumes, such as clover or alfalfa.
- Chemical scarification. Sulfuric acid is used to wear away the coat of hard seeds. Seeds are placed in an acid bath for a predetermined amount of time and then rinsed with water to remove any residual chemicals. Seeds are sown immediately after treatment.
- Hot water scarification. Seeds are placed in a large amount of 170°F to 212°F (77°C to 100°C) water. This softens the seed coat to make it pervious to water. In this process, the water is heated, and then the seeds are placed in the water. Finally, the heat source is removed. Seeds are kept in the water between 12 and 24 hours and planted shortly thereafter.

Chemical Dormancy

Natural chemicals that accumulate in fruit and seed coat tissues can inhibit germination or reduce gas exchange in seeds. Some fleshy fruits such as citrus, cucumbers, apples, and pears contain substances ranging from *phenols* (naturally occurring chemical compounds in seeds) to *abscisic acid* (a plant hormone) that inhibit germination. Some seeds are coated with a chemical inhibitor that is only removed through leaching after heavy rain, including iris and some desert plants. In other seeds, such as spinach and white mustard, a layer of mucilage (a thick, gluey substance) hampers gas exchange and limits seed germination, **Figure 13-8**.

Morphological Dormancy

Some seeds are not mature when they fall from the plant. In these seeds, the embryos have not fully developed before dropping. This condition can be found in a wide range of species, including anemone, poppy, ginseng, carrot, rhododendron, primrose, and others. To promote germination in these seeds, a temperature treatment (of either cold or warmth depending on the species) may promote germination.

Physiological Dormancy

The embryo controls physiological dormancy of seeds and delays germination. This increases the time for growth of the radicle to force open the seed covering. This dormancy can be overcome by manually removing the seed covering. Species such as lettuce, pepper, tomato, redbud, and lilac have demonstrated this kind of dormancy. *Photodormancy* occurs when seeds either require a period of light or dark conditions to germinate. Some seeds need after-ripening or time in storage to mature. This dormancy is not long lasting and can be broken with chilling treatments.

Eugene Gurkov/Shutterstock.com

Figure 13-8. The seeds of heirloom tomatoes are covered with a mucilage layer that must be removed before sowing.

> **Corner Question**
>
> What is the largest seed in the world?

Stratification

Intermediate and deep physiological dormancies require multiple months of moist, aerated cooling for seeds to germinate. This horticultural practice is called *stratification*. Seeds are stored in a moist, aerated medium at specific chilling temperatures for a designated amount of time. Depending on the species, stratification may last from one to five months. Many woody shrubs and tree species require stratification treatments. Some temperate, tropical, and semitropical species need a warm stratification treatment. In this practice, rather than providing cool temperatures, warm temperatures are paired with a moist, oxygen-rich medium to stratify the seed.

Seeds can be stratified outside in the fall. Seeds can be planted in a *seedbed* (a specially prepared space for seed germination) or in a cold frame. The wet, winter months will provide an environment that releases dormancy. However, seeds need to be protected from freezing, drying, and rodents. Many horticulturists will use refrigerated stratification for small batches of seed or high-value seed. Seeds are usually soaked for 12 to 24 hours and then planted in media such as sand, peat moss, vermiculite, or composted sawdust. A typical media mixture might have a 1:1 ratio of sand to peat or a 1:1 ratio of perlite to peat. Seeds can be stratified in layers to use space efficiently and placed in a container that limits drying and provides suitable aeration. Stratification temperatures usually fall within a range of 33°F–50°F (1°C–10°C). Once the specified time has passed, a few seeds are planted to see if germination occurs. When ready, seeds can be gently screened through a sieve and planted immediately.

Seed Propagation Techniques

Most agronomic, forestry, and vegetable and bedding plant growers propagate plants by seed. Depending on the crop, seeds can be sowed directly in the field, started in a greenhouse for plug production, or grown in nursery transplant beds to produce bare-root plants. Seed propagation requires an understanding of the germination requirements for each species grown. Four factors to consider for successful germination include:

- High-quality seed—needed for good germination percentage, rates, and uniformity.
- Seed selection of cultivars—select species that hold the desired characteristics for production.
- Seed dormancy—can be managed through treatments or sowing timing.
- Favorable environmental conditions—include the proper temperature, available water, oxygen, and light or dark conditions.

Field Seeding

Directly sowing seeds into the soil is the most common practice for commercially produced agronomic crops, many turfgrasses, vegetables, and some woody trees and shrubs. Many home vegetable and flower gardens are directly sown as well. Field seeding can be less expensive than using transplants, but variable environmental conditions can make it challenging to achieve uniform germination. Factors such as good seedbed preparation, high-quality seed, proper timing of planting, and seed treatments optimize the potential for good germination and seedling growth.

Seedbed Preparation

Good seedbeds have a soil texture and structure that provides adequate water contact with the seed, proper drainage and aeration, and minimal crusting. Large clumps or clods of soil should be removed. To prepare large seedbeds, use tractors with attachments to till and crumble the soil. In the home garden a tiller, spade, or rake can do the same thing, **Figure 13-9**. Building up the soil can provide nutrients and organic matter for soil structure.

vitaga/Shutterstock.com

Figure 13-9. A tiller can be used to till and crumble the soil in the same manner as a field tractor.

High-Quality Seed

Seed purchased from a reliable, certified seed dealer ensures commercial growers are planting seeds that have been tested for pureness of stock, have no weed seeds, and will uniformly germinate. Most seeds are hybrid seeds that offer high germination rates and vigorous, healthy seedling growth. Many home gardeners or hobbyists will save seeds, offering an inexpensive way to grow plants each year.

Timing of Planting

Each crop has its own environmental requirements, and proper soil temperature plays a key role in germination success. Research the need of each crop and create a planting calendar based on local conditions. For many crops, cool soil temperatures will result in uneven and poor germination and can cause stunted growth. High soil temperatures may induce *thermodormancy*, a secondary dormancy that inhibits germination.

Seed Treatments

Many diseases can impact seeds sown directly into the field. Some commercial growers use seeds that have been pretreated with a chemical fungicide to minimize disease risk. Priming is another practice that gives some seeds a head start toward faster germination rates and increased uniformity. Some seeds are small, so they are coated with a kind of clay through a process called *pelleting*. This process increases their size and allows them to flow through mechanical planters or be planted by hand, **Figure 13-10**.

Dwight Sipler, Stow, MA

Figure 13-10. Pelleting seeds changes their size and shape for improved mechanical planting. The lettuce seeds on the right are not coated. The coated seeds (on the left) are much easier to handle and may be planted by machine or by hand.

Planting Depth

The correct planting depth has been determined for most horticultural crops. Proper planting depth ensures that seeds will be able to emerge from the soil and establish themselves quickly. Seeds sown too deeply may rot in the soil or have staggered emergence. If the planting depth is too shallow, seeds may dry out or be washed away. A general guideline is to plant a seed to a depth about three to four times the diameter of the seed. Seeds that require light for germination are planted at a shallow depth and moisture levels are monitored.

Spacing

Appropriate seed spacing use the field area efficiently and maximizes yields. Seeds that are sown too closely can compete with each other for nutrients, water, and light, **Figure 13-11**. Seeds sown too far apart allow for weed growth and waste space. Many field-sown seeds are planted with a mechanical seeder that precisely plants seeds to optimal rates and depths.

li jianbing/Shutterstock.com

Figure 13-11. As seeds, these carrots were planted too closely. Crowding like this results in small roots for harvest.

Equipment

Mechanical seeders increase the efficiency of planting vegetable and agronomic crops. Seeders vary widely. They are selected based on the size and shapes of the seeds to be sown, soil properties, amount of land, and precision with which seeds need to be placed in a row.

Seeders have three main components. The seed hopper contains the seeds, a metering system measures out the number of seeds, and a drill places the seed in the soil. A seed drill might place seeds into an open furrow or punch individual holes in which to place the seeds. Random and precision seeders are available. Random seeders sow seeds without exact spacing but at a specified rate determined by the size of the seeding wheel and speed of the tractor, **Figure 13-12**. Precision seeders maintain a designated spacing and can use considerably less seed. Precision vacuum seeders use a vacuum to pick up a single seed and set it into the soil.

Watering

All seeds require adequate and consistent moisture for germination and early seedling growth. Many growers provide supplementary irrigation through overhead sprinklers, subsurface flooding, or drip irrigation. Natural rainfall also provides ample moisture in many areas of the country. Seeds are watered regularly to limit any soil crusting that might occur.

pryzmant/Shutterstock.com

Figure 13-12. This grass seed spreader is an example of a simple mechanical spreader.

Field Nurseries

Many nurseries produce their own woody transplants, fruit and nut tree rootstock plants, and vegetable transplants. Many of the practices employed in direct field seeding are used in field nurseries as well. However, woody transplants can initially be spaced very close together as they will be harvested and sold as bare-root plants for transplant into pots. Woody transplants require more controlled management to ensure their successful transplant into potted production.

STEM Connection
Squirting Cucumbers

The squirting cucumber (*Ecballium elaterium*) is related to pumpkins, squash, and gourds. This nonedible, poisonous cousin is a fascinating vegetable curiosity. As it grows along the ground, the vine forms a prickly, fleshy fruit. As the fruit matures, the slightest touch or tremor will launch a burst of mucilaginous liquid that contains its seeds. This dispersal travels at an amazingly fast speed. *Ecballium elaterium* is native to Europe, northern Africa, and some areas of Asia.

flaviano fabrizi/Shutterstock.com

Field Nursery Seedbeds

Nursery seedbeds have well-drained soil with a loamy texture. They are often incorporated with other crop rotations, including cover crops that provide nutrients and improve soil structure. Seedbeds are typically 3 1/2′–4′ wide with walkways in between. This spacing allows workers to reach easily into the middle of the bed to pull weeds.

Timing of Planting

The timing of crop planting is similar to field seeding. Sowing should occur based on the environmental requirements of the particular crop. In woody plant production, many seeds will need to be scarified and/or stratified for optimal germination. For woody plants that remain in the seedbed for a year or more, moisture, shade, nutrients, and weed and disease control must be carefully applied. Transplants that are ready to be harvested might be pulled mechanically and sold as bare-root plants. Many vegetable transplants used to be grown this way. However, they are now usually grown as plugs in a greenhouse.

Greenhouse Production

Greenhouses offer controlled conditions that are ideal for seed germination, proper temperatures that protect plants from freezing or chilling injuries, consistent watering, and soilless potting media for drainage and aeration.

Most commercial growers now specialize in producing plugs for transplants. The transplants are sold to retail nursery and greenhouse operations where they continue to grow prior to sale, **Figure 13-13**.

DutchScenery/Shutterstock.com

Figure 13-13. These geranium plugs will be shipped to a retail nursery where they will be potted into larger containers and sold to home gardeners.

Before the shift to plug production, growers would plant seeds in flats or germination trays. After a short period of seedling growth, they would then transplant the seedlings into cell packs, into individual containers, or directly into the ground. This method is still used by some small operations, botanic gardens, and home gardeners.

Plug Production

Millions of bedding plants, perennials, and vegetable plugs are produced each year in greenhouses that provide ideal germination and seedling growth conditions. Plug production has become a specialized industry. Growers invested in equipment that carefully manages germination conditions. Plugs can be grown in as little as six to eight weeks, allowing for multiple crops per season and fast shipping to clients. Plugs are transplanted to larger containers using the soil in which they were grown, reducing symptoms of transplant shock.

A plug seedling is grown in a small amount of soil in a small cell that is part of a large plug tray. A tray may have as many as 220–800 plugs. Much of soil building and seeding process is done by specialized machines. When using mechanical seeding, it is important to use high-quality pelleted or primed seeds to ensure even germination.

Seedlings have four distinct, observable stages of growth, **Figure 13-14**:
- Stage 1—radicle emergence or protrusion.
- Stage 2—cotyledon spread.
- Stage 3—three or four leaves unfolding.
- Stage 4—more than four leaves present.

At each of these stages, environmental conditions are controlled to maximize growth. Many plug producers use special germination chambers that provide proper lighting, temperature (70°F–75°F [21°C–24° C]), and humidity in stage 1.

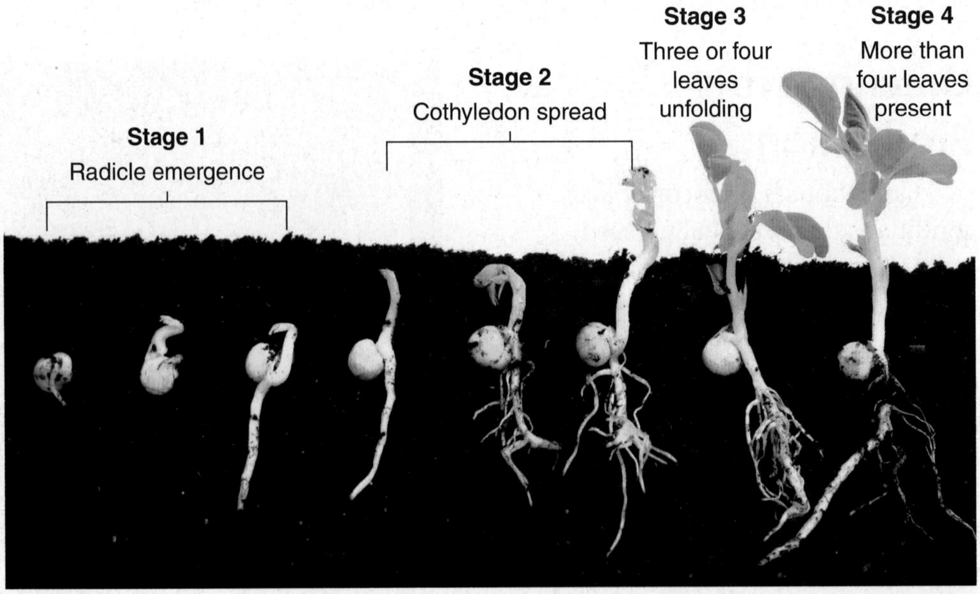

"With every deed you are sowing a seed, though the harvest you may not see."
—Ella Wheeler Wilcox

Figure 13-14. Seedlings have four distinct, observable stages of growth.

Plugs are often misted or fogged to maintain moisture and humidity. Humidity levels need to be carefully controlled to limit foliar disease. Light should be increased in stage 2 to prevent shoot (hypocotyl) elongation. Fluorescent or high-pressure sodium supplemental lighting in stage 2 and early in stage 3 provides sufficient intensity for growth. Nutrients should be available and then somewhat minimized to harden off seedlings for transplanting. Plugs are ready to be transplanted or shipped at stage 4. Many large-scale operations will have a mechanical transplanter that transfers plugs to bigger containers.

Media

Germination media is different than media used for other parts of potted plant production. Germination media has a fine texture to ensure soil-to-seed contact, moisture retention and drainage, adequate oxygen levels, and nutrients. The structure must also provide support for the seedling, **Figure 13-15**. Typical media components include fine-textured peat moss, perlite, ground bark, vermiculite, and slow-release fertilizers. The media should be sterile to prevent disease. Different textures are used depending on seed size. For example, a small petunia seed needs a finer textured germination media than a larger cockscomb seed.

Figure 13-15. This soilless potting media has very fine particles of peat moss and vermiculite to ensure soil-to-seed contact as well as adequate moisture retention/drainage, oxygen levels, and nutrients.

Mechanical Seeders

Mechanical seeders are a significant financial investment for greenhouse operations, but using them increases efficiency and reduces labor costs. Mechanical seeders may broadcast seeds across a flat or they may plant precisely, as in plug production. There are three types of seeders: template or plate, needle, and drum seeders. Template seeders use a metal plate with holes drilled into it, **Figure 13-16**. A vacuum in the seeder pulls seeds to attach to the template. Then when the vacuum is released, the seeds fall into the flat. Different seed sizes require different templates, making this a cumbersome option. It is, however, the least expensive option. A needle seeder has individual needles that lift seeds and place them into individual cells in a flat. The needles come in different sizes, are moderately priced, and can sow up to 100,000 seeds an hour. The drum seeder has a rotating cylinder that uses a vacuum to pick up seeds. As the drum rotates, seeds are deposited into the germination media. It is the fastest and most precise seeder, sowing as many as 800,000 seeds each hour. It is also the most expensive seeder.

Figure 13-16. This automatic needle seeder can be adjusted to sow different sized seeds.

Corner Question

What type of seeds do you eat?

Watering

Early stages of seedling growth need high levels of humidity and moisture. Humidity can be controlled through specialized germination chambers. On a smaller scale, germination tents may be used. These tents are made by placing seed flats on a greenhouse bench and covering them with polyethylene plastic. These chambers or tents prevent the seedlings from drying out in early stages of growth. Growers must provide ventilation in the chambers, however, to minimize risk of heat buildup.

As transplants begin to grow, they may be moved to a greenhouse. Here water can be delivered using a mist nozzle on a hose. Automatic boom misters have many nozzles that travel the length of a greenhouse, **Figure 13-17A**. Many greenhouses have subirrigation systems. Flats are placed directly on the concrete (which usually provides bottom heat). Water floods the floor and passively drains back into a water reservoir, **Figure 13-17B**. Nutrient solutions can be easily delivered through the water in this method. Capillary mat systems are made of a fabric that holds water when saturated and releases it to the flats as it is wicked up. These can limit evaporation and water loss, conserving water resources. Some operations use a float bed, similar to those used in the production of tobacco transplants. Styrofoam trays with individual cells are planted with seeds and placed on a nutrient-rich water solution, **Figure 13-17C**.

Transplanting

When stage 4 is reached, seedlings are ready to be hardened off and transplanted. Temperature and moisture are reduced to make seedlings ready to move and reduce symptoms of transplant shock. Transplant shock can result in premature stalk formation, increased susceptibility to disease, and diminished yields. Transplanting has historically been done by hand, but there has been a recent, significant shift to mechanized transplanting.

A — vallefrias/Shutterstock.com   B — T.W. van Urk/Shutterstock.com   C — Photograph by Matthew Vann

Figure 13-17. A—Automatic boom misters travel the length of the greenhouse. B—Flood floors in greenhouses conserve water, deliver bottom heat, and maximize space for young seedlings. C—These tobacco float trays sit in a nutrient solution that provides constant moisture to germinating seeds.

The container the plants are transplanted into will have holes made in the media, either manually or mechanically, using some form of a dibble, **Figure 13-18**. The roots are tucked into the hole in the media, and soil is packed gently around the plant. Transplants are allowed to continue growing to a more substantial size or are hardened off to be moved outside.

Seed Selection

Modern agricultural systems rely on the production of high-quality seeds. The majority of agronomic crops, annual bedding flowers and other ornamentals, and vegetables are propagated each season from seeds. Seeds hold the genetic traits incorporated by breeding and selection to create cultivars. These cultivars are adapted for specific environments and will yield a marketable and profitable crop. Customers demand that the products they buy represent the description that has been offered. Therefore, the genetic identity and purity of seeds forms the foundation for overall seed quality.

Using genetically pure seed, a grower or gardener can be assured that the seed is:

- True to name—seeds are labeled correctly with appropriate species, cultivar, and history.
- True to type—seeds grow into visual standards specific to the cultivar or species as it is labeled.
- Free of contaminants—little or no presence of weed seeds, no seedborne diseases, no seeds of other genetic material.

A variety of seed types are available to growers and gardeners. The seed that is used depends on the production goals and philosophy of the grower.

stockcreations/Shutterstock.com

Figure 13-18. Dibbles can be simple tools for home gardeners or mechanized tools used in greenhouse transplanting operations.

Landraces

Humans have long selected seeds from plants that showed promising characteristics. Seeds are saved from part of the harvested crop to be used the following growing season. A *landrace* is a locally adapted or traditional variety of a plant. Landraces maintain a tremendous genetic diversity that serves as an inherent buffer against environmental stresses, insects, or diseases that may afflict the crop. Landraces have adapted to local conditions and are still used in some parts of the world, **Figure 13-19**.

Landraces are open-pollinated, which means that pollination occurs by insect, bird, wind, humans, or other natural mechanisms and the flow of pollen between individuals occurs freely. Heirloom varieties are also open-pollinated species. These varieties have been selected and saved by gardeners over generations. Somewhat less diverse than a broad landrace, heirloom varieties have enjoyed resurgence in recent years.

D Pimborough/Shutterstock.com

Figure 13-19. The Irish government is dedicated to preserving landraces such as potatoes and other grains. Each of these plants has a stable set of unique genetic characteristics.

Figure 13-20. Heirloom varieties, such as Cherokee Long Ear corn, have been passed through generations and often have traits that cannot be found in hybrid seeds.

They have traits such as intense flavor, shapes, and colors that are not always present in hybrid seed, **Figure 13-20**. Heirloom seeds can be saved from one generation to the next without losing characteristics and performance of that variety.

Modern farms have shifted to hybrid varieties that tend to be uniform, produce high yields, and respond to the mechanization used on many farms. Many scientists are concerned that hybrids have led to a loss of genetic diversity. Conservation efforts are being made to maintain *germplasm* repositories (seeds or other materials from which plants are propagated). These repositories of landraces and other open-pollinated seeds can serve as raw genetic material for future use.

Wild Populations

Most native plant species evolve over time into populations with a fairly uniform phenotype (outward or observable properties of an organism) and genotype (inherited genetic properties), which are adapted to local environmental conditions. If a species has a broad geographic distribution, populations may look the same (have the same phenotype) but vary genetically to adapt to different areas. For example, wild common beans (*Phaseolus vulgaris*) are found throughout Central and South America. Different populations of wild beans show varying degrees of drought tolerance depending on their location. Plant breeders will sometimes draw from wild populations to improve traits in their breeding lines.

Some plants found in the wild may show a phenotypic difference in their morphology (size, shape, or structure). These plants are called botanical varieties, or simply varieties. For example, Alabama cherry (*Prunus serotina* var. *alabamensis*) is a variety of black cherry (*Prunus serotina*). Other botanical variations of black cherry also exist, including var. *eximia*, escarpment cherry; var. *rufula*, southwestern black cherry or Gila chokecherry; and var. *salicifolia*, the capulin black cherry. Most varieties can be produced by seed and are true to type, meaning they will display the characteristics of the parent plants. White flowering redbuds, a popular, woody, landscape ornamental, is a variety (*Cercis Canadensis* var. *alba*) and can be planted from seed, **Figure 13-21**.

Figure 13-21. White flowering redbud is a botanical variety of common redbud and produces seeds that will grow and also have white blooms.

The botanical form, or forma, is a secondary taxonomic rank. Form members usually have only slight physical differences, such as leaf color or leaf shape. They occur at any point within a species' range. Forms can be determined based on environmental factors rather than genetic factors. They can come from a sport or other mutation and are usually vegetatively propagated. Some forms have horticultural value, such as the thornless honey locust.

Gleditsia triacanthos forma *inermis* is a form of the usually very thorny honey locust, **Figure 13-22**.

Hybrids

Many annual bedding plants and other ornamentals, vegetables, and agronomic crops are hybrid varieties, produced by plant breeders and sold as seed. As discussed in Chapter 8, plant breeding is the process of cross-pollinating parent plants to create progeny that display desired characteristics. Commonly desired characteristics include increased disease resistance, novel flower colors, and improved vigor and yield. Hybrid plants have a quality called hybrid vigor, which results in improved growth performance. The seeds that result from hybrid plants are not saved because they are not true to type. Plants grown from these seeds may lose many of the traits of the cultivar.

Transgenic Cultivars

Transgenic describes an organism into which genetic material from an unrelated organism has been introduced. Transgenic cultivars have a genome that has been transformed by genetic engineering. Also known as a ***genetically modified organism (GMO)***, the plant has been transformed for a variety of reasons including resistance to disease, herbicides, or insects. Many agronomic crops, such as corn, soybean, and cotton, have been transformed to hold a Bt gene from the bacteria *Bacillus thurinigiensis* to produce a toxin that fights off pests. Other crops have genes that are resistant to an application of herbicide (glyphosate), making weed control more efficient on a large scale. In Florida, citrus greening is a severe and devastating disease. It can destroy entire orchards. Promising research shows that a gene from spinach inserted into an orange plant can result in disease resistance. Not all genetically modified plants have outside genes inserted into their tissues. In the case of the Arctic® apple, the apples' own genes have been modified to turn off the expression of a chemical that causes apples to turn brown.

There is considerable debate by consumers over transgenic plants. Concerns for gene contamination across fields, food safety, and other arguments have slowed widespread acceptance of this technology. Palmer amaranth is an aggressive weed that has been a serious problem in fields in the southeast United States. Herbicide resistant GMO cultivars managed to rid this weed from fields for many years. Many scientists believe that transgenic plants may be one tool used along with other strategies to manage pest problems. Others disagree or question the use of this biotechnology.

Sally Wallis/Shutterstock.com

Figure 13-22. The thornless honey locust is a form of the spiny honey locust and must be propagated vegetatively.

Did You Know?

The seed case that holds kapok tree seeds are filled with soft, fluffy material. This material was previously used as filler for life jackets or bedding because it is light, strong, and waterproof.

Seed Production

To ensure production of high-quality seeds, commercial seed producers follow a set of state and federal guidelines to produce seeds that are certified.

Certified seed have a specific legal definition set by the Federal Seed Act, which sets strict standards for seed growers selling nationally. Most states have additional seed laws that govern the labeling, sale, and transportation of agricultural, vegetable, and other seeds.

The seed certification process is intended to protect growers and gardeners by ensuring that they are purchasing seed that has the following characteristics:
- Seeds free of contaminating material, including noxious weed seeds, stones, fungi, soil, chaff, broken seed, and other debris.
- Seeds with genetic identity and purity. The cultivar or variety is true to the given description.
- Seeds with germination uniformity.

Seed producers practice strict field and equipment sanitation to minimize any seed contamination by weeds, disease, and other material. They usually grow seeds in isolated fields to minimize any cross-pollination by other genetic material. Throughout the growing process, seed inspectors from local governmental agencies visit the fields to ensure varietal purity, isolation, and freedom from weeds and diseases. After harvest, seed is properly cleaned, extracted, and stored in separate bins labeled with the type, variety, year produced, and field location. Seeds are sent to a seed testing lab and tested for the characteristics from the list of characteristics. This service is often performed by state departments of agriculture. If seeds meet these standards, they can be distributed and sold. The state regulatory agency has the authority to stop the sale of seeds that are found to be mislabeled or misrepresentative of the variety.

Plant Variety Protection Act

The Plant Variety Protection Act (PVPA) of 1970 (amended in 1994) provides plant breeders legal intellectual property rights protection to new varieties of plants that are sexually reproduced (by seed) or tuber-propagated. Breeders are issued a Certificate of Protection that gives them exclusive rights to sell, reproduce, import, export, or use the plant in producing a hybrid or different cultivar. The protection lasts up to 25 years and enables plant breeders to recover research costs. It also provides an incentive for grower to develop improved varieties.

A variety can be patented if it is:
- New material that has not been sold prior to the application.
- Distinguishable from any other existing variety.
- Uniform with any variation being describable, predictable, and commercially acceptable.
- Stable and when reproduced remains unchanged according to its distinctive characteristics.

According to the language written in the PVPA, the act is intended "to encourage the development of novel varieties of sexually reproduced plants and to make them available to the public, providing protection available to those who breed, develop, or discover them, and thereby promoting progress

> "Behold, my brothers, the spring has come; the earth has received the embraces of the sun, and we shall soon see the results of that love! Every seed has awakened, and so has all animal life."
> —Sitting Bull

in agriculture in the public interest." Varieties that are protected under the PVPA can be sold as seed stock only with the permission of the holder of the Certificate of Protection.

Seed Saving

Seeds should be harvested when the embryos are relatively mature. Immature seeds may result in little or no germination. If germination does occur, the seedling often emerges as weak or stunted. Immature seeds tend to have high moisture content, watery endosperm, and a light-colored seed coat. They tend to rot in storage and usually shrivel when dried.

Mature seeds come from fully ripe fruit and should be collected before the seed disperses, **Figure 13-23**. Ripeness appearances vary dramatically depending on species. Generally, ripe fruit will show color changes, chemical changes (such as in sugar or starch content), and shifts in specific gravity. Many gardeners know that seeds that float are usually not viable. Scientists use this same test to measure the specific gravity of seeds in comparison to the known specific gravity of water. Viable seeds are usually denser than nonviable seeds.

Przemyslaw Wasilewski/Shutterstock.com

Figure 13-23. Poppy seeds should be harvested when the fruit is mature, but before the seeds disperse.

Because seeds are harvested before dispersal, they must be extracted from their fruit. In some cases, the fruit is dried and the seeds fall out. Old flower heads of cockscomb or celosia can be placed upside down in a paper bag to dry. Once dried, the plant can be shaken, or threshed, to remove the seeds. The seeds are then sorted through sieves to separate them from the chaff. Seeds contained within fleshy fruit, such as raspberries and strawberries, can be macerated using a blender filled with water. The pulp is skimmed and rinsed to extract the seeds. In tomatoes, the pulp is separated from the juice and the seeds. The seeds are coated with a mucilage layer that must be removed through an acid treatment or fermentation. Once removed, the seeds can be rinsed and dried for storage.

Some seeds, such as a number of grasses, may require stronger mechanical force to extract the seeds. Hammer mills, threshers, and other machines are used to separate seed from fruit. Seeds such as some pines and conifers require high heat to remove them from the cone. They should be placed in a special kiln to dry the cone and coax it open.

After seeds have been removed from the fruit, they need to be separated from the chaff, insect parts, soil, and other foreign material. Most commercially harvested seeds use a set of sieves to isolate the seeds by blowing air through the screens. The sieve screens come in various sizes that are used to clean and grade the seeds, **Figure 13-24**.

ansem/Shutterstock.com

Figure 13-24. Seed sieves remove chaff, soil, rocks, and other unwanted material from harvested seeds.

Seed Storage

Seeds are living, respiring organisms. Placing them in the right environmental conditions will lengthen their shelf life. For most seeds, every 10% drop in moisture content (within limits) will double the length of viability. Every 50°F (10°C) decrease in temperature will also double the storage shelf life. Many species can be stored at temperatures of 0°F (–18°C) for a considerable amount of time. Some seeds require added moisture during storage. Citrus, walnut, oak, and chestnut species all prefer a chilled, moist environment. Cool, wet conditions can also encourage the growth of rot-causing fungi. Take precautions to minimize disease risks.

Careers in Seed Propagation

The seed propagation industry provides food and clothing to the world. The majority of human calorie consumption worldwide comes from seeds such as cereals, legumes, and nuts. Many cooking oils are made using seeds. Cotton fiber comes from cotton plant seeds. Seed propagation also provides crops used to feed livestock and beautify the environment. Seed sales representative and nursery propagator are just two careers in the seed propagation field.

Seed Sales Representative

Many seed companies employ sales representative to market their product to retailers, wholesalers, and growers. A seed sales representative is responsible for cultivating relationships with clients to facilitate the sale of seeds. Representatives must understand customer needs and develop trusting relationships with customers. Sales representatives manage forecasting and seed line demand issues. Sales representatives serve as a vital link for technical growing information and can handle any inquiries or complaints regarding a product. Sales representatives usually have an associate or bachelor's degree and significant experience in sales.

Nursery Propagator

Many nurseries specialize in the production of woody trees and shrubs that are produced primarily from seed. Nurseries have covered greenhouse space as well as outdoor growing space for container (plug) seedling production. A propagator is responsible for germinating and growing bare root seedlings and transplants plugs. Production responsibilities include soil mixing, sowing, thinning, irrigation, photoperiod manipulation, and pest control among other greenhouse production practices. Propagators may have to provide seed treatment such as stratification, scarification, and pre-germination techniques to enhance the germination of woody plants. Some propagators have a high school diploma, along with experience. Management and supervisory positions require a minimum of an associate or bachelor's degree.

CHAPTER 13 Review and Assessment

Chapter Summary

- Seeds are formed as a result of the fertilization process. All seeds have an embryo, a protective outer covering, and storage tissue.
- The three stages of seed of development are histodifferentiation, cell expansion, and maturation drying.
- Some plants are able to spontaneously produce seeds without going through the fertilization process. The process creates genetic clones of the parent.
- In order for germination to occur, the seed must be viable, favorable environmental conditions must be present, and dormancy conditions must be overcome.
- Seed germination has three phases starting with imbibition or the uptake of water, followed by lag phase where cellular activities prepare for seed growth, and then the emergence of the radicle or seed root.
- Temperature, water, oxygen, and light are critical environmental factors in seed germination.
- Dormancy allows a seed to germinate only when environmental conditions favor seedling survival. A hard seedcoat, chemical inhibitors, seed embryo maturity, or light or temperature requirements can cause seed dormancy.
- Four factors needed for successful cultivation include high-quality seed, seed selection of cultivars, seed dormancy, and favorable environmental conditions.
- Field seeding requires proper seedbed preparation, high quality seed, proper planting timing, possible seed treatments, appropriate planting depth and spacing, and adequate water.
- Greenhouses offer controlled conditions that are ideal for seed germination.
- Most commercial growers specialize in producing plugs for transplants. Plugs of bedding plants, perennials, and vegetables can be grown in 6–8 weeks.
- Seedlings can be characterized by four stages of growth: radicle emergence, cotyledon spread, unfolding of three or four leaves, and the presence of four or more leaves.
- Seeds hold the genetic traits incorporated through breeding and selection to create cultivars.
- Seeds come from different places and are selected based on their characteristics. Categories of seeds include landraces, heirloom varieties, wild populations, hybrids, and transgenic cultivars.
- To ensure production of high-quality seeds, commercial seed producers follow a set of state and federal guidelines to produce seeds that are certified.
- The Plant Variety Protection Act provides plant breeders legal intellectual property rights protection for new varieties of plants that are sexually reproduced (by seed) or tuber-propagated.
- Seeds should be properly harvested at maturity, extracted from the fruit, cleaned, and stored at proper temperatures for long shelf life.

Words to Know

Match the key terms from the chapter to the correct definition.

A. abscisic acid
B. apomixis
C. cell expansion
D. desiccation
E. genetically modified organism (GMO)
F. germplasm
G. histodifferentiation
H. imbibition
I. lag phase
J. landrace
K. maturation drying
L. phenol
M. photodormancy
N. plug
O. plumule
P. priming
Q. quiescent
R. radicle
S. radicle protrusion
T. scarification
U. seedbed
V. seedlot
W. thermodormancy
X. transgenic
Y. viable
Z. vivipary

1. A stage of seed development when the embryo and endosperm develop distinct characteristics.
2. A plant hormone that can inhibit germination.
3. The process of hydrating and then drying out a seed for greater germination uniformity.
4. A locally adapted or traditional variety of a plant.
5. A need of some seeds for either light or dark conditions in order to germinate.
6. A specially prepared space for seed germination.
7. A seed that is grown in a small container to transplantable size.
8. Drying out of a seed.
9. The rapid uptake of water by a seed.
10. The ability of a plant to produce seeds without going through the fertilization process.
11. The last period of early seed germination characterized by the emergence of the seed root.
12. Dormant or in a state of inactivity.
13. Alive; a living seed.
14. Seeds of a particular crop gathered at one time and likely to have similar germination rates and other characteristics.
15. A phenomenon in which seeds germinate inside their fruit without maturation drying.
16. A naturally occurring chemical compound in seeds that may prevent germination.
17. Seeds or other materials from which plants are propagated; serves as raw genetic material for future use.
18. A period during seed germination with little or no water uptake but with high cellular activities that prepares the seed to grow.
19. A plant that has had its genome transformed through genetic engineering.

20. A process of removing part of the seed coat to allow imbibition.
21. An organism into which genetic material from an unrelated organism has been introduced.
22. The seed root of a seed.
23. A secondary dormancy that inhibits germination.
24. The growing point for the developing shoot.
25. The phase of seed development when seeds have reached physiological maturity.
26. The second stage of seed development in which seed cells increase in size.

Know and Understand

Answer the following questions using the information provided in this chapter.

1. What parts do all seeds have?
2. What are the three stages of seed development? Briefly explain each stage.
3. What three conditions must be met in order for seed germination to occur?
4. Describe the process of seed germination.
5. What are three indicators of seed germination rates?
6. What is the optimal temperature ranges for plants in the following categories: cool temperature tolerant, cool temperature requiring, and warm temperature?
7. The rate of water movement into the seed depends on what factors?
8. How do light levels affect seedling growth?
9. List the ways seed coat dormancy can be removed.
10. Explain the process of stratification.
11. What are four basic factors to consider for successful germination?
12. Why is it important to use high-quality seed?
13. Why is it important to use the proper planting depth for seeds? What may happen if the planting depth is too deep or too shallow?
14. Why is it important to properly space seeds?
15. Why do plug producers need to know the stages of seedling growth?
16. Name three types of mechanical seeders used in greenhouse production and describe each.
17. What problems can transplant shock cause? What can be done to reduce transplant shock in seedlings?
18. Explain how seedlings are transplanted using mechanized methods.
19. What are landrace plants and why are they important?
20. What are some types of transgenic plants used today?
21. Explain the reasons for genetic modification of plants.
22. What general rights are growers provided under the Plant Variety Protection Act?
23. What are two important concepts for seed storage that increase the length of viability?

Thinking Critically

1. You have collected seeds from different tree species during a recent walk through an arboretum. You sow all the seeds, but not all of them germinate. What do you think is happening? What can you do to get the seeds to germinate?
2. You really love blueberries. You have a blueberry bush in your garden that produces many berries for harvest, but the flavor is not very strong. What could you do to create better tasting blueberries?

STEM and Academic Activities

1. **Science.** Most plant species have different germination requirements. Research a plant that you find interesting and write a report that summarizes your findings.
2. **Engineering.** As a greenhouse grower in the Southwest, water is a valuable resource. Design a watering system that would allow you to maximize your watering efficiency and use for plug production.
3. **Math.** You are a large commercial vegetable grower in the Southeast and need to maximize your space and production timing for optimal profits. Create a planting plan that accounts for growing seasons in the fall, spring, and summer, along with spacing and rotation of at least five crops.
4. **Math.** As a small nursery operator, you have a limited budget of $500 to purchase seeds that you will plant in your field seedbeds. You want the greatest diversity of seeds with enough of each species to be able to sell. You have five seedbeds and each is 25 yards long. After researching the cost of materials and the growing requirements, create a plan for how many seeds you can purchase and how many plants you can grow.
5. **Language Arts.** Imagine that you are the marketing director for an heirloom seed company that grows and sells flowers and vegetable seeds. One of your job responsibilities is to write articles for the grower's information center. Write an article about how to choose the best heirloom seeds. Focus on the characteristics of at least three different types of seeds.

Communicating about Horticulture

1. **Reading and Writing.** The ability to read and interpret information is an important workplace skill. Presume you work for a well-known, successful seed propagation company. Your employer is considering pitching a proposal to the local college to supply propagation needs for various research programs. He wants you to evaluate and interpret some research on innovative seed propagation techniques. Locate three reliable resources for the most current research on seed propagation of native plants in your region. Read and interpret the information. Write a report summarizing your findings in an organized manner.
2. **Reading and Listening.** In small groups, discuss the main topics in the chapter. Ask questions of other group members to clarify concepts or terms as needed.

SAE Opportunities

1. **Exploratory.** Visit a plug producer and observe each step of the process. Why do you think so many stages are mechanized? How does this help with germination uniformity?

2. **Exploratory.** Create a germination calendar for your garden plants. What seeds do you need to begin in the greenhouse? What seeds can you sow directly? Include planting timing, depth, and spacing for each plant. Why do you think growers develop planting calendars?

Peter Bernik/Shutterstock.com

3. **Experimental.** Conduct a germination experiment using your favorite tree or shrub. What environmental conditions will you use? Do you think your plant has any dormancy requirements? What do you expect to observe? How will you collect data? Share your findings with your class.

4. **Exploratory.** Job shadow someone from your state Crop Improvement Association. What are his or her daily tasks? What is the scope of his or her responsibilities? Why do you think the work is important for growers in your state?

5. **Placement.** Visit a local plant conservation society or a garden that collects seeds. Volunteer to harvest, extract, and clean seeds. Many organizations conserve native wildflower seeds. Why do you think this is important?

CHAPTER 14
Stem and Leaf Propagation

Chapter Outcomes

After studying this chapter, you will be able to:
- Understand the biological principles of stem and leaf propagation.
- Describe sources of plant material for stem cuttings and environmental conditions required for rooting.
- Describe the propagation methods for leaf cuttings and root cuttings.
- List materials used to mix rooting media.
- Explain the role of plant growth regulators in the propagation process.
- Identify careers related to stem and leaf propagation.

Words to Know

abscise	differentiate	preformed root
acclimatization	distal	proximal
adventitious root formation	hardwood	softwood
asexual propagation	herbaceous	stem cutting
auxin	leaf-bud cutting	stock plant
callus tissue	leaf cutting	suberin
cutting	nursery liner	vegetative propagation
cytokinins	phytohormone	wound-induced root
deciduous	plantlet	
dedifferentiate	polarity	

Before You Read

Before you read the chapter, read all of the table and photo captions. What do you know about the material covered in this chapter just from reading the captions?

While studying this chapter, look for the activity icon to:

- **Practice** vocabulary terms with Words to Know activities.
- **Expand** learning with identification activities.
- **Reinforce** what you learn by completing Know and Understand questions.

www.g-wlearning.com/agriculture

*H*ave you ever walked through a garden and pinched off a piece of a plant and later tucked it into a vase of water? Did you see roots begin to grow? This horticultural method of starting new plants from existing plant material is called *vegetative* or *asexual propagation*. It is an important clonal regeneration technique that produces a plant genetically identical to its parent. Many plants have been developed through plant breeding processes where seed propagation is not feasible, for sterile specimens, or simply for the joy of starting more plants.

A
Aggie 11/Shutterstock.com

B
saiko3p/Shutterstock.com

Figure 14-1. A—Cornstalks will naturally send adventitious brace roots toward the soil to stabilize the plant and provide additional support. B—Banyan trees have preformed root initials that, when given the right conditions, emerge as adventitious aerial roots.

Biological Principles of Leaf and Stem Propagation

The most widely applied method for vegetative propagation using a piece of leaf or stem is called a *cutting*. The cutting has the ability to adventitiously form roots, which allows subsequent normal growth and development of the new plant. The new plant is genetically identical to the plant from which the cutting was taken. *Adventitious root formation* is the process of roots forming from any plant part other than the root. Many adventitious roots form naturally, such as the brace roots on a cornstalk, or the aerial roots of a banyan tree, **Figure 14-1**.

The cellular tissue in the leaf or stem cutting actually *dedifferentiates*, meaning the cell regresses from a specialized function to a simpler state. For example, callus tissue forms from plant tissue in response to wounding. The *callus tissue* is a bundle of undifferentiated cells that cover the wound and begins to initiate new cellular divisions that *differentiate* (develop a specialized function) to form into meristematic growing regions or roots. There are two kinds of adventitious roots: preformed roots (with cells that have existing root initials) and wound-induced roots.

Preformed Roots

Preformed roots develop naturally on stems while they are still attached to the parent plant. These roots are underneath the stem tissue and may not emerge until the stem piece is cut and placed in environmental conditions conducive to the emergence of roots. For example, the easily rooted willow tree will form adventitious roots once it is placed in a potting medium and given a proper environment. Other species that have preformed roots include coleus (*Plectranthus scutellarioides*), *Hydrangea*, poplar (*Populus*), jasmine (*Jasminum*), currant (*Ribes*), and the houseplant pothos, **Figure 14-2**.

Wound-Induced Roots

Wound-induced roots only develop after a cutting is made. Roots develop as a direct response to the wounding that occurs when the stem piece or leaf from the parent plant is severed. Plants that are wounded (whether intentionally for the process of propagation or accidently as in when the lawn mower runs into a tree) exhibit a wounding response. The wounding response has three key steps:

1. The outer layer of injured cells die. The cambial layer forms a barrier zone that is rich in a waxy substance called *suberin* that seals the wound. This prevents plant tissue from drying out and protects the plant from pathogens.
2. Cells behind the wound begin to divide and form a callus.
3. Cells near the vascular cambium and phloem begin to divide and initiate adventitious roots.

amphaiwan/Shutterstock.com

Figure 14-2. Pothos are easy to propagate with preformed roots that emerge at the node.

Adventitious Shoot and Bud Formation

Plants started from leaf cuttings need to develop both roots and shoots. Adventitious shoot, or bud, formation arises from the active cellular regions of the primary or secondary meristems. Preformed, primary meristems are undifferentiated cells that are constantly dividing. Wound-induced, secondary meristems are cells that have already differentiated into some kind of tissue but change into new, undifferentiated meristem cells that allow bud tissue to grow.

Piggyback plant (*Tolmiea*), **Figure 14-3**, is a unique specimen that grows *plantlets* (small or young plants) that originate from the primary meristem. As the plantlets develop, they detach from the mother plant.

joto/Shutterstock.com

Figure 14-3. Piggyback plant forms adventitious shoot tissue from preformed shoot initials.

Figure 14-4. Mother of thousands develops foliar embryos from primary meristem tissue.

Figure 14-5. African violets will only begin shoot emergence after wounding.

Did You Know?

Piggyback begonia (*Begonia hispida* var. *cucullifera*) develops little plantlets along the veins of leaves, much like the piggyback plant, *Tolmiea*. An army of fuzzy little plantlets can separate from the mother plant and fall to the earth. Given the right conditions, they can begin to grow on their own.

If they are in a moist rooting medium, they can easily root and become independent plants. Mother of thousands plant (*Bryophyllum*) has preformed primary meristems that develop foliar embryos on the leaf margins and will easily root given the right environmental conditions, **Figure 14-4**.

Most leaf cuttings form adventitious buds through a wounding response that prompts the secondary meristem to dedifferentiate cells into new meristematic cells. The African violet is a classic example of this process, **Figure 14-5**. Leaves are cut from the stock plant and veins are wounded. New shoots emerge from tissue beneath the epidermis after adventitious roots have been established. Other species vary slightly in where the bud tissue arises. In the lily plant, buds originate in the parenchyma, and in the *Peperomia* plant, new shoots form from the callus tissue.

Plant Material Used for Stem and Leaf Cuttings

Along with stem and leaf cuttings, there are other methods of vegetative propagation. They will be explored in later chapters. The type of cutting used to propagate a plant depends on the species, how easily a species roots, and the most cost-effective option for propagation.

Most commercial propagators use stock plants as the source of material, **Figure 14-6**. *Stock plants* are plant material kept specifically for the purpose of propagation. Stock material is managed using techniques such as pruning and girdling (making incisions or bending the stem), which encourages growth that will have high rooting potential. Any stock material should be disease free, uniform, true to type, and fairly vigorous.

Figure 14-6. The stem cutting of this boxwood is typical for many woody ornamentals.

STEM Connection
Spiny or Spineless?

Honey locust (*Gleditsia triacanthos*) carries sharp thorns, some up to 1″, and could cause serious pain to any person or animal that draws too close. Interestingly, this native tree has no known animal predators from which to defend itself. The honey locust grows easily and quickly, has reasonably strong branches, provides dainty, dappled shade from pinnate leaves and is tough enough to withstand nearly any setting. Horticulturists found a thornless form, *Gleditsia triacanthos* var. *inermis*, that embodies the classic appeal of the thorned tree, but without any risk of bodily harm. It is vegetatively propagated to maintain thornlessness.

Sari ONeal/Shutterstock.com

Other sources of materials include trimmings from nursery plants, tissue culture liners, and plants growing in the landscape or in the wild. Cut plant material used for propagation may be a stem cutting, a leaf cutting, or a root cutting.

A *stem cutting* is a portion of the shoot and may include the tip or just a section of stem. Stem cuttings are classified by the plant's type of stem tissue. Stem tissue may be *hardwood* (mature, dormant, and woody plant material), semihardwood, *softwood* (soft, succulent, new growth on woody plants), *herbaceous* (nonwoody, soft-stemmed plants), or *deciduous* (trees or shrubs that lose their leaves annually). A *leaf cutting* is a leaf or a portion of a leaf used to propagate a new plant. Leaf cuttings may be just a leaf with no stem, or a *leaf-bud cutting* which includes a leaf blade, petiole, and a short piece of the stem with a bud attached.

Hardwood Cuttings

Hardwood refers to mature, dormant, and woody material. Hardwood cuttings tend to have a long shelf life and require no specialized equipment. They are the least expensive and easiest to use material for stem propagation.

Hardwood cuttings are taken from wood that has stopped actively growing and is dormant, generally in the late fall, winter, or early spring, **Figure 14-7**. The wood most often comes from the last year's growth. Some species can be propagated from wood that is two years old or older, such as figs, olives, and some plums. The material should be pathogen free, vigorous, and growing in full sunlight.

Chuck Wagner/Shutterstock.com

Figure 14-7. This dormant maple is a source of good hardwood cuttings.

It should have normal internode lengths and preferably be taken from the upper part of the plant. The material is ready to be cut when the wood is firm and does not bend, when the leaves can be removed without tearing the bark, or when the leaves have *abscised* (fallen off). Typically, the central and basal portions of the cutting are used for propagation because they contain the most stored carbohydrates needed for root and shoot growth. Avoid using material with flower buds, if possible, as flowers pull stored energy to flowering rather than rooting. Tip portions are generally discarded because of their low levels of carbohydrates. They also often have unwelcome flower buds.

Hardwood cutting lengths range from 4″ to 30″ (10 cm to 76 cm), depending on the species of plant and purpose of the cutting. For fruit trees, long cuttings allow for the insertion of a bud graft. Include at least two nodes in the cut, with the bottom cut immediately beneath a node and the top cut 1/2″ to 1″ (1.3 cm to 2.5 cm) above a node. The diameter of a cutting varies depending on species and may range from 1/4″ to 1″ (0.6 cm to 2.5 cm). Three types of cuts may be used in hardwood cuttings, **Figure 14-8**:

- Straight cutting—a straight cut across the stem. The most commonly used cutting.
- Mallet cutting—a cutting that includes a short section of the stem of older wood. Mallet cuttings are used on harder to propagate material.
- Heel cutting—a cutting that includes a small section of older wood at the base of the cutting. This type of cutting is also used on plants that may be difficult to root.

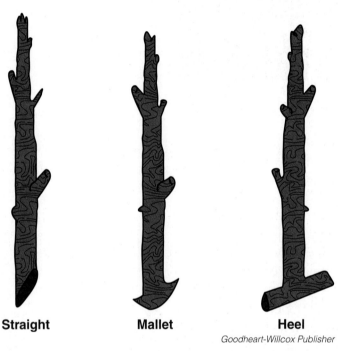

Straight   Mallet   Heel

Goodheart-Willcox Publisher

Figure 14-8. The diagram illustrates three ways to make hardwood cuts.

Early morning is the best time to take cuttings because the plant is fully turgid. Cuttings should be kept cool and moist until they can be stuck in a planting medium. An ice chest or plastic bag with wet paper towels can be used to store cuttings and avoid desiccation (drying up).

Hardwood cuttings are most commonly used to propagate deciduous woody species, but they can also be used for some evergreens, **Figure 14-9**. Many trees and shrubs readily and easily propagate, such as willow (*Salix*), *Forsythia*, privet, spirea, crape myrtle (*Lagerstroemia*), and dogwood (*Cornus*). Some commercial fruit trees are also started by hardwood cuttings and include fig (*Ficus carica*), quince (*Cydonia oblonga*), mulberry (*Morus*), grape (*Vitis*), pear (*Pyrus*), pomegranate (*Punica granatum*), and some plums. Other species require special considerations, such as hormone treatments, to encourage rooting and need finely regulated environmental conditions.

Corner Question

What is one of the most popular shrubs planted in the landscape?

Propagation Methods for Woody Ornamentals

Abelia (*Abelia* spp.**) SH, HW
Arborvitae, American (*Thuja occidentalis*) SH, HW
Azalea (deciduous) (*Rhododendron* spp.) SW
Azalea (evergreen & semi-evergreen) (*Rhododendron* spp.) SH
Barberry, Japanese (*Berberis thunbergii*) SH, HW
Barberry, wintergreen (*Berberis julianae*) SH
Basswood; American (*Tilia americana*) SW
Birch (*Betula* spp.) SW
Blueberry (*Vaccinium* spp.) SW, HW
Boxwood, common (*Buxus sempervirens*) SH, HW
Boxwood, littleleaf (*Buxus microphylla*) SH, HW
Camellia (*Camellia* spp.) SW, SH, HW
Cedar (*Cedrus* spp.) SH, HW
Chamaecyparis; false cypress (*Chamaecyparis* spp.) SH, HW
Cherry, flowering (*Prunus* spp.) SW, SH
Clematis (*Clematis* spp.) SW, SH
Cotoneaster (*Cotoneaster* spp.) SW, SH
Crabapple (*Malus* spp.) SW, SH
Crape myrtle (*Lagerstroemia indica*) SH
Daphne (*Daphne* spp.) SH
Dawn redwood (*Metasequoia glyptostroboides*) SW, SH
Deutzia (*Deutzia* spp.) SW, HW
Dogwood (*Cornus* spp.) SW, SH
Elderberry (*Sambucus* spp.) SW
Elm (*Ulmus* spp.) SW
English ivy (*Hedera helix*) SH, HW
Euonymus (*Euonymus* spp.) SH, HW
Fir (*Abies* spp.) SW, HW
Forsythia (*Forsythia* spp.) SW, SH, HW
Fringe tree (*Chionanthus* spp.) SW
Gardenia; Ellis Cape jasmine (*Gardenia jasminoides*) SW, SH
Ginkgo, maidenhair tree (*Ginkgo biloba*) SW
Goldenrain tree (*Koelreuteria* spp.) SW
Heath (*Erica* spp.) SW, SH
Hemlock (*Tsuga* spp.) SW, SH, HW
Holly, American (*Ilex opaca*) SH
Holly, Chinese (*Ilex cornuta*) SH, HW
Holly, English (*Ilex aquifolium*) SH

Holly, Japanese (*Ilex crenata*) SH, HW
Holly, Yaupon (*Ilex vomitoria* Aiton) SH, HW
Honeylocust (Gleditsia triacanthos) HW
Hydrangea (*Hydrangea* spp.) SW, HW
Japanese cedar (*Cryptomeria japonica*) SH
Jasmine (*Jasminum* spp.) SH
Juniper, Chinese (*Juniperus Chinensis*) SH, HW
Juniper, creeping (*Juniperus horizontalis*) SH, HW
Juniper, shore (*Juniperus conferta*) SH, HW
Larch (*Larix* spp.) SW
Leyland cypress (×*Hesperotropsis leylandii*) SH, HW
Lilac (*Syringa* spp.) SW
Magnolia (*Magnolia*) SH
Maple (*Acer* spp.) SW, SH
Mock orange (*Philadelphus* spp.) SW, HW
Mulberry (*Morus* spp.) SW
Photinia (*Photinia* spp.) SH, HW
Pine, eastern white (*Pinus strobus*) HW
Pine, mugo (*Pinus mugo*) SH
Pittosporum (*Pittosporum* spp.) SH
Poplar; aspen; cottonwood (*Populus* spp.) SW, HW
Quince, flowering (*Chaenomeles* spp.) SH
Redbud (*Cercis* spp.) SW
Rhododendron (*Rhododendron* spp.) SH, HW
Rose (*Rosa* spp.) SW, SH, HW
Rose of Sharon (*Hibiscus syriacus*) SW, HW
Serviceberry (*Amelanchier* spp.) SW
Shoeblackplant (*Hibiscus rosa-sinensis*) SW, SH
Smoketree (*Cotinus coggygria*) SW
Spirea (*Spiraea* spp.) SW
Spruce (*Picea* spp.) SW, HW
St. Johnswort (*Hypericum* spp.) SW
Sweetgum (*Liquidambar styraciflua*) SW
Trumpet creeper (*Campsis* spp.) SW, SH, HW
Tulip tree (*Liriodendron tulipifera*) SH
Viburnum (*Viburnum* spp.) SW, HW
Virginia creeper (*Parthenocissus quinquefolia*) SW, HW
Weigela (*Weigela* spp.) SW, HW
Willow (*Salix* spp.) SW, SH, HW
Wisteria (*Wisteria* spp.) SW
Yew (*Taxus* spp.) SH, HW

SW = softwood, SH = semihardwood, HW = hardwood   **spp = multiple species

Goodheart-Willcox Publisher

Figure 14-9. This table lists a number of woody ornamentals and the recommended method of propagation material.

There are four propagation processes that may be used to root hardwood cuttings:

- Direct fall planting. In regions with mild temperatures, cuttings can be taken in late fall and immediately planted in the nursery. Growers in California and Texas propagate hardwood cuttings in this manner.
- Warm temperature pretreatment. Hardwood cuttings are prepared in the fall while the buds are quiescent (or in a state of suspended growth). The cuttings are treated with *auxin*, a plant growth hormone that induces adventitious root formation. They are stored for three to five weeks in warm, moist conditions before planting or storing at a cooler temperature. This pretreatment allows root initiation while the buds are dormant, so there is little competition for carbohydrates.
- Bottom heat. For species that are difficult to root, a treatment of bottom heat can hasten rooting. Cuttings are collected during dormancy. Auxin, in the form of indole-3-butyric acid (IBA), is applied. After sticking cuttings in a soil medium, they are placed on a heating mat, **Figure 14-10**.

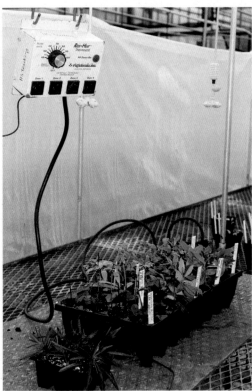

Elizabeth Driscoll/Goodheart-Willcox

Figure 14-10. Hard-to-root species benefit from bottom heat which hastens rooting in cuttings.

- Direct spring planting. Species that easily root can be taken as cuttings during dormancy and stored in cool (32°F–40°F or 0°C–4.5°C), moist conditions. In the spring, the cuttings are placed in propagation trays or in a field nursery bed and given intermittent misting until rooted. Most propagators use an intermittent mist system that delivers fine droplets of mist to reduce water loss.

The propagation methods used for hardwood cuttings of narrow-leaved evergreens are similar to the methods used for deciduous hardwood cuttings. However, they root much more slowly, sometimes taking a year to fully root. Because evergreen species retain their leaves, there is a risk of desiccation. The cuttings must be rooted in high humidity conditions or receive frequent, light misting. Some evergreen species, such as false cypress (*Chamaecyparis*), arborvitae (*Thuja*), and prostrate juniper, root easily. Upright junipers, spruces, hemlocks, firs, and pines are much more challenging to root. Each species has specific propagation needs. In general, most species respond well to high humidity, bottom heat, bright light, and a treatment of auxin to hasten rooting.

Semihardwood Cuttings

Semihardwood cuttings come from the partially matured wood of broadleaf evergreens and deciduous plants during the summer and early fall months. For the broadleaf evergreen shrubs, cuttings are taken from new shoots after a burst of growth has occurred. The timing of this growth varies depending on the species, usually occurring from late spring through early fall. Species such as *Camellia*, *Pittosporum*, *Rhododendron*, and holly propagate well as semihardwood cuttings, **Figure 14-11**.

Semihardwood cuttings range in length from 3″ to 6″ (7.5 cm to 15 cm). The lower leaves may be removed from the stem before inserting it into the propagation medium. These cuttings are more perishable than hardwood cuttings and will dry out quickly. Large leaves may need to be trimmed by as much as one-third to one-half to reduce the surface area. Minimizing the leaf surface reduces water loss through transpiration. To optimize rooting potential, semihardwood cuttings should be collected early in the morning when plants are turgid and then stored in cool, moist conditions. Cuttings are placed in a soilless medium and misted intermittently. Depending on the species, bottom heat, auxin, and wounding can be beneficial.

happykamill/Shutterstock.com

Figure 14-11. This camellia easily roots as a semihardwood cutting. It should be collected from new shoots after a flush of growth.

Softwood Cuttings

Collect softwood cuttings when soft, succulent, new growth on woody plants is beginning to harden (mature). The best softwood cuttings come from shoots that snap easily when bent and that have both mature leaves and small, young leaves. Most softwood cuttings are taken during growth flushes, which generally occur from April or May to August, depending on the region. Many woody ornamental species are started by softwood cuttings, including *Magnolia*, spirea, maple (*Acer*), lilac (*Syringa*), smoketree (*Cotinus*), *Hydrangea*, redbud (*Cercis*), and *Wisteria*.

Softwood cuttings typically root easily and more swiftly than other types of cuttings. Because the shoots are quite tender and stress easily, they must be prevented from drying out. Take cuttings in early morning when plants are turgid.

"Success is the result of perfection, hard work, learning from failure, loyalty, and persistence."
—Colin Powell

History Connection

J.C. Raulston

The late J.C. Raulston was a well-respected professor at North Carolina State University (NCSU) and strong advocate for the nursery industry. He was instrumental in introducing innovative plant material throughout the southeast and in establishing the NCSU arboretum (now the J.C. Raulston Arboretum). Mr. Raulston trialed hundreds of plants for growing, testing, and sharing with the industry and the public. His enthusiasm and promotion of new and interesting plant material shifted landscapes from a limited number of species to a broad range of species. He also pushed the plant palette to include a broad range of beautiful and functional woody trees and shrubs that flowered or had other interesting features in all months of the year. He invited nursery owners to the arboretum to take cuttings and he freely distributed rooted cuttings he propagated himself. Mr. Raulston worked tirelessly to enhance the diversity of trees and shrubs that could support the nursery trade.

Figure 14-12. This hydrangea cutting has two or more nodes and was collected from a lateral shoot.

Store the cuttings in moist, cool environments to minimize stress and desiccation. Take cuttings from a lateral or side portion of the stock plant. They should be 3″–5″ (7.5 cm–12.5 cm) long and have two or more nodes, **Figure 14-12**. Because softwood cuttings always have leaves, remove the lower portion to make it easier to stick. Larger leaves can be trimmed to diminish water loss. Cuttings are perishable and should not be stored for more that one or two days. To ensure successful rooting, temperatures should be between 75°F and 80°F (23°C to 27°C), light should be adequate but not excessive, and intermittent misting should be applied.

Herbaceous Cuttings

Herbaceous cuttings are made from succulent, fleshy, nonwoody plants. Species that are propagated as herbaceous cuttings include coleus, geranium, sweet potato, poinsettia, carnation, and many houseplants. Most herbaceous cuttings are taken from the tip of the stem where there is a naturally high concentration of auxin. These cuttings typically root more easily than cuttings from other parts of the stem. Cutting should be from 3″ to 5″ (8 cm to 13 cm) long. The lower leaves should be removed before inserting the cutting into a soilless medium for rooting. The conditions used for softwood cuttings, including bottom heat, encourage rapid and uniform rooting.

Did You Know?
Large, fibrous begonia leaves can be cut into pieces, with each piece containing a large vein. The piece is inserted upright into the rooting medium. The new plant develops from the large vein at the base of the section of leaf.

Leaf Cuttings

A leaf cutting is a leaf or a portion of a leaf used to propagate a new plant. Leaf cuttings use the leaf blade or leaf blade and petiole (stalk or area that attaches the leaf blade to the stem) to propagate new plants. Adventitious roots, buds, and shoots arise from the leaf, either at the base or on cuts made to the leaf. The original leaf dies and does not become part of the new plant. Not many plants can be started with leaf cuttings. Several plants that can be started this way include African violets, *Begonia*, *Peperomia*, snakeplants, pineapple lily, *Sinningia*, *Kalanchoe*, and some *Sedum* species, **Figure 14-13**.

Leaf cuttings of African violets and begonias often use the leaf blade or leaf blade and petiole. The cutting is placed in a potting medium, and new plants form at the base of the petiole, **Figure 14-14**. For some plants, such as rex begonia, the veins on the underside of the fleshy leaves of plants are cut. The leaf blade is laid flat in direct contact with the surface of the propagation medium. The leaf is secured with pins to keep it in constant contact with the medium surface. After spending time in a high humidity environment, new plants will emerge from where the cuts were made. The original leaf eventually shrivels and dies.

Figure 14-13. Peperomia propagates well as a leaf cutting.

Figure 14-14. Notice the new plants forming at the base of the begonia.

Some species, such as snakeplant (*Sansevieria*) and pineapple lily (*Eucomis*), are propagated using the entire leaf. The long leaf blades are cut in sections, **Figure 14-15**, and placed in propagation medium up to three-fourths of their length. The new plants arise from the base of the cutting.

All leaf cuttings require intermittent misting or high humidity conditions, such as in a propagation tent. Bottom heat may hasten root, bud, and shoot development. Ideal media temperatures are in the range of 65°F–77°F (18°C–25°C). An application of plant growth hormones can encourage bud formation and adventitious shoot formation.

Leaf-Bud Cuttings

Leaf-bud cuttings are similar to leaf cuttings, but they include a leaf blade, petiole, and a short piece of the stem with a bud attached. Because a bud already exists and can develop into the shoot, only the adventitious roots need to form. Many herbaceous greenhouse plants, some fruits (such as black raspberry, blackberry, boysenberry, and lemon), and some trees and shrubs (such as maple, camellia, and rhododendron) readily root as leaf-bud cuttings. Many plants can be started this way, and this method is useful if there is limited stock material, **Figure 14-16**. The cuttings are inserted into the soil medium with the bud under the surface. High humidity and bottom heat are applied to hasten growth.

Root Cuttings

Root cuttings are not a commonly used propagation method, but for some species this method is the easiest way to develop new plants. Pieces of a root are cut from stock plants. The pieces are placed in a rooting medium, covered with the medium, and often covered with plastic or glass to prevent the plant material from drying out. Correct *polarity* (spatial orientation within plants) should be maintained, with the top of the plant placed upward and the bottom of the root placed downward, **Figure 14-17**. Often, commercial propagators will make a diagonal cut on the *proximal* end (end nearest the point of attachment or crown) and a straight cut on the *distal* end (end farthest from the point of attachment, or crown). Cuttings that are inserted into the soil vertically should ensure the proximal end faces up. Cuttings are placed about 1″–2″ (2.5 cm–5 cm) deep, depending on root size.

Elizabeth Driscoll/Goodheart-Willcox
Figure 14-15. Snake plant leaves are cut into sections and will form new plants at the base of the cutting.

bepsy/Shutterstock.com
Figure 14-16. This lemon tree grew from a leaf-bud cutting because there was limited stock plant material.

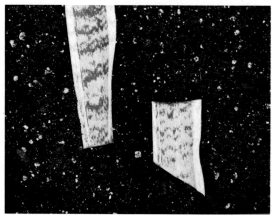
Elizabeth Driscoll/Goodheart-Willcox
Figure 14-17. Note the angled cut for the proximal end and the straight cut for the distal end to ensure proper polarity when placing the cutting in the soil.

Figure 14-18 lists a number of ornamental and edible specimens that can be propagated by root cuttings.

Hardening Off

Rooted cuttings need to adjust away from the high humidity environment of a propagation chamber. This hardening off process, also call *acclimatization*, gradually exposes plants to different environmental conditions. It enables the new plant to increase rates of photosynthesis and absorb water and nutrients through the root system. Cuttings that are left too long under the misting system can begin to deteriorate and growth will slow. This hampers production schedules and diminishes the quality of the new plant, with leaf drop and poor root growth. Most hardening off processes shorten the time of misting gradually until the cuttings are acclimated and mist is no longer needed.

Rooting Medium

The rooting medium used to propagate plants by stems or leaves should perform several functions. The rooting medium should:
- Hold the cutting in place during the rooting period.
- Provide moisture for the cutting.
- Allow gas exchange at the base of the cutting.
- Create a dark environment to encourage rooting.

> "Men are mortal. So are ideas. An idea needs propagation as much as a plant needs watering. Otherwise, both will wither and die."
> —B.R. Ambedkar

Specimens Commonly Propagated through Root Cuttings

Apple, crabapple (*Malus* spp.)	Kiwi (*Actinidia deliciosa*)
Barrenwort (*Epimedium* spp.)	Leadwort (*Plumbago* spp.)
Bayberry (*Morella pensylvanica*)	Summer ragwort (*Ligularia dentata*)
Bellflower (*Campanula* spp.)	Lilac (*Syringa vulgaris*)
Blackberry, raspberry (*Rubus* spp.)	Liriope (*Liriope* spp.)
Blanket flower (*Gaillardia* spp.)	Lungwort (*Pulmonaria* spp.)
Bleeding heart (*Dicentra* spp.)	Queen of the prairie (*Filipendula rubra*)
Bottlebrush buckeye (*Aesculus parviflora*)	Mint (*Mentha* spp.)
Bugbane (*Cimicifuga racemosa*)	Oriental poppy (*Papaver orientale*)
California poppy (*Eschscholzia californica*)	Plume poppy (*Macleaya cordata*)
Coneflower (*Echinacea purpurea*)	Poplar (*Populus* spp.)
Cranesbill (*Geranium* spp.)	Rose (*Rosa* spp.)
Daphne (*Daphne*)	Sage (*Salvia* spp.)
Evening primrose (*Oenothera* spp.)	Sassafras (*Sassafras albidum*)
Fall phlox (*Phlox paniculata*)	Sea holly (*Eryngium* spp.)
Gas plant (*Dictamnus*)	Siberian bugloss (*Brunnera macrophylla*)
Globethistle (*Echinops* spp.)	St. John's wort (*Hypericum calycinum*)
Glory-bower (*Clerodendrum trichotomum*)	Stokes' aster (*Stokesia laevis*)
Heartleaf saxifrage (*Saxifraga nelsoniana*)	Sweet gum (*Liquidambar styraciflua*)
Hop (*Humulus lupulus*)	Windflower (*Anemone hupehensis* var. *japonica*)

Goodheart-Willcox Publisher

Figure 14-18. These plants can be propagated as root cuttings.

As noted in Chapter 11, there are a number of different substrates that can be mixed to create a planting medium. For propagation mixes, growers use a mixture of organic materials, such as peat, sphagnum moss, and bark, and combine it with a mineral component, such as perlite, vermiculite, coarse sand, expanded shale, or rock wool, **Figure 14-19**. The mineral components provide aeration and drainage. Mineral soil (field soil) is rarely used, unless nursery liners are being grown in nursery beds. *Nursery liners* are young plants that will be grown in the nursery for an extended time until they are ready for sale.

Most commercial mixes are sterile to limit any introduction of pathogens. When mixing your own media, sterilize the soil with heat, using either an oven or steam.

Plant Growth Regulators

Plants naturally produce chemicals called *phytohormones* that are present in small amounts and regulate plant growth and other functions. The five major plant hormones include auxin, cytokinins, gibberellin, abscisic acid, and ethylene. Many of these hormones play a role in propagation by inducing or sometimes inhibiting the formation of adventitious roots and shoots. *Cytokinins* are a class of plant growth hormone that encourages bud formation and adventitious shoot formation. Auxin and cytokinins are the hormones most commonly used in propagation. These plant hormones are naturally and synthetically manufactured as plant growth regulators (PGRs) to promote successful propagation through root, bud, and shoot formation.

Elizabeth Driscoll/Goodheart-Willcox

Figure 14-19. Rock wool is commonly used as a medium to propagate herbaceous cuttings.

Auxin

Auxin plays a number of roles within the plant. Related to propagation, auxin can promote the growth of adventitious roots, particularly in species that are hard to propagate. Indole-3-acetic acid (IAA) is an auxin that is naturally produced by plants. It can be extracted and applied to cuttings to stimulate rooting. Indole-3-butyric acid (IBA) and 1-naphthalene acid (NAA) are synthetic auxins that encourage rooting. They remain the most widely used auxins for stem cuttings. Auxins can also be used in combination with each other. Common combinations include IBA and NAA or IAA and IBA.

AgEd Connection Agriscience Fair

Agriscience fairs provide opportunities for students to develop an experiment that illustrates their creativity, curiosity, knowledge, and science inquiry skills. Consider doing an experiment that determines the best methodology for propagating your favorite plant. Do you need to use a plant hormone? Which one? At what rate will you apply the hormone? Be sure to follow the science process, replicate your experiment, and share your findings in a professional way.

Elizabeth Driscoll/Goodheart-Willcox

Figure 14-20. Moistening the ends of the cuttings will increase the surface contact of the plant growth regulator to the base of the cut.

Apply auxin to fresh cuts. If you use cuttings taken earlier in the day or week, make new, fresh cuts. Auxins can be applied in powdered or liquid form. Cuttings are dipped into the substances and then stuck into the rooting medium. Generally, higher concentrations of auxin are used on plants that the grower expects to be hard to root. In most cases, recommended concentrations have been established. In powdered applications, the hormone is added to talcum powder. Cuttings are dipped individually or as a bundled group into the powder and then inserted into the rooting medium, **Figure 14-20**. If needed, cuttings can be moistened with water to encourage better contact with the powder.

Concentrated solutions of hormone can also be dissolved into a liquid, usually some kind of alcohol, such as isopropyl alcohol or ethanol. The basal ends of the cuttings are lowered 0.2″–4″ (0.5 cm–1 cm) into the solution for about three to five seconds and then inserted into the propagation medium. This process is called the *quick dip method*. Cuttings can be dipped in bundles or individually. The highest absorption of hormone occurs at the surface of the cut. Many commercial propagators prefer the quick dip method for uniformity of rooting and ease of application.

Cytokinins

Cytokinins play essential roles in cellular division. In propagation, the relationship between auxin and cytokinins determines whether root formation is promoted or shoots are developed. A high auxin-to-cytokinins ratio encourages root development. A high cytokinins-to-auxin ratio encourages adventitious shoot formation. A high concentration of both fosters callus formation. Cytokinins are an important factor in initiating buds and shoots from leaf cuttings.

Careers in Stem and Leaf Propagation

Stem and leaf propagation are important components of the horticulture industry. The work done in these areas guarantees a supply of new plants is always available for growers and consumers. Two careers in these areas are horticulture illustrator and nursery inspector.

Horticulture Illustrator

Horticulture illustration is the preparation of accurate renderings of botanical subjects and horticultural processes. These illustrations are used for textbooks, cooperative extension publications, journals, museum exhibits, websites, and many other applications. Illustrations may be hand or digitally drawn and communicate a visual explanation of complex information.

For example, an illustration of the leaf-cutting process helps gardeners understand the proper technique. Many illustrators have a background in fine arts and may have taken additional coursework specific to botanical or scientific illustration.

Nursery Inspector

A nursery inspector promotes the production and sale of nursery stock. Nursery inspectors conduct inspections of all plant material that will be sold or distributed to make sure they are free of diseases and pests. Most states require annual inspections of nurseries to ensure there are no damaging plant pests prior to the plants being shipped within and out of state. Inspectors are usually part of a state's agriculture department. They issue phytosanitary certificates by visiting nurseries, garden centers, chain stores, and landscape companies to monitor for any pests. Nursery inspectors should have a bachelor's degree in agriculture, horticulture, entomology, or a related field.

Career Connection
Mark Weathington
Arboretum Director

Mark Weathington

A very large, busy highway expansion is about to get started in the backyard of the J.C. Raulston Arboretum (JCRA). Mark Weathington, the arboretum's director, sees the increasing urbanization around the garden as an opportunity for innovation and to bring horticulture to the thousands of commuters that travel by each day. "We are trying to figure out a way to green the sound barrier walls through plants. We hope to provide a functional and beautiful space that minimizes noise and adds an attractive aesthetic."

Mark began his horticulture career while working in Virginia Tech's public garden. He learned nursery production and propagation while working in the public garden, and then worked at both the Atlanta Botanical Garden and Norfolk Botanical Garden. Today, Mark finds himself firmly rooted at the JCRA as the arboretum's director and continues to serve the needs of nurserymen in the southeastern United States by evaluating new landscape ornamentals. This work is a continuation of the JCRA mission to propagate and share new plants to increase the diversity of woody ornamentals available in the nursery trade.

The JCRA has grown into an inviting public garden where visitors can increase their appreciation and understanding of horticulture. Mark finds that when he arrives at work each morning, everyone is happy to be there, sharing the same goal and mission. He says, "There are artists that cannot live without art, I am passionate about plants in the same way. I love being around plants, talking about them, and sharing them." Mark takes deep personal satisfaction knowing the arboretum provides an interesting and innovative landscape that showcases plants that support the growth of the nursery trade across the United States and brings a spark to gardeners everywhere.

CHAPTER 14 Review and Assessment

Chapter Summary

- Plants have a unique ability to form new plants from a piece of stem, leaf, or root through a process called vegetative or asexual propagation. In this process, adventitious root, shoot, and bud formation occurs.
- There are two types of adventitious roots: preformed and wound-induced. Preformed roots develop naturally given the right environmental conditions. Wound-induced roots form in response to wounding or cutting of the plant.
- Stem cuttings can come from different types of tissue including hardwood, semihardwood, softwood, and herbaceous tissues.
- Hardwood cuttings are taken from dormant wood, usually from deciduous woody ornamentals and some narrow-leafed evergreens. Straight, mallet, and heel cuts are different techniques for making wood cuts.
- Semihardwood cuttings come from partially matured wood of broad-leaved evergreens and deciduous plants growing in the summer to fall months. Early morning collection of cuttings minimizes water loss and optimizes rooting potential.
- Softwood cuttings are taken from soft, succulent new growth of woody plants, usually from April through August. Softwood cuttings typically root easily.
- Herbaceous cuttings come from the succulent, fleshy growth of nonwoody plants. Bottom heat and intermittent misting encourage rooting of herbaceous cuttings.
- Leaf cuttings are derived from the leaf blade or leaf blade with the petiole attached. New plants form at the base of the leaf or cuts in the vein, with both adventitious root and shoot formation occurring. The original leaf dies; it does not become part of the new plant.
- Leaf-bud cuttings include the leaf blade, petiole, and small section of stem that includes a bud. This method of propagation is useful when there is limited stock material.
- Root cuttings use pieces of root cut from stock plants to propagate new material. The pieces are covered with rooting medium and often covered with plastic or glass to prevent them from drying out. Not all species can be propagated this way.
- Rooted cuttings need to acclimate to environments other than the high humidity environment of a propagation chamber. This hardening off process enables the new plant to increase rates of photosynthesis and absorb water and nutrients through the root system.
- Rooting medium should hold the cutting in place, provide adequate moisture, allow gas exchange at the base of the cutting, and create a dark environment favorable for rooting.
- Plants naturally produce chemicals called phytohormones. They are present in small amounts and regulate plant growth and other functions. The five major plant hormones include auxin, cytokinins, gibberellin, abscisic acid, and ethylene.

Words to Know

Match the key terms from the chapter to the correct definition.

> A. abscise
> B. acclimatization
> C. adventitious root formation
> D. auxin
> E. callus tissue
> F. cytokinins
> G. dedifferentiate
> H. differentiate
> I. distal
> J. hardwood
> K. herbaceous
> L. leaf-bud cutting
> M. phytohormone
> N. plantlet
> O. polarity
> P. preformed root
> Q. proximal
> R. softwood
> S. stock plant
> T. suberin
> U. vegetative propagation

1. The process of roots forming from any plant part other than the root.
2. A method of starting new plants from existing plant material.
3. A bundle of undifferentiated cells that cover the wound and begin to initiate new cellular divisions that differentiate to form meristematic growing regions.
4. Develop a specialized function, such as in shoot cells.
5. Spatial orientation within plants.
6. A single node and adjacent internode tissue with the leaf attached.
7. Mature, dormant, woody plant material.
8. Soft, succulent, new growth of woody plants.
9. Regress (for cells or tissue) from a specialized function to a simpler state.
10. To fall off or separate from.
11. A nonwoody, soft-stemmed plant.
12. A waxy substance that seals a wound on plants.
13. A plant hormone that induces adventitious root formation.
14. Having existing root initials.
15. A class of plant growth hormone that encourages bud formation and adventitious shoot formation.
16. A chemical naturally produced in a plant that regulates growth and other functions.
17. Located closest to the point of attachment.
18. Located farthest from the point of attachment.
19. A small or young plant.
20. Plant material kept specifically for the purpose of propagation.
21. Gradually exposing plants to different environmental conditions.

Know and Understand

Answer the following questions using the information provided in this chapter.

1. What is vegetative or asexual propagation of plants and why is it important?
2. What is the genetic makeup of a vegetatively propagated plant?
3. What are two kinds of adventitious roots?

4. What are the three steps in a wounding response?
5. Draw or give an oral explanation of how a plant started from leaf cuttings can develop adventitious shoots.
6. Explain how to select material for a hardwood cutting.
7. Describe three types of cuts that may be used for hardwood cuttings.
8. List and briefly describe the four propagation processes that may be used to root hardwood cuttings.
9. Why would you reduce the leaf area on semihardwood cuttings?
10. Describe the process of collecting and starting semihardwood cuttings.
11. Describe the process for propagating softwood cuttings.
12. What are some plants that are typically propagated as herbaceous cuttings?
13. Describe the technique used to propagate a begonia from a leaf cutting.
14. What environmental conditions are favorable for propagation using leaf cuttings?
15. How do leaf-bud cuttings differ from leaf cuttings?
16. Describe the process of propagation using root cuttings.
17. What is meant by polarity regarding root cuttings?
18. Why do rooted cuttings need to acclimate or harden off away from the propagation chamber?
19. What are the primary functions of a rooting medium?
20. Name five major plant hormones and what role they play in propagation.
21. What are some benefits of using plant growth regulators?
22. Give one example of a natural plant hormone and two examples of synthetic growth regulators.

Thinking Critically

1. You have collected several hardwood cuttings from a woody ornamental shrub species on a recent walk through an arboretum. You propagate the cuttings, but they do not form roots. What do you think is happening? What can you do to get the cuttings to root?
2. Imagine that you are a small greenhouse owner who primarily sells herbaceous material. The economy is slowing, and you are trying to keep your business afloat. What ideas do you have to keep your business thriving?

STEM and Academic Activities

1. **Science.** Design and implement an experiment to determine the rooting efficiency that occurs due to bottom heat. Document the method you use and your results.
2. **Science.** Most plant species have different propagation requirements. Research a plant that you find interesting and write a report that summarizes your findings.
3. **Math.** You are operating a small start-up retail nursery and need to determine the size of a greenhouse you want to build for propagation of bedding plants. You want to grow 2000 flats of bedding plants. If a bedding plant flat occupies 1 1/2 square feet per flat, how many square feet do you need? Additionally, 70% of the greenhouse space is production space, with the remainder in aisles. What size should the greenhouse be?

4. **Social Studies.** Research the work of J. C. Raulston, a famous horticulturist in the Southeast region of the United States. Find descriptions and images of his work. Select one plant that he propagated and introduced to the nursery trade. Does the plant introduction have any particular meaning to the nursery industry? What was his intent? How does this compare with plant introductions today?
5. **Language Arts.** Pretend that you are a commercial plant propagator. You keep a daily journal to help you remember information that you might need later. Today, you tried propagating five new species. Write a journal entry about these species and include all the information you might need in the future.

Communicating about Horticulture

1. **Reading and Speaking.** Research the propagation protocol for your favorite plant. Prepare a short demonstration for your classmates on this propagation technique, sharing information while actively showing your peers how to propagate your plant.
2. **Listening and Speaking.** Working with a partner, compare and contrast different propagation techniques. Consider the perspective of the manufacturer/grower, merchandiser, and consumer. In what situations would one technique be preferable to the other? Record the key points of your discussion. Hold a class discussion. Compare your responses to those of your classmates.

SAE Opportunities

1. **Exploratory.** Visit a commercial or home propagator and observe each step of the plant propagation process. What conditions does the grower create to encourage rooting and shoot development?
2. **Exploratory.** Create a propagation calendar for five of your favorite ornamental trees. When should you take cuttings? How should you prepare the cuttings? Include timing, depth, length, and other important details for each plant. Why do you think growers keep records of their propagations?

Jodi Riedel/Goodheart-Willcox Publisher

3. **Experimental.** Conduct an experiment using your favorite tree or shrub. Take cuttings at multiple points during the year and try to root them. What do you expect to observe? How will you collect data? Share your findings with your class.
4. **Exploratory.** Job shadow someone who commercially propagates plants. What are this person's daily tasks? What is the scope of his or her responsibilities? Why do you think the work is important for growers in your state?
5. **Exploratory.** Visit a local botanical garden or arboretum that propagates plants from stock in the garden. Volunteer to assist in propagation tasks and learn the techniques used for each plant.

CHAPTER 15
Layering and Division

Chapter Outcomes
After studying this chapter, you will be able to:
- Discuss the uses of layering in propagation.
- Describe the layering techniques used by growers.
- Describe the different types of natural layering.
- Explain the process of crown division.
- Identify types of geophytes and summarize their propagation methods.
- List careers related to layering and division.

Words to Know

air layering	mound (stool) layering	suberize
bulb	nontunicate	sucker
bulblet	offset	tuber
corm	pseudobulb	tubercle
cormel	rhizome	tuberize
geophyte	runner	tunicate
girdling	separation	
hilling	serpentine layering	
layering	stolon	

Before You Read
After reading each section (separated by main headings), stop and write a three- to four-sentence summary of what you just read. Be sure to paraphrase and use your own words.

While studying this chapter, look for the activity icon to:

- **Practice** vocabulary terms with Words to Know activities.
- **Expand** learning with identification activities.
- **Reinforce** what you learn by completing Know and Understand questions.

www.g-wlearning.com/agriculture

Irmantas Arnauskas/Shutterstock.com; warmer/Shutterstock.com; Denis and Yulia Pogostins/Shutterstock.com; i7do/Shutterstock.com; Photograph by Marko Penning, Penning Freesia B.V.

Have you ever walked by a neighbor's yard and noticed something odd about the rose bushes? Do some of the stems seem to be growing into the ground away from the plant and then out again? Your neighbor is probably using a traditional gardening technique to grow more roses. This propagation method is called layering.

Layering consists of several techniques used to propagate plants. Some layering techniques are intensive and must be done in very specific ways at certain times of the year. However, layering may also occur naturally, as is the case for several types of plants, such as strawberries and mint. Orchids also propagate naturally this way, **Figure 15-1**.

Another type of propagation is division. In the division method, one plant is divided to create multiple plants. This technique is commonly used for perennials such as bulbs, corms, and tubers. Crown division is used for some types of plants, whereas physical separation of geophytes is used for others. If you have ever removed a clove of garlic from a garlic bulb, then you have seen at least one step of geophyte separation.

aLittleSilhouetto/Shutterstock.com

Figure 15-1. Orchids can form a special kind of stem (called a *keiki*) that can be used to start new plants.

Layering in Propagation

Layering is a vegetative or asexual propagation technique that allows stems that are still attached to the parent plant to form roots. While the stem is initiating adventitious roots, it is simultaneously pulling in water, nutrients, sugars, and plant hormones, limiting any risk of water stress. The rooted stem is considered a layer following detachment (removal) from the parent plant. Some plants naturally layer, but most plants require specific layering techniques for roots to form.

The practice of layering is most successful when the following elements are considered:

- Attachment of stem to parent plant.
- Accumulation of sugars and auxin to rooting area.
- Exclusion of light in the rooting area.
- Stock plant invigoration.
- Seasonal timing.

The attachment of the stem to the mother plant provides a continual supply of carbohydrates from photosynthesis, water and minerals pulled from the soil solution, and the plant hormone, auxin, **Figure 15-2**. The accumulation of the sugars and auxin to the rooting area fosters root initiation. *Girdling* (making incisions or bending the stem) further encourages root formation. To girdle the plant, a sharp knife is pressed against the bark to remove a portion that exposes the cambial layers underneath. This triggers the release of hormones that encourage root formation. Auxin may be added as indole-3-butyric acid (IBA) to the girdled or cut area to enhance rooting.

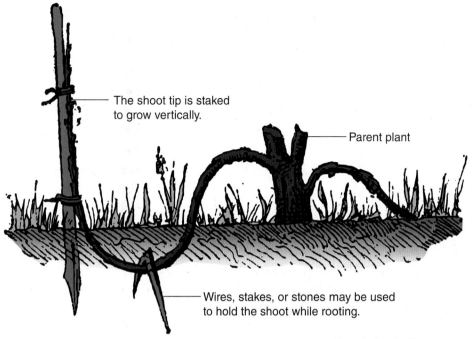

Figure 15-2. This diagram illustrates the general concept of layering, where a rooted plant or layer forms while still attached to the parent plant.

Most layers are covered with rooting medium, most often field or garden soil. This creates an environment devoid of light, which has been found to stimulate rooting.

Many commercially propagated layers come from stock plants. Growers will cut back stock plants to encourage the growth of vigorous new shoots in a process called invigoration. The newly sheared stock plants are covered in rooting medium and watered to provide a dark, moist environment favorable for rooting the newly emerged shoots. The timing of layering depends on the plant species. For most plants, using dormant stems and layering in the spring will take advantage of stored carbohydrates and foster root growth.

Commercial propagators who use layering will establish propagation beds that may last up to 20 years, **Figure 15-3**. Sites should have soil that is well drained and that provides adequate aeration. Access to continuous moisture should be provided, and temperature extremes should be avoided, perhaps through insulation. Field sites should be free of plant pathogens, insects, and weeds. Layering is labor intensive and expensive, but it may be the only propagation method available for species that are hard to propagate.

"Earth laughs in a flower."
—Ralph Waldo Emerson

Figure 15-3. A commercial propagation bed of apple rootstocks in Washington is used for layering.

Layering Techniques

There are a number of layering techniques, and their use depends on the plant species and the given situation. Types of artificial layering that growers use include:

- Simple layering.
- Compound layering.
- Air layering.
- Mound layering.
- Trench layering.

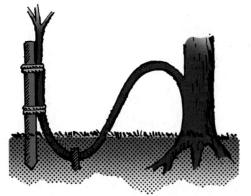

Pearson Scott Foresman/Shutterstock.com

Figure 15-4. Shoots are bent over in early spring and fall, and roots form on the buried part of the shoot.

Simple Layering

In simple layering, a flexible, low-growing stem is bent to the ground. Part of it is covered with soil with 6" to 12" of the tip remaining above the soil, **Figure 15-4**. The portion where the stem is bent will form adventitious roots. In many cases, wounding the underside of the stem can further enhance rooting. Many low-growing shrubs can be layered. Forsythia, azalea, and honeysuckle are examples of shrubs that can be layered. Commercial propagation of filberts or hazelnuts is done using simple layering.

The simple layering technique is typically done in early spring to allow the layer to grow through the season. It is then removed in the fall or spring before growth starts anew. Other simple layering will make use of mature branches in the late summer. The layer will be removed in the spring before growth begins or permitted to grow to the end of the season. Growers should occasionally check to ensure adequate moisture. Once roots have developed, the layer is removed from the parent plant and treated as a rooted cutting.

Compound Layering

The compound layering method can produce several layers (rather than just one) from a single stem. In compound layering, a long stem is bent and laid down horizontally. It is held in place with wire pegs and fully covered with soil, **Figure 15-5**.

Goodheart-Willcox Publisher

Figure 15-5. Compound layering produces several rooted layers compared to just one layer in simple layering.

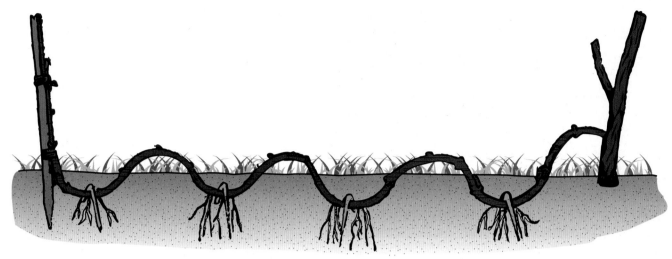

Morphart Creation/Shutterstock.com

Figure 15-6. Serpentine layering is most successful when using plants characteristic of vine-like growth.

Once new shoots begin to grow to at least 4″ (10 cm), a shallow trench is dug. The stem is laid in the trench and the shoots are held by pegs. As the shoots continue to grow, soil is filled in around them until they are rooted and can be cut.

A variation of compound layering is called *serpentine layering*. Shoots are laid horizontally to the ground. Some buds are exposed while others are covered with soil in an alternating pattern, **Figure 15-6**. Serpentine layering works well for vine-like plants.

Air Layering

Air layering involves wrapping a girdled stem with rooting medium and allowing adventitious roots to form. This layering practice has been used for centuries. It is useful for propagating houseplants, such as rubber plant, croton, monstera, philodendron, and dieffenbachia, **Figure 15-7**. Air layering can also help propagate woody ornamentals, including camellia and magnolia. Air layering helps propagate tropical fruits, such as lychee (*Litchi chinesis*), longan (*Dimocarpus longan*), and key lime (*Citrus aurantifolia*). For the best results, air layers are made in the spring on younger shoots. Growers have less success in rooting older stems.

To make the air layer, choose an area on the stem that is about 12″ (30 cm) from the tip of the shoot and just below a node. Ensure there are no leaves on the stem 3″–4″ in either direction. Girdle the stem, **Figure 15-8**. Expose the inner, woody tissue, and scrape it to eliminate the cambial tissue and prevent healing.

Tamara Kulikova/Shutterstock.com

Figure 15-7. Rubber plant is commonly propagated through air layering.

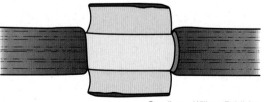

Goodheart-Willcox Publisher

Figure 15-8. In dicots, the bark is completely girdled to start the process of air layering.

Figure 15-9. Wrapping the wound with sphagnum moss covered in plastic provides a moist, dark space for roots to initiate and grow.

Application of IBA to the wound can be beneficial. This technique is used primarily on dicots. With monocots, propagators make a slanting upward 1"–1 1/2" (2.5 cm–4 cm) cut. The cut is held open with sphagnum moss.

For both methods, the exposed stem is surrounded with about two handfuls of soaked sphagnum moss. The moss is then securely wrapped with plastic. The moss retains moisture and prevents water from entering, **Figure 15-9**. Aluminum foil can also be used. It reflects sunlight and modifies temperatures.

Adventitious roots will begin to initiate. After the sphagnum moss fills with corky, thick early roots and fibrous secondary roots, the stem can be severed from the parent plant. Rooting generally occurs in two to three months or less. Spring and early summer layers can be harvested as the shoots become dormant in the fall. Removal of the air layer is most often successful when the plant is not actively growing. Transplanted layers should be acclimated with high humidity and low light conditions.

Mound Layering

Mound (stool) layering is a method in which soil is piled on the crown of the plant. The new shoots form adventitious roots that can be severed and transplanted. This layering technique is the most commercially important method of layering. It is used to produce millions of apple, pear, and other fruit tree rootstocks each year. Propagators cut back dormant stock plants severely to only 1" (2.5 cm) above the soil surface. As the stock plant begins to grow, soil, sawdust, bark, or a soil-sawdust mixture is mounded over the new shoots, **Figure 15-10**. *Hilling* is the piling of soil onto plants to promote desired growth. As shoots continue to grow, a second hilling of rooting medium covers the shoots to one-half of its height. A third and final hilling covers the shoots in mid-season when most shoots reach about 18" (45 cm). These shoots have been covered to a depth of 6"–8" (16 cm–20 cm). Layers are removed once the plant is inactive, after roots have formed. Apple rootstock, quince, currant, gooseberry, spirea, quince, magnolia, and cotoneaster respond well to mounding.

Did You Know?
The air layering technique was perfected in China. This technique is more than 4000 years old.

Figure 15-10. In mound (stool) layering, sawdust or soil covers the base of the stock plant to encourage the emergence of rooted layers.

Thinking Green

Garbage Can Garden

Many leftover scraps from the kitchen can actually be turned into a garbage can garden. If the remains of a ginger rhizome has a node, plant it in potting medium. Likewise, twist the top of a pineapple and tuck it into a pot filled with moistened medium. If the leafy remnants of carrots, beets, and turnips still have a small piece of root attached, place it on top of pebbles in a bowl with a little water and watch new leaves emerge. Avocado, mango, and papaya seeds can easily sprout. If you have nearly rotting potatoes, stick them in a pot, and watch them grow!

iofoto/Shutterstock.com

Trench Layering

For woody plant material that is difficult to root, trench layering may be the only technique that will produce clones. In trench layering, the mother plant is planted in a sloping position (30° to 45°) that allows shoots to be layered horizontally and pegged down in the base of a trench. Loose rooting medium, such as soil, bark, or sawdust, is filled around the new shoots as they develop. The rooting material excludes light, which causes the new shoots to etiolate, or elongate. Research suggests the etiolation is critical for initiating roots in species that are hard to propagate. Trench layering can be used with quince, apple, mulberry, walnut, and cherry plants.

Natural Layering

Many plants have modifications of their plant parts or growth habits that enable them to create layers through their natural biology. Tip layers, runners, stolons, offsets, and suckers are examples of natural layering.

Tip Layers

Blackberries, dewberries, and black raspberries have trailing growth habits and exhibit natural tip layering. When the shoot tip of new growth is inserted into the soil, it grows downward. The tip curves upward where roots will then form to become a new plant. The newly rooted tip layer is tender and prone to drying out. It should be dug and replanted immediately. The best time to replant is generally in late fall or early spring.

Runners

Runners are specialized stems that grow or "run" horizontally above the ground and form new plants. Strawberries have a runner growth habit. Stems arise from the axil of a leaf at the crown of the plant, and new plantlets (young or small plants) form along the nodes of the running stem, **Figure 15-11**.

Steve Longrove/Shutterstock.com

Figure 15-11. Strawberries form daughter plants or plantlets from runners that stem from the mother plant.

Other running plants include bugleweed (*Ajuga*), spider plant (*Chlorophytum comosum*), Boston fern, and the orchid genus *Dendrobium*.

In strawberries, runners begin to grow as temperature and day length increases (12 to 14 hours or more of light per day). Plantlets produced along the runner develop roots and remain attached to the mother plant. In turn, they send out their own runners to form more plantlets. As the weather turns cold, the runner stems die, and the plants are disconnected from each other. Each new plant can be dug when it is well rooted and transplanted.

Stolons

Similar to runners, *stolons* are horizontal stems, but they grow above the ground and produce plants or tubers. *Tubers* are swollen, underground stems that are storage organs for the plant. Potatoes are an example of a plant that produces tubers. The stolon can be considered a naturally rooted layer and can be severed from the parent plant and transplanted. Many plants that form stolons are considered weeds or a nuisance because of their prolific nature. Bermuda grass, mint, and lamb's ear are examples of plants that form stolons, **Figure 15-12**.

JoeyPhoto/Shutterstock.com

Figure 15-12. Bermuda grass rapidly grows stolons and is considered a weed in some gardens.

Offsets

Offsets are a type of lateral shoot or branch that forms at the base of a main stem. Offsets only occur in some plants, primarily in monocots. Agave is a monocot that commonly forms small plantlets (called pups) at the base of the parent plant. Similarly, the banana, pineapple, and date palm will form offsets, or offshoots, that can be cut from the main stem with a sharp knife, **Figure 15-13**. In many cases, the offsets will have sufficient roots and will only need to be transplanted and given care like a rooted cutting. If an offset does not have roots or has minimal roots, it can be propagated like a stem cutting. It should be placed in rooting medium in high humidity. Intermittent misting and possibly bottom heat should be applied.

Suckers

A *sucker* is a shoot that grows from an adventitious bud on a root. In horticultural practice, a sucker is widely considered to be any shoot that grows near the base of a plant whether it is from a root or a stem. Best practices in pruning call for the removal of suckers, a process called suckering. In propagation, however, root suckers are rooted layers and can be dug and cut

warmer/Shutterstock.com

Figure 15-13. Bananas form rooted offsets that can be removed with a sharp knife and transplanted.

from the parent plant during dormancy. Pineapples form both offsets and suckers, **Figure 15-14**.

Crown Division

The base of a plant where shoots arise is called the crown. In herbaceous perennials, the old stem from the previous season's growth dies after blooming. New lateral shoots initiate, and adventitious roots form along the base of the new shoots. As this process unfolds each season, the crown may expand significantly, and the base of the plant may become crowded. The crown of the plant can be divided to prevent such overcrowding. Some multibranched woody shrubs have a similar habit. New shoots grow from the crown. Although older shoots do not die, younger, more vigorous shoots may crowd them out.

Crown division is the process of dividing the crown of herbaceous perennials or woody, multibranched shrubs. Perennials that flower in the spring and summer are generally divided in the fall. Perennials that bloom in late summer and autumn should be divided in the early spring. Plants are dug from the soil. Using a sharp knife, ax, or handsaw, the crown is cut into sections and transplanted to new locations, **Figure 15-15**. A similar method can be used for potted plants. In some cases, the old section of the crown may not be very vigorous and may be discarded rather than replanted. Crown division is slow and little used by commercial growers, but it is a common practice for home gardeners.

Pugun and Photo Studio/Shutterstock.com
Figure 15-14. Some cultivars of pineapple are commercially propagated by suckers.

Sarycheva Olesia/Shutterstock.com
Figure 15-15. Hostas can be divided by digging up the plant and using a sharp knife or shovel to split the crown into sections that can then be replanted.

Division and Separation of Geophytes

Underground storage structures in plants are called *geophytes*. They include many commonly known modified plant parts, including bulbs, corms, tubers, rhizomes, and pseudobulbs. Most of these geophytes belong to herbaceous plants and have developed as a means for storing food, nutrients, and water. They also provide a means of vegetative or asexual reproduction, providing clonal regeneration of the plant species. Some of the geophyte structures, such as bulbs and corms, only need to be separated. Other structures, such as tubers and rhizomes, must be divided (cut into sections). Each structure has a specific propagation technique.

Corner Question

What perennials can be divided using crown division?

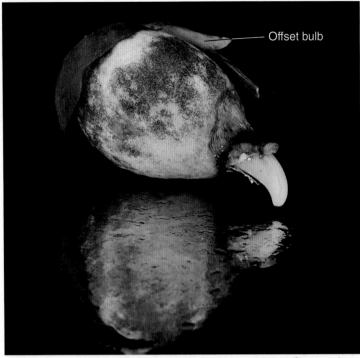

Figure 15-16. Tulip bulblets form from a main bulb and grow in size until they produce offset bulbs.

Bulbs

A *bulb* is a modified stem that contains a short, fleshy basal plate at the bottom from which roots grow. A bulb contains fleshy scales (primary storage tissue and modified leaves) and the shoot (consisting of developing flowers and leaf buds). Lateral buds in the fleshy scales develop into *bulblets* (a small bulb attached to the main bulb, which serves as a means of reproduction for the plant), **Figure 15-16.** The bulblets will increase in size until they become small bulbs that can be separated. *Separation* is a specific propagation term used with plants that produce bulbs or corms that can be easily pulled apart.

These small bulbs are planted and grown for several seasons until they flower and produce offset bulbs. Offset bulbs are daughter bulbs that are still attached to the parent bulb. If left undisturbed, they may remain attached for several years. The offset bulb is the commercial means of propagating new plant material. These bulbs are removed to replant into nursery rows where they grow into bulbs of sufficient size to produce flowers.

Most bulbs are monocots and have specialized modified leaves to provide a storage function as well as a means of asexual reproduction. There are two kinds of bulbs: tunicate and nontunicate.

Tunicate Bulbs

Think of the papery husk of a garlic bulb or the brown skin of a tulip bulb. These types of bulbs are *tunicate*, or laminate, bulbs that have outer bulb scales that are dry. Tunicate bulbs include onion, garlic, daffodils, tulips, and hyacinths. Plants in the Amaryllidaceae family are all tunicate bulbs, **Figure 15-17.** The papery covering or tunic protects the bulb from drying out and minimizes physical injury. The bulb scales are almost solid, concentric layers. They have adventitious root initials that will begin to grow when the environmental conditions and timing are right.

Figure 15-17. Amaryllis bulbs are tunicate, meaning they are wrapped in a papery sheath that provides protection from injury and drying out.

Nontunicate Bulbs

Nontunicate, or scaly, bulbs do not have a dry covering, but they have separate scales that are attached to the basal plate as shown in the lily bulb depicted in **Figure 15-18**. Without the covering, nontunicate bulbs can damage easily. They should be kept moist to prevent injury. Many bulbs have contractile roots that will physically pull the bulb to position it at a certain point in the ground.

Propagation

General propagation of bulbs begins with digging the bulb after the foliage has died back and dormancy is present. The bulb parts are separated, and the new bulbs are graded by size and stored at 65°F–68°F (18°C–20°C) until planting. Bulbs are planted according to size to help manage the harvest. For example, small tulip bulbs, less than 2″ (5 cm), will need at least three seasons to grow to a marketable size. A large bulb, 3.2″ (8 cm) may only need a single growing season to reach flowering size. The minimum size for tulip bulbs is 3.6″–4″ (9 cm–10 cm). This process is similar for most bulbs, with specific timing depending on plant species.

Lilies will increase naturally, but this takes considerable time. Commercial propagators of Easter lily (*Lilium longiflorum*) harvest underground stem bulblets in the late summer and plant them in the fall. The plants are allowed to grow for a season. After a single growing season, they are dug and replanted for one more year. Then they are harvested to be sold as commercial bulbs. Scaling is another method of propagation for lilies. With scaling, the individual bulb scales are removed and placed in a rooting medium in order for adventitious bulblets to form from each scale, **Figure 15-19**.

Hyacinth is one of the slowest bulbs to increase. A technique called basal cuttage was developed to hasten the development of bulblets. Using a scoring method, three deep, straight cuts are made across the basal plate of a mature bulb to reach the growing point. Bulblets form from the axils of the bulb scales. In another method, the entire basal plate is scooped out with a scalpel to stimulate adventitious bulblets to form on the exposed bulb scales. The bulbs are then placed in darkened conditions at 70°F (21°C) and later increased to 85°F–90°F (29.5°C–32°C) for two weeks. Humidity should be at 85% for two to three months. The mother bulb is then planted in the field in the fall. By spring, bulblets begin to rapidly develop. The bulblets are harvested and replanted to grow to marketable size.

Other methods of propagation that are less common but important for certain species include leaf cuttings and bulb cuttings. In bulb cuttings, the bulb is cut into 8 to 10 sections with basal plate attached. The sections contain three to four bulb scales. They are planted in rooting medium and warm temperatures to encourage adventitious bulblet formation.

Richard Griffin/Shutterstock.com

Figure 15-18. Lily bulbs are nontunicate and do not have a covering. They are more susceptible to injury or damage.

leungchopan/Shutterstock.com

Figure 15-19. Flowers, such as the narcissus, can be propagated by taking bulb scales and placing them in rooting medium to form individual bulblets on the base of the scale.

Corms

Gladiolus, freesia, and crocus are among the more commonly known plants with corms, **Figure 15-20**. A *corm* is a modified, enlarged stem that serves as a storage organ and method of reproduction. Corms are swollen stems that are surrounded by dry, scale-like leaves called the tunic. Corms have observable nodes and internodes and are composed of solid storage tissue. Corms naturally produce new corms as a means of reproduction for the plant and a source of propagation material for growers.

After flowering, the plant is dug. The mother corm, new corms, and cormels are separated. *Cormels* are small corms that are produced from a stolon on the base of the corm. The cormels are grown for one year until they produce a new corm. The new corms are planted for further growth for one to two seasons until they reach flowering size. Some large corms can be cut into sections with each section containing at least one bud. The section will develop a new corm.

Photograph by Marko Penning, Penning Freesia B.V.
Figure 15-20. Freesias develop storage organs known as corms.

Denis and Yulia Pogostins/Shutterstock.com
Figure 15-21. The edible portion of the potato is called a tuber, which develops from stolons.

rodimov/Shutterstock.com
Figure 15-22. Potatoes are propagated by taking a tuber and cutting it into sections, with each piece containing an eye. These are then planted in the field and develop into shoots that form new stolons and tubers.

Tubers

Tubers are modified, enlarged stems that have become specialized to be underground storage organs. Potato (*Solanum tuberosum*), caladium (*Caladium hortulanum*), and Jerusalem artichoke (*Helianthus tuberosus*) are all tubers. If you imagine a typical potato that has stayed too long in the pantry, the nodes (also called eyes) begin to develop shoots. The role of the tuber is to serve as a source of stored food to drive the development of shoots in the spring, **Figure 15-21**. After emergence and growth of shoots, the old tuber dies. As the primary shoots form, adventitious roots develop as well as lateral shoots or stolons. These shoots run underground and will eventually *tuberize*, or develop new tubers.

In potatoes, propagation occurs with a tuber being cut into sections that each contain one or more eyes, **Figure 15-22**. Generally referred to as seed potatoes, these pieces require warm temperatures and high humidity to allow the cut surfaces to heal and *suberize*. Suberize means to form a waxy substance (called suberin) that protects plants from desiccation and pathogens. Potatoes are then planted in the field.

Caladium tubers are also cut into segments with two buds per piece. They are planted in field nurseries until tubers are ready to be harvested. Harvested tubers are dried and stored at 60°F (16°C) until ready for market or planting.

Tubercles are small aerial tubers produced in the axils of leaves of hardy begonia (*Begonia grandis* subsp. *evansiana*) and cinnamon vine (*Dioscorea batatas*). The tubercles can be collected in the fall and planted in the spring.

Tuberous stems, rather than tubers, are produced by the enlargement of the hypocotyl of a seedling and may include the first nodes of the epicotyl and the upper section of the primary root. Tuberous begonias and cyclamens are tuberous stems, **Figure 15-23**. In cyclamens, an upper portion of the tuberous stem can be cut and planted for adventitious root formation. Tuberous begonias are propagated most successfully as stem or leaf cuttings.

Tuberous Roots

Tuberous roots are enlarged secondary roots and include sweet potato, dahlia, and cassava, **Figure 15-24**. Most tuberous roots are biennial, growing storage roots in the first season. After a dormant state, they begin to grow again in the spring, with shoots emerging from the root and using stored nutrients and sugars. The old root disintegrates, and new tuberous roots form to perpetuate the cycle.

To propagate dahlia, the crown is divided with each section containing a bud. In commercial sweet potato production, the roots are bedded in the field and covered with plastic to provide moisture and warm temperatures of 80°F (27°C) to encourage sprouting. The emerging sprouts are called slips. Slips can be hilled with more soil to encourage adventitious roots to form. Many slip producers simply cut the slips and plant them in the field where they will form roots.

Photograph by Stavroula Ventouri-ex www.srgc.net-

Figure 15-23. Cyclamens have tuberous stems that are commonly used in propagating new plants.

photowind/Shutterstock.com

Figure 15-24. Dahlias form tuberous roots that are the primary plant part for commercial propagation.

Rhizome

Iris, bamboo, sugar cane, banana, many grasses, ginger, and lily of the valley are important rhizomes. A *rhizome* is a modified stem structure that grows horizontally below or near the soil's surface. Rhizomes produce roots on the bottom and shoots on the top, **Figure 15-25**. The stem has nodes and internodes with a leaf-like sheath that surrounds the stem at each node. This sheath can become a leaf.

A *Izf/Shutterstock.com* **B** *Zigzag Mountain Art/Shutterstock.com*

Figure 15-25. A—Sugar cane is also a rhizome that forms a strong rooting system. B—Ginger plants form rhizomes with shoots emerging on the top of the modified stem structure and roots underneath.

Rhizomes are typically propagated by division. A plant is dug from the garden, the soil removed, and the rhizome cut into sections with each containing one or more buds. Adventitious roots and shoots will develop from the buds. Iris is a good example of this kind of plant. Plants are divided after flowering, usually in late summer. Some rhizomatous plants, such as bamboo, form aboveground shoots called culms, **Figure 15-26**. Culms may be cut and laid horizontally in the soil and covered. New shoots will emerge from the nodes.

Pseudobulb

Many orchid species will develop a swollen, fleshy stem structure called a *pseudobulb* (meaning false bulb). Serving as a specialized storage organ, the pseudobulb accumulates sugars, nutrients, and water, which enable survival during dormancy, **Figure 15-27**. Pseudobulb structures vary by species. In *Dendrobium*, the pseudobulb consists of many nodes that develop offshoots that form adventitious roots and can be cut and potted. *Cattleya*, *Laelia*, *Miltonia*, and *Odontoglossum* species have rhizomes that have pseudobulbs. Sections of rhizome are cut to include four to five pseudobulbs and planted in growing medium. New growth comes from the base of the pseudobulbs and nodes.

Boonsom/Shutterstock.com

Figure 15-26. Bamboo shoots emerge as culms. They can be cut and propagated to form new shoots and roots.

i7do/Shutterstock.com

Figure 15-27. Orchids form structures called pseudobulbs that can be separated into new plants.

Careers in Layering and Division

Layering and division techniques may be intensive and expensive for home gardeners. Professional propagators are skilled in managing the growth, health, and propagation of plants and trees. These professionals are key to ensuring healthy plants are sold commercially.

Fruit Nursery Propagator

All commercial apple rootstocks are grown through the process of mound layering, and many other fruits are grown through various layering processes. Responsibilities for a propagator for a fruit tree nursery may include managing the stock plants for quality and vigor and going through the appropriate propagation processes for each fruit type. The propagator is responsible for not only the propagating but care and maintenance of rooted propagules. Propagators need to keep detailed records and have a minimum of a high school diploma with experience in propagation.

Orchard Manager

Orchard managers play a key role in making decisions and managing the health of the fruit trees and orchard ecosystem. Managers are responsible for organization and supervision of pruning, planting, fertilizing, spraying, thinning, mowing, harvesting, and storage of fruit. Managers recruit, train, supervise, and evaluate the orchard employees and communicate with orchard owners. The position requires a bachelor's degree in agriculture or related field and/or at least five years of related horticultural and construction work experience.

Corner Question

What are some uses for bamboo?

CHAPTER 15 Review and Assessment

Chapter Summary

- Layering is vegetative or asexual propagation technique that allows stems that are still attached to the parent plant to form adventitious roots.
- With simple layering, a flexible, low-growing stem is bent to the ground and covered with soil. The bent portion of the stem will form adventitious roots.
- Compound and serpentine layering produce multiple layers from a single stem by covering multiple buds and stimulating multiple shoots to grow and initiate roots.
- Air layering is a technique in which the stem is girdled and the wood is covered with rooting medium for adventitious roots to grow into.
- Mound (stool) layering is a method in which soil is piled on the crown of the plant. The new shoots form adventitious roots that can be severed and transplanted.
- In trench layering, shoots are layered horizontally and pegged down in the base of a trench. Loose rooting medium is filled around the new shoots as they develop.
- Some plants have modifications that enable them to create rooted layers through their natural biology. This is called natural layering and includes tip layers, runners, stolons, offsets, and suckers.
- Crown division is the process of dividing the crown of an herbaceous perennial or multibranched woody shrub into multiple sections for replanting.
- Underground storage structures in plants are called geophytes. They include many commonly known modified plant parts, including bulbs, corms, tubers, rhizomes, and pseudobulbs. Plants with these parts are propagated by separation or division.
- There are two types of bulbs, tunicate and nontunicate. Both form little bulblets that in turn can produce bulbs that are grown to larger sizes for commercial use.

Words to Know

Match the key terms from the chapter to the correct definition.

A. air layering
B. bulb
C. bulblet
D. corm
E. cormel
F. geophyte
G. girdling
H. hilling
I. layering
J. mound (stool) layering
K. nontunicate
L. offset
M. pseudobulb
N. rhizome
O. runner
P. separation
Q. serpentine layering
R. stolon
S. suberize
T. sucker
U. tuber
V. tubercle
W. tuberize
X. tunicate

1. A modified, enlarged stem that serves as a storage organ and method of reproduction.
2. A propagation method that involves wounding the stem and wrapping it in rooting medium to encourage adventitious root formation.
3. A young, small corm produced from a stolon on the base of a mature corm.
4. The piling of soil onto plants to promote desired growth, as in the case of potatoes to produce more tubers.
5. A type of lateral shoot or branch that forms at the base of a main stem, which occurs only in some plants, primarily in monocots.
6. A storage organ that develops in some orchids and can be cut and potted to grow new plants.
7. To form a waxy substance (called suberin) that protects plants from desiccation and pathogens.
8. A modified stem structure that grows horizontally below or near the soil's surface, producing roots on the bottom and shoots on the top.
9. A propagation method in which soil is piled on the crown of the plant, and the new shoots form adventitious roots that can be severed and transplanted.
10. A modified stem that contains a short, fleshy basal plate at the bottom from which roots grow and which holds fleshy scales and the shoot.
11. A small bulb attached to the main bulb, which serves as a means of reproduction for the plant.
12. A swollen, underground stem that is a storage organ for the plant.
13. A small, aerial tuber produced in the axil of a leaf.
14. A variation of compound layering in which shoots are laid horizontally to the ground and some buds are exposed while others are covered with soil in an alternating pattern.
15. A specialized stem that grows horizontally aboveground and forms a new plant.

16. A type of bulb that has fleshy scales and does not have a dry covering, such as a lily bulb.
17. A general term for vegetative propagation techniques in which plants initiate roots while still attached to a parent plant.
18. An underground storage structure in a plant, such as a bulb, corm, or tuber.
19. A type of bulb that has a dry, papery covering or tunic.
20. A shoot that grows from an adventitious bud on a root or any shoot that grows near the base of a plant.
21. Horizontal stems that grow above ground and produce plants or tubers.
22. Wounding of a stem by cutting or bending the stem for the purposes of propagation.
23. The process of developing new tubers, which are swollen, underground stems that are storage organs for the plant.
24. A propagation term used to define plants that produce bulbs or corms that can be easily pulled apart.

Know and Understand

Answer the following questions using the information provided in this chapter.

1. Why is layering a successful asexual propagation technique for certain species?
2. Describe the conditions growers use to propagate layers from stock plants.
3. Identify the different types of artificial layering.
4. List five examples of shrubs that can be propagated using simple layering.
5. What time of year is simple layering typically done?
6. How does serpentine layering differ from regular compound layering?
7. Describe the specific steps to air layering of a dicot.
8. Which method of layering is the most commercially important?
9. Explain the role of hilling in mound layering.
10. Why would a grower use the trench layering method and how is it done?
11. What are five examples of natural layering?
12. What are three plants that exhibit natural tip layering?
13. What types of stems are runners? How are they a source of new plants?
14. What types of stems are stolons? How are they a source of new plants?
15. What type of plants (monocots or dicots) generally form offsets? What are four examples of plants that form offsets?
16. When are perennial crowns divided? Describe the process of crown division.
17. What are two types of geophytes that can be separated and used to grow new plants?
18. What is the role of the tunic on a tunicate bulb?
19. What is the function of contractile roots for bulbs?
20. What are some plants that have rhizomes?

Thinking Critically

1. You are trying to propagate a woody specimen by simple layering, and the layers are not rooting. What do you think is happening? Create a list of five possible reasons why the layers are not rooting.
2. A gardening friend approaches you with a propagation question and would like an answer. She has had an agave and would like more, but the plant is not forming any offsets. What questions would you ask your friend to determine the answer to the problem?

STEM and Academic Activities

1. **Science.** Research recent hyacinth cultivars. Choose five cultivars, and write a two-page report comparing the differences and similarities of the new varieties.
2. **Science.** Find out more about adventitious root formation. Why do some plants more readily initiate these types of roots? Write a report summarizing your findings.
3. **Math.** You are a wholesale greenhouse producer of lilies. You have a customer who would like to order 100 lily bulbs. What do you need to know about lily propagation to determine a cost that you can share with your customer?
4. **Social Science.** Role-play a situation in which you are working at your school's plant sale. How would you relay information to a homeowner about how you propagated many of the plants through layering? How would you convey the value of these plants?
5. **Language Arts.** Using the knowledge you acquired from this chapter, create an informational poster about plants that can be propagated through layering. Include the different types of layering techniques and give examples of each.

Communicating about Horticulture

1. **Writing and Speaking.** As the gardener for a public garden, create an informational pamphlet on the ways to propagate common garden plants. Include images in your pamphlet. Present the information you have written in the form of a presentation to your class and hand out the pamphlets as resources.
2. **Reading and Speaking.** In small groups, discuss the photographs and illustrations in chapters. Describe, in your own words, what is being shown in each image. Discuss the effectiveness of the illustrations compared to the text description.
3. **Listening and Speaking.** Interview a commercial propagator. Ask the person to describe a typical day at work. Here are some questions you might ask: What is the work environment like? What are the job duties? What kinds of management problems do you have to deal with? What other types of professionals do you work with? Report your findings to the class, giving reasons why you would or would not want to pursue a career similar to that of the person you interviewed.

SAE Opportunities

1. **Experimental.** Obtain some potatoes and cut them into sections for seed potatoes. Try planting them with and without suberization. What happens? Keep a record of your findings.

2. **Exploratory.** Create a propagation calendar for each of the different types of geophytes discussed in this chapter. When should you take cuttings? How should you prepare the cuttings? Include timing, depth, length, and other important details for each plant. Why do you think growers keep records of their propagation activities?

3. **Exploratory.** Create a list of fruit plants that grow on the campus of your school that could be sold in your school's plant sale. Determine which method of propagation is best suited for each fruit. Determine a time frame, how many layers from each plant you will need, and what the costs associated with this project will be.

4. **Exploratory.** Job shadow someone who professionally practices layering techniques. What kind of organization does she or he work for? How often does this person use layering in her or his work?

5. **Entrepreneurship.** Propagate bulbs, tubers, and corms in your school's garden and greenhouse and sell them to customers.

AlexeiLogvinovich/Shutterstock.com

CHAPTER 16
Grafting and Budding

Chapter Outcomes

After studying this chapter, you will be able to:
- Discuss the benefits of grafting for growers.
- Explain the requirements for successful grafting.
- Describe the techniques for different grafting methods.
- Explain how to propagate a plant asexually via budding.
- List careers related to grafting and budding.

Words to Know

approach grafting	cleft grafting	scion
bark grafting	grafting	side-veneer grafting
bark slipping	hole insertion grafting (HIG)	splice grafting
bench grafting	inarch grafting	topworking
bridge grafting	interspecific	wedge grafting
budding	interstock	whip-and-tongue grafting
budwood	saddle grafting	

Before You Read

Read the chapter title and tell a classmate what you have experienced or already know about the topic. Write a paragraph describing what you would like to learn about the topic. After reading the chapter, share two things you have learned with the classmate.

While studying this chapter, look for the activity icon to:

- **Practice** vocabulary terms with Words to Know activities.
- **Expand** learning with identification activities.
- **Reinforce** what you learn by completing Know and Understand questions.

www.g-wlearning.com/agriculture

ood crops have been developed from selecting high-performing plants and from breeding efforts to develop hybrid cultivars with traits desirable for growers. Woody species, however, are highly variable and rarely grow true to type from seed. Plant improvement by selection and breeding is slow and challenging. By propagating certain fruit crops, such as fig, grape, and pomegranate, using layering and through rooted offshoots, selection of genetically suitable plant material became more widely available. Many temperate fruit trees, however, do not root well. With the discovery of grafting, clonal propagation of these plants has made high-quality plant stock available throughout the world. Grafting originated as early as 2000 BCE in China and Mesopotamia.

There are several types of grafting, including splice, cleft, bark, and bridge grafting. These techniques join scions and rootstocks to propagate or heal plants. Budding is similar to other types of grafting, but it is unique in that it uses buds rather than scions to grow on other plants.

Benefits of Grafting

Grafting is a horticultural technique that joins together two different plants or plant parts so that they grow into one plant, **Figure 16-1**. A *scion* is a young shoot or twig of a plant, often used for the upper portion of a graft. Grafting joins the scion of one plant with the rootstock of another plant. *Budding* is a type of grafting in which a single bud is taken from one plant and grown on another plant.

Grafting and budding are asexual, or vegetative, propagation techniques. The plant that grows from the scion or bud is genetically identical to its plant of origin. Grafting is generally done on species that have a difficult time rooting. Grafting may be the only way to produce clonal plants for some species. In fruit trees, this is particularly true. Grafting provides a number of benefits by allowing grower to:

Swapan Photography/Shutterstock.com

Figure 16-1. This moon cactus is actually a graft between two different cactus species.

- Change varieties or cultivars. New cultivars with greater resistance to insects or diseases or ones that have enhanced stress tolerance or higher yields may be developed. However, replanting an orchard to use these new plants may be too costly. Grafting new scion material onto older, established orchards is called *topworking* and may help keep an orchard productive and profitable, **Figure 16-2**.

Corner Question

What kind of plants cannot be grafted?

- Optimize cross-pollination. Many fruit trees require cross-pollination by another fruit tree to guarantee fruit growth. Trees can be grafted to have multiple varieties and ensure that cross-pollination occurs.
- Use quality rootstock. Rootstocks are generally selected for superior growth habits, such as size, vigor, disease resistance, and insect and nematode resistance. They may also be selected for stress tolerances to drought, salinity, or temperature and for soil adaptation. For example, scions can be grafted onto dwarfing rootstock to produce a smaller or shorter tree, **Figure 16-3**.
- Increase clones. Many conifer species are difficult to propagate through vegetative cuttings. However, many species can be successfully grafted. For example, Colorado blue spruces are often grafted onto Norway or Sitka spruces.
- Produce certain plant forms. Numerous weeping, cascading, or dwarf horticultural specimens are grafted or budded onto a rootstock. Weeping flowering cherry is usually grafted as scion material onto a Mazzard cherry rootstock.
- Repair damaged plants. Damaged or injured specimens can sometimes be repaired through grafting. If the base of trunk is wounded, plant seedlings of that species around the trunk, and eventually you can graft the seedlings onto the tree above the injury. Injured branches may also be repaired through grafting.

Steven Winter

Figure 16-2. The art of grafting permits the creative combining of multiple cultivars in one tree, like this 'Bing' and 'Rainier' cherry which is perfect for a small urban backyard.

Ivonne Wierink/Shutterstock.com

Figure 16-3. Apple orchards often contain trees with scions grafted onto dwarfing rootstock to have a tree that is shorter and thus makes picking fruit easier.

History Connection
The Tree Circus

Axel Erlandson was a farmer in California. As a hobby, he shaped trees into fantastic sculptures through pruning, bending, and grafting. Calling his work *Circus Trees*, he opened a horticultural attraction in the late 1940s, claiming to have "the world's strangest trees." Erlandson's arbor sculpture is an artful application of the horticulture technique of grafting. The "Basket Tree" was formed by six sycamore trees planted in a circle. Then Erlandson grafted them together to form the diamond patterns. Over his lifetime, he created more than 70 living sculptures. About 25 of his trees remain today. They were moved and can be seen at Gilroy Gardens.

Photography by Pam Winegar

Successful Grafting

Grafting requires several factors for success:

- Rootstock and scion compatibility. Most grafting occurs within species, such as an apple to an apple. Even within closely related plant materials, incompatibility may occur and varies by cultivar. In some cases, different species may be compatible. For example, a tart cherry (*Prunus cerasus*) may be compatible to a sweet cherry (*Prunus avium*). Distantly related plants, such as a maple and a cherry, cannot be grafted. **Interstock** material may be used to bridge incompatible scion and rootstock material within species. Interstock material is grafted onto the rootstock and the scion material is grafted onto the interstock.

- Vascular cambia alignment. The vascular cambia of the scion and rootstock must make direct contact. Smooth cut surfaces are held together through various binding methods, such as wrapping tightly with film or tape, nailing, or wedging. Callus cells form and provide a bridge between the rootstock and scion material. A new vascular cambium forms across the callus bridge. New xylem and phloem tissue arise and form a functional conduit through the newly grafted plant. The scion material is provided with water and nutrients from the rootstock as the graft union forms.

- Proper timing. The species and technique used dictate the time of year to graft. For example, in bud grafting, the buds are usually dormant while the rootstock material can still produce callus tissue to heal the graft union.

- Graft union protection To prevent the graft union from drying out or water penetrating the graft, the cut surfaces should be covered with grafting wax, paint, tape, or film, **Figure 16-4**.

Stocked House Studio/Shutterstock.com

Figure 16-4. Covering all the cut surfaces of the graft to prevent drying out and water seeping in is important for graft healing.

- Post-graft care. Monitoring the new graft is important to ensure healing. For some potted plants, grafted material may need an environment with high humidity, low light, and proper temperature. In some cases, such as with grafted tomatoes, the scion material can root itself into the soil, losing the effectiveness of the disease resistance from the rootstock. Shoots that may emerge from the rootstock may also diminish scion growth. Sometimes staking may be needed for the rootstock to support the scion material, such as with grafted weeping ornamentals.

Timing Grafting

Graft during the inactive season, which is typically the winter and early spring. Plants growing in the field or orchard are grafted in place. Plants growing in containers may be moved indoors for the grafting process. Afterwards, they may be protected or placed in nurseries. This latter process is called *bench grafting*. Some trees can be grafted during the winter and stored until spring planting.

Selecting Scion Wood

Harvest high-quality scion wood from the previous season's shoots. Sharp tools, such as pruning shears or knives, should be used to cut scion material. Following cutting, immediately store scion materials in moistened burlap or plastic bags. Use flaming or sterilizing solution to clean grafting tools regularly throughout the grafting process. Sterilizing solutions may consist of isopropyl (rubbing) alcohol or a mix of 10% bleach and 90% water. Scion wood should be selected to be free from disease and damage. If possible, the scion should be used the same day it is cut. Rootstock plants should also be good quality, healthy, and true to type.

> "We marry A gentler scion to the wildest stock, And make conceive a bark of baser kind By bud of nobler race. This is an art Which does mend nature—change it rather; but The art itself is nature."
> —William Shakespeare

STEM Connection

Ketchup 'n' Fries™ Plant

Tomato or potato? How about both? For years, hobbyists have been able to graft a tomato onto a potato rootstock as an interesting novelty to show to family and friends. A simple splice graft that puts a tomato scion on top of a potato rootstock can yield cherry tomatoes for harvest. As the tomato winds down, the plant's energy is put toward developing the potato tubers. The grafted plants can be put into a healing chamber. They are planted in the garden after the graft has healed and the plant has been acclimatized.

A British mail order nursery has commercialized this novelty and dubbed it the TomTato, also known as Ketchup 'n' Fries. A US mail-order nursery has licensed the Ketchup 'n' Fries plant and made it available to any backyard gardener.

Types of Grafts

There are a number of different techniques for grafting, with each method having different uses. The most common grafting approaches are discussed below.

Bork/Shutterstock.com

Figure 16-5. The whip-and-tongue graft interlocks the scion and rootstock material together.

Whip-and-Tongue Grafting

Whip-and-tongue grafting is a method of joining plants in which the rootstock is cut with two notches and the scion material is cut to match and create a tight fit. This technique is popular for nursery crops and woody ornamentals. It is commonly used with relatively small material, 1/4″–1/2″ (6 mm–13 mm) in diameter, with the rootstock and scion being of equal size. The scion material should have at least two or three buds, **Figure 16-5**. Use the process to complete a whip-and-tongue graft:

1. Prepare the rootstock and scion. Cut the rootstock diagonally about 1″–1 1/2″ (2.5 cm–6 cm) long. Make the cut in one stroke of the knife to leave a smooth, even, flat surface. Make an identical cut at the scion's base. Next, make a reverse cut by placing the blade of the knife across the cut end of the rootstock halfway between the bark and pith. Cut at an angle through the bark and pith, running parallel to the diagonal cut and stopping at the base of the initial cut, **Figure 16-6A**.
2. Insert the scion. Match the cuts from scion and rootstock and fit them together so the cambial layers align, **Figure 16-6B**.
3. Secure the graft. Wrap the graft union with a grafting strip or tape, and seal it with a plastic paraffin film or grafting wax, **Figure 16-6C**.

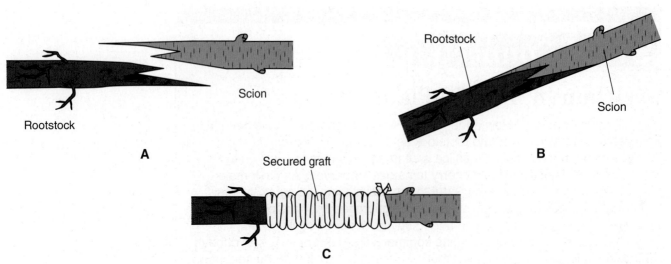

Goodheart-Willcox Publisher

Figure 16-6. Preparing a whip-and-tongue graft. A—Cut the rootstock and scion. B—Match the cuts from the scion and rootstock and fit the two plants together. C—Wrap the graft union.

Splice Grafting

Splice grafting is a joining method in which a simple diagonal cut of the same length and angle is made on the rootstock and scion. The scion and rootstock are placed on top of each other with the cambial layers aligned. This technique is also called *whip grafting*. This grafting method is most commonly used in herbaceous material that forms a callus easily or in woody material that is not flexible enough to allow a tight fit with a whip-and-tongue graft. Grafting of vegetable crops for disease resistance (such as tomatoes, potatoes, or squash) is a common use for this graft, **Figure 16-7**. Use the following process to complete a splice graft:

1. Prepare the rootstock and scion. The rootstock and scion are both cut on an angle, **Figure 16-8A**. The diameter of both materials should be as similar as possible. In squash, often just a cotyledon is used.
2. Insert the scion. The scion is placed onto the rootstock, **Figure 16-8B**. The vascular cambia should align.
3. Secure the graft. The scion and rootstock are held in place while the graft junction is wrapped or tied, **Figure 16-8C**. In herbaceous material, only a grafting clip may be needed to secure the material together. In woody material, the junction is sealed with grafting wax or grafting paint.

In herbaceous splice grafts, the plants are put into a healing chamber for a period of a few weeks to encourage graft healing and minimize drying out. The chamber has high humidity and little light for the first week or so. Gradually, light is added. Eventually, the grafted plants are hardened off.

vallefrias/Shutterstock.com

Figure 16-7. Tomato grafting is important for managing a number of soil-borne diseases, such as verticillium and fusarium wilt.

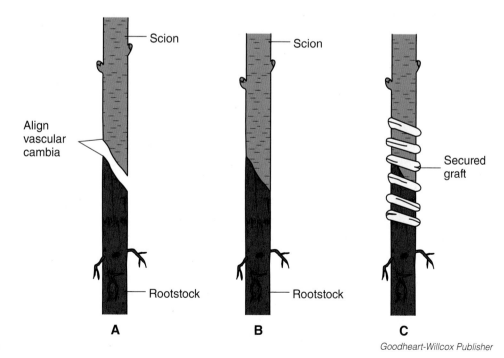

Goodheart-Willcox Publisher

Figure 16-8. Preparing a splice graft. A—Cut the scion and rootstock on an angle. B—Align vascular cambia.

Cleft Grafting

Cleft grafting is a plant-joining method that is used to topwork trees in the trunk or scaffold branches. This technique is also called splice grafting. It is a simple and widely used method to change varieties of fruit trees in an existing orchard through topworking. This method is typically used for apples, cherries, pears, and peaches. Cleft grafting can be used to propagate species that are difficult to root, such as some camellias, or to repair damaged plants. Graft during the inactive season, which is typically the winter and early spring. The rootstock is much larger than the scion piece.

The rootstock used for cleft grafting should be 1"–4" (2.5 cm–10 cm) in diameter, and the wood should be straight-grained. The scion should be 1/4"–1/2" (6 mm–13 mm) in diameter, straight, and 6"–8" (15 cm–20 cm) long. It should have at least three buds. Use the following process to complete a cleft graft:

1. Prepare the rootstock. Prepare the rootstock stem to be grafted by sawing through the branch to make a clean, smooth cut. Use a clefting tool and a mallet to make a split, or cleft, through the center of the stock 2"–3" (5 cm–7.5 cm), **Figure 16-9A**. Drive the tool into the center of the cleft to hold the rootstock open to insert the scion.

2. Prepare the scion. Use a sharp, clean grafting knife to make a smooth tapered cut about 1"–2" (2.5 cm–5 cm) on one side of the base of the scion near the base of the lowest bud. On the opposite side of the base, make a second smooth cut of the same length so it is slightly thicker than the other side. The wedge that is formed should be somewhat blunt, **Figure 16-9B**.

3. Insert the scion. Two scions are used in a cleft graft, and one is inserted at each end of the cleft, **Figure 16-9C**. With a grafting chisel or a small wedge, open the crack and insert the scion with the thicker side toward the outside. Align the cambia between the rootstock and the scion.

4. Secure the graft. After proper placing of the scion, use the clefting tool to close the rootstock to hold the scions in place. Use wax or paint to seal cut surfaces and to prevent desiccation, **Figure 16-9D**.

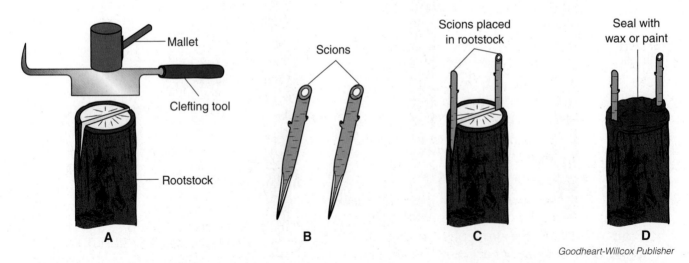

Goodheart-Willcox Publisher

Figure 16-9. Preparing a cleft graft. A—Cut the rootstock using a clefting tool and mallet. B—Prepare the scions by making a tapered cut on one end. C—Insert a scion at each end of the rootstock's cut. D—Seal cut surfaces with wax or paint.

If both scions in the cleft grow, one will usually be more vigorous and dominant than the other. After the first growing season, leave the stronger scion and remove the weaker scion.

Wedge Grafting

Wedge grafting is a plant-joining method in which the scion is cut into a V-shaped wedge and the rootstock is cut to form a V shape to hold the scion. Wedge grates are performed in late winter (in mild climates) or early spring before the bark begins to slip. *Bark slipping* refers to a condition in which the vascular cambium is actively growing and the bark can be easily detached or peeled in one even layer from the wood beneath it without wounding the plant material. The diameter of the rootstock should be 1"–4" (2.5 cm–10 cm). The scions should be 1/4"–1/2" (6 mm–13 mm) in diameter and 6"–8" (15 cm–20 cm) long. Use the following process for completing a wedge graft:

1. Prepare the rootstock. A sharp, sturdy, short-bladed knife is used to create a V-shaped wedge in the side of the rootstock, about 2" (5 cm) long. Two cuts are made, meeting at the bottom and as far apart at the top as the width of the scion. The cuts reach 3/4" (2 cm) long into the side of the rootstock. After these cuts are made, a screwdriver is pounded downward behind the wedge chip from the top of the rootstock to knock out a chip, leaving a V-shaped opening for insertion of the scion.
2. Prepare the scion. The base of the scion is shaped to a wedge, mirroring the same size and shape as the opening on the rootstock, **Figure 16-10A**.
3. Insert the scion. After the two vascular cambial layers are matched, the scion is tapped downward, firmly into place, and slanting outward slightly at the top so that the vascular cambial layers cross. If the cut on the rootstock is long enough and slightly narrowing, the scion should be so tightly held that it would be difficult to displace. In a rootstock that is 2" (5 cm), two scions should be inserted across from each other. In a 4" (10 cm) rootstock, three scions should be used that are evenly spaced apart, **Figure 16-10B**.
4. Secure the graft. After all scions are firmly tapped into place, all cut surfaces, including the tips of the scion, should be waxed thoroughly, **Figure 16-10C**. As the strongest scion begins to grow, the other scions are removed after a season or two.

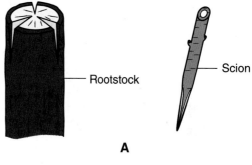

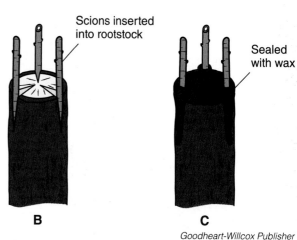

Goodheart-Willcox Publisher

Figure 16-10. Preparing a wedge graft. A—Use a knife to make V-shaped cuts in the top of the rootstock, and taper the ends of the scions to match. B—Insert scions into the rootstock cuts. C—Use wax to seal all cut surfaces.

Saddle Grafting

Saddle grafting is a plant-joining method in which the scion is cut in the shape of a saddle and sits on top of the rootstock. It can be easily done by hand or by using a grafting machine. It has often been used for grafting rhododendrons and for bench grafting of grapes. The diameters of the rootstock and scion should be the same. This type of graft is typically performed when the stock is not active in the winter. Use the following process to complete a saddle graft:

1. Prepare the rootstock. Use rootstock that is no larger than 1″ (2.5 cm) in diameter. Draw the grafting knife upward on both sides, making two cuts on each side to form an inverted V. Make the cuts between 1/2″–1″ (1.3 cm–2.5 cm) long.
2. Prepare the scion. Pull the grafting knife into the scion material from each side to make a V-shaped cut, **Figure 16-11A**. The cuts should be the same length on both the rootstock and scion to best match the cambial tissue when the materials are joined together.
3. Insert the scion. Insert the prepared scion into the saddle of the rootstock. Cambial alignment occurs easily if the rootstock and scion are the same diameter, **Figure 16-11B**. Adjustments may be needed otherwise.
4. Secure the graft. After the cambial layers are aligned, secure the graft with grafting twine, tape, or strip, **Figure 16-11C**. Then use a grafting compound to seal all cut surfaces.

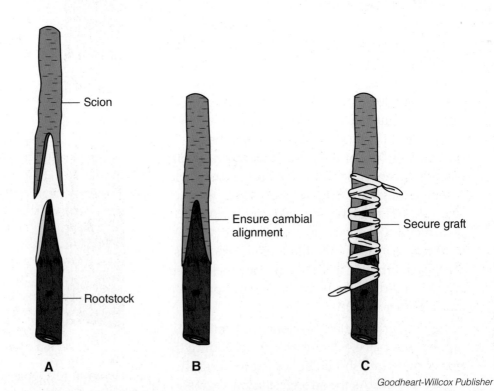

Goodheart-Willcox Publisher

Figure 16-11. Preparing a saddle graft. A—Cut the rootstock end to form an inverted V, and make a V-shaped cut on the scion's end. B—Insert the scion into the rootstock and ensure cambial layers are aligned. C—Tie the graft to secure it.

Hole Insertion Grafting (HIG)

Growers in China and Japan have pioneered processes in herbaceous vegetable transplant grafting. For example, watermelon has been grafted to squash rootstock to promote plant vigor and increase yield in the presence of disease. Tolerance to physical stresses and resistance to soil borne plant pathogens are also improved. *Hole insertion grafting (HIG)* is a plant-joining method used on herbaceous material in which the rootstock growing point is removed and a hole is inserted into the top of the plant. The scion material is cut in a wedge shape and inserted into the hole. HIG is the most popular grafting technique used on cucurbits (squash family members). This is because there are many suitable *interspecific* (occurring between different species) squash relatives that can be used as rootstocks and few materials are needed. Use the following process to complete a hole-insertion graft:

1. Prepare the rootstock. HIG is done on young transplants. The rootstock plant is ready for grafting once cotyledons and the first leaf start to develop. The growing point that includes the true leaves, the apical meristem, and axillary buds should be removed with a sharp probe. The upper portion of the rootstock hypocotyl is then pierced with the probe or a needle (1.4 mm in size) to make an opening for the scion.
2. Prepare and insert the scion. Cut the scion below the cotyledons with a sharp razor at a 45° angle on both sides to make a wedge. Insert the scion into the hole made in the rootstock, **Figure 16-12A**.
3. Secure the graft. Match the cut surfaces together, and transfer them to a humid room. See **Figure 16-12B**.

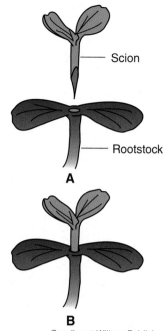

Goodheart-Willcox Publisher

Figure 16-12. Preparing a hole-insertion graft. A—Pierce the rootstock to make an opening for the scion, which is angled on both sides of the cut. B—Insert the scion into the hole of the rootstock.

STEM Connection — Healing Herbaceous Grafts

Courtesy of Dr. Suzanne O'Connell

The environment of a healing chamber generally has high humidity (80%–95%) and moderate temperatures between 75°F and 80°F (24°C and 27°C). The ability to adjust light levels is important. The important functions of a healing chamber are minimizing the transpirational water loss, allowing callus formation, and allowing vascular cambial layers to connect. The grafted plants should be slowly acclimatized.

Day 1: Perform the graft and place grafted plants in a pre-misted healing chamber. Close the plastic of the healing chamber and cover the chamber with black plastic.

Day 2: Keep the chamber closed and covered with black plastic.

Day 3: Open the chamber and mist the top and the sides. Close the chamber and partially move the black plastic away from the front of the chamber.

Day 4: Keep the chamber closed and fold the black plastic up on all sides of the chamber, but leave it on the top.

Day 5: Open the chamber and mist the inside. Leave the chamber open for thirty minutes. Mist the plants if any wilting occurs, and close the chamber. Remove the black plastic completely.

Day 6: Open the chamber, mist the inside, and leave the chamber open for an hour, misting the plants if any wilting occurs. Close the chamber.

Day 7: Open the chamber, and mist the inside. Leave it open for three hours, again misting the plants if any wilting occurs. Close the chamber.

Day 8: Open the chamber for six hours, misting the inside of the chamber and misting the plants if any wilting occurs. Close the chamber.

Day 9: Remove the plants from the chamber.

Bark Grafting

Bark grafting is a joining method in which a scion of one plant species is placed into the layer where the bark and wood separate in a rootstock plant. Similar to cleft grafting, bark grafting can be used to topwork flowering and fruiting trees, change varieties, or repair damaged wood. Bark grafting can be done on large diameter rootstock of 4"–12" (10 cm–30 cm). Bark grafting is typically done when the bark slips easily from the wood in early spring. Keep in mind that bark grafts are often weak, and you may have to stake them for the first few years. Use the following process to complete a bark graft:

1. Prepare the rootstock. Make a clean cut vertically through the bark about 2" (5 cm) long across the surface of the rootstock. Make slits 1"–2" (2.5 cm–5 cm) apart, **Figure 16-13A**.
2. Prepare the scion. Prepare several scions to insert around the cut in the rootstock. Scions should contain two or three buds and be about 4"–5" (10 cm–13 cm) long. On only one side, slice the base of each scion to a 1 1/2"–2" (4 cm–5 cm) tapered wedge, **Figure 16-13B**.
3. Insert the scion. Insert the scion under the loosened bark. Align the wedge-shaped tapered surface of the scion against the exposed wood under the flap of bark. Align the scion with the exposed wood under the bark to position it into place. Nail the scion in place using wire brads, **Figure 16-13C**. In some cases, the fit may be tight enough that nails are not needed. Insert a scion every 3"–4" (7.5 cm–10 cm) around the edge of the rootstock.
4. Secure the graft. Cover all cut surfaces with a sealant such as grafting wax or paint. As the scions begin to attach and grow, allow the strongest to grow and prune away all the others.

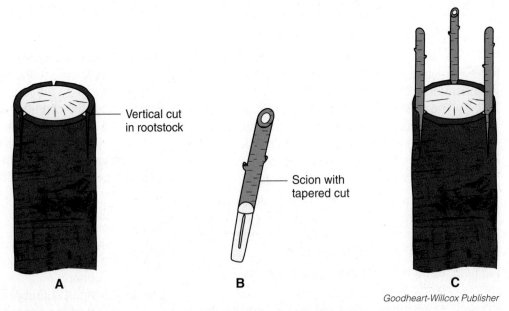

Goodheart-Willcox Publisher

Figure 16-13. Preparing a bark graft. A—Make vertical cuts in the bark of the rootstock. B—Cut the scions so the cut side is tapered. C—Insert the tapered scions into the cuts in the rootstock and seal.

Bridge Grafting

Bridge grafting is a joining method in which damaged tissue is removed from a plant and scions are inserted to provide support and reconnect the vascular tissue across the wounded area. It is used to repair injury done to the trunk, usually from lawnmowers, weed trimmers, animals, disease, or freezing, **Figure 16-14**. When the bark is significantly damaged, the phloem tissue that lies directly underneath may also be injured. This injury can limit the transport of sugars throughout the tree. Bridge grafting should be done in early spring as the tree begins actively growing and the bark slips easily. Use the following process to complete a bridge graft:

1. Prepare the rootstock. Remove the damaged tissue around the wounded area so the graft can be made on healthy stems. Make a cut in the bark of the rootstock above and below the injury in the form of flaps, **Figure 16-15A**. The cut should be the same width as the scion. Be careful to fold the flap away from the rootstock so the bark flap is not torn.
2. Prepare the scion. The scion material should come from wood about one year old that is dormant and 1/4″–1/2″ (6 mm–13 mm) in diameter. Ensure scions are straight and about twice the length of the rootstock's injury. Cut tapers 1 1/2″–2″ (3.8 cm–5 cm) on the same side at each end of the scion, **Figure 16-15B**.
3. Insert the scion. Insert a scion every 2″–4″ (5 cm–10 cm) around the wounded area. First, insert the scion firmly into a bark flap below the injured area in the live, undamaged bark. Then insert it into the upper flap, making sure that the scion material is right side up. The cut side of the scion should be against the wood of the rootstock. Tack the flap in place over the scion, **Figure 16-15C**.
4. Secure the graft. Once all of the scions have been inserted, cover the cut surfaces with grafting wax, being careful to work the wax around the scions. As the graft begins to attach and grow, remove any buds or shoots on the scions as they develop.

Liz Driscoll/Goodheart-Willcox Publisher

Figure 16-14. Placing mulch around the base of young trees will prevent injury from lawn mowers and trimmers.

"Graft good fruit all, or graft not at all."
—Benjamin Franklin

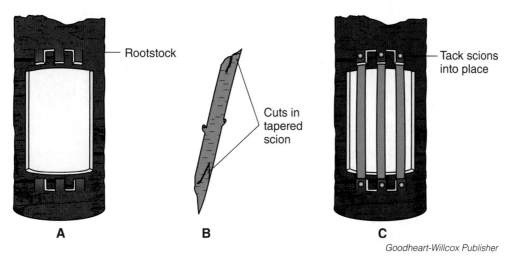

Goodheart-Willcox Publisher

Figure 16-15. Preparing a bridge graft. A—Cut the bark of the rootstock to create slots in the bark. B—Taper both ends of the rootstock and make cuts to fit in the rootstock's slots. C—Tacks are driven through the bark and scion into the rootstock.

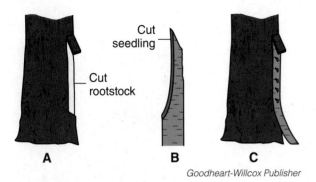

Figure 16-16. Preparing an inarch graft. A—Make vertical cuts in the rootstock, leaving a flap of bark at the top. B—Cut the seedling vertically on the side where it is to be grafted. C—Fit the seedling into the cut of the tree and nail in place.

Inarch Grafting

Similar to bridge grafting, *inarch grafting* is a joining method that supports a damaged plant stem. Inarch grafting uses existing shoots or suckers growing from the rootstock as scion material. Seedlings that are compatible may also be planted around the tree and grafted when active growth begins, **Figure 16-16**. Use the bark or bridge grafting methods to graft the tip of the scion above the injury. The upper end of the scion material should be 1/4″–1/2″ (6 mm–13 mm) thick and cut shallowly for 4″–6″ (10 cm–15 cm). The cut side should be facing the rootstock. Another smaller cut about 1/2″ (13 mm) should be done on the opposite side of the long cut, forming a sharp, wedge-shaped end on the scion.

Approach Grafting

Approach grafting is a method of joining two separate, rooted plants together. After a graft union has healed, remove the top of the rootstock plant from above the graft, and remove the bottom of the scion plant from below the graft. **Figure 16-17** illustrates spliced, tongue, and inlay approach grafting methods.

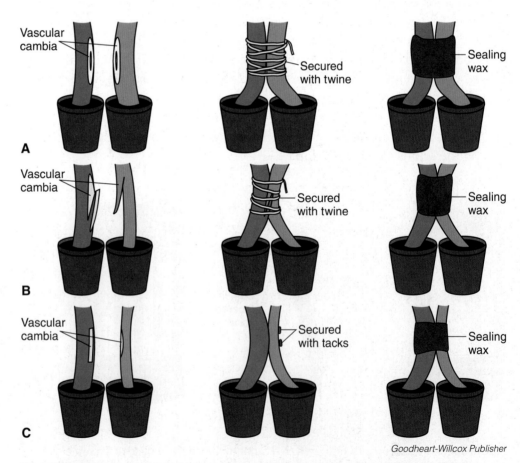

Figure 16-17. Three methods of approach grafting. A—Spliced approach grafting. B—Tongue approach grafting. C—Inlay approach grafting.

Approach grafting is used when establishing a graft union between plants that are difficult to graft. Often, approach grafting occurs between two plants growing in containers. For woody material, this type of grafting should be done when plants are actively growing and rapid healing of the graft union can occur. Approach grafting has been used to graft squash seedlings to promote disease resistance.

Side-Veneer Grafting

Side-veneer grafting is a plant-joining method in which a flap of bark is removed from the side of a plant stem (rather than the top of the rootstock as in other grafting methods) and a scion is attached to the cut area. Cut off the top of the rootstock plant after the graft has healed. In the nursery industry, a side-veneer graft was historically used to graft plants that are difficult to root, such as camellias and rhododendrons. Today, the side-veneer graft provides a means for grafting dwarf or compact conifers onto a rootstock. Follow these steps to complete a side-veneer graft:

1. Prepare the stock. Side-veneer grafting is typically done on potted plants. Rootstocks are selected from dormant plant material and brought into a cool greenhouse to foster root growth. Cut downward about 3/4"–1" (2 cm–2.5 cm) at the base of the stem to reveal a bark flap that still has some wood attached. Then remove the flap by cutting at its base, **Figure 16-18A**.
2. Prepare the scion. Select a scion that matches the diameter of the rootstock. Make a slanting cut 3/4"–1" (2 cm–2.5 cm) long at the base of the scion, **Figure 16-18B**.
3. Insert the scion. Align the cambial layers of the scion's cut surface with the cut surface of the rootstock.
4. Secure the graft. Secure the scion, **Figure 16-18C**. Cover the graft with grafting wax. Once the graft has healed, remove the bindings.

Did You Know?
Grafting is practiced in the craft of Bonsai. It is used to introduce a branch where it is needed, place smaller foliage into a tree that has large leaves that are difficult to reduce, and add "roots" to a specimen.

"Grafting is one of the oldest of the arts of plant craft."
—Liberty Hyde Bailey

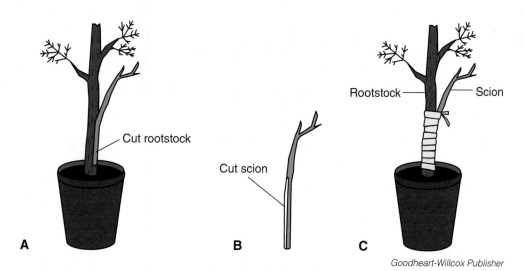

Goodheart-Willcox Publisher

Figure 16-18. Preparing a side-veneer graft. A—Make a vertical cut in the bark of the rootstock. B—Make a slanted cut at the base of the scion. C—Align the cambial layers of the rootstock and scion and secure the graft.

Budding

Budding is a type of grafting, often referred to as bud grafting. This method uses only a single bud from the chosen scion. Because only one bud is used from the scion material, many more plants can be created from the parent material. Budding will not scar the parent plant material (also known as a *stock plant*).

Unlike grafting (which includes a scion piece having several buds) this method can be done during the current growing season and during dormancy. Grafting success occurs in winter months during dormancy; however, budding takes place from early spring through late summer. Grafting is done when the bark is slipping.

Horticulturists who practice budding for some time find that this method is faster and that they have equal or higher rates of success than with other types of grafting. Like other types of grafting, budding is only successful when two compatible plants are united. Budding usually involves deciduous fruit trees, shade trees, or roses.

Three common methods of budding are patch budding, chip budding, and T-budding. These methods require **budwood** (short lengths of young branches used to secure vegetative buds as scion material). Only vegetative buds can be used successfully for budding. They are smaller and sharper than reproductive buds and are usually found near the terminal end of the stem. Budwood should be collected early in the morning from the interior of the plant. Buds that are one year old are best. Cut scions should not be allowed to dry out.

Patch Budding

Patch budding works best with ornamental trees, shrubs, and fruit trees that have thick bark. Follow these steps for patch budding:

1. Prepare the stock. Decide where you will place the graft on the rootstock. This area should be smooth and between two nodes. Using a grafting knife, make two cuts parallel to one another that run horizontally about one-third of the circumference of the stem. Make the first cut, and then make the second cut about 1" above the first cut. To connect the parallel cuts, make two vertical cuts. The four cuts should resemble a rectangle. Take the patch of bark off the rootstock.

2. Prepare the scion. Prepare the budwood by selecting a bud. Use the patch from the rootstock as a template and mark the dimensions around the bud. Cut around the bud and remove the bud sideways. This will ensure that cambial tissue from the wood core is removed and remains with the budwood.

3. Insert the scion. Set the patch into the rootstock immediately. All four sides should fit securely. Remove any excess tissue.

4. Secure the graft. Cover the union with grafting tape. Seal all four cuts, leaving a small hole for the bud to poke through.

Safety Note

Using a Grafting Knife

Keep safety in mind when using a grafting knife. Grafting knives should be sharp. A blunt knife leads to injuries. Keep your thumbs under the wood while cutting to avoid injury. Lock your thumbs together when holding onto the stems of plants while cutting. This provides controlled leverage.

Chip Budding

When bark is not slipping, chip budding works as the best budding vehicle. Viticulturists use this method on grapevines in the fall when bark is not slipping, **Figure 16-19**. Follow these steps for chip budding:

1. Prepare the stock. Select a site that is smooth on the rootstock. Find this area between two nodes. Using a grafting knife, make a 45° angle cut down and 1/4" into the rootstock. Make a second cut 1" above the first. Angle this cut down and into the wood until it joins the first cut, **Figure 16-20A**.
2. Prepare the scion. Find a bud stick (this contains the budwood) that is the same diameter as the rootstock. Select the bud and create the exact same cut as was achieved on the rootstock, **Figure 16-20B**.
3. Insert the scion. Place the chip from the budwood into the rootstock's notch that was made. Be sure to align cambial layers, **Figure 16-20C**.
4. Secure the graft. Cover the union with grafting tape. Pierce a hole in the tape where the bud can be exposed but keep all wounds covered, **Figure 16-20D**.

T-Budding

Countless fruit, citrus, rose, and dogwood plants are propagated in late summer when the bark is slipping via T-budding. The exact time of bark slipping depends on many environmental factors, such as soil moisture and temperature. Budding should not be attempted, however, when air temperatures are greater than 90°F (32°C).

Singkham/Shutterstock.com

Figure 16-19. Chip budding on grapes is a typical way to propagate new material onto established rootstock.

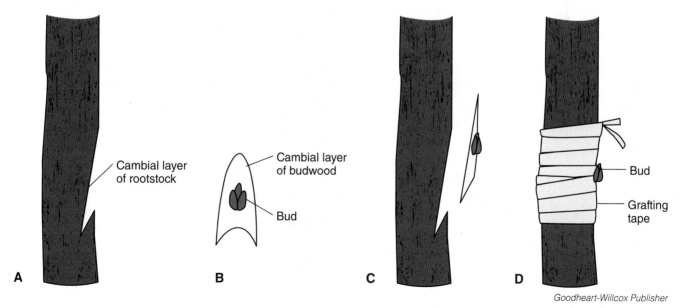

Goodheart-Willcox Publisher

Figure 16-20. Preparing stock for chip budding. A—Preparing the rootstock. B—Preparing the chip of budwood. C—Inserting the chip onto the rootstock. D—Secure with grafting tape and allow the bud to poke through.

Corner Question

Why do Honeycrisp apple seeds not grow Honeycrisp apple fruit?

The buds should be inserted on the cooler side of the plant (on the north or east side of the stems). Follow these steps for T-budding:

1. Prepare the stock. Select a site on the rootstock that is between nodes and is smooth (with approximately 2" of free space on the stem). Create a 1" vertical cut through the bark. At the top of the cut, create a horizontal cut that is about one-third of the stem's circumference, **Figure 16-21A**. These cuts should just break the stem's bark. At the intersection of the two cuts, use the knife to slightly lift the corners of the bark.

2. Prepare the scion. Find the bud to use on the scion. About 1/2" below the bud, cut the wood upward and go past the bud about 1". About 3/4" above the bud, make a horizontal cut through the bark. Make the cut deep enough to remove the bud, the bark, and a little bit of wood, **Figure 16-21B**.

3. Insert the scion. Insert the scion into the flaps on the rootstock and ensure there is contact of the cambia, **Figure 16-21C**.

4. Secure the graft. Wrap the union with grafting tape and create a hole to expose the bud, **Figure 16-21D**.

Successful Budding Tips

For successful budding, be sure to wash your hands, work tools, and work area. Do not touch the interior (wood or cambium). Do not put the buds down anywhere. Work quickly and efficiently to avoid contamination. Do not cover the bud with grafting tape. Use compatible plant specimens. Use budwood that is free of pests. Make sure the cambia align and make contact between the rootstock and budwood. Following these simple rules will help you succeed in budding.

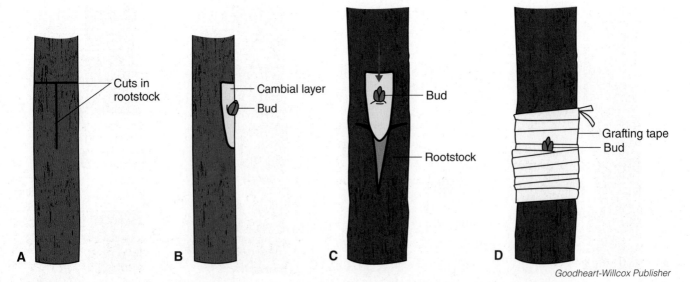

Goodheart-Willcox Publisher

Figure 16-21. Preparing stock for T-budding. A—Make vertical and horizontal cuts in rootstock. B—Make a similar cut in the budwood, removing the bark to expose the cambial layer. C—Insert the budwood into the cut rootstock. D—Secure with grafting tape and allow the bud to poke through.

Aftercare

Successful budding will be evident when the bud begins to grow, usually within 15 days. To ensure that the newly implanted bud will dominate the plant's growth, cut off the other branches above that bud. Remove any wrapping around the site of the union to ensure that there is no girdling (cutting through the bark all the way around the plant), which leads to serious damage or death of the plant.

Careers in Grafting and Budding

Grafting and budding techniques require skill and experience. It is important for professionals to be able to communicate about these topics. Professionals who are skilled in grafting and budding as well as other agricultural information may be in positions of leadership and management.

Agricultural Communications

Horticulture requires writers, television presenters, and radio media specialists to be able to communicate agricultural information to various groups of people, such as consumers, legislators, and agribusinesses. Positions in this field are broad and varied and could include sales and journalism. There is a strong need for people who possess good communication skills and knowledge of scientific agriculture. This position requires a bachelor's degree in communications or marketing, preferably with an emphasis on agriculture. At least 3 to 5 years of experience in the field is preferred.

Assistant Vineyard Manager

Under the direction of the vineyard manager, the assistant supervises and coordinates all of the various farming activities for assigned vineyards. Responsibilities may include general vineyard management work such as pruning, irrigation management, weed control, integrated pest management, cultivation, canopy management, and harvest. An assistant will also support vineyard development activities including site preparation, trellis and irrigation system installation, planting and grafting, and equipment selection. The assistant will supervise field crews and help coordinate activities of vineyard supervisors, foremen, and other related personnel. This position requires a bachelor's degree in horticulture or a related field, as well as 2 to 3 years of experience in a vineyard.

CHAPTER 16 Review and Assessment

Chapter Summary

- Grafting is a horticultural technique that joins together two different plants or plant parts so that they grow into one plant.
- The following factors must be present for a graft to be successful: rootstock and scion compatibility, alignment of vascular cambia, proper timing, protection of the graft union, and proper care after grafting.
- Most grafting occurs during the winter and early spring months while both scion and rootstock remain dormant.
- Rootstock and scion wood should be selected to be free from insect, disease, or winter damage. If possible, the material should be used the same day that it is cut.
- Grafting techniques include whip-and-tongue grafting, splice grafting, cleft grafting, wedge grafting, saddle grafting, hole insertion grafting, bark grafting, bridge grafting, inarch grafting, approach grafting, and side-veneer grafting.
- The whip-and-tongue grafting technique is a popular method used in nursery crops and woody ornamentals and is commonly used with relatively small material.
- Cleft grafting is a simple and widely used method to change varieties of fruit trees in an existing orchard through topworking.
- Hole insertion grafting (HIG) is the most popular grafting technique used on cucurbits because there are many suitable rootstocks and grafting promotes plant vigor.
- Bridge grafting is used to repair injury done to the trunk of a plant and to provide support and reconnect the vascular tissue across the wounded area.
- Budding is a type of grafting, often referred to as bud grafting. This method uses a single bud from the chosen scion.
- Budding techniques include patch budding, chip budding, and T-budding.

Words to Know

Match the key terms from the chapter to the correct definition.

A. approach grafting
B. bark grafting
C. bark slipping
D. bench grafting
E. bridge grafting
F. budding
G. budwood
H. cleft grafting
I. grafting
J. hole insertion grafting (HIG)
K. inarch grafting
L. interspecific
M. interstock
N. saddle grafting
O. scion
P. side-veneer grafting
Q. splice grafting
R. topworking
S. wedge grafting
T. whip-and-tongue grafting

1. Joining together two different plants or plant parts so that they grow into one plant.
2. A plant-joining method in which damaged tissue is removed from a plant and scions are inserted to provide support and reconnect the vascular tissue across the wounded area.
3. A plant-joining method used on herbaceous material in which the rootstock growing point is removed and a hole is inserted into the top of the plant. The scion material is cut into a wedge and inserted into the hole.
4. A plant-joining method that is used to topwork trees in the trunk or scaffold branches.
5. A plant-joining method in which a scion of one plant species is placed into the layer where the bark and wood separate in a rootstock plant.
6. A method of joining two separate, rooted plants together.
7. Any grafting that occurs in potted plants in the nursery (typically with pots sitting on a bench or table).
8. A joining method used to support a damaged plant stem in which existing shoots or suckers growing from the rootstock or compatible rooted plant material planted around the tree is used as scions.
9. A plant-joining method in which the scion is cut in the shape of a saddle and sits on top of the rootstock.
10. A method of joining plants in which a simple diagonal cut of the same length and angle is made on the rootstock and scion. They are placed on top of each other with the cambial layers aligned.
11. A plant-joining method in which a flap of bark is removed from the side of a plant stem and a scion is attached to the cut area. Once the graft has healed, the top portion of the rootstock is cut off.
12. A plant-joining method in which the scion is cut into a V-shaped wedge and the rootstock is cut to form a V-shape to hold the scion.
13. A method of joining plants in which the rootstock is cut with two notches and the scion material is cut to match and create a tight fit.
14. The process of grafting new scion material onto a mature tree.
15. A young shoot or twig of a plant, often used for the upper portion of a graft.
16. A piece of plant material that is grafted between the rootstock and scion to allow joining and growth of incompatible varieties.
17. Short lengths of young branches used to secure vegetative buds as scion material.

18. Occurring between different species.
19. A type of grafting that uses a single bud from the chosen scion, which is inserted into the rootstock material.
20. A condition in which the vascular cambium is actively growing and the bark can be easily detached or peeled in one even layer from the wood beneath it without wounding the plant material.

Know and Understand

Answer the following questions using the information provided in this chapter.

1. What is grafting?
2. Why is grafting an important propagation technique for certain species?
3. What are some benefits of grafting for growers?
4. What conditions are needed for successful grafting?
5. What time of year do most grafts occur and where is grafting done?
6. What factors should be considered in selecting scion wood?
7. Whip-and-tongue grafting is popular for what type of plants?
8. Explain the process of splice grafting. What plants are commonly grafted using this method?
9. What are four examples of fruit trees that can be propagated using cleft grafting? What are two other uses for cleft grafting?
10. What are two examples of uses of saddle grafting? When is saddle grafting typically done?
11. Describe the steps to making a hole insertion graft.
12. What are three important functions of a healing chamber for grafted plants?
13. What are some uses for bark grafting and when is it typically done?
14. Why would a grower use bridge grafting and how is it done?
15. Explain how side-veneer grafting is different from most other methods of grafting.
16. Why would a grower use budding as a propagation technique?
17. What safety considerations should be followed when using a grafting knife?
18. Describe the process for chip budding.
19. What environmental factors should be considered in T-budding?
20. What practices will help ensure successful budding?

Thinking Critically

1. You are a pecan grower in Texas. A late winter ice storm has hit your orchard, and there is significant damage to your trees. Describe the resources, information, and competencies you will need to repair your trees. Will you need any unusual mechanics or systems of operation?
2. A friend just bought a new house with a very small yard in a downtown area. She really wants to grow fruit trees. What kind of recommendations do you have for her?

STEM and Academic Activities

1. **Science.** Investigate the science behind the callus formation and vascular cambium development in a graft union. What cellular changes are occurring? What might be some reasons why some grafts do not attach and grow?

2. **Science.** Your agriculture teacher has asked you to do a demonstration of tomato grafting for your peers. Using at least three resources, research the process of tomato grafting. Write a report that includes how, where, and why tomatoes are grafted. Present your report to the class and give a demonstration of the graft.
3. **Math.** You have an apple orchard and want to wedge graft your fruit trees. To prepare, you need to take cuttings of scion material. If you have 125 rootstocks that are 2″ (5 cm) and 130 that are 4″ (10 cm), how many scions do you need to cut?
4. **Language Arts.** In small groups, discuss the illustrations in this chapter. Describe in your own words what is being shown in each illustration. Discuss the effectiveness of the illustrations compared to the text descriptions.
5. **Language Arts.** Working in groups of two or three, read the procedure on how to make a cleft graft and practice making cleft grafts. Once each member of the group is able to make the graft, compare and discuss issues or problems you each encountered.

Communicating about Horticulture

1. **Reading and Speaking.** Read one of the grafting techniques described in the chapter. Obtain materials to actually demonstrate the technique. Using a digital video recorder, film yourself performing the graft while you are speaking about what you are doing. Show the video to your class and have them give you constructive feedback.
2. **Reading and Writing.** Research one of the budding strategies. Draw a step-by-step diagram to show someone how to do the technique. Write detailed descriptions with each drawing to clarify what is occurring. In what way do diagrams help readers understand concepts more fully?

SAE Opportunities

1. **Experimental.** Select one of the grafting or budding methods discussed in this chapter. What kind of plants could you use with this approach? Select a plant. With the supervision of your teacher, try your grafting technique. What happens? Keep a record of your findings.

pixinoo/Shutterstock.com

2. **Exploratory.** Create a grafting schedule for five fruit trees. When should you perform the graft? How should you prepare the rootstocks and scions? Include information about the size of material, length, timing, and other important details for each plant. Why do you think growers keep records of their grafting?
3. **Exploratory.** Job shadow someone who professionally practices grafting techniques. What kind of organization does this person work for? How often does he or she use grafting in the work?
4. **Entrepreneurship.** Use bud grafting to create a number of different woody ornamentals using stock plants from your school landscape or in partnership with a local nursery. Sell your budded plants to customers at a school plant sale.

CHAPTER 17
Tissue Culture: Micropropagation

Chapter Outcomes

After studying this chapter, you will be able to:
- Understand micropropagation history.
- Describe the advantages and disadvantages of micropropagation.
- Recognize the tools and equipment used in micropropagation.
- Describe materials used to make growth media for micropropagation.
- Identify the stages of micropropagation culture.
- Discuss the future of tissue culture and micropropagation.
- Explore careers related to micropropagation.

Words to Know

acclimatization	endogenous contamination	recultured
apical meristem	explant	somaclonal variation
aseptic	growth medium	subculturing
callus	*in vitro*	tissue culture
clean room	laminar flow hood	totipotency
clone	micropropagation	
cryopreservation	protoplast	

Before You Read

Review the list of terms at the beginning of the chapter. Write what you think each term means. Then look up the term in the glossary and write the textbook definition.

While studying this chapter, look for the activity icon to:
- **Practice** vocabulary terms with Words to Know activities.
- **Expand** learning with identification activities.
- **Reinforce** what you learn by completing Know and Understand questions.

www.g-wlearning.com/agriculture

mihalec/Shutterstock.com; tony4urban/Shutterstock.com; KentaStudio/Shutterstock.com; MgPL/Shutterstock.com; Mihai Simonia/Shutterstock.com

Tissue culture is a collection of techniques used to grow plants in a nutrient medium under sterile conditions. ***Micropropagation***, also referred to as tissue culture, is an asexual propagation method in which plants are manipulated on a cellular level, causing them to duplicate themselves repeatedly and rapidly. This process is important in the agriculture and horticulture industries because it quickly produces identical plants (*clones*). Using this method of asexual propagation, very small pieces of plant tissue are placed in a sterile culture or test tube that contains a distinctive ***growth medium*** (a substance containing nutrients and hormones used for plant growth), **Figure 17-1**. Ultimately, cells divide to create new tissues and organs. The final result is a new organism: a cloned plant.

When a plant cannot be produced from seeds (sexual propagation), breeders can use micropropagation to quickly produce the plant. (Keep in mind that sexual propagation is still the least expensive and fastest way to disseminate new varieties if the plants can be produced by seeds.) Micropropagation is a form of biotechnology in which plants are being manipulated using science, and cells and entire plants are cloned. Micropropagation is an essential tool used by horticulturists, mostly professionals, in today's thriving green industry.

borzywojo Studio/Shutterstock.com

Figure 17-1. Micropropagation is accomplished using a sterile technique by technicians who specialize in micropropagation.

History

Although micropropagation celebrated its 100th anniversary in 2004, many horticulture professionals consider this propagation technique as still in its infancy. However, this technique is considered to be one of the most technologically advanced methods of propagation. The earliest advances in tissue culture began in the early part of the nineteenth century and the majority of micropropagation protocol was created in the 1940s through the 1960s.

- In 1839, scientists Matthias Schleiden and Theodor Schwan proposed that all life could be grown from a single cell. They determined that a single cell contains all of an organism's genetic information and the ability to differentiate into any type of cell. This is known as ***totipotency***.
- Gottlieb Haberlandt conceptualized the idea of *in vitro* cell culture in 1902. *In vitro* is Latin for "in glass." For scientific purposes, ***in vitro*** refers to the growth of an organism outside the body, such as in a test tube or petri dish. Haberlandt was able to isolate cells and put them in culture, but never successfully had the cells divide.
- Plants embryos were cultured by E. Hannig in 1904.

- Lewis Knudson germinated an orchid embryo *in vitro* in 1922, **Figure 17-2**.
- The first plant hormone, indole acetic acid (IAA), was discovered in 1926 by Fritz W. Went.
- In 1939, constant regeneration and proliferation of plants was achieved by scientists Roger J. Gautheret, P.R. White, and Dr. P. Nobecourt.
- Coconut milk was added to the growth medium and plants began to thrive with the addition of this compound, as discovered by J. Van Overbeek in 1941.
- In 1946, Professor Ernest Ball raised whole plants of lupine *in vitro*. This was the first entire plant grown via tissue culture.
- In 1962, Toshio Murashige and Folke Skoog developed the first reliable standardized artificial medium for micropropagation.

nakorn/Shutterstock.com

Figure 17-2. These orchids are the result of micropropagation.

Advantages and Disadvantages

As with other propagation methods, micropropagation has advantages and disadvantages. Horticulturists continue to use micropropagation for the many advantages it offers, including:
- The uniformity of clones in age, quality, and desired characteristics.
- The ability to grow large numbers of plants in small spaces in a short amount of time.
- The ability to produce a viable plant when breeding plants that are not usually compatible.
- The ability to generate vast quantities of clones from one plant.
- The ability to produce pest-free plants (at the time of distribution from the laboratory).
- The ability to aid in the conservation and replication of rare or endangered plant species.

Did You Know?

There are nearly 150,000 orchid hybrids. Orchids were the first plant to be micropropagated and are currently the most commonly micropropagated.

Thinking Green

Micropropagation Uses a Sustainable Product

Technicians, or culturists, typically use agarose to create the medium of growth. Agar, derived from a red algae's cells, primarily comes from the genera *Gelidium* and *Gracilaria*. This seaweed substance is plentiful and easily replenishes itself. Some forms of these algae are being researched as potential biofuel sources.

Gelidium sesquipedale

Joxemai/Wikipedia

Labor and Equipment Costs

The main disadvantage of micropropagation is the cost of labor and equipment. Micropropagation does not happen by chance. Micropropagation requires the employment of technicians skilled in the art of micropropagation as well as specialized equipment. Technicians must be trained to use *aseptic* (free from contamination) techniques to prepare and maintain work areas because micropropagation must be conducted in a *clean room* (a space that is completely sterile and void of any contaminants).

In addition to trained technicians and clean room requirements, special tools, such as a *laminar flow hood* (a machine that filters and purifies the air), are needed. A laminar flow hood may cost several thousand dollars and the filter, which must be replaced regularly, can also be costly, **Figure 17-3**. Micropropagation also requires a growing medium comprised of various chemicals, nutrients, and hormones. This medium must be carefully prepared in a sterile environment with the proper formulation for each species, variety, or cultivar of plant being micropropagated.

Supply and Demand

Skilled labor and equipment expenses are reflected in the high price of micropropagated materials. Plants that are micropropagated must be those in demand and desired in large quantities, **Figure 17-4**. For example, if there is high demand for a new, unique plant, a grower can meet that demand in the shortest amount of time possible through micropropagation. As with all things that are bought and sold, supply and demand or economics play a major role in deciding whether to micropropagate plant material for the open market.

Figure 17-3. A technician from Agri-Starts works to prepare plant material for micropropagation with the use of a laminar flow hood.

Figure 17-4. Popular plants, such as the (A) calathea and (B) Venus flytrap, are the result of micropropagation. Venus flytraps have difficulty propagating in nature and the use of micropropagation ensures future generations of this unique carnivorous plant.

Genetic Diversity

As explained earlier, plants reproduced via micropropagation are clones with the same genetics. This can be of great benefit when a grower expects to replicate the exact same traits of the parent plant. However, the lack of genetic diversity may also be detrimental. Imagine if everyone in your class had the exact same genetics. Not only would you all look exactly the same, you would have the same immunities or susceptibilities to diseases and illnesses. If one person in the class became ill, everyone would likely follow suit. This is true in the plant world as well. If a viticulturist has a vineyard with only one variety of vines and one plant suffers because of a fungus, it is likely they will all be susceptible. For this reason, it is important for growers to maintain genetic diversity in their crops.

Corner Question

Can plants be bred if they are incompatible?

Environmental Requirements

Supplying plant material with the proper environmental conditions is very important for successful micropropagation. When using micropropagation techniques, the *explant* (cells, tissues, or entire plants that are in an *in vitro* culture), is removed from the parent plant and placed in an environment similar to its native habitat. The micropropagated plants require light, nutrition, hormones, moisture, and the appropriate temperature as well as a sanitized area in which to grow, **Figure 17-5**. The items needed to achieve and maintain an aseptic environment include a laminar flow hood, test tubes or petri dishes, an artificial light supply, a growth medium, ethanol or isopropyl alcohol solution (70%), bleach solution (10%), forceps, scalpel, plastic tape for sealing test tubes and petri dishes, and hand soap.

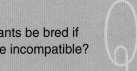

Photography by Agri-Starts, Inc.

Figure 17-5. A technician at Agri-Starts inspects plant material that has been cultured via micropropagation. The growing area is maintained as a sterile environment in which the temperature and humidity is carefully controlled.

To prevent the introduction of disease or pests, the explants' surface must also be sanitized of all bacteria, fungi, viruses, and pests before being introduced to the sterile environment. The explants may need to be subcultured if bacteria, fungus, or viruses arise from within the surface of the sanitized tissue. This type of contamination is called *endogenous contamination*.

While working, the technician (culturist) must maintain sterility in the entire work area. Actions that should be taken to ensure sterility include washing and sterilizing the work area (tabletops, floors, and tools), proper hand washing, and keeping doors or windows near the work area closed. Traffic flow through the area should also be kept to a minimum.

Did You Know?
Certain types of agar can be used as a vegetarian alternative to gelatin. Agar is used in countless foods, such as ice cream, preserves, and soups.

Growth Media

Growth media are combinations of nutrients, growth regulators (hormones), and water used to grow and sustain the explants or cells *in vitro*. These substances, or the concentration of the substances, are manipulated and changed for various stages of growth. The medium must be sterile and may be either a liquid or a gel. Inorganic salts, vitamins, sugars, and trace elements are also included in various amounts in the growth medium.

The medium will always have a water base. A gelling agent is added to the medium to form a gelatin-type texture. Gelling agents vary, but they are often derived from algae (agar). Many commercial micropropagation labs use alternative gelling agents derived from elements such as corn or potato starch. As the explants develop, they are *recultured* or transferred into a different test tube or petri dish containing nutrients and growth regulators needed for the next growth stage, **Figure 17-6**.

luchunyu/Shutterstock.com

Figure 17-6. Plants can be recultured in growth medium. These tobacco plants have been recultured to secure more tobacco plants.

Stages of Micropropagation

There are five stages of micropropagation. Each stage plays a vital role in the production and transitioning of a plant from an *in vitro* environment into a natural environment.

Safety Note
Micropropagation

Fire, ethanol, alcohol, and bleach are used to sterilize work areas, equipment, and plant materials. It is imperative that these elements be stored properly and used following a strict protocol to prevent worker injury and equipment damage. Always wear the appropriate personal protective equipment (PPE) and keep hair tied back and sleeves and clothing snug when working with fire. Know where the closest eyewash station and fire extinguishers are before beginning the micropropagation process. You should also review pertinent safety data sheets (SDS) before working with chemicals.

Safety Note

Hand Washing

Do not wash your hands with hotter water than is comfortable. Also, gently wash your hands. Do not scrub them to the point of discomfort. Harshly scrubbing your hands may cause abrasions and this, coupled with bacterial or fungal contaminants, may cause serious health problems. If your hands have any cuts or abrasions, the alcohol solution will cause a great deal of discomfort.

Alexander Raths/Shutterstock.com

Stage 0: Selection and Cultivation of Stock Plants

This stage requires the meticulous cultivation of stock plants (parent plant material kept specifically for the purpose of propagation). Particular attention is paid to the plants to ensure there are no diseases or insects impacting the plant's growth, **Figure 17-7**. This stage is in place to limit or prevent contamination in the following stages.

Stage 1: Initiation or Establishment

In stage 1, an aseptic environment is established and the explant is sterilized and transferred to an *in vitro* culture. To prevent contamination, technicians must keep their hair tied back or covered and they must wash their hands with warm soapy water for 30 seconds before handling plant material and tools. All jewelry should be removed. Wearing latex or nitrile gloves and a simple face mask at this time is also recommended. All work areas should be cleaned with a 10% bleach solution and hot soapy water. Finally, everything must be sprayed with a 70% solution of ethanol or isopropyl alcohol, **Figure 17-8**.

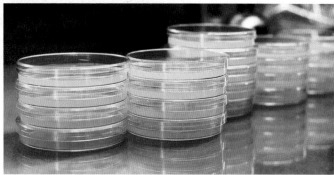
masa_damon/Shutterstock.com

Figure 17-7. Micropropagation equipment and materials include sterile petri dishes filled with growth medium. This medium includes sugars, agar, hormones, and salts to aid in plant development.

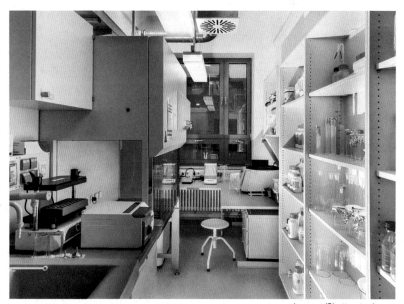
anyaivanova/Shutterstock.com

Figure 17-8. A sterile lab is essential for tissue culture. This is one of the larger expenses associated with micropropagation.

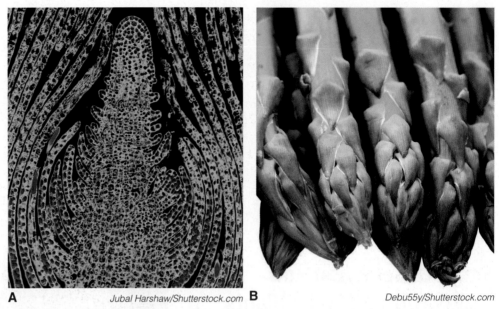

Figure 17-9. A—This apical meristem of a plant has been dyed and magnified under a microscope. B—The apical meristem of asparagus is beneath the terminal leaf buds.

Explant material may be obtained from several sources, including:

- Single cells from plants.
- Small pieces of plant tissues from leaves, stems, shoot tips, flowers, or embryos.
- The *apical meristem* (a swelling of cells at the furthest tip of a plant shoot or the bud), **Figure 17-9**.
- Undifferentiated masses of cells that are from parenchyma and lack a defined function, called a *callus*, found on other explants.

Figure 17-10. Extreme caution must be made when taking an explant to another stage of micropropagation. Using sterile tools is essential.

Just as the work surfaces, tools, and existing environment must be sanitized, the explant material must be completely sterile. To sterilize the explant, technicians wash the plant material in a bleach or peroxide solution. In addition, explants may be dipped in ethanol or isopropyl alcohol. The explant is placed on the growth medium in a test tube or petri dish with sterilized forceps or tweezers, **Figure 17-10**. Once the explant is established, it is ready for reculturing in stage 2.

Stage 2: Multiplication

In stage 2, the explants established in stage 1 are transferred (with sterilized tools) to a second medium containing nutrients needed for the plant material to multiply. This medium is referred to as the multiplication medium. To prevent contamination, technicians use standard aseptic techniques (cleaning and PPE) as well as a laminar flow hood and sterilized forceps during this procedure. Forceps may be sterilized by dipping them in alcohol and "flaming" them with a small flame, and dipping them into two separate sterile water containers.

STEM Connection
Cincinnati Zoo's CREW CryoBioBank®

LianeM/Shutterstock.com

The Carl H. Lindner Jr. Family Center for Conservation and Research of Endangered Wildlife (CREW) is a research facility dedicated to saving endangered plants and animals from extinction. The CryoBioBank, a genome database, is a division of this facility. The CryoBioBank stores thousands of plant and animal tissue samples through a process known as **cryopreservation**. Cryopreservation uses liquid nitrogen to preserve the samples at subzero temperatures. Tissues can be stored and remain viable for decades under these unique environmental conditions.

Scientists and horticulturists work together at this facility to preserve various species, especially those in danger of extinction. Plants such as the four-petal pawpaw (*Asimina tetramera*) have been given a new lease on life through the CryoBioBank. This plant, native to and found only in southeastern Florida, has been nearly driven to extinction by loss of habitat and the difficulty of propagating by seed. The scientists at CREW developed a means of propagation by tissue culturing the meristems and preserving these tips in liquid nitrogen. Additionally, the four-petal pawpaw has been micropropagated and prepared for transplant to its native Florida.

Medium Composition

The sterile multiplication medium is typically composed of a gel containing vitamins, sugars (basal medium), and a cytokinin growth regulator. Cytokinins are a class of growth regulators used to encourage shoot production. Using a high cytokinin to auxin ratio in the multiplication stage encourages axillary or adventitious shoot formation. A high number of adventitious shoots that do not elongate may result if the medium has too high a concentration of cytokinin.

Subculturing

The plantlets developing in the multiplication medium remain in stage 2 for about 5–7 weeks (depending on the plants' particular needs and growth) before reculturing. Plantlets at this stage may be divided, detached, and cultured again in another stage 1 or 2 medium, depending on the technician's motive. When plantlets are recultured in stage 2 medium, the process is referred to as *subculturing* (establishing a new culture by moving some cells from an existing culture). There is a limit to how many times plantlets can be subcultured. Undesirable or unintentional changes may occur if a plantlet is subcultured too many times. This occurrence is called *somaclonal variation*. Somaclonal variations can be intentional, too. Breeding specialists often welcome unique changes in genetic material that could possibly yield a new cultivar.

Corner Question
What is the actual size of a meristem?

Kitto Studio/Shutterstock.com

Figure 17-11. In stage 2, plantlets are multiplied. The tissue has no roots and only has stem and leaf tissue.

jaboo2photo/Shutterstock.com

Figure 17-12. A rhizome is an underground or subsurface stem. The rhizome appears black in micropropagation.

Environmental Conditions

As every plant has exclusive growth requirements, the environmental conditions needed during stage 2 depend on the culture being used and the plant material being propagated. Technicians must determine light quantity and quality, the amount of moisture needed, and the type and amount of nutrition required based on the explants' needs.

When plantlets are ready for transfer to stage 3, they have shoots and leaves that are rich in green pigment. The plantlets do not have any roots in stage 2, **Figure 17-11**.

Stage 3: Rooting

In stage 3, plants are being prepared for transplant. The plantlets must develop roots to absorb water and nutrients once they are transplanted in soil. In stage 3, the plantlets are carefully transferred to a sterile rooting medium containing nutrients and growth regulators such as cytokinins, cytoninins, and auxin. These hormones control cell growth, tissue formation, and ultimately how plants develop. This medium is often called the transplant medium as it prepares the plantlet for potting or planting in soil in stage 4.

Roots often develop below the surface of the medium. Sometimes, rhizomes (underground lateral stems) may appear and serve as the origin for other roots, **Figure 17-12**. Root development in this stage helps signify when plants are ready for transplanting to soil. The amount of time required for plants to develop roots adequate for transplanting varies (ranging from weeks to months) from species to species.

Did You Know?

If a plantlet is subcultured in stage 2 every six weeks, it potentially can yield anywhere from 5 to 120 plants each time, or a little less than 1000 each year.

Stage 4: Acclimatization

In stage 4, plantlets are transferred to a sterile potting medium and acclimated for the transition to a nonregulated environment. *Acclimatization* is the gradual exposure of plants to different environmental conditions. This process is also known as hardening-off. The primary reason micropropagated plants must be acclimated is because they have been kept in an ideal environment and may have poor control of water loss and nutrient uptake.

When plantlets are removed from the *in vitro* rooting culture of stage 3, the roots are lightly washed to remove any medium, which could provide the perfect habitat for bacteria or fungi in a high-humidity environment. Plantlets are placed in a sterile potting mix that is considered optimal for that particular plant's growth. Initially, plantlets are exposed to high amounts of humidity and very little sunlight.

During the first weeks after transplant, plantlets may be covered with plastic or humidity domes to increase the amount of moisture. Stage 4 may require up to several weeks. Just before plants are ready to be transplanted outdoors or in a greenhouse, they are subjected to drier air and greater quantities and intensities of light. These conditions help harden the plants for survival in their new environment, **Figure 17-13**.

Corner Question

What can be done to improve the growing conditions while plants are *in vitro*?

The Future of Tissue Culture and Micropropagation

Micropropagation is a biotechnology method used in forestry, agronomy, and horticulture to preserve plant and tree species; create new and improved fruits, vegetables, and ornamentals; and to expedite the development of some genetically modified crops. Although there is some controversy regarding the use of biotechnology techniques in plant production, micropropagation shows great promise to help producers meet the food and fiber needs of the world's growing population. Some of the benefits of micropropagation include:

- Massive replication of plant materials that can be used for beneficial production of biopharmaceuticals (plant-derived compounds used to treat illnesses).
- Use of explants to determine if cells (rather than entire plants) are sensitive to various chemicals (particularly pesticides).

"Use plants to bring life."
—Douglas Wilson.

A *Photography by Agri-Starts, Inc.* B *smcfeeters/Shutterstock.com*

Figure 17-13. A—These plants survived the first three stages of tissue culture and are being acclimated to their new environment in stage 4. B—This potato plant has been placed into a container with a new potting soil medium.

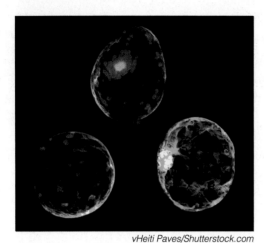

Figure 17-14. The protoplast of a tobacco plant.

- Use of a meristem from a contaminated stock or stock with a virus to produce clean stock after micropropagation.
- Chromosome doubling by using antimitotic agents during tissue culture. This technique can change a sterile seed into a fertile one. This technique can also be used to create new and unique ornamental cultivars of plant material.
- Somatic hybridization, which occurs when plants lose their cell walls due to the introduction of cellulose enzyme. This produces a *protoplast* (also known as a somatic cell), **Figure 17-14**. Once this occurs, cells tend to merge or fuse together regardless of whether they are similar plants. Cells are grown into callus phase, then into plantlets, and eventually into full plants via tissue culture. Unlike with traditional breeding techniques, plants cells can be combined regardless of their compatibility.
- Creation of plants that have changes in their genetics that are not expressed physically but rather in the environment in which they are grown. Examples include growing a southern magnolia in New York, a rhododendron in alkaline conditions, or a tomato in soil rich with unusually toxic salts, **Figure 17-15**.

Careers in Micropropagation

A career in micropropagation and tissue culture ensures a future filled with opportunities for growth. Technicians find many career opportunities in research and development. These jobs are found in the private sector, with government agencies, and with colleges and universities (especially those that are agriculturally related or land grant universities).

Corner Question
What is one plant that you eat almost everyday that is a result of somatic fusion (hybridization)?

Figure 17-15. A—The future of micropropagation and tissue culture involves somaclonal hybridization. A plant could be fused with another plant to develop new traits. Imagine the beauty of the southern magnolia withstanding the killer winters of Maine. B—What about an acid-loving rhododendron withstanding the alkaline soils of your state? Their beauty could be enjoyed around the world via somaclonal hybridization techniques.

Micropropagation Lab Technician

Technicians, sometimes called culturists, must be proficient in plant tissue culture techniques. This includes being able to prepare media, micropropagate materials, and index (create databases) plants. Technicians work in a laboratory and must take notes, make charts and graphs, and have the ability to synthesize information into a written document. Attention to detail in this position is extremely important. Technicians must also maintain accurate daily records and ensure that all plant material, laboratory supplies, and equipment are tracked in an inventory system or database. An associate's degree or equivalent work experience in the same field is usually required.

Biotechnology positions, such as a micropropagation technician, often have policies that demand ethical and professional behaviors from their employees. Technicians are often prohibited from discussing their research and data with anyone outside of the company.

Did You Know?
Frogs can survive freezing and thawing throughout the winter months. As long as no more than 65% of their total body water freezes, they will survive. New cells will be regenerated to replace damaged cells.

Cryopreservation Scientist

Cryopreservation scientists use cryogenics to preserve and store botanical germplasm (a collection of genetic resources for an organism), plant tissue culture, and seeds in facilities around the United States. One such facility is the USDA's Agriculture Research Service (ARS) National Center for Genetic Resources Preservation (NCGRP) cryogenic storage vault in Ft. Collins, Colorado, **Figure 17-16**.

The NCGRP facility, which operates more than 45 cryogenic tanks, is the largest agricultural gene bank facility in the United States. The staff conducts research to acquire more efficient and effective methods for preserving plant germplasm. The research findings, preservation techniques, and specialized technology developed by the NCGRP have been embraced and implemented by gene banks around the world. Scientists from around the world travel to the NCGRP for research and training. (The NCGRP also researches and stores genetic materials from animals.)

A *USDA/Lance Cheung*

B *USDA/Lance Cheung*

Figure 17-16. A—Specialized tanks at a USDA/ARS cryopreservation bank in Colorado are used to preserve plant germplasm for future use. B—A technician at a cryopreservation plant exams plant germplasm that was stored in the cryopreservation tanks at more than –300°F (–184°C) and will remain viable for decades.

People who work at this facility have varied educational training related to the study of plants. Many of the employees at this facility are professional researchers with doctoral degrees. However, there are many employment opportunities for laboratory technicians with plant science backgrounds. In these positions, employees must have problem-solving skills, be inquisitive, employ good communication skills, and pay attention to detail.

Career Connection

Ty Strode
Vice President and Marketing Director

Ty Strode is part owner of his family-run company in Apopka, Florida, called Agri-Starts Inc., and is also one of the most influential young growers in the green industry. In 2013, the *Greenhouse Product News* named Strode to their *40 under 40* list. This honor highlights Strode as one of the premier horticultural growers in the world.

Agri-Starts, founded in 1984 by Ty's father Randy Strode, began by growing foliage items for Florida growers. Today, Agri-Starts micropropagates 500 different types of plants annually and ships up to 300,000 micropropagated plants each week. Agri-Starts is currently focused on propagating edible plants, such as fig, blueberry, and blackberry, and has partnered research efforts with several universities to micropropagate the best and newest edible varieties. In addition, the company has invested in green technologies, such as water reclamation and integrated pest management techniques. Agri-Starts was also recently recognized by the Florida Department of Agriculture as the environmental leader of the year for their land ethic and conservation efforts.

Photography by Agri-Starts, Inc

Ty Strode encourages students to join the green industry where opportunities are limitless. He believes the "industry (is filled) with incredible people and even the greatest competitor is still respected." Positions in his company and others in the industry include jobs in sales, marketing, finance, accounting, and freight logistics.

CHAPTER 17 Review and Assessment

Chapter Summary

- Tissue culture is used to grow or maintain plants in a nutrient medium under sterile conditions. Micropropagation, also sometimes referred to as tissue culture, is an asexual propagation method in which plants are manipulated on a cellular level.
- Since the 1970s, many advances have been made in tissue culture. Current techniques are possible because of the stabilization of tissue culture medium made by Murashige and Skoog in 1962.
- Micropropagation offers benefits or advantages, such as providing identical plant material at the quickest rate of replication and providing a breeding method for plants that are not usually compatible.
- Micropropagation has some disadvantages. It is a costly endeavor that can only be achieved through sterile techniques performed by skilled technicians.
- When using micropropagation techniques, it is essential to have a clean room with a laminar flow hood to reduce contamination. Scalpels, forceps, and tweezers must be made sterile through the use of bleach solutions, ethanol, isopropyl alcohol, or flaming.
- Growth media are combinations of chemicals, vitamins, nutrients, growth regulators (hormones), and water used to grow and sustain cells *in vitro*. The growth medium must be sterile and can be a liquid or a gel.
- There are five stages of micropropagation. Stage 0 includes selection and cultivation of stock plants. Stage 1 includes establishing an aseptic environment, sterilizing explants, and transferring explants to an *in vitro* culture. Stage 2 involves multiplication of plantlets. Stage 3 involves rooting plants in preparation for transplanting. Stage 4 involves hardening off plants.
- Micropropagation is used in forestry, agronomy, and horticulture.
- The process of micropropagation, which manipulates plants at a cellular level, shows great promise. Micropropagation can be used to create new and improved fruits, vegetables, and ornamental plants.
- Several technical careers are available in micropropagation, including laboratory technician or cryopreservation scientist.

Words to Know

Match the key terms from the chapter to the correct definition.

A. acclimatization
B. apical meristem
C. aseptic
D. callus
E. clean room
F. clone
G. cryopreservation
H. endogenous contamination
I. explant
J. growth medium
K. *in vitro*
L. laminar flow hood
M. micropropagation
N. protoplast
O. recultured
P. somaclonal variation
Q. subculturing
R. tissue culture
S. totipotency

1. An asexual propagation method in which plants are manipulated on a cellular level causing them to duplicate themselves repeatedly and rapidly.
2. A collection of techniques used to grow or maintain plants in a nutrient medium under sterile conditions.
3. The protoplasm of a living cell from which the cell wall has been lost or removed.
4. Growth of an organism outside the body, typically in a glass test tube or petri dish.
5. A swelling of cells at the furthest tip of a plant shoot or the bud.
6. The plant cell or tissue removed from the parent plant and placed in an *in vitro* culture.
7. A substance containing nutrients and hormones that is used for plant growth.
8. Sterile or free of contaminants.
9. Changes seen in plants that have been produced by tissue culture or those that have been subcultured.
10. Establishing a new culture by moving some cells from an existing culture.
11. The ability of a cell to differentiate into any type of other cell.
12. Gradually exposing a plant to different environmental conditions.
13. When the explant is transferred to a different test tube or petri dish containing nutrients and growth regulators necessary for a certain stage of plant growth.
14. A piece of equipment that filters and purifies the air and is used in tissue culture.
15. A sterile area that is void of contaminants.
16. A cluster of cells that are from parenchyma and lack a defined function.
17. Freezing organs, cells, or other biological materials in subzero temperatures using liquid nitrogen.
18. When bacteria, fungus, or a virus comes from within the surface of sanitized tissue.
19. A plant that is identical to the original plant and every other plant produced from the original plant.

Know and Understand

Answer the following questions using the information provided in this chapter.

1. What is micropropagation and why is it important?
2. How are cloned plants produced using micropropagation?
3. Why is micropropagation considered an application of biotechnology?
4. List two important micropropagation dates in history and explain what took place.
5. List three benefits or advantages and three risks or disadvantages of using micropropagation.
6. What type of plant material is commonly micropropagated?
7. Describe the conditions that are needed for micropropagation of plant material.
8. What are some safety tips to follow for micropropagation?
9. What are the three main ingredients in growth media used for micropropagation?
10. What happens in stage 0 of micropropagation?
11. What is one tip for effective hand washing?
12. What are the three main activities in stage 1 (initiation) for micropropagation?
13. What are four sources of explant material used for micropropagation?
14. How does a technician successfully transfer an explant using aseptic techniques?
15. Describe what happens to plantlets in stage 2 of micropropagation.
16. What is acclimatization and why do plants need to be acclimatized?
17. What are three benefits of tissue culture and micropropagation?
18. When does somatic hybridization occur and what does it produce? What does this mean for plant breeding?
19. What are some skills needed by a micropropagation lab technician and what are some of the duties for this job?
20. How is cryopreservation being used to foster preservation of plant material and species at the National Center for Genetic Resources Preservation (NCGRP)?

Thinking Critically

1. A rare plant is nearing extinction in your state due to loss of habitat. What approach would you suggest to solve or prevent this plant catastrophe?
2. You work in a facility that practices micropropagation. Recently, your business has been attacked by the news media for using water for irrigation during a drought. What could your business do in the future to lessen consumption of water for plants, leaving more for drinking water for people?

STEM and Academic Activities

1. **Science.** Use a common plant, such as an African violet. Propagate this plant via tissue culture. Ask your instructor for support and guidance in this endeavor.
2. **Science.** If one tissue can create 20 plantlets at the end of an eight-week cycle and each of those plantlets is then placed in a test tube and the cycle is repeated continuously for a year, how many plantlets could be potentially micropropagated?
3. **Engineering.** Engineer a do-it-yourself or at-home tissue culture facility that simulates a laminar flow hood.
4. **Language Arts.** Research information about a current tissue culture endeavor involving plants. Write an editorial article in response to this current biotechnology practice or research. What are the benefits and risks of this practice? State your opinions about the practice.
5. **Language Arts.** Contact a local nursery. Interview an employee and determine if the company uses micropropagated plants. Whether the company does or does not practice micropropagation, ask why and how it impacts the business.

Communicating about Horticulture

1. **Reading and Speaking.** Make a time line of biotechnological advancements in the last 20–30 years. Focus on those relating the most to agriculture. Research advancements and the scientists who brought about these changes. Create a 10- to 15-point time line showing significant milestones that have led to new discoveries and theories. Describe your findings to the class.
2. **Listening and Speaking.** In a group, create a presentation sharing specific examples of how agriculture interacts with one of the following fields of science: natural, formal, or social. Present two to three examples of how this field of science is used in a specific sector of agriculture. Be prepared to share your examples with the class.

SAE Opportunities

1. **Exploratory.** Visit a botanical garden, zoo, university, or private corporation that has a facility which micropropagates plants. Observe their protocol for micropropagation and investigate what plant materials they use for micropropagation.

2. **Improvement.** Create a tissue culture facility or clean room at your school. Research what you can do to successfully accomplish tissue culture without a laminar flow hood.

3. **Experimental.** Use various media to prepare a plant via tissue culture.

4. **Exploratory.** Create a list of native plants in your state that are endangered. Discuss how tissue culture and micropropagation could improve the chances for these plants to survive in the future. Summarize this information in a video documentary, blog, or virtual poster.

5. **Placement.** Find a job (paid or unpaid) at a laboratory facility that works with micropropagation or tissue culture techniques.

KentaStudio/Shutterstock.com

CHAPTER 18
Greenhouse Operation and Maintenance

Chapter Outcomes
After studying this chapter, you will be able to:
- Describe the factors to consider when planning a greenhouse.
- List and describe greenhouse structures.
- Recognize greenhouse components.
- Understand how to maintain greenhouse structures and equipment.
- Identify careers related to greenhouse operations and maintenance.

Words to Know

aspirated thermostat	greenhouse range	shutter
cold frame	high tunnel	static pressure
cool cell	horizontal air fan (HAF)	sump pump
eave	hotbed	sump tank
emitter	louver	thermostat
fogger	manometer	topography
gable	perforated convection tube	vent
glazing	photocell	
greenhouse orientation	ridge	

Before You Read
Before reading this chapter, review the highlighted terms within the body. Determine the meaning of each term.

While studying this chapter, look for the activity icon to:

- **Practice** vocabulary terms with Words to Know activities.
- **Expand** learning with identification activities.
- **Reinforce** what you learn by completing Know and Understand questions.

www.g-wlearning.com/agriculture

Ruud Morijn Photographer/Shutterstock.com; CreativeNature R.Zwerver/Shutterstock.com; oleandra/Shutterstock.com; gtfour/Shutterstock.com; PerseoMedusa/Shutterstock.com

The Roman emperor Tiberius Caesar ailed from an unknown illness. His medical team advised him to eat a cucumber each day as treatment. In 30 BCE, there were no supermarkets, electricity, or running water. To grow a cucumber year-round in Rome's climate seemed impossible. Tiberius rallied his team to create a structure that could grow a cucumber for his daily consumption. These ancient Romans enclosed a stone-walled structure with sheets of mica (a nearly translucent mineral that breaks into thin pieces). This house, known as a specularium, was surrounded by constantly burning fires that provided the additional heat needed for growing cucumbers. This was the birth of the greenhouse.

Today, greenhouses around the world vary from small backyard models to beautiful, grand structures, **Figure 18-1**.

> "The young plants may be defended from cold and boisterous windes, yea, frosts, the cold aire, and hot Sunne, if Glasses made for the only purpose, be set over them, which on such wise bestowed on the beds, yeelded in a manner to Tiberius Caesar, Cucumbers all the year, in which he took great delight."
> —Thomas Hill

Greenhouse Planning

Planning is the most important step in the purchase or construction process. Numerous factors must be considered before construction or purchase of a greenhouse. These factors include site location, crops to be cultivated, climate, greenhouse orientation, operating costs, and market opportunities. These aspects of greenhouse construction, operation, and management all influence the final decision regarding the selection or construction of a greenhouse facility.

Site Location

Selecting the location for a greenhouse is not as simple as finding a piece of land to buy or rent. Significant thought must go into determining where a greenhouse will be erected. Physical characteristics of the site must be considered.

MKDK/Shutterstock.com

Figure 18-1. The royal greenhouses in Laeken (Belgium) were built in the late 1800s. Progress in construction methods and materials allowed builders to create elaborate greenhouses that housed fruit-bearing trees as well as ornamental plants. Do you think that greenhouses are as popular now as they were in earlier centuries?

The features of the land include *topography* (the elevations and slope of a specific land area), drainage, and soil quality. Obstructions on the site and access to utilities and adequate roads must also be considered. These characteristics dictate whether a site can be considered as a possibility for a greenhouse.

Corner Question

Where is the highest mountain peak in North America?

Level Site

Sites should be nearly level to reduce the costs of excavation. Land that has a significant slope must be leveled. This grading requires the use of excavation equipment and labor and adds significant costs to the construction of the greenhouse. Level sites provide convenient access for customers and workers and make it easier to use equipment and machinery, **Figure 18-2**. Several greenhouses or a greenhouse range (connected greenhouses) can be constructed on a level site. This arrangement may help to control construction costs.

T.W. van Urk/Shutterstock.com

Figure 18-2. A large, flat greenhouse floor allows room for the placement of irrigation equipment.

Greenhouses should not be placed on peaks or in valleys. Higher elevations are often windier than lower elevations. Greenhouses placed in low areas may not have proper air circulation. As cold air settles in low areas, greenhouses in low areas may require more heat, costing more money. Low areas also are at risk for standing water or flooding after heavy rains.

Drainage and Soil Quality

Adequate drainage and soil quality are important considerations when choosing a greenhouse site. Greenhouses that are housed on gravel must have soil (below the gravel) that drains. If they do not, water will stand on the greenhouse floors. A site that has a loam or sandy soil will offer better drainage.

Soil quality is extremely important for field-grown nursery operations. If there is any possibility of a greenhouse operation using this growing method, then the soil quality must be determined. A soil pH of five to seven is ideal for growing nursery plants in the ground. Although soil can be amended, avoid soils with extreme pH levels. Soils should have the ability to hold water and nutrients. Good soil structure contributes to better plant growth. In addition, a site with proper soil structure helps support foot and equipment traffic.

Did You Know?

The oldest greenhouse in the United States is on the Lyman Estate in Waltham, Massachusetts. This glass greenhouse was designed by English gardener William Pell. The house was erected in 1793, and a complex of four greenhouses was constructed from 1798 to 1930. Two hundred years later, these greenhouses stand strong and are open to the public.

Obstructions

Greenhouse locations should be free of obstructions. Obstructions include:
- Large stones and outcroppings of rocks. These obstacles can restrict the use of equipment and machinery.
- Trees. The area should not be shaded by trees, especially evergreens. Deciduous trees may provide some beneficial coverage during summer months while leaving the site exposed in winter months.
- Buildings. Structures can shade the area and make it difficult for machinery to maneuver.
- Recreational sites. Thrown objects can damage greenhouses.

> **Corner Question**
>
> How many miles of roads are in the United States?

Although wooded areas should be avoided, a windbreak or nearby hills can be beneficial. A natural tree windbreak or a hill located to the north and northwest of the site help to reduce wind speed in winter months. This deflects cold air from the site.

Nearby Utilities

Electricity, fuel, and water must all be available in a greenhouse. Check on the availability of utilities to the site and determine what types of fuel can be supplied (natural gas or propane, for example). Water must be supplied to this site by a well, a retaining pond, or the city water system. Each of these methods has its own set of advantages and disadvantages. Some growers use multiple water sources. Well water should be analyzed to ensure it has the proper pH, low levels of salts, and average or minimum levels of mineral impurities.

Road Access

The greenhouse site must have ready access to roads. Many growers prefer to build near a major roadway or interstate highway. Some growers prefer access to nearby airports as well. Whether a grower sells directly to the public or to wholesalers, easy access to roads is important.

Crops to Be Cultivated

The crop a grower will produce may determine the location and type of greenhouse that will be constructed. Finding the ideal location for a specific crop helps to alleviate future problems. The economic value of a crop also affects how much money will be used to construct a greenhouse. Low-value crops do not need a state-of-the-art greenhouse. Crops must provide profits to help pay for the greenhouse. Therefore, a greenhouse must be selected to match the value of the crop.

Climate

Climate refers to the general weather pattern for a geographical area. The crop selected for growth should match the climate of the proposed greenhouse location. Factors to consider include the maximum and minimum temperatures, precipitation, and wind. Weather history related to violent storms (hurricane, tornadoes), extreme droughts, or heavy rains should also be considered. Geographical features of the area, such as a large body of water, can also affect climate conditions.

A feature that is often forgotten is the number of hours of daylight in an area. The number of hours of natural sunlight has an enormous impact on plant growth. Alaska has nearly 24 hours of daylight during certain times of the year. At other times, this same location is cloaked in darkness for more than 20 hours a day, **Figure 18-3**.

Cheryl A. Meyer/Shutterstock.com

Figure 18-3. Day length impacts the growth of plant material. These gigantic cabbages were grown in Alaska, where they have nearly 20 hours of daylight each day during the summer months.

Greenhouse Orientation

Greenhouse orientation is the way a greenhouse is positioned on a site. Greenhouse orientation affects the amount of sunlight that enters the structure. It also affects heating and cooling efficiency. Greenhouses should be positioned in a way that maximizes sunlight throughout the year. This orientation will increase plant growth and help create higher profits. General rules regarding greenhouse orientation include:

- Greenhouses above 40°N latitude in the Northern Hemisphere should run east to west.
- Greenhouses below 40°N latitude should run north to south.
- Ridge and furrow greenhouses (described later in the chapter) at all latitudes should run north to south. This orientation balances a shadowing effect that happens from the roof and gutter of each neighboring greenhouse, allowing the shadow to pass over each house throughout the day.
- Head houses (solid structures normally used for transplanting or storage) should be placed at the north end of north-to-south oriented greenhouses.

Greenhouse doors should not open into prevailing winds. Ventilation equipment, such as exhaust fans, should face away from the prevailing wind. Cooling equipment should face the prevailing wind, **Figure 18-4**. All these factors contribute to an energy-efficient greenhouse.

Figure 18-4. Greenhouse growers must consider the direction of prevailing winds when determining the structure's orientation. Ventilation facing prevailing winds will struggle to exhaust air from inside the greenhouse.

Operating Costs

Various factors contribute to the costs of running a greenhouse business. Some of these factors include property taxes, fuel, labor, and transportation. Operating costs vary by greenhouse site. Property taxes are higher near urban areas, fuel costs vary almost hourly, and labor may not be plentiful or dependable in some areas, **Figure 18-5**. The amount of available workers dictates what can be grown and maintained at a greenhouse operation. Labor costs are an important consideration when selecting a location for a greenhouse business. Transportation facilities must be accessible. Growers deal with transportation and delivery logistics every day of the year. Together, these economic factors contribute to the selection of a greenhouse site.

Figure 18-5. Greenhouse labor is needed for a greenhouse operation to survive.

Market Opportunities

A greenhouse located close to its target market improves the opportunity for sales, reduces shipping costs for customers, and protects crops from damage before customer delivery.

Market opportunities are also affected by the number of competitors in the area. Geographic areas with a small population cannot sustain a large number of greenhouses. This point is null, however, for growers intending to sell to a regional or national market. In this case, growers may want to position themselves near other regional or national growers. Growers often cooperate with one another on the use of equipment, labor, and shipping or by providing a needed supply to other growers.

Prospective greenhouse locations must also assess the future market and the opportunity for expansion. Growers cannot truly predict market trends. However, growers can create a plan to direct the operation in reaching the desired size. There must be enough space to accommodate additional greenhouses, buildings, storage sites, parking areas, shipping facilities, and any other facilities that will be needed for the growth of the company.

Greenhouse Structures

Greenhouse structure styles vary with each plant production facility. Each type of structure meets the needs of a specific crop. Types of greenhouse structures include even span, uneven span, lean-to, Quonset, Gothic arch, ridge and furrow, and sawtooth, **Figure 18-6**. In addition to greenhouses, other structures used for growing horticultural plants include cold frames, high tunnels, and hotbeds.

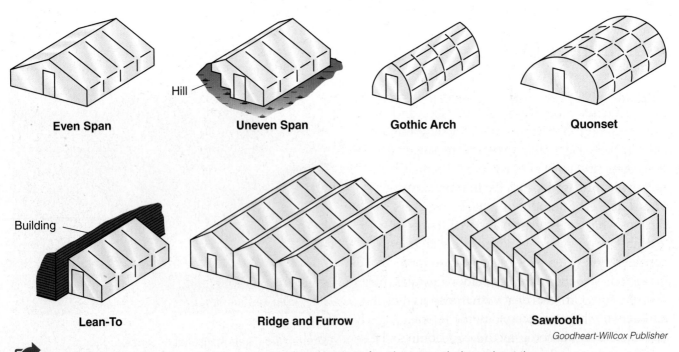

Goodheart-Willcox Publisher

Figure 18-6. Greenhouses provide a protected environment for plant growth throughout the year.

Even Span

When asked what a greenhouse looks like, most people describe an even-span greenhouse. On an even-span greenhouse, the distance on the roof from the eave to the ridge is the same on both sides of the greenhouse. The *eave* is the part of the greenhouse roof that meets the wall of the greenhouse. The *ridge* is highest point of a greenhouse roof. This type of structure allows even light distribution throughout all parts of the greenhouse.

Uneven Span

This greenhouse structure is commonly used on a slope or area with reduced daylight. On hills or slopes, the short side of the roof is on the upper side of the grade. When used in locations that have less daylight, the longer side of the roof faces the direct sun rays. In both of these situations, uneven span greenhouses are a wise choice.

Lean-To

A lean-to greenhouse is one-half of an even-span greenhouse connected to a wall of another existing structure. This type of greenhouse is usually placed against a south wall of the existing building. Shade from the existing building poses a problem. Therefore, the greenhouse should be erected to run east to west to make the most of the available sunlight. In these greenhouses, summer overheating and excess moisture may be problems.

Quonset

The roof of a Quonset greenhouse is a semicircular arch that begins at the ground level. Since there is no *gable* (the part of a wall that encloses the end of a pitched roof), there are no supporting posts or roof pitch. Quonset greenhouses can be difficult to ventilate properly. Many growers use Quonset houses because of the inexpensive construction costs. However, if low-quality or weak materials are used in the construction process, the structure may be damaged by high winds. In areas that receive a great deal of snow, these structures may not be able to hold the load and can collapse.

Gothic Arch

A Gothic arch greenhouse has a similar structure to a Quonset greenhouse, but with a gable down the center of the roof. This structure can withstand greater snow loads than a Quonset greenhouse. This structure allows condensation to run off, reducing the amount of water that would drip onto plants or people inside the greenhouse.

Ridge and Furrow

A ridge and furrow greenhouse encloses a large area under one roof. Several even-span greenhouses are joined at the eaves. There are no inside walls where the eaves join. Gutters are placed on top of posts where the eaves meet. This construction method is used in larger commercial ventures. This is one type of *greenhouse range* (a series of greenhouses that are attached to one another).

> "Who loves a garden loves a greenhouse too."
> —William Cowper

Sawtooth

A sawtooth greenhouse covers a large area and consists of a series of roof spans, sloped in the same direction, with a vertical opening between each roof. The sawtooth greenhouse structure was very popular in the 1970s and 1980s in regions where snow is not an issue. This design is commonly used in tropical regions because of the increased ability to circulate air and the ease of cooling provided by the vertical openings between roof levels.

Other Growing Structures

In addition to greenhouses, cold frames, high tunnels, and hotbeds provide protection or enhanced growing conditions for plants.

Cold Frames

A *cold frame* is a structure used to grow plants that is passively heated (heated by the sun). It is a small structure that has a vent for releasing built-up heat, **Figure 18-7**. Cold frame walls can be made from wood, concrete, straw bales, or other materials. Glass, plastic, or cloth covers the top and allows light to reach the plants. During the day, the cover of the cold frame is opened to prevent overheating. At night, the cover is closed.

Alison Hancock/Shutterstock.com

Figure 18-7. Cold frames can be used to grow young seedlings and bedding plants. What types of plants grow best in a cold frame?

High Tunnels

A *high tunnel* is a square or semicircular structure made of a frame and a covering, such as a plastic film, **Figure 18-8**. It is heated by the sun and is used to extend the growing season for plants. High tunnels are commonly used to cultivate edible crops. These structures offer protection from frost and modify the environment enough to enhance the production of plants. High tunnel structures insulate plant material and also provide protection from heavy winds and rain. High tunnels are used by vegetable and fruit growers around the world.

USDA Natural Resources Conservation Service

Figure 18-8. High tunnels can be used to protect crops from excessive rain, wind, and frost.

Hotbeds

Similar to cold frames, a *hotbed* is a structure consisting of four walls and a glass or plastic top. The main difference between hotbeds and cold frames is that hotbeds include an additional source of heat. Heat is supplied via a heating cable, hot water, steam, or a mass of decomposing organic material placed beneath the plants.

Thinking Green

High Tunnels in Organic Gardening

An organic vegetable farm is not complete without a high tunnel. This growing structure provides the ideal environment for extending growing seasons or providing a closely controlled system for the cultivation of leafy greens, strawberries, and other plants. Crops are planted directly into well-amended soils or into straw bales. High tunnel systems are used to cultivate vine crops such as tomatoes and cucumbers. These crops can be grown vertically and easily supported with a system of clips and string. High tunnels offer protection from extreme temperatures and climates. They help prevent the invasion of pests and pathogens, and allow intensive crop production on a small area of land.

USDA

Greenhouse Components

Greenhouses work because all the parts function together in a system. These components are part of an overall system that is used to heat and cool a structure intended to grow plant material efficiently.

Covering Materials

Several types of materials can be used to cover a greenhouse structure. *Glazing* is the covering material that is placed on the outside of the structure. The glazing forms the weatherproof seal that permits adequate quantities of sunlight into the building for plant growth. Selecting which covering material to use depends on durability, cost, light transmission, rate of heat loss, lifespan of material, and maintenance costs of each material. Four common types of glazing are glass, polyethylene film, polycarbonate fiberglass, and Solexx paneling.

Glass

Glass is the most transparent of the listed materials and allows the most light to reach the plants. Glass allows a great deal of heat loss, however, and is initially expensive. Glass has the longest lifespan of any covering material, but it is fragile and heavy. Adequate support must be built to carry the load.

Polyethylene Film

Polyethylene film is inexpensive, lightweight, and easy to install. Polyethylene is a plastic film that has only moderate durability (lasting usually two to three years) and light transfer. Solexx paneling is a polyethylene material that is durable, flexible, and easily installed. A thin wall of insulation between two outside layers provides additional energy savings. The material is relatively inexpensive and lasts for at least 10 years, **Figure 18-9**.

Courtesy of Solexx and the Greenhouse Catalog

Figure 18-9. Flexible Solexx panels allow for a variety of installations.

Did You Know?

Polyethylene film greenhouses usually consist of two layers. The double-layer film reduces heat requirements by 40%. The inside layer is 4 mil and the outside layer is 6 mil. Between the two layers of polyethylene film is a 2" space. A fan blows air through the layers and creates an insulation barrier between the outside air and the inside air.

Polycarbonate Fiberglass

Polycarbonate fiberglass is a corrugated, multiwall material that is almost indestructible initially. Light transfer is high but diminishes over time. This product usually must be replaced every 10 to 20 years as it becomes brittle from weathering.

Fans

Fans aid in air circulation, ventilation, and temperature regulation of a greenhouse. Fans turn on based on a temperature reading that is determined by the use of a thermostat. A *thermostat* is a device that measures temperature. The thermostat probe is placed at plant level and shielded from direct sunlight (to prevent a misread). Two types of fans used in greenhouses are exhaust fans and circulation fans.

Exhaust Fans

Exhaust fans are large fans that pull air out of the greenhouse and work in conjunction with vents and cooling systems. These devices work to force air out of the greenhouse, removing stagnant air and reducing temperatures and plant disease. These fans help create a cooler environment for plants and for those working in the greenhouse during hot summer months, **Figure 18-10**.

Safety Note

Fan Safety

All fans must have a protective cage surrounding the blades to ensure that nothing (hair, clothing, or body parts) can get entangled. Fan blades move as quickly as the blades of a mower and can cause just as much damage.

Circulation Fans

Circulation fans help create constant temperatures throughout a greenhouse or high tunnel. Mounted fans hang above plant material and force heated air to circulate evenly to all plants. These fans improve plant production, reduce heat stress, eliminate condensation on walls and ceilings, keep greenhouses fresher, and improve working conditions. Several types of circulation fans are available. A common type is a *horizontal air fan (HAF)*. These fans are mounted overhead around the greenhouse to gently guide warm air in a circulating pattern.

Chris Hill/Shutterstock.com

Figure 18-10. The circulating fan near the ridge of this greenhouse helps distribute warm air.

Corner Question

What is a hydrothermal vent?

STEM Connection

Static Pressure

Static pressure is the force exerted by a still liquid or gas, especially water or air. When exhaust fans are running, there is usually a minor air pressure decrease in the building compared to outside air pressure. Using exhaust fans creates negative pressure, drawing air into the building. This method controls air exchange and is efficient and simple.

Proper negative pressure pulls air into the greenhouse at the appropriate speed and direction for mixing with native air. Cold air passes into the greenhouse through vents and is denser than warm air inside. The cold air must be forced in at the proper speed or this air will fall down toward the floor.

vvoe/Shutterstock.com

(This is why drafts feel cooler in poorly ventilated buildings.) Greenhouses with adequate static pressure will have fewer cold spots and no drafts at the level where plants are located.

A *manometer* is a device used to measure static pressure by determining the difference between inside and outside air pressure. This is expressed by inches of water column (wc).

Louvers, Shutters, and Vents

Regardless of the time of year or temperature, greenhouses need fresh air to reduce disease and supply the carbon dioxide needed for optimum plant growth. Greenhouse features that help supply fresh air include louvers, shutters, and vents, **Figure 18-11**.

- *Louvers*—Vertical or horizontal slats placed in a frame that can be angled to allow air and light exchange between the greenhouse and the outside environment.
- *Shutters*—Areas of the greenhouse that can be opened or closed to allow or prevent air exchange between the greenhouse and the outside environment.
- *Vents*—Parts of the greenhouse that can be opened to release built-up heat to the outside environment, reduce condensation, and improve airflow.

A   LYphoto/Shutterstock.com

B   marekuliasz/Shutterstock.com

C   Robert Schneider/Shutterstock.com

Figure 18-11. Greenhouses have several features to provide fresh air. A—Cooling pad louvers help to bring in air that then goes through the cooling cells and evaporates. The exhaust fans help to pull in cooled air and then move that air down the length of the greenhouse. B—Shutters can be opened or closed, depending on the environmental requirements of the greenhouse. C—Greenhouse vents are often located near the ridges of the building. They release air that has been heated inside of the greenhouse due to radiant energy.

STEM Connection

Retractable Roof Greenhouse

Retractable roof greenhouses provide basic protection to crops using stationary or retractable insect screening. They have curtain systems for shading, heat retention, and light blocking. They also have perimeter walls with roll-up curtains. Roofs allow maximum light and infrared radiation during the early morning and late afternoon to warm plants and maximize photosynthesis. Retractable roofs can provide natural sunlight and allow a great deal of air exchange. Retractable roofs are expensive and the investment must be carefully considered.

UMB-O/Shutterstock.com

Evaporative Cooling

Evaporative cooling systems create cool air through the process of evaporation coupled with hot air removal. This commonly used cooling method is considered the most energy-efficient method for a greenhouse.

The process begins in a *sump tank*. This holding container stores several gallons of water. A *sump pump* (a motor that pulls water from one location to another) forces the water through pipes that saturate *cool cells* (cellulose pads or panels) with water, **Figure 18-12**. As the water evaporates into the greenhouse, the water changes from a liquid to a gas. This is called a *phase change*. Energy is lost from the air, resulting in a temperature reduction. To improve the cooling process, an exhaust fan is placed at the opposite end of the greenhouse to pull warm air through the greenhouse. On the other side of the evaporative cooling wall is a cool cell louver, **Figure 18-13**.

Kavee Vivii/Shutterstock.com

Figure 18-12. A cooling cell made of cellulose is dampened with water. The water evaporates off the cells to cause a cooling effect in the surrounding environment.

zhu difeng/Shutterstock.com

Figure 18-13. The brown wall at the far end of the greenhouse, past the plant material, is the evaporative cooling system.

The air is coldest near the evaporative cooling pad and warms from radiant energy as it passes through the greenhouse to the exhaust fans.

Heating Systems

Several types of heating systems are used in greenhouses. These systems include hot water heating, infrared heating, steam heating, and forced hot air heating. The overall goal of the system is to heat a space that quickly loses heat in the most efficient and economical manner. An accurate thermostat reading is essential for efficient heating.

Hot Water Heating

In a hot water heating system, a boiler is used to heat water. The water circulates through heating pipes throughout the greenhouse. Heat is released from pipes that are in the vicinity of the plants. These pipes can be under growing tables or on the floor. This method distributes heat evenly throughout the greenhouse. This method takes a long time to heat a greenhouse and is expensive to install.

Infrared Heating Systems

An infrared heating system uses propane or natural gas to heat a steel radiant tube to temperatures between 500°F and 800°F (260°C to 427°C). A flue discharges heat between 150°F and 200°F (66°C to 93°C). In this system, plants and soil are heated rather than the air as in traditional circulating air systems. These systems are hung in the peaks of the greenhouse, **Figure 18-14.**

jeka84/Shutterstock.com

Figure 18-14. Infrared heating systems became popular in the 1990s, especially in retail settings. Do you recognize this type of heater?

Steam Heating

With steam heating, pressurized steam is forced through pipes. As steam condenses to water, it gives off heat. The condensed water returns to the boiler, is reheated, and the process begins again. Steam can be transported efficiently, and there are fewer pipes needed than for hot water systems. Overheating can be a problem. Maintenance costs are higher, and steam nozzles may clog. There is also a danger of severe burns to greenhouse workers at the site of line breaks or near steam nozzles.

Safety Note

Carbon Monoxide

Unit heaters create a lethal gas by-product known as carbon monoxide. To ensure that this gas does not poison people or plants in the greenhouse, a heating unit must be coupled with a heater ventilation system. This vent is usually a tube that runs to the outside and allows the gas to dissipate into the surrounding environment and not within the greenhouse walls.

Corner Question

What produces 5% of all carbon dioxide (CO_2) created in the world each year?

Thinking Green

Burning Biomass to Fuel Greenhouses

Greenhouses feel the pressures of the fuel crunch. Rising costs of fuel impact profits. Some greenhouses turn to alternative energy sources. Burning wood chips, wood pellets, corn, and other grains have helped some greenhouses become more energy efficient. This green technology is sustainable and much more cost efficient than using energy sources such as propane.

Stocksnapper/Shutterstock.com

Forced Hot Air Heating

In a forced hot air heating system, unit heaters distribute hot air evenly throughout a greenhouse with the help of circulating fans. Perforated tubing also may be used. Tubes made of polyethylene line the length of the greenhouse above the benches. Plastic tubes with small holes (called *perforated convection tubes*) release the air that is blown down the tube. The heaters are often fueled via steam, hot water, propane, or natural gas, **Figure 18-15**.

Jodi Riedel/Goodheart-Willcox Publisher

Figure 18-15. A unit heater is important for heating the greenhouse during colder temperatures.

Floors

Flooring is an important component of a greenhouse. Several choices are available, each with its own advantages and disadvantages.

Standard Concrete

Standard concrete is extremely durable and can withstand heavy loads. This material is best suited for areas with heavy foot traffic or where equipment and machinery will be used. This material does not drain and should not be placed in greenhouse aisles or under benches, **Figure 18-16**.

Porous Concrete

Porous concrete permits drainage, prevents puddling, and prohibits weed growth. Unlike standard concrete, no sand is included in this mix. Porous concrete has only about one-fourth of the strength of standard concrete. It is well-suited for aisles and walkways or areas for equipment not weighing more than 600 pounds.

Corepics VOF/Shutterstock.com

Figure 18-16. This greenhouse floor is made of a nonporous concrete. The floor can be flooded to water the plants.

Gravel

Gravel flooring material cannot be sanitized and so can allow weeds, disease, insects, and pests to travel into the greenhouse, **Figure 18-17**. While gravel is an economical choice initially, costs associated with disease prevention and loss can surpass the savings. In addition, floors made of gravel (or soil) may not adequately drain. This can cause puddles and slippery floors that create safety hazards for workers or customers. An economical compromise is to use porous concrete in aisles and walkways and to use gravel beneath benches.

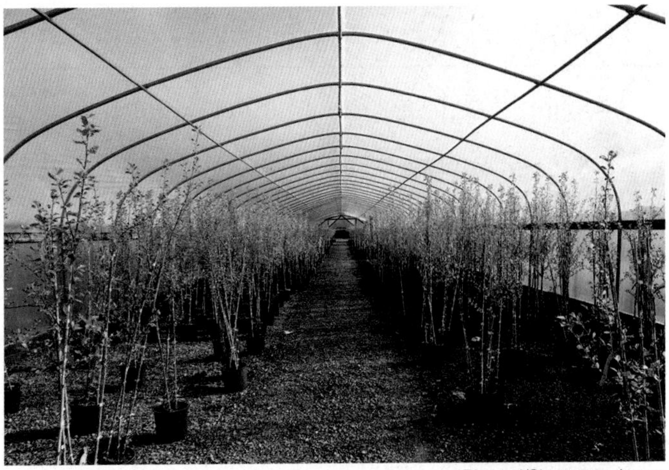

Figure 18-17. Gravel floors can be problematic because they cannot be thoroughly cleaned and are breeding grounds for insects and disease.

Benches

Benches are structures that place plants above the floor in a greenhouse. Considerable thought and care must be invested before selecting a bench material. Wood, concrete, and plastic are materials commonly used for greenhouse benches.

Wood

Hardy wood can be selected from locust, redwood, cedar, and cypress trees. These woods should be treated with a preservative to diminish decay caused by the high humidity of a greenhouse environment. Wood tends to warp and absorbs chemicals, disease, or soils found in the greenhouse. Removing these contaminants can be a difficult task. Wood also requires more maintenance than other materials used for benches.

Concrete

Some growers pour an entire bench structure, legs and all, from concrete. This process creates permanent benches. It is important that drainage be provided within these bench systems. They are easily sterilized and little maintenance is required, **Figure 18-18**.

Figure 18-18. Concrete benches are permanent. Once they are placed, they stay in that location.

Plastic

Plastic benches, often prefabricated, are lightweight and durable. They can be moved easily and can aid in creating a customer-friendly environment in the greenhouse. They are easier to maintain and sanitize than wood benches. However, plastic benches may not be as durable as other bench materials.

Corner Question

How do scientists determine the hardness of a wood?

Thinking Green

Recycling Pallets for Greenhouse Benches

A portable, inexpensive, and earth-friendly bench system is one made of reused pallets. The pallets are placed on top of cement blocks. These benches are very inexpensive and easy to assemble or disassemble. This system also keeps pallets out of the local landfill or from being burned. The pallets will decompose over time so they are not suitable for a retail greenhouse. Wood pallets may also be engineered into unique structures, such as a mobile greenhouse growing system.

Baloncici/Shutterstock.com

BW Folsom/Shutterstock.com

Irrigation

Irrigation applies water to the growing media of greenhouse plants using a number of methods. Two basic methods of greenhouse irrigation are overhead watering and subirrigation.

Overhead Watering

Overhead watering involves wetting the foliage of plants in addition to wetting the media in which the plants are growing. There are several methods of overhead watering:

- Hand watering. Workers use a watering wand and breaker (head of the watering wand that has small holes to emit water) to distribute water over the plant. This system is an inefficient use of labor. Water is often wasted and splashes on leaves, leading to the spread of plant disease.
- Overhead sprinklers. These systems have a tendency to waste water and wet the foliage, which can lead to plant injury or disease. The nozzles from this system often have a circular pattern that overlaps. Plants may receive too much or too little water. Sprinklers and nozzles can provide various sizes of droplets from a fog to a mist to a normal raindrop.

Safety Note

Irrigation

Keep hoses off the ground so that workers and customers do not trip on them. This is also safer for the plants. Fewer diseases will be spread when hoses do not touch plants or the floor. A trolley system may be used to prevent hoses from being contaminated with insects or pathogens living on the greenhouse floors.

gtfour/Shutterstock.com

- Drip irrigation. This system saves water and labor while promoting overall plant health and vigor. Tubes are placed within each plant's growing medium, **Figure 18-19**. Plants are slowly watered via an *emitter*, a device that releases water in an irrigation system.
- Water trays and saucers. Underneath each of the plants is a system of saucers that collects water that did not go into the pot during overhead watering. This reduces runoff and leaching by allowing later water absorption by the plant. These inexpensive products also are reusable.
- Boom irrigation. Pipes fitted with water nozzles move over the tops of plants and water them. This system is also used on seedlings, cuttings, and plugs. An overhead rail that is powered by a motor or a cart moves the watering system above the plant material.

jacglad/Shutterstock.com

Figure 18-19. Drip irrigation is an effective and efficient way of watering evenly. It helps to conserve water and money.

Subirrigation

Subirrigation, also called zero runoff irrigation, delivers water to the roots of plants below the soil surface. Subirrigation is an environmentally responsible irrigation option. Benefits include water and fertilizer conservation (by at least 50% compared to overhead systems); improved plant health; vigorous, uniform plant growth; and reduced labor costs. Subirrigation is done using flood floors, trough systems, ebb and flood systems, or capillary mats.

- Flood floor—a perfectly sloped floor made of nonporous concrete is flooded with a nutrient solution. Berms are installed along the floor to create zones. PVC pipes under the concrete supply water and quickly remove nutrient solution from the area. The nutrient solution returns to a holding tank where it is treated to maintain the proper pH and nutrients and to ensure that it is free of disease-causing pathogens, **Figure 18-20**.

Ruud Morijn Photographer/Shutterstock.com

Figure 18-20. This greenhouse floor is flooded with water. Water submerges all the pots for a few minutes, and then the water drains to be recycled for plant use.

Thinking Green

Subirrigation Reduces Water Consumption

Agriculture is the number one consumer of fresh water in the world. In the United States, agriculture consumption is 80%; it is more than 90% in the western states. Subirrigation systems can significantly reduce consumption through recirculation of water and reduction in evaporation or runoff. Greenhouses also receive an economic return from improved plant growth and quality when using subirrigation systems.

- Trough system—a trough is placed at a slight grade over existing benches (about a 3% slope). Pots are spaced throughout the trough and tubes emit a nutrient solution that passes the base of the pots. Water collects at the low end and flows into a storage tank for recycling.
- Ebb and flood system—completely level trays or benches hold plants. A nutrient solution floods these watertight benches approximately 1″ deep for 10 minutes. A valve is opened and the solution drains through a filter to a holding tank. The solution is used again after it is treated against disease and the pH and nutrients are adjusted.
- Capillary mat—pots are set on a mat material saturated with a nutrient solution. The plant takes up the nutrient solution through the holes in the pot. The mat material is placed on top of a drip tube that supplies the nutrient solution, and a plastic tarp holds the mat and prevents water loss, **Figure 18-21**.

PRILL/Shutterstock.com

Figure 18-21. Capillary mats include a water pipe under a geotextile mat. The plants absorb water through a process called capillary action.

Additional Greenhouse Equipment

Several types of devices and equipment available in commercial greenhouses help growers maintain an optimal growing environment for plants. This equipment includes controls, thermostats, humidistats, photocells, lights, shade cloths, fertilizer injectors, and weather stations. These items may be considered a luxury in hobby greenhouses.

Controls

Many types of control systems are available for use in greenhouses. Some simple control systems automatically turn on equipment, such as heaters, fans, and evaporative cooling fans. These controls work with sensing devices, such as a thermostat, to relay information to a computer program. These systems are programmed to react to the data that is presented. System costs vary widely, depending on the technology. Some systems can be controlled remotely by a smartphone or other devices. This allows plant growing conditions to be monitored and changed if needed.

Thermostats

Thermostats and aspirated thermostats are used to measure temperatures and work with control systems. An *aspirated thermostat* is a device that measures temperature and has a small fan attached to blow the air across the temperature sensor. An aspirated thermostat provides more accurate temperature readings to relay to the control system.

Thinking Green

Energy Efficient Poinsettia Production

Poinsettia production can be extremely costly because these tropical plants require warmer temperatures in November and December. This increase in temperature means higher use of energy resources as well. Cold growing involves using an early flowering cultivar that tolerates colder temperatures. Growers can reduce the temperature of the greenhouse by up to 15°F (–9.4°C) and still grow beautiful poinsettia plants. Several advantages of this method include better shipping capabilities and hardiness of the poinsettia, fewer chemical growth hormone applications, and less frequent irrigation (since plant media will stay saturated longer in cooler temperatures).

Leena Robinson/Shutterstock.com

Humidistats

A humidistat monitors the relative humidity within a greenhouse. Humidity plays a major role in plant health, and excessive humidity can lead to disease. Additional humidity can be supplied with a *fogger*. Foggers atomize water into tiny droplets, smaller than mist. This can help increase humidity and also decrease the greenhouse temperature.

Photocell

Some control systems (especially those with fewer digital parts) may use a photocell to determine the time of day. A *photocell* is a device that reads the amount of light that is present in an area in a measurement known as foot-candles, **Figure 18-22**.

Lights

Many production greenhouses provide supplemental light during periods of low light to extend day length or to break up a night to promote flowering of photoperiodic plants. Supplemental lighting options include:

- Incandescent lighting—common home lights that light up instantly, give off warm light, and are inexpensive but inefficient. They have a short life span of only 750–2500 hours. These lights are not useful for growing quality plant material.
- Compact fluorescent lighting—bulbs last 7000–24,000 hours, depending on bulb type. Bulbs provide a full spectrum of fluorescent light that resembles natural light.
- High-pressure sodium lighting—recommended for supplemental lighting where natural lighting is already present. These lighting fixtures produce a warm white light and have a life span of 16,0000–24,000 hours.
- Light-emitting diode (LED) lighting—popular lights with life spans of 35,000–50,000 hours. This lighting uses 75% less energy than incandescent bulbs.
- Metal halide lighting—recommended for areas with no natural lighting. These light fixtures provide crisp, white light and have a life span of 5000–20,000 hours.

Carlos Yudica/Shutterstock.com

Figure 18-22. A photocell measures the amount of light in a greenhouse in foot-candles.

Corner Question

What is a foot-candle of light?

Thinking Green

Greenhouse Curtains

Energy-saving curtains reduce heating and cooling costs. The curtains (also called screens, thermal blankets, and night curtains) cover the roof and sometimes the sides of the greenhouse to reduce nighttime heat loss and daytime heat gain by forming a thermal barrier. The curtains can reduce nighttime heat loss by 20%–75%, reducing heating costs by 30%–50%. Summer shading (blackout material) energy savings range from 15% to 99%, reducing temperatures by an average of 10°F (–12.2°C) when using a semiporous aluminum material. These savings decrease costs for growers and conserve resources.

mihalec/Shutterstock.com

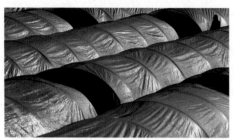
CNRN/Shutterstock.com

Figure 18-23. Shade cloths help stop light from getting into a greenhouse, keeping greenhouses cooler.

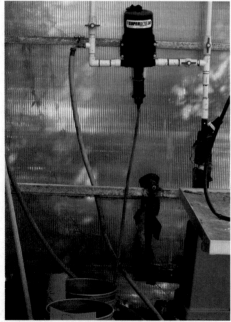
Jodi Riedel/Goodheart-Willcox Publisher

Figure 18-24. A fertilizer injector adds a small amount of concentrated fertilizer to the water that is being supplied to plants.

Shade Cloth

Shade cloth is a material that stops some sunlight from reaching the greenhouse. This material can be placed on top of the greenhouse or inside of the greenhouse, **Figure 18-23**. Some greenhouses have automated shade cloths that cover crops for portions of the day based on temperature or photocell readings relayed to a control system. This reduction in light prevents some photosensitive crops from being damaged by the light and also reduces temperature during periods of intense sunlight and high outdoor temperatures.

Fertilizer Injectors

Fertilizer injectors are machines that inject concentrated water-soluble fertilizers into the irrigation system used in the greenhouse. The water that is released into the hose or irrigation system is a diluted concentration of fertilizer and water. Rates of fertilizer are applied in parts per million (PPM) of nitrogen. The grower must formulate a stock solution according to the type of injector, the ratio of fertilizer injected to water, and the amount of nutrients in the fertilizer being used. Several types of injectors are available for growers and they vary in precision and price, **Figure 18-24**.

Weather Stations

A weather station is a device that measures the humidity, sunlight, precipitation, and temperature. It provides valuable information about outdoor weather conditions and the greenhouse environment. This data is used by control equipment and computer programs.

It helps growers make decisions about how to change a greenhouse's growing environment to promote optimum plant growth. A weather station may be mounted on a rooftop or on a freestanding pole, **Figure 18-25**.

Maintaining Structures and Equipment

To make sure that a well-constructed greenhouse has a long, useful life, greenhouse owners must follow certain maintenance procedures regularly. A greenhouse is a major investment that needs structural and equipment maintenance to keep it in good condition.

Structural Maintenance

Greenhouses are structures that house plants and employees. This area must provide conditions made for plant growth and workers' safety.

Greenhouse Structures

Greenhouses can be made from a variety of materials, including wood, metal, and plastic. Each type of material requires periodic inspection and maintenance.

- Wood—check for rot and insect infestation. Inspect areas for diseases that include algae and fungi growth. Wood must be cleaned and painted or stained to prevent decay.
- Metal—inspect for rust. Remove rust with a metal brush and repaint.
- Plastic—has a very long life span, but can degrade due to photosensitivity. Inspect the material to see if it is sturdy and can support materials. If the plastic material is broken, repair it or dispose of it appropriately.

Check all doors, flooring, supports, and framing to identify needs for structural repairs or replacement.

Greenhouse Covering Materials

Greenhouse coverings or glazing materials may need to be repaired or replaced over the course of the greenhouse life.

- Glass—replace broken panels. To provide shading in summer months, glass is sometimes whitewashed with a semipermanent paint to make the panels opaque.
- Polycarbonate—has a useful life of 10–20 years. As the material ages, repair or replace broken or brittle panels.
- Polyethylene plastic film—has a life span of 2–3 years. This material will become brittle and discolor. Once brittle, it must be replaced.

Igors Jefimovs/Shutterstock.com

Figure 18-25. A weather station can help growers make decisions about how to change the greenhouse environment. **Where else might you see a weather station?**

Corner Question

What color should greenhouse structures and equipment be painted?

Did You Know?

The greenhouse effect happens not only on our planet, but also in the greenhouse. Short wavelengths of light enter the greenhouse and are absorbed. The energy converts into heat energy and is radiated in the form of longer wavelengths that cannot pass through the greenhouse covering material. The radiant energy builds up. As a result, the temperature of the greenhouse increases significantly.

Greenhouse Equipment Maintenance

All the equipment inside the greenhouse works together to make the greenhouse operate and function as an artificial growing space. If just one item does not work properly, this can lead to lost crops and income.

Fans

Exhaust and circulating fan safeguards must be continuously monitored for worker protection and efficiency of environmental controls. Maintain a schedule to monitor protection guards, fan bearings, motors, and fan belts. Fan belt tension must be checked monthly. Replace fan belts when worn, weak, or frayed.

Heating Equipment

Heating units and all heating-related equipment must be maintained. A heating unit that does not work efficiently can harm crops, especially by contaminating the air. Follow these maintenance guidelines:

- Provide adequate levels of oxygen to gas-fired units.
- Ensure heating units are vented to the outside.
- Follow the maintenance manual for heating units.
- Have all servicing of heating units and equipment done by a certified technician.

Cooling Pad Systems

Several items work together to create a cooling effect in the greenhouse. The cooling pad system works with louvers, exhaust fans, and a control system to perform evaporative cooling. Water quality contributes to the effectiveness of a cooling pad system. Cooling pads, if properly maintained, can last for up to 10 years. They are most effective, however, in the first five years of service, **Figure 18-26**. Follow these procedures to maintain the cooling system:

- Clean the sump tank of particulates or impurities that settle at the bottom of the tank.
- Ensure that the water source for all outdoor sump tanks is covered to keep the tanks free of insects, diseases, and debris. Before cold weather arrives, drain the sump pump, the pipes, and the sump tank to avoid damaging the equipment.

nature10/Shutterstock.com

Figure 18-26. These cooling pads are being cleaned using an organic peroxide solution that is effective in controlling bacteria, fungus, algae, and many other pests.

- Check for clogs in the PVC line that delivers water to the cool cells. Make sure that any foreign materials are expelled.
- When the cool cells are dried, gently brush off any algae or dirt.
- During the summer months, algae grows on the pads, reducing their effectiveness. Use an algaecide or a peroxide product to reduce algae growth in the recirculating tank.

Planning Maintenance Procedures

A state-of-the-art greenhouse has many pieces of equipment that are key to successful plant growth. Overlooking any of these maintenance procedures could negatively impact the greenhouse. Follow these guidelines when planning maintenance procedures:

- Read all owner's manuals and determine the maintenance needs of each piece of equipment.
- Create a spreadsheet that outlines what must be maintained on each piece of equipment and how often.
- At least once a week (daily is preferred), walk through the greenhouse and check equipment to determine if any item is working improperly.

> **Safety Note**
>
> **Unplug Equipment Before Servicing**
>
> Before you start to work on any item in the greenhouse, be sure that the power source has been shut off. If you are working on anything electrical in the greenhouse, power down first to ensure your safety and the safety of others.

Greenhouse Structure Careers

The greenhouse construction industry offers a number of rewarding careers. Each of these positions has specific educational requirements and various abilities. The careers highlighted in this section focus on jobs requiring some plant knowledge. Along with horticultural information, each career has a unique skill set that is required of the professional.

Greenhouse Engineer

Greenhouse engineers have an educational background that focuses on engineering (specifically agricultural engineering). A greenhouse engineer focuses on engineering greenhouse structures for commercial, independent, academic, and research facilities. The services provided by greenhouse engineers include planning, consulting, designing, building, and producing feasibility studies for greenhouses. They design greenhouses and the equipment that helps the greenhouse to function. Greenhouse engineers may perform inspections, project management, or evaluations as well. Greenhouse engineers should be critical thinkers, analysts, and problem solvers. They should also have communication skills as they work often with other professionals or team members. This position requires a strong foundation in science, technology, engineering, and math.

> **Did You Know?**
>
> The Netherlands has over 9000 greenhouses that account for 0.25% of its total land use.

Greenhouse Construction Worker

Greenhouse construction workers are part of both the construction and agricultural industries. They use heavy equipment, work with electricity, and deal with fuel sources. They also use ladders and work in elevated areas. Safety is the number one priority while working. Greenhouse construction workers should be physically fit and enjoy working outdoors. This job is also mentally demanding. Individuals must be able read plans created by an engineer and then create a structure from those plans. Workers must also communicate effectively and have strong math skills.

Career Connection

Neil Devaney
Account Executive, Greenhouse Sales

Neil Devaney

As an account executive at Stuppy Greenhouses of North Kansas City, Missouri, Neil Devaney's experience in the greenhouse industry helps him with his diverse customer base. Neil and his team design and build structures that meet their customers' needs. He is responsible for making initial contact with garden centers, schools, commercial growers, and nurseries along the East Coast of the United States. He understands greenhouses and customizes designs and equipment for each grower's unique environment and crops.

Devaney graduated from Clemson University with a degree in agricultural economics. He has more than 21 years of experience in the greenhouse industry. Challenges he faces in the industry include price pressures for raw materials and selling structures that are expensive. He creates lasting relationships with greenhouse growers by ensuring that the products that the customers use are valuable to the grower's mission.

A trend that Devaney has noticed in greenhouse sales has been a change in culture for greenhouse growers. Today's growers are steering toward supplying food. Aquaponics, hydroponics, herbs, fruits, and vegetables are being grown in greenhouses. Growers are filling a need for fresh food. In addition, Devaney has noticed that growers with fewer than three acres in production are consolidating. These growers on smaller operations are often family-run businesses, but younger family members are not always interested in sustaining the companies.

The greenhouse industry has changed significantly over the past two decades. The future of this industry will continue to evolve to serve customers seeking beauty and food from plants.

CHAPTER 18 Review and Assessment

Chapter Summary

- Planning for greenhouse construction or purchase depends on the site location, crops to be cultivated, climate, greenhouse orientation, operating costs, and market opportunities.
- Structures used for growing horticultural plants include greenhouses, cold frames, high tunnels, and hotbeds. Types of greenhouse structures include even span, uneven span, lean-to, Quonset, Gothic arch, ridge and furrow, and sawtooth.
- Greenhouse components contribute to the overall function of the greenhouse. For optimum plant growth, growers use various coverings, fans, louvers, shutters, and vents. Cooling and heating systems, flooring, benches, and irrigation systems are also important parts of the greenhouse system.
- Several types of devices and equipment are available in most commercial greenhouses. These items help growers maintain an optimal growing environment for plants and may include controls, thermostats, humidistats, photocells, lights, shade cloths, fertilizer injectors, and weather stations.
- To make sure that a well-constructed greenhouse has a long, useful life, greenhouse owners must follow proper maintenance procedures regularly. Keeping the greenhouse structure and the related equipment in good repair is important for successful operation of the greenhouse.
- Many careers are related to the design, construction, maintenance, and sales of greenhouses. Greenhouse engineer, greenhouse construction employee, and greenhouse sales account executive are three examples of these careers.

Words to Know

Match the key terms from the chapter to the correct definition.

A. aspirated thermostat
B. cold frame
C. cool cell
D. eave
E. emitter
F. fogger
G. gable
H. glazing
I. greenhouse orientation
J. greenhouse range
K. high tunnel
L. horizontal air fan (HAF)
M. hotbed
N. louver
O. manometer
P. perforated convection tube
Q. photocell
R. ridge
S. shutter
T. static pressure
U. sump pump
V. sump tank
W. thermostat
X. topography
Y. vent

1. The part of the greenhouse roof that meets the wall of the greenhouse.
2. The part of a wall that encloses the end of a pitched roof greenhouse.
3. The highest point of a greenhouse roof.
4. A device used to measure static pressure.
5. A device that measures temperature.
6. A device that measures temperature and has a small fan attached to blow the air across the temperature sensor.
7. A machine that emits tiny water droplets in order to cool a greenhouse or add humidity.
8. A device that releases something, such as water, in an irrigation system.
9. The way a greenhouse is positioned on a site.
10. A structure consisting of four walls and a glass or plastic top that uses a heat source to encourage plant growth.
11. A plastic pipe with holes that distributes heated air in a greenhouse.
12. A motor that pulls water from one location to another.
13. A part of the greenhouse that can be opened to release built-up heat, reduce condensation, and improve airflow.
14. A device that reads the amount of light in an area, measured in foot-candles.
15. A holding container or reservoir for water.
16. A structure made of a frame and a covering, such as plastic film, in a semicircular or square shape that is heated by the sun and used to extend the growing season for plants.
17. The covering material of a greenhouse.
18. A small structure used to grow plants that is passively heated by the sun and has a vent for releasing radiant and built-up heat.
19. A series of greenhouses that are attached to one another.
20. A pad or panel made of cellulose that is used in an evaporative cooling system for a greenhouse.

21. A fan that is suspended above the growing environment of a greenhouse and circulates air.
22. The force exerted by a still liquid or gas.
23. An area of the greenhouse that can be opened or closed to allow or prevent air exchange between the greenhouse and the outside environment.
24. An angled slat that allows air and light exchange between the greenhouse and the outside environment.
25. The elevation and slope of a specific land area.

Know and Understand 🔗

Answer the following questions using the information provided in this chapter.

1. List factors to consider when planning for a greenhouse.
2. What are some characteristics that should be considered when selecting a greenhouse location?
3. Why is greenhouse orientation important? Which way should a greenhouse be oriented?
4. What are some factors that contribute to the costs of running a greenhouse business?
5. List seven types of greenhouse structures.
6. What are four types of growing structures used for horticultural plants?
7. What is the main difference between hotbeds and cold frames?
8. Three types of greenhouse glazes include glass, polyethylene film, and polycarbonate fiberglass. List the advantages and disadvantages of each.
9. Describe the function of an exhaust fan in a greenhouse.
10. Describe the function of circulation fans in a greenhouse.
11. What are three greenhouse features that are used to help supply fresh air in the greenhouse?
12. Explain how an evaporative cooling system works in the greenhouse.
13. What are some types of heating systems used in greenhouses?
14. What procedures are followed to ensure that carbon monoxide gas from unit heaters does not poison people or plants in the greenhouse?
15. How do standard concrete and porous concrete differ?
16. Explain the function of greenhouse benches and list three materials commonly used to build them.
17. What are five methods of overhead watering for plants in the greenhouse?
18. What are some benefits of using subirrigation for plants in a greenhouse?
19. What are some types of devices and equipment found in commercial greenhouses that help growers maintain an optimal growing environment for plants?
20. What can greenhouse owners do to help ensure that a well-constructed greenhouse has a long, useful life?
21. What education is needed to prepare for a career as a greenhouse engineer? What are some duties or activities of this job?

Thinking Critically

1. During your summer break, your teacher asked you to take care of the school's plants while she is on vacation for a week. Just after the teacher leaves, you learn that the person who promised to drive you to school every day is sick and will not be able to provide transportation. What do you do to remedy this situation?
2. Your greenhouse has overhead irrigation and hoses for hand watering. Hand watering is so time-consuming that a student would have to miss class to water the greenhouse plants. The overhead sprinklers could be used, but you are not sure how long the sprinklers will need to run to ensure that everything is saturated. How would you determine how long to irrigate the plants with the overhead sprinklers?

STEM and Academic Activities

1. **Engineering.** Design your school's greenhouse to scale using one of the following: a ruler, an architect or engineer's scale, or an online or virtual design program.
2. **Math.** Determine the area in square feet and volume in cubic feet of your school's greenhouse. Next, determine the total exposed surface area. If your school does not have a greenhouse, determine the area and volume of your classroom.
3. **Math.** Determine how much concrete or gravel would be needed to cover the floors of your greenhouse or classroom to a depth of six inches.
4. **Social Science.** Write an editorial for your school or community newspaper that outlines how your school's greenhouse uses energy. Describe what could be done to improve your school's growing environment and make it more sustainable.
5. **Language Arts.** Visit a local garden center or greenhouse and make a list of its equipment. Talk to the employees. Ask them to explain any equipment you are not familiar with.

Communicating about Horticulture

1. **Reading and Speaking.** Select a historical era that interests you. Using at least three resources, research the history of greenhouses in that era. Include topics such as structural advances, environmental controls, and technological advances. Using the information gathered through your research, write a report. Present your report to the class using a presentation graphics package.
2. **Reading and Speaking.** Water rights trigger heated debates among people from different parts of the country. Identify a water-based agricultural issue either in your state or your region of the country. Research that issue and follow the National FFA Organization's career development event guidelines for the Agricultural Issues Forum to prepare an analysis and presentation of the issue for local civic groups.

SAE Opportunities

1. **Exploratory.** Job shadow a greenhouse sales employee.
2. **Exploratory.** Create a design for a greenhouse and determine how much the greenhouse would cost to build.
3. **Experimental.** Create a series of small hoop houses made of different transparent materials, such as polyethylene plastic, polycarbonate, and another product of your choice. Put the same type of plant under each type of glazing and compare the growth of the plants under each material.
4. **Improvement.** Work with your instructor to develop a plan to improve your greenhouse in one way. Find the funding for this project through grants or fund-raising efforts and make this project become a reality.
5. **Entrepreneurship.** Create a service for cleaning greenhouses and sanitizing them after each planting season.

Alison Hancock/Shutterstock.com

CHAPTER 19
Greenhouse Production

Chapter Outcomes

After studying this chapter, you will be able to:
- Identify the environmental requirements for plants grown in greenhouses.
- Describe the crop inputs for greenhouse grown crops.
- Recognize plant materials used in greenhouse production.
- Describe types of greenhouse crops.
- Identify greenhouse grown plant material.
- List careers related to the greenhouse industry.

Words to Know

bedding plant	cyclic photoperiodic lighting	liner
bio-stimulant	evergreen	night interruption (NI)
container capacity	flagging	plant growth regulator
critical day length (CDL)	incomplete fertilizer	(PGR)
critical night interval (CNI)	insoluble fertilizer	

Before You Read

Arrange a study session to read the chapter with a classmate. After you read each section independently, stop and tell each other what you think the main points are in the section. Continue with each section until you finish the chapter.

While studying this chapter, look for the activity icon to:
- **Practice** vocabulary terms with Words to Know activities.
- **Expand** learning with identification activities.
- **Reinforce** what you learn by completing Know and Understand questions.

www.g-wlearning.com/agriculture

A grower dictates what crops are grown in a greenhouse and how the greenhouse is maintained. As outlined in Chapter 18, *Greenhouse Operation and Maintenance*, many factors guide growers in their decisions regarding greenhouse culture. Each plant crop produced in a greenhouse requires a unique growing environment. No two crops are the same, and no two seasons are replicated. Every year, new inputs and variables play a role in the crop's success.

Growers cultivate herbs, vegetables, microgreens (edible seedlings), cut flowers, and medicinal plants. Greenhouse growers manage the greenhouse environment to meet the needs of the crop and to achieve optimal growing conditions. Optimum growing conditions mean healthier plants, and healthier plants mean higher profits and business success.

Environmental Requirements

Whether a greenhouse grower is producing a plant for flowers, foliage, or food, he or she must provide all of the same elements necessary for plant growth in the outside world. These elements include a container or other place to grow, light, carbon dioxide, nutrients, water, pest control, and appropriate temperature. The environment for plant growth must be provided in a greenhouse space or a plant will not be able to perform its essential life functions, such as photosynthesis and respiration.

Light

Several methods of delivering artificial light were outlined in Chapter 18, *Greenhouse Operation and Maintenance*. These sources of light include incandescent lighting, compact fluorescent lighting, high pressure sodium lighting, LED lighting, and metal halide lighting. Each of these sources provides additional light to increase intensity (amount of light) or duration (length of light). Growers can use their choice of light and timing to supplement natural daylight and achieve desired growing conditions, **Figure 19-1**.

Supplemental Lighting

Supplemental lighting involves increasing the intensity of light that reaches plants. This is especially important in northern climates in the winter, or for those climates with many overcast days. The daily light integral (DLI) is the total quantity of photosynthetic light received by the plant. If the DLI is too low, then additional lighting must be applied to promote uniform and quality plant growth. A high light intensity of 400 foot-candles to 600 foot-candles, often provided by high pressure sodium or LED lights, can be used to achieve the necessary DLI required by a specific crop.

A.Krotov/Shutterstock.com

Figure 19-1. A greenhouse can provide supplemental light or light that will modify a plant's photoperiod.

> ## STEM Connection | Moles of Light
>
> Today's growers often measure DLI in $mol \cdot m^{-2} \cdot d^{-1}$, which means the number of moles of light (mol) per square meter (m^{-2}) per day (d^{-1}). What is a mole of light? A mole is a very large constant number (6.022×10^{23}, which equals 602,200,000,000,000,000,000,000). What DLI is needed to grow high-quality plants? The amount of light needed depends on the crop. However, an average target minimum DLI inside a greenhouse is $10-12 \; mol \cdot m^{-2} \cdot d^{-1}$. Plant quality generally improves as the average DLI increases. As DLI increases, the rooting, branching, flower number, and stem thickness increases; sometimes plant height decreases.

Photoperiodic Lighting

Photoperiodic lighting involves creating a long- or short-day effect to impact plant growth or flowering. The length of uninterrupted darkness is known as the photoperiod. A plant's response to the number of uninterrupted hours of darkness is called *photoperiodism*, and this is what induces flowering. Each plant has its own **critical day length** (the number of hours of light needed to initiate flowering in a plant). This concept is also expressed as **critical night interval (CNI)** because plants respond to dark, and not to light, for flowering. CNI is the length of dark period needed to initiate flowering in a plant. Plants are referred to as short-day, long-day, or day-neutral plants. A long-day plant is also known as obligate or facultative.

- Short-day plants—require short days and long nights to flower. Examples of short-day plants include poinsettias, chrysanthemums, and holiday cacti, **Figure 19-2**.
- Long-day plants—require long days and short nights (or interrupted long nights) to flower. Examples of long-day plants include coneflowers, spinach, and dill.
- Day-neutral plants—will flower regardless of day length. Examples of day-neutral plants include impatiens, geraniums, tomatoes, African violets, and roses.

A *Neirfy/Shutterstock.com*

B *Muskoka Stock Photos/Shutterstock.com*

C *joloei/Shutterstock.com*

Figure 19-2. A—A poinsettia is a short-day flower that blooms with short days and long nights. B—Mums must have short days to bloom. In the summer, a black cloth can be placed over the plants for 13 hours to create a short-day effect and induce flowering. C—Holiday cacti, such as this Thanksgiving cactus, often do not rebloom in people's homes because they are given additional lighting at night. This makes the plant react as if it is a long day and short night, inhibiting flower formation.

History Connection

Percy Julian
Pioneer Research Chemist

Kentaro Foto/Shutterstock.com

Percy Julian was an American chemist who researched the synthesis of medicinal drugs from plants. He pioneered the synthesis of physostigmine, which comes from a bean plant. As an innovator in the industrial large-scale chemical synthesis of the human hormones testosterone and progesterone, Julian synthesized hormones from plant sterols. His work laid the foundation for the production of drugs such as cortisone. Later, Julian founded a company to manufacture steroids from the wild Mexican yam. His work greatly reduced the cost of steroids, making them available to all people in need of these medicines. With more than 130 chemical patents to his name, he was one the first African Americans to earn a doctorate degree in chemistry.

There are a number of ways growers can artificially create the needed photoperiod for plants regardless of what is occurring in the natural outdoor environment.

- Long days—achieved by adding hours of light to extend the day length with artificial light. Another way to simulate long days is by night interruption (NI). *Night interruption (NI)* is a method of creating a long-day effect for plants by disrupting the dark period with light. NI is achieved by turning on lights for 10 minutes and off for 20 minutes between 10:00 pm and 2:00 am or just leaving the lights on for the entire four-hour duration. This intermittent lighting method is also called *cyclic photoperiodic lighting.* This method does not require high-intensity lighting and can be achieved with as little as 10 foot-candles or an incandescent bulb.
- Short days—accomplished by covering the plants from light with a black cloth from 4:00 pm until 8:00 am, **Figure 19-3**.

Using proper lighting is very important for greenhouse growers with photoperiodic plants. Some growers have had poinsettias not flower due to the light supplied at night from a soda machine or a nearby parking lot lamp. These unintentional interruptions of the dark period for plants can lead to a great loss in profits.

Air

The air surrounding a plant can be controlled to promote better quality. Air contains pollutants in a greenhouse just as it does outdoors. A faulty heater can release poisonous carbon monoxide that can be harmful to people and plants.

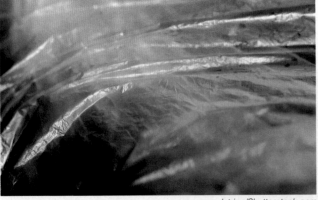
roundstripe/Shutterstock.com

Figure 19-3. Shade cloth, which is a black cloth or plastic, should be used to cover plants to create a short-day and long-night effect. Just after dusk, the cloth should be placed. It should be removed the following morning before heat is trapped. A period of night interruption can create a different photoperiodic effect of long days and short nights.

Sulfur dioxide is released from the burning of fossil fuels and can be present due to the type of energy source used to heat a greenhouse.

Other gases that are already present in the atmosphere may not be present in quantities that will limit plant growth. Carbon dioxide (CO_2) is 0.03% of air by volume naturally. CO_2 is necessary for photosynthesis.

In a greenhouse where ventilation may be limited at some points of the year, there may not be enough gaseous exchange with the native atmosphere. Nighttime is especially problematic as greenhouse vents usually are closed. Plants are no longer photosynthesizing, but they are still respiring.

Carbon dioxide (CO_2) can be supplemented to a greenhouse environment. Providing supplemental CO_2 is also called carbon dioxide fertilization. CO_2 can be supplemented to a greenhouse through the burning of natural gas or propane. Large greenhouses may also use liquid CO_2 available in a pressurized tank or solid CO_2 found as a dry ice that is put into a cylinder and distributed. All methods of supplementing CO_2 require the proper equipment, distribution, and management of this gas. Excessive levels of carbon dioxide can be just as harmful as too little.

Did You Know?
Pure carbon dioxide is heavier than air. When adding this gas to a greenhouse, it must be distributed with circulating fans or other methods of delivery.

Nutrients

Greenhouse growers achieve plant growth with the help of supplemented nutrients. A plant in a container cannot grow or flourish with water alone. There are 17 essential nutrients that must be available to plants, **Figure 19-4**. Growers use several nutrient delivery methods and nutrient formulas to encourage optimum plant growth.

Complete and Incomplete Fertilizers

The three essential mineral nutrients that a plant requires in the greatest amount are called primary macronutrients. These nutrients include nitrogen (N), phosphorus (P, supplied in the form of phosphate), and potassium (K, supplied in the form of potash). Nitrogen is an important part of amino acids, which make proteins, so it is needed for every protein and enzyme a plant produces. It is also part of the chlorophyll molecule. Nitrogen is responsible for the green growth of a plant while phosphorus promotes the growth of roots, flowers, fruits, and seeds. It is essential for protein synthesis, the formation of membranes, and is part of ATP, the energy used by plants.

17 Essential Plant Nutrients	
Element	**Chemical Forms Absorbed**
Hydrogen (H)	H_2O (water)
Carbon (C)	CO_2 (carbon dioxide)
Oxygen (O)	O_2, H_2O (oxygen, water)
Macronutrients (Primary Nutrients)	
Nitrogen (N)	NO_3^-, NH_4^+ (nitrate, ammonium)
Potassium (K)	K^+ (potassium ion)
Phosphorus (P)	$H_2PO_4^-$, HPO_4^{2-}, PO_4^{3-} (orthophosphate, monohydrogen phosphate, phosphate ion)
Macronutrients (Secondary Nutrients)	
Calcium (Ca)	Ca^{2+} (calcium ion)
Magnesium (Mg)	Mg^{2+} (magnesium ion)
Sulfur (S)	SO_4^{2-} (sulfate)
Micronutrients	
Boron (B)	H_3BO_3 (boric acid)
Chlorine (Cl)	Cl^- (chloride)
Copper (Cu)	Cu^+, Cu^{2+} (cuprous ion, cupric ion)
Iron (Fe)	Fe^{3+} (ferrous ion)
Manganese (Mn)	Mn^{2+} (manganous ion)
Molybdenum (Mo)	MoO_4^{2-} (molybdate ion)
Nickel (Ni)	Ni^{2+} (nickelous ion)
Zinc (Zn)	Zn^{2+} (zinc ion)

Goodheart-Willcox Publisher

Figure 19-4. These 17 essential elements must be available to plants.

Corner Question

There are 17 essential nutrients, but one more element is now being added to some fertilizers. What is this element? How might it aid or hinder plant growth?

Potassium helps increase overall hardiness and vigor of the plant. It is essential for the synthesis of sugars, starch, and proteins, and is important for the growth of meristematic tissues.

Fertilizer labels always highlight the analysis or percentages of nitrogen, phosphorus, and potassium in every container near the fertilizer name. This is known as the guaranteed N-P-K analysis and is written like this: 10-10-10 or 12-2-12. From these numbers, growers calculate how much of this specific fertilizer to apply to plants based on the results of a soil test. A complete fertilizer has all three primary macronutrients (nitrogen, phosphorus, and potassium). An *incomplete fertilizer* is one that lacks one or more of these three primary macronutrients.

Why would a grower ever apply an incomplete fertilizer? A plant may be lacking an essential nutrient or may not need one of these for the type of growth that the grower desires. Ornamental grasses or foliage plants are cultivated only for their leaves. Thus, only a small amount of phosphorus is needed for root growth. Adding phosphorus to a mature turfgrass lawn only helps with the germination of weed seeds, as turfgrass roots have no problem accessing phosphorus in the soil (assuming the soil is not deficient in phosphorus).

Growers often rotate fertilizers to achieve their expected results. A grower may use a complete fertilizer, such as 20-10-20, for a period of time and then rotate to a different fertilizer, such as 0-15-0. This will apply various nutrients in different concentrations and may also help regulate pH of the growing medium.

Did You Know?

Scouring rushes (horsetails) have so much silicon that pioneers used them to scrub pots and pans.

Soluble and Insoluble Fertilizers

Fertilizers come in two different forms. Soluble fertilizers can be dissolved in water and delivered through fertigation (adding fertilizer to irrigation water), **Figure 19-5**. Soluble fertilizers reach the plants in specific concentrations via fertilizer injectors and irrigation systems.

Insoluble fertilizers are those that do not readily mix with or dissolve in water. Slow-release fertilizers (SRF) and controlled-release fertilizers (CRF) allow growers to mix a fertilizer into the growing medium at the time of planting and help in reducing fertilizer runoff. These insoluble products slowly release nutrients to plants over time. Growers usually do not rely on SRF or CRF products alone. They supplement these products with soluble fertilizers.

mihalec/Shutterstock.com

Figure 19-5. A fertigation system in a greenhouse stores various nutritional solutions (fertilizers) that are injected into irrigation lines.

SRFs are less predictable and release their nutrients based on multiple factors. SRFs have one or more layers of sulfur, wax, urea, or other material on the outside of the fertilizer. As water comes in contact with the material, it seeps through the coating's imperfections or pores and releases fertilizer. This process is impacted by the temperature and pH of the growing medium, microbial activity, and other factors.

CRFs are water-soluble fertilizers protected by a membrane (made of plastic, polymer, or a resin base) that limits the solubility of fertilizer, **Figure 19-6**. The amount of nutrients that are released depends only on the soil temperature and the coating that is used. Higher media temperatures cause the products to release more quickly, and the product may need to be applied more than once.

Organic and Inorganic Fertilizers

Another decision a grower must make in fertilizer selection is whether to grow organically, which means growing plants without synthetic fertilizers, genetically modified organisms, or antibiotics. Organic and sustainable growers use organic fertilizer because of the belief that they are less harmful to the environment and produce quality plants. Changing to an organic fertilizer program is not a simple switch. Several changes must be made to growing protocols. Growers who have short-term crops, such as vegetable seedlings and transplants, have a much easier time changing to organic fertilizers. Factors to consider when growing organically are:

- Organic fertilizers can be higher in salt concentrations.
- Organic granular fertilizers, such as compost, can be inconsistent.
- Nutrient availability in organic fertilizer solutions is temperature dependent.

Today, several fertilizer companies offer organic fertilizers that are complete or incomplete and soluble or insoluble. Many of these products may be more expensive for growers, but the perceived benefits to the environment may make them worth the extra expense.

A *Bildagentur Zoonar GmbH/Shutterstock.com* B *Photographee.eu/Shutterstock.com*

Figure 19-6. A—A polymer resin is used to create a semipermeable membrane around controlled- or slow-release fertilizers. These resin-coated fertilizers are a tool recognized in some career development events. B—A sulfur-coated or urea-coated fertilizer is a slow-release type of fertilizer.

Temperature

A greenhouse grower controls temperature through a series of equipment and systems. These tools include:
- Ventilation systems.
- Evaporative cooling systems.
- Heating units.
- Foggers.
- Shade cloths.
- Energy-saving curtains, **Figure 19-7**.

Growers operate these tools with control systems that range from simple dials to computers.

mihalec/Shutterstock.com

Figure 19-7. An energy-saving blanket or curtain is used to conserve energy and maintain temperatures in the greenhouse.

DIF

DIF is a term that refers to the difference between day and night temperatures. (Subtract the night temperature from the day temperature.) A +DIF occurs when the daytime temperature is warmer, and a –DIF occurs when the nighttime temperature is warmer. A +DIF results in normal internode (stem growth between nodes) elongation, but a –DIF produces shorter internodes.

Growers can create a –DIF by cooling the greenhouse at dawn for two to three hours. An optimal –DIF is –10°F (–23.3°C). This treatment has led to growers using fewer chemical plant growth regulators (PGRs). *Plant growth regulators (PGRs)* are hormones that speed or slow growth of plants. Many plants in greenhouse production respond to –DIF; however, not all plants react with shorter internodal growth, **Figure 19-8**.

Ailisa/Shutterstock.com

Figure 19-8. The tomato plant on the left has a –DIF. –DIF leads to shorter and less leggy plants.

Heat and Cold Stress

Most greenhouse plants flourish in temperature ranges between 50°F and 77°F (10°C and 21.1°C). Above and below this temperature range, plants become distressed and physiological changes occur. Heat stress causes reductions in growth due to damage of photosynthetic machinery in the plant. Symptoms of this include chlorosis; bud, flower, or fruit abortion; and lack of growth, **Figure 19-9A**.

Cold stress causes plants to slow down growth. Plant roots grow poorly, respiration slows, plants become nutrient deficient, and plants look stunted, **Figure 19-9B**. Plants may have sunken or water-soaked lesions on leaves from cold damage. Wet soil and cool temperatures also can lead to disease problems for plants. Growers should attempt to prevent both heat and cold temperature stresses in plants.

Water

Plants in a greenhouse require watering. Freshwater is a finite natural resource and conservative use can alleviate water shortages while saving money at the same time. Subirrigation and drip irrigation methods help direct water to exactly where it is needed, which reduces waste. Water can also be reclaimed, treated, and used again.

A   *Alena Brozova/Shutterstock.com* B   *Teresa Prendusi/USDA*

Figure 19-9. A—A plant that was exposed to high temperatures will have yellowing or chlorosis. B—These geranium leaves were in a greenhouse where the heaters failed and the leaves were damaged by low or freezing temperatures.

Thinking Green

Rainwater Catchment

A greenhouse has a large roof area. When it rains, the water from the roof can be collected and used for irrigation. Water travels down the roof to gutters and is collected in a reservoir or tank, often called a *cistern*. A pump is then used to deliver water to irrigation systems in the greenhouse.

woodygraphs/Shutterstock.com

Water Quality

Water quality is a limiting factor that growers around the country struggle with every day. Water that has high or low pH can play a major role in common nutrient deficiencies, since nutrients are only available to the plant within narrow pH ranges. Growers should learn the pH of their water and find fertilizers that will counteract any pH issues. Consistent water testing (including pH levels) is key to a greenhouse nutrition program.

Water may also have suspended solids or pollutants that can contribute to greenhouse problems. Solids include organic matter, sand, and algae. Filters should be used to ensure these solids do not clog valves, nozzles, and emitters in irrigation systems. Pollutants can include excessive amounts of nutrients. Some water may have too much chloride, sodium, or other elements.

Solids and pollutants can be removed or treated using water treatment methods. Water treatment includes deionization and reverse osmosis, **Figure 19-10**. These methods are very expensive. Other methods of water treatment can involve blending reclaimed, collected, or nonproblem water with the current water source.

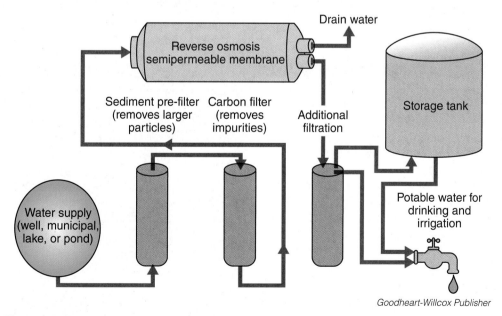

Goodheart-Willcox Publisher

Figure 19-10. Reverse osmosis is a high-priced method for purifying water. Although it works well, it is often cost prohibitive. **Why is reverse osmosis expensive?**

The Watering Decision

Growers make several critical decisions every day. One of these decisions is when and how much to water. Underwatering can be from not watering often enough or not saturating the entire growing medium. Underwatering results in shorter internodes, brittle stems, flower loss, leaf drop, burnt leaf margins, and root loss in plants, **Figure 19-11A**.

Overwatering has its own set of problems. Overwatering is usually accomplished by watering too often. Quality rooting media have significant drainage. A single time of excess watering does not usually hurt plants. However, long-term overwatering leads to wilting, nutrient deficiencies, stunting, disease, and root death in plants. Plants respire and need oxygen for this necessary life function. Most plants that are sitting in water will eventually die, **Figure 19-11B**.

Remember these guidelines for successful watering:

- Use a well-drained growing medium. The medium should hold water and also allow adequate aeration.
- Water thoroughly every time. Make sure that the medium is adequately saturated.
- Water when the plant is flagging. *Flagging* is the condition of a plant when it is just about to wilt. The plant tips will begin to point downward and foliage will have a gray or lackluster appearance.
- Water in the morning. Morning is the best time to water because plants need water for the heat of the day, and the water should be available to them early.
- Do not water a greenhouse in the evening. Water will not evaporate as quickly as during the day, and sitting water can lead to disease and insect problems.

A *Axente Vlad/Shutterstock.com*

B *Jodi Riedel/Goodheart-Willcox Publisher*

Figure 19-11. A—These pepper plants have not had enough water, and both leaves and the fruit show symptoms of inadequate watering. B—A poinsettia that had been overwatered for several days actually shows signs of wilting. Wilting is not always associated with underwatering or disease.

STEM Connection

Automated Irrigation Sensors

Growing medium (also called *substrate*) can be monitored by sensors that determine when a plant needs to be irrigated. This control takes the guesswork out of when to water and saves energy and money. The latest sensors gauge substrate moisture, electrical conductivity, and temperature. These sensors allow growers the possibility of incorporating irrigation and fertigation measurement with automation. Water sensors primarily are coupled with drip irrigation systems or capillary mats.

USDA Forest Service

Pest Control

Pest control is covered in great detail in later chapters. Greenhouse growers constantly battle insects, disease, mollusks (such as slugs), animals (such as rodents), and weeds. These factors pressure growers and plants. A grower must implement integrated pest management (IPM). This is a series of steps that includes scouting, identification of pests, cultural controls, natural controls, and chemical controls. A grower will have a threshold of tolerance and must be very vigilant to ensure that a pest does not overpopulate and negatively impact plant health.

> "Plant diseases are shifty enemies."
> —E. C. Stakman

Crop Inputs

An input is something that is taken in or used by a process. Several inputs to produce crops are at work in a greenhouse. Inputs discussed here include growing media, plant growth regulators, containers, trays, tags, and labels.

Media

Growers need growing media, also called *substrate*, to act as a balance and support for plant roots. The growing medium should also hold water, allow air circulation, and operate as a nutrient reservoir. Some characteristics of growing media that growers should consider include:

- Texture, impacted by the components of the media, may include peat, perlite, vermiculite, bark, coconut coir, and wood products.
- The soil pore spaces (voids in the media, or all of the space not made up of a solid material) are important for healthy root growth, as well as water and air infiltration.
- *Container capacity* is the maximum amount of water that can be held by a growing medium against the pull of gravity. (This is similar to field capacity for garden soil.)
- The bulk density is the weight of the medium relative to volume. Various components of a medium have different bulk densities and impact the behavior of the medium, **Figure 19-12**.

Greenhouse growers purchase or make their media based on the plant's needs and what works for their growing system. Growers can modify or purchase media with additives as well. Fertilizers and bio-stimulants are two common media additives. A *bio-stimulant* is a microorganism that fixes nitrogen into a form plants can use and creates a healthier environment for root growth, boosting plant growth. For example, using mycorrhizae (fungus that grows with the roots of a plant in a symbiotic relationship) is a natural way to provide nitrogen in a form plants can consume.

> **Did You Know?**
> A good grower takes plants out of containers regularly to inspect the roots. Healthy roots make a healthy plant. Most plants exhibit a root that should touch the edges of the pot and have white, fuzzy hairs growing behind the root tips.

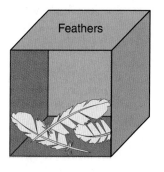 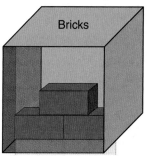

Goodheart-Willcox Publisher

Figure 19-12. Shown here are two containers of equal size. If one box is full of feathers and one box is full of bricks, the box with the bricks has greater bulk density, measured in pounds per cubic foot (lb/ft^3). Greenhouse media typically have a bulk density of 10 lb/ft^3–30 lb/ft^3. Sand has a greater density than other media, such as peat moss and vermiculite.

Hands-On Horticulture

Greenhouse Medium Guidelines

Growers using greenhouse medium can follow some simple guidelines to make a positive difference in their cultivation practices.
- Use the growing medium shortly after receiving it.
- If you switch to a different medium, be prepared to water differently.
- Use medium with the proper pH for the plant. Peat moss is the main component of most commercial medium mixes and is acidic with a pH of 3.5–4.5.
- Use the right kind of medium for your crop.
- Allow the medium to dry between waterings to prevent algae and fungal growth, as well as insect infestations.
- If the growing medium does not dry out, switch products.
- Once you have crops growing, occasionally test the growing medium for pH and electric conductivity (EC), also known as measured salt levels.

Did You Know?

According to the US Census of Agriculture, there are currently more than 1.1 billion square feet of commercial greenhouses or other protective structures in the United States where plants are grown. That is about 50 square miles or more than 25,000 acres of sheltered areas used for plant production.

Plant Growth Regulators

Plant growth regulators (PGRs) increase or inhibit plant growth. Hormones that are in the plants, just like those in other living organisms, control the type of growth a plant experiences. Most PGRs either slow or hasten growing. However, some PGRs will break dormancy or increase growth of organs, such as roots or leaves. Using a product that inhibits growth can make a shorter, bushier plant by shortening internodes. Several active ingredients in products control the plant processes and can be included in various concentrations. PGRs are available to growers as powders and liquids and have several modes of application.

Benefits of Using PGRs

There are several benefits to using plant growth regulators (PGRs). After application of PGRs, plants become darker green, use less water, resist fungi, and are overall healthier plants. Using a PGR can also help in integrated pest management (IPM).

Common Mistakes of PGR Applications

Growers who are unfamiliar with PGR applications may be somewhat timid about using them. Some experiences involve applications of PGRs where plants never grew again. This can happen if a grower applies too much of a PGR. Correct math is important in creating the correct formulation. Growers must be sure to put the decimal point in the right place and have a system to recheck their math. Some PGR errors are correctable. If you have a problem related to PGRs, talk to a PGR expert. Learn from other growers' experiences and record your findings. Even though the exact scenario will not be repeated, and plant cultivation changes based on environmental factors, the information may be helpful in the future.

Containers, Trays, Tags, and Labels

If you have ever flipped through a greenhouse supply catalog, you know that it can be a challenge to find what you are looking for. Growers are unique and have their own needs for cultivation. Moreover, many growers want to stand out from the rest. They demand products that will help them to grow superior plants and aid in marketing. A profusion of colorful, uniquely shaped, company-branded, and decomposable pots are available. Growers must discover products that work best for them and their customers.

Containers

Many traditional sizes and types of plant containers produce great plants, **Figure 19-13**. There are countless decorative pots, hanging baskets, and everything in between to give the perfect plant a perfect home. These containers are commonly made of plastic. Clay and metal pots can be used for plants, but they are significantly more expensive. Growers incorporate many biodegradable container products into their growing systems as well.

The diameter of the top of the pot, whether circle, square, hexagon, or another shape, is usually measured in inches. Pots are referred to as 3″, 4″, 5″, and so on. These pots can fit into what is called a flat (tray) for easier transport.

Another method of measuring container size is by the volume of material it holds. Examples are by quarts and gallons. A gallon pot holds one gallon of soil. These pots also have a diameter, but may be described by volume as well.

A *Arina P Habich/Shutterstock.com*

B *apple2499/Shutterstock.com*

C *Sukpaiboonwat/Shutterstock.com*

Figure 19-13. A—Decomposable pots are being used by growers as a way to prevent horticultural materials going to the landfill. B—Containers are often measured by their diameter or volume. C—This grower uses a variety of container sizes, colors, and shapes for greenhouse materials.

Thinking Green

Container Recycling Programs

Every year Americans discard millions of agricultural plastic items into landfills. These plastic containers can be put to good use instead of going to landfills. Find out if a local greenhouse retailer offers container recycling. Contact the cooperative extension agency in your county and ask about container recycling. You can also contact local garden groups and schools, which often accept plant containers to be reused. Every agricultural plastic item that does not end up in the landfill is money, material, and space saved. You can also check local garden centers to see if a program exists, or talk to your teacher about accepting or starting a recycling program at your school.

jtairat/Shutterstock.com

Inserts

Cell pack inserts are plastic units (typically four to six units per pack) that fit into trays and in which small plants grow. People often refer to these cell pack inserts as four-packs or six-packs. Inserts vary in number, size, depth, and shape just as pots do. Traditionally, cell packs were meant to be put into a tray known as a 1020 (to accommodate cell packs 10″ wide by 20″ long in total). Inserts for these trays could be purchased based on a numbering system. Examples of cell pack insert configurations include:

- 606—6 packs, 6 cells.
- 804—8 packs, 4 cells.
- 1206—12 packs, 6 cells.
- 1801—18 packs, 1 cell.

The first number represents how many packs, and the second number represents the number of cells in a pack. Tray sizes are changing and are now available in many other sizes, **Figure 19-14**.

Plug Trays

Plug trays are rigid sheets of plastic with multiple cells. They are rigid and strong enough to be used without being placed in another tray. The cells often are shaped as squares, circles, or hexagons. Plug trays house seeds, seedlings, cuttings, and liners. A *liner* is a plant used for transplanting that is in a container with a 1″ to 3″ diameter. These trays are available in different sizes, depending on how many cells are requested.

Figure 19-14. A—This insert of pansies is known as a six-pack. These inserts come in a variety of sizes and fit into trays for easier transport. B—This tray shows multiple inserts of cabbage seedlings.

STEM Connection

Calculating Growing Space

Growers must determine how much space can be allocated for a particular crop. They must first determine the size of their benches. Then they must decide what size plant will be grown before it is sold. For example, if a greenhouse bench is 4′ wide and 40′ long, that is equal to 160 ft^2 of growing space. To determine how many standard flats can fit in this space, the grower must understand that a flat is approximately 11″ wide by 21″ long.

For example, imagine a bench is 40′ long, or 480″ long, and can fit nearly 24 flats lengthwise (480/21). The width of the bench is 4′, or 48″, and can fit 4 flats (48/11). Therefore, 24 flats lengthwise by 4 flats wide is 96 flats (24 × 4), the number that can be grown in this space.

Common sizes are 26, 52, 144, 288, and 512. These are sizes of plug trays that are represented by the number of cells within that tray.

Tags and Labels

Owners need to be able to identify plants, especially when they are newly planted seedlings or immature plants. Growers have much plant material to track and often use tags and labels to properly identify plants. Growers usually will include the scientific name followed by the cultivar. An example is the cultivar of *Juncus effusus* known as 'Big Twister.' Additionally, the tag may include the date the plant was propagated and the size of container it is in.

As plants change hands from a grower to a retailer, the plant identification is changed or updated. Plants are now given tags or labels that will include the name, identifying characteristics, cultural requirements, and the USDA hardiness zones for which the plant is suited. Customers need information to make decisions about whether to purchase a plant. Having a picture is especially important for a plant that is not in flower.

Some businesses focus on tag and label production. These companies work to create attractive labels that are functional and meet the advertising or marketing needs of the grower, **Figure 19-15**. Many plants are now registered or trademarked. The label for such a plant may state its unique characteristics.

> **Corner Question**
>
> What is branding?

Jodi Riedel/Goodheart-Willcox Publisher

Figure 19-15. Tags contain critical information for consumers. The most helpful tags include a photograph of the plant at maturity as well as cultural information.

Plant Materials

Plant materials vary based on the time of year, how they are best shipped, how much they cost, and the preference of growers. Plants may start in the greenhouse as a seed, put down roots in a propagation greenhouse, or be shipped from the Netherlands as a bulb. However the grower receives the plant material, there is a niche for each product available. Plant materials used by growers include seeds, unrooted cuttings, plugs, liners, bare root plants, bulbs, and tubers.

Seeds

Not all plants develop viable seeds. Some plants do not produce seeds. Some plants produce plants from seed that look nothing like the parent plant. (These plants are referred to as not true to type.) Other plants have special requirements, such as vernalization (cold treatment) or scarification (injuring the seed coat), that make them difficult to germinate. However, many plants are easy to cultivate from seeds, **Figure 19-16**.

Surkov Dimitri/Shutterstock.com

Figure 19-16. These tomato seedlings are approaching a time of transplant. Seedlings must be mature enough for transplant to avoid injury.

Jodi Riedel/Goodheart-Willcox Publisher

Figure 19-17. Automation, such as this plug tray device, makes greenhouse growing more efficient and less laborious.

Several annual bedding plants, herbaceous perennials, vegetables, and herbs are started in a greenhouse by seed. *Bedding plants* are plants used mainly for ornamental purposes. They are planted in beds and often last just one season. They are also called annuals. Seeds are usually sown into plug trays for commercial production. In large-scale production, automated seeders sow seeds into plug trays of various sizes. Using automated seeders helps to reduce labor costs and time, **Figure 19-17**. Can you imagine planting 522 seeds on just one plug tray? Better yet, imagine doing that for eight hours straight. Automation has revolutionized horticulture just as it has other industries. Smaller growers often do not use automated seeders because of their limited growing space. They can afford to pay the labor costs for sowing hundreds of trays (rather than tens of thousands of plug trays as used by large operations).

Seeds have unique germination requirements. Some seeds require light, while others need darkness; some prefer moist conditions, and some can rot in too much water. Growers pay special attention and keep records regarding seed germination for future reference.

Unrooted Cuttings

Cuttings are tips of stems that have been removed (cut) from the plant. Typically, these stem cuttings will be rooted by another grower. Growers who produce cuttings have stock plants from which tip cuttings are taken. Unrooted cuttings can be shipped all over the world using unique packing techniques and air travel. Cuttings that are taken in Costa Rica on a Monday can arrive at another grower in Maine by Tuesday. Acquiring unrooted cuttings is a common practice today for growers of all sizes as unique material from around the globe can be easily transported.

Did You Know?

Some seedlings do not transplant well, and seed should be direct sown (sown directly in the place where they will remain). Examples are carrots, radishes, and turnips. Transplanting causes the apical meristem of the root tip to be injured, which leads to plant damage and shock.

Hands-On Horticulture

Planting Seeds by Hand

Sow seeds in a medium that is very fine and holds a great deal of moisture. Use a flat or a plug tray, and sow seeds two to three times as deep as the seed is big. Cover with potting soil and a thin layer of vermiculite. Gently mist until thoroughly saturated or set the flat or tray into a vessel that will allow water to be absorbed from the bottom (capillary action). Place in the greenhouse and begin cultivating the plant according to its needs.

Hands-On Horticulture

Planting Bare Root Plants

Bare root material is different than a seed, and it must be planted differently as well. After receiving the material, lay it out and allow to air out for two to three hours. Fill the growing container one-third full with potting soil. Fan out the roots and radiate them over the growing medium. Hold the crown of the plant and fill the container the rest of the way with potting soil. Ensure the crown is visible above the soil line. Water the plant to saturate the growing medium thoroughly.

Corner Question

What is an All-American Selection plant?

Plugs and Liners

Some growers purchase plugs and liners ready for transplant to larger containers or inserts.

Plugs

Plug plants are young plants with well-established, complete, and independent root systems. These plants are easy to transport and transplant into containers or inserts, **Figure 19-18**. Plugs are grown from seeds or asexually propagated material.

Purchasing plugs from another grower specializing in this type of plant material can help reduce costs and time invested. Growers can order plugs from all over the world. Agriculture is a global business. Plug growers in Central America, South America, and Asia ship plugs around the world.

Jeeranan Thongpan/Shutterstock.com

Figure 19-18. These plug trays hold hundreds of plants each.

Liners

Liners are defined by the USDA as plants with an established root system that are produced in standard nursery containers that are equal to or greater than 5/8" in diameter but less than 3" in diameter at the widest point of the container or cell interior. Like plugs, liners include seedlings and plants from tissue culture and other asexual methods of propagation, **Figure 19-19**. Liners are slightly larger than plugs and often include tray sizes of 72 cells or more per tray. Liners are very sturdy and provide a quick way to grow saleable plants.

Lipsett Photography Group/Shutterstock.com

Figure 19-19. These plants are larger than plugs and are known as liners.

Bare Root

Bare root plants are free of soil and are shipped while dormant. The bare root material includes the roots, a crown (area where roots and stem meet), and some stem material, **Figure 19-20**. Roots are sold by grade and often include #1 and #2. A #1 root material is large and usually takes six to eight weeks of production time in a two-gallon pot.

FPWing/Shutterstock.com

Figure 19-20. These are bare root plants that can be planted in containers.

Bulbs and Tubers

Many different plants produce what is known as a bulb or a tuber. A bulb is a modified stem that contains a basal plate at the bottom from which roots grow. It holds storage tissue, modified leaves, and a flower bud(s) that is in a dormant state. A tuber is an underground stem that is a storage structure that can be used for plant propagation. Bulbs and tubers can be grown throughout the year; however, they are often thought of as having a season, **Figure 19-21**.

- Spring blooming bulbs, such as daffodils, tulips, croci, and hyacinths, can be planted in the fall and overwintered in containers (outdoors).
- Summer blooming bulbs, such as amaryllises, lilies, and ranunculuses, are planted in late winter or early spring in the greenhouse.
- Summer blooming tubers, such as begonias, caladiums, and dahlias, do best with long summer days and can be planted in containers of all sizes, shapes, and forms.

Growers can order bulbs, tubers, and other dormant storage structures of plants (corms, rhizomes, etc.) and plant them just as they would seeds.

Paulina Grunwald/Shutterstock.com

photowind/Shutterstock.com

Figure 19-21. A—Bulbs are an asexual structure that can be planted similarly to a seed. B—A tuber, such as this one found on a dahlia, is another plant organ that can be removed and asexually propagated. A tuber can be planted similarly to seeds or bulbs.

Greenhouse Crops

Many plants can be grown in the greenhouse environment. Growers usually study market trends to determine which plants will be profitable. There must be a want or need from customers or there is no need to grow the plants. A hobby grower can try whatever he or she may wish to grow in the greenhouse. However, commercial wholesale and retail growers are in the business to make a profit. They must grow plants that customers are willing to purchase. Types of crops grown in a greenhouse include container plants, foliage plants, vegetables and herbs, plants for cut flowers, bedding plants, and perennial plants.

Container Plants

Gardeners are demanding containers or planters for their landscapes. As the number of urban dwellers increases and garden customers lose access to land to cultivate, gardeners are using more containers. Growers are supplying mixed hanging baskets with lush annuals, decorative containers filled with mixes of tropical and bedding plants, and ready-to-plant containers that can be slipped out of a pot (or planted directly if in a compostable pot) into a gardener's container of choice. There are many possibilities for growing plants in containers in greenhouses and landscapes across the country.

Several plants often are grown in containers as a blooming potted plant for specific holidays.

- Poinsettias, Christmas cacti, paperwhites, amaryllises, and cyclamens are grown for Christmas, **Figure 19-22A**.
- Mums are grown for Halloween.
- Thanksgiving cacti are grown for Thanksgiving.
- Cyclamens and orchids are grown for Valentine's Day, **Figure 19-22B**.
- Chrysanthemums, kalanchoes, Easter cacti, and Easter lilies are grown for Easter, **Figure 19-22C** and **Figure 19-22D**.
- Geraniums and hydrangeas are traditionally given on Mother's Day.

> "Gardening is a medicine that does not need a prescription...and with no limit on dosage."
> —Author unknown

Foliage Plants

Foliage plants are often tropical in nature and provide beautiful leaves in and around the garden, in residences, and in buildings. Lush ferns, philodendrons, golden pothos, dracaenas, peace lilies, and spider plants have lovely leaves of every shape, size, and color that can enhance the interior or

A —Photology1971/Shutterstock.com
B —A.Ryser/Shutterstock.com
C —margostock/Shutterstock.com
D —Olga Popova/Shutterstock.com

Figure 19-22. A—A cyclamen is a flowering potted plant often grown for holidays such as Christmas and Valentine's Day. B—A lady slipper orchid is a beautiful and unique potted flowering plant. C—A kalanchoe is a potted flowering plant that consumers can often find at florists, gift shops, and even grocery stores. D—This Easter cactus is similar to Christmas and Thanksgiving cacti, but it has a unique flower shape.

Tatiana Volgutova/Shutterstock.com

Figure 19-23. A *Chlorophytum comosum* cultivar, known as a spider plant, is an easily cared for foliage plant. This cultivar has twisted leaves.

exterior environment, **Figure 19-23**. These tropical plants can remain alive for years in environments that are between 50°F to 77°F (10°C to 25°C) with adequate light levels and proper care. These plants change the character of spaces instantly and are welcomed equally by gardening novices and experts.

Vegetables and Herbs

Recently, there has been an increase in people who want to grow their own food and know every input in its production. Growers can provide vegetables and herbs to start a garden or simply produce the vegetables and herbs in the greenhouse. Greenhouses were originally constructed with the objective of growing vegetables. Today, the production of greenhouse vegetables is a way for people to access locally grown produce year-round. Common vegetables and herbs grown in a greenhouse include tomatoes, peppers, eggplants, cucumbers, lettuce, basil, cilantro, parsley, chives, and dill.

Cut Flowers

Cut flowers include plants that are grown for aesthetic purposes, **Figure 19-24**. A floral designer uses the flowers as a medium for his or her art. Cut flower arrangements include flowers that are harvested from all over the globe. The US cut flower production in greenhouses has reduced in size. This is due to competitors in other countries producing quality plants at low costs. Once air transportation of cut flowers became a suitable means of delivery, the cut flower industry in the United States began to wane. Today, the United States imports 64% of its cut flower products.

Ruud Morijn Photographer/Shutterstock.com

Figure 19-24. Gerbera daisies are beautiful cut flowers grown in greenhouses.

STEM Connection

What Are Microgreens and Sprouts?

A new market in sprouted or germinated plants has exploded in the horticulture industry. Growers can take seeds and germinate them for consumers to eat. They can harvest seedlings (microgreens) for customers as well. Microgreens and sprouts are packed with nutrients. Microgreens are edible seedlings from plants, such as beets, broccoli, cabbage, turnips, chard, and grains. They are often used in the food industry as garnish, in salads, and for sandwiches. Growers use lentils, alfalfa, kale, radishes, broccoli, peas, and countless other seeds to meet the market's demands for this unique greenhouse product.

Andi Berger/Shutterstock.com

Cut flowers include not just ornamental flowers but also unique foliage and even edible flowers. Foliage is used in floral arrangements and can range in size, shape, and color. Chefs use flowers for several dishes or as a way to decorate foods and desserts. Flower petals can be used in salads, and orchids may ornament a soufflé.

Almost any plant can be used in cut flower arrangements. Cut flowers are included in bouquets, holiday arrangements, wedding decorations, and staging for special occasions. Cut flowers are not just roses. Heirloom plants, such as zinnias and daisies, make a beautiful arrangement. However, unique varieties of succulents paired with dusty miller foliage can make an arrangement a little more interesting.

Corner Question

Which state in the United States grows the most cut flowers?

Bedding Plants

A plant that will be set into a garden bed or container when it is about to bloom is referred to as a bedding plant. These plants, often called annuals because of their life span of less than one year, are used for display for one or two seasons. Gardeners replace these annuals with another plant, usually another bedding plant, at the end of the season.

Some people have considered bedding plants boring; but in the past two decades, this sector of the green industry has been growing. Geraniums, African marigolds, petunias, and impatiens are now considered to be rich in color, shape, texture, and size. What was once a simple red geranium may now be scented, variegated, or splashed with brilliant oranges or pinks, **Figure 19-25**. Ornamental grasses, papyrus, and even ornamental corn provide height and texture. Torenia, diascia, and Million Bells® blanket the borders of gardens. The list of unique flowers and the millions of cultivars that researchers develop provide continuous color and flare for the landscape.

Horticulturists now cover garden beds with mixes of bedding plants, tropicals, flowers for cutting, vegetables, herbs, and perennials. A garden is never truly finished, and it must be continuously cultivated and improved. Bedding plants add unique flavors to the landscape every season. Growers produce these plants to provide engaging and fresh plant material to garden enthusiasts everywhere. Some growers still offer old favorites for those with traditional plant tastes.

Did You Know?

Eyeball plant (*Acmella oleracea*) is also known as the toothache plant. If you chew on the leaves, your mouth and tongue get a numbing effect for a few minutes. Be careful; excessive drool is a side effect of this zesty green.

A hans engbers/Shutterstock.com

B Lennard Janson/Shutterstock.com

C In Tune/Shutterstock.com

Figure 19-25. A—Geraniums are a common flowering annual (also known as a bedding plant). B—Gazania flowers are very drought and heat tolerant. C—An eyeball plant is a unique flowering bedding plant. It lacks petals, and when eaten, it gives a person's mouth a numbing effect for a few minutes.

Perennial Plants

Plants that require more than two years to complete their life cycle or live for many years are called perennials. Perennials grown in greenhouses are usually herbaceous or evergreen. Herbaceous plants die down to the ground and become dormant. They grow again the following season. *Evergreen* plants are those that remain green all year.

Consumers want perennials because they only need to be planted once, and they will remain in the garden for many years. People consider perennial gardens to be easier to maintain than other gardens. Many perennials do not have flowers that bloom for an entire season or provide lasting color like bedding plants do. Perennials commonly offer a window of bloom, perhaps a week or two, and then provide only foliage for color and texture.

Perennials can provide unique foliar color or textures to the landscape. Coral bells (*Heuchera*) have beautifully colored leaves that range from chartreuse to burgundy, **Figure 19-26**. Other perennials with striking foliage are ornamental grasses, such as zebra grass (*Miscanthus*).

Customers who want a perennial that provides a longer showing of flowers have several choices to consider. Shasta daisies (*Leucanthemum*) include cultivars such as 'Banana Cream' and 'Crazy Daisy' that offer unique traits compared to a typical Shasta daisy. Daylilies and *Coreopsis* cultivars can include a range of colors and flowering times to meet the customer's needs. For a long-blooming winter flower, try Hellebores (*Helleborus*), which are beautiful while they bloom and have attractive spent seed heads as well.

Monika Pa/Shutterstock.com

Figure 19-26. This unique planting shows coral bells (*Heuchera*), which is an evergreen perennial in many parts of the United States.

Careers in Greenhouse Production

Career opportunities are plentiful in the greenhouse industry. Aside from growers, many other personnel assist in the growing operation. Sales and marketing, customer service, pest management technicians, greenhouse technicians, researchers, and public representatives operate at most wholesale and retail nurseries. In small greenhouse businesses, a few individuals may wear many of those hats, while in larger operations, employees specialize in one field.

Plant Tag Technician

Companies that offer plant tags, labels, and other packaging materials for the greenhouse industry hire technicians to manufacture the materials, **Figure 19-27**. Technicians require a high school diploma or GED. A technician works on a factory floor producing materials. A work week will include rotations to various machines, presses, and other equipment that

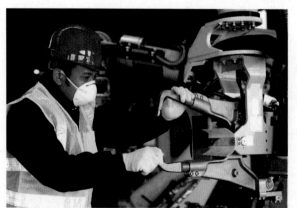

ndoeljindoel/Shutterstock.com

Figure 19-27. This technician works on a press that makes plant tags and labels. Some companies manufacture the plant tags and labels that many growers use.

create the packaging, advertising, and marketing materials. Technicians must pay particular attention to safety, as the machines that are in use can cause serious injury to workers. This position allows workers to mature and apply for other job opportunities within the company. Once a technician has exhibited a solid work ethic and performed the duties assigned satisfactorily, then the employee can move up to management and supervision of other employees.

Greenhouse Customer Service

Many greenhouses are wholesalers and do not allow the public to enter their doors. Regardless of whether a greenhouse is a wholesaler or a retailer, there are still customers, and there is a need to sell to the customers. Customer service involves taking orders and helping customers meet their purchasing goals. Customers can reach service agents via phone, on the Internet, or in person.

A person in customer service must be knowledgeable about the plant materials that are for sale. He or she must be friendly and work hard to satisfy clients. Someone in customer service should also be a problem solver and critical thinker as he or she often responds to issues associated with product delivery. A positive attitude, attention to detail, and professional behavior can help a customer service agent thrive in the greenhouse industry.

Career Connection

Denise Etheridge
Homewood Nursery

A grower's tasks include the monitoring and control of the greenhouse environment, nutrition management and fertilizer application, superior knowledge and understanding of crop and crop health issues, and crop record keeping. Denise Etheridge, an experienced grower of 22 years at Homewood Nursery in Raleigh, North Carolina, has many additional skills she uses every day.

Like most growers, in addition to cultivating greenhouse crops, Denise manages a crew of 16 people during peak season. She looks for individuals with horticultural passion who are dependable, responsible, honest, and have good communication skills. Homewood Nursery is successful not only because of its superior retail greenhouse plant material, but also because of the people who work with customers and their superior knowledge of plants.

As a grower with an arsenal of plant knowledge, Denise loves a good challenge. No two

Denise Etheridge

growing seasons are the same, and the challenges that arise may be different, or new problems may arise that require creative solutions. Denise's job is to "figure it out and fix it."

In college at North Carolina State University, Denise was a freshman envisioning her ideal job in computers. She was inspired in an introductory horticulture class taught by Bryce Lane. She changed her major and realized that she wanted to grow flowers. One of her favorite parts of growing is seeing the first brightly colored geranium that peaks through the sea of green foliage. She says she "always forgets how brilliant the blooms really are."

CHAPTER 19 Review and Assessment

Chapter Summary

- Plants require a container or other place to grow, light, carbon dioxide, nutrients, the appropriate temperature, water, and pest control.
- Growers use a number of inputs or tools for the greenhouse plant production process. These crop inputs include media, plant growth regulators (PGRs), containers, trays, tags, and labels.
- Plant materials used by growers include seeds, unrooted cuttings, plugs, liners, bare root plants, bulbs, and tubers.
- Types of crops grown in a greenhouse include container plants, foliage plants, vegetables and herbs, plants for cut flowers, bedding plants, and perennial plants.
- Career opportunities are plentiful in the greenhouse industry. Aside from growers, many other personnel assist in the growing operation. Sales and marketing, customer service, pest management technicians, greenhouse technicians, researchers, and public representatives operate at most wholesale and retail nurseries.

Words to Know

Match the key terms from the chapter to the correct definition.

A. bedding plant
B. bio-stimulant
C. container capacity
D. critical day length (CDL)
E. critical night interval (CNI)
F. cyclic photoperiodic lighting
G. evergreen
H. flagging
I. incomplete fertilizer
J. insoluble fertilizer
K. liner
L. night interruption (NI)
M. plant growth regulator (PGR)

1. A lighting method that uses a pattern of light and darkness during the night from 10:00 pm to 2:00 am to create a long-day effect for plants.
2. A fertilizer that is lacking one of the three primary macronutrients: nitrogen, phosphorus, or potassium.
3. A plant that is used mainly for ornamental purposes in beds and often is used for a particular season or two of the year.
4. A hormone (synthetic or natural) that changes plant growth rates, often as an inhibitor or accelerator.
5. A fertilizer that does not readily mix with water.
6. The number of hours of light needed to initiate flowering in a plant.
7. A method of creating a long-day effect for plants by disrupting the dark period with light.
8. A plant that remains green throughout its life cycle and all seasons.
9. A microorganism that fixes nitrogen into a form plants can use and creates a healthier soil environment for root growth, boosting plant growth.
10. The condition of a plant when it is just about to wilt.
11. The maximum amount of water that can be held by a substrate against the pull of gravity.
12. The length of dark period needed to initiate flowering in a plant.
13. A plant used for transplanting that is in a 1" to 3" diameter pot.

Know and Understand

Answer the following questions using the information provided in this chapter.

1. What types of plants do growers cultivate in greenhouses?
2. What are the necessary elements for plant growth in a greenhouse?
3. How are supplemental and photoperiodic lighting different?
4. Carbon dioxide is needed for what essential process for plants? What are ways that carbon dioxide can be added to greenhouse air?
5. Describe the contributions of pioneer research chemist Percy Julian.
6. What is the primary difference between a complete fertilizer and an incomplete fertilizer?

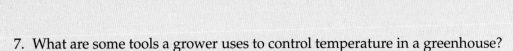

7. What are some tools a grower uses to control temperature in a greenhouse?
8. What are three characteristics of water that impact plant growth?
9. What are some guidelines for successful watering?
10. What are some materials that particles in growing media may be made of?
11. What is the role of bio-stimulants in plant production?
12. What are some benefits of using plant growth regulators (PGRs)?
13. What are some materials that containers for growing plants are made of?
14. What information is typically found on plant labels at retailers?
15. What are some types of plant materials used by growers?
16. How are liners and plugs alike and how are they different?
17. How is a bare root plant planted?
18. What types of crops are typically grown in a greenhouse?
19. What are three examples of foliage plants?
20. What types of plants are used as cut flowers? What are some uses of cut flowers?
21. What do consumers like about using perennial plants?
22. What are some careers associated with greenhouse industry?

Thinking Critically

1. You find that the media you use for bedding plants is holding too much water. You have 20 bags left of the material. What could you do to remedy this situation? Create a list of five possibilities.
2. A teacher approaches you with a plant problem and would like an answer. He has had a Thanksgiving cactus for the past four years. The first three years it bloomed, but this year it did not. What questions would you ask the teacher to determine the answer to the problem?

STEM and Academic Activities

1. **Science.** Determine how many foot-candles of light your school's greenhouse has at different points of the day, different places in the greenhouse, or different times of the year. Create a spreadsheet to enter the data. Use the data to create a graphical representation of your findings.
2. **Technology.** In your school's greenhouse, develop a system to turn on lights in the night to express long days and short nights. Create a list of materials that will be needed and determine a cost for the project.
3. **Math.** If a greenhouse has 50 benches that are 50′ long and 6′ wide, how many 1020 flats could fit in this space?
4. **Social Science.** Role-play being a customer service agent for a retail greenhouse. Include another student who is portraying a difficult customer in the scenario.

5. **Language Arts.** Choose five of the plants mentioned in this chapter. For each plant, write an advertising description that would be appropriate for a plant catalog. Be sure to highlight all the selling points.
6. **Language Arts.** Using the knowledge you acquired from this chapter, create an informational poster outlining the crops grown in greenhouses. Include the different types of crops and give examples of each.

Communicating about Horticulture

1. **Reading and Speaking.** Many school agriculture departments operate a greenhouse. What happens to the wastewater from irrigation in the greenhouse? What is the quality of that water? What happens to the wastewater? How could that water be reused? Research ways in which the wastewater could be reclaimed. Present your findings to the class.
2. **Reading and Writing.** Is there less oxygen in the air in winter? Research this topic and write a 1- to 2-page report explaining if and how cold winter months affect oxygen levels on Earth.

SAE Opportunities

1. **Exploratory.** Job shadow a customer service representative for a greenhouse or related company.
2. **Experimental.** Use one crop and test different methods of PGRs on the crop. Analyze your results and determine conclusions to share.
3. **Experimental.** Research which cut flowers will grow well in a greenhouse in your region. Try growing a few of the different types and determine which cut flower yielded the best results in your greenhouse.
4. **Exploratory.** Create a list of annual bedding plants that could be sold in your school's greenhouse. Find a company that sells plugs for the crops you have chosen. Create a schedule for cultivation of these plants after receiving them from the plug producer. Include cultural information and size of material that will be sown. Determine how many of each plant you will need to purchase and what the cost will be to fill your greenhouse.
5. **Entrepreneurship.** Grow vegetables, herbs, or microgreens in your school's greenhouse and sell them to customers.

KENG MERR/Shutterstock.com

Chapter 20: Twenty-First Century Horticulture

Chapter Outcomes

After studying this chapter, you will be able to:
- Explain how various hydroponic systems work.
- Describe aquaponic systems.
- Describe rooftop gardens.
- Discuss vertical gardening systems.
- Discuss raised bed gardening.
- List careers related to nontraditional gardening systems.

Words to Know

aeroponic system	drip system	intensive green roof system
aggregate	ebb and flow system	lasagna composting
aquaponics	effluent	nutrient film technique (NFT)
biofilm	extensive green roof system	square foot gardening
biopharming	ground level ozone	vertical gardening
culling	heat island effect	water culture system
deep water culture (DWC)	hydroponics	

Before You Read

After reading each section (separated by main headings), stop and write a three- to four-sentence summary of what you just read. Be sure to paraphrase and use your own words.

While studying this chapter, look for the activity icon to:

- **Practice** vocabulary terms with Words to Know activities.
- **Expand** learning with identification activities.
- **Reinforce** what you learn by completing Know and Understand questions.

www.g-wlearning.com/agriculture

As our world population continues to grow and more land is lost to development, the agriculture industry is faced with the task of producing more food and fiber in less space and on poorer quality land. In order to meet these needs and overcome so many obstacles, growers have turned to new, innovative growing systems, often located in the most unlikely places. Crops are currently being cultivated in places such as abandoned mines, WWII bomb shelters, and growth chambers located deep in the earth. These nontraditional growing areas require alternative growing methods such as hydroponics with water and artificial lighting in place of soil and natural daylight. Other nontraditional growing systems that have become increasingly popular include rooftop gardening, vertical gardening, green roofs or walls, and aquaponics.

In this chapter, you will learn about nontraditional growing systems as well as biotechnology applications being used to help growers meet the needs of an increasing population and decreasing land availability.

Hydroponics

Hydroponics, a water-based, soilless growing method, is the fastest-growing sector of the green industry. This technology provides a means of growing plants in places with poor soils or limited land availability. It also allows food to be cultivated near consumers and delivered fresh on a daily basis—often the same day it is harvested. This is especially true in urban areas where growers have installed hydroponic systems on rooftops and in repurposed buildings, **Figure 20-1**.

ssguy/Shutterstock.com

Figure 20-1. Plants can be grown indoors, hydroponically, using various light systems.

The initial investment for a hydroponic system is high because the grower must acquire a location, equipment, power source, and possibly a water treatment system. If water quality is low, the water must be treated (which is extremely expensive) or brought in. The grower must also provide aeration, nutrients, light, temperature control, appropriate pH balance, and pest control. Growers have control over every facet of the growing process and must fully understand their crops' needs and carefully monitor their system to keep plants healthy and productive. If not properly monitored, diseases can spread quickly from sick plants to healthy plants through contaminated water.

Did You Know?

London's underground bomb shelters are being used to house a subterranean hydroponic farm that will produce 11,000–44,000 lb of crops annually.

History of Hydroponics

Hydroponic gardening is not a new concept. Early forms of hydroponics date to around 600 BCE. For example, the Hanging Gardens of Babylon (one of the seven wonders of the ancient world) were elaborately engineered and contained many types of trees, shrubs, and vines. These gardens are said to have flourished in the arid, desert-like climate due to a hydroponic-type system supplied with water from the Euphrates River. During the tenth and eleventh centuries, the Aztecs developed a floating garden that resembled hydroponic systems used today. They built rafts with roots and reeds, topped them with soil, and floated them on Lake Texcoco. Crops were sown in the soil and grew as the rafts drifted around the lake, **Figure 20-2**.

During the early 1600s, British scientist Sir Francis Bacon formally researched hydroponic, or soilless, gardening. His research was published in 1627 and propelled a surge of hydroponic research and advancement. Another British scientist, John Woodward, conducted hydroponic experiments in the late 1600s. Many of his experiments focused on determining the source of plant nutrition.

topten22photo/Shutterstock.com

Figure 20-2. Floating mats of plants, like these in Asia, were considered some of the first hydroponic growth systems.

The twentieth century experienced a wax and wane of hydroponic research and advances.

- In 1925, researchers at agricultural experiment stations began to search for solutions to the problem of replacing soils within greenhouses. The idea of using a nutrient or water solution began to be studied.
- Between 1925 and 1935, university experiment stations in states such as New Jersey, Indiana, and California began to test sand and gravel culture.
- William Gericke, a University of California Berkeley scientist, promoted hydroponics using the term *aquaculture*. Later, he found this term already in use to describe the study of aquatic organisms and coined the term *hydroponics*.
- In 1940, two other scientists from Berkeley, Dennis Hoagland and Daniel Arnon, published *The Water Culture Method for Growing Plants without Soil*. This text, considered one of the most important hydroponic resources, is still used today.
- During World War II, the armed forces grew plants hydroponically on barren Pacific Islands to feed the troops.
- Starting in 1973, the rising cost of petroleum led to increased costs of plastic. This inflation in price led many growers to abandon the idea of hydroponics.
- Disney's EPCOT Center Land Pavilion showcased various methods of hydroponics as the way to grow crops in the future. In 1982, innovators at Disney forecasted a need for this growing technique as a way to solve agricultural problems associated with conventional farming practices.
- A former Sony semiconductor factory now houses the largest hydroponic facility in Japan. Mirai grows 10,000 heads of lettuce under 17,500 LED lights in the 25,000 ft^2 building. Plants grow 2 1/2 times faster and use 1% of the water used in conventional outdoor lettuce cultivation.

"Gardening is learning, learning, learning. That's the fun of them. You're always learning."
—Helen Mirren

History Connection

Hydroponics on Wake Island

In the 1930s, Pan American Airways (PAA) expanded its flight service to include flights from the United States to China. The flight distance required various points for refueling and restocking food supplies for its crews and passengers. With little land to choose from, PAA chose Wake Island as one of its stopping points. As there was no arable land and it was expensive to airlift in fresh vegetables, the airline installed a hydroponic growing system. The airline continued operations on the island until the beginning of World War II.

KC-135_Stratotanker_boom/US Air Force

Did You Know?
There are between 20,000 and 25,000 hectares of commercial hydroponic production globally. They supply about $6 to $8 billion worth of produce.

Hydroponic Systems

There are seven main types of hydroponic systems. Each method of hydroponics has advantages and disadvantages. Growers must evaluate the system's setup, maintenance, yield, cost, and reliability to determine which would work best for them. Some methods use a misting system, some use a growing substrate, and others use only liquids. However, all methods grow plants without the use of soil. Each method must provide oxygen to plant roots, a nutrient solution, and the proper pH to ensure optimum plant growth and quality. Types of hydroponic systems include wick, ebb and flow, drip, nutrient film technique (NFT), water culture, deep water culture, and aeroponics.

Wick System

This simple, passive system uses a wick that is fed into a reservoir filled with a nutrient solution. The plants are suspended in a tray filled with a medium or aggregate of the grower's choice. The nutrient solution transfers from the reservoir to the plant material suspended in an aggregate through the wick, **Figure 20-3A**. A wick works by capillary action (the ability of a liquid to flow in opposition to external forces such as gravity). The water is pulled up against the force of gravity through a process called adhesion. The edges of the droplets of water are forced up the surface of the wicking material (typically made from cotton). Growers may use an air pump to add oxygen to the nutrient solution.

Ebb and Flow System

Ebb and flow is a method of hydroponics in which the growing chamber or tray holding plants is periodically flooded with a nutrient solution and then drained, **Figure 20-3B**. The nutrient solution contains vital elements such as nitrogen, phosphorus, and potassium. A reservoir tank holds the nutrient solution and a growing chamber perches above the reservoir. The nutrient solution is pumped into this space. The nutrient solution rests in the tray, where it can be absorbed by the roots, and then drains back into the reservoir. This method, also known as flood and drain, is the most popular method used by both hobbyists and professional growers.

Drip System

A *drip system* uses a reservoir of liquid nutrient solution that is pumped into a series of thin tubes (driplines) that slowly release the solution onto the base of each plant, **Figure 20-3C**. Recovery drip systems designed to collect and reuse the excess nutrient solution require more maintenance than nonrecovery systems due to variance in pH and nutrient strength levels in the reservoir. Drip systems are used in hydroponic systems as well as in other growing applications, including greenhouses, gardens, nurseries, and landscapes. Plants can be in growing trays, buckets, or baskets filled with an aggregate.

Nutrient Film Technique (NFT) System

The *nutrient film technique (NFT)* is a versatile hydroponic method that circulates a thin, constant stream of liquid nutrient solution over the plant roots. This system is commonly made with PVC pipe that has a series of holes across the top of the pipe. The plant material is suspended in baskets in this series of holes, **Figure 20-3D**. The baskets may be filled with aggregate, such as clay pellets or rockwool. (In hydroponics, *aggregate* is a solid, inert material that supports plant life. It is often referred to as a growing medium or substrate.) The roots of the plants hang down in the channel where the nutrient solution flows over the root tips. The nutrient solution must flow continuously or the roots will quickly dry out.

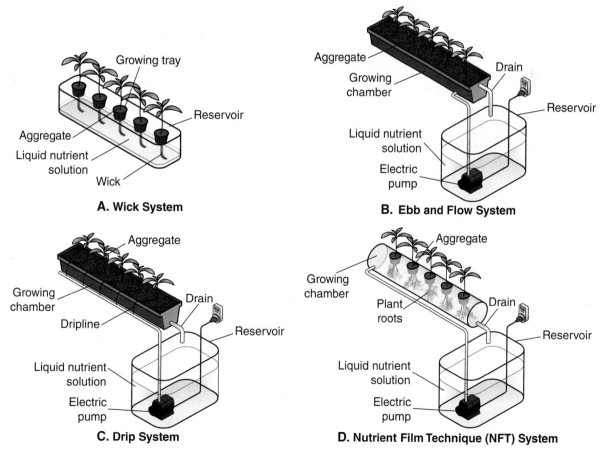

Figure 20-3. The various hydroponic systems

Zern Liew/Shutterstock.com

KAMONRAT/Shutterstock.com

Figure 20-4. Lettuce are fast-growing, water-loving plants that grow well in water culture systems. This simply designed system does not work well with large or long-term plants.

Sura Nualpradid/Shutterstock.com

Figure 20-5. These plants in an aeroponic system are being misted with a nutrient solution.

Water Culture System

In *water culture systems*, light plastic or foam trays containing plants float on a nutrient solution. The plant roots hang below the tray and are submerged in the nutrient solution, **Figure 20-4**. An air pump is used to supply oxygen to the roots through the nutrient solution. This system requires few mechanisms and is often a starter system for individuals attempting to grow hydroponically. Water culture may also be called floating hydroponics.

Deep Water Culture (DWC) System

Deep water culture (DWC) systems consist of a main reservoir and a series of buckets. The nutrient solution is pumped from the main reservoir to the buckets through tubing. The lid of each bucket holds a plant or plants and may or may not contain aggregate. The roots are suspended in the nutrient solution below the lid. An air pump or bubblers may be used to provide oxygen to the nutrient solution.

Aeroponic System

In an *aeroponic system*, plant roots are suspended in the air and misted intermittently with a nutrient solution, **Figure 20-5**. This system does not use any type of growing medium. This method requires a great deal of supervision by a grower and constant suspension or training of new plant growth.

System Components

Hydroponic systems use various components to create a functioning, soil-free system for cultivation. Equipment is needed to circulate water, turn equipment on or off, maintain water temperature, and add oxygen to the nutrient solution. Other important components include the aggregate used to support plants as well as the containers used to hold the plants and contain the nutrient solution. Some of the pieces of equipment that work together to hold, circulate, or aerate the liquid solution are listed and described in **Figure 20-6**. (This equipment will vary by individual systems and grower preferences.)

Equipment Used to Contain, Aerate, and Transport Nutrient Solution	
Air pump	An electric pump used to pump air from the surrounding environment into an air line connected to an air stone or bubbler.
Air line	Tubing used to deliver oxygen (ambient air) to the nutrient solution, often to an air stone from an air pump.
Air stone (bubbler)	A piece of porous stone used to diffuse oxygen into the nutrient solution.
Chiller	An electric device used to cool water to the appropriate temperature for plant growth. The constant cooler temperature also helps reduce fungal populations.
Dripline	Tubing used to deliver nutrient solution to a nozzle that emits drops or a trickle of solution to individual plants.
Drip manifold	A component from which multiple driplines are connected and through which the liquid nutrient solution is first pumped.
Grow tray or growing chamber	A container that holds the plant material and usually a nutrient solution.
Mist nozzle	A component that emits fine droplets of water in an aeroponic system.
Nutrient pump	An electric pump used to pull water from the reservoir to the growing tray, drip manifold, or individual lines.
Reservoir	A container or storage tank used to hold the liquid nutrient solution.

Goodheart-Willcox Publisher

Figure 20-6. These components make it possible to adequately control and monitor the liquid nutrient solution used in hydroponic systems.

Additional devices and systems that make a hydroponic system work more efficiently or effectively include:
- Climate controls—used to maintain growing area temperatures at optimum levels for plant growth. Climate controls may be used to add and/or remove heat from the growing area.
- Carbon dioxide (CO_2) systems—used to increase the level of carbon dioxide in growing areas to supplement and promote photosynthetic output.
- Lighting systems—artificial light source used to provide sufficient or supplemental light at optimal quality and in appropriate quantity.
- Nutrient solutions—countless solutions available that may be used to create a nutritional program suited to specific systems and crops.
- Pest control—integrated pest management (IPM) that may include organic and synthetic forms of pest controls using biological, cultural, and chemical means.
- pH meters—monitoring and adjustment of nutrient solution/water pH is vital to ensure plants receive sufficient nutrients.
- Timers—automatically turn controls on and off at appropriate intervals to maintain such factors as temperature (water and air), oxygen, and water levels.

Without these implements, growing hydroponically could be quite cumbersome and much less profitable.

Corner Question

Why does a hydroponic system benefit from a chiller?

Aggregate Forms

Hydroponic systems do not use soil but may use an aggregate (growing medium) to support the plant roots and help hold up the plants, **Figure 20-7**. The medium should not amend or change the chemical makeup of the nutrient solution. It should be coarse and porous, which will allow plant roots to access nutrients and oxygen easily. The growing media may be organic or inorganic and can come from a multitude of sources. Some growing aggregates include:

- Coconut coir—a lightweight organic material made of coconut husks.
- Hydroton™—a pelletized expanded clay material.
- Oasis cubes—a lightweight, synthetic material often used as a floral design substrate.
- Perlite—a white, volcanic rock material.
- Pine bark—an organic material made of chipped tree bark.
- River rock—rocks with rounded edges.
- Rockwool—an expanded basaltic rock fiber material.
- Vermiculite—a lightweight, metallic, mica material.

Tortoon Thodsapol/Shutterstock.com

Figure 20-7. The coconut coir will provide support for both the cucumber plant and its roots.

Safety Note
Rockwool Safety
Some people are allergic to rockwool. It can be irritating to your skin. It is best to wear gloves and protective clothing when working with this material.

"Garden as though you will live forever."
—William Kent

Crops

The cost of establishing a hydroponic system makes it important to choose crops with high returns. Crops commonly grown using hydroponics include tomatoes, peppers, herbs, cucumbers, lettuce, and microgreens. Many of these crops have short life spans and produce a significant yield, which allows growers to quickly recoup the costs of implementing the hydroponic systems. Growers are continuously expanding crop selections and attempting to grow new crops for consumers.

With a highly controlled environment, hydroponics is also an ideal method of growing plants for biopharming. *Biopharming* is growing plants that have been altered, often via genetic engineering, for medicinal uses. These biotechnological strategies are not without controversy. Although the growth of these crops takes place in enclosed environments to decrease the possibility of contaminating other crops, critics are still very wary of these agricultural methods.

Aquaponics

Hydroponics is growing plants without the use of soil while aquaculture is growing fish. Aquaponics is the combination of these two systems. *Aquaponics* is a system of growing plants in water that has been or is being used to grow fish, snails, crayfish, or other aquatic creatures. Growers harvest the aquatic creatures, most commonly fish, and the plants.

The *effluent* (waste) produced by fish can be toxic to the fish, especially if they are being raised in a confined space. Aquaponics uses the effluent to contribute to the plant's nitrogen cycle. Nitrogen-fixing bacteria fix the ammonia into nitrates and then nitrites (which are used by plants as nutrients). Once nutrients have been used by the plants, the water is recirculated to the fish tank and the cycle begins again, **Figure 20-8**.

Chapter 20 Twenty-First Century Horticulture 515

Corner Question

What is a Brix scale?

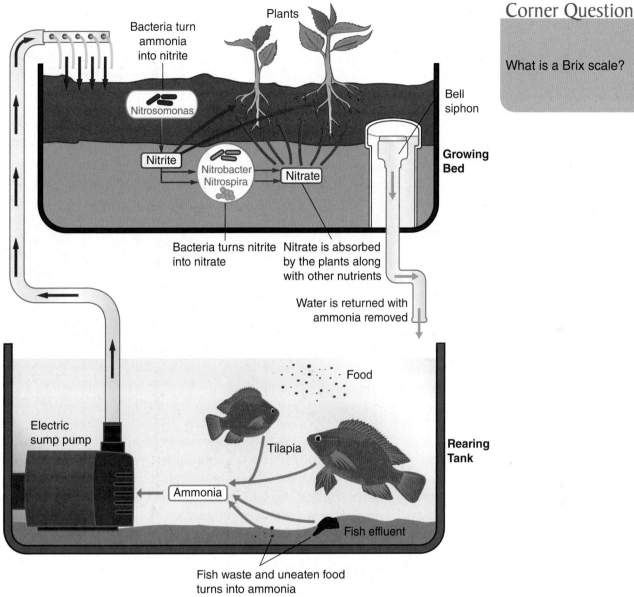

Moriz/Shutterstock.com; Chalintra.B/Shutterstock.com

Figure 20-8. An aquaponic system uses five main elements in the production of food. Fish and plants are the products while oxygen, water, and bacteria (along with fish effluent) act to produce the two crops.

Aquaculture farming systems range in size, complexity, and crops grown just as any other farming ventures do. Many growers consider aquaculture one of the solutions to dwindling land availability and an increasing need for food by a growing population.

History of Aquaponics

Aquaculture has been around for centuries. However, aquaponics and the cultivation practices used today rest on the research of the 1970s. Several institutions pioneered science and practices that enable aquaponics to thrive today.

Thinking Green

AgVets Hydroponics

Mike Walker, a US Marine Corps veteran, and his partners formed a company called AgVets to support veterans' return to civilian life and work. The objective of this company is to help veterans and their families see the opportunities for agricultural careers.

AgVets latest venture involves opening and operating greenhouses with hydroponic systems. Military veterans and their family members are given the opportunity to enter internships and learn how to grow crops, such as lettuce, basil, miniature cucumbers, and microgreens, hydroponically. There will also be opportunities to introduce disadvantaged youth to greenhouse farming. A university-based training program helps train workers through a mentoring program. Eventually, selected workers will manage large greenhouse operations. Mike says they are "growing their own growers."

Currently, AgVets' workforce provides flavorful produce to nearby markets with delivery the same day as harvest. The workers reap the therapeutic impacts of their work and also find a fulfilling career in hydroponic production. AgVets fosters economic, environmental, and social benefits for a growing world and population.

A.Krotov/Shutterstock.com

New Alchemy Institute

The New Alchemy Institute was a research center in Hatchville, Massachusetts. The center was a dairy farm converted to an aquaculture facility. The institute experimented with growing fish in bioshelters and reported on the use of aboveground, translucent tanks. The tanks were operated using solar power. The pond water, rich with fish waste, was used to irrigate crops in greenhouses.

North Carolina State University

In the 1980s, a graduate student (Mark McMurtry) at North Carolina State University in Raleigh, North Carolina, developed a research project to grow tilapia fish and vegetables. Mark McMurtry's experiment resulted in a new system with high fish protein production, water recirculation, water waste reduction, and high yields of vegetable crops. The S & S Aqua Farm near West Plains, Missouri, modified the North Carolina State University system to create a more efficient method called the Speraneo System. Tilapia fish were raised in 500-gallon tanks from which the water was circulated to vegetables grown in gravel inside a greenhouse.

The University of Virgin Islands System

A research team at the University of Virgin Islands team revolutionized the deep water culture (DWC) system of hydroponics, merging the growth of tilapia in rearing tanks as a fertilizer source. The fish tanks are linked together with a floating raft that acts as a stage for basil, Swiss chard, lettuce, and other crops. This system recirculates water through tubing with pumps. It has operated continuously for over a decade, allowing for harvests of fish every six weeks and lettuce and other crops weekly.

System Components

The same components used in hydroponic operations are used for the plant-growing portion of an aquaponic operation. These components, referred to as the hydroponic subsystem, will vary depending on the method being used and the grower's preferences. Aquaponic operations also use equipment specifically for the fish or other aquatic creatures being raised. Equipment and devices used for the aquaculture portion of an aquaponic system include:

- Electric sump pump—the mechanism that pumps water from one point to another in the system.
- Biofilter—an element that provides a habitat for bacteria to grow and convert ammonium (fish waste) into nitrates (a form usable by plants).
- Settling basin—a unit that catches waste and filters suspended solids.
- Rearing tank—the location where the fish are raised.

Growing in an Aquaponic System

The three main classes of organisms grown in an aquaponic system are the plant crop, the fish (or other aquatic creatures), and bacteria. Each of the organisms is an integral part of the system and each contributes and uses various elements to continue the cycle.

Plants

Leafy greens and herbs are commonly cultivated by aquaponic growers using the effluent water that is rich with nutrients from fish and bacteria. Plants grown using aquaponics include kohlrabi, cabbage, basil, dill, parsley, lettuce, and cilantro, **Figure 20-9**. Other plants that growers choose vary, but may include tomatoes, cucumbers, peppers, eggplant, radishes, taro, melons, and peas. The hydroponic method used varies by operation, but is usually limited to the following:

- Ebb and flow.
- Drip.
- Nutrient film technique (NFT).
- Deep water culture (DWC).

After the water passes through the hydroponic portion of the system, it is filtered and oxygenated before returning to the fish-rearing tanks.

Aquatic Creatures

Not all aquatic creatures thrive in an aquaponic system. For example, saltwater fish do not work with plant systems due to the high concentrations of salt used in the water. There are several freshwater fish that can be grown. However, some fish statistically grow faster and are healthier in aquaculture facilities than others. Tilapia is the number one fish used in commercial aquaponics, **Figure 20-10**. Other fish used include perch, catfish, cod, and barramundi. Some systems also grow snails, prawns, or crayfish.

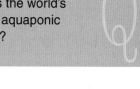

Corner Question

What is the world's largest aquaponic system?

Binh Thanh Bui/Shutterstock.com

Figure 20-9. Cilantro is a fast-growing, water-loving plant that grows well in aquaponic systems.

Ammit Jack/Shutterstock.com

Figure 20-10. Tilapia is the number one fish grown in commercial aquaponic operations because it is fast-growing, resistant to disease and parasites, and can handle a wide range of water quality and temperature.

Career Connection — Rebecca Nelson

Nelson and Pade Aquaponics

In 1984, Rebecca Nelson and John Pade constructed a hobby greenhouse in California. They followed with the construction of two commercial tomato hydroponic greenhouses. Later that decade, the company expanded into aquaponics. Today, Nelson and Pade, Inc.®, is housed in Montello, Wisconsin, and hobbyists and experts consider Nelson and Pade, Inc., a premiere aquaponic design, engineering, and supply company.

The company designs and engineers aquaponic systems, supplies aquaponic equipment, and helps hobbyists and professionals develop their knowledge in aquaponics. In 1997, the company began publishing *Aquaponics Journal* and began directing all of its efforts into this sector of agriculture. Today, the company is still growing. It currently has a 14,000 square foot demonstration greenhouse and 17,000 square feet of shop and warehouse space housed on 12 acres. The company offers courses affiliated with the University of Wisconsin-Stevens Point.

Nelson and Pade Aquaponics

Rebecca Nelson, one of the founders of Nelson and Pade, shows some of her tilapia and crops in this highly efficient aquaponic system.

Nelson and Pade have pioneered the aquaponics movement and are still key players in these growing endeavors.

Bacteria

Have you ever had a goldfish and wondered what happened when you found it floating in its glass bowl? The goldfish may have died from a high level of ammonium. Fish constantly release ammonium into the water as a by-product of their metabolism. In an aquaponic system, the ammonia is converted into a usable form of nitrogen by various bacteria and then absorbed by the plants.

- Nitrosomonas is a bacteria that converts ammonia into nitrites.
- Nitrobacter and nitrospira further the nitrogen cycle by converting nitrites into nitrates, a nutrient easily used by plants.

Bacteria housed in the aquaponic system form a biofilm over surfaces. A *biofilm* is a group of microorganisms that stick together on a surface. The coating formed by these groups of beneficial microorganisms is the thin, slimy coating found on the rocks, filters, and roots in an aquaponic system. Biofiltering units in aquaponic systems promote the growth of these microorganisms. Solid waste (uneaten food and waste) is removed from the tank containing the bacteria and is often treated for use as plant fertilizer. Once the solids are removed and effluent has been converted to nitrate, the water containing the nutrients is pumped to the hydroponic system.

Safety Note

Biofilms can grow on the vegetables harvested from an aquaponic operation. Biofilms can contain beneficial bacteria as well as harmful bacteria, such as listeria and E. coli. Therefore, vegetables should always be washed before consumption. Careful sanitation procedures must be part of good agricultural practices at aquaponic facilities to prevent food contamination.

Rooftop Gardening

Rooftop gardens are yet another form of gardening that has been in use for centuries. Evidence of rooftop gardens and elevated terraces has been found in ruins dating back to earlier than 600 BCE. These ancient gardens provided food, recreational areas, and shade to the inhabitants of many urban areas, much in the same manner that they do today.

Corner Question

What is the weight of a green rooftop system?

Techniques

A variety of gardening techniques can be situated on top of buildings. The type of gardening technique used for a rooftop garden may range from simple container plantings to an elaborate, intensive roof system. The type of gardening technique used often depends on the structural capability of the building.

Container Gardening

Container gardens are the simplest form of rooftop gardening. Container plantings are usually low in cost and fairly easy to maintain. The main elements of container gardens are the containers, growing medium, plants, a method for watering the plants, and a method of adding nutrients for plants. Container plantings allow the gardener flexibility in choice of containers, plant materials, and planning or arrangement. Growers may also more easily move seasonal plants indoors when needed.

Green Roof Systems

A green roof system is a carefully designed rooftop garden that requires professional design and installation. The garden is designed not only to cultivate plant materials, but to function as the roofing material for the structure. There are two methods of green roof systems:

- *Extensive green roof system*—a live, planted garden that forms a matting and helps to increase insulation, reduce runoff, and increase a roof's lifes pan. It has vigorous and hardy plants that require little maintenance, such as sedums and other succulents or prairie plants.
- *Intensive green roof system*—a landscape on top of a roof similar to a typical garden planted at ground level. Intensive roof systems may have complex or elaborate plantings. They usually require more extensive maintenance that involves fertilization, irrigation, pruning, and harvesting, **Figure 20-11**.

These systems significantly increase insulation for the building, which reduces energy needs and costs. The green roof system commissioned depends on the objectives of the building owner, projected maintenance needs, and the structural capability of the building.

Marius GODOI/Shutterstock.com

Figure 20-11. Small trees may be grown with an assortment of other plant materials in an intensive green roof system.

Thinking Green

Riverbend Nursery

Riverbend Nursery in Riner, Virginia, grows many plants and prides itself on cultivating plant systems specifically meant for green roofs. The company is a licensed grower for a company called LiveRoof®. The nursery produces sedums and other hardy materials meant for green roof systems.

Kelly Connoley-Phillips, the Sales and Marketing Manager at Riverbend Nursery, approaches architects, landscape architects, and engineers to introduce them to Riverbend's green roof products. Once LiveRoof is specified as the green roof system on a project, she works with the nursery staff to select a plant mix for the project based on the client's needs. Once the selection is finalized, growers set the 12- to 16-week growing process in motion. The green roof materials are shipped to the certified installer for same-day installation.

Kelly emphasizes the benefits of incorporating a green roof into new construction. There are several benefits aside from the aesthetic beauty. One of the most important financial benefits is that this will "extend the roof's life by 200%, skipping two replacement cycles." It can also help to decrease a building's heating and cooling costs. These benefits, coupled with the impact of decreasing the heat island effect and reducing runoff, should help architects and business owners realize the importance of going green, especially when constructing a roof.

Riverbend Nursery

Planning a Rooftop Garden

Several factors must be considered when planning a rooftop garden.
- Consider the engineering of the roof. Is it stable enough to hold a garden? What load can be supported? What is the overall condition of the roof? What access is there to the roof?
- Determine the objectives for this garden. What types of plants will be used? Will this space be used for leisure or food production? What is the budget of the project? What shadows are cast by surrounding buildings?
- Design the system to meet the needs of the objectives. What material will be used as a growing medium? What plant species will be grown? What irrigation system will be installed? What maintenance will be associated with the plant species used?
- Obtain the proper permits. Will there be a need for new electrical, building, or other permits? What windbreaks will be needed?

Structural Capability

The best time to install a rooftop garden is during new construction. Working with an existing roof presents several obstacles that can be costly, although not impossible, to overcome. Installing a green roof system on new construction is easier and more cost effective because all structural requirements and access points can be included in the original design. If an existing structure is to be modified to accommodate a rooftop garden, a licensed structural engineer or architect must first evaluate the structure.

The engineer or architect will determine the structure's load-bearing capability and whether additional supports are needed. The total load (weight) of the garden includes heating and cooling equipment, plants, media, water (including snow or ice), any equipment or materials, and the people who may visit or work on the roof.

Prior to construction, permits from city or other local governments must be obtained. Plans created by licensed engineers or architects must also be submitted for review.

Access

Access to the rooftop garden should be a primary concern during the planning stage. If the space is open to the public, two or more ways to access the site may be required. Multiple access points may be included in new construction and may include both indoor and outdoor entry points. Access usually includes stairs or a fire escape.

STEM Connection
Urban Heat Island Effect

During warmer months, cities often experience a phenomenon known as heat island effect. **Heat island effect** is an increase in temperature of urban areas due to a variety of environmental factors, including the radiant energy trapped by dark surfaces (roofs and asphalt pavement primarily) and the heat produced by increased air conditioner use. The higher temperatures, combined with air pollution, contribute to higher levels of smog and ground level ozone. **Ground level ozone** is a harmful air pollutant that is formed from chemical reactions of other pollutants near Earth's surface. Smog and ground level ozone are both harmful to human health and may cause problems such as eye irritation, asthma aggravation, and permanent lung damage. The heat island effect also negatively impacts the quality of the environment and building operating costs.

Increasing the amount of growing vegetation in a city is an excellent means of reducing the heat island effect and improving air quality. Plants provide heat absorption while also cooling the surrounding air through evapotranspiration. With limited land space available, rooftop gardens are a logical and practical solution. Rooftop gardens have additional benefits, including:
- Reducing energy costs as the plants and soil act as insulators for both heating and cooling.

Lena Serditova/Shutterstock.com

- Reducing rainfall runoff from buildings. Rainwater would otherwise run off the building, collect additional pollutants, and add polluted water to sewer systems.
- Filtering and cooling rainwater before releasing excess water into the sewer systems.
- Filtering the air by taking in carbon dioxide and producing oxygen through photosynthesis.
- Providing recreational areas for a building's inhabitants.
- Providing space for food production and enhancing the city skyline.

Cost

The initial cost of a rooftop garden depends on the type of garden and whether structural modifications must be made. The cost of a container garden may be the least expensive, but will vary depending on the materials (plant, media, and containers) used. Green roof systems cost about 50% more than conventional roofs. An extensive green roof system usually costs less than an intensive green roof system. Both systems, however, can increase the life of a roof by more than 50%, offsetting the initial investment.

Materials

The materials used in a green roof system must be installed to create an environment that will foster plant growth and maintain the roof's integrity. The materials are installed in layers in the following order (**Figure 20-12**):

- The bottom layer is composed of a lightweight, but sturdy insulation.
- A waterproof membrane is placed on the insulation layer. The waterproof membrane protects the building from moisture. This membrane is designed to withstand acidity released by roots.
- A root barrier is placed on the waterproof membrane to prevent roots from penetrating the membrane and insulation.
- A drainage layer made of a gravel, clay, or plastic comes next. This layer diverts excess water and aerates growing media. Additional drainage points or tiles should also be incorporated into the garden design. Some products used in the drainage layer are also designed to store water for later plant use.

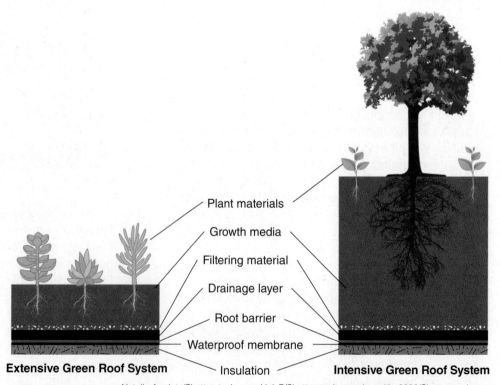

Natalia Aggiato/Shutterstock.com; Hal_P/Shutterstock.com; kostolom3000/Shutterstock.com

Figure 20-12. Green roof systems use layers of materials to create a growing environment conducive to plant growth.

- A geotextile or filtering material rests on top of the drainage layer. This mat permits water penetration while preventing erosion of the media.
- The last layer contains the media, plants, and any wind blankets or breakers. The media used must be lightweight, but must also promote water retention and drainage.

Irrigation and Drainage

When planning a rooftop garden, the water supply and storage system must be an integral part of the design. The garden design must include an irrigation system for water distribution, a drainage system to control excess water flow, and a means of storing excess water for later use. Although water from the city supply system may be used for irrigation, rainwater should be the primary source of water. Green roof systems are designed to take in rainwater as it falls and may also retain excess water for later use by plants. Excess rainwater may also be stored in rooftop cisterns, **Figure 20-13**. A drip system should also be installed to distribute water throughout the garden when needed.

Rooftop gardens reduce runoff by absorbing rainwater and diverting the excess to another system. The excess water must be dealt with using a drainage system. Water not absorbed by plants or media can be diverted to a cistern, and additional water directed to the sewer system. If too much water stays on the rooftop, there will be excessive weight added and plant root systems may suffer. Drainage systems can include gutters, downspouts, tiles, and drains. Screens or barriers may be used to prevent obstruction of the drainage system.

Corner Question

What is the weight of water per gallon?

sunipix55/Shutterstock.com

Figure 20-13. Cisterns are commonly used to store rainwater for irrigation purposes. Elevated tanks enable the grower to use gravity to help distribute water throughout the garden.

Growing Media

Natural soils made of clay, sand, and silt are extremely heavy. Growing media well suited for rooftop gardens include compost, organic matter, and recycled materials. The media must allow water to be held while also being permeable. In addition, the media ingredients must resist heat, tolerate frosts and thaws, provide a space for rooting and holding nutrients, and be resistant to fire. The media must secure plants in a manner that will prevent them from falling over due to wind. (Wind speed doubles for every ten stories of building height.)

Plants

Plants suitable for rooftop gardening vary, but they must be hardy. Generally, plants that do well in this environment also perform well in poor soils. Plants must be able to withstand harsh environmental conditions, as extreme weather conditions are magnified on rooftops. Insulation for roots helps with continuous freezing and thawing cycles that are encountered on many rooftops in the United States. Windy conditions also increase desiccation, or extreme drying, of plant material. Landscape architects and gardeners typically choose drought-tolerant and native species for successful rooftop gardens.

Maintenance

Rooftop gardens require the same degree of maintenance as other garden spaces. Rooftop gardens require irrigation, weeding, pruning, fertilizing, harvesting, *culling* (removal of dead or old plants), replanting, and the addition of soil amendments. Additional maintenance tasks include supporting windbreaks, cleaning drainage pipes, and winterizing irrigation equipment.

Vertical Gardening

As space for gardening becomes more limited, gardeners search for new ways or places to grow plants. Many people live in urban areas or take up residency in apartments and townhomes with little or no green space. Gardeners can grow plants without a great deal of space by planting a vertical garden. *Vertical gardening* is a method of growing plants on a vertical surface, such as a wall or trellis, rather than on a horizontal surface, **Figure 20-14**. Growers might also use vertical garden elements to add beauty or variety to a space. Methods of vertical gardening that people find successful for ornamental plants or food production include using pockets, trays, pot hangers, found objects, planters, and green walls.

stocker1970/Shutterstock.com

Figure 20-14. A vertical garden is one that is upright, like this one that blankets the side of a building.

Pockets

Several companies make products with a pocket design to hold plants. Pockets composed of a recycled felt material can be constructed as individual pockets or in rows of pockets. The pockets are filled with a potting soil and then plants are added. Pockets work especially well for growing edibles, such as herbs and strawberries. They are also excellent for houseplants and can add a touch of green to any wall in a building or home. Grommets or other methods are used to hang the material on a vertical surface. The pockets can be watered by hand or a drip tube or soaker hose can be attached to irrigate the plants.

Trays

Trays used for vertical gardening are similar to greenhouse or nursery trays. They are typically constructed of plastic or a decay-resistant wood, such as cedar. The trays are usually divided into planting cells. The planting cells range in size, depending on the type of plant material used. For these systems, it is best to use shallow-rooted material. The trays are mounted with a bracket and rest at a 30° angle to promote drainage and aeration, **Figure 20-15**.

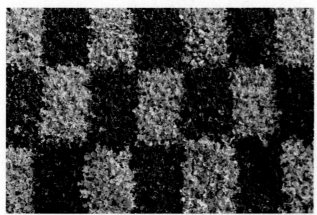

Piyachok Thawornmat/Shutterstock.com

Figure 20-15. Plant materials for vertical gardens can be grown in trays before installation. This lettuce, grown in two cultivars, covers a vertical space.

Thinking Green

Woolly Pockets Vertical School Gardens

Santee Education Complex, Alex De Cordoba

Miguel Nelson is a sculptor who combines his artistic abilities and love of plants to decorate the walls of event spaces. Inspired by compliments regarding the way he used plants to decorate plain walls for special events, Nelson used his decorating ideas to create a vertical garden system and helped found a company called Woolly Pocket. Wooly Pocket, with offices in Los Angeles, California, and manufacturing in Phoenix, Arizona, makes pocket systems using durable, 100% recycled milk jugs.

To get students involved in gardening and eating fresh, healthful food, Woolly Pockets offers a program for schools to get their own vertical growing system. Schools can sign up on the Woolly Pocket website and register for a Woolly School Garden. Woolly Pockets helps schools find funding for their vertical planting system. The program is a prime opportunity for FFA chapters in search of community projects. An FFA chapter can partner with elementary or preschools for a PALS project to help obtain and install a vertical system with younger children.

Pot Hangers

A pot hanger is a device made of polypropylene that supports containers. The hidden hanger can support up to 100 pounds and tolerate high winds. The device supports any device that has a lip for it to clamp onto. It can then be attached to any surface.

Found Objects

Salvaged or recycled materials can be used for growing plants and make attractive additions to the landscape, **Figure 20-16**. Gutters, troughs, pallets, shutters, old shoes, and soda bottles can all be modified or fitted to suit a vertical garden. Gardeners must ensure that these planters have adequate drainage and are near a readily available source of water. The plants can be watered by hand or a drip tube or soaker hose can be attached to irrigate the plants.

Imfoto/Shutterstock.com

Figure 20-16. Reusing materials to create growing areas is a creative and resourceful way to start a small garden.

Planters

Several types of planters adapt well to vertical planting systems. These planters include planting tubes, stacked containers, and barrels.
- Planting tubes are made of plastic or another fiber. They act as a hanging basket with holes distributed evenly around the tube. Soil fills the container and plants are placed in the holes.

526 Horticulture Today

Figure 20-17. Strawberries planted in these stackable containers are easily maintained as a vertical garden. They conserve space and offer an easy harvest of delicious strawberries.

Figure 20-18. Vertical wall gardens may be planted outdoors or indoors.

- Stacked containers can create a tower effect. Some are sold as a system; others are created by gardeners. These towers are easy to maintain, use less horizontal space, and are aesthetically pleasing, **Figure 20-17**.
- Barrels can be modified to provide spaces for plants to grow. Slots are made in these upright containers where plants can be inserted. Usually, up to 40 plants can be grown in a fifty-gallon barrel.

Green Walls

Green walls (often referred to as living walls) blanket the interior or exterior sides of buildings and structures with plant materials, **Figure 20-18**. Most walls incorporate an irrigation system into their design. A landscape architect professor, named Stanley Hart, patented the idea in 1938.

Green walls offer several benefits:
- Reduction in temperature surrounding the living wall. This helps to counteract the heat island effect found in urban areas.
- Reuse of non-potable water (water that is not fit for drinking). In recirculating systems, the plants can help to absorb some contaminants from the water.
- Use of a surface that would otherwise go unused. Areas that lack horizontal space for plants or are arid and cannot support plant life can benefit from green walls.
- Beautify the exterior and interior of buildings. Studies have shown that green spaces indoors promote a more productive work environment.

Before installing a green wall space, several factors should be considered. The factors include:
- How the wall will be constructed.
- The direction of the sun's rays and where they travel throughout the day.
- The media that will be used for this mode of vertical gardening: loose media, mat media, or structural media.

Loose Media

Systems that use loose media have potting soil placed in a package and attached to a wall, **Figure 20-19**. These systems must have media replaced annually when planted outdoors and biennially when planted indoors. These systems tend to erode and should not be used in applications over eight feet high. Additionally, these systems should not be used in areas where there is a great deal of public interaction as watering can be difficult and untidy.

Mat Media

A mat made of coconut coir or a felt-like product can be used to line a wall. This system is best for the interior of a building. These systems do not support mature or aggressive root growth. The mat will need to be replaced about every three to five years because the material will be covered with roots and plants will overwhelm the material.

Structural Media

Structural media systems have a life span of 10 to 15 years. Blocks of material with precise pH levels, water holding capacity, and aeration are manufactured into various sizes, shapes, and thicknesses. This strong material withstands continual watering, fertilization, heat, seismic activity, and wind.

Mrs_ya/Shutterstock.com

Figure 20-19. Loose media systems do not last as long as other vertical systems and can be difficult to water or maintain.

Raised Bed Gardening

Raised bed gardens elevate the surface of the cultivation area 12″ to 30″ above ground level. A raised bed is the perfect option for a gardener faced with a poor soil situation or for a container-type roof garden, **Figure 20-20**. The raised bed may be built from new or recycled products and can be filled with topsoil or compost to create optimal growing conditions. Soil in the beds is typically deep, loose, and fertile and has good aeration, drainage, and permeability.

Gardeners with physical disabilities or difficulty bending can garden more easily with raised bed gardens. Raised bed gardens can be easily weeded, irrigated, mulched, and generally maintained. To accommodate those with physical challenges, raised bed gardens should be constructed at least 30″ tall and no more than 4′ across. Raised bed gardening works well in urban and suburban situations where soil surrounding buildings is of poor quality and in areas where construction once took place or heavy foot traffic has impeded the soil's health.

Raised Bed Media

Gardeners can choose how to compose the growing media in their raised bed garden. The quickest and simplest method is to simply add organic matter to the needed depth, width, and length. Adding compost, aged manures, and leaves to the soil can help quickly fill the bed. Gardeners may use lasagna or sheet composting to fill raised beds.

Alison Hancock/Shutterstock.com

Figure 20-20. Raised beds with drip irrigation are situated on this intensive rooftop garden.

Lasagna composting is a system of building layers of organic matter to construct the growing media in a raised bed. The following steps are used in lasagna composting.

1. Cut the grass in the area to be gardened as close to the soil level as possible.
2. Cover the area with 8 to 10 layers of newspaper and/or a layer of cardboard and apply water until the material is saturated. Be sure to overlap the edges of the paper and/or cardboard.
3. If you are using a frame, place it on top of the newspaper/cardboard layer.
4. Pile layers of chopped grass, vegetation, mulch, soil, or compost on top of the newspaper or cardboard. Apply water to saturate the material.
5. Cover with more newspaper and/or cardboard and then again with organic matter, watering each time. Use a mixture of organic brown and green materials, such as grass clippings, kitchen scraps, sawdust, seaweed, and used potting soil.
6. Allow the material to decompose so that the material looks like earth and smells like good, sweet soil. This process will take several months, depending on the temperature and moisture level of the pile. Continue the process until the raised bed garden has the appropriate dimensions.

Intensive Gardening

Intensive gardening uses several horticultural techniques to yield the greatest harvests for a raised bed. Plants are carefully placed to promote growth. An example may include planting corn, beans, and squash together. The tall corn stalks act as a trellis for bean vines to grow on. The squash leaves provide shade around the base of the corn and bean plants to help cool the soil and prevent weeds from growing. Gardeners spend a great deal of time planning for every season's planting. Plants that are harvested are quickly replaced. Gardeners often exercise a method known as *square foot gardening*, **Figure 20-21**. Plants are placed every foot and are planted in a staggered pattern where leaves overlap to prohibit weed growth. Gardeners use certain compact varieties and species of plants to conserve valuable growing space. Harvests from this method are usually 4 to 10 times higher than conventional gardening methods.

S. M. Beagle/Shutterstock.com

Figure 20-21. Square foot gardening is an efficient and well-planned method of gardening.

Straw Bale Gardening

A raised bed garden does not have to be permanent nor use a frame. Straw bale gardens use composted straw bales as a planting medium. Straw bales are usually about 4′ long and 18″ to 24″ wide. When the straw bale is placed on the ground, it is nearly 24″ tall. A bale will turn into a block of compost over time and can act as a platform for plant growth, **Figure 20-22**.

Jodie Riedel/Goodheart-Willcox Publisher

Figure 20-22. Decomposed straw bales can be used to create nutrient-rich garden beds.

> **Did You Know?**
> The decomposition of a straw bale is an exothermic reaction. This means that as the microbes decompose the straw bale, heat is given off. A decomposing straw bale can reach temperatures in excess of 140°F (60°C).

Gardeners can speed the process of decomposition by using high nitrogen fertilizers or blood meal and applying water. To check whether the bale has decomposed sufficiently, insert a trowel into the bale to examine the material. It should be cool to the touch and smell like sweet soil. The interior of the bale should also look darkened and no longer resemble straw.

The decomposed bale will house between three to four vegetable plants. Some plants, such as tomatoes, should be planted in a bush variety. Two tomato plants may be housed in each bale. The plants must be adequately watered. Soaker hoses or a drip irrigation system work well. Additionally, plants must be heavily fertilized to ensure that they are receiving the appropriate amount of nitrogen. The carbon from the straw bales will lock up a great deal of the fertilizer, so additional nitrogen fertilizer must be applied. A good organic source is blood meal that is 12% nitrogen.

Careers

Each of the gardening techniques discussed in this chapter offers countless career and entrepreneurial opportunities. Aside from growers, many other personnel are needed to design, install, and maintain equipment.

Sundraw Photography/Shutterstock.com

Figure 20-23. An aquaponic manager cares for the fish in his operation.

Aquaponic System Manager

An aquaponic system manager possesses agricultural training and education—at least a bachelor's degree. This person develops crop and fish management programs at an aquaponic facility, **Figure 20-23**. He or she ensures proper planting, cultivation, and harvesting of fish and plants. The aquaponic manager provides technical expertise for daily crop production operations. He or she determines the types, quantities, projected sales volumes, and budgets for all crops. In addition, the manager may hire, train, and supervise all other workers.

Living Wall Designer

When envisioning an artist, most people picture a painter or a sculptor, not a horticulturist. A living wall or green wall designer is an artist who uses plants as a medium and walls or other vertical spaces as his or her canvas. Vertical landscape designers create beautiful, living art that also provides environmental benefits. A living wall designer has a plant science background coupled with an emphasis in art. Living wall designers may also be landscape architects or designers, botanists, or agronomists.

Career Connection: Jennifer Nelkin Frymark

Gotham Greens

In New York City and Chicago, Gotham Greens grows hydroponic crops on the rooftops of warehouses and grocery stores. Gotham Greens designs, builds, and operates urban greenhouses for hydroponic vegetable and herb production. They cultivate their plants in a pesticide-free, ecologically sustainable environment year-round to supply reliable, safe, and wholesome produce to the local community.

The chief agricultural officer, Jennifer Nelkin Frymark, learned how to grow plants hydroponically at Arizona State University. She has also worked in places such as Antarctica growing fresh produce for scientists in a continent covered with ice. She has a strong background in plant physiology and expertise in greenhouse system design, hydroponic systems and controls, integrated pest management, plant nutrition, and staff training.

Frymark is passionate about sustaining the environment and bettering her world. She uses hydroponic systems for her business for several reasons. Hydroponic systems recirculate water and produce crops with 20 times less water than conventional agriculture. Hydroponics also yields 20 to 30 times more than conventional agriculture. There is very little usable land in metropolitan areas, and hydroponics provides a more efficient avenue for plant growth. With urban locations,

Gotham Greens

Gotham Greens sells locally and eliminates long-distance travel for crops. The shipping reduction provides fresher produce, reduces fuel consumption, and prevents additional carbon emissions. In addition, Gotham Greens employs local community members.

Gotham Greens was recently named to the "Coolest New Businesses in America" list by Business Insider. This did not happen by chance. Frymark and her partners' efforts have made her company stand out and expand. Their work with companies such as Whole Foods has also helped Gotham Greens gain recognition in urban markets.

CHAPTER 20 Review and Assessment

Chapter Summary

- Hydroponic growers cultivate plants using water and nutrients without soil.
- British scientist Sir Francis Bacon formally researched hydroponic gardening in the seventeenth century. Advances in hydroponics have continued to the present day.
- Types of hydroponic systems include ebb and flow, water culture, deep water culture, aeroponics, nutrient film technique, drip system, and wick system.
- Crops grown using hydroponics include tomatoes, peppers, herbs, cucumbers, lettuce, and microgreens.
- Aquaponics is a system of growing plants with water that has been used to grow fish, snails, or other aquatic creatures.
- An aquaponic system uses a hydroponic subsystem for the plant portion of its operation as well as specific equipment (sump pump, biofilter, settling basin, rearing tank) for the aquatic creature portion.
- Plants grown using aquaponics include lettuce, kohlrabi, cabbage, basil, dill, parsley, and cilantro.
- Rooftop gardens are situated on top of buildings. They are an excellent way to reduce the heat island effect in urban areas. Plants provide heat absorption while also cooling the surrounding air.
- Rooftop gardens come in a variety of forms. Both container plantings and entire rooftop systems can add beauty to an area and provide cost savings for heating and cooling.
- Factors that should be considered when planning a rooftop garden include the load-bearing ability of the structure, the objectives for the garden, and the permits needed for construction.
- Vertical gardening is a method of growing plants on an upright surface, such as a wall or trellis. Methods of vertical gardening include using pockets, trays, pot hangers, found objects, planters, and green walls to grow ornamental plants or food.
- Raised bed gardens create an elevated gardening area.
- Career opportunities associated with nontraditional gardening systems include chief agricultural officer for a hydroponic facility, an aquaponic system manager, and vertical garden wall designer.

Words to Know

Match the key terms from the chapter to the correct definition.

A. aeroponic system
B. aggregate
C. aquaponics
D. biofilm
E. biopharming
F. culling
G. deep water culture (DWC)
H. drip system
I. ebb and flow system
J. effluent
K. extensive green roof system
L. ground level ozone
M. heat island effect
N. hydroponics
O. intensive green roof system
P. lasagna composting
Q. nutrient film technique (NFT)
R. square foot gardening
S. vertical gardening
T. water culture system

1. A harmful air pollutant that is formed from chemical reactions of other pollutants near Earth's surface.
2. A hydroponic growing method in which plant roots are suspended in the air and misted intermittently with a nutrient solution.
3. A method of hydroponics in which the nutrient solution is pumped from a main reservoir and circulated through an attached system of buckets containing plants.
4. A landscape on top of a roof similar to a typical garden planted at ground level.
5. A media that is used to cultivate plants in a hydroponic system.
6. A method of hydroponics in which the containers holding plants are periodically flooded with a nutrient solution and then drained.
7. A method of hydroponics in which plants in trays float with their roots submerged in a nutrient solution.
8. A hydroponic method in which a liquid nutrient solution is slowly released onto the base of each plant through a series of thin tubes.
9. The increase in temperature of urban areas due to a variety of environmental factors.
10. A system of growing plants in water that has been used to grow fish, snails, or other aquatic creatures.
11. A hydroponic method that circulates a thin stream of water containing nutrients over the roots of plants.
12. A method of growing plants on an upright surface, such as a wall or trellis.
13. A system of gardening in which every plant is allotted a specific amount of space in a garden.
14. Growing plants that have been altered, often via genetic engineering, for medicinal uses.
15. The waste from living creatures.
16. A group of microorganisms that stick together on a surface and form a thin, slimy coating.
17. Removing dead or old plants.
18. A method of growing plants in a water-based system.
19. A live, planted garden on top of a structure that forms a matting and helps to increase insulation, reduce runoff, and increase a roof's life span.
20. A system of building layers of organic matter to construct the growing media in a raised bed.

Know and Understand

Answer the following questions using the information provided in this chapter.

1. What are some problems associated with growing hydroponically?
2. What was Sir Francis Bacon's contribution to hydroponics?
3. What are the types of hydroponic growing systems discussed in this chapter?
4. What are some types of aggregates used in hydroponic systems?
5. What are some crops that are grown using hydroponics?
6. Why are some people wary of biopharming methods?
7. How does aquaponics differ from hydroponics?
8. What role do bacteria play in an aquaponic system?
9. What are the main elements used in an aquaponic system for the production of plants and fish?
10. What are the main elements of container gardens?
11. What is a green roof system?
12. What are some benefits of a rooftop garden?
13. What are some factors that should be considered when designing a rooftop garden?
14. What are the elements or materials that make up a green roof system for a rooftop garden?
15. Why might growers choose to plant a vertical garden?
16. What are some methods of vertical gardening that people find successful for ornamental plants or food production?
17. What is a green wall and what are some benefits offered by green walls?
18. Describe a raised bed garden.
19. Describe the education needed for and job duties of an aquaponic system manager.

Thinking Critically

1. You recently started an aquaponic system to grow tilapia. You are using city water. Within the first 48 hours, 75% of your fish died. Hypothesize the reasons for their sudden death.
2. You visited a vertical wall that was constructed almost three years ago using pockets on the interior of a building. The plants did very well for the first year and a half. They have been watered, fertilized, and exposed to the correct amount of light. Now, these once vibrant plants look sick and weak. What could be the reason for their decline?

STEM and Academic Activities

1. **Science.** Find a building that has a green roof system and compare the engineering required of this building to your home or school.
2. **Science.** Design a living wall for a space in your school. Select the appropriate plant materials.
3. **Engineering.** Design and construct a vertical garden using a recycled plastic barrel.
4. **Social Science.** Contact a company that offers hydroponic systems for urban areas. Ask questions about its workforce and who it employs to operate the business. Create a short video or marketing announcement about this company and what it provides for its employees or consumers.
5. **Language Arts.** Write a position paper outlining the advantages and disadvantages of green roofs.

Communicating about Horticulture

1. **Reading and Writing.** Compare and contrast three types of hydroponic growing aggregates. Create a chart with the pros and cons of each type.
2. **Reading and Writing.** Research aquaponics and write a two-page report explaining the benefits of using this technology.

SAE Opportunities

1. **Exploratory.** Job shadow a hydroponic grower.
2. **Experimental.** Compare different growing media on one hydroponic growing system that uses an aggregate culture.
3. **Exploratory.** Research how hydroponic or aquaponic systems can positively impact urban communities around the world.
4. **Entrepreneurship.** Design and construct vertical garden systems using recycled materials such as pallets and sell them to the public.
5. **Entrepreneurship.** Grow plants hydroponically and sell the produce.

I love photo/Shutterstock.com

CHAPTER 21
Nursery Production

Chapter Outcomes

After studying this chapter, you will be able to:
- Describe elements related to the market niche for nursery operations.
- Discuss production methods of nursery operations.
- Describe sustainable nursery production practices.
- Identify nursery-grown plant materials.
- List careers related to nursery production.

Words to Know

capillary mat	mechanical transplanter	socket pot
circling	oscillating sprinkler	substrate
crown	phytosanitary	suction lysometer
drip irrigation	certificate	sustainable nursery
electrical conductivity (EC)	pot bound	water sprout
leachate	root pruning	winterization

Before You Read

Take two-column notes as you read the chapter. Fold a piece of notebook paper in half lengthwise. On the left side of the column, write main ideas. On the right side, write subtopics and detailed information. After reading the chapter, use the notes as a study guide. Fold the paper in half so you see only the main ideas. Quiz yourself on the details and subtopics.

While studying this chapter, look for the activity icon **to:**

- **Practice** vocabulary terms with Words to Know activities.
- **Expand** learning with identification activities.
- **Reinforce** what you learn by completing Know and Understand questions.

www.g-wlearning.com/agriculture

A plant nursery facilitates growth of plant material under a number of different conditions. Nurseries grow seedlings, transplants, stock plants for propagation, container plants, and field-grown plants. These plants may include ornamental or edible trees, shrubs, bulbs, grasses, groundcovers, herbaceous perennials, and annuals, **Figure 21-1**. Nurseries may incorporate growing structures into their production that can include greenhouses, high tunnels, and shade or lath houses.

Market Niche

"Effective marketing is the key to profitability."
—Author unknown

Nursery businesses are found in the United States and around the globe. Like other businesses, a nursery tries to identify one or more specific areas of a broad market for which it will produce products. This specific area is called a market niche. Nurseries find their market niche based on the nursery site location, the types of plants that are cultivated, and production methods.

Like all businesspeople, entrepreneurs in the plant industry must work hard. An extensive knowledge of plants, marketing, and management are essential for success.

Site Selection

When selecting a nursery site, several characteristics should be considered. These include the location, growing conditions, soil type, water and utility availability, and existing structures or barriers. The topography of the site should have a slight grade of less than 5%. A site that is very low lying or has high hills should be avoided due to frost pockets or high winds.

Elena Pavlovich/Shutterstock.com

Figure 21-1. These plants, a number of differently shaped conifers, are growing in a field nursery.

538 Horticulture Today   Copyright Goodheart-Willcox Co., Inc.

Crops

Crops depend on market demands and the types of plants suited for the site. Specialty crops suited for a niche market are best produced in small nursery operations. Greenhouse production can increase what can be grown in a native environment, but a greenhouse may be costly to heat and cool. Outdoor facilities can grow material that is suited only for that USDA hardiness zone, **Figure 21-2**.

Types of Nurseries

Nursery crops are marketed (advertised and sold) via several avenues, which include wholesalers, re-wholesalers, mail-order companies, retailers, and landscapers.

Ruud Morijn Photographer/Shutterstock.com

Figure 21-2. The winter means a reduction in temperature. Nursery managers must ensure that the plant material is protected and that winterization takes place. **What materials could be used for proper winterization of a nursery?**

Wholesalers

Wholesalers sell plants at low prices to retailers, landscapers, and re-wholesalers. The plants are sold at significantly reduced prices compared to what consumers pay for plants. This is possible because the wholesaler sells fewer types of plants but in greater quantity. They may specialize in a particular type of plant, such as conifers, evergreens, roses, or herbaceous perennials, to help them keep cultivation costs low.

Re-wholesalers

Re-wholesalers are businesses that purchase diverse crops from wholesalers and sell them to landscapers. The landscapers may require more diverse plant material and in smaller quantities than a typical wholesaler may wish to provide.

Mail-Order Companies

Mail-order companies sell directly to consumers through print catalogs and websites. Orders are delivered rather than picked up by a customer at a retail outlet. Mail-order nurseries sell plants in containers or bare root because of high shipping costs. Specialty nurseries thrive doing mail-order sales because a customer can order plants regardless of the customer's location.

Retailers

Retailers sell directly to the final customer. Businesses must be conveniently located for their customer base. These businesses can grow their own plant material or purchase and resell plants from a wholesaler, **Figure 21-3**.

Blend Images/Shutterstock.com

Figure 21-3. A retail nursery worker should be friendly to customers and be knowledgeable about plant material.

Career Connection: Tony Avent

Plant Delights Nursery

Photograph by Seny Norishingh

Almost 30 years ago, Plant Delights Nursery in Raleigh, North Carolina, was founded to serve and help fund the Juniper Level Botanic Garden. Tony Avent, a horticulturist who studied at North Carolina State University, wanted to explore the world and collect plants. He was driven to acquire unique plants and display them at his garden.

Juniper Level Botanic Garden also served as a facility for propagation and research for Avent. Through his selection, propagation, and cultivation of *Hosta*, *Colocasia*, *Epimedium*, and numerous other ornamental perennials, Avent created a mail-order business, supplying customers with specimens from the garden collection. Throughout the year, dedicated nursery workers ship plant materials all over the world to anxiously awaiting customers. Plant Delights Nursery now devotes 10% of its income to Juniper Level Botanic Garden.

The myth that a nursery cannot be lucrative has been disproven by Plant Delights Nursery. Tony Avent also authored a book called *So, You Want to Start a Nursery*. If you are interested in starting your own mail-order or other type of nursery, reading his book would be a perfect starting point.

Did You Know?

In the United States, nursery-grown crops are valued at nearly $7 billion and account for nearly 40% of all horticulture crops grown.

Landscapers

Landscapers are businesses that design and install plants and garden features or structures. Landscape nurseries grow products for their own use. They may also have some plants to sell to their landscape clients or other customers. Landscapers want plants in large sizes to give a finished look to the gardens they install. They also use balled and burlapped wood plant materials. Landscapers want plant material that is identified by cultivar and of high quality.

Licensing and Shipping Regulations

Government agencies such as the USDA regulate nursery operations. Nurseries must obtain permits or licenses to grow plants through their state government. Nurseries must have a dealer license to sell plants and must obtain additional licensing to sell plants out of state. When plants cross a state or international line, they require a special *phytosanitary certificate*. This official document states that the plant material has been inspected and appears to be free of pathogens, insects, and weeds.

Market Outlook

Since 2006, the nursery industry and green industry have suffered from the housing market crash and economic downturn. When building of housing developments nearly ceased, so did the demand for green products, such as trees and shrubs. In addition to a lethargic housing market, sluggish consumer spending, high costs of inputs, and increased labor expenses led to a downturn in the nursery industry.

In 2013, however, the housing market increased, and the need for plant material swelled. Economic recovery in the United States allowed nursery growers to begin to see profits increase. According to some cooperative extension agencies, about 35% of nursery growers are now ready to make increases in capital investment and prepare for future growth.

Production Methods

Nursery operations use one of three production methods: the container-grown method, the field-grown method, or the pot-in-pot (PNP) method. Each method of cultivation has its own advantages and disadvantages. A nursery operator must decide which method meets the needs of its customers, plant materials, and growing environment.

> **Did You Know?**
> Container-grown crops generate about 10 times more sales than field-grown crops.

Container-Grown Production

Almost 80% of all nursery crops are planted in containers because these plants are popular with customers. Container-grown plants are cultivated with soilless media in a variety of container types. The plants are then grown in greenhouses, tunnels, or in the ground in nursery operations, **Figure 21-4**.

Rigucci/Shutterstock.com

Figure 21-4. Container-grown nursery material provides a number of advantages over other methods of nursery production, such as ease in transportation and lower weight than field-grown plant material.

Hands-On Horticulture

Phytosanitary Certification

Who authorizes phytosanitary certificates in the United States? Several government agencies work together to regulate phytosanitary certificates. These organizations include:
- United States Department of Agriculture (USDA)
- Animal Plant Health Inspection Service (APHIS)
- APHIS Plant Protection and Quarantine (PPQ)
- National Plant Protection Organization (NPPO)

These agencies work together to regulate importation and exportation of plant material. Now, imagine the process and paperwork that is needed for plant material to be secured through these systems. Nursery workers must understand how to navigate the system and have patience in dealing with this task.

Types of Containers

Several factors influence the type of containers a nursery grower uses. The cost, size, shape, shipping capacity, durability, appearance, insulation value, features that control root growth, and sustainability of the material all play a role in which container is used.

- Cost. The cost needs to be low. Thousands to millions of pots are used annually by small and large nurseries alike. Containers can be a major expense in production.
- Size. Nursery containers range in size. They are often sold based on volume of media held by the container (1 gallon, 3 gallons, 5 gallons).
- Shape. The shape of the container plays a role in ease of stacking and storage as well as the ability to withstand tipping in the greenhouse.
- Shipping capacity. This determines how many of a particular size and shape of pot can be shipped for delivery.
- Durability. The length of time that a plant will be housed in a container should match the life span of a container. Plastic, fiber, metal, and wood are useful for different amounts of time.
- Appearance. In retail nurseries, customers want a container that is attractive. Nurseries also have a chance to advertise or brand their product on containers to entice customers. This is less important to landscapers, who typically use black, round pots, **Figure 21-5**.
- Insulation value. Growers in places with extreme cold temperatures or cycles of frosts and thaws must pay close attention to how well the container insulates roots.
- Features that control root growth. Copper can control root growth. Plants grow taller, are less pot bound, and leach less water in these pots. Fiber pots treated with copper do not decompose as quickly as other fiber pots. When roots come in contact with the copper, they die back. This helps prevent *circling* (roots growing around the inside of the pot) or *pot bound* (roots filling the pot) root formation. See **Figure 21-6**.
- Sustainability of material. Containers that are recycled, biodegradable, or sourced from renewable resources are popular with many nursery growers. These materials are effective for plant production and friendly to the environment.

weerayut ranmai/Shutterstock.com

Figure 21-5. Paper pots or bag pots have been used to produce plants. Many nursery workers feel that this method of cultivation is a more viable option for twenty-first century horticultural production.

Scott Latham/Shutterstock.com

Figure 21-6. A problem associated with container production is that roots may circle the pot or become pot bound. There are ways to prevent this irregular root growth. **What is one way to reduce pot circling?**

STEM Connection

Container Engineering

The nursery industry has discovered that modifications made to plant containers can control plant root growth.
- Bottomless pots. These containers lack a bottom. When roots grow to the bottom and are exposed to air, they die back. This is called air pruning. This causes significant branching and a more robust root system within the pot. These containers are made of plastic or paper.
- Ribbed containers. Dr. Carl Whitcomb invented a unique container of this type called the RootMaker. RootMaker containers have staggered walls and a staggered bottom. This container design prevents root circling and forces roots to grow downward and outward to holes that direct the path of the roots.
- Copper-coated containers. Growers may use a commercial paint with copper embedded in the product to coat nursery pots. They may also purchase nursery containers with copper embedded in the container walls. These products can help nursery growers avoid pot bound plants.

Corner Question

How many acres does a container nursery operation need to be profitable?

Planting

To be successful, growers must plant crops that are appropriate for the nursery's growing environment. Countless plants and cultivars can be selected, but they must be suitable for the site (greenhouse, tunnel, or outdoor area). Once a crop is selected, the nursery manager must choose the appropriate container and *substrate* (growing medium or material in which a plant's roots grow).

Aged pine bark is the most commonly used growing medium in the nursery industry, **Figure 21-7**. Other amendments, such as peat and sand, can be added to create a better growing medium. The medium must be well drained and able to hold some moisture. It must also allow for resaturation (holding liquid again after drying out).

Plants must be potted either manually or with some automated help. Automation can include potting machines that physically fill up containers with media in place of employees. Both labor and automation are significant costs. Automation, unlike labor, has higher initial costs, but returns on the investment will likely be seen in time, **Figure 21-8**. Wholesale operations often use automation to reduce costs of production in the long run.

©iStock.com/naruedom

Figure 21-7. Pine bark is a major component of most nursery substrates. It is a sustainable component and a welcome alternative to peat moss.

HildaWeges Photography/Shutterstock.com

Figure 21-8. An automated planter increases the speed of nursery planting.

Figure 21-9. A method of monitoring pest populations rests with insect strips. These traps collect insects and allow nursery workers to identify and count pest populations.

hans engbers/Shutterstock.com

Maintenance

Plants in containers must be heavily irrigated. Growers may choose to use drip systems, overhead irrigation, or capillary mats for most nursery-grown plant material. Growers must cluster similar plants together to ensure the plants are watered similarly.

Nutrient applications can be controlled-release fertilizers or fertigation applications. Growers need to determine which method best meets the needs of the crop. Additional maintenance of container-grown crops includes pruning, *winterization* (protection from cold and harsh conditions), weed control, disease control, algae control, and insect control, **Figure 21-9**. These maintenance factors all must be closely monitored and regulated by the nursery manager.

Harvest

Nursery-grown plants can be harvested at various stages, which can include:
- Liners—plants that will be transplanted and grown into a saleable size plant.
- Whips—straight stems with some branching. Whips are often seedlings that are one to two years old.
- Finished materials—plants that have all the characteristics of a mature plant.

These crops can be harvested any day of the year. The labor that is required for harvesting is very light. Harvest may involve simply picking up and moving the plants in containers.

Field-Grown Production

Until the mid-1900s, field-grown production was nearly the only method of production in the United States. Unlike container production, growers pay particular attention to the soil's profile and overall health. Success in field-grown production depends on the site location, crop selection, and planting methods as well as proper maintenance and harvesting of crops. According to some cooperative extension agencies, at least 200 acres is needed for a field-grown nursery to be profitable.

Corner Question

How much labor is required for each acre of container nursery production?

Site Location

Site characteristics for field-grown nurseries are the same as many of those needed for container-grown operations. A slight slope of less than 5% is ideal. There should be few low valleys or high points where temperature variances are significant. Few spots, if any, should support standing water or flooding.

Accessible water and utilities are essential for production. Nursery managers should be aware of the site's history. Was there crop production? Were pesticides or chemicals used?

The site's soil is of the utmost importance to a field-grown nursery grower. If the nursery plans to grow bare root crops, the soil must be sandy and easily removed from roots after harvest. For balled and burlapped (B&B) production, the soil must stay firmly attached to roots, which means it should have some clay content, **Figure 21-10**. The soil must be fertile, well drained, and free of some significant pests. Once soil is analyzed, amendments of organic matter and nutrients must be applied to increase the productivity of the soil for field-grown crops.

J. Bicking/Shutterstock.com

Figure 21-10. A balled and burlapped plant should be planted quickly after harvest. Unlike container plants that can be held in a pot for some time, B&B plants can suffer from delayed transplant.

Crops must be protected from wind, which can dry plant material. A barrier of trees or shrubs may already exist along the perimeter of the nursery grounds. If there is nothing to help slow down the velocity of the winds, then the careful planting of trees and shrubs to create a living fence is strongly recommended for nursery operations of all types.

Did You Know?

Burlap is a coarse fabric that uses a large woven pattern made of jute, hemp, or a similar plant-based fiber.

Crop Selection

A nursery grower must select the appropriate plant materials that will tolerate the climatic conditions of the nursery and the soil conditions. Growers can choose from plants highly demanded by a given market, grow crops that are part of a niche market, or try uncommon species for specialized customers. Whatever the choice of the grower, the goal is to make profits. This requires that there be customers willing to purchase the plant products, **Figure 21-11**.

Planting

When and how to plant will depend on the type of plant material cultivated. Nursery managers often plant rows of liners between late fall and early spring. This process can be accomplished manually or by using *mechanical transplanters* (machines that move plants from one place to another).

senee sriyota/Shutterstock.com

Figure 21-11. Plants in high demand, such as these drought-tolerant succulents, are constantly changing, depending on market changes. Ten years ago, these plants were not readily available, but can be found globally today.

Figure 21-12. Planting a bare root plant may be easier for nursery workers. This method is also good for shipping. *Why would bare root production save nurseries on shipping costs?*

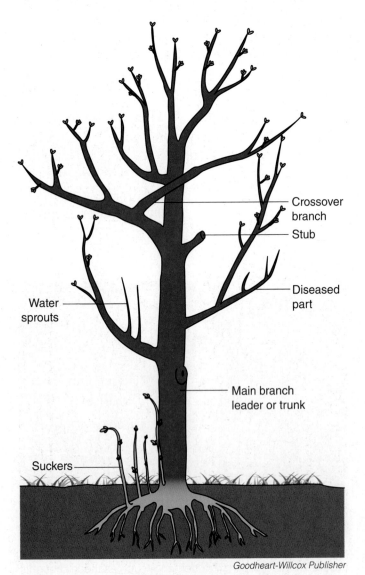

Figure 21-13. Suckers and water sprouts should be removed from plant material. Pruning can take place year round, especially on plant structures like these.

The *crown* of the plant (where the stem and roots meet) should never be planted below the soil level. The crown should be at the same depth at which the plant was previously grown, **Figure 21-12**.

Plant material should be spaced in a row spacing that permits optimum growth. Plants should not crowd one another and should be spaced according to the size desired upon sale. During the first year after planting, plants may require staking to help support straight growth and prevent breakage during wind or rain.

Maintenance

Immediately after plants are placed in the soil, irrigation or hand watering must take place to saturate the soil. New plants are likely to wilt from unestablished roots and will need supplemental water for the first two years of growth. After this, plants will only need water supplied during periods of drought. Field nurseries can be irrigated with driplines or overhead irrigation.

Fertilizer is applied before planting and then as needed based on the results of soil tests. Nutrients can be applied using equipment such as a tractor when many acres need fertilizing. Different soils, different crops, and different times of the year will require unique fertilizer applications that must be accurately calculated, distributed, and recorded.

Pruning is important for field-grown crops. Pruning is done to remove dead, diseased, or damaged stems. Stems that are crossed, rubbing other stems, *water sprouts* (stems that grow vertically from other stems or branches), or suckers (plant stems that grow from the base of the plant vertically, near the trunk) must also be removed, **Figure 21-13**. Successful pruning should maintain a natural shape with compact growth. In another method called *root pruning*, the roots of a plant are severed to promote growth of some kind. Root pruning can be done regularly or just before harvest to ensure a better transplant. Root pruning can create an output of more fibrous roots or induce flowering or fruiting.

Thinking Green

Human Waste as Fertilizer

The human effluent that is produced and put into sewage treatment plants can have an incredible value for plants. Milorganite® is a rich, slow-release fertilizer produced from a waste treatment plant in Milwaukee, Wisconsin. Milorganite is produced by removing all water content and adding heat-dried microbes to digest the organic matter in the wastewater. The remaining water is cleaned and returned to Lake Michigan. The resulting microbes are then dried, bagged, and shipped all over the country as Milorganite fertilizer. Gardeners can use the waste product to fertilize all kinds of plants while bettering the environment. Milorganite is one of the world's largest recycling efforts.

Corner Question

What pest lives in the soil and can make it impossible for nurseries to ship their product to other states?

Harvesting

In comparison to container-grown nursery plants, harvesting field-grown crops is much more labor intensive. Plants can be dug by hand or a tree spade (a mechanical digger) can be used. Ideally, harvesting takes place in the months when plants are dormant and less transplant shock can occur. However, the demand for field-grown stock may be year round, and harvesting must take place during any season.

Plants are usually harvested for B&B within three to five years after initial planting. The caliper (diameter) of the tree's trunk determines the size of the B&B. Once the plant is dug, a laborer immediately places burlap around the root mass, which is still covered in soil, **Figure 21-14**. Often, a wire cage is then placed around the burlap to promote easier transport and durability. The wire cage and burlap are tied and secured. The ideal time from harvest to transplanting should not be more than two weeks. If transplanting must be delayed, then careful attention must be paid to watering and temperatures (ensuring that the roots do not desiccate or freeze).

Zigzag Mountain Art/Shutterstock.com

Figure 21-14. The size of root ball that is dug or harvested depends on the diameter (caliper) of the plant's trunk. The size of a B&B plant is determined by measuring the trunk of the tree using a tree caliper.

Pot-in-Pot (PNP) Production

Pot-in-pot (PNP) nursery production involves planting material in containers (production pots) and then placing those pots into *socket pots* (permanent containers placed in the ground). This method of nursery production is used for shade trees, ornamental trees, and large shrubs. This system crosses container production with field-grown production to give benefits of both techniques.

Did You Know?

Although less labor is generally required for field-grown production, there are still many tasks that must be accomplished by workers. These tasks include planting, pruning, pest control, fertilizer application, irrigation, harvesting, and packing for shipment. One worker needs to be employed for every seven to eight acres of field-grown nursery production.

Figure 21-15. PNP nurseries must use fabric and irrigation, usually drip, for successful growth.

Figure 21-16. A socket pot is the pot that the production container is placed into and stabilized.

PNP production provides a number of advantages to nursery growers, **Figure 21-15**. Container-grown plants can be harvested and shipped at any time of the year with little advance notice. Field-grown plants require digging and shipment of very heavy material. Unlike a container system, PNP allows plants to be secured in ground so that they do not tip over or suffer from extreme cold.

Site Selection

For a PNP site, soil drainage is the number one issue. The socket pots can easily fill with water. If the surrounding soil does not permit drainage, then the socket pot becomes a bucket filled with water. When soil is not permeable, the water fills all the spaces in the growing medium and the plant cannot respire. This situation leads to poor growth or death for the plant. A drainage system can be constructed around and beneath the pots, however, to ensure that the water can drain from the socket pot, **Figure 21-16**.

Quality water and utilities must be readily accessible to the PNP nursery. A retaining pond or a site to reclaim water runoff will help to improve the environmental impacts while reducing costs associated with water.

Unlike other nursery systems, the slope or grade of the nursery should be minimal. Only a 1%–2% grade is desired for PNP operations. As for other nursery operations, extreme low areas can be pockets for frost, and sites with hills would not be suitable for PNP operations.

Installation

Planning and preparation contribute to the successful installation of socket pots. One of the most important decisions for a PNP nursery is the socket pots used. The socket pot decides the size of the production pot and the size of the marketable plant material the nursery offers.

The drainage of the existing soil will determine how the socket pot is installed. Soils with good drainage can use an auger to dig holes for each socket pot. Soils with poor drainage can use excavation equipment to dig trenches for placement of a drainage tile, followed by the socket pot.

Once the socket pot is placed, the soil is then backfilled around the pot. The rim or lip of the lid extends just above the existing soil surface.

Corner Question

How much does the average PNP operation cost to establish?

A weed barrier fabric or geotextile can then be laid over the area and cuts made over the socket pot's opening, **Figure 21-17**.

Crop Selection

PNP nursery growers select valuable trees and shrubs for their production. Since PNP operations require high capital investments and have high labor costs, they must select crops that will yield high profits. The selected species should be suited for the site's climatic growing conditions.

Planting and Maintenance

The planting of the production pots is similar to planting for container-grown production. Growers select media to meet the needs of the plants being produced. This often starts with a pine bark mix with various amendments. The production pots receive the same treatments as container production, including irrigation, fertilizer application, pruning, staking, and pest control.

Jodi Riedel/Goodheart-Willcox Publisher

Figure 21-17. Nursery workers lay a geotextile over the socket pot holes. Slices in the fabric are then made and the socket pots are put into place.

STEM Connection

Aboveground System for PNP

An aboveground system (AGS) for PNP provides versatility for growers. Nursery managers can easily move the socket pot from one area to another. The idea is just like a PNP system with a few differences that can make AGS an easier alternative to PNP. Unlike PNP (where socket pots are embedded in the ground), AGS have socket pots fixed to the pavement or concrete with screws or nails. The screws or nails secure the socket pots in place and prevent the AGS from tipping over and allow for more security than a traditional container system in a nursery. The real difference between a traditional PNP and an AGS is that this system can be installed anywhere, regardless of soil type or drainage. The AGS can be situated on top of a parking lot, gravel, or any other flat surface. This is the perfect choice for an area of a nursery that may not have permanence. An AGS still provides some insulation (provided by the additional pot) and many of the other benefits associated with traditional PNP systems.

Jodi Riedel/Goodheart-Willcox Publisher

Did You Know?

Labor required for PNP is somewhere between container production and field-grown production. The average PNP tree requires 10 minutes for planting, 30 minutes for pruning, 30 minutes for maintenance, and 5 minutes for harvest. That is 1 hour and 15 minutes of labor required per plant.

Since the plants' containers are not exposed to the elements, there is very little need to winterize the PNP system. Careful attention must be paid to ensure that there is a tight seal between the two pots and between the socket pot and the ground. Growers reap the benefits of PNP production during the colder months of winter when less work has to be completed than for container-grown plants.

Plant material moves through the PNP system usually through a three-year rotation. Plants are not potted into their final container. They usually rotate through three varying sizes. Plants begin in one size of pot, and then are usually transplanted twice. Each new production pot has a corresponding home with a socket pot. The plant is allowed to grow and is maintained until its final or marketable size.

Harvest

Plants will grow to maturity and be ready for harvest at varying times based on the size of the production container, the type of plant grown, and the cultivation practices. Once the plant is mature by market standards, it can easily be harvested. Production pots are simply lifted from socket pots.

Unlike field-grown plant material, PNP plants can be harvested year round and with very little lead time needed from customers. Unlike field-grown plants, weather does not impact the harvest—saturated or frozen soil is not a factor. However, the location of plant roots can present a problem in harvesting.

Plant roots can creep out of production pots and into socket pots. If a plant's roots do escape into a socket pot, the harvest (removal from the socket pot) can damage the plant roots. To prevent this from occurring, growers can use pots lined with copper. They also monitor the plant's root growth and transplant the plant to a larger pot when needed. They can use squares of fabric (made with copper or a product that prevents root growth woven into the fabric) to line the bottom of the socket pot, **Figure 21-18**.

Jodi Riedel/Goodheart-Willcox Publisher

Figure 21-18. Alan Erwin from Panther Creek Nursery shows a root shield product that is made of a geotextile. The fabric includes copper and helps to ensure that the roots of the plants will not pass through the socket pot and into the native soil.

Production Method Advantages and Disadvantages

As stated earlier, a nursery operator must decide which production method to use. There are many factors to take into account, including resources, environment, and customer needs. There are also different advantages and disadvantages of each production method. These factors include amount of land, amount of labor need, and harvesting requirements, **Figure 21-19**.

Container-Grown Production	
Advantages	**Disadvantages**
• Requires less land • Easier to transport • Easier to harvest, done year round • More customers want containerized plants • Return on an investment is realized in one-half to one-third the time of a field nursery	• Container areas require more money to construct and maintain • At least one employee needed per acre of production • A lot of gravel needed for drainage • Needs winter protection • Higher degree of management • Plants will blow over during every storm and wind • Requires specialized equipment for mixing media • More transportation required • 27,000–40,000 gallons of water are required per acre (for 200 irrigations a year) • Intensity of management is greater • Plants are more susceptible to physiological stresses (heat, cold)
Field-Grown Production	
Advantages	**Disadvantages**
• Less labor per acre needed • Larger plants can be more easily produced • Less irrigation is required • Lower start-up costs • Requires less year-round attention • Soil provides insulation or buffer to heat and cold stress • Some plants can be produced more economically in the field than in containers	• More land required • Good soils needed • Harvesting equipment needed • Must have a hard-working labor pool • Greater shock to plant during transplant • Harvesting usually takes place only when plants are dormant • May have to harvest regardless of need, "sacrifice" plants to shift or make space for growing trees • Rootballs from B&B are very heavy and costly to ship, requires mechanical loaders for transport
Pot-in-Pot (PNP) Production	
Advantages	**Disadvantages**
• Less water • No winter protection needed • Can harvest year round • No blown over containers • Greater root growth than container-grown plants • Cooler root system • Healthier rootball than traditional container plants	• Very well drained soil • Time to check irrigation • High start-up costs • Two containers must be purchased in the beginning for each plant • Susceptible to lethal high-temperature damage during post-production

Goodheart-Willcox Publisher

Figure 21-19. This table shows the advantages and disadvantages for different types of nursery production methods.

Sustainable Nursery Production

It seems unimaginable that any nursery would choose to grow plants using only conventional growing techniques, but many nurseries do so. Conventional nursery production depends on using plastic containers, chemical pesticides, synthetic fertilizers, and imported growing substrate components. Some growers hesitate to add sustainable practices in their nurseries for a number of reasons, which may include lack of training, resistance to newer techniques, and higher costs. Today, many nurseries are transitioning to sustainable growing methods because of the economic and environmental benefits that result from these growing practices.

AgEd Connection

Equipment and Supplies Identification

The following materials and equipment are included in the identification portion of some career development events. Students must identify materials and equipment used in the industry. The tools and materials may be presented physically or as photographs. Use the illustrated glossary (page 913) and the e-flashcards on your textbook's student companion website at www.g-wlearning.com/agriculture to help you study.

anvil-and-blade pruner	galvanized pipe	mower blade balancer	sand
architect's scale	garden (spading) fork	nursery container	scoop shovel
ball cart (B&B truck)	garden (bow) rake	oscillating sprinkler	shade fabric
bark medium	gas mask	peat moss	sharpening stone
bark mulch	grafting band	pick axe	siphon proportioner
bow saw	grafting tool	planting/earth/soil auger	soaker hose
brick paver	granular fertilizer	planting bar	soil sampling tube
broadcast (cyclone) spreader	grass shears	pole pruner	solenoid valve
bubbler head, irrigation	gravity (drop) spreader	polyethylene film	spade
bulb planter	ground/pelleted limestone	polyethylene pipe	spark plug gap gauge
burlap	hearing protection	pop-up irrigation head	sphagnum moss
chain saw	hedge shears	posthole digger	spray suit
chaps	hoe	pot-in-pot units	square point (flat) shovel
compressed air sprayer	hook-and-blade pruners	power blower	string trimmer
core aerifier	hose-end repair fitting	power hedge trimmer	tape measure
cut-off machine	hose-end sprayer	propagation mat	time clock
drip emitter, irrigation	hose-end washer	pruning saw	topsoil
dry-lock wall block	hose repair coupling	PVC (polyvinylchloride) pipe	tree caliper
duster	impulse sprinkler	reel mower	tree wrap
dust mask	landscape fabric	resin-coated fertilizer	trowel
edger (power or hand)	leaf rake	respirator	T-square
edging	loppers	rotary mower	vermiculite
engineer's scale	mattock	rototiller	vertical mower
erosion netting	measuring wheel	round point shovel	water breaker
fertilizer tablet	mist nozzle (mist bed)	safety goggles	wire tree basket

A *sustainable nursery* is one that incorporates growing practices that will maintain or better the environment for today and the future. Many growers attempt to produce plants with as little negative impact on the environment as possible. Sustainable production should include actions, products, and services that satisfy the needs of customers and help protect the environment. Sustainable nursery production will reduce the waste of natural resources, raw materials, and energy by increasing efficiency and improving the quality of its operations. A sustainable nursery manages irrigation, nutrients, substrate, and the growing environment to reduce its impact on Earth while producing quality plants for its customers.

"Always do your best. What you plant now you will harvest later."
—Og Mandino

Water Management

Water is a critical factor in plant growth. The amount of water, the quality of water, and the timing of irrigation contribute to crop quality. Additionally, poor water management and application contributes to pest problems and pollution of surrounding water for other uses, such as drinking, **Figure 21-20**.

Proper water management delivers several benefits to growers including:

- Quality plant growth.
- Reduced nutrient loss.
- Reduced rate of disease.
- Reduced frequency of insect pests.
- Reduced fertilizer and pesticide applications.
- Reduced levels of runoff from agricultural chemicals.

Rigucci/Shutterstock.com
Figure 21-20. A retention pond can collect excess water and provide an irrigation source for the nursery operation.

All these benefits combine to save growers money by reducing production costs.

Irrigation (also known as watering) is a critical factor in most nursery operations, especially those that produce container-grown plants. A plant requires water only once the growing medium is dry. Overwatering can be more detrimental to plant health than underwatering in many situations. Plants require irrigation when the medium feels dry to the touch or when the plant first begins to show signs of water stress (referred to as flagging). Nursery workers need to monitor plants, or irrigation sensing equipment must be used.

Plants cannot be watered on a schedule without considering several factors that contribute to a plant's water use. These factors include:

- Weather—rain, humidity, wind, and temperature all impact water consumption.
- Type of plant—whether or not the plant is drought tolerant can play a major role in how often watering is required, **Figure 21-21**.

Zvonimir Atletic/Shutterstock.com
Figure 21-21. Succulents are excellent plants to cultivate in sustainable nursery production. These plants require very little water.

Corner Question

How many gallons does the state of Georgia estimate that nurseries use every day and every year?

- Type and size of container—the larger the container the more water it holds. Certain containers allow more water to evaporate than others. Fiber containers hold less moisture that some other types of containers.
- Planting substrate—growing media with more sand or bark dry more quickly than those with peat moss.
- Surface mulch—placing a mulching material on top of the substrate can help to reduce watering intervals.
- Location—plants in shade will use less water than those in sun.
- Plant growth—plants that are dormant or growing very little will require less water than those plants that are actively growing.

A nursery grower must determine the most appropriate method of irrigation to satisfy the needs of the plants. Nursery workers apply irrigation through hand watering, overhead sprinklers, drip irrigation, and capillary mats.

Hand Watering

Hand watering requires an incredible amount of labor. A person who waters by hand must recognize which plants need watering. This takes training and attention to detail. Workers need to pay attention to plant signals and the way that the substrate looks. They must also water a plant until the container is completely saturated.

Paying people to water plants is not an expense that most nurseries care to have. Hand watering takes place on initial planting or to water in places that were missed by automated irrigation systems.

"If there is magic on the planet, it is contained in water."
—Loren Eisely

Overhead Irrigation

Overhead irrigation is an uneven method of watering plants that uses a great deal of water. Overhead sprinklers apply water above the plants in a circular pattern while *oscillating sprinkler* heads apply water in a rectangular pattern. A container nursery using overhead irrigation can use between 15,000 and 45,000 gallons of water per acre per day in the summer. An overhead irrigation system is the least expensive to install, but it has some disadvantages. Determining irrigation times for overhead systems is the same as for any irrigation system, **Figure 21-22**. Containers must be monitored to determine how much time passes before they are thoroughly saturated. Water is also applied to areas where there may not be plants present. To prevent this, water valves can be used to turn on and off parts of the water system.

sakhorn/Shutterstock.com

Figure 21-22. Overhead irrigation is commonly used in nursery production. This method can waste water or overwater in places where there is overlap of sprinkler heads.

Drip Irrigation

Drip irrigation (also known as trickle feed) delivers water to plants using small tubes that slowly release water directly to the plant's roots.

Each container or plant has a trickle tube that can supply water or a water and fertilizer solution. Weighted drip tubes are placed in the substrate of containers. Water should remain on until water comes out of the bottom of the pot's drain holes. This method of irrigation uses 60%–70% less water than an overhead irrigation system.

Drip irrigation systems have a high initial cost, but they have excellent precision when delivering water, **Figure 21-23**. These systems also allow workers to continue working while irrigation is running. Using drip irrigation produces less runoff than other methods. This system is less impacted by wind or foliage cover as well. Aside from the initial cost of the system, there is little maintenance. The drip tubes can get clogged and will need some upkeep.

vallefrias/Shutterstock.com

Figure 21-23. Drip irrigation has many forms. Tubes with premade holes or emitters can be placed into tubing. Drip irrigation saves on water consumption and can improve plant growth.

Capillary Mats

Some nurseries manage capillary mats as part of their irrigation system. A *capillary mat* is a material that is covered with a coating and that absorbs and releases water to plants that rest on it. These mats are constructed with absorbent fabric. The bottom is lined with an impermeable plastic layer and then covered on top with a perforated plastic (or similar) material to minimize evaporation and water loss. A pot that has holes in the bottom sits on top of the mat, and the substrate pulls up the water through capillary action. Capillary mats can work well with plants of various sizes and are especially well suited for retail nurseries. This process will not over- or under water plant material. Runoff is not an issue with this method. However, the initial cost is much more costly than overhead or drip irrigation systems. Capillary mats must also have a level surface to work and require regular cleaning and maintenance.

"When the well's dry, we know the worth of water."
—Benjamin Franklin

Runoff Management

The essential resource for nursery production is water. All water that does not go into the pot, is not used by the plants, or leaches through the pot must be caught and contained by a runoff system created at the nursery. The design of the nursery should include a water catchment system.

The nursery beds can be graded so that runoff water is directed to strategically placed ditches, canals, and retention ponds. The collected water can be remediated (treated and cleaned) by plant material. Plant roots absorb the water that has pesticides and fertilizers and actually anchors many of the chemicals within the roots. The plant roots harbor many of the chemicals that would otherwise linger in the irrigation water. Any additional pesticides and pathogens can be treated before water is used again. This water reuse can save nurseries significant costs.

Did You Know?

Small lab and field systems have been capable of 90% nitrate removal from water in only 30 minutes with a flow rate of 2.5 gallons per minute. These systems have unlimited abilities to help small nurseries remove nitrogen from runoff waters.

Water Treatment

Population growth and increased demands for agricultural irrigation has resulted in changes to aquifers and the chemistry of some water. Water conservation and sustainable management of water resources are needed to protect this natural resource. Nurseries must implement sustainable irrigation practices that use conservation and recycling of water.

Corner Question

How much less water do capillary mats use than overhead sprinklers?

Corner Question

How many acres of wetland are in the United States?

Surface water and groundwater quality are impacted by nursery irrigation practices. The runoff that is created by irrigation should be considered a vital resource that cannot be wasted. A water and nutrient management plan changes runoff into a valuable resource that is reapplied to crops. The water must be treated to ensure its safety before being reused. These treatment plans can include chemical, thermal, or radiation treatments to remove contaminants and pathogens. Runoff contaminants of concern include nutrients, pathogens, and pesticides.

Nurseries can include a number of sustainable and nonchemical methods to treat water. These methods can include:

- Constructed wetlands—mimic natural wetlands. They contain vegetation, soils, microbes, and other organisms and can remove contaminants. They are tailored to meet the needs of the nursery and are low maintenance, **Figure 21-24**.

FCG/Shutterstock.com

Figure 21-24. A constructed wetland is an excellent way for nurseries to capture and clean water on site.

- Floating wetlands—are used by nurseries with existing irrigation, retention ponds, or storm water pools to clean contaminants. Plants create a column with roots that provide habitat for microorganisms, filtrate particulates, and uptake contaminants in root matter.
- Bioreactors—use microorganisms to clean polluted or contaminated waters. This is an excellent way to remove fertilizers and other carbon-based pesticides.
- Planted vegetation (buffer) strips—can also use plants and other organisms to help clean the water. Plants absorb the excess nutrients and contaminants, slow moving water, trap particulate matter, and reduce general contamination to other areas.

Unlike chemical means that may also require additional energy sources, these methods of water treatment are cost-effective for producers.

Nutrient Management

Fertilizer application varies based on the way that plants are cultivated, whether in the ground or in containers. The applications of fertilizer can take place before planting or after planting, **Figure 21-25**. A grower can use soluble, slow-release, organic, or synthetic fertilizers to supply nutrients plants need.

Jodi Riedel/Goodheart-Willcox Publisher

Figure 21-25. Fertilizer can be applied before or after planting. Fertilizer injectors may be computer controlled and designed to monitor application amount and timing.

Fertilizer Pre-Plant Application

A grower can apply fertilizers before planting to meet a number of objectives. Fertilizers can help to adjust the pH of the growing media, supply macronutrients, and supply micronutrients. Dolomitic limestone can help to increase the pH while iron sulfate can reduce the pH. When fertilizer nutrients are added, they can change the pH of the substrate. Careful monitoring must take place, and changes to create the proper pH will need to be made.

Fertilizer Post-Plant Application

Applications of fertilizer after planting are done using controlled-release fertilizer and/or fertigation. In the first method, nutrients are delivered to plants slowly over a period of time. Release rates depend on the formulation of fertilizer, temperature, irrigation rates, and other environmental factors. Growers have access to many formulations for specific plant needs.

Fertigation uses water-soluble fertilizers to deliver nutrients to plants. You may be familiar with these fertilizers as the "blue stuff." These fertilizers mix readily with a water solution. They deliver readily usable nutrients to the plants. The nutrient-rich runoff water must be collected and treated accordingly.

Nutrient Monitoring

As with any integrated approach to management, a sustainable system requires continual monitoring of all components, especially nutrients. Having too few nutrients leads to poor plant health, as does overapplication of fertilizer. Overapplication, however, has an additional negative impact on the environment.

> "Fertilizer does no good in a heap, but a little spread around works miracles all over."
> —Richard Brinsley Sheridan

STEM Connection
Nutrient Analysis

The results of water and substrate tests are essential in managing any nursery facility. Using an EC/pH meter, collect a sample of irrigation water. Select plants from each type of crop for analysis. Collect samples from three or four pots in a section of the same production area. Solution for testing can be *leachate* (water that is collected after it has percolated through a soil). Using the pour-through method, pots are irrigated and water that flows through the bottom is collected. Another method uses a **suction lysometer** (a device used to pull a solution out of a given material). These cylindrical probes are imbedded in the substrate. A vacuum action draws up the solution. EC tests are conducted on samples collected. Analysis for nitrogen, phosphorus, potassium, calcium, magnesium, and other nutrients can be performed by soil scientists, agronomists, and plant pathologists.

Jodi Riedel/Goodheart-Willcox Publisher

Electrical conductivity (EC) is a measure of the amount of electrical current a material can carry. EC can be used to indicate the nutrient status of a substrate. High EC levels mean that there are dangerous levels of salts in the substrate, which may be a result of overfertilization. Monthly or biweekly monitoring of EC and pH levels should be recorded and action should be taken. Adjustments to fertilizer applications should be made as needed.

Substrate Management

A key component of sustainable nursery operation rests in the type of substrate that growers use. Several ingredients make up formulations of media used in the nursery industry. The responsibility of choosing the appropriate substrate rests on the lead grower at the nursery, and the decision for which formula to choose depends on several factors.

The traits of the substrate used affect other parts of the sustainable nursery system. Substrate must hold water, but it should also allow air movement in the pores near the root zone. Oxygen in the root zone is required for many plant processes, **Figure 21-26**. Substrates that do not hold much water will also have difficulty retaining nutrients. This can lead to excessive fertilizer and irrigation applications, which are not part of sustainable production methods.

Most medias used by nursery growers include a combination of the following ingredients:

- Sand.
- Pine bark.
- Peat moss.
- Vermiculite.
- Perlite.
- Controlled-release fertilizer.

Did You Know?
It can take from five to nine years before a nursery operation is profitable.

"The ultimate test of man's conscience may be his willingness to sacrifice something today for future generations whose words of thanks will not be heard."
—Gaylord Nelson

Jodi Riedel/Goodheart-Willcox Publisher

Figure 21-26. This student is checking the plant's root zone. Usually healthy white, fuzzy roots illustrate root vigor.

Some of these ingredients, such as peat moss, are not considered sustainable resources. Peat moss is harvested from bogs that can take hundreds to thousands of years to replace. Several ingredients that are considered more sustainable have been recently investigated and analyzed by many land grant universities and businesses. These renewable and sustainable substrate ingredients include:

- Coconut coir.
- Calcined clay (similar to that found in cat litter).
- Peanut hulls, **Figure 21-27**.
- Pecan shells.
- Cotton gin waste.
- Sawdust.
- Manure.
- Compost mix.

Vladimir Shulenin/Shutterstock.com

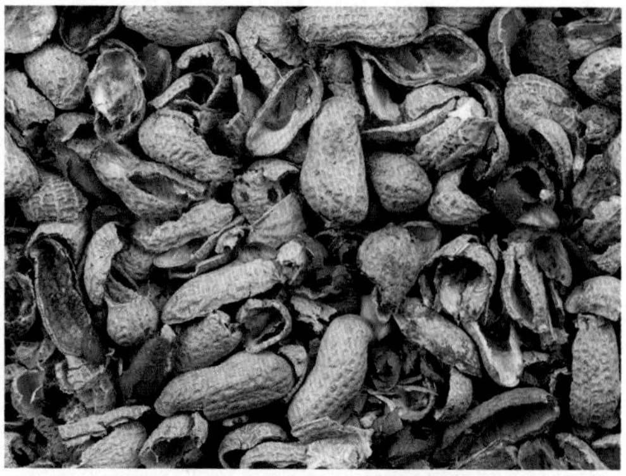

Figure 21-27. Nurseries may use potting soils that include sustainable and organic ingredients such as peanut shells.

The idea of incorporating these ingredients into soilless mixes is not new. The incorporation of these blends into everyday nursery production is becoming less novel and more the norm. Several issues impact the use of these sustainable ingredients. Costs for the material, transport costs, uniformity of product, and consistent availability of the product are some of these issues.

Environmental Management

The environment that surrounds a plant can be manipulated to some extent to benefit the health and growth of a plant. Careful attention must be paid to identify unwanted organisms in the environment, such as pathogens, insects, and weeds.

Growing Structures

Various growing structures can be found at different nurseries around the country. Common examples of these structures that are to help control the growing environment include:

- Greenhouse—temperature, ventilation, and light are controlled.
- High tunnel—high framed walls covered in a plastic material to protect plants from extreme cold. Irrigation and ventilation are provided.
- Low tunnel—usually no taller than 6′ high; used to protect plants from cold by simply covering the plants during cold temperatures.
- Shade house—provides protection from

absolut/Shutterstock.com

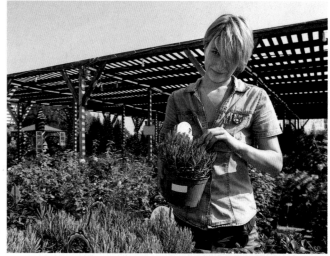

Figure 21-28. A lath or shade house provides filtered light for plants requiring fewer foot candles of light for proper growth.

sun for shade-loving plants, **Figure 21-28**.

Plant Spacing

Whether plants are planted in the ground or in containers, there must be appropriate space provided for the plants to ensure optimum growing conditions. Plants cultivated in containers should be spaced so that the edges of the plants barely touch one another, resulting in compact growth and higher yields, **Figure 21-29**. As plants increase in size, they should be spaced out, or some containers should be removed and placed elsewhere to ensure adequate room for growth. Plants grown in the ground must be spaced according to their final size at the time of transplant or sale.

Plants spaced unsuitably can have poor growth. Spacing that is too tight results in plants shading one another. This practice can cause leaf and branch death due to the lack of photosynthesis. Spacing that is too wide wastes irrigation water and valuable space that could be used to produce a saleable plant.

Praisaeng/Shutterstock.com

Figure 21-29. At their current stage, these orchid liners are small and appropriately spaced for optimum growth. As they mature, they will each require more space to reach maturity and blossom.

Winterization

Plants grown in containers need winterization or protection from extreme cold. The roots of many plants suffer from root injury when temperatures drop below 20°F (–6°C). To ensure that plants survive in these temperatures, growers must change the surrounding environment. Using greenhouses and both high and low tunnels are a great way to protect plants. If these structures are unavailable, growers can use a frost fabric, also known as a floating row cover. This material is draped over the plants and secured to the ground. Other insulation materials can be used, such as straw, burlap, or plastic, to protect plants. If nothing else can be done, growers can pack plants tightly and ensure that they are protected from severe winds, **Figure 21-30**.

pan demin/Shutterstock.com

Figure 21-30. Providing winter protection in nurseries is crucial in cold climates.

Pruning

Corner Question

What can be done to a plant if it begins to outgrow the container?

Growers prune plant material to achieve several objectives. Container-grown plants are usually pruned to achieve a certain shape or to create a more compact plant. Pruning is also used to remove dead, diseased, or damaged branches in the field. Stems may be removed to ensure the safety of workers where low-lying branches could be hazardous. Field-grown plants must also be pruned to create aesthetically pleasing shapes.

Creative pruning involves both art and science. Horticulturists must know the plant material and realize how their pruning techniques will impact plant growth. The pruner must be able to identify key parts of plant material, such as the central leader, lateral buds, and flower buds. Additionally, the pruner must know whether the plant is a multi-stemmed or single-stemmed plant. The pruner must then make precise cuts to achieve the expected results.

Pruning can take place at various times of the year. Dormant pruning can take place during slow work times for employees, and this is the time when most nursery pruning takes place. Many nurseries prune immediately after planting or during the growing season so that the horticulturist can better envision how the cuts will impact growth. Summer pruning involves removal of water sprouts and suckers, maintenance of a central leader, and corrective shaping. Winter pruning involves removal of no more than one-third of the plant material. Extra-long branches at the bases should be pruned to one-half of their length. Pruning should be done to remove crossing, rubbing, parallel, or competing branches.

"Nature provides exceptions to every rule."
—Margaret Fuller

Integrated Pest Management

Integrated pest management (IPM) is an approach to managing pests that uses common sense and economical practices and results in the least possible hazard to people, property, and the environment. Any sustainable nursery operation includes IPM as part of its system. IPM incorporates various monitoring, prevention, and control methods to combat pests. Multiple strategies are combined to optimize pest control, including:

- Determining thresholds for pests (population levels).
- Monitoring, scouting, and correctly identifying pests, **Figure 21-31**.
- Recording data.
- Using cultural controls such as resistant cultivars, cleaning facilities, removing breeding grounds, planting at appropriate times, and correct irrigation.
- Employing biological controls such as cats to hunt rodents or beneficial insects such as ladybugs to control harmful pests.
- Quarantining plants that are infected or infested from healthy plants.
- Controlling pests with pesticides.

Decha Thapanya/Shutterstock.com

Figure 21-31. Many insects impact the growth of nurseries, like these scale insects. This type of insect, scale, has a piercing and sucking mouthpart that can also spread disease.

Careers in Nursery Production

Nursery operations include careers for people with many interests. Finance, management, business, horticulture, engineering, public relations, advertising, and customer service are some examples. Rewarding job positions include equipment salesperson and plant buyer.

Thinking Green

A Ring of Fire to Control Weeds

Fire can kill anything, especially weeds. Flame weeding is a sustainable way to control weed growth. Flaming does not actually burn the plant cells, but it sears them with the heat. This is an effective control against broadleaf weeds. However, stubborn grassy weeds may require additional applications. Weeds can resprout, and additional applications need to be made about two to three weeks after the first application.

Irina Fischer/Shutterstock.com

Equipment Salesperson

Nursery operations use many types of equipment. Tractors, attachments, augers, and tree spades are just a few of the items that nursery managers need to make their operations efficient. Many of the machines that wholesale nursery workers operate may be custom designed and manufactured to meet their specific needs.

People interested in equipment sales should have familiarity with how machinery works. Additionally, individuals dealing with equipment sales should be personable and knowledgeable about business relations. These employees should have some education in an agriculturally related field.

Plant Buyer

Can you imagine buying plants but never having to pay for them? Plant buyers experience this luxury every day. Landscaping, re-wholesaling, and retail nurseries usually employ a plant buyer. This person purchases the plant materials needed by the business.

Plant buyers should be horticultural enthusiasts. This person should have a wealth of plant knowledge and be aware of unique plant materials and their specific characteristics. Knowing about current and future trends in horticulture guides buyers in their plant purchases. Experience in working with unique plant material or a nursery operation can be valuable for a plant buyer, **Figure 21-32**.

Dustin Hardin/Shutterstock.com

Figure 21-32. A buyer at a nursery purchases plant material and pays close attention to the market.

Career Connection | Alan Erwin

Panther Creek Nursery

Alan Erwin leads a team of 21 nursery employees at a premiere PNP operation in Willow Springs, North Carolina. A member of the FFA and a graduate of North Carolina State University, Alan has worked in horticulture his entire life. For the last 15 years, Alan has managed Panther Creek Nursery. The 58-acre nursery prides itself on its sustainable practices, which include a runoff program where all water is recaptured, water recycling, nutrient monitoring, and soil conservation. When clients walk through the nursery, they see pristine plant material with very few weeds. Alan ensures that his nursery is extremely clean and weed free (which also helps to reduce insects and disease). Alan says there is "never a second chance for a first impression."

Panther Creek provides top-quality crops, in part due to the PNP system. Although the system is expensive to install, the plant material has superior rooting compared to plants in traditional container systems. A drip irrigation system that is closely monitored and a specialized controlled-release fertilizer contribute to superior plants. The plants are harvested year round and are available for customer pickup or delivery within two hours after the sale. Alan monitors every facet of production.

Jodi Riedel/Goodheart-Willcox Publisher

From monitoring plant material to pulling orders and loading trucks, he ensures that Panther Creek offers excellent nursery products.

AgEd Connection

Plant Identification

As part of a career development event, students must identify plants commonly raised in nurseries and used in landscaping. Plants range from smaller flowering specimens, to shrubs, and trees. Plants may be presented as intact, live specimens or as photographs. Use the illustrated glossary (page 900) and the e-flashcards on your textbook's student companion website at www.g-wlearning.com/agriculture to help you study and identify these common plants.

Chapter 21
Review and Assessment

Chapter Summary

- A plant nursery facilitates growth of plant material under a number of different environmental conditions. Nurseries grow seedlings, transplants, stock plants for propagation, container-grown plants, and field-grown plants.
- A nursery tries to identify specific areas of a broad market for which it will produce products. Nurseries find their market niche based on the nursery site location, types of plants that are cultivated, and production methods.
- Nursery crops are marketed via several avenues, which include wholesalers, re-wholesalers, mail-order companies, retailers, and landscapers.
- Government agencies regulate nursery operations. Nurseries must obtain permits or licenses to grow plants and must have a dealer license to sell plants.
- Nursery operations employ one of three production methods: the container-grown method, the field-grown method, or the pot-in-pot (PNP) method.
- Container-grown plants are cultivated with soilless media in a variety of container types. The plants are then grown in greenhouses, tunnels, or on the ground.
- The site's soil is of the utmost importance to a field-grown nursery grower. The soil must be fertile, well drained, and free of significant pests.
- Pot-in-pot (PNP) production involves planting material in containers and then placing those pots into socket pots placed in the ground. This method is used for shade trees, ornamental trees, and large shrubs.
- Many nurseries are transitioning to sustainable growing methods because of the economic and environmental benefits. Sustainable practices include management of water, nutrients, substrate, and the growing environment.
- The plant material that nursery workers cultivate in nurseries is nearly limitless and includes trees, shrubs, vines, groundcovers, fruit, and herbaceous perennials.
- Nursery operations offer many career opportunities, including equipment sales and plant buying.

Words to Know

Match the key terms from the chapter to the correct definition.

A. capillary mat
B. circling
C. crown
D. drip irrigation
E. electrical conductivity (EC)
F. leachate
G. mechanical transplanter
H. oscillating sprinkler
I. phytosanitary certificate
J. pot bound
K. root pruning
L. socket pot
M. substrate
N. suction lysometer
O. sustainable nursery
P. water sprout
Q. winterization

1. A material that is covered with a coating and that absorbs and releases water to plants that rest on it.
2. Protecting plants from cold and harsh winter conditions by covering with material, packing tightly, or placing in a sheltered place.
3. Severing the roots of a plant to promote growth of some kind, such as flowering or fruiting.
4. A condition in which the roots of a plant are tangled and wrap around the interior of the pot.
5. A machine that moves plants from one place to another.
6. A growing medium or the material in which a plant's roots grow.
7. A nursery that incorporates growing practices that will maintain or better the environment for today and the future.
8. A method of watering plants that uses small tubes that slowly release water directly to the plant's roots.
9. An official document that shows a plant has been inspected and appears to be free of pathogens, insects, and weeds.
10. The area on a plant where the roots meet the stem tissue.
11. A condition in which the roots of a plant fill the interior of the pot.
12. Plant stems that grow vertically from other stems or branches.
13. A container used in pot-in-pot production that holds another container with plant material and is located in the ground.
14. A device used to pull a solution out of a given material.
15. A measure of the amount of electrical current a material can carry.
16. A device that delivers irrigation water in a rectangular pattern.
17. Water that is collected after it has percolated through a soil.

Know and Understand

Answer the following questions using the information provided in this chapter.

1. The market niche for a nursery is based on what factors?
2. What factors should be considered when choosing a site for a nursery?
3. List the five types of nurseries described in this chapter.
4. What is the purpose of a phytosanitary certification for plants and when is one needed?
5. What are three types of production methods used by nurseries?
6. What are some factors that determine the type of container to use in a nursery?
7. What is the most commonly used growing medium in the nursery industry?
8. What are some activities involved in maintenance for container-grown crops?
9. What are the soil characteristics of a site that is suited for a field-grown nursery?
10. When are field-grown plants usually harvested for B&B and what steps are involved?
11. Why is soil drainage important for PNP production and what can be done to ensure good drainage?
12. In PNP production, what determines harvest time and how is harvesting done?
13. Why might some growers hesitate to add sustainable practices in their nurseries?
14. List 10 tools used in nursery production that may be included in the identification portion of a career development event.
15. What are four methods of irrigation a nursery can use?
16. How much water is saved using drip irrigation instead of overhead irrigation on a typical nursery?
17. What are four methods used by sustainable nurseries to treat water?
18. When is fertilizer applied and what are some types of fertilizers growers can use to supply nutrients plants need?
19. What are some sustainable or renewable ingredients used in nursery substrates?
20. What are some structures that are used to help control the growing environment for plants?
21. What is integrated pest management (IPM) and what are some strategies used in IPM?

Thinking Critically

1. You have recently moved to Florida where your family has purchased some very flat land with very sandy soils. Describe what you think would be the best nursery for this site. Justify your decision.
2. You recently realized the financial opportunity to sell alpine plants. What do you think would be the best way to market these plants? What alpine plants could be cultivated in your region? How do you cultivate these plants (greenhouse, nursery, PNP)? Defend your response.

STEM and Academic Activities

1. **Science.** Design a rotation for a PNP system that would include a three-year rotation for plants from one-gallon to three-gallon to five-gallon containers.
2. **Math.** A plant that is grown in a container may need to grow several years before it is sold. Do the math: A plant is grown from seed and takes four weeks to germinate. The plant then grows for four months and is transplanted to a small container. The plant is now grown for one year in the container. It is then transplanted to its final container where it will stay for 30 months until it is sold to a retail nursery. At the retail nursery, it sits on a shelf for three weeks and then is sold. What is the age of the plant at sale?
3. **Social Science.** Contact a nursery in another city, state, region, or country. Ask five specific questions about the nursery operation and report the answers to the class.
4. **Language Arts.** Choose a nursery plant that is available on the market today that is the result of current breeding practices. Determine who the breeder was in this plant's production. Research how this plant was bred and determine its parents. What characteristics resulted from the crosses that were made? How does this plant differ from the parent material? Write a one-page summary about this plant, its history, and the breeder.

Communicating about Horticulture

1. **Reading and Speaking.** Create and present an informative, prepared speech for an audience unaware of the nursery industry in your state. Take this time to educate them about this vital industry and its numerous impacts.
2. **Reading and Writing.** Develop an infographic about the nursery industry. Post this information in a space where people can see the information.

SAE Opportunities

1. **Exploratory.** Job shadow a nursery worker.
2. **Experimental.** Grow nursery plants in various sustainable medias and examine their growth.
3. **Exploratory.** Create a marketing logo or campaign for the nursery industry in the United States. Survey consumers to determine their feelings about the logo or advertising campaign created. Analyze your data.
4. **Entrepreneurship.** Develop a nursery business selling plants of your choice with a nursery system of your choice.
5. **Placement.** Get a job working at a nursery (wholesale, retail, mail order, or other).

kurhan/Shutterstock.com

CHAPTER 22 Vegetable Production

Chapter Outcomes

After studying this chapter, you will be able to:
- List the reasons vegetables are important for good health.
- Describe the market opportunities for vegetable production.
- Explain the environmental requirements for vegetable production.
- Describe different production methods.
- Summarize postharvest handling and storage practices.
- List careers related to the olericulture industry.

Words to Know

crop rotation	interplanting	perched water table
dripline	leaching	row cover
good agricultural practice (GAP)	low tunnel	sanitation
	olericulture	

Before You Read

Write all the chapter terms on a sheet of paper. Highlight the words that you do not know. Before you begin reading, look up the highlighted words in the glossary and write the definitions.

While studying this chapter, look for the activity icon to:
- **Practice** vocabulary terms with Words to Know activities.
- **Expand** learning with identification activities.
- **Reinforce** what you learn by completing Know and Understand questions.

www.g-wlearning.com/agriculture

Kale chips, radish candy, and kohlrabi slaw are vegetables that are fun to eat and to grow. Rich in healthful nutrients, vegetables play a key role in growing healthy bodies and minds. Digging deep into the soil and watching a seed sprout and develop into something edible remains one of the principal joys of gardening.

Vegetables for Health

Eating vegetables is an essential part of a healthy diet because vegetables contain minerals and nutrients vital for health and maintenance of your body. The United States Department of Agriculture (USDA) recommends that half of your plate should be filled with fruits and vegetables, **Figure 22-1**. An overall healthy diet includes fruits, whole grains, low-fat dairy, and varied proteins (not just meat). Be aware of the number of calories you eat. Select foods that are low in fats, sugars, and salts. A healthy lifestyle also includes physical activity.

protonz/Shutterstock.com

Figure 22-1. A diet rich in vegetables is delicious and healthy for your body.

Health Benefits

A healthy lifestyle that includes eating vegetables and fruits can help reduce the risk of heart disease, heart attacks, and strokes. Other benefits of eating vegetables include:

- Vegetables such as broccoli and other brassicas, dark leafy greens, garlic, tomatoes, and winter squash may protect against some types of cancers.
- Fiber-rich vegetables and foods may lessen the risk of heart disease, obesity, and type 2 diabetes.
- Vegetables and fruits that are rich in potassium may lower blood pressure. They may also reduce the risk of developing kidney stones and help decrease bone loss.
- Higher consumption of low-calorie foods such as vegetables can help to lower total calorie intake.

Nutrients Found in Vegetables

Plants take in nutrients from the soil solution. Those nutrients are the same nutrients that people consume when eating a raw carrot or sautéed leafy greens. Nutrients are essential for the processes our bodies perform every day. No single vegetable contains all the nutrients needed

IPM evaluation should include a regular review of records to identify any changes in pest activity (increase or decrease) that will determine whether the management techniques are working. The review should examine the correlation between methods used and changes in pest populations. The following questions can be useful in evaluating an IPM program:

- What problems have been identified?
- Is the monitoring program adequate?
- Are all pest populations below unacceptable levels that require correction actions (action thresholds)?
- Do additional or different actions need to be taken?
- Can time and effort to control pests be reduced?
- What changes are necessary?

Action Thresholds

Identifying action thresholds is an essential part of any integrated pest management program. An *action threshold* is the point at which the pest reaches an unacceptable level, and where some type of corrective action to reduce its numbers is economically justified, **Figure 29-17**. Action thresholds and the corrective action taken vary depending on the pest, site, geographical location, and time of year. For example, even a few cockroaches are not tolerated in food service areas (very low action threshold). On the other hand, clover in a regularly mowed area will not crowd out the desired turf (high or no action threshold). Some pests can have an action threshold of one sighting. Other predictable and unacceptable pests may require no sighting at all to take action.

Figure 29-17. Clover in the lawn may not be aesthetically pleasing to some people, but it rarely impacts the growth of turf. Therefore, it has a high action threshold.

An *economic injury level (EIL)* is the lowest population density of a pest that will cause economic damage and justify the costs of pest control. Corrective action (pest control) may result in the highest yields, but it may not necessarily be the most cost effective. Treatment at an action threshold is the most cost-effective management approach, but it may or may not result in the highest yields. Most agronomic and horticultural crops have thresholds that have been established by research. For example, codling moths in apples can cause significant economic damage if a certain population size is reached. To determine the infestation level, traps containing lures are hung in the orchard. If more than five codling moths are caught in a trap per week, there is a high risk for problems from future generations of codling moths. By applying a pesticide within five to seven days, growers can keep populations to an acceptable level.

> "I do some of my best thinking while pulling weeds."
> —Martha Smith

Corrective Actions

When there is a pest problem, growers consider the available cost-effective control strategies and use a combination that eliminates unacceptable pests and helps prevent future infestations. Monitoring is always necessary unless the action threshold is zero. Corrective actions include physical, mechanical, biological, and chemical controls.

Physical and Mechanical Controls

Physical and mechanical control techniques directly remove or kill pests. They may also prevent insect pests from reaching their hosts by means of a barrier or trap.

yuris/Shutterstock.com

Figure 29-18. Hoeing a garden is an easy, inexpensive way to remove weeds.

Insect Removal

Physical removal techniques are labor intensive and include hand picking, spraying with water, and cutting. For example, some aphids can be knocked off plant foliage by spraying a stream of water on the leaves. Japanese beetles can be knocked off by hand into a bucket of soapy water. Bagworm eggs overwinter inside bags. These bags can be cut from the plants in the winter months and destroyed.

Weed Removal

On a small scale, weeds can be removed through hand pulling, with hoes or string trimmers, or through tillage, **Figure 29-18**. Hand removal of weeds can be time and labor intensive, which can be expensive. Techniques known as flame cultivation or thermal weeding use propane flame torches or infrared torches to heat the plant until cellular rupture occurs and kills the plant, **Figure 29-19**. Grit blasting (shooting small and targeted amounts of grit onto plants) using ground peach pits or other material has shown early evidence as an effective method to kill weeds.

Irina Fischer/Shutterstock.com

Figure 29-19. Flame torches are used often in organic production to eradicate weeds in a quick, less labor-intensive manner.

Pest Barriers

Some pests can be prevented from reaching their host by installing a barrier. For example, a simple row cover can provide protection to solanaceous crops, such as eggplants, from the ravages of flea beetles. Putting screens in greenhouses and buildings will bar entry to many pests. Fine netting prevents birds from eating or damaging ripening fruit, **Figure 29-20**. Paper collars placed around stems of plants can prevent access by cutworms. Barriers may also be used to prevent weeds from spreading. For example, a metal border can stop the spread of bermuda grass into a garden. Landscape fabric can be laid down prior to planting to keep weeds to a minimum. An *anti-transpirant* spray (chemical compounds applied to the leaves of plants to reduce transpiration) can be used for some woody landscape ornamentals. This spray creates a protective barrier against infection by some plant pathogens that could cause disease.

Serkan Ogdum/Shutterstock.com

Figure 29-20. Netting placed over plants or trees prevents birds from eating or damaging fruit, but allows airflow and water to pass through.

Traps and Attractants

Traps and attractants are not only used to monitor populations, but also to trap pests to lower plant damage. For example, a sticky barrier wrapped around the trunks of trees and woody shrubs will capture crawling insects, **Figure 29-21**. Pheromone traps are primarily used in monitoring; but in some cases, such as in codling moths, the traps can be used to reduce population numbers. Traps may also be used to provide an artificial breeding or resting site. For example, small numbers of gypsy moth larvae have been effectively controlled by placing a band of folded burlap around the tree trunk to provide an artificial resting site for the caterpillars. Once caterpillars have gathered, they can be collected and destroyed.

Biological Controls

Many natural predators of pests exist in agricultural and horticultural ecosystems. These natural predators are called *beneficials* and may be insects, arachnids, mammals, and even microbes, **Figure 29-22**.

Gypsy Moth Larva
Peter Waters/Shutterstock.com

Adult Gypsy Moth
D. Kucharski K. Kucharska/Shutterstock.com

Figure 29-21. Gypsy moths are devastating, exotic pests that feed on many tree species. They may be controlled by trapping the larvae in burlap wrappings.

Pests	Natural Enemies	
Aphids	aphid midge aphid parasites bigeyed bugs damsel bugs lacewings lady beetles leather-winged beetles	minute pirate bugs parasitic flies parasitic wasps predatory mites soldier beetles spiders syrphid flies
Beetles	Anaphes species damsel bugs elm leaf beetle parasite	leather-winged beetles soldier beetles spiders tachinid flies
Bugs such as lygus, plant, and stink bugs	Anaphes species assassin bugs bigeyed bugs	damsel bugs lacewings
Caterpillars	alfalfa butterfly parasite assassin bugs *Bacillus thuringiensis* bigeyed bugs birds caterpillar parasite damsel bugs egg parasitic wasps	grape leaffolder parasite lacewings lady beetles minute pirate bugs parasitic flies predaceous bugs and wasps spiders tachinid flies
Elm leaf beetle	parasitic flies	parasitic wasps
Lace bugs	assassin bugs lacewings lady beetles	parasitic wasps pirate bugs spiders
Leafhoppers	assassin bugs bigeyed bugs	damsel bugs spiders
Mealybugs	citrus mealybug parasite dustywings lacewings lady beetles	mealybug destroyer lady beetle minute pirate bugs parasitic wasps

Pests	Natural Enemies	
Mites	bigeyed bugs damsel bugs dustywings lacewings lady beetles minute pirate bugs parasitic mites	predatory fly larvae sixspotted thrips spider mite destroyer lady beetle western predatory mite
Psyllids	lacewings lady beetles parasitic wasps	pirate bugs spiders
Scale, cottony cushion	lady beetles parasitic flies	vedalia beetle
Scales	lacewings lady beetles minute pirate bugs	parasitic wasps predatory mites
Slugs, snails	birds decollate predatory snail parasitic flies	predaceous ground beetles snakes toads
Thrips	damsel bugs lacewings minute pirate bugs parasitic wasps	predatory mites predatory thrips spiders
Weevils, root or soil-dwelling	entomopathogenic nematodes	parasitic wasps
Whiteflies	bigeyed bugs dustywings lacewings lady beetles	minute pirate bugs parasitic wasps spiders whitefly parasite
Whitefly, giant	lacewings lady beetles parasitic wasps	syrphid fly larvae

Goodheart-Willcox Publisher

Figure 29-22. This table lists common natural enemies used in biological control, including predators and parasites. Natural enemies that limit pests are essential components of integrated pest management programs.

Biological control promotes and protects beneficial predators and parasites that help control pests. When scouting for pest monitoring, scouts should also monitor the presence of beneficial organisms. The presence of beneficials will impact IPM decisions. For example, tomato hornworm can be parasitized by a type of wasp. Avoiding the use of an insecticide that could kill these wasps allows predation to occur, **Figure 29-23**.

Growers may purchase and introduce parasitic beneficials as part of their IPM. One of the most common examples is the use of a ladybug or ladybird beetle as an aphid predator, **Figure 29-24**. Both the larvae and adult beetles eat aphids as well as other soft-bodied insects. Other important predators include lacewings, soldier bugs, and spiders as well as a variety of others. Parasitic beneficials are organisms that live on or inside the pest host, such as the braconid wasp. The host often dies after the parasite has completed its development. Most agricultural beneficial parasites are small wasps and flies that use caterpillars, whiteflies, aphids, and other pests as hosts.

Microbial Control

Bacteria, viruses, fungi, and nematodes can also be used as biological control measures. The naturally occurring, soil-dwelling bacterium *Bacillus thuringiensis* (Bt) is commonly used to control pests. Bt produces a protein crystal that is toxic for lepidopteran insects (butterflies and moths). The ingested Bt destroys the larvae's intestinal lining and they die within days.

The Bt toxin can be sprayed on plant foliage or genetically inserted into plant material, **Figure 29-25**. It specifically affects caterpillars and will not harm other insects in the landscape. Microbial insecticides are fairly slow acting and are most effective when applied when pest numbers are low and in early stages of their development. Bt has been used as an organic-certified pesticide.

Elizabeth O. Weller/Shutterstock.com

Figure 29-23. Tomato hornworms are often parasitized by a natural enemy, the braconid wasp.

Pavel Mikoska/Shutterstock.com

Figure 29-24. Ladybug adults and larvae are voracious predators of aphids.

AlissalaKerr/Shutterstock.com

Figure 29-25. Cotton is commonly transformed with the Bt gene, which limits damage by any lepidopteran pest.

Chemical Controls

Pesticides are an integral part of many IPM programs, but they are not a substitute for other effective measures. When action thresholds are exceeded and nonchemical control techniques are insufficient or not practical, chemical control plays a critical role in managing pest populations. Pesticides should be used only when needed, but not only as a last resort. Small infestations of some highly-detrimental pests can quickly get out of hand. The application of a pesticide may be needed before there is time to try other methods. For example, timing an application to match pest life cycles will provide more efficient control.

Chemical methods to manage pests include many types of compounds. Some will repel or confuse pests. For example, bug spray repels mosquitoes, but it does not kill them. Some pesticides interfere with pest biology, such as interrupting weed photosynthesis or insect molting processes. Others, including some botanical and most conventional insecticides, are broadly toxic to living systems. The term *pesticide* is an inclusive term for many chemicals. Specific pesticides include insecticides (to control insects), herbicides (to control weeds), fungicides (to control fungi), rodenticides (to control rodents), and others.

Pesticide Application

Pesticides are required by law to have a detailed label and to be used according to the label to ensure maximum benefits and minimal hazard to human health and the environment. Pesticide labels describe how to use the pesticide correctly, including the best timing for application. Pesticide labels also specify what types of personal protective gear should be worn by the applicator.

Although pesticides are valuable tools in pest management, overuse or misuse has led in some cases to resistance to the pesticide, outbreaks of secondary pests, adverse effects on nontarget organisms, unwanted pesticide residues, and direct hazards to the user. Following best practices in pesticide use is a critical component of IPM. These practices include rotating types of pesticides, proper timing, safe use, and using pesticides in combination with other control tactics. Pesticides and their application are covered in more detail in Chapter 33, *Pesticide Management and Safety*.

Careers in Integrated Pest Management

Many interesting careers are related to integrated pest management. IPM methods must be researched and developed, users must be educated in their application, and chemical as well as biological components must be marketed, sold, and distributed. Three careers in IPM you may want to consider include research scientist, pesticide sales, and management in a pesticide company.

Research Scientist

Research scientists in IPM work for universities, state departments of agriculture, federal agencies such as the EPA or USDA, and other collaborating organizations, **Figure 29-26.** Researchers work to develop crop-specific IPM systems. These IPM systems include best practices and new control methods with the reduction of pesticide impact while still supporting profitable farming. Researchers often build relationships with farmers, crop associations, and farmer networks to encourage the widespread adoption of IPM systems. Researchers may write grant proposals and research papers to support their work. Higher-education degrees (master's or doctorate) are required along with at least five years of experience.

avemario/Shutterstock.com

Figure 29-26. Research scientists working on IPM methods work in labs, greenhouses, and fields. They also work with farmers and other growers to implement monitoring and control methods.

Pesticide Sales

If you have excellent communication and organization skills, a position as a sales associate for a pesticide company may be a good fit for you. Job responsibilities include explaining the company's services to existing customers as well as acquiring new customers. Sales associates must understand pesticide products, target pests, application methods, and what product would best suit a grower's operation. Most companies require some prior sales experience with a history of demonstrated sales results.

Manager

After being a sales associate, you might advance to an administrative position as a sales manager or director. In this position, you will have the opportunity to train, coach, and motivate others. You must have strong decision-making and problem-solving skills for this job. Computer skills, as well as excellent communication, interpersonal, and organizational skills, are also needed. Supervisory experience is preferred. This position generally requires a relevant college degree or equivalent experience.

CHAPTER 29 Review and Assessment

Chapter Summary

- A pest is any organism that damages or disrupts plant growth. Integrated pest management (IPM) is an environmentally sensitive approach that includes a variety of strategies (cultural, physical, biological, and chemical) to most effectively control pests.
- Pest management decisions are based on a variety of factors, such as the size of the affected area, estimated number of pests, life-cycle stage of the majority of pests, type of crop, urban or rural setting, time of year, and costs.
- Pests are organisms that damage plants or disrupt crop development. Pests include weeds, birds, rodents, mammals, insects, snails, nematodes, and plant pathogens, such as bacteria, virus, fungi, and oomycetes
- Pest prevention creates an initial environment that is unfavorable to pests and includes sanitation, habit modification, and selecting plant materials that are free of viruses, bacteria, fungi, or weed seeds.
- Inspection and monitoring of sites through regular observation for the appearance of insects, weeds, diseases, and other pests will allow growers to locate, identify, and rank the severity of any pest outbreaks that occur.
- Inspection and monitoring involves scouting sites using sweep nets, insect traps, soil samples, weed counts, and visual observations. Pest identification is important in selecting the correct control measures.
- Recordkeeping is a vital component of an IPM program. Maintaining good records will help solve pest problems and offer a historical perspective of pests.
- An action threshold is the point at which the pest reaches an unacceptable level and where corrective action to reduce its numbers is economically justified.
- Growers often take corrective measures to eliminate unacceptable pests and help prevent future infestations. Corrective actions include physical, mechanical, biological, and chemical controls.
- Physical and mechanical control techniques directly remove or kill pests or exclude insect pests from reaching their hosts by means of a barrier or trap.
- Biological control is a practice that promotes, protects, and introduces beneficial predators and parasites to help control pest populations.
- Chemical controls using pesticides are an integral part of many IPM programs, but not a substitute for other effective measures. Pesticides are required by law to have a label and to be used according to the label.
- Many interesting careers are related to integrated pest management. Three of these careers are research scientist, and sales associate or manager for a pesticide company.

Words to Know

Match the key terms from the chapter to the correct definition.

- A. action threshold
- B. anti-transpirant
- C. beneficial
- D. economic injury level (EIL)
- E. integrated pest management (IPM)
- F. oomycete
- G. pest
- H. pesticide
- I. pheromone
- J. plant pathogen
- K. scouting
- L. vector

1. A natural predator or parasite that controls the population of pests, such as insects, mammals, bacteria, and other microbes.
2. The regular checking of a field, garden, orchard, lawn, greenhouse, or other area that identifies potential pest problems early while they can still be managed and thereby reduce crop loss and control costs.
3. The lowest population density of a pest that will cause economic damage and justify the costs of pest control.
4. An organism that causes disease in plants.
5. A chemical compound that can be applied to the leaves of plants to reduce transpiration and provide protection from plant pathogens.
6. An organism that transmits a disease or parasite from one plant to another.
7. The point at which the volume of a pest reaches an unacceptable level, and where some type of corrective action to reduce its numbers is needed and economically justified.
8. A chemical that is emitted by insects to attract mates and which can be synthetically produced and used in lures to trap insects for monitoring.
9. An environmentally-sensitive approach to controlling pests that uses a combination of cultural, physical, biological, and chemical controls.
10. A chemical that is used to control insects, weeds, fungi, rodents, and other pests.
11. Filamentous protists, including water molds and downy mildews that can cause diseases, such as blights and rots.
12. Any organism that damages or disrupts plant growth.

Know and Understand

Answer the following questions using the information provided in this chapter.

1. What is integrated pest management (IPM)?
2. What are two main benefits of using IPM?
3. What factors are considered in making decisions for IPM?
4. What are agricultural/horticultural pests? Give several examples.
5. What impacts do weeds have in farms and gardens?
6. How do nematodes damage plants?
7. What types of damage can harmful bacteria cause in plants?

8. What are two examples of oomycetes that can cause plants to rot and how can incidents of these pathogens be minimized?
9. What risk do wildlife and other vertebrates pose as pests in the garden and what are some examples of vertebrates that can be pests?
10. Describe sanitation techniques that prevent pest infestations.
11. What are three examples of habitat modification that may prevent or reduce pests?
12. What are three common examples of plants that have been genetically modified to resist glyphosate and how does this affect pest management?
13. How does having genes from the soil-dwelling bacterium *Bacillus thuringiensis* (Bt) inserted into plants aid in pest management?
14. What is scouting (as it relates to IPM) and what are some methods for scouting?
15. Why is it important to properly identify a pest?
16. What are some questions that can be useful in evaluating an IPM program?
17. Describe physical and mechanical control techniques and two examples of these controls.
18. Explain biological control as part of IPM and give an example of a biological control.
19. Explain how pesticides can be an important part of IPM.
20. What duties are involved in a pesticide sales associate job and what skills are needed for this job?

Thinking Critically

1. One of your fields has a serious outbreak of a disease. You decide to use a pesticide. As you are getting ready to spray, the weather changes. It is windy and possibly will rain. What course of action will you take?
2. You are scouting your field and find a large area with a high population of an insect that you have never seen before. The insects are causing significant damage to your crop. What steps will you take?

STEM and Academic Activities

1. **Science.** Choose four different pests in your neighborhood that interest you. Find both the common and scientific name of each, using binomial nomenclature. Make a table with three columns: Common Name, Scientific Name, and Example. In the Example column, either draw a picture or attach a photograph of the pest.
2. **Technology.** Research recordkeeping or inventory control software for small businesses. Determine which software might be best suited for a farmer managing pests. What data fields are critical? How can this be adapted for use with field monitoring and pest management?
3. **Technology.** Investigate technologies that allow digital mapping of pest incidences. Prepare a summary that describes the process, methods, and usefulness of this application.
4. **Math.** Scouting requires observing fields in several spots. Suppose a field that is one acre in size should be scouted in 10 different locations. Create a map with marks indicating the locations for scouting that represents all parts of the field.

5. **Language Arts.** Pretend that you are a crop consultant, and you keep a daily journal to help you remember information that you might need later. Today, you scouted several different fields and made pest management decisions. Write a journal entry about these visits that includes all the information you might need in the future.

Communicating about Horticulture

1. **Reading and Speaking.** Working in groups of five students, draw a poster that illustrates the key concepts in this chapter. Use your textbook and research online for images. Take turns quizzing one another on the fundamental ideas within the chapter.
2. **Reading and Speaking.** Debate the topic of different ways to manage pests. Divide into three groups: physical and mechanical control, biological control, and chemical control. Each group should gather information to support a pro argument for their control method. Use definitions and descriptions from this chapter, as well as other resources, to support your side of the debate and to clarify word meanings as necessary. Do additional research to find expert opinions, costs associated with each method, and other relevant information.
3. **Writing.** Pick a plant pest that causes significant economic damage in your state. Use research from your local cooperative extension or land-grant university to learn about the biology and control strategies for the pest. Write a one-page summary that describes your findings.

SAE Opportunities

Jari Hindstroem/Shutterstock.com

1. **Exploratory.** Job shadow a crop consultant. What are the daily responsibilities for this job? What do you like or not like about this position? What education and experience are required to have this position?
2. **Exploratory.** Find a plant on your school campus that has an obvious pest problem. What are the options to manage the pest? How will your decision impact other plants, living organisms, and your fellow students?
3. **Exploratory.** Research what beneficials could be used to control insect pests in your school greenhouse. How does the control by beneficials compare to other methods classes may have used in the past?
4. **Exploratory.** Create a pest management plan for your school greenhouse. What preventive measures will you implement at the start? What protocols will you have in place that will allow you to make informed decisions when pest problems do arise?
5. **Exploratory.** Visit a local pesticide dealership and ask to participate in a sales call. How is this job (pesticide sales associate) different than a crop consultant job? What are the responsibilities of a sales representative? If you wanted this kind of position, what would you need to do to be marketable in the future?

CHAPTER 30
Insects

Chapter Outcomes
After studying this chapter, you will be able to:
- Describe the anatomical structures of insects and their functions.
- Summarize each step of an insect's life cycle.
- List the ways pheromones and allomones influence insect behavior.
- Explain the taxonomic hierarchy of insects.
- Describe the methods for collecting and pinning insects.
- Describe careers related to entomology.

Words to Know

allomone	instar	phytophagus
complete metamorphosis	mandible	polyphagus
diapause	micropyle	proboscis
elytra	monophagus	sclerite
embryogenesis	ocelli	setae
entomology	oligophagus	spermatheca
exoskeleton	ommatidia	stylet
hemelytra	ovipositor	tegmina
incomplete metamorphosis	parthenogenesis	vivipary

Before You Read
Write all the chapter terms on a sheet of paper. Highlight the words that you do not know. Before you begin reading, look up the highlighted words in the glossary and write the definitions.

Female praying mantids that eat the heads of their male friends, assassin bugs that pierce and suck the innards of their prey, and ants that harvest honeydew from aphids and in return offer plants protection, are just a few examples of fascinating insects and their interesting behaviors. ***Entomology*** is the branch of biology that is the study of insects. Entomology is important in horticulture because growers can use their knowledge of insect biology and behavior to help manage insect pests and encourage pollinators and other beneficial insects.

Anatomy

Insects belong to the Animalia kingdom in the phylum Arthropoda and specifically fall within the class Insecta. Insects are characterized by having:
- An exoskeleton.
- A segmented body divided into three regions (head, thorax, and abdomen).
- One pair of antennae.
- Three pairs of legs.
- Wings (usually, but not always).

Exoskeleton

Insects, as well as other arthropods (spiders, mites, scorpions, and crustaceans), are supported by a hardened external skeleton called the *exoskeleton*. The muscles and organs are on the inside of the skeleton. The exoskeleton serves as armor, shielding the insect from harm. It protects against physical injury and provides a barrier against the environment, including predators, pathogens, and chemicals. The exoskeleton also provides a watertight layer against desiccation and a sensory interface with the environment. Many *setae* (hairs) cover some insects, sending signals to the insect to react in some way.

The exoskeleton consists of a noncellular covering called the cuticle. The epidermis, which is a single outer cell layer of the body, forms the cuticle. Small, rigid plates of the exoskeleton called *sclerites* are joined together by soft, flexible membranes, allowing for movement and flight, **Figure 30-1**. Sclerites do not grow. As the insect develops, it will molt to reveal a new cuticle underneath.

enterphoto/Shutterstock.com   anat chant/Shutterstock.com

Figure 30-1. An insect's exoskeleton provides protection, prevents desiccation, and allows for sensory interaction with the environment. The exoskeleton consists of small, rigid plates called sclerites. The ventral sclerites underneath the insect's body are called *sternites*.

The Head

An insect's head contains many critical components, including the mouth, eyes, and antennae, **Figure 30-2**.

Antennae

The antennae are a pair of sense organs located near the front of an insect's head. Antennae are primarily used to detect smell and are covered with tiny hairs that can capture odor molecules. Insects also use their antennae to touch the environment to determine location, food, and predators. Some insects can perceive changes in humidity and temperature through their antennae. Others use their antennae to identify sounds, and some can use their antennae to determine air speed during flight. There are a number of antennae types, **Figure 30-3**.

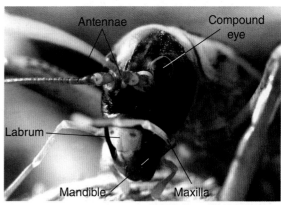

Attila Fodemesi/Shutterstock.com

Figure 30-2. The basic components of an insect's head.

Antennae Types

Name	Appearance	Examples	Name	Appearance	Examples
Aristate (pouch-like with lateral bristle)		House flies	Moniliform (bead-like)		Termites
Clavate (gradually clubbed at the end)		Carrion beetles	Pectinate (comb-like)		Male glowworms and fire-colored beetles
			Plumose (brush- or feather-like)	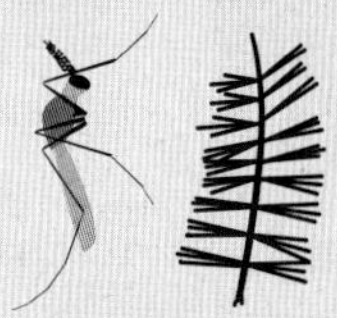	Mosquitoes, moths
Flabellate (fan-like)		Lined June beetle	Serrate (sawtoothed)	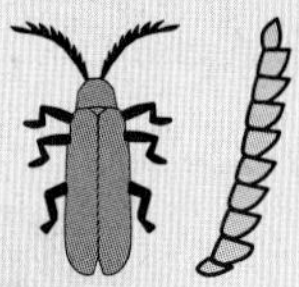	Click beetles
Geniculate (hinged or bent like an elbow)		Weevils and ants	Setaceous (bristle-like)		Dragonflies
Lamellate (nested plates)		Scarab beetles	Stylate (ending in a long, slender point)	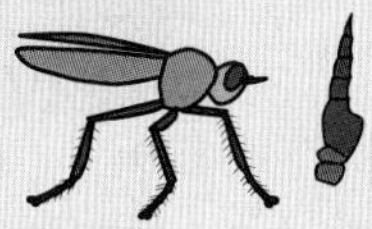	Horse and deer flies

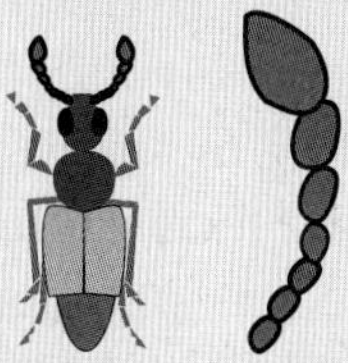

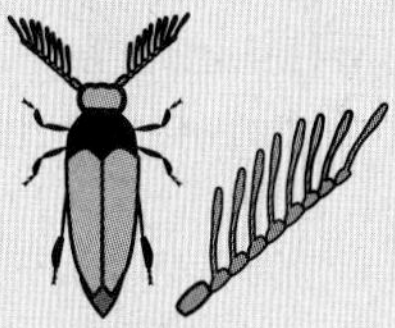

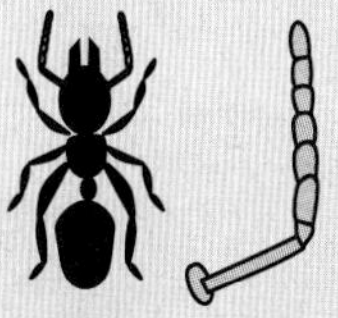

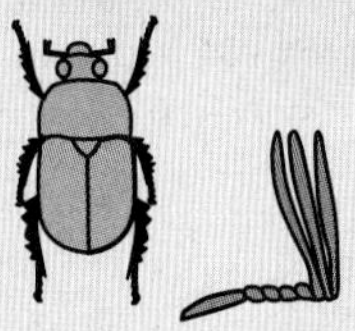

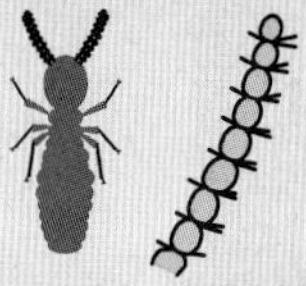

L. Shyamal

Figure 30-3. Observe the differences between the antennae.

Compound Eyes and Ocelli

Most insects have two compound eyes, which are made up of *ommatidia*, visual sensing facets that collectively make up the entire compound eye. The number of ommatidia varies considerably between species. For example, some worker ants have fewer than six while some dragonflies may have more than 25,000. Some insects have reduced or absent compound eyes. Examples include parasitic insects, many soil-dwelling insects, and those species that inhabit dark places such as caves. Compound eyes allow for a wide field of vision. Dragonflies, for example, have incredibly large compound eyes and can see in nearly every direction, **Figure 30-4**. Most insects also have *ocelli*, modified eye structures that can detect changes in light intensity.

Mouthparts

All insects have the same basic five mouthparts: the labrum, the mandibles, maxillae, hypopharynx, and the labium, **Figure 30-5**. However, mouthparts vary by species. The mouthparts of each species have evolved, depending on the feeding behavior and food in its diet. Insect feeding behaviors are divided into the following categories:

- Biting and chewing (grasshoppers, dragonflies).
- Sipping (nectar) and chewing (pollen) (honeybees).
- Probing and sipping (butterflies, moths).
- Sponging and lapping (flies).
- Piercing and sucking (mosquitoes, assassin bugs, aphids).

Mandibles

Grasshoppers have biting mouthparts that allow them to crush and chew their food, **Figure 30-6A**. Their *mandibles* (parts of the insect mouth that are often modified for specialized feeding) are hinged to allow them to eat grasses, leaves, or smaller insects. Some biting insects, such as ant lion larvae, have mandibles that are longer and sharply pointed so they can impale their prey, **Figure 30-6B**.

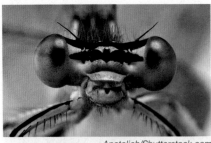
Anatolich/Shutterstock.com

NERYX.COM/Shutterstock.com

Figure 30-4. Dragonflies are effective predators because of their large compound eyes that offer a nearly 360° view.

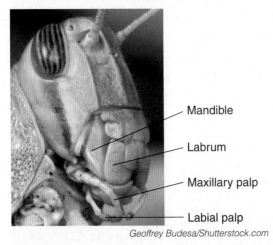
Geoffrey Budesa/Shutterstock.com

Figure 30-5. The basic structure of an insect mouthpart.

A *kurt_G/Shutterstock.com*

B *Manfred Ruckszio/Shutterstock.com*

Figure 30-6. A—The mandibles of a chewing insect are well-defined and strong, allowing them to eat through leaves and even bark. B—Ant lion larva have pointed mandibles to impale their prey.

Proboscis

Many insects eat liquid food, such as nectar, blood, sap, or cellular contents of plants. The honeybee has a modified mouth part that folds together to form a *proboscis*. The proboscis is used to sip nectar, **Figure 30-7**. Additionally, it has functional mandibles that can manipulate wax and aid in pollen consumption. The proboscis is jointed and can be folded up when not in use. Butterflies and moths have a proboscis (but without a functional mandible) that is long enough to reach the nectar deep in narrow tubular flowers. Their proboscis can be coiled up toward their body. These are probing and sipping mouthparts. Houseflies have a proboscis that is designed for lapping food.

Stylets

Many insects have specialized mouthparts that form *stylets*. Stylets allow an insect to pierce the epidermal layer of a plant or animal and withdraw liquids. Mosquitoes have stylets that can pierce human skin and draw blood. Plant pests, such as aphids, thrips, stinkbugs, leafhoppers, and others, use a stylet to puncture a plant's epidermis and suck up the cellular contents, **Figure 30-8**. Many of these insects also transmit plant pathogens through their feeding behavior. The mouthparts of some insects may change, depending on the insect's life cycle stage. Butterfly larvae, for example, have a chewing mouthpart that is more or less nonexistent in its adult form.

Dmitri Gomon/Shutterstock.com

Figure 30-7. Honeybees have a modified mouthpart called a proboscis that allows them to suck up nectar from flowers.

Nathanael Siders/Shutterstock.com

Figure 30-8. The stylet mouthpart enables insects to pierce the epidermal layer of a plant and suck nutrients, proteins, carbohydrates, and water from the cells within.

The Thorax

The thorax is the middle segment of an insect's body. It holds the legs and wings that have been specialized for locomotion. The thorax is divided into three segments, with each segment housing a pair of legs, **Figure 30-9**.

Did You Know?
Flies predigest their food by spitting on it. The spit has enzymes that break down the food material to be lapped up by flies.

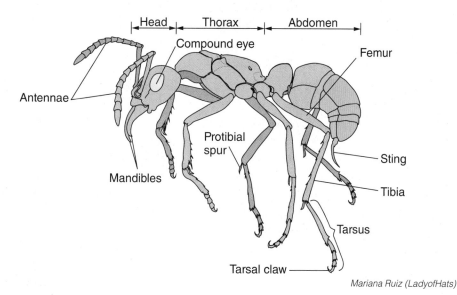

Mariana Ruiz (LadyofHats)

Figure 30-9. Basic worker ant anatomy.

"Happiness is a butterfly, which, when pursued, is always just beyond your grasp, but which, if you will sit down quietly, may alight upon you."
—Nathaniel Hawthorne

Insect Legs

Insect legs have evolved for specialized functions. For example, the digging legs of a mole cricket allow it to burrow through soil, and the grasping forelegs of a praying mantis permit it to capture and consume its prey. Many insects, such as grasshoppers, flea beetles, and leafhoppers, have a jumping ability. This jumping ability is enabled through adaptations in the hind legs, **Figure 30-10**. Aquatic insects, such as the giant water bug, have hind legs that are flattened like a paddle for swimming. Honeybees have special compartments on their hind legs called pollen baskets, **Figure 30-11**. Pollen is placed and stored in the pollen baskets during foraging. Some insects, especially larvae with limited need for transportation, have no legs. The feet of insects are flexible and have claws that allow many to climb vertical surfaces.

Tyler Fox/Shutterstock.com

Figure 30-10. The hind legs of jumping insects are enlarged to allow for great leaps.

Ikordela/Shutterstock.com

Figure 30-11. The pollen basket on a honeybee provides storage space to hold pollen while foraging.

The Wings

Insects are the only invertebrates to have wings. With the exception of mayflies, only adult insects have functional wings. Wings have a network of veins that provide support and rigidity and enclose transparent cuticles. Venation patterns are unique to species and can be used in identification, **Figure 30-12**.

EtiAmmos/Shutterstock.com

Figure 30-12. Many venation patterns in the wings are identifying clues for different species.

STEM Connection
Butterfly Scales

The word *lepidoptera* means scaled wings in Greek. This name is well suited to the butterflies and moths that belong to this order because their wings are covered with thousands of tiny scales overlapping in rows. The scales, arranged in designs and colors unique to each species, provide beauty to an observer and help the insect perform critical functions. Scales provide insulation and thermoregulation, produce pheromones (in males), and aid in guiding flight. The patterns also provide passive defense through camouflage or mimicry.

Many insects have two pairs of wings. Others have one pair, and some have no wings. In a number of insects, the front wings serve as protection for the delicate hind wings. In beetles, the front wings are hardened and are called *elytra*. When a ladybird beetle flies, the front elytra wings open to reveal the soft flight wings, **Figure 30-13A**. In grasshoppers and cockroaches, the front wings are like leathery parchment and are called *tegmina*, **Figure 30-13B**. Many true bugs (order Hemiptera) have thickened parts of their front wings called *hemelytra*. They are thick and leathery at the base but become more membranous toward the tip of the wing.

Insects use muscle contractions to power their wings. As with insect antennae and legs, insect wings vary in design and function by species. Dragonflies can move each of their four wings independently, giving them precise control and mobility when capturing prey, laying eggs, or mating. Many other flying insects rely primarily on one pair of wings to drive flight. For example, insects such as flies, moths, butterflies, bees, and wasps have one pair of wings that is well developed. Their wings have hooks or hairs that link the forewings to the hind wings, functionally acting as one wing on each side.

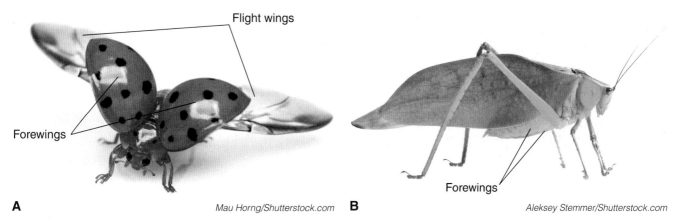

A   *Mau Horng/Shutterstock.com*   B   *Aleksey Stemmer/Shutterstock.com*

Figure 30-13. A—Beetles have hardened forewings (elytra) that protect the delicate flight wings underneath. B—The forewings of a katydid are called tegmina and are thin like parchment.

> **Corner Question**
> How many descendants might one female aphid have over the course of a growing season?

The Abdomen

The role of the abdomen is to store and process nutrients, circulate blood, pump oxygen, develop and deposit eggs, produce sperm, and play a role in mating functions. Unlike the rigid structures of the head and thorax, the abdomen is flexible and can expand and contract. Most female insects have an *ovipositor* on their abdomens, used to deposit eggs, **Figure 30-14**.

Ovipositors differ by species. Some insects can use their ovipositor to make a hole in which to lay eggs. For example, grasshoppers dig holes in the soil, cicadas make holes in wood to lay eggs, and parasitic wasps may spear a larva of another insect species and insert their eggs inside. In stinging insects (bees, wasps, and ants), the ovipositor forms a stinging shaft.

Figure 30-14. Cicadas (A) use ovipositors to deposit insect eggs, sometimes inside twigs (B).

The abdomens of immature insects are often very soft and allow for feeding and molting. These larvae also have prolegs that allow for locomotion and grasping of material. For example, the movement of a caterpillar climbing a stem is aided by the prolegs, **Figure 30-15**. Soft-bodied aquatic immature insects may have external gills, as found in mayflies (Ephemeroptera), damselflies (Odonata), and stoneflies (Plecoptera).

Internal Systems

Insects have similar functions and needs as other animals. *Breathing* in insects occurs through a waterproof cuticle called the spiracle, **Figure 30-16**. Oxygen enters the insect's body and travels very close to cells for gas exchange. Blood travels through insects in an open system. The blood travels through the dorsal blood vessel (the insect equivalence to a heart), through the body cavity, and through specialized structures that circulate it through the wings, legs, antennae, and abdomen. Insects have a gut that serves as the digestive unit, absorbing nutrients for growth and development. Insects also have reproductive organs. The nervous and endocrine systems of an insect serve as the communication system, coordinating actions and responding to stimuli and controlling development.

Figure 30-15. Prolegs aid insect larva in locomotion and climbing to reach food sources.

Figure 30-16. Insects breathe through special organs on the sides of their bodies called spiracles. This silkworm's spiracles can be seen through high-powered magnification.

Growth and Development

Insect development encompasses changes in form and growth in size. The development of every insect includes three primary stages: the embryo, the immature insect, and the adult.

The Egg and the Embryo

Life begins as an egg for most insects, **Figure 30-17**. Each egg, or female gamete, is created in the female's reproductive system, and needs to be fertilized by a male gamete to form an embryo. Through the act of mating, a female receives sperm from her male partner. The female can store the sperm for long periods of time in a special structure called the *spermatheca*. In fact, a honeybee queen will mate with dozens of male drones to increase the genetic diversity among her brood. As a developing egg travels past the opening to the spermatheca, a few sperm are released onto its surface. The sperm swim toward the *micropyle*, a small opening in the eggs, and fuse their nuclei into the eggs to form diploid zygotes. Honeybee queens lay unfertilized eggs that become drones or fertilized eggs that become female. Most of the female eggs will develop into worker bees and a few may be raised as a queen.

happykamill/Shutterstock.com

Figure 30-17. The eggs of a butterfly are laid on a host plant that the larvae can eat once hatched.

After the egg is fertilized, it progresses through a phase of rapid growth and development known as *embryogenesis*. During this stage, the embryo goes through cellular division and differentiation into body segments, appendages, and specialized structures. Once the immature insect emerges from the egg, it is considered a first instar nymph (or larva). (*Instar* is the phase between two periods of molting.)

STEM Connection

Viviparous Aphids

Some insects, such as aphids, will lay fertilized eggs as well as give birth to live young that have been asexually produced. The process of asexual reproduction in insects is called **parthenogenesis**. Aphids combine parthenogenesis with **vivipary** (giving birth to living offspring that develop in the mother's body) in which embryos can start development even before the birth of the mother. In considering managing pests, this results in a shortening of time between aphid generations, overlapping generations, increased reproductive potential, and a possibility of increased rate of development of resistance to insecticides.

Figure 30-18. The grub of a beetle is in an immature phase and will molt several times before pupating into an adult.

Figure 30-19. Silverfish are ametabolous, meaning the immature and adult stages are very similar in appearance and the reproductive organs are functional only in the adult.

Figure 30-20. The immature and adult forms of a stinkbug are rather similar.

Metamorphosis

The newly hatched, immature insect will begin to feed, grow, and mature, **Figure 30-18**. This stage is divided into instars. As the insect develops from one instar to the next, it makes a new cuticle and sheds the old in a process called molting. During the different instars, the insect is changing forms through a phenomenon termed metamorphosis. There are three kinds of metamorphoses:

- Ametabolous.
- Hemimetabolous (incomplete).
- Holometabolous (complete).

Ametabolous Metamorphosis

Only a few primitive insects, such as silverfish and bristletails, change very little as they molt from one instar to the next. Insects in the first instar look very much like those in the adult stage, except the adult has mature reproductive organs, **Figure 30-19**.

Hemimetabolous Metamorphosis

Insects such as mayflies (Ephemeroptera); dragonflies and damselflies (Odonata); grasshoppers (Orthoptera); termites, roaches, and mantids (Dictyoptera); and true bugs (Hemiptera) undergo an *incomplete metamorphosis*. Incomplete metamorphosis insects begin as an egg, transition into an immature larva (sometimes called a nymph), and change into an adult. Their appearance, feeding patterns, and habitat are similar throughout the larval and adult stages. For example, the immature form of the stinkbug is fairly comparable to that of the adult, **Figure 30-20**.

Holometabolous Metamorphosis

In holometabolous, or *complete metamorphosis*, an insect goes through four phases: egg, larva, pupa, and adult, **Figure 30-21**. The pupal stage seems inactive, as there is little outward movement. Physiologically, however, the tissues and organs are changing into adult organs. In some cases, an immature organ will shift to become the functional adult organ. Other organs and tissues may be present but inactive in the larval stage. During the process of pupation, they differentiate into their final structures. Many insect orders have complete metamorphosis, including beetles (Coleoptera), flies (Diptera), butterflies and moths (Lepidoptera), and bees and wasps (Hymenoptera).

Nicolas Primola/Shutterstock.com

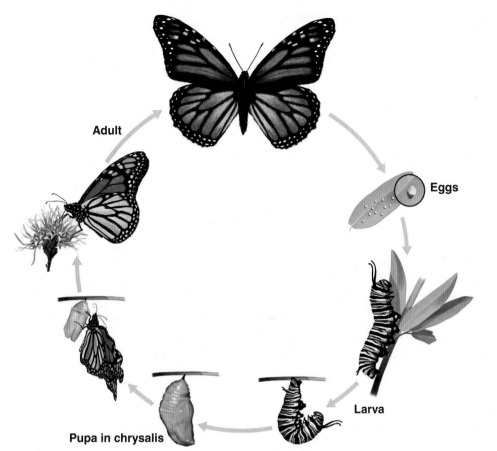

Figure 30-21. Butterflies have holometabolous metamorphosis and go through four distinct life stages: egg, larva, pupa, and adult.

Did You Know?

Caterpillars have developed interesting defense characteristics. Many caterpillars resemble plant parts or bird droppings. Others may just look frightening to scare off predators. Some swallowtail caterpillars have huge "eye spots" on their thorax regions and are able to inflate their bodies so they appear snake-like.

Some insects go through a resting period, similar to a plant's dormant stage, called ***diapause***. Diapause allows insects to enter a state of arrested development and survive unfavorable climactic conditions, such as cold temperatures. Depending on a species, diapause may occur in the egg, larva, pupa, or adult stage. For example, the mourning cloak butterfly overwinters in northern regions as an adult, **Figure 30-22**. This behavior is fairly unique for butterflies.

MarkMirror/Shutterstock.com

Figure 30-22. The mourning cloak butterfly spends its winters as an adult in diapause to survive cold conditions.

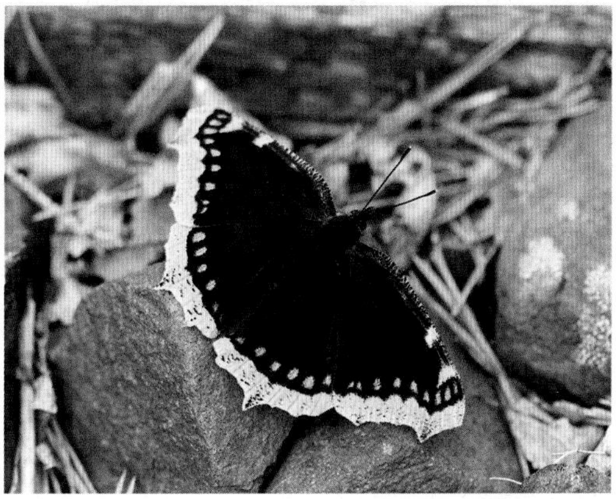

Chemical Signals

Insects constantly receive chemical signals that verify the presence of food, enemies, suitable mates, and nesting sites. Many chemicals elicit a behavioral response, such as scurrying away in the face of a predator or following a trail of food. Pheromones and allomones are two chemical signals important to insects. Pheromones act as chemical signals between individuals of the same species. ***Allomones*** act as chemical signals between different species to the benefit of the originator.

AgEd Connection: Pests and Disorders Identification

As part of a career development event, students must identify certain common pests and/or the symptoms and damage they cause. The pest may be presented as an intact specimen, with a photograph, or as a preserved specimen (insect mount). Some of the common greenhouse and nursery pests that may be included on the identification list include aphids, bagworms, borers, leafhoppers, leaf miners, scale, spider mites, snails, slugs, whiteflies, and white grubs. Use the illustrated glossary (page 924) and the e-flashcards on your textbook's student companion website at www.g-wlearning.com/agriculture to help you study and identify these common pests.

Aphid

Shipher Wu (photograph) and Gee-way Lin (aphid provision), National Taiwan University

Bagworm case

Steve Heap/Shutterstock.com

Borer and wood damage

Henrik Larsson/Shutterstock.com

Leafhopper

YapAhock/Shutterstock.com

Leaf miner

topimages/Shutterstock.com

Scale

Decha Thapanya/Shutterstock.com

Spider mite damage

D. Kucharski K. Kucharska/Shutterstock.com

Snail

Alexander Raths/Shutterstock.com

Slug

Lisa S./Shutterstock.com

Snail/slug damage

Savo Ilic/Shutterstock.com

Whitefly

D. Kucharski K. Kucharska/Shutterstock.com

White grub

fotosav/Shutterstock.com

Copyright Goodheart-Willcox Co., Inc.

Sex Pheromones

Pheromones are used primarily to attract male and female insects of the same species. A female may release a pheromone to signal she is ready to mate, or a male may release a pheromone to announce his presence. Sex pheromones are used to attract insects to pheromone traps, which are used to monitor and control pest populations.

Aggregation Pheromones

Many insects cluster together for feeding, mating, shelter, attack, or other reasons, **Figure 30-23**. Aggregation pheromones encourage insects to come together for a common purpose. For example, bark beetles will produce an aggregation pheromone to increase their population levels at a tree to overcome the tree's defenses. Aggregation pheromones are commonly placed in insect traps being used to monitor the presence and concentration of pest populations.

pan demin/Shutterstock.com

Figure 30-23. Pheromones drive aphids to aggregate for a common purpose of feeding.

Alarm Pheromones

Alarm pheromones are triggered as a predator approaches and may prompt an attack, as in the case of many social insects. Fire ants, for example, are social insects that will quickly swarm and attack a perceived threat, **Figure 30-24**. Honeybees will also send out an alarm pheromone in the presence of an enemy. The bees may gather into a swarm and sting the aggressor. Insect pests such as aphids also use alarm pheromones to warn each other of a perceived threat. If a predator, such as a ladybug, draws near, they secrete a fluid indicating the aphids should drop, leap, or fly to safety.

SweetCrisis/Shutterstock.com

Figure 30-24. Fire ants release an alarm pheromone when attacked.

Territorial Pheromones

Territorial pheromones communicate to other insects that a particular area is inhabited or occupied. A female may lay her eggs in a location and mark it with a pheromone to warn other insects away. These same pheromones may be used to map out a pathway to a food source or habitat. For example, ants may mark the way home or the pathway to food.

Allomones

Allomones play two primary roles in the insect world:
- Defense.
- Beneficial associations between insect species.

Defensive Allomones

Defensive allomones produce an objectionable taste that allows some insects to avoid being eaten. Monarch butterflies have a blood-borne allomone that causes a predatory bird to vomit after ingestion. Predatory birds associate the objectionable taste (and vomiting) with the monarch's bright wing coloration, **Figure 30-25**.

James Laurie/Shutterstock.com

Figure 30-25. Monarch butterflies have a defensive toxin that is a blood-borne allomone that makes it unpalatable to predators.

> "I do believe that an intimacy with the world of crickets and their kind can be salutary—not for what they are likely to teach us about ourselves but because they remind us, if we will let them, that there are other voices, other rhythms, other strivings and fulfillments than our own."
> —Howard E. Evans

Some insects will spit or regurgitate an unpleasant chemical that is highly offensive to many potential predators. For example, bombardier beetles will discharge a hot secretion that is followed by an audible popping when they are disturbed.

Beneficial Allomones

Some insects use allomones to encourage beneficial relationships with other species. For example, some leaf-cutter ants, certain termites, and bark beetles form associations with fungal species. The insects will *farm* the fungi, clearing any contamination that may compete with the fungi. In return, insects harvest the fungi, presumably to aid in digestion.

There is still much to be learned about the chemical ecology of insects, particularly about its use in managing pests and potentially promoting beneficial insect populations.

Taxonomy

Organizing the diversity of insect species is the science of taxonomy. As with plants, insects are classified into a hierarchical structure. The first (and broadest) group is the domain Eukaryota. Other categories in the structure include kingdom, phylum, subphylum, class, order, family, genus, and species (the narrowest grouping). For example, the classification of the Monarch butterfly is as follows:

Domain: Eukaryota
Kingdom: Animalia
Phylum: Arthropoda
Subphylum: Hexapoda
Class: Insecta
Order: Lepidoptera
Family: Nymphalidae
Genus: *Danaus*
Species: *D. plexippus*
Common Name: Monarch butterfly

Insect taxonomy uses binomial nomenclature similar to that used for plants, with the species consisting of the genus and a specific epithet, such as the honeybee, *Apis mellifera*, **Figure 30-26**. Characteristics that are used to classify insects include morphological data (such as structure of mouthparts), body shape and size, antennae, and wing venation as well as genetic information that can isolate an insect's DNA.

The sheer volume of species in entomology is overwhelming. It is important that students be familiar with the major groups of insects, especially the insect orders and, in many cases, insect families.

Protasov AN/Shutterstock.com

Figure 30-26. The European honeybee has a scientific name of *Apis mellifera*, meaning "bee" and "honey-bearing" in Latin.

Agricultural Pests and Beneficials

Phytophagus insects (those that eat plants) eat plant tissues to acquire sugars, proteins, fats, vitamins, salts, and water. Depending on the species, insects will feed on every plant part, including roots, stems, leaves, fruits, seeds, and flowers.

Some insects will feed on multiple plant parts; others specialize the location of their feeding. Insects may be categorized by their choice of food and its location.

- Insects that feed only on a specific host plant for all of their nutritional needs are called *monophagus*. For example, the grape phylloxera is a tiny aphid-like insect that causes stunting or death to wine grapes (*Vitis vinifera*) by feeding on their roots.
- Insects that feed on a broader range of plant material are considered *polyphagus*.
- Insects that will feed on closely related plant species are referred to as *oligophagus*. For example, the Colorado potato beetle does not eat just the leaves of a potato; it will also dine on many solanaceous crops, including eggplant, pepper, and tomato, **Figure 30-27**. Most insects fall into this category.

Anton Vakulenko/Shutterstock.com

Figure 30-27. The Colorado potato beetle will feed indiscriminately on many solanaceous crops, such as potatoes, tomatoes, peppers, and eggplants.

Feeding Behaviors and Plant Damage

Agricultural insect pests are insects that damage or disrupt the development of plants or crops. Pest insects affect crops in different ways and at different stages of plant growth. The insects' feeding behavior (biting, crushing, boring, chewing, piercing, sucking) determines what type of damage will be caused to the plant. This behavior also determines whether the damage will be merely aesthetic, or if it will reduce yields or kill the plants.

Biting and Chewing

Insects with strong mandibles will bite, crush, or chew the plant materials. Infestations of biting and chewing insects can be devastating to a crop, often resulting in plant death.

Boring

Boring insects also chew through plant materials with powerful mandibles. Some boring insects bore into tree bark, whereas others prefer the soft tissue in fruits, nuts, or seeds, **Figure 30-28**.

Piercing and Sucking

Insects use stylets to penetrate plant material and extract fluids containing nutrients and water. Some stylets are strong enough to penetrate tree bark. Plant pests with stylets often transmit plant pathogens through their feeding behavior, further traumatizing the plant.

"You will catch more flies with a spoonful of honey than with a gallon of vinegar."
—Romanian Proverb

Henrik Larsson/Shutterstock.com

Figure 30-28. Bark beetles, such as the European oak bark beetle, can use their mandibles to bore through wood.

Figure 30-29. Marble oak galls are formed due to an insect laying eggs inside the plant tissue.

Gall-Producing Insects

Gall-producing insects deposit their eggs into plant tissue, and the plant responds by abnormally swelling into a gall, **Figure 30-29**. Galls can also occur when insects are feeding. Damage caused to the plant can be insignificant or serious.

Pathogen Transmission

Damage by insects can go beyond the actual eating of plant tissues. Transmission of plant pathogens, as well as toxicity to the plant, can cause irreparable harm. Sucking insects may transmit a toxin that causes leaf spotting, streaking and browning on leaf blades, leaf curling, yellowing, wilting, growth reduction, and even plant death. Leafhoppers, for example, can cause "hopperburn," or a browning and curling of leaf edges.

Vectors

Insects are also vectors for bacterial, viral, and fungal diseases of plants. (A vector is an organism that transmits diseases or parasites from one plant or animal to another.) As the mouthpart is inserted into the plant tissue, microbes from the surface of the stylet are transferred directly into the plant. For example, fire blight is a bacterial disease that seriously impacts certain species in the rose family (Rosaceae) and is especially destructive to apples (*Malus* spp.), pears (*Pyrus* spp.), and crabapples (*Malus* spp.). Fire blight is caused by the bacterium *Erwinia amylovora*, which overwinters in diseased branches and rapidly multiplies in the spring. Large numbers of bacteria will ooze to the surface of the plant and form a sweet, sticky substance. Insects such as aphids, ants, bees, beetles, and flies are attracted to this ooze, pick up the bacteria on their bodies, inadvertently carry the bacteria to opening blossoms, and transmit the disease to healthy plants.

Reproduction Rate

Some agricultural pests can produce multiple generations within a growing season. This rapid reproduction ability makes it challenging to manage populations. With high reproductive rates, insects are also constantly diversifying their genetics. In turn, this allows for the increased development of pesticide resistance.

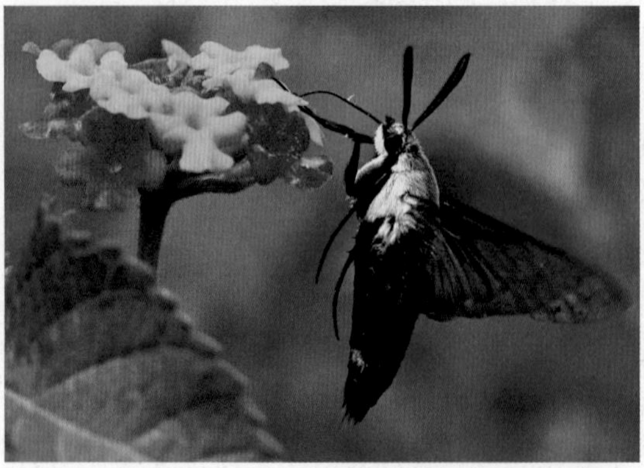

Beneficial Insects

Not every insect is a plant pest, and many insects play a critical role in the reproduction and perpetuation of many flowering plant species. Others may serve as natural predators of insect pests. Many flowering plants have coevolved with a beneficial insect to have a specialized association with that insect. These flowering plants have special shapes, structures, colors, nectars, and odors that attract the insect. The beneficial insect may be the preferred, or the only, pollinator for the plant, **Figure 30-30**.

Did You Know?
Insects are among the most abundant organisms in the world. It is estimated that less than 3% are pests of agriculture, humans, and animals.

Figure 30-30. This hummingbird moth has a long proboscis that can reach the nectar at the bottom of the narrow tubular lantana flower.

Commercial Pollination

Many fruit crops require an insect pollinator for pollination to occur. Apples, beans, blueberries, blackberries, cherries, cranberries, cucumbers, pears, plums, raspberries, strawberries, and tomatoes are a few major crop examples. Having enough pollinators during bloom is essential to produce a harvestable crop. Proper pollination quickens maturity, increases fruit size at yield, and produces a more desirable fruit shape. To ensure proper pollination occurs, growers may rent beehives from commercial beekeepers. These beehives are placed in and around crop fields or orchards during critical bloom periods, **Figure 30-31**.

It is estimated that honeybees are worth $14.6 billion to the agricultural industry and contribute to one-third of the food we eat. Native bees are also important pollinators. In some cases, they are more efficient than honeybees at the individual level.

Corner Question

How many trips would a honeybee have to make to collect enough nectar for a pound of honey?

branislavpudar/Shutterstock.com

Figure 30-31. Honeybee hives may be rented by fruit tree growers and placed in orchards to optimize pollination and fruit set.

Natural Predators

Many insects are natural enemies of insect pests, and their presence can be essential to a successful IPM system. Some predator insects prey on organisms by attacking or killing them. The green lacewing is a predator insect, **Figure 30-32**. Green lacewings are generalists, feeding on a wide variety of small insects such as thrips, mites, whiteflies, aphids, leafhoppers, and mealybugs. All green lacewing larvae are predaceous, and some species have predaceous adults. Green lacewings are commercially available and may be introduced in greenhouse and nursery settings. Ladybugs, or rather ladybird beetles, are another example of a natural predator whose presence is encouraged and often introduced to crop fields, nurseries, and greenhouses.

A Henrik Larsson/Shutterstock.com **B** Evgeniy Ayupov/Shutterstock.com

Figure 30-32. Green lacewings are predaceous as larva (A) and adults (B) and are natural enemies of many insect pests.

STEM Connection

The Yucca Moth

All species of yucca east of the Rocky Mountains are pollinated by one species of moth, *Tegeticula yuccasella*. After mating, the female moth prepares to lay her eggs. The preparation process begins with the female removing pollen from a yucca plant's flower anthers. She transports the pollen to another flower on a different yucca plant using specialized structures (curved tentacles in her mouth). She deposits the pollen on the flower's stigma, effectively fertilizing it. She then uses her ovipositor to pierce the flower ovary and lays her eggs. When the eggs hatch, the fertilized ovaries will have produced seeds for the larvae to eat. The larvae crawl to the soil and overwinter as pupa. The female lays only a few eggs, ensuring that the plant will have enough seeds and not abort the fruit. This relationship provides protection and food for developing larvae as well as plant pollination and seed production.

Ikordela/Shutterstock.com

Parasitic Predators

Other insects are parasites that feed on, and kill, their host insect. Most of these insects are parasites as larvae. Tachinid fly larvae are internal parasites of immature beetles, butterflies, moths, sawflies, earwigs, grasshoppers, or true bugs. Insects lay eggs on foliage near the host insect. The host insect ingests the maggots after they hatch. In other species, the adult fly adheres eggs to the body of the host, and the maggots burrow into the host's body after the eggs hatch. Some species can pierce the host's body with an ovipositor and deposit eggs inside. For example, the braconid wasp lays eggs inside the tobacco hornworm. The eggs hatch inside the body of the hornworm (the larva of a moth) and begin to feed. They eventually pupate outside the body of their host, ultimately killing it.

Growers and gardeners can incorporate natural insect predators as part of a pest management program. Some natural insect predators may be purchased commercially. Others may need to be cultivated by creating beneficial habitats that encourage their presence.

Collecting Insects

Insects can be collected from almost anywhere: school grounds, parks, gardens, and along sidewalks. When collecting insects, observe plants in the garden, especially flowers and leaves, and closely examine tree branches, bark, and stems. Many decomposing insects can be found under rocks and logs. A walk around the edges of ponds, lakes, streams, rivers, and bogs may reveal an abundance of aquatic insects. Use bait, a sweep net, pitfall traps, light traps, or killing jars to collect insects.

Bait

Bait, such as a piece of fruit or other food, can be used to lure insects. Different baits attract different insects. Traps may also be baited with synthetic pheromones or allomones.

Sweep Nets

Sweeping is done by moving a net back and forth through tall plants to collect small insects, **Figure 30-33**. Similarly, using an aerial net allows for the capture of flying insects. The net can be swung through the air or around plant material.

Kirsanov Valeriy Vladimirovich/Shutterstock.com

Figure 30-33. Insect nets allow for the capture of numerous insects to be collected and identified.

Pitfall Traps

Another common collecting method is using pitfall traps. Pitfall traps are made by first digging a hole deep enough to keep the container lip level with the soil surface. Fill the container with a preservation agent such as soapy water or ethyl alcohol. Walking or crawling insects, many of which are active at night, fall into the trap.

Light Traps

A light trap uses a light source (usually a black light) to attract nocturnal insects. Light traps vary in design and may have a funnel or net with a container from which the insects cannot escape.

Killing Jars

Once insects are collected, they can be placed in killing jars. Killing jars may simply be a mason jar with plaster of Paris at the bottom to hold a small amount of ethyl acetate. The gases from the ethyl acetate kill the insects after several hours. Insects may also be put in the freezer for one day (or longer) to kill them. Separate jars should be used for specimens with delicate wings, such as moths and butterflies, to avoid wing damage.

> **Safety Note**
> **Ethyl Acetate Hazard**
> Do not ingest or inhale ethyl acetate, as it may be hazardous to your health. Also avoid skin and eye contact to avoid irritation. Refresh used kill jars in a well-ventilated area.

Preserving Insects

Preserving insects allows for an intimate observation and study of an organism that can be elusive in natural environments. Collecting and pinning insects preserves specimens for learning and identification, **Figure 30-34**. Insects can be stored and displayed in special collection boxes or in simple plastic containers. When handled carefully, a collection can be a learning tool for many years.

Nathan B. Dappen/Shutterstock.com

Figure 30-34. Insect collections are important for identifying insects and learning more about how their structures help the insect function.

Pinning

When examining a pinned specimen, you should be able to view the insect's structures without interference from the pins. Use the following steps to practice pinning insects.

1. Rest the insect specimen on a pinning block (foam or cardboard).
2. Secure the insect by holding it in place with your fingers or a pair of forceps.
3. Insert the insect pin at the proper location, **Figure 30-35**. Entomologists use specialized insect pins that do not damage the insect body or rust. The pins are available in sizes that relate to the thickness of the pin and are used based on insect characteristics such as thickness, hardness, and size.
4. The pin must go through the insect. Keep approximately 3/8" (.95 cm) of the pin showing above the insect body. This will enable you to hold the pin without touching or damaging the insect.

Some insects can be pinned by spreading their wings. This creates a specimen that looks realistic and natural. Lepidopteran species' wings are always spread. For others, such as Orthoptera (grasshoppers), it is not necessary but can be interesting to observe. Using a spreading board, pin the insect in the proper location on the thorax and gently spread the wings until they are level with the top of the board.

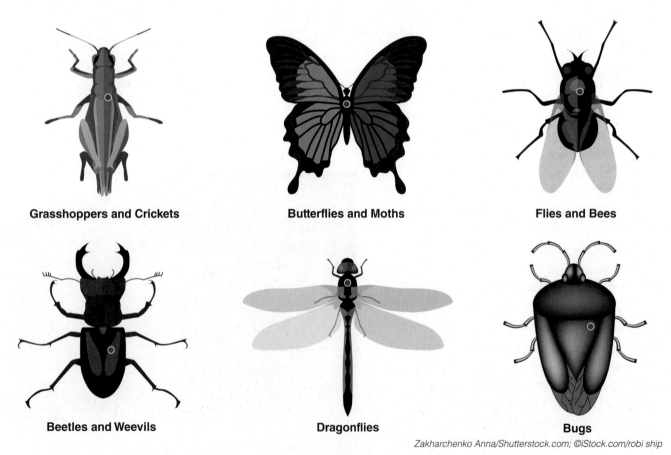

Zakharchenko Anna/Shutterstock.com; ©iStock.com/robi ship

Figure 30-35. Each type of insect is pinned in such a way as to cause the least amount of damage and allow the best view of the insect's anatomy.

Position the wings and use strips of paper to anchor the wings down, **Figure 30-36**. The paper strips are held in place by insect pins.

Gluing

Insects that are too small to be pinned are glued on a small paper triangle called a *paper point*. Using stiff paper, such as an index card or card stock, cut a triangle. Using alcohol-soluble glue, attach the point firmly to the insect's right side on the middle of the thorax. Place an insect pin through the point on the wide end. The insect may have to be held in place with forceps until the glue dries.

grafvision/Shutterstock.com

Figure 30-36. Strips of paper can be used to anchor the wings until the insect is pinned in place.

Labeling

Two labels should be placed on the pin below each insect specimen. Both labels should be the same size and lined up parallel to the length of the body of the insect. The top label should have the county and state in which the insect was collected, the collection date, and the name of the collector. The bottom label should show the order of the insect, family, scientific name, and common name.

Careers Related to Insects

Entomologists are people who study insects. Entomologists may be employed in the area of agricultural research or by zoos, botanical gardens, or other organizations that have an interest in studying or controlling insects.

Agricultural Research Entomologist

Many research entomologists find employment in the agricultural industry. They may evaluate new lines of agricultural and horticultural crops for susceptibility to major insect pests (both field and laboratory research). Other entomologists may develop control strategies for managing insect pests in the field. An expertise in insect rearing is often needed to supply the insect pests for research. Many entomologists work for agricultural chemical companies, cooperative extensions, state departments of agriculture, or the United States Department of Agriculture. Educational requirements for a research entomologist may include a bachelor's, master's, or doctorate degree in entomology or a related field.

Education Entomologist

Many entomologists are dedicated to educating the public about insects. These entomologists may work for organizations such as zoos, botanical gardens, and butterfly houses. The popularity of live insect exhibits (such as butterfly or bug houses) has created a need for entomologists trained in insect husbandry. These entomologists may design and prepare insect exhibits for public viewing. A college degree and experience in education is necessary for this type of position.

Career Connection: The Bug Chicks

Kristie Reddick and Jessica Honaker

Kristie Reddick and Jessica Honaker are two quirky scientists with master's degrees in entomology from Texas A&M University who share a love of bugs. Reddick and Honaker describe their business, The Bug Chicks, as a fun, accurate science media designed for parents, teachers, and bug dorks. Reddick and Honaker use their business to share interesting stories of insects, spiders, and other arthropods, as well as advocate for positive role models of women in science. The two entomologists are also determined to increase science literacy in youth and adults alike.

Reddick began her studies in acting and dancing and became hooked on entomology after being introduced to blood-curdling arachnids and poop-scouring dung beetles in Kenya. Honaker began her college years with the intent of becoming a physical therapist, only to discover that, while dissecting mosquito larvae, invertebrates had stolen her heart, too. While pursuing their master's degrees at Texas A&M, they filmed videos that captured their mutual excitement and passion for communicating about insects with others. Their work inspired the creation of "The Bug Chicks."

Photography by The Bug Chicks Kristie Reddick and Jessica Honaker of The Bug Chicks

The Bug Chicks is Kristie and Jessica's business that artfully weaves together their desire to inspire young people into science through different media formats. They blog, give presentations, hold workshops, and produce funny and educational videos. Most recently, they have taken their ideas across the country, traveling with a green couch to explore backyard wildlife and sending a message that science is cool and relevant, and anyone can do it. They both suggest if you find something that you really love and want to do, go for it with everything you have!

CHAPTER 30 Review and Assessment

Chapter Summary

- Entomology is a branch of biology that is the study of insects. Entomology is important in horticulture because it helps people to understand how to manage insect pests and encourage pollinators and other beneficial insects.
- Insects can be characterized by having an exoskeleton, a segmented body, three pairs of legs, one pair of antennae, and usually (but not always) wings.
- Insects may have one of three kinds of metamorphosis: ametabolous, hemimetabolous (incomplete), or holometabolous (complete). All insects have stages as an egg, an immature insect, and an adult.
- Pheromones are chemical signals and may be used to attract a mate, call insects to aggregate, give an alarm, or mark a territory or trail.
- Allomones are chemical signals that act between different species. Allomones may be used to allow insects to take food and shelter from social insects or as a defensive mechanism to warn predators.
- Insects are classified into a hierarchical structure. Characteristics that are used to classify insects include morphological data, such as structure of mouthparts, body shape and size, antennae, wing venation, and genetic information.
- Insect pests can be monophagus, oligophagus, or polyphagus and may eat the plant and/or vector plant pathogens.
- Not every insect is a plant pest, and many insects play a critical role in the reproduction of many flowering plant species. Others may serve as natural predators of insect pests.
- Preserving insects allows for an intimate observation and study of these organisms. Pinning or gluing insects preserves specimens for studying and as an identification tool.
- Entomologists study insects for various reasons. Entomologists may be employed in agricultural research or by zoos, botanical gardens, or other organizations that have an interest in studying or controlling insects.

Words to Know

Match the key terms from the chapter to the correct definition.

A. allomone
B. complete metamorphosis
C. diapause
D. elytra
E. embryogenesis
F. entomology
G. exoskeleton
H. hemelytra
I. incomplete metamorphosis
J. instar
K. mandible
L. micropyle
M. monophagus
N. ocelli
O. oligophagus
P. ommatidia
Q. ovipositor
R. parthenogenesis
S. phytophagus
T. polyphagus
U. proboscis
V. sclerite
W. setae
X. spermatheca
Y. stylet
Z. tegmina

1. A rapid stage of growth and development after the egg is fertilized during which the embryo goes through cellular division and differentiation.
2. A process in which insects pass through the life cycle stages of egg, larva or nymph, and adult.
3. A process in which an insect goes through four phases: egg, larva, pupa, and adult.
4. A structure on the abdomen of a female insect that deposits eggs.
5. The insect development phase between two periods of molting.
6. A chemical signal that acts between different species to the benefit of the originator.
7. A special reproductive structure found in female insects that stores sperm.
8. Parchment-like forewings found in grasshoppers and cockroaches.
9. The process of asexual reproduction in insects.
10. A modified mouth part that allows insects to suck up a liquid.
11. Modified eye structures that detect changes in light intensity.
12. A branch of biology that is the study of insects.
13. A small opening in the egg by which sperm enters.
14. Visual sensory facets that make up an insect's compound eyes.
15. A modified mouthpart that allows insects to pierce the epidermal layer of a plant or animal and draw liquids into the mouth.
16. The hairs that cover some insects; used to send signals to the insect to react in some way.
17. A type of insect that eats plants.
18. A small, rigid plate that is part of the exoskeleton of an insect.
19. A type of insect that feeds on multiple plant species that are somewhat closely related.
20. A type of insect that feeds on a wide range of plants.
21. One of the forewings of Hemiptera (true bugs) that is slightly thickened at the base.
22. A type of insect that feeds on only one plant species.

23. A state of arrested development (similar to plant's dormancy) that allows an insect to survive unfavorable conditions.
24. A hardened external skeleton found in all insects.
25. The hardened forewings of beetles.
26. A part of the insect mouth that is often modified for specialized feeding.

Know and Understand

Answer the following questions using the information provided in this chapter.

1. What is entomology, and why is it important for horticulture?
2. List the key characteristics that define an insect.
3. Describe the insect exoskeleton.
4. What purpose do antennae serve?
5. Describe the mouthparts of a honeybee.
6. What are some examples of plant pests that have stylets used to feed on plants?
7. What are some specialized functions insect legs may have?
8. List three different kinds of forewings that may be found in some insects.
9. Describe the role of the abdomen in insects.
10. What structure allows insects to "breathe"?
11. How does a honeybee queen create genetic diversity within her brood?
12. Summarize the key steps of holometabolous metamorphosis.
13. Why is understanding pheromones important for pest management?
14. Explain the taxonomy (classification system) for insects.
15. How do insects cause harm as agricultural pests?
16. What are some plants affected by fire blight, and how is this disease spread?
17. Summarize the behavior of insect predators.
18. How can growers incorporate natural insect predators as part of a pest management program?
19. What are some methods that can be used to collect insects?
20. What are some job activities for research entomologists in the agricultural sector, and what education is required for this job?

Thinking Critically

1. Biomimicry is a field that draws from examples in nature to solve real-world problems. For example, the reflective scales of the blue morpho butterfly have inspired designs in iridescent fabrics and cosmetics as well as in digital displays. Examine the insects around you, and try to determine how you might solve a problem or create a technology based on your observations of insect structure and function.
2. Why is it important for growers to understand the anatomy and physiology of the insects they are trying to control or encourage?

STEM and Academic Activities

1. **Science.** Investigate the science behind insect pheromones. Write a three-page report explaining the chemical reasons for insect behavior and why different species use different pheromones.
2. **Science.** Investigate current research programs in entomology. Choose a research program that interests you. Prepare a report on the scientific methods used in this program and how the results affect the horticultural industry.
3. **Technology.** Agricultural pests are a critical problem for the productivity of a farm. What technologies are available today that allow growers to efficiently manage pest populations? How are these technologies different from those that were available 20 years ago?
4. **Social Science.** Insects have played a significant role in history, from uses in fabric dyes and foods to roles in human disease epidemics. Research how insects have influenced events throughout history and share your findings with your peers.
5. **Language Arts.** Write a short chapter for a book that will be used to teach entomology to third-grade students. The chapter should contain basic information about insects. Keeping the audience in mind, develop the topic thoroughly. Use concrete details and extended definitions to help the students understand the concepts.

Communicating about Horticulture

1. **Reading and Speaking.** Working in groups of three students, create flash cards for the key terms in this chapter. On the front of the card, write the term. On the back of the card, write the pronunciation and a brief definition. Use your textbook and a dictionary for guidance. Then take turns quizzing one another on the pronunciations and definitions of the key terms.
2. **Reading.** Plan an insect collection route. Work with two or three peers on this activity. Each member of the group should choose five local addresses and various kinds of insects to be collected. Determine the amount of allotted time to spend in each area. Using a map of the local area, plan the most efficient and fastest route.
3. **Writing and Speaking.** Use line drawings on poster board to detail the basic structures of an insect. In the margins, create a diagram that shows the different types of metamorphosis of insects. Display the drawings as reference tools for the class. Be prepared to explain to the class how and why each type of body part contributes to the function of the insect.

SAE Opportunities

1. **Exploratory.** Job shadow an entomologist. What are the daily responsibilities for this job? What do you like or not like about this position? What education and experience are required to have this position?
2. **Exploratory.** Inventory the number of different pollinator species on your school campus and their populations. What is an optimal number of pollinators? What could you do to increase the numbers and diversity of pollinating insects on your school grounds?
3. **Exploratory.** Research what parasitic insects could be used in your school greenhouse. Are they easily available? How effective are they? What are the benefits and costs of using this method to control pests?
4. **Exploratory.** Visit a local butterfly house. What butterfly species are present? How are the insects reared and managed? What kinds of foods and plants are available for feeding and mating?
5. **Exploratory.** Create an insect collection for your agriculture teacher. Collect and pin as many insect orders and families as you can find. How will this be a useful tool in learning about horticulture?

Evgeniy Ayupov/Shutterstock.com

CHAPTER 31 Disease Management

Chapter Outcomes
After studying this chapter, you will be able to:
- Describe disease development.
- Identify organisms that cause disease.
- Identify abiotic and biotic diseases.
- Describe the disease cycle.
- List signs and symptoms of disease.
- Describe the management of plant diseases.
- Identify plant diseases.
- List careers related to plant disease management.

Words to Know

abiotic	gall	pathologist
bacteria	haustorium	resistant host
biotic	host	sign
disease	immune	susceptible host
disease triangle	injury	symptom
entomologist	inoculum	toxicity
environment	parasitic plant	virus
fungus	pathogen	

Before You Read
Before reading this chapter, flip through the pages and make notes of the major headings. Analyze the structure of the relationships of the headings with the concepts in the chapter.

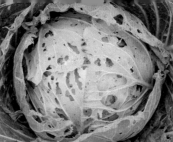

While studying this chapter, look for the activity icon to:

- **Practice** vocabulary terms with Words to Know activities.
- **Expand** learning with identification activities.
- **Reinforce** what you learn by completing Know and Understand questions.

www.g-wlearning.com/agriculture

Does your city or town have a street named Elm? If not, odds are that there is an Elm Street somewhere near you. Much of North America was blanketed with elm trees until the introduction of Dutch elm disease (DED) in 1931, **Figure 31-1**. A *disease* is a disorder in the structure or function of a living organism, such as a tree. The devastation from this disease changed the landscape of most of North America forever.

A furniture company in Ohio imported some elm trees from France, not knowing they were infected with Dutch elm disease. To make matters worse, the exotic and invasive European elm bark beetle that accompanied the logs spread the disease more efficiently than native bark beetles in the United States. By 1977, the city of Minneapolis marked 31,475 elm trees with Dutch elm disease. Most of these trees were over 50 years old and had trunk diameters of nearly 3′ (.91 m). The loss of these trees cannot be overstated as it devastated urban streetscapes, especially in major cities such as New York, Chicago, and Detroit. By the 1970s, many scientists believed that the American elm (*Ulmus americana*) would be extinct by the turn of the century because more than 75% of all elm trees lost their lives to the disease.

Did You Know?
The American Arbor Society estimates that only about one in every 100,000 American elm trees is resistant to the Dutch elm disease. This makes selecting and breeding from that one genetically resistant tree like finding a needle in a haystack.

A *fungus* is a eukaryotic organism that attaches itself to a host and decomposes the organism and absorbs nutrients from it. A bark beetle infects the elm with a fungus (*Ophiostoma*) that causes Dutch elm disease. The fungus spreads via the tree's vascular system. The tree tries to stop the spread of the fungus by plugging up its xylem, the tubes that transport water and nutrients throughout the tree. Infected trees exhibit withering and yellowing of leaves, premature leaf drop, and eventual root death and overall system shutdown. Suckers, or shoots from the original tree, may emerge and grow for up to 15 years, but they too will succumb to Dutch elm disease.

Many methods of prevention, treatment, and even breeding have been attempted to overcome the effects of Dutch elm disease. Initially, insecticides were used to combat the vector of the disease, the beetle. This method of control proved to be ineffective. Later, a fungicide injected into the vascular system of the tree prevented the fungus spores from inhabiting the tree, **Figure 31-2**. This method of control has helped to save thousands of trees throughout North America. Geneticists work to select resistant varieties and continue research in an effort to find a cure to this devastating tree disease.

Joseph O'Brien, USDA Forest Service, Bugwood.org

Figure 31-1. This picture from the late 1970s shows a typical American street lined with American elm trees.

A Joseph O'Brien, USDA Forest Service, Bugwood.org   B Whitney Cranshaw, Bugwood.org

Figure 31-2. A—Damage and symptoms from a European elm beetle. B—Foresters have tried a number of different techniques to stop the infection of the American elm with Dutch elm disease. Here, a tree was girdled and a fungicide was put into the tree's vascular tissue.

Corner Question

Why have American elm breeding efforts failed?

Disease Development

Plant diseases impact the global population every day. Diseases lead to food crop loss before, during, and after harvests. In tropical regions where the conditions constantly promote disease, up to two-thirds of some crops can be lost, contributing to hunger problems. Other horticultural crops, such as timber, flowers, ornamental plants, and turf, are also significantly impacted by plant disease. Estimates of crop losses (food, fiber, and ornamental plants) due to plant pests and diseases are in the hundreds of billions of dollars annually.

Professionals who study and work to combat diseases play a critical role in disease management. A *pathologist* studies diseases, and an *entomologist* studies insects. People in these professions work together because insects are often the vector (transmitter) of diseases in plants. Together they create a system of disease diagnosis and management.

Did You Know?

Some pathogens that infect plants can also infect people. For example, *Pseudomonas aeruginosa* causes rot in soft vegetables, such as lettuce. The bacterium can cause complications in the lungs, urinary tract, blood, and existing wounds in humans.

Disease Triangle

Three factors impact the development of a disease. For a disease to develop there must be a *pathogen* (an organism that causes disease), a *host* (an organism that can allow a pathogen to survive), and the appropriate *environment* (surroundings or conditions in which something lives or exists). These factors make up what is known as the *disease triangle*, **Figure 31-3**. These variables will determine whether and how a disease is expressed. If any one of these factors is not present, then the disease cannot develop.

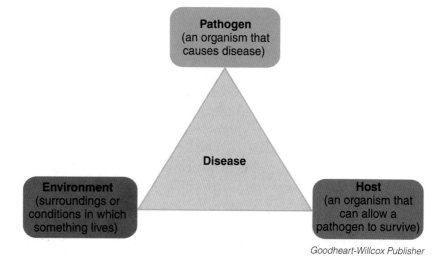

Figure 31-3. The disease triangle shows how pathogens, the environment, and a host interact to determine the amount of disease evident.

STEM Connection

Rose Picker's Disease

Sporothrix schenckii is a fungus that lives in soil, plants, and decaying vegetation. The fungus can enter the human body through a cut caused by a rose prickle, a piece of hay, or pine seedlings—anything that can cause a break in the skin. The fungal spores can also be inhaled. This fungus causes "rose picker's disease" (sporotrichosis). An open sore appears on the skin at the site of infection. In very rare cases, the central nervous system, bones, or joints can be affected.

HHelene/Shutterstock.com

Susceptible Host

A susceptible host plant acts as a nursery for the pathogen, providing food resources and a substrate for the pathogen's growth. Some plants are more susceptible to disease and are better hosts for pathogens than others. A particular plant species may or may not act as a suitable host for a pathogen. Some plants are *immune* (resistant) to certain diseases and are known as **resistant hosts**. However, all plants are susceptible to at least one pathogen. Some plants are more prone to infection by pathogens. Growers may want to avoid using such plants in the landscape or garden.

The general health and growth stage of the plant also contribute to the susceptibility (ability to be infected) of a host of a pathogen. Plant density, or how closely plants are spaced, affects how quickly the disease spreads from one host plant to another, **Figure 31-4**. Healthy, properly spaced plants are less likely to acquire an infection. A *susceptible host* is a plant that is prone to disease.

Vlad Teodor/Shutterstock.com

Figure 31-4. Disease can spread easily between plants that are too tightly spaced.

Pathogen

A pathogen can be bacteria, a fungus, a virus, a disease-carrying nematode, or a parasitic plant. *Bacteria* are microscopic, unicellular, prokaryotic (without a nucleus or membrane-bound organelles) organisms that reproduce asexually by dividing. A *virus* is a nonliving entity that consists solely of nucleic acid (DNA or RNA) and a protein coat. Pathogens may infect all parts of the plant, including the stems, leaves, roots, flowers, fruits, and seeds. These pathogens cause one or more plant systems to not function properly. As a result, plants may show distress, fail to grow or fruit, and even die. A number of different pathways in which a pathogen can infect a plant are discussed later in this chapter.

The progression of a disease in a host is determined by how virulent (infectious) and aggressive the pathogen is. Pathogens can adapt to survive in even extreme environmental conditions. Some pathogens are able to quickly reproduce and disperse over a distance.

Corner Question

What organism is the vector for the bacteria *Yersinia pestis* that causes bubonic plague?

Environment

Several factors contribute to a favorable disease environment. Moisture in and temperature of the air and soil are key factors. Warmer temperatures and higher humidity often increase the speed of the disease cycle. Additional factors, such as poor air circulation, little sunlight, and stagnant water, can contribute to the environment for a disease.

During periods of rain, the greenhouse's humidity may stay near 100%. This high humidity along with warm temperatures is the perfect environment for breeding pathogens. Decreasing the temperature and reducing the humidity can greatly decrease the diseases caused by pathogens, **Figure 31-5**.

> **Corner Question**
>
> What is rose rosette?

Organisms That Cause Disease

Pathogens transmit disease. As in humans, when plants are infected with a disease, there is a causal agent responsible for the disease. Pathogens that are responsible for plant disease include viruses, bacteria, fungi, and other infectious agents.

Vlad Teodor/Shutterstock.com

Figure 31-5. A horticulturist monitors the health of her plants on a humid and cloudy day, when pathogens can cause more problems.

Viruses

Viruses are pieces of either DNA or RNA (genetic material) wrapped inside a protein coat. They do not metabolize, maintain homeostasis, or reproduce on their own, and they are not made of cells, so they are not considered living, **Figure 31-6**. These intracellular parasites must use a host cell to form more virus particles. Viruses can enter a plant through a wound site or be injected by a vector such as an aphid.

Bacteria

Bacteria are unicellular organisms with no nucleus or membrane-bound organelles. These unicellular organisms reproduce asexually. Asexual reproduction in bacteria occurs through a process called binary fission, in which the cells split in two.

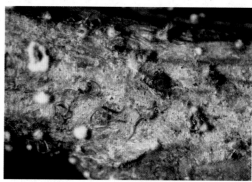

Mary Ann Hansen, Virginia Polytechnic Institute and State University, Bugwood.org

Figure 31-6. A virus can spread throughout a host, such as this pine.

Fungi and Fungal-like Organisms

Fungi and fungal-like organisms (FLO) are heterotrophic, which means they feed off something else and cannot make their own food (like autotrophs can). *Pythium* and *Phytophthora* are examples of FLOs that cause plant rot. These organisms have filamentous growth and can produce spores. Fungi and FLOs cause the most plant disease of all the pathogen groups, **Figure 31-7**.

Julie Vader/Shutterstock.com

Figure 31-7. Powdery mildew attacks the leaves of many squash plants.

> **Did You Know?**
> The largest nematode ever recorded was found inside the placenta of a sperm whale. It was over 26′ (8 m) in length.

Organisms Detrimental to Plants

In addition to viruses, bacteria, and fungi, parasites also pose a threat to plants. These parasites include nematodes as well as parasitic plants. These parasites harm plants by feeding off them.

Nematodes

Nematodes are microscopic worm-like organisms, **Figure 31-8**. These simple, multicellular organisms have fewer than 1000 cells each. Nematodes have a tubular digestive system with an opening at both ends. Although worm-like, they are classified differently from earthworms or any other worm. The total number of identified nematode species is estimated to be one million. Nematodes inhabit environments like salt and fresh water, soil, and other organisms.

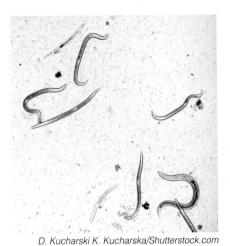

D. Kucharski K. Kucharska/Shutterstock.com

Figure 31-8. Nematodes are microscopic worms that cause various injuries and diseases to plants.

Parasitic Higher Plants

Parasitic plants feed off other plants. They feed off the nutrients being transported throughout the vascular system of the plant. Parasitic plants have a modified root called a *haustorium*, which anchors itself into the vascular tissues of the host plant. Examples of parasitic plants include:

- Mistletoe, **Figure 31-9A**.
- Dodder, **Figure 31-9B**.
- Indian paintbrush, **Figure 31-9C**.

Mistletoe can be found all over the United States and is most easily seen in deciduous tree canopies during the dormant season (when leaves are not present). Mistletoe is often used to decorate for winter holidays. The plant is poisonous to people, but the fruits and seeds are a source of nutrients for wildlife.

A Ottochka/Shutterstock.com
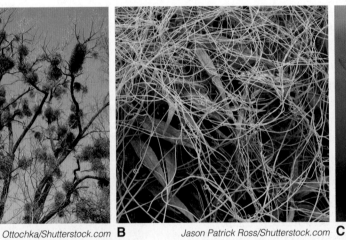
B Jason Patrick Ross/Shutterstock.com

C Tom Reichner/Shutterstock.com

Figure 31-9. A—Mistletoe can be seen in the canopies of deciduous trees when they are dormant. B—Dodder is a leafless parasitic plant that anchors into a host plant's vascular system. C—Indian paintbrush is a parasitic plant as well as a state flower.

Thinking Green

Using Baking Soda to Treat Plant Fungi

Many organic proponents suggest spraying a sodium bicarbonate (baking soda) solution on plants before or at the time a fungus appears on a plant. Baking soda solutions can help and are not bad for the environment. However, they are not 100% helpful for plants either. Baking soda is alkaline. Most plants prefer acidity. Over time, using baking soda can change the pH of your soil, making it more alkaline. In addition, the solution or concentration may cause plant damage.

Types of Disease

Diseases negatively impact the way a plant performs. Diseases cause the plant to show abnormalities, **Figure 31-10**. Just because a plant looks sick does not mean that it is diseased. There are a number of cultural disorders that may also appear to be diseases. Wilting can be caused by a disease, the result of overwatering or under watering, or even a nutrient deficiency. To determine the exact cause of the plant abnormality, a horticulturist must investigate every angle of culture.

absolutimages/Shutterstock.com

Figure 31-10. These diseased tomatoes can show symptoms of numerous types of diseases.

Testing for Disease

When in doubt about what is causing a plant problem, a horticulturist can send a plant tissue sample to a lab to be examined by a plant pathologist. The pathologist can then study the tissue on a molecular level and identify the exact organism that is causing the disease. If a pathogen is identified, the horticulturist can then begin to treat the disease or prevent it from spreading to uninfected plants.

Most land-grant universities have a plant pathology lab or clinic that can identify pest problems. Whether it is a weed, insect, cultural disorder, or disease, the team of pathologists, entomologists, and other scientists work together to correctly identify the problem and help determine a solution. These services are usually not free, but they are much less expensive than at a private lab. Contact your local cooperative extension agency or your land-grant university if you need to submit plant material for testing.

Did You Know?

In the year 2000, apple growers in Michigan suffered devastation to their orchards from fire blight, a disease caused by bacteria. The state estimated a loss of 450,000 apple trees resulting in a total economic loss of $42 million.

Two Types of Plant Disorders

Two types of plant disorders, referred to as abiotic and biotic, cause plants to act abnormally.

> **Corner Question**
>
> What are some common air pollutants that can impact plant growth?

Abiotic

Abiotic disorders are caused by some factor other than living organisms. (The prefix *a-* means without.) They are not infectious and do not spread between plants (are not contagious). Abiotic disorders may result from:

- Reactions to pollutants.
- Nutritional deficiencies and toxicities.
- Temperature conditions.
- Underexposure or overexposure to light.
- Water stress.

Abiotic damage does not show disease symptoms, such as oozing, cysts, or fungal growth. In addition, abiotic damage is not the result of animals, insects, or any other living organism.

Biotic

An infectious disease that is caused by a living organism and that can be spread between plants is known as a *biotic* disorder. Biotic causal agents include pathogens, insects, mites, and animals. Vectors (organisms that spread disease) transmit pathogens. A mosquito or a flea is a common vector for disease in humans, and an aphid or a whitefly is a common vector for plant diseases, **Figure 31-11**.

corlaffra/Shutterstock.com

Figure 31-11. Aphids can infect plants with pathogens and also cause other damage to plants.

Sick Plants versus Injured Plants

Plants can be sick as the result of disease, but they can also be injured in various ways. A diseased plant performs suboptimally as the result of a pathogen. A plant with an abiotic disorder is underperforming due to nutrient deficiencies, water stress, or some other abiotic factor. *Injury*, however, is damage caused as a result of an *immediate* action or event within a plant's environment. Examples of injury include:

- Ice, hail, or wind damage.
- Road salt damage (after snow).
- Damage from chemical burns.
- Pesticide damage.
- Mechanical damage.
- Lightning strike damage.

An injured plant is usually easy to diagnose because the cause and effect situation is obvious. After a significant storm, for example, there may be broken tree limbs scattered on the ground, or shredded leaves in fields of crops. Once injured, plants are more susceptible to disease through open wounds, **Figure 31-12**.

> **Corner Question**
>
> How much electricity is in one bolt of lightning?

Corner Question

Are humans ever intentionally inoculated with disease?

Figure 31-12. A—A lightning strike can cause damage to plant material. B—Pesticide burn or toxicity impacts overall plant health. C—The leaf margins will burn from road salt applied during winter snowfall. D—Mechanical damage by a lawn mower blade can cause life-threatening or just aesthetic damage to plants.

Disease Cycle

The disease cycle is the series of events in the development of a disease. The parts of the disease cycle include inoculation, entrance, establishment, growth, reproduction, and dissemination, **Figure 31-13.** The extent of the disease is measured by visual symptoms. *Symptoms* are a plant's reactions to a disease, such as wilting. If any part of the disease cycle is interrupted, the disease will not progress. The disease can be interrupted intentionally by the grower in an effort to control the disease or prevent future infection.

Inoculation

A pathogen comes in contact with its host in the inoculation part of the disease cycle. An *inoculum* is a pathogen (or its parts) that is capable of causing infection and disease when transferred to the appropriate host and environment.

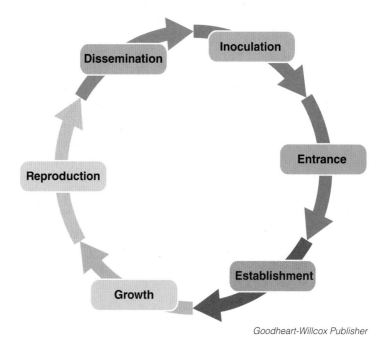

Figure 31-13. The life cycle of a disease begins with inoculation and ends with dissemination.

Entrance

After inoculation, a pathogen gains entrance into its host via several avenues, such as:
- Open wounds.
- Stomata (pores in the epidermis of the leaf or stem), **Figure 31-14**.
- Lenticels (pores in stems).
- Hydathodes (pores in stems).
- Vectors.

At this stage, the plant is considered infected by the pathogen. The surrounding environmental conditions, the general health of the plant, and the concentration of pathogens will determine how quickly and to what extent the disease will become established.

Establishment, Growth, and Reproduction

At this point in the disease life cycle, the pathogen must establish itself within the host and begin replicating. The host acts as a substrate for the replication of the pathogen. The pathogen infects cells, tissues, and organs. Now, during growth and reproduction, the disease becomes apparent.

Bacteria can double in population every 20–30 minutes, depending on the species. One cell becomes two, then four, then eight, and eventually millions within a few hours. Bacteria and viruses reproduce more quickly with warmer temperatures and high humidity. This is why food should be stored properly. Foodborne illnesses are caused by bacteria that quickly reproduce when not properly stored in a refrigerator. Cold air in a refrigerator slows down, but does not stop bacterial reproduction.

Dissemination

During dissemination, the pathogen can infect surrounding plants and the disease can continue to spread. When a pathogen infects another plant host,

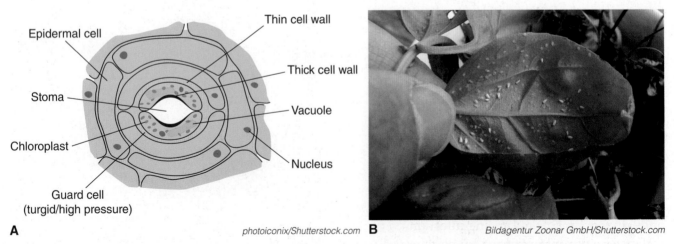

Figure 31-14. A—Disease can enter a plant's system through structures such as the stomata. B—Insects, such as whiteflies, that have a piercing and sucking mouthpart transmit a variety of pathogens to plants.

it begins a secondary disease cycle. The pathogen can also remain in plant tissues of live or dead plants. The pathogen can overwinter in plant tissues and inoculate a host again at a later time.

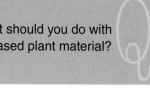

Corner Question

What should you do with diseased plant material?

Signs and Symptoms of Disease

One of the first steps of managing a disease is to identify the signs and symptoms and to diagnose the actual type of disease. Signs and symptoms of disease differ from one another. A *sign* is when the pathogen or part of the pathogen is observed. Examples of signs are:

- Hyphae (branching filaments of a fungus), **Figure 31-15**.
- Mycelium (the vegetative parts of a fungus).
- Spores.
- Bacterial secretions.
- Fruiting bodies.
- Nematodes.

Signs may be identified in the field; however, many signs must be observed and diagnosed by a pathologist in a laboratory setting because of their microscopic size. Symptoms are a plant's reactions to a disease, such as leaf distortion or blight. Symptoms differ from signs in that they can be visually recognized as an abnormality different from an injury. Initial symptoms often are invisible or nondescript. The pathogen may cause a *toxicity* (poisonous effect). Vital plant processes may be interrupted, which may lead to death. As the symptoms progress, they can be classified into the following categories: wilting, abnormal tissue color, defoliation, abnormal increase in tissue size, dwarfing, replacement of host tissue, and necrosis.

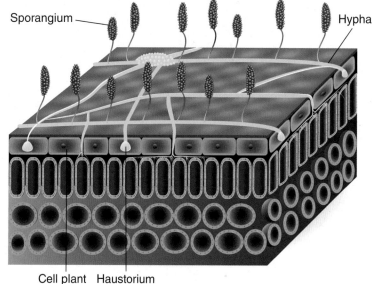

Designua/Shutterstock.com

Figure 31-15. Fungi act as parasites on plants. They need the host plant for survival and a food source.

Wilting

Various bacterial and fungal pathogens can block the xylem of a plant's vascular system. The plant attempts to translocate water and nutrients, but the xylem is blocked by the colonization of the bacteria or fungi. Parts of the plant or the entire plant will die, **Figure 31-16**.

Miyuki Satake/Shutterstock.com

Figure 31-16. Plants that have a disease called wilt look just like those that have suffered from being under or over watered.

Figure 31-17. *Abutilon striatum* 'Thompsonii' (also called a flowering maple) was infected with a virus that caused a mosaic appearance. This disease, however, only makes the foliage appear unique, and there are no other ill effects. Horticulturists covet this unique plant.

Abnormal Tissue Color

Stems, leaves, and flowers can change their pigment in response to a pathogen. Mosaic (mottling), chlorosis (yellowing), necrosis (browning), and reddening or purpling of tissue is common.

Although mosaics are a disease and cause an abnormal appearance in leaf tissue, some mosaics appear to be desirable. *Abutilon striatum* 'Thompsonsii' (flowering maple) suffers from a mosaic virus; however, this virus is actually beneficial. The mosaic disease makes the leaf unique and beautiful and does not cause the plant to perform poorly, **Figure 31-17**. *Abutilon mosaic bigeminivirus* (AbMV) is transmitted by whitefly and causes the leaves to be heavily mottled. This effect is something desired by some plant collectors and enthusiasts.

> "If you would know strength and patience, welcome the company of the trees."
> —Har Borland

Defoliation

Plants may lose leaves or drop fruit in response to a disease. The loss of leaves is known as defoliation. If plants are losing leaves out of season, it may be an indication of disease.

Deciduous plants naturally lose their leaves each fall in North America due to shortening days. Temperature is not a factor in determining when leaves are dropped. The plant measures the hours of darkness. As the length of night increases, the plant sends signals to its chloroplasts, and they begin to die. Other pigments from the leaves are then revealed, showing bright fall colors such as yellow, red, and purple, **Figure 31-18**. Eventually, the leaf dies and falls from the plant.

Abnormal Increase in Tissue Size

A plant's response to a disease may be an increase in the number of cells, larger cell sizes, or a distortion of plant tissues. Plant tissues can twist and curl in the stems, leaves, and roots. *Galls*, tumor-like growths, can develop in response to an insect laying eggs within the plant, **Figure 31-19A**.

Galls can be unique and colorful deformations in plant tissues. Various insects can cause galls on plants. Some insects responsible for galls include moths, flies, aphids, and beetles, **Figure 31-19B**.

Figure 31-18. The deciduous plants of the fall defoliate (lose leaves) according to how many hours of darkness are measured by the plant's system.

A *Jaclyn Schreiner/Shutterstock.com* B *Ariene Studio/Shutterstock.com*

Figure 31-19. A—A gall is a tumor-like symptom that is often associated with a vector. B—This oak gall hosts small epidermal tubes that harbor immature insects.

Did You Know?
Corn smut is a delicacy in many South American cultures. Cuitlacoche (weet-la-ko-che) is a fungus and has a rich mushroom or truffle taste. Aztecs enjoyed cuitlacoche as part of their diet, and it was included in tamales and stews.

Galls have been used throughout history for medical purposes. In the 1700s, the French used galls as a way to treat fevers. Today, galls are used in exotic and gourmet cuisine.

Dwarfing

Plants that appear stunted in size are known as dwarfed. Parts of the plant or the entire plant can become dwarfed as a result of disease.

Replacement of Host Plant Tissue

Disease-causing organisms can replace the tissues of a healthy plant. This can often be seen on flowers and fruit tissues. Corn smut, caused by the pathogen *Ustilago maydis*, replaces entire kernels of corn with large sac-like structures filled with black spores, **Figure 31-20**.

Necrosis

Tissues of leaves, stems, and roots may die as a result of a pathogen. This is known as necrosis, **Figure 31-21**. A fungal disease, called damping-off, results in the death of stem tissues. The seedling's stem tissues die, and eventually, the entire plant suffers the same fate.

Managing Plant Diseases

Plant diseases can be managed; however, prevention of disease is always best. If a disease sign or symptom is identified, then the disease must be diagnosed. After the appropriate diagnosis, the disease can be treated. Using integrated pest management (IPM) practices is essential for combatting plant disease.

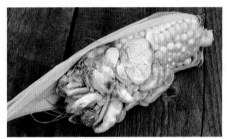

Nataliia Melnychuk/Shutterstock.com

Figure 31-20. Smut is the replacement of plant tissue with a pathogen.

Rocky Mountain Research Station/Forest Pathology, USDA Forest Service, Bugwood.org

Figure 31-21. Rotting is exhibited in the saguaro cactus.

University of Arkansas Forest Entomology Lab, University of Arkansas, Bugwood.org

Figure 31-22. Devices such as these traps can help to monitor and catch pests that spread disease.

Parts of IPM include cultural practices, using disease-resistant plants, incorporating beneficial organisms, quarantining diseased plants, and employing chemicals when necessary, **Figure 31-22**.

Cultural Practices

Effective cultural practices promote healthy plant growth and prevent plant diseases. Plants need appropriate amounts of water, light, nutrients, and space as well as temperatures that allow optimal growth. Proper spacing of plants to allow air circulation is critical in disease prevention. Plants that have optimal growth are more likely to resist diseases.

Genetically Resistant Plants

Plants that can resist pathogens should be used whenever possible. Some plants may naturally resist a disease. Other plants have been genetically engineered or selectively hybridized (bred) to withstand diseases. For example, tomatoes have been bred to resist a variety of diseases. Use disease-resistant plants to make prevention of diseases much easier.

Beneficial Organisms

Another method of suppressing plant diseases is to use beneficial organisms. Some organisms can act as parasites or predators to other pathogens. Some bacteria and fungi have proven to combat a wide range of pathogens. Some products have biological control agents that target pests, such as mildew. Ladybugs and praying mantises are often released to eat disease vectors such as aphids and beetles.

STEM Connection: Papaya Ringspot Virus

Genetically engineered papayas were created by the University of Hawaii in the late 1990s to resist the devastating disease called papaya ringspot virus (PRSV). Papaya ring spot virus causes the foliage to appear abnormally small and skinny.

Hawaii grows, ships, and exports the majority of the papayas throughout the United States and parts of Asia. As a major commodity for this island state, Hawaii depends on healthy papaya plants. When these plants were developing a devastating disease, action had to be taken.

Researchers at the University of Hawaii joined forces to create a genetically engineered papaya that could withstand the virus. Since they are not living, viruses are not treatable and one of the only ways to deal with the disease is prevention.

After a great deal of research, Hawaiian growers were able to trial this new genetically modified organism, but not without controversy. It took a great deal of education and persuasion to convince consumers and government agencies that the new GMO papaya was safe and that the flavor would remain the same.

Alberto Pantoja, USDA Agricultural Research Service, Bugwood.org

STEM Connection: The Beagle Brigade

Dog enthusiasts around the world realize the intensity of the beagle breed's keen sense of smell. Over the years, breeders have trained beagles to be man's best friend and a resource for sniffing. Trainers work with the breed to track animals that will be hunted, find missing persons, and even search for lost possessions.

The USDA and APHIS also employ beagles for another purpose. Each year, the Beagle Brigade (housed at airports and customs points throughout the United States) finds and helps prevent over 75,000 prohibited agricultural items from entering through the borders of the United States. Trainers must use proven animal science principles and the latest technology to properly train these animals to protect the people and the environment of the United States.

Quarantines

Separating infected plants from those that are not infected is a good way to control the spread of a disease. Many plant diseases never enter the United States because of quarantining laws established by the United States Department of Agriculture (USDA) and the Animal and Plant Health Inspection Service (APHIS). These organizations mandate how plant materials are trafficked into the country. Through their vigilant service, the United States is protected from many exotic diseases that could otherwise cross our borders.

Chemical Applications

Many of the diseases that impact plant growth involve fungi. Fungicides treat fungus and are the most-used worldwide pesticide.

Thinking Green

Natural Fungicides

Several oils that are derived from plants can help control fungus without negatively impacting the environment. These oils include:
- Tea tree oil.
- Cinnamon essential oil.
- Rosemary oil.
- Oregano oil.
- Jojoba oil.
- Neem oil.
- Citronella oil.

The oils are applied to the plant's surface. Special care must be made when applying the oils as they may burn the plant's surface. Personal protective equipment (PPE) must be worn when applying these oils (or any pesticide).

Dipak Shelare/Shutterstock.com

Disease Index

Entire textbooks and websites are dedicated to the identification and management of plant diseases. Important plant diseases in greenhouses, ornamental gardens, edible gardens, and turfgrass sites impact growers every day. Diseases that affect plants include anthracnose, blights, cankers, club root, damping off, galls, leaf blister, leaf curl, leaf spot, mildew, mold, root-knot nematode, rot, rust, scab, smut, viruses, and wilts.

Anthracnose

Anthracnose is a fungal disease that causes small dead spots on the leaves, stems, fruits, or flowers, with raised edges and a sunken center, **Figure 31-23**. There may be circles of brown and pink exhibited on the leaves associated with this disease. Avoid touching plants when they are wet to prevent spreading disease. If disease is found, cull (remove) the diseased plant and discard it. Do not add to a compost pile.

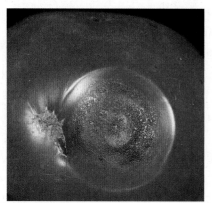

Clemson University - USDA Cooperative Extension Slide Series, Bugwood.org

Figure 31-23. Anthracnose has caused a sunken center in this fruit.

Blights

Blights appear as a scorching on stems and leaves. Leaves will suddenly wither, plant growth will cease, and the plant will die. Cut below where blight occurs to remove diseased parts. Burn diseased material that has been cut from the plant. Use resistant cultivars of plants when available.

Fire Blight

Fire blight is a destructive bacterial disease that impacts pears, apples, and roses. It causes the plants to appear scorched or burned, **Figure 31-24**, and impacts fruit production.

Early Blight

Early blight, also known as *Alternaria*, is a fungal blight that attacks vegetables, ornamentals, and fruit and shade trees around the world. It is caused by the *Alternaria* fungus. Tomatoes, peppers, eggplants, and potatoes are susceptible. Infected plants exhibit symptoms of leaves having concentric rings of brown or black spots. Leaves display symptoms first, but the fruits of these plants will also have dark, sunken spots. To combat this disease, apply beneficial fungi to the soil before planting, use resistant cultivars, encourage air circulation, rotate crops between families, and apply a baking soda spray two weeks before symptoms typically appear on the plant.

Grandpa/Shutterstock.com

Figure 31-24. Fire blight causes millions of dollars of damage to rose, apple, and pear crops around the country.

AgEd Connection

Disease Identification

As part a career development event, students must identify certain common diseases and the damage they cause. Disease symptoms and damage may be presented with a photograph, or as a preserved specimen. Some of the common plant and crop diseases that may be included on the identification list include anthracnose, apple scab, black spot, blights, botrytis, canker, cedar-apple rust, crown gall, fire blight, powdery mildew, and root rot. Use the illustrated glossary (page 926) and the e-flashcards on your textbook's student companion website at www.g-wlearning.com/agriculture to help you study and identify these common disease symptoms and damage.

Late Blight

Phytophthora, also known as late blight, is a water mold that infects azaleas, lilacs, hollies, and solanaceous crops (potato, tomato, eggplant, and pepper). Dark lesions, surrounded by yellow or brown tissue, are found on leaves. The presence of white fuzz indicates that the water mold is producing spores. If spores wash into the soil from rain or irrigation and come in contact with growing potatoes, the potatoes will rot. Tomatoes will have greasy-looking lesions and may show white fuzz also. Plants usually will die during a wet season accompanied by late blight. The disease overwinters in plant material; therefore, all infected material should be disposed of and burned. The best way to control late blight is to prevent it in the first place. Do not save potato seed to use the following year. Plant resistant varieties. Be vigilant, especially during wet weather. Apply a bacteria called *Bacillus subtilis* to the soil to help combat the disease.

Bacterial Blight

Bacterial blight attacks many plants, but it is especially troublesome to legumes. Foliage and the fruit exhibit water-soaked lesions followed by drying, and they eventually drop off. Some spots may have yellow bacterial secretions. Do not work with the legumes when they are wet to avoid bacterial blight problems. To combat this disease, rotate crops, use resistant cultivars, and remove and destroy any infected plant materials.

Cankers

Cankers commonly form on woody stems of plant material. Symptoms include cracks or splitting of wood, sunken areas, or other abnormal tissue, **Figure 31-25**. Cankers may girdle stems, resulting in the loss of branches or the entire plant.

phomphan/Shutterstock.com

Figure 31-25. Citrus fruits, such as this lime, can have canker or lesions on the fruit.

Cytospora Canker

Cytospora canker is a fungus that attacks stone fruits (peaches, cherries, nectarines, and plums), spruces, and poplars. Symptoms appear as discolored, circular lesions on the bark. To combat this disease, cut out the canker areas with clean pruning shears. Use resistant cultivars to avoid this disease.

Nectria Canker

Nectria canker is a fungus that attacks vines, shrubs, and some hardwood species, including maples. Small sunken areas form on the stem and may be accompanied by a white, red, or orange spore-producing structure. If plants are infected, remove dead and diseased branches and burn them. Clean pruning shears between cuts and especially between uses on different trees.

> **Did You Know?**
> Canker does not only appear on woody stems; it can also be found on some citrus fruit. A bacterial canker can infect citrus fruit, such as limes, and cause lesions on the lime peel.

Club Root

Club root is a disease that impacts plants in the mustard family, such as cabbage and broccoli. This fungus causes wilting of the leaves during the day and may lead to yellowing in leaves. Defoliation may also occur. Roots will swell and appear distorted. The fungus can live for many years in the soil, even if host crops are not present. It is important to use cabbage cultivars that are resistant to the disease. Avoid growing cabbages in the same garden space each year.

Damping Off

Damping off is a disease caused by soilborne fungi that may cause seeds to rot before germination, or cause seedlings to rot off at the soil line and fall over. Keeping the soil moist but not waterlogged, increasing air circulation, and using fungicides to treat seeds or seedlings can help prevent damping off.

History Connection: Irish Potato Famine

No other people suffered as greatly as the Irish during what became known as the Irish Potato Famine of the 1840s. Much of Ireland had become dependent on the potato as a critical source of calories and nutrition. Ireland grew a monoculture of potatoes and focused their cultivation on one cultivar known as the 'Irish Lumper.'

The pathogen, *Phytophthora infestans*, is believed to have originated in the Andes Mountains, along with the potato. Some seed potatoes that were sent to Europe were infected, and the pathogen was spread throughout Europe via wind patterns. In 1845, about 40% of the potato crop was ruined; by the next year the entire potato crop was ruined. Because Ireland depended on one variety of potato that was not resistant to *Phytophthora infestans*, the effects were devastating.

More than one million people in Ireland died from starvation, or from diseases such as cholera and typhus (resulting from people in poor health moving into close proximity while waiting to leave Ireland for England or other countries), and more than two million people emigrated because of the potato disease in the 1840s. The disease was not contained in Europe, however. The United States and Canada also recorded instances of late blight during 1845 and 1846, but because their populations did not depend on one crop, the effects were not devastating.

Galls

A gall is a tumor-like growth on plant tissue. Galls are often the result of an insect vector, mite, nematode, bacteria, virus, or fungus. If the gall is dissected and no insect is found inside, then the culprit for the disease is a pathogen.

Crown gall is caused by bacteria and it infects and kills many types of plants, including brambles, grapes, vegetables, and ornamental plants. Galls are rounded like a ball and have a cork-like appearance. They appear at the soil line, on roots, or on branches, **Figure 31-26**. To control this pest, buy resistant plants that appear healthy and avoid planting in soil that once had plants with crown gall.

University of Georgia Plant Pathology, University of Georgia, Bugwood.org

Figure 31-26. The crown gall on this branch is a tumor-like growth.

Leaf Blister and Leaf Curl

Leaf blister and leaf curl cause distorted and curled leaves on many tree species. These fungal diseases cause blisters and yellow bumps. The pathogen may cause the plant to lose leaves and ultimately to die.

Peach curl is a disease that attacks peach and almond trees. The disease symptoms include new leaves that are pale or reddish with a midrib that does not grow within the leaf, causing the leaf to appear puckered. Fruit also exhibits damage, and the tree may eventually die. A dormant oil application prior to the time that buds begin to swell (during the late winter or early spring) can control the disease.

Did You Know?

Georgia ranks third in peach production. The state, however, boasts that its peaches are the juiciest and sweetest. Georgia grows nearly 140 million pounds of peaches during four months of each year.

Leaf Spot

Leaf spot is a fungal disease that causes spots on the leaves. It can lead to serious plant problems. One type, called black spot, is a serious rose pest that causes black spots with yellow margins on leaves, **Figure 31-27**. An infestation can result in leaf drop. Control mechanisms include destroying all leaves from the roses when they fall, mulching underneath rose bushes to prevent spores of the pathogen from splashing on uninfected leaves, and using resistant cultivars of roses.

Mildew

Mildew is easily recognized as a white, gray, or powdery substance that covers leaf and stem tissues. This fungal growth leads to the browning and shriveling of leaves, and it may cause fruit to be of poor quality. Using disease-resistant cultivars when available and applying appropriate fungicides to combat mildew can control this aggressive plant disease.

Yuan-Min Shen, Taichung District Agricultural Research and Extension Station, Bugwood.org

Figure 31-27. Black spot has caused the discoloration on the leaves of this plant.

Mold

Many types of fungal diseases are molds. These diseases appear as woolly threads on the infected plants. Also known as gray mold, *Botrytis* impairs the performance of many flowers and fruits. In high humidity, this disease thrives. It appears on mostly dead or dying tissues. Removing spent flowers and ripened fruit combats *Botrytis*, as does increasing air circulation and decreasing humidity. You may have noticed *Botrytis* on strawberries that you have purchased from the grocery store and left on your counter for a day or two, **Figure 31-28**.

Scott Bauer, USDA Agriculture Research Service, Bugwood.org
Figure 31-28. *Botrytis* can occur when fruit is improperly stored.

Root-Knot Nematode

Root-knot nematode disease is the result of a nematode feeding on the roots of the plant. Large tumor-like galls, or knots, form throughout the root system. The infection results in lower yields and makes edible plants unacceptable to consumers. Plants display additional symptoms aboveground, such as wilting, chlorosis, and stunting. Because the roots of the plant are impacted, the ability to absorb and translocate essential nutrients is limited. Eventually, severely diseased plants may die. Cultural controls, such as rotating crops, planting cover crops, and using resistant cultivars, work well against root-knot nematodes. An infestation may require the use of chemical controls (such as soil fumigants) to control root-knot nematodes.

> "When you're green you're growing, when you're ripe you rot."
> —Ray Kroc

Rot

Rots result in the decomposition of plant tissues in the stems, roots, and fruits, **Figure 31-29**. Rot is often seen in stored plant tissues, such as bulbs and tubers. Rots can be soft and wet or hard and dry, and they can be caused by either bacteria or fungi. To combat root rot, use beneficial fungi and bacteria in the soil. To control stem and fruit rot, increase air circulation and use appropriate chemicals as a last resort. To store bulbs, tubers, and fruits, regulate the environment to keep these perishable items according to scientific recommendations. Requirements may include a certain temperature, relative humidity, and light exposure.

Mary Burrows, Montana State University, Bugwood.org
Figure 31-29. This plant has been affected by root rot, which has caused its roots to decompose.

Corner Question
What is that smell in the gym locker room?

Rust

Rust is a fungus that causes powdery, orange-brown or rust-colored spores to appear on the leaves and stems of plant material. The powdery spores can be carried by the wind and spread the disease. Remove infected plants and burn the material. Spray plants with a neem oil to help control rust disease.

STEM Connection Cedar-Apple Rust

Cedar-apple rust is caused by *Gymnosporangium juniperi-virginianae*. This disease disfigures apples, pears, and cedars on the leaves, stems, and fruit. For apples, the disease cycle begins in April or May with pale yellow or orange spots appearing on the foliage. The spots grow larger and appear as concentric rings that are rust colored. A sign of the disease may include drops of orange liquid. Later in the season, black dots may develop on the orange spots. In the summer, tube-like structures form on the undersides of the leaves where the spots are located.

Cedar trees exhibit a reddish-brown gall that is 1″ (2.54 cm) in diameter. The galls appear to have pitting or look a bit like chicken skin. The pits then evolve into long, jelly-like orange horns in the spring. These horn-like arrangements swell during April and May rains. Go outside after a rainstorm in late spring and see if you can spot these unique galls. Eventually the galls dry up, so they will not always be seen.

Melinda Fawver/Shutterstock.com

Scab

Scab is a fungal disease that causes scabby lesions on fruits, leaves, and tubers. Apples, peaches, and potatoes are commonly infected, **Figure 31-30**. Controlling scab can be accomplished by increasing air circulation through pruning. Additionally, using a sulfur fungicide on a spray schedule suggested by your local cooperative extension agency or land-grant university can help to control scab.

Smut

Smuts are fungal growths that are unique in appearance. Normal organ tissues, usually on fruit, are replaced with the fruiting bodies of these disease-causing organisms. Large, sac-like blisters are filled with spores. Refer to **Figure 31-20**.

Corn smut can cause kernels, tassels, stalks, and leaves to exhibit smut galls. The galls mature and burst, releasing spores. The spores disperse and lay in the soil, waiting to infect future crops. Remove infected galls and burn them before they burst.

University of Georgia Plant Pathology, University of Georgia, Bugwood.org

Figure 31-30. Apple scab is a fungal disease that appears as scabby lesions.

Virus

A plant infected with a virus may function improperly and can exhibit a range of symptoms. Leaves may curl, streak, mottle, or develop ring-shaped spots. Using virus-resistant cultivars and plant material that has been certified is the best way to prevent viral diseases.

Tobacco mosaic virus (TMV) can be transmitted from tobacco to many susceptible crops. Individuals who smoke cigarettes can transmit and spread

Corner Question

What happens to your jack-o'-lantern after halloween?

the disease by handling plants. Growers often ask employees who smoke to wash their hands and wear gloves when working with plants. TMV is expressed by mottling and some puckering of plant leaves. Infected plants do not photosynthesize efficiently, and overall growth is slowed.

Wilts

Bacteria and fungi can attack the vascular system of a plant, causing the plant to look permanently wilted. Part or all of the plant will die when infected with wilt. These pathogens can live in the soil for up to 10 years after infecting plants.

Stewart's Wilt

Stewart's wilt disease is spread by flea beetles and attacks sweet corn. Beetles chew the leaves and then transmit the bacteria, which causes symptoms of wilt and streaking in the leaves. When the stalks are cut, bacterial ooze will appear. The plant will eventually die or no corn will be produced. To combat this disease, use resistant cultivars and control flea beetles with an appropriate insecticide.

Fusarium and *Verticillium* Wilt

Fusarium and *Verticillium* fungal wilts cause yellowing and wilting in a wide variety of plants. Rotate crops to avoid planting the same family of plants in an area where this disease has occurred. Remove and burn infected plants. Use resistant cultivars of plants when available.

Careers in Disease Management

Careers in plant pathology, entomology, and horticulture can focus on the diseases of plants. People in these careers may do research, work in a lab or clinic, or treat plants in a greenhouse or nursery operation.

Plant Pathologist

A plant pathologist specializes in plant health. Plant health management requires proper identification of disease-causing organisms, an understanding of disease-causing agents, a knowledge of plant growth, and an understanding of disease impacts.

In preparing to be a plant pathologist, a person may take college courses in areas such as botany, microbiology, crop science, soil science, ecology, genetics, biochemistry, molecular biology, and plant physiology. Although plant pathology is an interdisciplinary science, most pathologists specialize and take courses for master's and doctoral degrees. Plant pathologists are employed by colleges and universities, government agencies, international institutes, and industrial firms, **Figure 31-31**. They may also work as private practitioners.

Stephen Coburn/Shutterstock.com

Figure 31-31. A pathologist uses a microscope to examine plant tissues that have been sent to the lab.

Farm Advisor

A farm advisor provides expertise in areas such as soil, irrigation, nutrition, and overall plant health. The advisor must usually possess at least a master's degree in a related agricultural field. In addition, he or she must have extensive knowledge and experience in plant production (usually within a specific commodity). The farm advisor should be able to not only recognize pests and plant problems, but also thoroughly understand how to prevent and treat these pests. A career as a farm advisor requires a great deal of general knowledge, but especially some specialty in plant pathology.

Farm advisors can have a limitless list of responsibilities, including conducting and reporting needs assessments, designing and conducting on-site research, monitoring crops, and producing profitable yields on the farm. These duties require a great deal of mental and physical stamina. The farm advisor should be a good communicator, an excellent record keeper, a problem solver, and have a good work ethic. A farm advisor can be self-employed, contracted, or employed by a company. These professionals work all over the world wherever agricultural production occurs, **Figure 31-32**.

Edler von Rabenstein/Shutterstock.com

Figure 31-32. A farm advisor helps a client with his crop. She gives advice based on sound agricultural knowledge, especially in plant pathology.

Career Connection — Tabitha West

Cedar Valley Nursery

Tabitha and Mark West are part owners of a nursery in Ada, Oklahoma, called Cedar Valley Nursery. Mark is a trained and educated horticulturist. Tabitha has a microbiology background with a focus on medical technology. She began by helping Mark at the tree farm. In May 1999, she started working at their 40-acre nursery facility that houses 55 greenhouses.

Tabitha's background in microbiology prepared her for a part of the job where she is valued most. Her strong science background and knowledge of fungi and viruses that affect humans easily transferred to working with plants. The integrated pest management (IPM) program that she helped to develop at Cedar Valley Nursery aided in her award of the "40 Under 40" most influential greenhouse growers by *Greenhouse Product News* in 2012.

Tabitha West

Cedar Valley Nursery propagates and produces mostly tree liners that are shipped to 35 states. Tabitha works with her team to monitor seedlings, cuttings, and liners for diseases every day. Her plants have what is known as a "grower's shadow." Tabitha claims, "The more time the plants see the grower's shadow, the healthier the plants will be." This undoubtedly true statement simply means that monitoring and scouting for pests is essential in the IPM program. Practicing a strong preventive disease program will help produce healthier plants.

CHAPTER 31 Review and Assessment

Chapter Summary

- Plant diseases lead to food crop loss before, during, and after harvests. Estimates of crop losses due to plant pests and diseases are in the hundreds of billions of dollars annually.
- Professionals who study and work to combat diseases play a critical role in disease management. A pathologist studies diseases, and an entomologist studies insects.
- Diseases develop based on three factors: host, pathogen, and the environment. These three elements make up what is known as the disease triangle.
- Organisms that cause diseases are known as pathogens and may include viruses, bacteria, fungi, nematodes, and parasitic plants.
- A plant that looks sick may not be diseased. A number of cultural disorders may appear to be diseases.
- Abiotic disorders are caused by some factor other than living organisms. An infectious disease that is caused by a living organism is known as a biotic disorder.
- Injury is damage caused as a result of an immediate action or event within a plant's environment, such as hail or wind damage.
- The disease cycle involves inoculation, entrance, establishment, growth, reproduction, and dissemination. If any part of the disease cycle is interrupted, the disease will not progress.
- Disease signs are actual parts of the disease organism that can be observed. Disease symptoms are a plant's reactions to a disease, such as wilting, tissue abnormalities, defoliation, tissue size changes, dwarfing, tissue replacement, and necrosis.
- Integrated pest management (IPM) includes cultural practices, using disease-resistant plants, incorporating beneficial organisms, quarantining diseased plants, and using chemicals to control plant diseases.
- Diseases that affect plants include anthracnose, blights, cankers, club root, damping off, galls, leaf blister, leaf curl, leaf spot, mildew, mold, root-knot nematode, rot, rust, scab, smut, viruses, and wilts.
- Careers in plant pathology, entomology, and horticulture can focus on the diseases of plants. People in these careers may do research, work in a lab or clinic, or treat plants in a greenhouse or nursery operation.

Words to Know

Match the key terms from the chapter to the correct definition.

A. abiotic
B. bacteria
C. biotic
D. disease
E. disease triangle
F. entomologist
G. environment
H. fungus
I. gall
J. haustorium
K. host
L. immune
M. injury
N. inoculum
O. parasitic plant
P. pathogen
Q. pathologist
R. resistant host
S. sign
T. susceptible host
U. symptom
V. toxicity
W. virus

1. Microscopic, unicellular organisms that reproduce asexually, some of which cause disease.
2. The three factors needed for a disease to develop: a host, a pathogen, and an appropriate environment.
3. A eukaryotic organism that attaches itself to a host and decomposes the organism and absorbs nutrients from it.
4. An organism that is capable of being infected by a pathogen and exhibiting symptoms of disease.
5. A pathogen (or its parts) that is capable of causing infection and disease when transferred to the appropriate host and environment.
6. A person who studies diseases.
7. A host that is prone to disease.
8. A microscopic organism that infects and lives off other organisms.
9. A type of disease caused by living organisms.
10. A person who studies insects.
11. A tumor-like growth that can develop in response to an insect laying eggs within a plant.
12. Damage caused as a result of an immediate action or event within an environment, such as wind damage to a plant.
13. A plant that anchors itself into another plant and feeds from its vascular system.
14. A host that is unable to be infected by a disease-causing organism.
15. A plant's reaction, such as wilt or blight, that is an indication of a disease.
16. A disorder in the structure or function of a living organism, such as a person, animal, or plant.
17. The surroundings or conditions in which something lives or exists.
18. A modified root of a parasitic plant that anchors itself into the vascular tissues of the host plant.
19. Resistant to or not affected by something, such as a disease.
20. An organism that causes disease.

21. An indication of a disease given by the observable presence of a pathogen or its parts.
22. A level or degree to which something is poisonous.
23. A type of disease caused by some factor other than living organisms.

Know and Understand ↗

Answer the following questions using the information provided in this chapter.

1. What was the cause of the Dutch elm disease when it was introduced to the United States?
2. Name and briefly describe the three parts of the disease triangle.
3. How does the spacing of plants affect their susceptibility to disease?
4. What are the five types of pathogens that affect plants?
5. What effects do plant pathogens have on plants?
6. How does a parasitic plant feed from a host plant, and what are three examples of parasitic plants?
7. Why might a grower need to send a plant tissue sample to a lab for testing?
8. How do abiotic and biotic diseases differ?
9. What is a plant injury? List one example.
10. What are the six parts of the disease cycle?
11. What are five ways a pathogen can enter a plant?
12. What are seven categories of disease symptoms?
13. Before a disease can be treated, what step must take place?
14. What parts of IPM must be included in disease management?
15. How can using certain genetically engineered plants help prevent plant diseases?
16. Why were genetically engineered papayas created by the University of Hawaii?
17. What are some diseases that affect plants?
18. What steps can be taken to combat early blight disease?
19. What causes damping off and how can it be prevented?
20. Which plant disease can be transmitted by cigarette smokers?
21. What are two examples of careers related to plant disease management?

Thinking Critically

1. You ordered a load of topsoil for your garden. You planted some cucumbers this year. Recently, you find that all of your plants have died of wilt. You know that this must be a disease because you watered appropriately and fertilized according to your soil test report. What would you do?
2. You recently culled your garden after experiencing some serious disease problems this past season. You have gathered all the dead plant material and are ready to burn the plants. However, you find out there is a city ordinance against burning this year due to a drought. What can you do to get rid of the diseased plant material?

STEM and Academic Activities

1. **Science.** Last winter was very long, and now the plants are finally starting to emerge from their dormancy. The hostas that are planted near your mailbox have scorched edges. There has been plenty of rain this spring, and there are no symptoms of pest damage. What do you think could have caused the young leaves to emerge burnt looking?
2. **Science.** Identify 10 common diseases within your school's greenhouse or garden. Create a dichotomous key to help others identify the diseases. Create a system that includes yes and no answers and helps people understand the symptoms they see.
3. **Math.** Calculate how much bacteria will exist after 24 hours if one bacterium cell splits every 30 minutes into two bacteria.
4. **Science.** Contact an agent with your local cooperative extension agency or a garden center employee and ask him or her to tell you about the most common plant diseases noted in your growing area. Discuss how to manage these diseases with this person.
5. **Language Arts.** Write a letter to a horticultural magazine or newspaper with a question about a disease you identify on one of your plants. Describe all the symptoms and see if the author can respond with the correct identification of the problem. Be sure to include cultural information about the plant's environment (watering, location, sun exposure, and more).

Communicating about Horticulture

1. **Reading and Writing.** Research a current disease that is impacting your region. Create a bulletin for growers to help them understand how to identify the disease as well as understand the life cycle, symptoms, prevention and treatment of the disease, and where to find more information about this problematic disease.
2. **Reading and Speaking.** Develop a presentation about disease prevention in greenhouses and gardens. Highlight how commercial and home horticulturists can change their current practices to prevent disease rather than treating disease.

SAE Opportunities

1. **Exploratory.** Job shadow a plant pathologist.
2. **Experimental.** Use a variety of different oils to treat a fungus and see which one is the most effective against the disease.
3. **Exploratory.** Research insects that are vectors for disease.
4. **Entrepreneurship.** Develop a line of products used to combat fungus using essential oils.
5. **Placement.** Get a job working at a garden center and focus on learning about plant disease control.

Vlad Teodor/Shutterstock.com

CHAPTER 32

Weeds

Chapter Outcomes

After studying this chapter, you will be able to:
- Describe what constitutes a weed and how weeds impact horticultural systems.
- List the characteristics of weeds.
- List examples of different weed biology.
- Describe the categories for identifying weeds.
- Compare and contrast different weed management methods.

Words to Know

broadleaf weed	post-emergent herbicide	self-compatible weed
bunch-type weed	pre-emergent herbicide	solarization
contact herbicide	rosette	systemic herbicide
nonselective herbicide	sedge	tillering
no-till	selective herbicide	

Before You Read

Skim the chapter by reading the first sentence of each paragraph. Use this information to create an outline for the chapter before you read it.

While studying this chapter, look for the activity icon to:

- **Practice** vocabulary terms with Words to Know activities.
- **Expand** learning with identification activities.
- **Reinforce** what you learn by completing Know and Understand questions.

www.g-wlearning.com/agriculture

Since the rise of agriculture nearly 10,000 years ago, men and women have had to wrangle with unwanted plants competing for critical resources, **Figure 32-1**. Weeds have an adverse outcome on the economy by negatively impacting crop yields, species diversity of the natural environment, or even human health. Billions of dollars each year are spent on managing weeds and preventing their introduction. Prior to applying different control methods, a grower must be able to identify the weed species and understand its biology. Only then can the best management strategy be implemented.

Sever 180/Shutterstock.com

Figure 32-1. Weeds have been present since the early days of agriculture, nearly 10,000 years ago. One of the most tenacious and well-known is the dandelion. What are common uses for dandelions?

Definition of a Weed

There are numerous definitions of a weed, including a plant that is out of place or a plant growing where it is not wanted. In commercial horticultural production and gardening, weeds compete with plants for water, nutrients, and light and can diminish plant quality and yield. Weed seeds can reduce crop quality by contaminating a harvest with their presence. Some weeds may serve as hosts for crop diseases or provide shelter for insects to overwinter. Other weeds, such as quackgrass, have allelopathic qualities, producing chemical substances that are toxic to crop plants.

Impact of Weeds

Weeds impact agriculture and, more largely, human health and productivity in the following ways:

- Increase crop production and processing costs by higher equipment wear and fuel costs.
- Reduce product/crop quality.
- Add to the amount of water and nutrients required for crop production.
- Act as alternate hosts for insects and diseases.
- Increase animal production costs and product quality and diminishes land values.
- Affect human and animal health (allergies, poisonings), **Figure 32-2**.
- Decrease wildlife habitat.
- Decrease water quality and damage watersheds and systems, which decreases recreational opportunities.

Elena Elisseeva/Shutterstock.com

Figure 32-2. Some weeds, such as ragweed, can impact human health by aggravating allergies. What other weeds are common allergens?

- Displace native, threatened, and endangered species (plants, animals, and insects), **Figure 32-3**.
- Increase costs at industrial and utility sites (to control weeds).

Costs of Weeds

Cropland and rangeland acreage in the United States constitute nearly 1 billion acres across the country, with each farm needing to manage weeds in some way. Crop production is diminished through the competitive nature of weeds for light, nutrients, and water. As much as a 100% crop failure can occur if weeds are not controlled.

natalia deryabina/Shutterstock.com

Figure 32-3. Purple loosestrife is a nonnative, invasive species that has displaced wildlife habitat and food sources. How was purple loosestrife introduced to the United States?

Weed Characteristics

Weeds have a number of characteristics that permit them to survive and allow them to flourish in many different environmental conditions. Weeds often have some or all of the following characteristics:

- Aggressive establishment through rapid growth (deep root system).
- Prolific seed production.
- Vegetative reproductive structures.
- Seed dispersal mechanisms.
- Seed dormancy.
- Staggered germination.
- Long-term survival of buried seeds.
- Adaptation for spreading.

Many weeds grow rapidly as seedlings and have the ability to reproduce very quickly. For example, pigweeds have a C4 photosynthetic pathway. This gives them the capacity to grow and develop rapidly in high temperature and high light conditions, as much as 2″–4″ (5 cm–10 cm) in as little as two to three weeks. A number of weeds, such as quackgrass, can reproduce both sexually through seed production and asexually through the vegetative structures of rhizomes.

"Even the richest soil, if left uncultivated will produce the rankest weeds."
—Leonardo da Vinci

Weed Seeds

Many weeds, such as the Canadian thistle, have mature seed as soon as two weeks after flowering. Some weed seeds have very broad germination temperature ranges and germinate much more quickly than their crop counterparts. *Self-compatible weeds* are those that do not need to cross-pollinate in order to set seed; they are self-pollinating. Even if there is only one self-compatible weed in a garden, it can still be detrimental through its ability to disperse weed seeds. Weeds that do require cross-pollinators often do not need the aid of a specific pollinator or are wind-pollinated, as in the many grasses that are considered weeds.

Multiple species of weeds have the potential to produce large numbers of seeds and have seed dispersal mechanisms that can fling seeds far and near, **Figure 32-4**.

msnobody/Shutterstock.com

Figure 32-4. Weeds grow rapidly, such as this thistle that can disperse mature seed as soon as two weeks after flowering.

Corner Question

How fast can mile-a-minute weed grow?

Some annual weeds can produce multiple seed crops in a year. Seeds can hold several kinds of dormancy, permitting them to weather unfavorable conditions and germinate at optimal times. Many weeds are viable for considerable periods of time, maintaining a long dormancy and resisting decay.

Environmental Conditions for Weeds

Weeds thrive in a wide range of environmental conditions. Some weed roots can grow deeply into the soil. For example, bindweed roots have reached a depth of 10′ (3 m), accessing water and nutrients beyond the reach of crops. Perennial weeds have roots and storage organs that contain food reserves, allowing them to tolerate environmental stress and cultivation, **Figure 32-5A**. Through other adaptive mechanisms, weeds can tolerate stresses, such as low or excessive levels of certain nutrient elements in the soil; drought; waterlogging; temperature extremes; and repeated grazing, mowing, or tillage. Many weeds have modified structures that thwart the efforts of grazing animals or insects. For example, nettles are covered with small spines along the stems to detract unwanted visitors, **Figure 32-5B**. In some cases, weeds have modified growth habits to compete for resources, such as having a climbing habit, having allelopathic chemicals, or growing as a rosette. A *rosette* is a growth habit of many biennials in their first year in which leaves attach in a circle around the base of a stem. Allelopathic plants have the ability to release a chemical into the environment that inhibits growth by other plants. Weeds can be found everywhere in agriculture. They are easily spread and adapt to new areas and habitats. In some cases, invasive species that have been unintentionally introduced to the United States have no natural predators and aggressively populate wild and cultivated lands. Common examples are purple loosestrife and kudzu, **Figure 32-5C**.

Safety Note

Toxic Weeds

Some weeds are edible whereas others are toxic. Entire plants may be poisonous, or the toxicity may be confined only to seeds, roots, berries, or leaves and stems.

A   Richard Griffin/Shutterstock.com

B   Manfred Ruckszio/Shutterstock.com

C   Scott Ehardt

Figure 32-5. A—Dandelions have large, fleshy taproots that enable them to reach and store water. B—Nettles will thwart efforts of predators with their stinging hairs along the stems and leaves. C—Kudzu was originally introduced to stabilize steep banks, but it has rapidly taken over the Southeast.

STEM Connection — Seed Viability Experiment

Dr. William James Beal, a professor at Michigan Agricultural College (now Michigan State University) in East Lansing, Michigan, initiated a seed viability experiment in 1879. He started the study "with the view of learning something more in regard to the length of time seeds of some of our most common plants would remain dormant in the soil and yet germinate when exposed to favorable conditions" (Beal, 1886). He selected 50 seeds of 23 different kinds of plants. The seeds were mixed together in moderately moist sand, placed in different jars, and buried on a sandy knoll in hidden locations on campus. One jar would be dug up every five years. Later, this time period was extended to 10 years. The experiment is still going today, and the most recent jar was uncovered in 2000. The next jar will be dug up in 2020, with five more jars remaining.

After 120 years, seedlings emerged within a week after being planted in a seed-sand mixture with seeds continuing to germinate over the next 39 days. After flowering, the plants were positively identified as *Verbascum blattaria*, *Verbascum thapsus*, and *Malva neglecta*. This experiment illustrates the sheer longevity that some seeds have to remain viable. Some seeds were still germinating nearly 1.5 months later. Weeds behave similarly, with long viabilities and staggered germination. The implications for weed control are persistence in management and continual monitoring.

Benefits of Weeds

Although weeds can wreak devastating havoc in farms and gardens, they can also provide significant benefits. Some of the benefits weeds can provide include:

- Soil stabilization. Weeds readily populate bare soil, such as after a wildfire, construction, or other land-disturbing event.
- Habitat and food for wildlife. Many honey bees and other native bees will forage among weeds, such as clover, mustards, bindweeds, and other weeds.
- Soil enrichment. Weeds decompose and add organic matter to the soil as well as fix nitrogen (in the case of vetches and clovers).
- Genetic reservoir. Many modern-day cultivars are crossed with wild-type plants, including some weeds.
- Medicinal uses. Traditional healing draughts made from weeds have managed illnesses that range from mild depression to rheumatism to bleeding.
- Food for people. Foragers can find edible weeds, including dandelions, nettles, lambs quarters, henbit, and many others.
- Aesthetic qualities. Goldenrod and Queen Anne's lace are often considered weeds, but both are also used in floral design, **Figure 32-6**.
- Employment opportunities. Annually, weed control costs billions of dollars for the United States. Weed management offers job opportunities with chemical companies and as chemical applicators and crop consultants.

"A weed is but an unloved flower."
—Ella Wheeler Wilcox

A — Madlen/Shutterstock.com   B — iofoto/Shutterstock.com

Figure 32-6. A—Goldenrod may be considered a weed by some, but others enjoy it as a cut flower. B—Queen Anne's lace may also be used in floral designs.

Weed Biology

The life cycle of a weed influences how it may be a problem and which methods may be effective for management. Weeds can be classified by their biology as an annual—either winter or summer, biennial, or perennial.

Annuals

An annual plant will complete its life cycle from seed germination, flowering and fruit set, and seed maturity in one growing season. Annual weeds tend to grow quickly and produce large numbers of seeds. They can be easier to control than perennial weeds. Summer annual weeds will germinate in the spring with warmer soil temperatures and grow throughout the spring and summer, **Figure 32-7A**. They will flower and set seeds in mid- to late summer and die in the fall. Winter annual weeds will germinate in the fall or early winter and grow during the spring. They flower and mature seeds in the late spring or early summer before dying, **Figure 32-7B**.

A www.ansci.cornell.edu/plants/medicinal/portula.html

B Severyn Bogdana/Shutterstock.com

Figure 32-7. A—Purslane is a summer annual that germinates as soil temperatures warm in the late spring and early summer. B—Chickweed is a common winter annual weed that prefers cooler temperatures.

Biennials

A biennial is a plant that completes its life cycle over two growing seasons. Seeds germinate and grow vegetatively throughout the season. Plants overwinter in a rosette stage. They complete life the following year, which is an important consideration for control. Biennials can be managed prior to flowering.

Perennials

Perennial weeds can persist and grow for many years. They are divided into two groups: simple and creeping. Simple weeds normally spread by seeds. If a shoot is injured, simple perennial weeds may grow a new plant through vegetative means. Creeping perennials generally reproduce by stolons, rhizomes, tubers, aerial bulblets, and bulbs. They also reproduce by seed, **Figure 32-8**.

"What would the world be, once bereft, Of wet and wildness? Let them be left. O let them be left; wildness and wet; Long live the weeds and the wilderness yet."
—Gerald Manley Hopkins

Corner Question

How many seeds can pigweed produce?

A Olivier Pichard

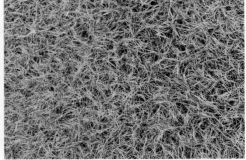
B komkrit Preechachanwate/Shutterstock.com

Figure 32-8. A—Curly dock is a simple perennial weed that can generate a new plant from a very small piece of cut root. B—Bermuda grass is a creeping weed that readily propagates by both stolons and rhizomes.

Parasitic Weeds

Parasitic weeds use host plants as sources for nutrients and water, **Figure 32-9**. A parasitic plant directly attaches to another plant through a haustorium, or modified root that forms a link between the parasite and host plant. The parasitic plant draws nutrients and water from the host plant through the haustorium. There are two main types of parasitic plants: stem parasites and root parasites. Parasitic weeds are a challenge to manage. Chemicals and extended fallow periods are the primary means of control.

luckytonyom/Shutterstock.com

Figure 32-9. Dodder is a parasitic weed that absorbs nutrients and water from its host plant.

Weed Identification

Weeds are generally categorized into three large groupings: grassy weeds, sedges, and broadleaf weeds. A *sedge* is a type of weed that is characterized by parallel venation and a triangular, solid stem. A *broadleaf weed* is a dicot plant with net venation and multiple leaf arrangements and structures. These groupings have implications for control. Effective weed management requires proper identification of weeds. If a weed is wrongly identified, the methods for control may not work and may result in wasted money and time. For example, a chemical control that works on a broadleaf weed does not work on a grassy weed. Many books, cooperative extension agency publications, and websites feature photo galleries of weeds at various stages to aid in correct identification.

STEM Connection | Mistletoe

Mistletoe is an evergreen parasite that can damage and even kill its host plant. It grows on a number of ornamental landscape tree species, including alder, ash, birch, box elder, cottonwood, locust, silver maple, walnut, oaks, and zelkova.

Female flowering mistletoe plants produce berries that are small, sticky, and white. They are an attractive food source for a number of birds. The birds feed on berries, digesting the pulp and excreting the seeds, which stick tightly to any branch on which they land. The seed begins to germinate and eventually grows through the bark and into the cambial layers. There the haustoria develop. The haustoria slowly extend up and down within the branch as the mistletoe grows. If the visible portion of the mistletoe is removed, new plants often resprout from the haustoria.

Mistletoe takes both water and mineral nutrients from its host trees. Healthy trees can withstand a few mistletoe infections, but affected branches may weaken or die. Profoundly infested trees may be reduced in vigor, stunted, or even killed, often in combination with other stresses such as drought or disease.

The most effective treatment is mechanical removal of mistletoe before it produces seed and spreads to other limbs or trees. Pruning infested branches can stop the spread in healthy trees. Severely infested trees should be removed and replaced with less susceptible species to protect surrounding trees.

TwilightArtPictures/Shutterstock.com Martin Fowler/Shutterstock.com

Grassy Weeds

Grassy weeds are monocots. They have long, narrow leaves with veins running parallel to each other and similar leaf shapes among species. (Leaves do have a netlike pattern.) Grasses are wind-pollinated and do not have showy or colorful flowers. The ability to identify grasses depends on recognizing growth habits, certain vegetative features, and seed heads. Grasses have round, oval, or flat stems with hollow internodes.

The growth habit of grasses can be divided into three different classes: *bunch-type weeds*, rhizomatous weeds, and stoloniferous weeds. Grasses with a bunch-type growth habit produce new stems through *tillering* (sending out of new stems by a mother plant), **Figure 32-10A**. A tiller is a stem that arises from a bud in the crown. Although all grasses produce tillers, those that spread by tillering are categorized as bunch-type grasses. Plants considered rhizomatous grow through horizontal creeping underground stems called *rhizomes*, **Figure 32-10B**. Stoloniferous grasses are similar to rhizomatous grasses except that lateral growth occurs by horizontal creeping of aboveground stems called *stolons*.

A *s74/Shutterstock.com* B *igor.stevanovic/Shutterstock.com*

Figure 32-10. A—Tillers are new stems produced by a grass that allows it to spread. B—Johnsongrass is a major weed that spreads through underground rhizomes.

Sedges

Sedges are annual or perennial grass-like plants with aerial flower-bearing stems, **Figure 32-11**. In annual forms, the stem is solitary and has mostly basal leaves. Perennial forms have a thick rootstock or an underground rhizome, usually with shortened internodes. Sedges usually have triangular stems with leaves arranged in groups of three. Root systems are fibrous in some plants, including species such as yellow and purple nutsedge, and produce rhizomes and tubers. Sedges can be very competitive with a desired crop for water, nutrients, and space.

Tamara Kulikova/Shutterstock.com

Figure 32-11. Sedges are perennial grass-like plants that have triangular stems.

Broadleaf Weeds

Broadleaf weeds have noticeably different physical features from grasses. They have distinct leaf shapes, netlike venation, and branching stems, **Figure 32-12**. Leaf structure, arrangement, and other surface characteristics, such as hairs or spines, can be used in identification. Leaves are alternately or oppositely arranged. Some broadleaf weeds grow in a rosette, with leaves in a circular pattern from a central growing point located at or beneath the soil surface. Others grow and spread by means of creeping above stolons or below rhizomes. Broadleaf weeds can produce a fibrous root system or a root system dominated by a large, fleshy taproot. Broadleaf weeds often bear colorful flowers of different sizes and shapes. At certain times of the year, flowers can be very useful identification aids.

mimohe/Shutterstock.com

Figure 32-12. There are many broadleaf weeds, including these lamb's quarters.

Weed Management

Managing weeds involves a combination of methods that prevent an initial introduction of weeds and additional techniques that control weed populations. These control methods for weeds used in integrated pest management (IPM) include mechanical, cultural, biological, and chemical controls.

Prevention

Weeds are constantly moving by means of a plant's own dispersal mechanisms and with the aid of animals or birds. Human practices, such as moving contaminated seed or machinery, also contribute to weed dispersal. By implementing effective preventative measures, weeds may be stopped before they become a problem. Some methods for prevention include:

- Purchasing seed free of weed seeds.
- Cleaning equipment before moving it among fields and farms.
- Preventing weed seed production by removing flowers.
- Buying clean hay for animals. (Many animal feeds are contaminated with weed seeds.)
- Preventing vegetative spread of perennial weeds.
- Scouting for new weeds.
- Treating small plots to prevent weeds from spreading.
- Identifying weeds properly, **Figure 32-13**.

Nadezhda Kulikova/Shutterstock.com

Figure 32-13. Properly identifying a weed is the first step in determining control methods.

Mechanical Control

Mechanical control of weeds requires a knowledge of weed identification and is often most effective at a particular stage of weed growth. It can be costly with labor, time, equipment, and fuel inputs, but it is an essential component of IPM. Mechanical methods for control include:

- Tillage.
- No-till.
- Mowing.
- Hand pulling.
- Weeding tools.
- Flaming.
- Solarization.
- Mulching.

"A weed is a plant that has mastered every survival skill except for learning how to grow in rows."
—Doug Larson

Tillage

Tillage involves cultivating the soil with a tractor or tiller to bury weeds, separate roots from shoots, dry out vegetation, and exhaust storage reserves. It is especially effective for perennial weeds. Tillage may also bring up dormant seeds.

No-Till

No-till involves the planting of new plants in the crop residue from the previous season's growth. This form of conservation tillage is primarily practiced to limit soil erosion. In fields with low weed populations, research has shown that this can further reduce weed populations over time.

Mowing

Mowing removes shoot growth and prevents seed production. In some cases, mowing can deplete perennial storage reserves over time, **Figure 32-14**.

Hand Pulling

Hand pulling and weeding tools are particularly efficient in gardens and landscape beds. Weeding tools, such as hoes, weed whackers, and cultivators, are especially useful in organic production, **Figure 32-15**.

Flaming

Flaming uses high temperatures from a flame weeder to disrupt cellular membranes and causes dehydration, which results in plant death.

Solarization

Solarization uses the radiant heat from the sun to kill weed populations. Plastic sheets are placed on the beds, trapping solar radiation to raise temperatures.

Mulching

Mulching excludes light, preventing shoot growth by weeds. A number of different mulches are used, depending on the situation. Examples include plastic, paper, shredded bark, decomposed leaves, and aged manure, **Figure 32-16**.

Cultural Control

Understanding how a crop is grown, the soil, weed history, and environmental factors can influence how to manage weeds. Some methods of cultural control for weeds include:

- Crop competition. Many crops can outgrow their weed competitors. Cropping patterns, such as high-density planting, intercropping, soil amendments, and no-till, give crops a boost over weeds.
- Planting date. Early crop planting or delayed crop planting may provide a competitive advantage for the crop over weeds.
- Cover crops. Cover crop plantings can be killed or used as a living mulch. A new crop can be planted into the cover crop, **Figure 32-17**.
- Crop rotation. Certain weeds associate with certain crops more than others, and rotating fields can reduce weed populations.

alexkich/Shutterstock.com
Figure 32-14. Mowing can remove the flowers before weed seeds have the opportunity to mature.

Alexander Lukatskiy/Shutterstock.com
Figure 32-15. Hand weeding or hoeing remain viable strategies in weed management, especially in organic production and home gardening.

Linda Hughes/Shutterstock.com
Figure 32-16. Mulch will prevent weeds from growing and provide a pleasing appearance in landscape beds.

Figure 32-17. Cover crops, such as a winter rye, may suppress weed growth.

Corner Question

What is the oldest weeding tool?

Biological Control

Biological control involves using living organisms to reduce the population of weeds. For example, a natural enemy of the weed may be introduced into the weed's environment. Biocontrol agents may be insects, fungi, bacteria, or even animals. For example, kudzu has successfully been controlled in certain areas by the use of grazing animals, such as goats and sheep. Much of the research on biocontrol with insects has shown success in control of invasive weed species rather than in agricultural applications.

Chemical Control

Herbicides are chemicals used to kill plants. They play a significant role in managing weeds in agricultural settings. Herbicides have a number of advantages and disadvantages just as any other weed control strategy. Herbicides require much less time and labor to broadcast or spray across a crop than removing weeds by hand. They can be selectively used to target certain weeds. For example, a broadleaf weed in turfgrass can be controlled without harming the turf. They can provide comprehensive and effective weed control.

Herbicides can also be costly to purchase and use, requiring specialized equipment to properly apply. Herbicides can cause toxicity in humans if the proper protective gear is not used or worn to protect the applicator, **Figure 32-18**. Historically, some herbicides were persistent in nature, which means they stay in the environment a long time. This is true less often with newer formulations. Weeds can develop resistance to herbicides, especially if chemical control is the primary method of weed management.

Figure 32-18. Always practice safe use and wear personal protective equipment when using herbicides.

Herbicides are classified as selective or nonselective. *Selective herbicides* will kill some weeds, but they will not impact other plants. Herbicides labeled for control of weeds in lawns, for example, will kill broadleaf weeds (dandelions and thistles) but will not kill grasses. Other herbicides are specifically formulated to kill grassy weeds and would not be appropriate to use in a lawn area. A *nonselective herbicide* is a chemical that kills or damages every type of plant. These are often used for killing weeds growing in sidewalk cracks and driveways. When a weed and an ornamental plant are growing side by side, nonselective herbicides should be applied very carefully to avoid damaging the ornamental plant.

Herbicides can also be divided into pre-emergent herbicides or post-emergent herbicides. A *pre-emergent herbicide* is a chemical applied (usually to the soil) prior to germination to kill any weeds that start to grow. It is used mostly for controlling annual weeds. A *post-emergent herbicide* is a chemical applied to actively growing weeds to kill the plants. They can be contact herbicides or systemic herbicides. *Contact herbicides* kill only the parts of the plant on which they are sprayed. *Systemic herbicides*, however, translocate from the point of contact throughout the plant to kill it. They are most effective in controlling perennial weeds where new growth can emerge from underground shoots. For optimal weed control, herbicides should be applied at the time of year when weeds are most susceptible.

Whenever chemicals are used, the herbicide label should be strictly followed. These labels show legal requirements for use and contain detailed information on how to use the product correctly. Labels share information on potential hazards associated with the herbicide and instructions you should follow in the event of a poisoning or spill. Following label instructions will allow you to minimize the risks and maximize the benefits. These instructions relate to protective gear and equipment, application rate, proper timing, storage, and disposal.

Careers in Weed Management

As mentioned previously, due to the constant need to control weeds, various job opportunities are available in weed management. These careers include chemical applicators and crop consultants.

AgEd Connection
Weeds Identification

As part of a career development event, students must identify major local weeds in horticultural systems. Weeds may be presented as intact specimens, with a photograph, or as preserved specimens. The list of weeds includes annual bluegrass, broadleaf plantain, buckhorn plantain, chickweed, crabgrass, dandelion, henbit, nutsedge, oxalis, purslane, and white clover. Use the illustrated glossary (page 928) and the e-flashcards on your textbook's student companion website at www.g-wlearning.com/agriculture to help you study and identify these weeds.

Chemical Applicators

Chemical applicators work with herbicides that help control or kill weeds. Chemical applicators are responsible for identifying where chemical treatment is needed, and then they complete this task. They mix the chemicals needed before performing the application. Chemical applicators need to have knowledge of these chemicals as well as an understanding of the safe handling of these materials. To become a chemical applicator, you must have an associate degree, followed by training resulting in certification.

Crop Consultants

Crop consultants must have a great breadth of knowledge to advise growers about efficient crop growth. They must understand the different types of growing media and recognize various plant diseases to be able to advise appropriately. Efficient crop growth includes prevention and control of weeds. Consultants must be able to collect data, write reports, and present on their findings. They also need to be able to problem solve and instruct the growers on what changes to make to their techniques. Crop consultants are typically certified and require a bachelor degree as well as continuing education.

Career Connection | Dr. Carol Somody

Senior Stewardship Manager, Syngenta

Carol Somody grew up in a crowded city and never saw a farm until she was a teenager. When she was young, her grandfather planted the seed for her love of growing things. Carol eventually went to the land grant university in her state, Rutgers University, to pursue a career in agriculture. As a young woman studying agronomy in the 1970s, Carol was continually discouraged by some school advisors and urged to pursue something "more appropriate" for her gender and upbringing. A key turning point occurred in her college career when she met a professor who mentored and encouraged her. This professor, who was a weed scientist, encouraged Carol to reach her goal by becoming a weed scientist herself. Building on a foundation of a bachelor's in agronomy, Carol then focused her energies solely on weed science and ultimately obtained a PhD.

Fieldwork, spending long hot hours behind a hoe, gave Carol critical experience and insight into weed biology. Her work also gave her the understanding that weed management requires the consideration of every tool in the IPM toolbox. Since 1982, Carol has worked for Syngenta and its legacy companies that invest in developing pesticides, seeds, and other technologies that support growers in maximizing crop yields in an environmentally sound and economically efficient manner.

Carol focuses on education of all audiences on the safe use of pesticides within the framework of IPM. She is an advocate for the science behind pesticides and their use, tirelessly promoting the need for proper management of pesticides from purchase to disposal, and protection of applicators, farm workers, and the environment.

Dr. Carol Somody

CHAPTER 32
Review and Assessment

Chapter Summary

- A weed is a plant that is out of place or a plant growing where it is not wanted. In horticultural production and in gardening, weeds compete with desired plants and crops for water, nutrients, and light, resulting in diminished plant quality and yield.
- Weeds have a number of characteristics that permit them to grow in many different environmental conditions. Some of these characteristics include rapid growth, prolific seed production, reproduction vegetatively, long seed dormancy, staggered germination, and adaptations for spreading.
- Weeds provide a number of benefits, such as soil stabilization, habitat and food for wildlife, aesthetic qualities, soil enrichment, genetic diversity, medicinal uses, and food for people. Many job opportunities are associated with weed control.
- Weeds vary in their biology. They can be characterized by having an annual, biennial, or perennial life cycle. Some weeds are parasitic.
- Weeds are categorized as grassy weeds, sedges, or broadleaf weeds. They can be controlled based on their grouping.
- Weed management starts with prevention, which includes using seeds and hay that are free of weed seeds, scouting, and treating small populations of weeds. Cleaning equipment before moving it among fields and farms also helps in prevention.
- Mechanical weed control might include tillage, no-till, mowing, hand pulling, flaming, solarization, and mulching.
- Cultural control of weeds involves using crop competition, planting dates that are early or delayed, cover crops, and crop rotation.
- Biological control involves using living organisms to reduce the population of weeds. Biocontrol agents may be insects, fungi, bacteria, or even animals.
- Chemical control involves using herbicides to prevent or kill weeds. Herbicides can be selective or nonselective. Whenever chemicals are used, the herbicide label should be strictly followed to avoid hazards to people, animals, and the environment.

Words to Know

Match the key terms from the chapter to the correct definition.

A. broadleaf weed
B. bunch-type weed
C. contact herbicide
D. nonselective herbicide
E. no-till
F. post-emergent herbicide
G. pre-emergent herbicide
H. rosette
I. sedge
J. selective herbicide
K. self-compatible weed
L. solarization
M. systemic herbicide
N. tillering

1. A chemical that is applied (usually to the soil) prior to weed germination to kill any weeds that start to grow.
2. The process of new stems or tillers being created by a mother plant.
3. A type of weed that is characterized by parallel venation and a triangular, solid stem.
4. A type of grassy weed that produces new stems through tillering.
5. A chemical that is applied to actively growing weeds to kill the plants.
6. A weed that does not need to cross-pollinate in order to set seed; a self-pollinating weed.
7. A chemical that will kill all kinds of plant.
8. A dicot weed with net venation and multiple leaf arrangements and structures.
9. A growth habit of many biennials in their first year in which leaves attach in a circle around the base of a stem.
10. A weed control technique that uses the radiant heat from the sun to kill weed populations.
11. A chemical that is sprayed on the weed foliage and kills only the plant tissue it touches.
12. The practice of planting new plants in the crop residue from the previous season's growth.
13. A chemical that can kill some types of plants but not others.
14. A chemical that can translocate or move within the plant from the point of entry to kill the plant.

Know and Understand

Answer the following questions using the information provided in this chapter.

1. What is a weed, and how do weeds affect desirable plants and crops?
2. What nine impacts do weeds have on agriculture?
3. What eight key characteristics do weeds have that permit them to survive in many environments?
4. How can just a few self-compatible weeds be detrimental to a garden?
5. What are some stresses that weeds can tolerate due to their adaptive mechanisms?
6. How does allelopathy benefit weeds?
7. What are four benefits that weeds can bring?

8. What are three ways weeds are classified by their biology?
9. Describe the life cycle of annual weeds.
10. Describe the life cycle for biennial weeds.
11. How does a parasitic weed feed on other plants?
12. Weeds are generally categorized into what three large groupings?
13. List the three categories of grassy weeds and describe the growth habit of each type.
14. What are the characteristics of a broadleaf weed?
15. What types of control methods for weeds are used in integrated pest management?
16. What are eight methods of weed prevention?
17. Describe the tillage mechanical weed control method.
18. Describe four methods of cultural control for weeds.
19. Describe biological control methods for weeds.
20. Describe safe procedures for using herbicides.
21. What are some of the weeds you might be asked to identify in the FFA CDE for Nursery and Landscape?

Thinking Critically

1. You find a population of weeds growing in your garden and take appropriate measures to eliminate the weeds. A few weeks later, the weeds return, and you have to start over. What do you think happened? What could you do differently?
2. Why is it important for growers to understand the biology of the weeds they are trying to control?

STEM and Academic Activities

1. **Science.** Investigate the science behind allelopathy. Write a three-page report explaining the reasons plants exude certain chemicals and how different species may use different chemicals.
2. **Science.** Investigate current research programs in biological control of weeds. Choose a research program that interests you. Prepare a report on the scientific methods used in this program and how the results affect the horticultural industry.
3. **Technology.** Visit a local commercial nursery and ask about the technology used to provide weed control. If there are no commercial nurseries in your area, research this topic on the Internet. Write a short report of your findings.
4. **Engineering.** Mechanical control of weeds may require the use of a weeding tool. Research examples of different weed tools, such as a hoe, rake, or other implement. Design a new weeding tool that would require less physical strength and be as equally as effective.
5. **Language Arts.** Imagine that you are the marketing director for a large company that sells herbicides. One of your job responsibilities is to write articles for the grower's information center. Write an article about how to choose the best herbicide. Focus on the characteristics of at least three different types of herbicides.

Communicating about Horticulture

1. **Listening and Speaking.** In small groups, discuss with your classmates—in basic, everyday language—your knowledge and awareness of the weeds in your everyday surroundings. Take notes on the observations expressed. Review the points discussed, factoring in your new knowledge of weeds. Develop a summary of what you have learned about weed biology and their presence in our everyday surroundings. Present your findings to the class, using the terms that you have learned in this chapter.

2. **Listening and Speaking.** Visit a local company that works in weed science. Ask to interview their experts on weeds. Prepare a list of questions before your interview. Here are some questions you might ask: What is your work environment like? What are your job duties? What type of research are you currently doing? What type of facilities do you use for your research? What impact will your research have on the weed industry? Ask if you can have a tour of their facilities. Report your findings to the class, giving reasons why you would or would not want to pursue a career similar to that of the person you interviewed.

SAE Opportunities

1. **Exploratory.** Job shadow a scientist who studies weeds. What are the daily responsibilities of his or her job? What do you like or not like about this position? What education and experiences are required to have this position?

Jari Hindstroem/Shutterstock.com

2. **Exploratory.** Inventory the number of different weed species on your school campus and their populations. What could you do to decrease the numbers of weeds?

3. **Exploratory.** Research what herbicides could be used in your school landscape. Are they easily available? How effective are they? What are the benefits and costs of using these methods to control weeds?

4. **Exploratory.** Visit a local produce farm. Ask the farmer to share his or her weed management practices. Go into the fields and see what weeds you can identify. Think through possibilities for control. Write a short report and share it with your teacher.

5. **Exploratory.** Create a weed collection for your teacher. Collect, press, and create herbarium specimens for as many weed species as you can find. How will this be a useful tool in learning about horticulture?

CHAPTER 33
Pesticide Management and Safety

Chapter Outcomes
After studying this chapter, you will be able to:
- List types and formulations of pesticides.
- Explain how to read a pesticide label.
- Describe methods of safe pesticide application.
- Identify pesticide toxicities, poisoning, and first-aid treatment for poisoned persons.
- Explain how to store and to dispose of a pesticide safely.
- List careers related to pesticide management and safety.

Words to Know

active ingredient	contact pesticide	molluscicide
acute toxicity	EPA registration number	nematicide
agricultural pest	fungicide	pesticide formulation
algaecide	insecticide	restricted entry interval (REI)
biochemical pesticide	LC_{50}	rodenticide
biopesticide	LD_{50}	signal word
chronic toxicity	miticide	systemic pesticide

Before You Read
Before you read the chapter, read all of the table and photo captions. What do you know about the material covered in this chapter just from reading the captions?

While studying this chapter, look for the activity icon to:

- **Practice** vocabulary terms with Words to Know activities.
- **Expand** learning with identification activities.
- **Reinforce** what you learn by completing Know and Understand questions.

www.g-wlearning.com/agriculture

In general, a pest is something that is a nuisance. However, an *agricultural pest* is an insect, disease, weed, or animal that attacks a crop or food source and causes damage. Pests should be controlled using a strategic plan called integrated pest management (IPM). IPM is an approach to managing pests that uses commonsense, economical practices and results in the least possible hazard to people, property, and the environment.

Sometimes, when all other methods of IPM have been exhausted, pesticides (chemicals) are used to control pests that damage or attack plants, animals, and other organisms. Pesticides destroy pests that attack plants, animals, and other organisms, **Figure 33-1**. Before using a pesticide as part of the IPM program, the applicator must identify the pest being targeted and the best pesticide to use. The applicator should also know how to safely apply, store, and dispose of the pesticide in a manner that will ensure the safety of people, animals, plants, and the environment.

Ron Rowan Photography/Shutterstock.com

Figure 33-1. Various insects, such as adult Japanese beetles, can cause damage to plants, including holes in the leaves. A large infestation will leave plants entirely defoliated.

Types of Pesticides

Pesticides may be synthetic or organic. Synthetic pesticides are created with manufactured chemicals. Organic pesticides are derived from natural ingredients and do not contain manufactured chemicals. Both types of pesticides are toxic to the targeted pest(s), and may have detrimental effects on the environment and human beings when improperly applied or overused.

Pesticides (synthetic and organic) are divided into the following categories: insecticides, miticides, herbicides, fungicides, nematicides, molluscicides, biopesticides, rodenticides, and algaecides. The pesticides in each of these categories are used to control specific problems.

Insecticides

An *insecticide* is a chemical used to prevent, control, or decimate insect populations. Insecticides are categorized by the method in which they are taken in by pests.

- *Contact pesticides* are insecticides that kill insects through touch, or by entering the insect's system through ingestion. Contact pesticides are attached to the plant surface that is consumed by the pest. Contact pesticides mainly target insects with chewing mouthparts. Poisons that are ingested are known as stomach poisons.
- *Systemic pesticides* are translocated through the plant's vascular system. Insects with piercing and sucking mouthparts will take in the insecticide when they feed on the sap. Systemic poisons target insects with piercing sucking mouthparts.

Corner Question

Are ticks a pest to plants?

STEM Connection
Chemicals from Flowers

A powerful chemical used to control many different insects can be pyrethrin, or a pyrethroid. Pyrethrins are a chemical derived from chrysanthemum flowers. Chemists, however, have synthetically made a chemical called a pyrethroid. The synthetic form of pyrethrin combats insects in the same manner as the naturally occurring pyrethrin. An organic pesticide can contain pyrethrins but not pyrethroids (since they are a synthetic chemical).

val lawless/Shutterstock.com

Corner Question
What famous herbicide was used as a method of combat during the Vietnam War?

Miticides

Tiny, spider-like organisms are known as mites, **Figure 33-2**. Mites are also closely related to ticks. *Miticides* kill insects on contact or through the mite's ingestion of the poison. For example, a flea (insect) and tick control that is applied to a dog or cat enters the animal's bloodstream. When the flea or tick feeds on the animal, it takes up the poison and is killed. This method of delivering the poison is systemic.

Herbicides

An herbicide is a weed killer. A weed is a plant that grows in a place where it is unwanted, **Figure 33-3**. Nonselective herbicides kill all plants. Selective herbicides target specific types or spectrums of weeds. Pre-emergent herbicides are applied to a site to create a chemical barrier at the soil level before seeds germinate. New weed seedlings are killed by this chemical as the seeds germinate. A post-emergent herbicide controls weeds after they are growing.

Fungicides

Fungicides, the most widely used and applied type of pesticide, control or prevent fungal growth. Fungicides come in contact with the fungus that feeds on the plant material.

D. Kucharski K. Kucharska/Shutterstock.com
Figure 33-2. Mites, such as these spider mites, have a piercing sucking mouthpart and can damage plants.

Bildagentur Zoonar GmbH/Shutterstock.com
Figure 33-3. Weeds, such as this chickweed, are controlled by several chemicals used in herbicides.

Fungal pathogens cause various diseases, **Figure 33-4**. Fungicides are usually sprayed preventively rather than to control a fungus that is actively growing (showing signs and symptoms of disease).

Nematicides

Microscopic, multicellular worm-like organisms that inhabit soil and water are known as nematodes. Nematodes often feed on the roots of plants. A *nematicide* can be applied to soils (often in the form of a gas, known as a fumigant) to control nematode populations.

Molluscicides

Molluscicides control types of mollusks. Slugs and snails are two forms of mollusks that cause extensive plant damage, **Figure 33-5**. Molluscicides are available in granular form and are applied as bait. The slugs and snails eat the bait, which is poison.

Art Phaneuf Photography/Shutterstock.com

Figure 33-4. Various fungi, such as mildews found on this squash plant, can be prevented and sometimes treated with fungicides.

Biopesticides

Biopesticides are pesticides derived from natural products, such as plants, animals, and microorganisms. According to the Environmental Protection Agency (EPA), there are more than 195 registered biopesticides and more than 780 biopesticide products on the market today. Three types of biopesticides are microbial pesticides, plant incorporated protectants (PIPs), and biochemical pesticides.

Microbial Pesticides

Microbial pesticides include microorganisms such as bacteria and fungi. An example is a fungus that controls specific insects or weeds.

TwilightArtPictures/Shutterstock.com

Figure 33-5. A slug is a common pest that chews on the foliage of numerous horticultural crops. Slugs can be controlled with traps and molluscicides. Some gardeners use copper strips to shock these pests.

Thinking Green

Organic Controls for Slugs

Several methods can be used to control slugs that may damage garden plants.

- Diatomaceous earth—This powdery substance, made from the crushed remains (skeletons) of diatoms, contains a large amount of sharp silicon. Growers place the diatomaceous earth around the plants to create a barrier. As the slugs cross the barrier, the silicon causes cuts that lead to their dehydration and eventual death.
- Copper barrier—A copper wire or barrier can be placed into the soil and used as fencing. As the slug passes over the copper with its body, it reacts with the copper and is shocked.
- Clover—Plant a patch of sacrificial clover. The slugs will be lured from your prized plants and will feast on the clover instead.
- Beer trap—Place a small container in the ground with the lip at ground level. Fill the container with beer. Slugs are attracted to the sweet, fermented malt and will drown when they fall into the trap.
- Hand picking—Slugs can be picked off plants by hand and disposed of in a closed container. This type of removal is best done at night, when slugs are most active.

Plant Incorporated Protectants (PIPs)

Plant incorporated protectants (PIPs) come from plants that produce pesticidal substances from within the plant due to transgenic modifications. For example, through the use of biotechnology, Bt corn produces a nerve toxin called *Bacillus thuringiensis* within the plant. The toxin attacks larvae (such as the European corn borer) as they try to feed on the genetically modified corn.

Biochemical Pesticides

A nontoxic, naturally occurring mechanism used to control pests is known as a *biochemical pesticide*. An example is using pest pheromones to attract and capture pests in pheromone traps, **Figure 33-6**. Pheromone traps have proven to be effective in controlling pests such as the Japanese beetle.

Rodenticides

Not only are mice and rats a problem in homes and businesses, they are also a problem in greenhouses, gardens, storage facilities, and farms. Additional rodent pests common in agricultural settings include gophers, woodchucks (also known as groundhogs), moles, and voles, **Figure 33-7**. Applications of *rodenticides* (chemicals used to control rodents) are often in the form of poisonous bait pellets, packs, or blocks.

Algaecides

Algaecides control algae that can grow anywhere there is water. An evaporative cooling pad or a concrete floor in a greenhouse is the perfect environment for algae growth. Algaecide forms and application methods vary.

Pesticide Formulations

Pesticides must be formulated to be effective at controlling pests. A *pesticide formulation* is a stable mixture of active and inert ingredients used to create a product that controls pests. The formulation makes the final product easier and safer to use, and more effective in combatting a target pest.

Dale Spurgeon/USDA ARS

Figure 33-6. Pheromone traps, such as this boll weevil trap, can be used to help lure and capture insects to limit crop damage.

Peter Trimming/Flickr

Figure 33-7. A vole is a type of rodent that lives in underground burrows. Voles create runways and burrows that destroy lawns and eat the roots of plants. These organisms can be controlled with rodenticides.

Corner Question

Can algae be farmed?

Ingredients

A pesticide formulation may consist of:
- *Active ingredients*—the chemicals that control the target pest population.
- Carriers—something to help deliver or carry the active ingredients.
- Surfactants—surface-active ingredients that help ingredients adhere or spread to the targeted area.
- Adjuvants—other ingredients such as dyes, stabilizers, or other substances to enhance the effectiveness of the pesticide.

Pesticide manufacturing companies provide pesticides in different formulations to make products safer and easier to apply, and more effective at controlling pests. Using certain formulations of pesticides can also help prevent contamination of the environment. Formulations for pesticides include aerosol sprays, dust, wettable powders, granular pellets, liquid concentrates, and emulsifiable concentrates.

Aerosol Sprays

Aerosol sprays are applied through a spray can device, **Figure 33-8**. This formulation is convenient and easy to apply because all the applicator must do is press a tab to spray the pesticide. There is no mixing because the product is already formulated for release. This method of delivery is very efficient. Aerosols are expensive and are best used only for small areas. Aerosol formulations are identified on the SDS and pesticide label by the letter A.

Dust

Dust formulations are very fine particles that are applied by shaking the dust from a canister or a duster. A duster is an application device that forces the dust through a tube for dispersal. Clay, or another fine powder, may be used to bind to the active ingredient of the pesticide to create the dust. However, some pesticides may be purely made of the active chemical. Spreading the dust and getting even coverage can be a challenge with this type of product. Dust formulations are identified on the SDS and pesticide label by the letter D.

Robert Rozbora/Shutterstock.com

Figure 33-8. Pesticides are available in spray cans that are useful for smaller applications. Gloves should be worn when handling and applying aerosol pesticides.

Safety Note

Aerosol Cans Are Explosive

Never use an aerosol can near an open flame, and never try to puncture a can of aerosol spray. These canisters are highly pressurized. They can explode and cause serious damage if punctured or heated by fire or another heat source. *Always read and follow the label when using aerosol can pesticides.*

Wettable Powders

Wettable powders are dust-like formulations that are mixed with water or oil and sprayed through a sprayer. Wettable powders are economical and solve the application problems that are characteristic of dust formulations. Wettable powders provide even coverage and delivery of the pesticide. Wettable powder formulations are identified on the SDS and pesticide label by the letters WP.

Granular Pellets

Dry, coarse pellets that are applied using a spreading device (broadcast or drop-type spreader) are known as granular formulations. Baits and turf products are often granular formulations. Granular formulations are identified on the SDS and pesticide label by the letters GR.

Liquid Concentrates

Liquid concentrates are diluted with water and applied through a spraying device, **Figure 33-9**. Liquid concentrates are economical and generally easy to apply. However, mixing the product and using a sprayer require manual labor. Home gardeners may use a small, portable sprayer. Growers use large sprayers placed in a truck bed or behind a tractor to treat large areas. Liquid concentrations are identified on the SDS and pesticide label by the letters LC.

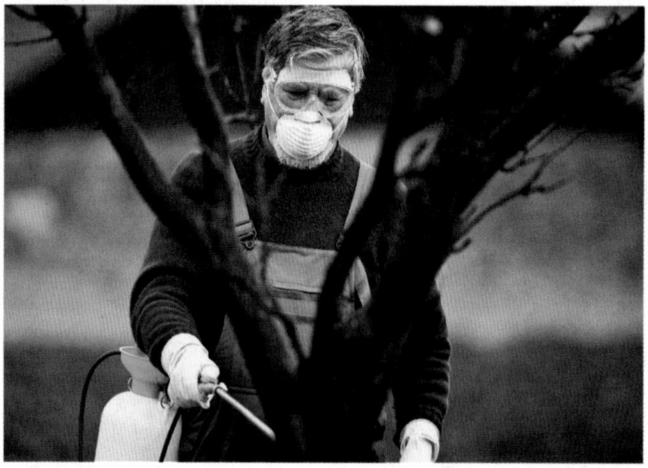

lightpoet/Shutterstock.com

Figure 33-9. Applicators must always wear the appropriate PPE, regardless of the size of the area or location being treated.

Emulsifiable Concentrates

Emulsifiable concentrates are pesticide solutions with emulsifying agents in a water-insoluble organic solvent. The pesticide solution is suspended in the emulsifying agent, much in the same way oil and vinegar do not mix in a salad dressing. When added to water, this formulation has a milky appearance. Emulsifiable concentrates are identified on the SDS and pesticide label by the letters EC.

Pesticide Labels

The pesticide label is a lengthy document created by scientists, the government, and lawyers. The pesticide label's objective is to ensure maximum benefits to users while reducing safety and environmental risks. When you are considering the application of a pesticide, it is very important that you read the label to understand how, when, and where to apply the pesticide. Before you buy a pesticide, read the label to make sure this product is appropriate for the particular pest you wish to control. Read the label for each step of use to ensure the safest and most effective use of the pesticide. Read the label before you purchase, mix, apply, store, and dispose of the pesticide.

Corner Question

What is FIFRA?

Not following the label is dangerous for several reasons, including environmental risks, safety risks, and legal implications. Failure to follow the label instructions may also result in less effective pest control with the product.

A Legal Document

The pesticide label is a legal and binding agreement between the applicator and the pesticide manufacturer. Pesticide manufacturers are under strict laws governed by the Environmental Protection Agency (EPA). The chemical undergoes years of research and testing before it is released to the public. The label contains explicit instructions and information based on this research and testing. Failure to comply with the directions of the label can have legal repercussions. If someone knowingly does not follow the directions of the pesticide label, he or she can be criminally prosecuted.

Sections of a Pesticide Label

The information found on a pesticide label is very detailed and meant to cover many issues associated with its application. Some of the most important information on the pesticide label includes:

- EPA registration number.
- Active ingredients list.
- Signal words.
- Precautionary statements.
- Environmental hazards section.
- First-aid instructions.
- Storage and disposal information.

EPA Registration Number

The *EPA registration number* is a number assigned to a pesticide after it has been reviewed and verified by the EPA. The number provides certification that all information and data found on the label has been reviewed by the EPA. It also indicates the product has been reviewed, and has been determined to have minimal or low risk when the label's directions are followed. The EPA registration number does not mean that the EPA supports the product or guarantees it to be effective. The label simply indicates that the EPA has reviewed the product.

STEM Connection: Insecticides and Bumblebees

What pesticide misuse resulted in the death of more than 25,000 bumblebees? In 2013, in a parking lot in Wilsonville, Oregon, more than 25,000 bumblebees and other pollinators were found dead or dying. A local landscape company had not followed the label directions of an insecticide called Safari. They sprayed the insecticide on linden trees while they were blooming. Because the trees were blooming, pollinators were working and collecting pollen that was laced with the poison. The misuse of the pesticide resulted in a loss of at least 150 colonies of bees in the local ecosystem.

Sergey Lavrentev/Shutterstock.com

History Connection: Skull and Crossbones

The symbol of the skull and crossbones originates with the medieval *Danse Macabre* (Dance of Death) symbol. The symbol took on its current form by the fifteenth century and accompanied war ships, military flags, and other insignia expressing recklessness and ferocity. In the eighteenth century, this symbol came to signify piracy. The symbol was used to mark the entrances of Spanish cemeteries. Since the nineteenth century, the skull and crossbones has been used as a symbol of warning on containers, such as those filled with poison.

Jess Kraft/Shutterstock.com

Active Ingredients

The active ingredients are those that provide control against the target pest. An active ingredient can be a synthetic or natural chemical.

Signal Words

Signal words are language on a pesticide label used to call attention to potential threats to human health. The words *caution*, *warning*, and *danger* are the signal words indicated by the EPA. The signal words have specific meanings:

- Caution—these pesticides are the least harmful to human health.
- Warning—this is a more dangerous pesticide and has more potential to negatively impact human health.
- Danger—this is the most dangerous type of pesticide. A pesticide with the word *danger* is only available for use by licensed or certified pesticide applicators. A label with the word *danger* will also have an illustration of a skull and crossbones.

Precautionary Statements

Precautionary statements describe the personal protective equipment (PPE) that should be worn by the applicator, protection for children and pets, and requirements for a treated area. PPE can include goggles, masks, respirators, gloves, shoes, socks, long sleeves, pants, and other protective gear for the applicator.

Environmental Hazards

The environmental hazards section outlines potential environmental damage that may result from using the product. The label will discuss the possible outcomes to wildlife, aquatic life, plants, animals, and water resources.

Safety Note

Agricultural Worker Protection Standard

The EPA's Agricultural Worker Protection Standard (WPS) was published in 1992 and is a regulation intended to protect agricultural workers from injury and poisoning associated with pesticides. The WPS offers protection to more than two million workers and pesticide handlers that work at more than 600,000 agricultural work sites. The WPS requires employers to provide workers with proper education, safety, and notification, and to provide mitigation when exposure does occur.

The directions for use help the applicator understand the purpose of the product and what pests it is designed to control, **Figure 33-10**. The product label will describe where it can be applied (nursery, greenhouse, outdoors, or indoors) and what pests it controls. The pesticide cannot be used against a nontarget pest or in a location that is not described in the directions for use section of the label.

First-Aid Instructions

In case someone is poisoned, the first-aid instructions define how to handle the situation. The *Statement of Practical Treatment* outlines first-aid protocol specific to that pesticide or poison. In case of poisoning, call 911 or a poison control center with the name of the pesticide. Administer the first-aid treatment that is described. Take the pesticide container or label to the hospital.

Storage and Disposal

Safe storage and disposal of a pesticide is equally as important as safe application. Products must always remain in their original container and away from children and pets. Pesticides should not be stored at extreme temperatures. All pesticides should be kept in a locked cabinet or storage facility.

mrfiza/Shutterstock.com

Figure 33-10. A health worker is using a fog in a tropical region to control insects that spread dengue virus. This same application may not be labeled for use inside a building.

Pesticide Application

Several factors should be considered in applying pesticides safely. These factors include:
- Gaining applicator certification.
- Selecting personal protective equipment.
- Determining the correct amount to use.
- Mixing properly.
- Applying correctly.
- Enforcing restricted entry intervals.

Pesticide Applicator Certification

Only trained and certified applicators may apply restricted pesticides. In accordance with national standards determined by the EPA and USDA, states, territories, and tribes are permitted to provide pesticide applicator certification and training programs. Certified applicators must undergo training, pass an exam, pay annual certification fees, and periodically renew their certification through education and/or testing.

Funding for Safety Programs

The EPA provides funding to review the competency of restricted-use pesticide applicators through the pesticide safety education program (PSEP). Since 1975, the EPA has had an interagency agreement (IAG) with the USDA to distribute funds to the state cooperative extension services for the purpose of training restricted use pesticide applicators. The joint efforts of the EPA, USDA, and cooperative extension services have helped educate individuals who work with these powerful pesticides. The applicators learn about appropriate use, storage, disposal, and safety for people and the environment.

Selecting Personal Protective Equipment

Figure 33-11. Read the pesticide label to understand which type of personal protective equipment must be worn.

Whether an applicator uses personal protective equipment (PPE) is not a choice. The label describes whatever necessary PPE must be worn during pesticide application. Personal protective equipment that should be worn during pesticide application includes long-sleeved shirts, goggles, long pants, shoes, socks, and nonpermeable gloves, **Figure 33-11**. When working with and applying pesticides, it is best to cover as much bare skin as possible. The less skin that is exposed, the less likely an applicator is to be poisoned through skin contact. It is also necessary to cover your head with a hat or hood as well.

Determining the Correct Amount to Use

Many pesticide products can be purchased in a form that is ready to use; however, others must be mixed in a quantity specific to the job at hand. Mixing too much, or too little, product can cause problems. Mixing too much pesticide may mean additional storage, waste, or disposal concerns. Mixing too little pesticide means more time is required to mix another batch (losing time and money).

Safety Note

Personal Protective Equipment

Personal protective equipment (PPE) varies, depending on the job at hand. The Occupational Safety Health Association (OSHA) regulates the proper use of PPE on job sites. PPE for a landscape worker differs greatly from the PPE for someone who is applying a pesticide. Not all pieces of PPE have the same effectiveness for every job. Know what you need to wear to protect yourself, and wear the PPE even if it is uncomfortable.

Determining exactly how much pesticide product is needed to treat the targeted pest for the specified area is critical. The pesticide label will include ratios you can use to determine how much pesticide you will need. Good measurements and careful calculations result in precise and responsible pesticide applications. Follow these steps to determine the amount of pesticide needed for an application:

- Determine the size of the area to be treated: length × width = area.
- Calculate how much pesticide is needed for the target area: X ounces of pesticide per 1000 ft^2 (304.8 m^2).
- Calculate dilutions of the pesticide product if the formula is not ready-to-use: X ounces of pesticide per gallon of water.
- Example:
  Area is 100′ (30.48 m) × 20′ (6.1 m) = 2000 ft^2 (609.6 m^2)

  Pesticide needed is 1 ounce of pesticide per 1000 ft^2 (304.8 m^2). Thus, 2 ounces of pesticide is needed. The pesticide must be diluted to 0.5 ounces per 1 gallon of water. Thus, if 2 ounces are needed, then 4 gallons of water would be needed for the 2000 ft^2 (609.6 m^2).

Mixing a Pesticide

If pesticides must be mixed to create the appropriate concentration, the applicator must use the appropriate measurements and follow safety protocols when mixing. Some safety protocols include:

- Never eat or drink when mixing pesticides.
- Wear appropriate PPE, including goggles, gloves, long sleeves, long pants, socks, and shoes.
- Mix in a well-ventilated area, preferably outdoors in adequate light.
- Mix only the amount that was calculated at the concentration recommended. Doubling the strength of a pesticide will not make it more effective, and may make it more dangerous. Do not make more than you need.
- Never use measuring equipment (teaspoons, cups, or jars) that will be used for anything other than pesticide measurement and mixing.
- Keep children, pets, and any other sensitive materials away from the area where mixing occurs.
- If mixing a concentrate, add water first and add the pesticide second. This will prevent splashing of the pesticide or possible exposure by adding water to the pesticide.
- Keep pesticides in their original containers. Use clearly marked containers to hold mixed pesticides. The mixed pesticide should be used immediately.
- If a spill occurs, clean it up immediately. Sprinkle the spill with vermiculite, sawdust, or cat litter (refer to the cleanup section of the label). Sweep the pesticide-soaked material into a garbage bag, and dispose of it according to the pesticide storage and disposal section of the label.

Applying Pesticides

It is important to assess the surrounding environment before application of a pesticide begins. Thoroughly read the label to understand how, when, and where the pesticide should be applied. Keep the following general guidelines in mind when preparing to apply pesticides:

- Check the surrounding area for water, people, pets, livestock, and other elements or organisms that are in the targeted site and may be affected by the pesticide.
- Check the weather forecast to see if rain or wind may be an issue. Pesticides should never be applied on windy or rainy days. The pesticide may have restrictions regarding how soon a pesticide may be applied before rain is expected.
- Check the label for the appropriate temperature at which a pesticide may be applied. Extreme high or low temperatures should be avoided.
- Use coarse droplets from spray equipment to prevent pesticide from drifting off target.
- Apply pesticides in the garden around dusk. This is when pollinators, such as honeybees, will not be pollinating.
- Never apply pesticides near a well or other water source.
- Use pesticides indoors only when absolutely necessary (interiorscapes). Ventilate the area and remove all food sources from the site before application.
- Triple-rinse all spraying equipment once application is completed.
- Store and dispose of all pesticide material according to the pesticide label.
- Properly remove, wash, or dispose of PPE. Follow proper washing techniques for PPE that can be reused.

Safety Note
Windy Days
Never apply pesticides on windy days. The small droplets can easily be swept away with the wind and treat or contaminate areas that were not meant to be exposed. This can lead to unintentional toxicities of nontarget populations. Unintentional contamination through pesticide application is the fault of the applicator and is punishable by law.

Applying insecticides during the appropriate part of an insect's life cycle is important for the chemical to work effectively. Correctly identifying the pest and determining what part of the life cycle the pest is in (adult, nymph, pupa, larva, or egg) are critical for selecting the best chemical control, **Figure 33-12**. Knowing what type of life cycle (complete or incomplete metamorphosis) the insect has is also important.

Restricted Entry Interval (REI)

The *restricted entry interval (REI)* denotes how much time must pass before a person can enter an area that has been treated with a pesticide. The pesticide label indicates the REI in hours. Depending on the pesticide's potential for toxicity, some pesticides labeled *caution* may have a REI of zero hours while others may have an REI of up to 48 hours.

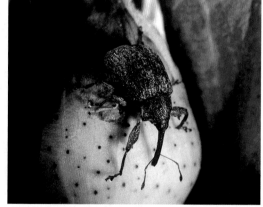

USDA Agricultural Research Service

Figure 33-12. Some pesticides are effective only during certain life cycle stages of the targeted pest. Read the pesticide label to ensure you are using the correct pesticide at the correct stage of the targeted pest.

TFoxFoto/Shutterstock.com

Figure 33-13. Signs to state that an area has been treated with a pesticide must be in a language that is understood by most people in the area. This sign, written in English and Spanish, is appropriate for many areas of the United States.

REIs must be posted on treated areas in a language that individuals in the area can understand. Usually, in the United States, REIs are posted in English and Spanish, **Figure 33-13**. The signs that are used for REI are very noticeable.

Toxicity

Through the EPA, the United States regulates pesticides that are considered toxic to human health and the environment. The toxicity of a pesticide is its ability to poison organisms. Poisons enter an organism through:

- Dermal contact (through skin).
- Inhalation (through respiration).
- Oral contact (through ingestion).
- Eye contact (through eye membranes).

Applicators are most likely to be affected through dermal contact or inhalation. Skin rapidly absorbs substances through cuts, the back of hands and necks, armpits, and the groin area. Inhalation of fine mists, such as aerosols or dusts, can also lead to exposure. Children, pets, and wildlife may ingest pesticides in granular forms. Thus, granular forms of pesticides, such as bait pellets, must be properly stored and applied only when pets or children will not be using the area.

Types of Toxicity

Toxicity of a poison may be described as acute or chronic. *Acute toxicity* is a measure of how poisonous a pesticide is after a single exposure. *Chronic toxicity* is a measure of how poisonous a pesticide is after repeated exposures, over a length of time. A very small amount of a toxin that is continually stored in the fat of an organism will build to toxic levels that can prove harmful or deadly.

Lethal Dose

A measurement used to determine the amount of acute oral and dermal toxicity is LD_{50}. LD stands for lethal dose (the amount of a substance needed to cause death). The *50* signifies that 50% of a test population of animals died when exposed to this quantity. The lower the LD_{50} of a toxin, the higher the toxicity.

STEM Connection | **Lethal Dose**

The acute LD_{50} of many pesticides is much lower than common household items that humans use in their everyday life. Numbers are expressed as milligrams (mg) of poison per kilograms (kg) of body weight.

- LD_{50} of table salt is 3000 mg/kg for rats, glyphosate (active ingredient in herbicides) is 5600 mg/kg for rats.
- LD_{50} of aspirin is 200 mg/kg for rats, malathion is 1375 mg/kg for rats.
- LD_{50} of nicotine is 10 mg/kg for rats, Sevin dust is 650 mg/kg for rats.

LD_{50} values are specified in milligrams of a substance per kilogram of a test animal's body weight.

Lethal Concentration

The measurement for acute inhalation toxicity is measured by *LC_{50}* values. *LC* stands for lethal concentration. The values are measured in milligrams per liter. (Liter is a volume measurement.) The lower the LC_{50} number, the more toxic the pesticide is by volume. Pets and children should be kept from areas where pesticides have been used, **Figure 33-14**.

Stone 36/Shutterstock.com

Figure 33-14. Children and pets are extremely susceptible to pesticides. Extreme caution and attention should be paid when applying pesticides near them.

Toxicity Categories

The EPA has established guidelines and educational materials for handling pesticides. In addition to the toxicity categories listed below, the EPA has published guidelines for REI (restricted entry intervals). Pesticides with greater toxicity have a greater REI.

- Toxicity I chemicals—the signal word *danger*, skull and crossbones, and *danger-poison* are included on the pesticide label. When the chemical is applied, warning signs must be posted and no one may enter the treated area for 48 hours after application.
- Toxicity II chemicals—the signal word *warning* is included on the pesticide label. Warning signs must be posted in the treated area, and no one may enter the area within 24 hours after application.
- Toxicity III chemicals—the signal word *caution* is included on the pesticide label. People may enter the area as soon as the mist or dust settles.
- Toxicity IV chemicals—the signal word *caution*, or no warning, may appear on the label. People may enter the treated area immediately or work within the targeted area as the treatment is applied.

Pesticide Poisoning

Recognizing pesticide poisoning is critical in preventing serious injury or death. If poisoning occurs, immediately contact a poison control center and dial 911 for an emergency. Find the pesticide label and have that ready for professionals. Symptoms of pesticide poisoning include:

- Redness, swelling, blistering, or pimpling of skin.
- Redness, swelling, or blistering of eyes, nose, mouth, and throat.
- Shortness of breath.
- Rapidness of breath.
- Drooling.
- Nausea, vomiting, abdominal cramps, and diarrhea.
- Headache, muscle twitching, and numbness.

If someone develops symptoms of poisoning after exposure to these chemicals, seek medical attention immediately to determine if the symptoms are pesticide related. Blood or urine analysis may be needed to determine pesticide toxicity.

Corner Question

What was the largest recorded case of pesticide poisoning in the United States?

First Aid

First aid should precede, but never replace, professional medical assessments and treatment. Once first aid has been administered, call both 911 and the poison center at 1-800-222-1222. Have the pesticide label available when calling the poison center.

Spills

Keep the following guidelines in mind when taking care of someone who has spilled a pesticide on his or her skin or clothing.

- Implement first-aid practices based on the Statement of Practical Treatment when pesticide poisoning occurs.
- Remove the exposed clothing immediately.
- Wash the exposed area immediately with freshwater and soap to dilute the chemical. Dilution of the poison is imperative.
- Cover any chemical burns with a loose, clean cloth. If the situation permits, and emergency personnel are not present, take the victim to an emergency treatment center.

Eye Exposure

For eye exposure, hold the eye open and flush with clean water (or saline) for a minimum of 15 minutes. Flushing the eyes will rinse and dilute the poison. Do not use drops or ointments to flush eyes. Seek professional medical help as soon as possible after flushing the eyes. The eye membrane absorbs poisons faster than any other external part of the body. Eye damage can occur in a few minutes with many types of chemicals.

Inhalation

If a person has inhaled a poison, immediately move that person to a fresh air environment, and call 911. Do *not* expose yourself! If the victim is unable to stand or unconscious, and you do not feel safe entering the area to retrieve the victim, immediately call 911. If it is safe to enter the area, retrieve the victim and help him or her to fresh air. (If you cannot move the person, open windows and doors to ventilate an enclosed space.) Keep the victim stationary, and loosen any clothing that would restrict breathing. If the victim is unresponsive, and you are trained to do so, administer artificial respiration (CPR) while waiting for emergency personnel to arrive, **Figure 33-15**.

Remain calm when helping someone, and remember to keep your health a priority. If you are exposed to the chemical in your first-aid efforts, it is important that you also receive medical attention.

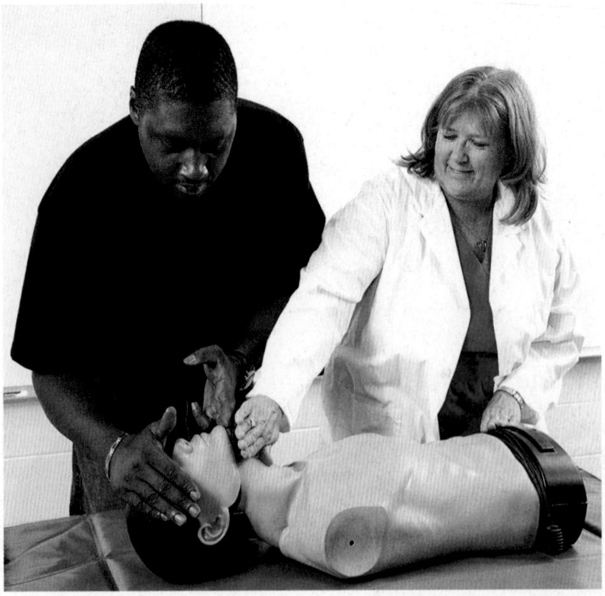

Lisa F. Young/Shutterstock.com

Figure 33-15. Learning and knowing CPR and other first-aid procedures can help in treating a person who has been poisoned.

Safety Note

Handling a Pesticide Emergency

You may need to help someone who has been poisoned by a pesticide. If the person is unconscious, having trouble breathing, or having convulsions, act quickly:
- Give first aid immediately.
- Call 911, or ask someone else can to do so while you begin first-aid treatment.

If the person is awake or conscious, not having labored breathing, and not having convulsions:
- Contact your local poison center (1-800-222-1222).
- Read and follow Statement of Practical Treatment on the pesticide label.
- Give first aid.

Storage and Disposal

The storage and disposal of a pesticide is as potentially harmful to human and environmental health as improper mixing or application. When considering how to safely store a pesticide, keep these guidelines in mind:
- Purchase only the amount of pesticide that will be used in the near future. This will reduce or eliminate the need to store leftover pesticides.
- Follow all storage instructions on the pesticide label.
- Store pesticides in a temperature-regulated facility.
- Store pesticides in a locked cabinet or storage facility.
- Keep pesticides contents in the original container.
- Keep pesticides out of reach of children and pets.
- Store pesticides away from an ignition source.
- Do not store pesticides in a location where flooding is possible.

Disposal

Proper disposal of pesticides is important for safety. If you purchase only the amount of pesticide you need for an application, you will not have to store or dispose of leftover pesticides. If you cannot use your supply, contact other growers or gardeners who may have the same pest problem, and may have use for the leftover pesticides. Always keep the pesticide label with the pesticide in its original container. If you must dispose of a pesticide, consider the following:
- Do *not* burn leftover pesticides, pour them down a drain, or throw them into the garbage. Pesticides are toxic and, when disposed of improperly, may cause damage to the environment, people, and other living organisms.
- Contact your local cooperative extension service agent and ask for suggestions for disposal of the pesticide.
- Contact your local solid waste agency, health department, or the EPA to learn about hazardous waste collection programs in your community.
- Contact Earth 911 (contact information available online). This agency, and others like it, can help direct you to the appropriate disposal of your leftover pesticide.
- Adhere to state and local laws when applying, storing, and disposing of pesticides. State and local pesticide disposal laws may be harsher than federal requirements found on the pesticide label.

Safety Note

Rinsing Pesticide Containers Safely

Follow these guidelines to rinse pesticide containers safely:

1. While wearing personal protective equipment, pour any excess pesticide into the sprayer.
2. Fill the pesticide container one-fourth full of clean water, recap, and shake the container for 30 seconds. Pour the rinse water into sprayer.
3. Repeat two additional times, shaking the container each time.
4. Carefully rinse the outside of the container and the cap over the sprayer (or a bucket) to catch the rinse water.
5. Dispose of the pesticide container according to local regulations.
6. Apply the diluted rinse material according to label directions onto targeted pests.

Container Reuse

Never reuse empty pesticide containers. An empty pesticide container has as much potential to be hazardous as a full container of pesticide. Residues that are left inside a pesticide container have the potential to be combustible. When empty, rinse the container at least three times and replace the cap securely. Dispose of the container according to the pesticide label instructions.

Careers

When you think about a career in pesticides, you most likely picture someone spraying a house for pests, such as cockroaches and termites. Yes, this is one career associated with pesticides; however, many other careers that involve science, math, technology, engineering, marketing, advertising, communications, and law are related to pesticides. Consider a career as a pesticide chemist, a lawyer for an agricultural chemical company, or the manager of a bee care facility.

Pesticide Chemist

A pesticide chemist is someone who researches the use and development of safer chemicals to combat insects, diseases, weeds, and other pests. A pesticide chemist may be employed by a college or university, government agency, or a private company. A pesticide chemist may be a laboratory technician, research assistant, or scientist, depending on the amount of education and training he or she has received. Chemists can create formulations, analyze chemicals, and be involved in quality control. There are many opportunities for chemists in the pesticide industry, **Figure 33-16**.

PhotoSky/Shutterstock.com

Figure 33-16. Chemists have many opportunities in the pesticide industry.

Lawyer for an Agricultural Chemical Company

Lawyers for agricultural chemical companies must have knowledge of the agriculture industry as well as knowledge of local, national, and even international laws. These lawyers usually have a working understanding of government agencies such as the EPA, USDA, US Fisheries and Wildlife Service, and the Department of Justice. Lawyers deal with regulatory issues, compliance to regulations, environment and chemical exposure, and claims. Lawyers may represent companies in litigation dealing with regulation and practices involving chemicals. Some attorneys are responsible for the creation of legal language on pesticide labels. They review the labels to ensure all legal aspects are covered fully, and that the company creating the label and chemicals has created a legal and binding document in accordance with the EPA and other relevant government agencies.

In addition to a bachelor's degree and a license to practice law, lawyers working in the agricultural industry may also have an education or background in agricultural studies.

Career Connection: Dr. Rebecca Langer-Curry

Bayer Bee Care

Dr. Rebecca Langer-Curry works for Bayer CropScience and is head of the North American Bayer Bee Care Program. Together, Dr. Langer-Curry and her colleagues use the Bayer Bee Care Program, and the Bayer Bee Care Center, to promote and protect pollinator health in North America and around the world.

Bayer has more than 25 years invested in research related to promoting bee health. To further their efforts, Bayer opened the Bayer Bee Care Center in North Carolina. Dr. Langer-Curry is the project manager at the center, which focuses on promoting cooperative efforts between apiculturists and agriculturists to establish sustainable solutions for bee care problems. The center is open to the public and provides a platform for education, research, and demonstration for scientists and the public.

Dr. Langer-Curry has a doctorate in pathobiology. Her career in science includes management of biosafety programs for labs and greenhouses. She has also worked in academia. Dr. Langer-Curry encourages students interested in apiculture to study a science and not to "fall into habit, but rather ask new questions and seek answers to those questions." Dr. Langer-Curry also advises students to explore business and marketing because knowledge in these areas will be helpful in the business world.

Dr. Rebecca Langer-Curry

CHAPTER 33

Review and Assessment

Chapter Summary

- An agricultural pest is an insect, disease, weed, or animal that attacks a crop or food source and causes damage. Pests should be controlled using a strategic plan called integrated pest management (IPM).
- When all other methods of IPM have been exhausted, pesticides (chemicals) may be used to control pests. Pesticides destroy pests that attack plants, animals, and other organisms.
- Many types of pesticides are available, including insecticides, miticides, herbicides, fungicides, nematicidies, molluscicides, biopesticides, rodenticides, and algaecides.
- A pesticide formulation is a stable mixture of active and inert ingredients used to create a product that controls pests. Formulations for pesticides include aerosol sprays, dust, wettable powders, granular pellets, liquid concentrates, and emulsifiable concentrates.
- Reading the pesticide label is very important before selecting, purchasing, mixing, applying, storing, or disposing of a pesticide. The pesticide label is a legal and binding agreement between the applicator and the pesticide manufacturer.
- The information found on a pesticide label is very detailed and can be dozens of pages long. This information is meant to cover many issues associated with the pesticide.
- Signal words on a pesticide label are used to call attention to potential threats to human health. The words *caution*, *warning*, and *danger* are the signal words indicated by the EPA.
- Factors to consider in applying pesticides safely include gaining applicator certification, selecting personal protective equipment, determining the correct amount to use, mixing properly, applying correctly, and enforcing restricted entry intervals.
- The toxicity of a pesticide is its ability to poison organisms. Poisons enter an organism through dermal contact, inhalation, or oral contact (ingestion).
- Recognizing pesticide poisoning is critical in preventing serious injury or death. Individuals must be able to recognize a pesticide poisoning and provide appropriate first aid to the victim.
- Storage and disposal of pesticides are just as important as proper application. Read the label and follow the directions for storage and disposal.
- Several careers that involve science, math, technology, engineering, marketing, advertising, communications, and law are related to pesticides. Three of these careers include, a pesticide chemist, a lawyer for an agricultural chemical company, and a project manager for a bee care facility.

Words to Know

Match the key terms from the chapter to the correct definition.

A. active ingredient
B. acute toxicity
C. agricultural pest
D. algaecide
E. biochemical pesticide
F. biopesticide
G. chronic toxicity
H. contact pesticide
I. EPA registration number
J. fungicide
K. insecticide
L. LC_{50}
M. LD_{50}
N. miticide
O. molluscicide
P. nematicide
Q. pesticide formulation
R. restricted entry interval (REI)
S. rodenticide
T. signal word
U. systemic pesticide

1. A chemical used to prevent, control, or decimate insect populations.
2. An insect, disease, weed, or animal that attacks a crop or food source and causes damage.
3. An insecticide that kills insects through touch, or by entering the insect's system through ingestion.
4. A chemical that is translocated through a plant's vascular system; targets insects with piercing and sucking mouthparts.
5. A product used to control or prevent mites.
6. A chemical used to control or prevent fungal growth.
7. A chemical product used to control nematodes.
8. A chemical product used to control mollusks (snails and slugs).
9. A pesticide that is derived from natural products, such as other plants, animals, and microorganisms.
10. A non-toxic, naturally occurring mechanism used to control pests.
11. A chemical substance used to control rodents.
12. A chemical used to control algae.
13. A mixture of active and inert ingredients (adjuvants, surfactants, and carriers) used to create a product that controls pests.
14. A chemical in a pesticide that works to control the targeted pest.
15. A number given to a pesticide once it has been reviewed and verified by the Environmental Protection Agency (EPA).
16. Language, such as *caution*, *warning*, and *danger* on a pesticide label used to call attention to potential threats to human health.
17. The time that must elapse before someone can enter an area after it has been treated with a pesticide.
18. A measure of how poisonous a pesticide is after a single exposure.
19. A measure of acute oral and dermal toxicity needed to kill 50% of a test population of animals.
20. A measure of how poisonous a pesticide is after repeated exposures, over a length of time.
21. A measure of acute inhalation toxicity needed to kill 50% of a test population of animals.

Know and Understand

Answer the following questions using the information provided in this chapter.

1. What is an agricultural pest?
2. What is integrated pest management?
3. What are pesticides?
4. What are some types of pesticides used to control selected or target pest populations?
5. What is an herbicide and when are post-emergent herbicides used?
6. What are four ways to control slugs besides using a molluscicide?
7. What are three types of biopesticides?
8. Describe the type of biopesticide called a plant incorporated protectant.
9. What are four types of substances that can be included in a pesticide formulation?
10. What are six formulations in which pesticides are available?
11. What type of legal repercussion may result from failure to comply with the directions of a pesticide label?
12. What are some types of important information found on a pesticide label?
13. What do the signal words *caution*, *warning*, and *danger* mean when printed on a pesticide label?
14. What must a person do to qualify as a certified applicator?
15. Why should a pesticide applicator cover as much of his or her bare skin as possible?
16. What are some safety protocols that should be followed when mixing pesticides?
17. When is the best time of day to apply pesticides in a garden and why?
18. What are four ways poison can enter an organism?
19. What should you do in the event of a pesticide poisoning?
20. What guidelines should you keep in mind when storing pesticides?

Thinking Critically

1. You recently noticed your neighbor spraying what appeared to be a pesticide in his lawn right next to a stream that leads to a river and, eventually, the ocean. What would you do?
2. On a visit to a relative's home, you notice that pesticides in the garage are not stored properly, and they are within the reach of children. The situation appears to be dangerous. What would you do to address the situation?

STEM and Academic Activities

1. **Science.** You have a small hobby greenhouse that has been infested with thrips. What would be the most effective method of covering all plant material with an insecticide? Justify your answer.

2. **Science.** Identify a biopesticide and research how it is engineered. Describe the process through an illustrated diagram.
3. **Math.** Calculate how much insecticide will be needed to treat a lawn that is 0.25 acres if the pesticide used should be applied at a rate of 0.25 fluid ounces per gallon and 1 gallon of pesticide should be applied per 100 ft^2 (30.48 m^2).
4. **Language Arts.** Write a position paper on whether you think neonicotinoids are contributing to the loss of bee colonies. Include facts and statements from scientific research that is well cited. Use MLA format for the report.
5. **Language Arts.** Contact a Cooperative Extension Service agent to determine the requirements in your state for a pesticide license. Create a poster or pamphlet that outlines the process to inform people in your school or community about pesticide certifications in your state.

Communicating about Horticulture

1. **Reading and Speaking.** Research two pesticides you have at your home and determine how to store and dispose of each pesticide properly. Create a five-minute presentation about the appropriate methods of storage and disposal for each.
2. **Writing and Speaking.** Visit your local extension office and interview them about any recent pesticide accidents within your community, region or state. Create an informative poster about your experience to tell the story of pesticide problems where you live.

SAE Opportunities

1. **Exploratory.** Job shadow a pesticide sales representative.
2. **Experimental.** Use an organic and a synthetic product to control weeds in your lawn. Compare the results of both treatments.
3. **Exploratory.** Research pesticide safety and worker protection standards within the local area, the state, and the nation. Create a video to demonstrate your understanding of these standards that could be shown to people working with pesticides in various settings.
4. **Improvement.** Work with your local Cooperative Extension Service office and develop a pesticide recycling program for your community. Make your school or another point in your community a drop-off point for these chemical containers.
5. **Exploratory.** Volunteer with the EPA, USDA, or a Cooperative Extension Service office to learn more about pesticide safety.

science photo/Shutterstock.com

Plant Identification

The following plant identification glossary contains more than 100 plants (ranging from smaller flowering specimens, to shrubs, and trees) commonly grown and used in landscaping applications. This illustrated glossary has been provided to help you familiarize yourself with these plants, and as a means of studying for career development events in which plant identification is a major component. To help you identify the plants, each entry includes the botanical/scientific name and at least one common name. This glossary is by no means all-inclusive, as there are innumerable varieties and cultivars available to growers everywhere. However, it contains a good variety for you to begin your studies. Use this glossary, as well as the e-flashcards on your textbook's companion website at www.g-wlearning.com/agriculture to help you study and identify these common plants.

alybaba/Shutterstock.com
Abelia x grandiflora (Glossy Abelia)

Bildagentur Zoonar GmbH/Shutterstock.com
Abies concolor (White Fir)

ukmooney/Shutterstock.com; Frank11/Shutterstock.com
Acer palmatum cv. (Japanese Maple)

Jonathon Billinger; Anatoly Vlasov/Shutterstock.com
Acer platanoides cv. (Norway Maple)

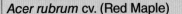
Ftlombardo; Richard A. McQuirk/Shutterstock.com
Acer rubrum cv. (Red Maple)

Melinda Fawver/Shutterstock.com
Acer saccharum cv. (Sugar Maple)

900 Horticulture Today

Plant Identification

F. D. Richards; Arina P. Habich/Shutterstock.com

Crataegus phaenopyrum (Washington Hawthorn)

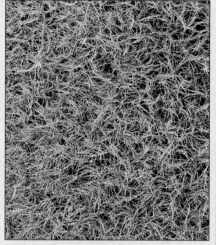
komkrit Preechanwate/Shutterstock.com

Cynodon dactylon cv. (Bermudagrass)

Forest & Kim Starr

Dieffenbachia maculata cv. (Spotted Dumb Cane)

LucaLuca

Dracaena deremensis 'Warneckii' (Striped Dracaena)

moritorus/Shutterstock.com

Dracaena fragens 'Massangeana' (Corn Plant)

jaroslava V/Shutterstock.com

Echinace a purpurea (Purple Coneflower)

Bondarenko/Shutterstock.com

Epipremnum spp. (Pothos)

Matt Lavin; Kees Zwanenburg Shutterstock.com

Euonymus alatus (Winged Euonymus)

Agnes Kantaruk/Shutterstock.com

Euonymus fortunei cv. (Wintercreeper)

Fagus sylvatica cv. (European Beech)

Ficus benjamina (Benjamin Fig)

Ficus elastica 'Decora' (Decora Rubber Plant)

Forsythia x intermedia cv. (Border Forsythia)

Fraxinus americana cv. (White Ash)

Gaillardia aristata cv. (Common Blanketflower)

Gardenia jasminoides 'Fortuniana' (Common Gardenia)

Ginkgo biloba (Ginkgo, Maidenhair Tree)

Gleditsia triacanthos inermis cv. (Thornless Honeylocust)

 *Torti Tude*	 *Daimond Shutter/Shutterstock.com*	 *val lawless/ Shutterstock.com*
Juniperus horizontalis cv. (Creeping Juniper)	*Lagerstroemia indica* cv. (Crape Myrtle)	*Leucanthemum x superbum* cv. (Shasta Daisy)
 *Lilyana Vynogradova/Shutterstock.com*	 *Bruce Marlin*	 *BONNIE WATTON/Shutterstock.com*
Liquidambar styraciflua (Sweet Gum)	*Liriodendron tulipifera* (Tuliptree)	*Liriope* spp. cv. (Lily-Turf)
 *Thirteen/Shutterstock.com*	 *Calin Tatu/Shutterstock.com*	 *Stephen Farhall/Shutterstock.com*
Lobularia maritima (Sweet Alyssum)	*Lonicera japonica* 'Halliana' (Hall's Japanese Honeysuckle)	*Magnolia grandiflora* cv. (Southern Magnolia)

Copyright Goodheart-Willcox Co., Inc.

Vladimira/Shutterstock.com	Forest & Kim Starr	©iStock.com/Achim Prill
Parthenocissus tricuspidata (Boston Ivy)	*Pelargonium x hortorum* cv. (Zonal Geranium)	*Pennisetum setaceum* (Fountain Grass)
Vahan Abrahamyan/Shutterstock.com	KENPEI	Jukka Palm/Shutterstock.com
Petunia x hybrida cv. (Petunia)	*Philodendron scandens oxycardium* (Heartleaf Philodendron)	*Picea abies* (Norway Spruce)
USDA National Resources Conservation Service	Stan Shebs	islavicek/Shutterstock.com
Picea pungens cv. (Colorado (Blue) Spruce)	*Pieris japonica* (Lily-of-the-Valley Bush)	*Pinus mugo* (Mugo Pine)

Equipment and Supplies Identification

The following equipment and supplies identification glossary contains more than 90 images of tools and materials used in the horticulture industry. This illustrated glossary has been provided to help you familiarize yourself with these implements and materials, as well as a means of studying for career development events in which their identification is a major component. This glossary is by no means all-inclusive, but provides a good starting point for your studies. Use this glossary, as well as the e-flashcards on your textbook's companion website at www.g-wlearning.com/agriculture to help you study and identify these tools and materials.

Anvil-and-blade pruner	Ball cart (B&B truck)	Bark medium
A.M. Leonard Inc.	*A.M. Leonard Inc.*	*Jodi Riedel/Goodheart-Willcox Publisher*

Bark mulch	Bow saw and bow saw blade	Brick paver
bofotolux/Shutterstock.com	*A.M. Leonard Inc.*	*safakcakir/Shutterstock.com*

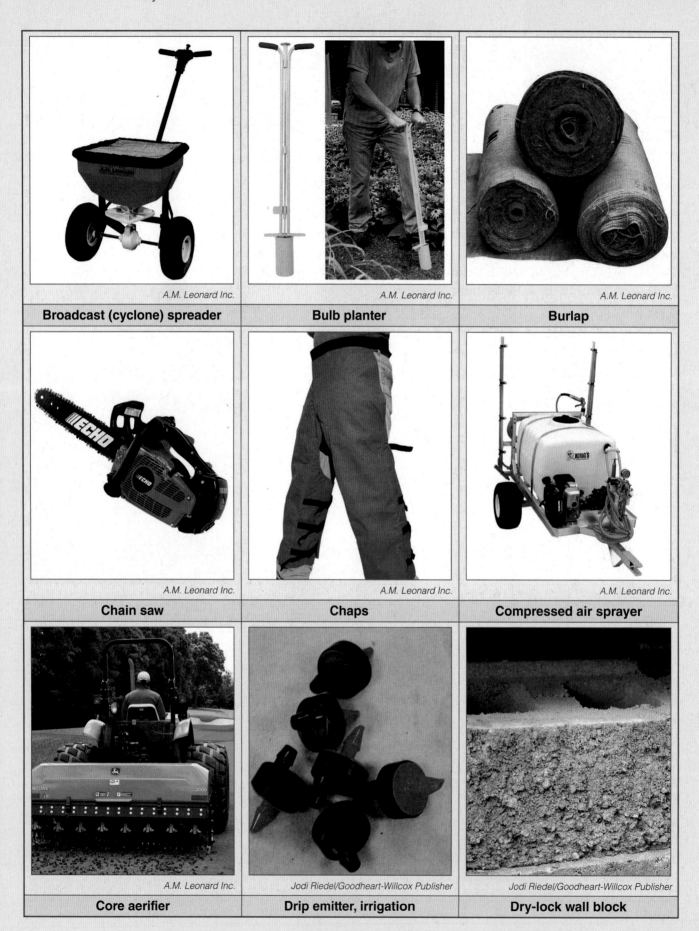

Equipment and Supplies Identification 915

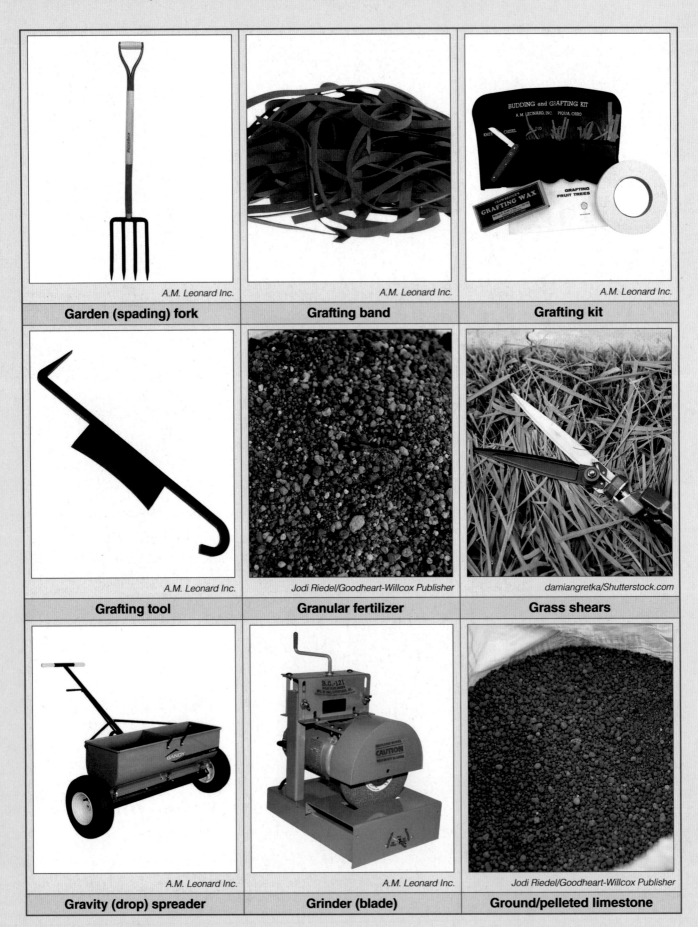

Equipment and Supplies Identification 917

A.M. Leonard Inc.
Half-mask respirator

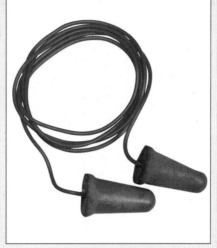
A.M. Leonard Inc.
Hearing protection

A.M. Leonard Inc.
Hearing protection

A.M. Leonard Inc.
Hedge shears

A.M. Leonard Inc.
Hoe

A.M. Leonard Inc.
Hose-end repair fitting (female)

A.M. Leonard Inc.
Hose-end repair fitting (male)

Jodi Riedel/Goodheart-Willcox Publisher
Hose-end sprayer

A.M. Leonard Inc.
Hose-end washer

A.M. Leonard Inc.
Hose repair coupling

A.M. Leonard Inc.
Impulse sprinkler

A.M. Leonard Inc.
Landscape fabric

A.M. Leonard Inc.
Leaf rake

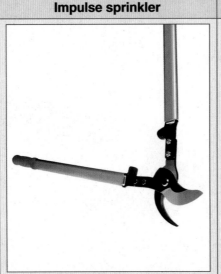
A.M. Leonard Inc.
Loppers

A.M. Leonard Inc.
Mattock

A.M. Leonard Inc.
Measuring wheel

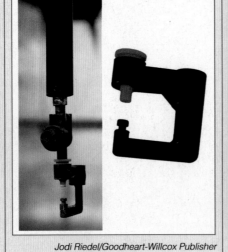
Jodi Riedel/Goodheart-Willcox Publisher
Mist nozzle

Jodi Riedel/Goodheart-Willcox Publisher
Nursery container

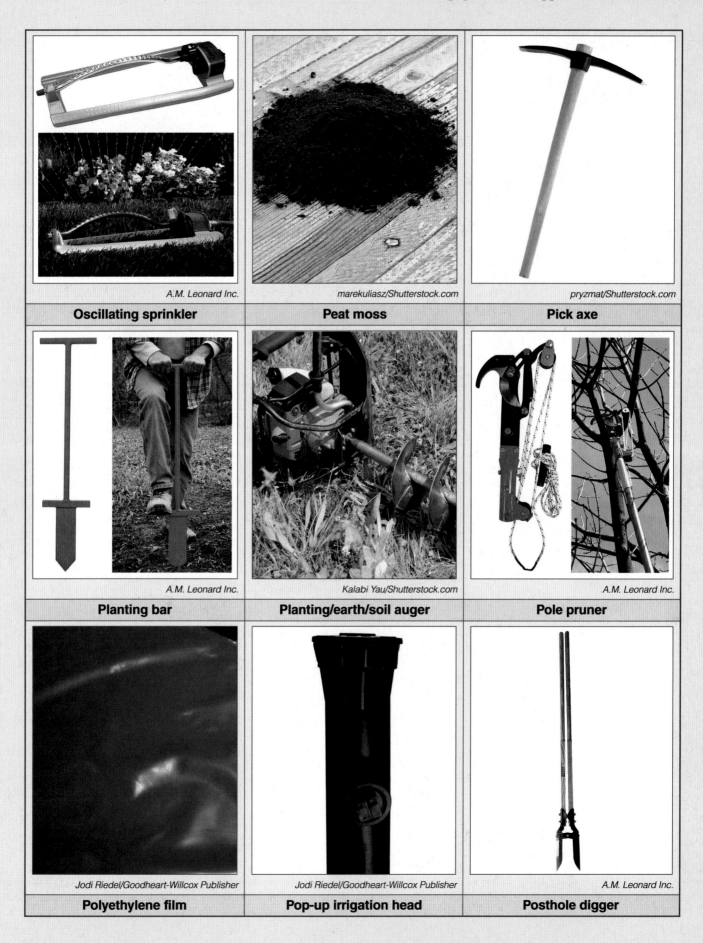

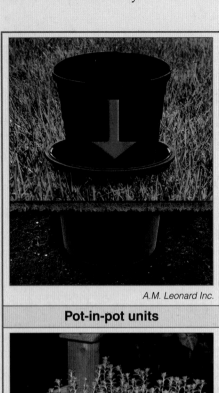 *A.M. Leonard Inc.* **Pot-in-pot units**	 *A.M. Leonard Inc.* **Power blower**	 *Le Do/Shutterstock.com* **Power hedge trimmer**
 *A.M. Leonard Inc.* **Propagation mat**	 *A.M. Leonard Inc.* **Pruning saw**	 *Toa55/Shutterstock.com* **PVC (polyvinylchloride) pipe**
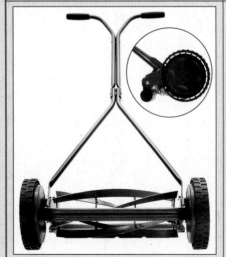 *Kitch Bain/Shutterstock.com* **Reel mower**	 *Jodi Riedel/Goodheart-Willcox Publisher* **Resin-coated fertilizer**	 *A.M. Leonard Inc.* **Rototiller**

922 Horticulture Today

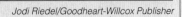 *Jodi Riedel/Goodheart-Willcox Publisher* **Solenoid valve**	 *A.M. Leonard Inc.* **Spade**	 *Scott Gauthier* **Spark plug gap gauge**
 *Videowokart/Shutterstock.com* **Sphagnum moss**	 *A.M. Leonard Inc.* **Spray suit**	 *A.M. Leonard Inc.* **Square point (flat) shovel**
 *Duplass/Shutterstock.com* **String trimmer**	 *Hurst Photo/Shutterstock.com* **Time clock**	 *Sandra van der Steen/Shutterstock.com* **Topsoil**

Copyright Goodheart-Willcox Co., Inc.

Equipment and Supplies Identification 923

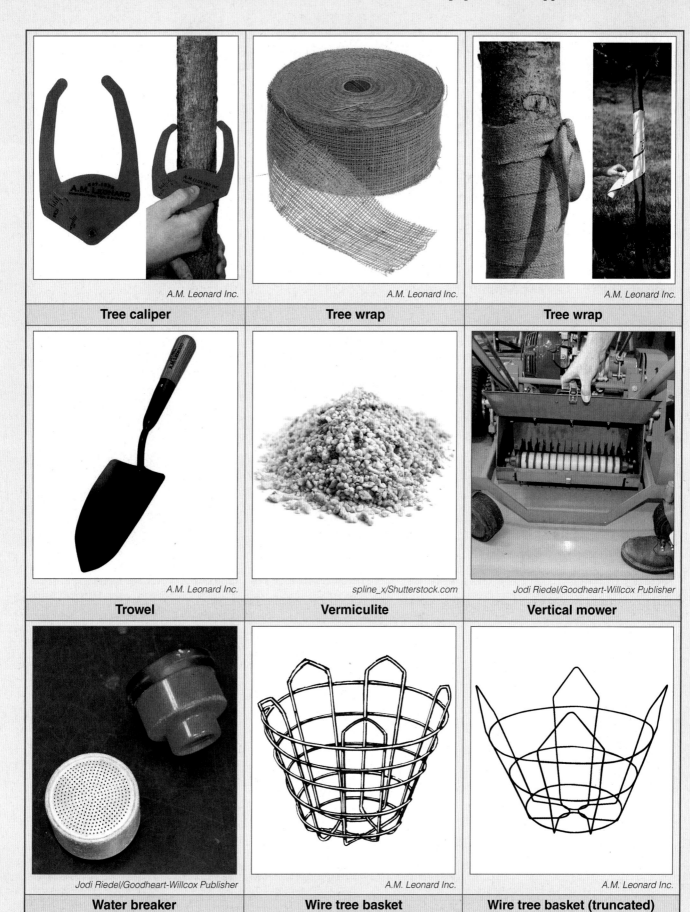

Pests and Disorders Identification

The following pests and disorders identification glossary contains a small sampling of common agricultural pests and the damage they cause to plants and trees. This illustrated glossary has been provided to help you familiarize yourself with these pests and damage, as well as a means of studying for career development events in which their identification is a major component. This glossary is by no means all-inclusive, but provides a good starting point for your studies. Use this glossary, as well as the e-flashcards on your textbook's companion website at www.g-wlearning.com/agriculture to help you study and identify these pests and disorders.

Aphid	Bagworm	Borer
corlaffra/Shutterstock.com	*Steve Heap/Shutterstock.com*	*Henrik Larsson/Shutterstock.com*

Chlorosis	Leafhopper	Leaf Miner
Alena Brozova/Shutterstock.com	*YapAhock/Shutterstock.com*	*topimages/Shutterstock.com*

Disease Identification

The following disease identification glossary contains a small sampling of common plant diseases that affect horticultural and agricultural crops worldwide. This illustrated glossary has been provided to help you familiarize yourself with these diseases, as well as a means of studying for career development events in which their identification is a major component. This glossary is by no means all-inclusive, but provides a good starting point for your studies. Use this glossary, as well as the e-flashcards on your textbook's companion website at www.g-wlearning.com/agriculture to help you study and identify these diseases.

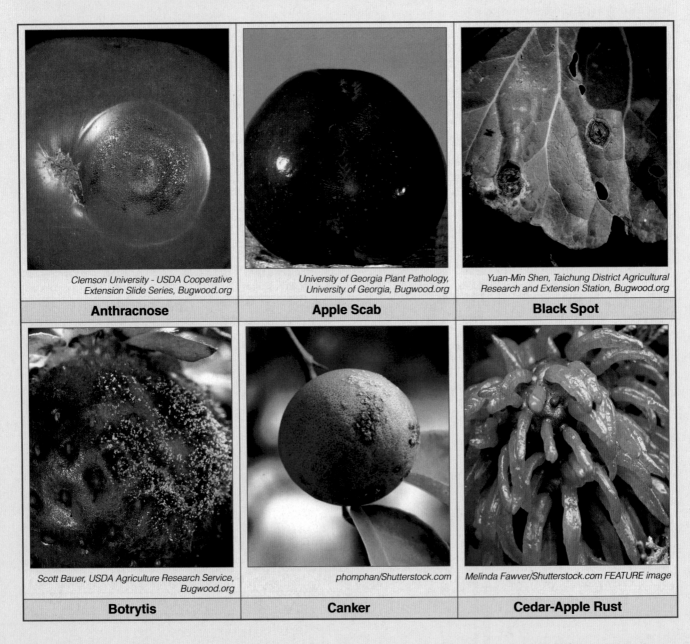

Clemson University - USDA Cooperative Extension Slide Series, Bugwood.org	University of Georgia Plant Pathology, University of Georgia, Bugwood.org	Yuan-Min Shen, Taichung District Agricultural Research and Extension Station, Bugwood.org
Anthracnose	**Apple Scab**	**Black Spot**
Scott Bauer, USDA Agriculture Research Service, Bugwood.org	phomphan/Shutterstock.com	Melinda Fawver/Shutterstock.com FEATURE image
Botrytis	**Canker**	**Cedar-Apple Rust**

926   Horticulture Today

Disease Identification 927

Weeds Identification

The following weeds identification glossary contains a small sampling of common weeds. This illustrated glossary has been provided to help you familiarize yourself with these plants, as well as a means of studying for career development events in which their identification is a major component. This glossary is by no means all-inclusive, but provides a good starting point for your studies. Use this glossary, as well as the e-flashcards on your textbook's companion website at www.g-wlearning.com/agriculture to help you study and identify these weeds.

Annual Bluegrass	Bermuda Grass	Broadleaf Plantain
Rasbak	komkrit Preechachanwate/Shutterstock.com	JonasS at Lithuanian Wikipedia

Buckhorn Plantain	Chickweed	Cocklebur
Harry Rose, South West Rocks, Australia	Bildagentur Zoonar GmbH/Shutterstock.com	NY State IPM Program at Cornell University

928   Horticulture Today   Copyright Goodheart-Willcox Co., Inc.

Glossary

A

abiotic. A type of disorder caused by some factor other than living organisms. (31)

abscise. To fall off or separate from. (14)

abscisic acid. A plant hormone that can inhibit germination. (13)

absolute scale. The true size of an object. (24)

acclimatization. Gradually exposing plants to different environmental conditions. A hardening off process. (14, 17)

achene. A one-seeded dry *indehiscent* fruit that is held freely within the pericarp. (8)

action threshold. The point at which the volume of a *pest* reaches an unacceptable level, and where some type of corrective action to reduce its numbers is needed and economically justified. (29)

active ingredient. A chemical in a *pesticide* that works to control the targeted *pest*. (33)

acute toxicity. A measure of how poisonous a *pesticide* is after a single exposure. (33)

adenosine triphosphate (ATP). A nucleotide found in the *mitochondria* of a *cell* and the principle source of energy for cellular reactions. (8, 9)

adhesion. The binding of water molecules to soil particles. (11)

adsorbed. Taken up and held by a soil particle. (11)

adventitious. Growing from an unusual place, as in plant roots that grow from stems or leaves. (8)

adventitious root formation. The process of roots forming from any plant part other than the root. (14)

advertising. The act of calling public attention to a product or service offered by a company. (5)

aeration. The process by which air is circulated, for example, in soil to provide oxygen for root respiration. (11)

aeroponic system. A *hydroponic* growing method in which plant roots are suspended in the air and misted intermittently with a nutrient solution. (20)

aggregate. The binding together of soil particles to form a soil structure. A solid, inert material that supports plant life in a *hydroponic* system. (11, 20)

agricultural literacy. Having the knowledge necessary to synthesize, analyze, and communicate basic information about agriculture. (2)

agricultural pest. An insect, *disease*, weed, or animal that attacks a crop or food source and causes damage. (33)

Agricultural Proficiency Award. A prize or recognition given by FFA for an exemplary SAE project. (2)

agriscience internship. A job placement working in a school's agricultural education program. (2)

agronomist. A person who studies or practices cultivating crops for food, fiber, and fuel. May specialize in areas such as biotechnology, soil science, weed science, or plant breeding. (4)

agronomy. The science, study, and technology of cultivating crops for food, fiber, and fuel, including the management of soil. (4)

A horizon. The fertile, upper, outermost layer of soil. Also referred to as the *surface* horizon or *topsoil*. (11)

air drainage. The process by which cold air sinks, flowing downhill to the lowest available point where it accumulates until dispersed by heat or wind. (10)

air layering. A propagation method that involves wounding the stem and wrapping it in rooting medium to encourage *adventitious root formation*. (15)

algaecide. A chemical used to control algae. (33)

allele. One of a number of variant forms of the same gene. (9)

allomone. A chemical signal that acts between different *species* to the benefit of the originator. (30)

ammonification. The process by which ammonium ions (NH_4^+) are made by saprophytic *bacteria* and fungi by incorporating nitrogenous compounds into amino acids and proteins with excess nitrogen (in the form of ammonium ions) being released as a by-product of their metabolism. (12)

angiosperm. A flowering plant that has seeds enclosed in fruit. (7, 8)

anion. A negatively-charged *ion*. (11)

annotate. To write questions or comments on a document as you read it. (3)

annotated bibliography. A document that lists citations for sources used and briefly describes each work. (3)

anther. The part of the *stamen* of a flower that produces pollen. (8)

anti-transpirant. A chemical compound that can be applied to the leaves of plants to reduce *transpiration* and provide protection from *plant pathogens*. (29)

apical meristem. A swelling of *cells* at the furthest tip of a plant shoot or the bud. (17)

apiculturist. A person who studies and maintains bees. (2)

apomixis. The ability of a plant to produce seeds without going through the *fertilization* process. (13)

approach grafting. A method of joining two separate, rooted plants together. (16)

aquaponics. A system of growing plants in water that has been used to grow fish, snails, or other aquatic creatures. (20)

aquascaping. The craft of arranging aquatic plants and *hardscape* materials in a pleasing manner. (24)

aseptic. Sterile or free of contaminants. (17)

asexual propagation. Horticultural method of starting new plants from existing plant material. (14)

aspirated thermostat. A device that measures temperature. It has a small fan attached to it that blows air across the temperature sensor for a more accurate reading. (18)

assimilation. The formation of organic nitrogen compounds from inorganic nitrogen compounds present in the *environment*. (12)

asymmetrical design. A design in which the visual weight is equal on both sides of the central vertical axis (CVA), but the sides are not mirror images. (25)

audience. The people who will read, hear, or see the message. (3)

auricle. A small, claw-like appendage that projects from the *collar* of a grass blade. (28)

autotroph. An organism that uses the process of *photosynthesis* to make its own food. (8)

auxin. A plant hormone that induces *adventitious root formation*. (14)

B

bacteria. Microscopic, unicellular organisms that reproduce asexually, some of which cause *disease*. (31)

balled-and-burlapped (B&B). A harvesting method in which the root system of a plant or tree is wrapped in burlap and tied with twine, which serves to protect the roots during transport. (4)

banding. A fertilizer application method in which the fertilizer is spread in bands or lines where developing roots will easily reach it; either to the side and below the seed rows, slightly below the seeds, or in between rows. (12)

bare root (BR). A harvesting method in which all soil is removed from the plant's root system. The roots may be covered with a plastic bag for shipping purposes. (4)

bark. Pieces of the outermost layer of trees or other plants. (11)

bark grafting. A plant joining method in which a *scion* of one plant *species* is placed into the layer where the bark and wood separate in a rootstock plant. (16)

bark slipping. A condition in which the vascular cambium is actively growing and the bark can be easily detached or peeled in one even layer from the wood beneath it without wounding the plant material. (16)

base plan. A map showing a site with the details of the landscape and its structures. (24)

basket weave. A design pattern in which two bricks are laid horizontally next to two vertically laid bricks to look as though they are weaving in and out of each other. (27)

bedding plant. A plant that is used mainly for ornamental purposes in beds and often is used for a particular season or two of the year. Also referred to as an annual. (19)

bench cut. A cut that removes upright limbs back to side branches. (23)

bench grafting. Any *grafting* that occurs in potted plants in the *nursery* (typically with pots sitting on a bench or table). (16)

beneficial. A natural predator or parasite that controls the population of *pests*, such as insects, mammals, *bacteria*, and other microbes. (29)

berry. A fruit with fleshy walls (pericarps), without a stone, and often with a number of seeds, such as grapes or oranges. (8)

B horizon. The subsoil layer that contains much of the original parent material that has been weathered. (11)

bibliography. A list of all research and sources that are used to create a text or presentation. (3)

biennial. A plant with a two-year life cycle, with flowering occurring in the second year followed by plant death. (10)

bine. A long, flexible climbing stem. (23)

binomial nomenclature. A two-word naming system, such as that used for plant *species*. (7)

biochar. A type of charcoal used for agricultural purposes with high nutritional content. (11)

biochemical pesticide. A nontoxic, naturally occurring mechanism used to control *pests*. (33)

biodiesel. A type of fuel made from materials such as vegetable oils or animal fats. (2)

biofilm. A group of microorganisms that stick together on a surface and form a thin, slimy coating. (20)

biological hazard. An organism that can cause harm to another living organism. Examples include mold, blood, plants, insects, *pests*, and *diseases*. (6)

biopesticide. A *pesticide* that is derived from natural products, such as other plants, animals, and microorganisms. (33)

biopharming. Growing plants that have been altered, often via genetic engineering, for medicinal uses. (20)

bioplastic. A type of biodegradable plastic made from organic components. (11)

bio-stimulant. A microorganism that fixes nitrogen into a form plants can use and creates a healthier soil *environment* for root growth, boosting plant growth. (19)

biotic. A type of disorder caused by living organisms. (31)

blade. The flat, leaf part of the grass. (28)

blanching. Keeping plants from light to prevent *photosynthesis* and to cause white tissue growth. (10)

bonsai. The Japanese art of pruning and manipulating plants into miniature form. (26)

botanist. A scientist who studies plants, including their structure, genetics, ecology, *classification*, and economic importance; often called a plant scientist. (4)

botany. The scientific study of plants, including their structure, genetics, ecology, *classification*, and economic importance. (4)

boutonniere. A small floral arrangement that is worn on the lapel area of clothing. (25)

bramble. A prickly, rambling vine or shrub. (4)

brand. A name, label, logo, or image under which a product is sold. (5)

bridge grafting. A plant joining method in which damaged tissue is removed from a plant and *scions* are inserted to provide support and reconnect the vascular tissue across the wounded area. (16)

broadcasting. A fertilizer application method in which fertilizer is spread uniformly over a field and tilled or incorporated into the soil prior to planting. (12)

broadleaf weed. A *dicot* weed with net venation and multiple leaf arrangements and structures. (32)

brown waste. Carbon-based materials, such as sawdust and cardboard, that are a common ingredient used to create *compost*. (11)

budding. A type of *grafting* that uses a single bud from the chosen *scion*, which is inserted into the rootstock material. (16)

budwood. Short lengths of young branches used to secure *vegetative* buds as *scion* material. (16)

bulb. A modified stem that contains a short, fleshy basal plate at the bottom from which roots grow and which holds fleshy scales (primary storage tissue and modified leaves) and the shoot (the developing flower and leaf buds). (15)

bulblet. A small *bulb* attached to the main *bulb*, which serves as a means of reproduction for the plant. (15)

bulk density. The mass (weight) of a given volume of dry soil and the solid and pore spaces of a soil. (11)

bunch-type weed. A type of grassy weed that produces new stems through *tillering*. (32)

C

callus. A cluster of *cells* that are from *parenchyma* and lack a defined function. (17)

callus tissue. Tissue that forms to cover a wound. The bundle of undifferentiated *cells* begins to initiate new cellular divisions that develop a specialized function to form into *meristematic* growing regions or roots. (14)

capillary mat. A material that is covered with a coating and that absorbs and releases water to plants that rest on it. (21)

capillary water. Water that is held in the soil against gravitational pull and is typically available to plants. (11)

capsule. A dry *dehiscent* fruit that splits open to release seeds and comes from flowers that had many *pistils*. (8)

carbon fixation. A process which occurs in the last stage of *photosynthesis* in which carbon atoms from carbon dioxide are used to make organic compounds in which chemical energy is stored. (9)

career exploration. The investigation of occupations. (2)

carotenoid. A plant pigment that reflects yellow, orange, and red light and assists in capturing light energy. (9)

caryopsis. A dry *indehiscent* fruit in which seeds are firmly attached to the fruit wall. (8)

Casparian strip. A waxy barrier that rings the endodermal *cells* in the roots. (9)

cation. A positively charged *ion*. (11)

cation exchange capacity (CEC). The amount of *cations* that a soil can hold. (11)

cell. The basic unit of a plant that contains many different organelles (structures) that drive plant processes. (8)

cell expansion. The second stage of seed development in which seeds undergo a phase of swift cell enlargement due to the accumulation of food reserves in the form of carbohydrates, fats, oils, and proteins. (13)

cellulose. Polysaccharide material that makes *cell* walls. (8)

Centers for Disease Control and Prevention (CDC). An operating unit of the Department of Health and Human Services that monitors safety risks and health hazards in the United States and abroad. (6)

central leader system. A training method for fruit trees that is characterized by one main, upright trunk called the leader. (23)

chelate. A chemical compound composed of a metal *ion* and a large organic molecule that can prevent *cations* from reacting with *anions* to form insoluble compounds. (12)

chemical hazard. A toxic substance that may be encountered in the workplace and that may cause a wide range of effects. (6)

chilling injury. A condition in which plants are damaged by low temperatures in the field or in storage. (10)

chimera. A type of *mutation* that allows two genetically distinct tissues to coexist. (9)

chlorine. A harsh chemical that can be used as a bleach or disinfectant for water purification. (26)

chlorophyll. A green pigment located in the *chloroplasts* of plant *cells* that is a receptor of light energy in the red and blue wavelengths. (8, 9)

chloroplast. A *plastid* that contains *chlorophyll* and is the site where *photosynthesis* occurs. (8)

chlorosis. Yellowing of plant leaves. (11)

C horizon. The subsoil layer below the *A horizon* and *B horizon* that contains much of the original parent material that has not been weathered. (11)

chromatin. A filament-like structure that contains *DNA* and proteins that makes up *chromosomes*. (9)

chromoplast. A pigmented *plastid* that contains *carotenoids*. The chromoplasts are often responsible for the saturated yellow, orange, and red colors found in flowers, aging leaves, fruit, and some roots such as carrots. (8)

chromosome. Cellular structures that carry the genetic information of a plant. (9)

chronic toxicity. A measure of how poisonous a *pesticide* is after repeated exposures, over a length of time. (33)

circling. A condition in which the roots of a plant are tangled and wrap around the interior of the pot. (21)

class. The taxonomic rank that separates or identifies plants in a *phylum*. Class is above *order* in taxonomic ranking. (7)

classification. The system of identifying and grouping like organisms together. (7)

clayey soil. Easily compacted soil with the finest particle size (less than 0.002mm), which contributes to slow drainage and poor *aeration*. (11)

clean room. A sterile area that is void of contaminants. (17)

cleft grafting. A plant joining method that is used to topwork trees in the trunk or scaffold branches. (16)

clone. A plant produced through *micropropagation* that is identical to the original plant and every other plant produced from the same original plant. (17)

coconut coir. A brown fibrous material that is the result of shredding coconut husks. (11)

cohesion. A property of water in which hydrogen bonding between adjacent water molecules allows the water to be pulled upward through the plant. (9, 11)

cold frame. A small structure used to grow plants that is passively heated by the sun and has a vent for releasing radiant and built-up heat. (18)

collar. The band at the backside of a leaf where the leaf blade and sheath meet. (28)

collenchyma. Plant tissue that has thickened *cell* walls made of *cellulose* and that provides support and minimizes breakage. (8)

color harmony. Guidelines for combining colors and color values in a design. Also referred to as *color scheme*. (25)

color scheme. Guidelines for combining colors and color values in a design. Also referred to as *color harmony*. (25)

color theory. A body of practical guidance for using and mixing colors. (24)

common name. A word or term for a plant that is used in everyday language. (7)

communication. A process in which a message is sent by one person and received and understood by another person(s). (3)

community supported agriculture (CSA). A farming practice in which people pay in advance for shares of the produce that is delivered at harvest. (4)

complete fertilizer. A fertilizer containing all three primary *macronutrients*: nitrogen, phosphorus, and potassium. (12)

complete metamorphosis. A process in which an insect goes through four phases: egg, larva, pupa, and adult. (30)

compost. Decomposed *organic matter* that may include animal manures, food wastes, and *vermicompost*. (11)

concept plan. A map or design that shows the general layout and possibilities for a landscape. (24)

cone. The fruit of a hop vine. (23)

confidence. A feeling of self-assurance and belief in one's ability to be successful. (1)

contact herbicide. A chemical that is sprayed on the weed foliage that kills only the plant tissue it touches. (32)

contact pesticide. An *insecticide* that kills insects through touch, or by entering the insect's system through ingestion. (33)

container capacity. The maximum amount of water that can be held by a *substrate* (growing medium) against the pull of gravity. (19)

container water garden. A water element that is housed in a vessel and can contain plants and aquatic animals. (24)

cool cell. A *cellulose* pad or panel that is used in an evaporative cooling system for a greenhouse. (18)

cordon. The major branch that comes from the trunk of a vine. (23)

core ideology. Basic ideas, standards, and principles. (5)

corm. A modified, enlarged stem that serves as a storage organ and method of reproduction. (15)

cormel. A young, small *corm* produced from a *stolon* on the base of a mature corm. (15)

corsage. An arrangement of flowers pinned to clothing or worn on the wrist. (25)

cortex. Plant tissue of the outer layers of plant stems and roots. (8)

cotyledon. First leaf to emerge from a seed. (7)

cover crop. A crop planted to add nutrients to the soil and manage erosion. (11)

critical day length (CDL). The number of hours of light needed to initiate flowering in a plant. (19)

critical night interval (CNI). The length of dark period needed to initiate flowering in a plant. (19)

critical thinking. Using objective reasoning or consideration before forming a judgment or taking action. (3)

crop rotation. A production practice in which vegetables of the same plant family are planted in different locations every year. (22)

crown. The area on a plant where the roots meet the stem tissue. (21)

cryopreservation. Freezing organs, *cells*, or other biological materials in subzero temperatures using liquid nitrogen. (17)

cull. An undesirable, low-quality fruit that is composted or thrown away. (23)

culling. Removing dead or old plants. (20)

cultivar. A name for a plant that has been bred or selected for horticultural purposes. (7)

cutting. A widely applied method for *vegetative propagation* using a piece of leaf or stem. The cutting adventitiously forms roots, which allows subsequent normal growth and development of the new plant that is genetically identical to the plant from which the cutting was taken. (14)

cyclic photoperiodic lighting. A lighting method that uses a pattern of light and darkness during the night from 10 pm to 2 am to create a long day effect for plants. (19)

cytochrome. A protein that carries electrons during *respiration*. (12)

cytokinesis. The process in which a single *cell* divides to form two daughter cells and the cell plate is formed. (9)

cytokinins. A *class* of plant growth hormone that encourages bud formation and *adventitious* shoot formation. (14)

cytoplasm. Part of a *cell* that surrounds the organelles and *nucleus* and contains amino acids, sugars, enzymes, and waste products (on their way to disposal). (8)

cytoskeleton. Protein *filaments* and motor proteins in the *cell*, composed of three major structural fibers: microtubules, microfilaments (also called actin *filaments*), and intermediate *filaments*. (8)

D

daily light integral (DLI). The amount of light intensity a plant receives throughout a given day. (10)

deciduous. Trees or shrubs that lose their leaves annually. (14)

dedifferentiate. A process in which the *cell* regresses from a specialized function to a simpler state. (14)

deep water culture (DWC). A method of *hydroponics* in which the nutrient solution is pumped from a main reservoir and circulated through an attached system of buckets containing plants. (20)

deficiency. A lack or shortage of something, such as plant essential nutrients. (12)

degree day. The required number of heat units that a plant needs to have to reach a certain point of development, usually flowering or harvest. (10)

dehiscent. A type of fruit that will split open at maturity and freely release seeds for dispersal. (8)

delegate. To assign or entrust a task or responsibility to another person. (1)

demographics. Characteristics or traits of a group. (3)

denitrification. A process in which *bacteria* reduce nitrate to a volatile form of nitrogen, such as nitrogen gas (N_2) or nitrous oxide (N_2O), which then return to the atmosphere. (12)

deoxyribonucleic acid (DNA). A substance in *cells* that contains a plant's genetic material and provides genetic information to *ribonucleic acid (RNA)*. (8)

desiccation. Drying out of a seed. (13)

desorbed. When nutrients are released by the soil particles to the soil water solution for availability to plants for uptake. (11)

diapause. A state of arrested development (similar to plant's *dormancy*) that allows an insect to survive unfavorable conditions. (30)

dichotomous key. A tool used to identify plants by pairing choices against each other until all choices have been exhausted and the plant *species* remains. (7)

Dicotyledoneae (dicots). A seed with two *cotyledons* (first leaves), net-veined leaves, and flower parts in multiples of fours and fives. (7)

DIF. A term used in *horticulture* to mean the day temperature minus the night temperature. (10)

differentiate. A process in which a *cell* develops a specialized function. (14)

diploid. A *cell* that contains two complete sets of *chromosomes*. (9)

direct sales. A distribution method in which a company sells its products to customers without another party involved. (5)

discrimination. Unfair treatment of another person. (6)

disease. A disorder in the structure or function of a living organism, such as a person, animal, or plant. (31)

disease triangle. The three factors needed for a disease to develop: a host, a pathogen, and an appropriate *environment*. (31)

distal. Located farthest from the point of attachment. (14)

domain. The highest and most inclusive taxonomic ranking for all living organisms. (7)

dominant. The relationship between one *allele* that is expressed over a second allele. For example in flower color, the dominant color will be expressed even in the presence of a recessive allele. (9)

dormancy. A condition in which buds and seeds are inhibited from growing until a certain environmental requirement is met. (10)

double cut. A pruning cut that starts with a cut going halfway underneath the branch followed by a cut from on top of the branch that meets the undercut.

drawing scale. The relationship between the distances on a plan to actual distances on a site. (27)

drip irrigation. A method of watering plants that uses small tubes that slowly release water directly to the plant's roots. (21)

dripline. The outermost circumference of a tree's branches where water drips from the leaves and onto the ground. (22)

drip system. A *hydroponic* method in which a liquid nutrient solution is slowly released onto the base of each plant through a series of thin tubes. (20)

drupe. A fruit with a fleshy, soft mesocarp and a seed enclosed by a hard, stony endocarp. (8)

E

eave. The part of the greenhouse roof that meets the wall of the greenhouse. (18)

ebb and flow system. A method of *hydroponics* in which the containers holding plants are periodically flooded with a nutrient solution and then drained. Also referred to as flood and drain. (20)

ecologist. Scientist focused with understanding ecosystems as a whole. (7)

economic injury level (EIL). The lowest population density of a *pest* that will cause economic damage and justify the costs of pest control. (29)

effluent. The waste from living creatures. (20)

electrical conductivity (EC). A measure of the amount of electrical current a material can carry. (21)

electrical conductivity (EC) meter. A tool that measures the amount of *soluble* salts in a sample of soil or media. (11)

elytra. The hardened forewings of beetles. (30)

embryogenesis. A rapid stage of insect growth and development during which the embryo goes through cellular division and differentiation. (30)

emitter. A device that releases something, such as water, in an *irrigation* system. (18)

empathy. The ability to understand and share feelings of others. (1)

emphasis. The prominence given to something, such as a focal point in a design. (24)

endogenous contamination. When *bacteria*, *fungus*, or a *virus* comes from within the surface of sanitized tissue. (17)

endoplasmic reticulum (ER). A complex, folded membrane system that provides a channel for transporting proteins and lipids in the *cell*. (8)

enologist. A person who uses grapes to make wine using chemistry and food science knowledge. (4)

entomologist. A person who studies insects. (31)

entomology. A branch of biology that is the study of insects. (30)

entrepreneur. A person who organizes and operates a business. (5)

entrepreneurship SAE. A hands-on learning project in which the student operates a business and is responsible for all financial risks. (2)

environment. The surroundings or conditions in which something lives or exists. (31)

envisioned future. What a company or other organization plans to achieve. (5)

EPA registration number. A number given to a *pesticide* once it has been reviewed and verified by the Environmental Protection Agency (EPA). (33)

epidermis. *Cells* that cover the outside areas of a plant and protect the plant from environmental stresses, minimize water loss, and provide a site for gas exchange. (8)

ergonomic hazard. A repetitious movement or pose that may lead to physical stress. (6)

espalier. A training method in which a plant grows flat against a wall, lattice, or other support. (23)

ethylene. A natural chemical produced and emitted in varying quantities by fruits and vegetables and by decaying plant materials. (4)

etiolation. A plant growth response in absence of light. (10)

evapotranspiration. The release of water through plant leaves, which then evaporates and helps cool the air. (4)

evergreen. A plant that remains green throughout its life cycle and all seasons. (19)

evidence. Facts or information about a topic indicating whether a belief or position is correct or valid. (3)

exoskeleton. A hardened external skeleton found in all insects. (30)

explant. The plant *cell* or tissue removed from the parent plant and placed in an *in vitro* culture. (17)

exploratory SAE. A hands-on learning project in which the student explores agricultural careers and subjects. (2)

extensive green roof system. A live, planted garden on top of a structure that forms a matting and helps to increase insulation, reduce runoff, and increase a roof's lifespan. (20)

F

family. The taxonomic rank that separates or identifies plants in an *order*. Family is above *genus* in taxonomic ranking. (7)

fertigation. The process of adding fertilizer to irrigation water. (12)

fertilization. The process in which the male gamete combines with the female gamete to create a genetically unique organism. (9)

fertilizer grade. A listing of nutrients contained in a fertilizer by weight. (12)

field capacity. The maximum amount of water that a soil can hold against the pull of gravity. (11)

filament. A thin stalk in the *stamen* of a flower that supports the *anther*. (8)

filler flower. Fine or airy materials used to fill in voids and often made of very small clusters of blooms or branched *inflorescence*. (25)

flagging. The condition of a plant when it is just about to wilt. The plant tips will begin to point downward and foliage will have a gray or lackluster appearance. (19)

Flemish bond. A brick design with bricks laid along the wall showing their sides and those laid across the wall showing their ends. (27)

floricane-fruiting. *Brambles* that produce fruit only on second-year canes. (23)

floriculture. The study, cultivation, and marketing of flowers and ornamental plants. (4)

focal point. The area of dominance or emphasis, as within a floral design, to which the eyes naturally travel. (25)

fogger. A machine that emits tiny water droplets in order to cool a greenhouse or add humidity. (18)

foliar application. A method of providing *micronutrients* to plants by spraying fertilizer directly onto the leaves. (12)

follicle. A dry *dehiscent* fruit that splits along the length of the fruit releasing seeds for dispersal. (8)

forester. A person who plants, manages, and cares for forests. (4)

forest stand. Trees or other growth occupying a specific area that are uniform in *species*, size, age, arrangement, and condition. (4)

forestry. The science or practice of planting, managing, repairing, and caring for forests. Also referred to as *silviculture*. (4)

form. The overall shape of a floral arrangement or the shapes of the individual materials used in an arrangement. (25)

form flower. A flower that is unique in some manner (color, shape, texture, or size) and does not necessarily fit into the other *form* categories. (25)

freezing injury. A condition in which plants are damaged when low temperatures freeze the water in plant tissue. (10)

functional diagram. A series of loosely drawn circles that divide spaces into areas that will be decided on by the designer. (24)
fungicide. A chemical used to control or prevent fungal growth. (33)
fungus. A eukaryotic organism which attaches itself to a host and decomposes the organism and absorbs nutrients from it. (31)

G

gable. The part of a wall that encloses the end of a pitched roof greenhouse. (18)
gall. A tumor-like growth that can develop in response to an insect laying eggs within a plant. (31)
general safety hazard. A common danger that most employees encounter at work, such as slipping on wet floors, falling from heights, injuries from machinery, or electrical shocks. (6)
genes. Specific sequences of nucleotide pairs on a *DNA* molecule that hold the information to build and maintain cellular processes and pass genetic traits to offspring. Different genes are responsible for any number of physical traits and drive biochemical processes and responses in plants. (9)
genetically modified organism (GMO). A plant that has had its *genome* transformed through genetic engineering. (13)
genome. The entire set of *chromosomes* of a plant. (9)
genotype. The genetic makeup of an organism, often in reference to a particular trait. (9)
genus. A subset of organisms within a *family* that share similar characteristics. Genus is above *species* in taxonomic ranking. (7)
geocarpy. A rare means of plant reproduction in which the flower stalk (after pollination) elongates and pushes into the ground where the fruit matures. (4)
geophyte. An underground storage structure in a plant, such as a *bulb*, *corm*, or *tuber*. (15)

geotextile. A permeable, inorganic mulch material made from a plastic. (11)
germplasm. Seeds or other materials from which plants are propagated; serve as raw genetic material for future use. (13)
girdling. Wounding of a stem by cutting or bending the stem for the purpose of propagation. (15)
glazing. The covering material placed on the outside of a greenhouse. (18)
glycolysis. The process in which a glucose molecule is split into two molecules of a compound called *pyruvate* and produces a small amount of *adenosine triphosphate (ATP)*. (9)
goal. An objective to be achieved. (5)
Golgi bodies. Cell organelles that control the flow of molecules in the *cell*, modifying some before packaging them into vesicles for transport to other parts of the cell or for excretion outside the cell. (8)
good agricultural practice (GAP). A farm-level production method used to ensure that fresh produce is safe for human consumption. (22)
grafting. Joining together two different plants or plant parts so that they grow into one plant. (16)
grana. A stack of *thylakoid* disks found within the *chloroplast*. (9)
graphic. A pictorial representation of an item in a landscape plan. (24)
gravitational water. Free water that moves through the soil by the force of gravity. (11)
greenhouse orientation. The way a greenhouse is positioned on a site. (18)
greenhouse range. A series of greenhouses that are attached to one another. (18)
green industry. The portion of the *horticulture* industry that cultivates and arranges outdoor plant materials to create spaces that are inviting, beautiful, and useful to people and the ecosystem. (4)
green waste. Organic materials rich in nitrogen, such as leaves, fresh manures, grass clippings, and coffee grounds, used to create compost. (11)

ground level ozone. A harmful air pollutant that is formed from chemical reactions of other pollutants near Earth's surface. (20)
growth medium. A substance containing nutrients and hormones that is used for plant growth. (17)
guard cell. A specialized epidermal plant *cell* located around pores that opens to allow for gas exchange and control water loss. (8)
guys. Heavy wires or ropes that are attached to an object (such as a tree) and anchored in the ground to provide stability (27)
gymnosperm. A nonflowering plant with seeds that develop without an ovary. (7, 8)

H

haploid. A *cell* that contains a single set of *chromosomes* from one parent. A sperm and an egg cell are *haploids*. (9)
harassment. Repeated treatment that bothers or annoys another person. (6)
hardscape. The constructed areas around a building or in a landscape, such as pavers, patios, sidewalks, and retaining walls. (4)
hardwood. Mature, dormant, woody plant material. (14)
harmony. In floral design, a pleasing interaction or blending among the elements of a design. (25)
haustorium. A modified root of a *parasitic plant* that anchors itself into the vascular tissues of the host plant. (31)
header. A brick that is laid in a design with its end or short side exposed on the design. (27)
heading cut. A cut that removes the terminal portion of a branch to just above a bud or side branch, invigorating growth and promoting lateral branch emergence. (23)
heat island effect. The increase in temperature of urban areas due to a variety of environmental factors. (20)
heat stress. A condition in which plants are damaged due to high temperatures. (10)

hedgerow design. A floral design that is patterned after a garden hedgerow. (25)

hemelytra. One of the forewings of Hemiptera (true bugs) that is slightly thickened at the base. (30)

herbaceous. Nonwoody, soft-stemmed plant. (14)

herbarium. A repository of collected plant specimens. (7)

herbicide. A chemical that will kill a plant. (28)

herringbone. An interlocking pattern of bricks set at 45° or 90° angles. (27)

heterotroph. An organism that uses external sources of food or energy. (8)

high-density orchard. A fruit tree orchard with between 150 and 180 trees per acre that comes into bearing within 2 to 3 years of planting. (4)

high tunnel. A square or semicircular structure made of a frame and a covering, such as plastic film. It is heated by the sun and used to extend the growing season for plants, especially edible crops. (18)

hilling. The piling of soil onto plants to promote desired growth, as in the case of potatoes to produce more *tubers*. (15)

histodifferentiation. A stage of seed development when the embryo and endosperm develop distinct characteristics. (13)

hole insertion grafting (HIG). A plant joining method used on herbaceous material in which the rootstock growing point is removed and a hole is inserted into the top of the plant. The *scion* material is cut into a wedge and inserted into the hole. (16)

homozygous. Having identical pairs of *genes* for a pair of hereditary characteristics. (9)

horizon. A layer of soil distinguished by properties and characteristics developed through the five factors of soil formation. (11)

horizontal air fan (HAF). A fan that is mounted above the growing *environment* of a greenhouse and circulates air. (18)

horticultural oil. A *pesticide* treatment consisting of different types of oils used to kill insect *pests*. (26)

horticulture. The science, art, technology, and business of plant cultivation. (4)

horticulturist. A person who specializes in the science, art, technology, and business of plant cultivation. (4)

host. An organism that is capable of being infected by a pathogen and exhibiting symptoms of *disease*. (31)

hotbed. A structure consisting of four walls and a glass or plastic top that uses a heat source to encourage plant growth. (18)

hydrometer. An instrument used to measure the percentages of sand, silt, and clay in a sample to determine the soil textural class. (11)

hydroponics. A method of growing plants in a water-based system. (20)

I

imbibition. During early germination, the rapid uptake of water by a seed. (13)

immune. Resistant to or not affected by something, such as a *disease*. (31)

improvement SAE. A hands-on learning project in which the student makes something related to agriculture better in some manner. (2)

inarch grafting. A joining method used to support a damaged plant stem in which existing shoots or suckers growing from the rootstock or compatible rooted plant material planted around the tree are used as *scions*. (16)

incomplete dominance. A phenomenon where a plant displays characteristics of both parents. (9)

incomplete fertilizer. A fertilizer that is lacking one of the three primary *macronutrients*: nitrogen, phosphorus, or potassium. (19)

incomplete metamorphosis. A process in which insects pass through the life cycle stages of egg, larva or nymph, and adult. (30)

indehiscent. A type of fruit that keeps the seeds within the fruit walls after leaving the parent plant. (8)

infiltration. The manner in which water moves through the soil. (11)

inflorescence. Clusters of single flowers gathered on a stem. (8)

information literacy. The ability to recognize when information is needed and to find, evaluate, use, and communicate information. (3)

injury. Damage caused as a result of an immediate action or event within an *environment*, such as wind damage to a plant. (31)

inoculum. A pathogen (or its parts) that is capable of causing infection and *disease* when transferred to the appropriate host and *environment*. (31)

insecticidal soap. A synthetic soap solution used to kill insect *pests*. (26)

insecticide. A chemical used to prevent, control, or decimate insect populations. (33)

insoluble fertilizer. A fertilizer that does not readily mix with water. (19)

instar. The insect development phase between two periods of molting. (30)

integrated pest management (IPM). An environmentally sensitive approach to controlling *pests* that uses a combination of cultural, physical, biological, and chemical controls as well as commonsense and economical practices. The goal is to effectively control *pests* with the least possible hazards to people, property, and the *environment*. (4, 29)

intensive green roof system. A landscape on top of a roof similar to a typical garden planted at ground level. (20)

intensive market coverage. A distribution strategy in which a company attempts to sell its products using all available outlets. (5)

interconnection. A seamless linkage between diverse features to create blending, such as the way plants and other elements are related to one another in a landscape. (24)

interior plantscaping. The practice of designing, installing, and maintaining indoor spaces with plants. (26)

interiorscape. A landscape designed for and installed in an interior space. (4, 26)

interiorscaper. A person who designs, installs, and maintains interior landscapes. (4)

interiorscaping. The design, installation, and maintenance of plants inside buildings. (4)

International Code of Botanical Nomenclature. A set of rules that guides the naming or renaming of plant *species*. (7)

internode. A space between *nodes*. (8)

interplanting. A method of planting early maturing crops next to later maturing crops to maximize space efficiency. (22)

interspecific. Occurring between different *species*. (16)

interstock. A piece of plant material that is being grafted between the rootstock and *scion* to allow joining and growth of incompatible varieties. (16)

interveinal area. The space between veins in a leaf. (12)

intuition. The ability to understand or know something without the use of reason. (1)

in vitro. Growth of an organism outside the body, typically in a glass test tube or petri dish; Latin for "in glass." (17)

ion. An atom or molecule that has lost or gained one or more valence electrons and become a positively charged *cation* or negatively charged *anion*. (11)

irrigation. The practice of applying water to land or soil to assist in growing plants. (10)

J

job interview. A meeting where an employer and a job applicant discuss a job and the applicant's qualifications for the job. (5)

juvenile stage. A period of growth that occurs before plants reach adult form. (10)

L

lag phase. A period during seed germination with little or no water uptake, but with high cellular activities that prepare the seed to grow. (13)

laminar flow hood. A piece of equipment used in tissue culture that filters and purifies the air. (17)

landrace. A locally adapted or traditional variety of a plant with genetic diversity that acts as a buffer against environmental stresses, insects, or *diseases* that may afflict the crop. (13)

landscape architect. A professionally trained and educated individual with a knowledge of plant material and engineering. (24)

landscape contractor. An individual who installs and maintains landscape elements. (24)

landscape designer. A person with horticultural expertise, formally or informally educated, who often works on small landscape projects. (24)

lasagna composting. A system of building layers of *organic matter* to construct the growing media in a raised bed. Also referred to as sheet composting. (20)

lawn. An expanse of *turfgrass* that is used for recreation or beauty. (28)

layering. A general term for *vegetative propagation* techniques in which plants initiate roots while still attached to a parent plant. (15)

LC_{50}. A measure of acute inhalation *toxicity* needed to kill 50% of a test population of animals. (33)

LD_{50}. A measure of acute oral and dermal *toxicity* needed to kill 50% of a test population of animals. (33)

leachate. Water that is collected after it has percolated through a soil. (21)

leaching. The movement of nutrients out of the soil into the groundwater. (22)

leadership. The act of directing, guiding, or motivating a person or group of people. (1)

leadership path. A plan or map showing steps and activities to complete to develop and exhibit leadership skills. (1)

leaf bud cutting. A single *node* and adjacent *internode* tissue with the leaf attached used to *propagate* a new plant. (14)

leaf cutting. A leaf or a portion of a leaf used to *propagate* a new plant. (14)

leaf mesophyll. Plant tissue that makes up the internal layers of leaves. (8)

legend. An element that describes symbols or abbreviations used in a document, such as a design plan. (27)

legume. A dry, *dehiscent* pod that opens along two seams at maturity. (8)

letter of application. A document that requests that the sender be considered for a job opening and introduces the writer's *résumé*. (5)

leucoplast. A colorless *plastid* that performs functions such as synthesizing starch and forming oils and proteins. (8)

light quality. The wavelengths of light that a plant receives. (10)

light quantity. The amount and duration of light emitted by the light source. (10)

ligule. A thin structure that clasps the top of the leaf sheath. (28)

line. The visual path that creates the foundation, shape, and *form* of a floral arrangement. (25)

line design. A floral arrangement that emphasizes lines rather than the plant material. (25)

line flower. A flower that has a long stem, a spike, or linear *form*. (25)

line-mass design. A floral arrangement in which a mixture of line and mass designs is used. (25)

liner. A plant that is used for transplanting that is 1″–3″ in diameter. (19)

loamy soil. Soil with a mixture of sand, silt, and clay particles; moderate water-holding capacity; and strong ability to store plant nutrients. (11)

locavore. A consumer who is interested in eating locally grown foods. (4)

louver. An angled slat that allows air and light exchange between the greenhouse and the outside *environment*. (18)

low tunnel. A season extension technique in which small hoops are placed over rows of vegetables and covered in fabric or plastic. (22)

loyalty. The quality of being dependable and showing support or allegiance to others. (1)

lysosome. Digestive system in the *cell*, using enzymes to break down large molecules like proteins, polysaccharides, lipids, and nucleic acids. (8)

M

macronutrient. A nutrient that plants need in high quantities for normal growth and development. (12)

macropore. A large pore space in the soil. (11)

mandible. A part of the insect mouth that is often modified for specialized feeding. (30)

manometer. A device used to measure *static pressure*. (18)

marketing. The total system of business activities designed to plan, price, promote, and distribute products. (5)

mass design. A floral arrangement in which *form* and mass are more important than the individual elements or lines. (25)

mass flower. A closed-form, single flower that has a dense, round shape. (25)

material safety data sheet (MSDS). A document that contains information on the potential health impacts of a chemical or other dangerous substance. Also referred to as a *safety data sheet (SDS)*. (6)

maturation drying. The phase of seed development when seeds have reached physiological maturity. (13)

mechanical transplanter. A machine that moves plants from one place to another. (21)

mechanics. The devices and techniques used to keep floral placements secure and stable in a design. (25)

meiosis. The process of sexual reproduction of *cells* that includes nuclear division and in which the cell's *chromosomes* are divided in half. (9)

meristematic. Least differentiated cells that are responsible for *cell* formation and growth. (8)

microclimate. A small area with different environmental conditions than the surrounding area. (10)

microgreen. A recently germinated plant (sprout) that is edible and used for food. (2)

micronutrient. A nutrient that is essential for plant growth, but in small or minute amounts. (12)

micropore. Small pore space in soil through which *capillary water* moves. (11)

micropropagation. An *asexual propagation* method in which plants are manipulated on a cellular level, causing them to duplicate themselves repeatedly and rapidly. (17)

micropyle. A small opening in an insect egg through which sperm enters. (30)

migrant worker. A person who moves from place to place to do seasonal work. (6)

mission statement. A passage that identifies the purpose or reason for the existence of a company or organization. (5)

miticide. A product used to control or prevent mites. (33)

mitochondria. The parts of a *cell* that drive the process of *respiration* and energy transfer. (8)

mitosis. In *asexual reproduction*, the steps a *cell* undertakes to duplicate and divide a complete set of *chromosomes*. (9)

modified central leader system. A variation of the central leader form in which the central leader is removed after five to nine main lateral branches develop. (23)

moisture stress. A condition in *turfgrass* caused by lack of water. Signs of stress include changing leaf blade color; wilted, folded, or curled leaf tips; or depression marks that remain in the turf from foot traffic. (28)

molluscicide. A chemical product used to control mollusks (snails and slugs). (33)

Monocotyledoneae (monocot). A seed with only one *cotyledon* (first leaf) in the seed, parallel-veined leaves, and floral structures in multiples of three. (7)

monophagus. A type of insect that feeds on only one plant *species*. (30)

morphology. The physical form and structure of an organism. (7)

mound (stool) layering. A propagation method in which soil is piled on the *crown* of the plant, and the new shoots form *adventitious* roots that can be severed and transplanted. (15)

muck. A soil composed of primarily decomposed *organic matter*. (23)

mulching. Adding a layer of material (organic or inorganic) over the garden soil for aesthetic value, weed suppression, soil temperature moderation, water retention, or as a platform for traffic. (11)

mulchmat. A nonwoven wool or cotton matting that is used to hold down *topsoil*, prevent erosion, moderate soil temperature, reduce weed activity, and retain moisture for soils. (11)

mutation. A naturally occurring genetic change that affects the appearance and functions of a plant. *Mutations* can also be induced through chemicals or radiation. (9)

mycorrhiza. A *fungus* that grows in association with the roots of a plant in a symbiotic or mildly pathogenic relationship. (11)

N

nanometer (nm). The unit of measurement used to quantify light wavelengths. (10)

National Institute of Occupational Safety and Health (NIOSH). A government agency that conducts research and makes recommendations dealing with workplace safety, injury, and illness. (6)

necrosis. Death of plant tissue, usually resulting in dark brown or black coloration. (10)

necrotic lesion. Tissue death that often occurs in plant leaves as spots along the margins or in entire leaves as a result of a nutrient deficiency or *plant pathogen*. (12)

nematicide. A chemical product used to control *nematodes*. (33)

nematode. A roundworm that lives in the soil. Some are plant parasitic and considered significant *pests*. (23)

nicotinamide adenine dinucleotide phosphate (NADPH). A molecule that acts as a carrier for electrons in *photosynthesis*. (9)

night interruption (NI). A method of creating a long day effect for plants by disrupting the dark period with light. (19)

nitrification. The process by which several *species* of soil-dwelling *bacteria* will oxidize ammonium *ions* into nitrite *ions*. (12)

node. The place on a plant stem where a leaf develops. (8)

nonselective herbicide. A chemical that will kill all kinds of plants. (32)

nontunicate. A type of *bulb* that has fleshy scales and does not have a dry covering, such as a lily *bulb*. (15)

north arrow. An orientation point on a landscape design that shows the direction north. (27)

no-till. The practice of planting new plants in the crop residue from the previous season's growth. (32)

nucleolus. A part of the *cell nucleus* that is responsible for the formation of *ribosomes*. (8)

nucleus. A large organelle found in a plant *cell* that contains the plant's genetic material. Two primary functions of the *nucleus* are to control cellular activities by determining when and which proteins are produced and to store the cell's genetic information and pass this information on through cellular division. The nucleus contains *deoxyribonucleic acid (DNA)* and *ribonucleic acid (RNA)*. (8)

nursery. A place where young plants and trees are cultivated for sale and for planting elsewhere. (4)

nursery and landscape industry. The portion of the *horticulture* industry that cultivates and arranges outdoor plant materials to create spaces that are inviting, beautiful, and useful to people and the ecosystem. (4)

nursery liner. A young plant that will be grown in the *nursery* for an extended time until it is ready for sale. (14)

nut. A dry, *indehiscent* fruit that has a hardened pericarp with a loose seed inside. (8)

nutrient film technique (NFT). A *hydroponic* method that circulates a thin stream of water containing nutrients over the roots of plants. (20)

O

Occupational Safety and Health Administration (OSHA). A government agency created to ensure safe and healthy working conditions for Americans. (6)

ocelli. A modified eye structure that detects changes in light intensity. (30)

offset. A type of lateral shoot or branch that forms at the base of a main stem, which occurs only in some plants, primarily in monocots. (15)

oil bodies. Lipid droplets, spherical in nature, that are found throughout *cells* in the plant and are most concentrated in fruits and seeds. (8)

olericulture. The science, cultivation, processing, storage, and marketing of herbs and vegetables. The practice of producing edible vegetable crops. (4, 22)

oligophagus. A type of insect that feeds on multiple plant *species* that are somewhat closely related. (30)

ommatidia. Visual sensory facets that make up an insect's compound eyes. (30)

oomycete. Filamentous protists, including water molds and downy mildews that can cause *diseases*, such as blights and rots. (29)

open center system. A fruit tree training method that keeps the center of the tree free of large branches, allowing sunlight to penetrate the lower fruiting wood. (23)

order. A taxonomic ranking that separates or identifies plants in a *class*. *Order* is above *family* in taxonomic ranking. (7)

organic food. Food that has been raised naturally, without chemicals or genetic modifications of any kind. (4)

organic matter. Living and dead organisms. (11)

ornamental horticulture. The study, cultivation, and marketing of flowers and ornamental plants. (4)

oscillating sprinkler. A device that delivers *irrigation* water in a rectangular pattern. (21)

overhead. The ongoing expenses of operating a business, such as utilities and rent. (5)

overseeding. Sowing grass seed over the *thatch* of dormant grass. (28)

ovipositor. A structure on a female insect's abdomen that is used to deposit eggs. (30)

P

paraphrase. To use your own words, sentences, and organization to summarize something written (or spoken) by someone else. (3)

parasitic plant. A plant that anchors itself into another plant and feeds from its vascular system. (31)

parenchyma. Plant tissue that is made up of the vascular tissue of the *cortex* (outer layers), *pith* (inner layers) of stems and roots, and *leaf mesophyll* (internal layers of leaves). (8)

parent material. The material (bedrock, sediment, or organic material) that is weathered to form soil. (11)

parliamentary procedure. A democratic and efficient way of conducting business based on guidelines in *Robert's Rules of Order*. (1)

parthenogenesis. The process of *asexual reproduction* in insects. (30)

pathogen. An organism that causes *disease*. (31)

pathologist. A person who studies *diseases*. (31)

pedologist. A scientist who studies soil. (11)

pedology. The study of soils in their natural *environment*. (11)

peds. *Aggregates* produced through the binding of sand, silt, and clay particles. (11)

perched water table. Groundwater that is temporarily located above unsaturated soil due to compaction or some other soil formation factor. (22)

perennial. A plant that returns year after year. (4)

perforated convection tube. A plastic pipe with small holes that distributes heated air in a greenhouse. (18)

periderm. The outside *cell* layer in woody plants that provides protection and prevents water loss. (8)

perlite. A very lightweight, pea-sized (or smaller) rock that is white and comes from volcanoes. (11)

perseverance. The ability to continue trying to do something regardless of setbacks or difficulties. (1)

personal protective equipment (PPE). Materials or devices worn to provide a shield or defense from dangers. (6)

perspective. The size and placement of objects to create depth and distance in a design, such as the foreground, midground, and background in a landscape. (24)

pest. Any organism that damages or disrupts plant growth, such as weeds, birds, rodents, mammals, insects, snails, *nematodes,* and *plant pathogens* such as *bacteria*, *virus*, fungi, and *oomycetes*. (29)

pesticide. A chemical that is used to control insects, weeds, fungi, rodents, and other *pests*. (29)

pesticide formulation. A mixture of active and inert ingredients (adjuvants, surfactants, and carriers) used to create a product that controls *pests*. (33)

petal. A modified leaf that surrounds the female and male reproductive parts of a flower. (8)

phenol. A naturally occurring chemical compound in seeds that may prevent germination. (13)

phenotype. The interaction between a plant's genetic makeup and the *environment*; a visible expression of an organism's observable characteristics. (9)

pheromone. A chemical that is emitted by insects to attract mates and which can be synthetically produced and used in lures to trap insects for monitoring. (29)

phloem. Plant tissue that carries photosynthetic products synthesized in the leaves down the plant and provides structural support. Also called vascular tissue. (8)

photoblastic. A characteristic of seeds that have a germination response to the presence or absence of light. (10)

photocell. A device that reads the amount of light in an area. Measured in foot-candles. (18)

photodormancy. An occurrence in which seeds require a period of either light or dark conditions to germinate. (13)

photon. A light particle and a measure of *light quantity*. (10)

photoperiod. The duration of day length (the amount that light is present) and the relationship between the dark and lighted periods. (10)

photosynthesis. The process by which plants capture energy from the sun to convert simple molecules of carbon dioxide (CO_2) and water (H_2O) into complex carbohydrate molecules that can be used by plants and animals as sources of energy and building blocks for other molecules. (9)

phototropism. The physical movement of a plant or its parts toward or away from a light source. (10)

pH paper. Paper test strips that change color to indicate the pH level (acidity or alkalinity) of soil. (11)

phylum. A taxonomic ranking that separates or identifies plants in a *kingdom*. *Phylum* is above *order* in taxonomic ranking. (7)

physical hazard. A condition or substance within the work *environment* that may cause a person harm. (6)

phytohormone. A chemical naturally produced in plants that regulates growth and other functions. (14)

phytophagus. A type of insect that eats plants. (30)

phytoremediation. The use of living plants to remove organic and inorganic contaminants from soil. (12)

phytosanitary certificate. An official document that shows a plant has been inspected and appears to be free of pathogens, insects, and weeds. (21)

phytotoxicity. A poisonous effect by a substance on plant growth. (12)

pistil. The female organs of a flower that include the stigma, style, and ovary. (8)

pith. The inner layer of plant tissue of stems and roots. (8)

placement SAE. A hands-on learning project in which a student has an internship that is paid or unpaid within the agriculture and natural resources industry. (2)

plagiarism. Using work (such as ideas, writing, or images) created by someone else without permission and presenting it as one's own work. (3)

plant growth regulator (PGR). A hormone (synthetic or natural) that changes plant growth rates; often an inhibitor or accelerator. (19)

plant hardiness zone. One of several areas identified across the United States based on average annual minimum winter temperatures. (10)

plantlet. A small or young plant. (14)

plant pathogen. An organism that causes *disease* in plants, including *bacteria*, fungi, *viruses*, and *oomycetes*. (29)

plant science. The study of plant growth, reproduction, and adaptation as well as the use of plants for food, fiber, and ornamental purposes. (4)

plasmalemma. A cellular membrane that is permeable to water, but impermeable to *ions* because of their charge. (9)

plasma membrane. Permeable layer inside the cell wall that controls what substances enter or leave the *cell*. (8)

plasticulture. An impermeable, plastic, inorganic mulch material that is placed over a planting area after a *drip irrigation* system is installed; commonly used in vegetable production. (11)

plastid. An *organelle* that contains food or pigment. (8)

plug. A seed that is grown in a small container to transplantable size. (13)

plugging device. Equipment that pulls pieces of leaf, stem, and root material from the turf. (28)

plumule. The growing point for a developing shoot on a seedling. (13)

polarity. Spatial orientation within plants. (14)

polycarbonate. A type of thermoplastic polymer that can be used as a greenhouse covering. (10)

polyethylene. A type of plastic that can be used as a greenhouse covering. (10)

polyphagus. A type of insect that feeds on a wide range of plants. (30)

pome. A fruit with a tough endocarp that encloses the seeds. (8)

pomologist. A professional involved in the cultivation, processing, storing, and marketing of fruits and nuts. (4)

pomology. The cultivation, processing, storing, and marketing of fruits and nuts. (4)

porosity. The state of having space or gaps. (11)

post-emergent herbicide. A chemical that is applied to actively growing weeds to kill the plants. (32)

postharvest. The cooling, cleaning, sorting, storing, packing, and shipping of produce, flowers, and other plant materials. (4)

pot bound. A condition in which the roots of a plant fill the interior of the pot. (21)

pot-in-pot (PNP). A method in which a plant is grown in a pot and that pot is placed in another pot that has been sunk into the ground. The system makes watering and moving the plant easier than growing the plant in the ground. (4)

pre-emergent herbicide. A chemical that is applied (usually to the soil) prior to weed germination to kill any weeds that start to grow. (32)

preformed root. Roots with *cells* that have existing root initials. (14)

preliminary design. A landscape map that breaks down the concept plan and includes more thorough details. (24)

priming. The process of hydrating and then drying out a seed for greater germination rate and uniformity. (13)

primocane-fruiting. *Brambles* that produce fruit at the top of first-year canes in late summer. (23)

priority. A determination or ranking of the importance or urgency of a task or activity. (1)

proboscis. A modified mouth part that allows insects to sip liquids. (30)

procrastination. Delaying or postponing a task or activity. (1)

professionalism. The exercise of judgment, skill, and polite behavior that is exhibited by someone who is trained to do a job well. (5)

profit margin. The amount by which income exceeds costs of doing business. (5)

propagate. To grow plants from seeds or other methods, such as rootings or cuttings. (4)

proportion. The relative sizes and amounts of elements in a design. (25)

protoplast. The protoplasm of a living *cell* from which the cell wall has been removed. (17)

proximal. Located closest to the point of attachment. (14)

pseudobulb. A storage organ that develops in some orchids and can be cut and potted to grow new plants. (15)

pyrolysis. Using heat to decompose organic material in the absence of oxygen. (11)

pyruvate. A compound that supplies energy to plant *cells* as part of the Kreb's cycle in *respiration*. (9)

Q

Q10. A term used to describe the doubling of reactions as a biological response to increases of temperature in 18°F (10°C) increments. (10)

quiescent. A state of inactivity. (13)

R

radicle. The seed root of a seed. (13)

radicle protrusion. The last period of early seed germination characterized by the emergence of the seed root. (13)

rain garden. A landscape that collects runoff water and pollutants, filters out pollutants, and slowly releases the water. (24)

ratio. The relative value of one thing compared to another. In reference to fertilizer, a ratio is the comparative proportion of elements in the mix. (12)

recessive. The relationship where one *allele* is only expressed when the second, dominant *allele* is not present. (9)

recultured. When the *explant* is transferred to a different test tube or petri dish containing nutrients and growth regulators necessary for a certain stage of plant growth. (17)

reel mower. A piece of equipment that uses scissor-like action to cut grass blades. (28)

relative humidity (RH). The amount of water vapor in the air compared to the amount of water vapor that air can hold at a given temperature. (9)

relative scale. The size of an element of the landscape in relation to another object. (24)

relief. An *environment* where the rocks and soil sit on the landscape. (11)

renewal pruning. The process of removing one-third of a plant's old mature stems per season from the ground level in order to stimulate new growth. (27)

repetition. The recurrence of an event or thing, such as repeated elements within the landscape. (24)

research and experimentation SAE. A hands-on learning project in which the student conducts research or uses the scientific method to solve a problem related to agriculture. (2)

reseller sales. A distribution method in which other parties buy a company's products and then sell them to customers. (5)

resistant host. A host that is unable to be infected by a disease-causing organism. (31)

respiration. The process in which glucose (the product of *photosynthesis*) combines with oxygen to produce energy in a form that can be used by plants. (9)

respire. The process in which a plant gives off carbon dioxide and takes in oxygen. (11)

restricted entry interval (REI). The time that must elapse before someone can enter an area after it has been treated with a *pesticide*. (33)

résumé. A document that contains a concise summary of a person's education, skills, work experience, and other qualifications for a job. (5)

rhizobia. *Bacteria* that converts elements such as nitrogen into a usable form that can be taken up by a plant. (11)

rhizomatous. A plant that produces *rhizomes*, or underground lateral stems. (28)

rhizome. A modified stem structure that grows horizontally below or near the soil's surface, producing roots on the bottom and shoots on the top. (15)

rhythm. A pattern of repeating or alternating elements in a design that suggests a sense of movement. (25)

ribonucleic acid (RNA). A substance that manufactures proteins based on the genetic information provided by *DNA* and plays a significant role in gene expression. (8)

ribosome. Consists of ribosomal RNA (rRNA) and protein and is actively involved in the synthesis of proteins. (8)

ridge. The highest point of a greenhouse roof. (18)

rodenticide. A chemical substance used to control rodents. (33)

root pruning. Severing the roots of a plant to promote growth of some kind, such as flowering or fruiting. (21)

root zone. The soil surrounding the plant's roots that serves as a natural reservoir from which the plant draws moisture and nutrients. (10)

rosette. A growth habit of many *biennials* in their first year in which leaves attach in a circle around the base of a stem. (32)

rotary mower. A piece of equipment that uses a blade that cuts grass by sucking and tearing the grass blades. (28)

row cover. A lightweight fabric that is laid on top of vegetables to provide protection from frost or insects. (22)

runner. A specialized stem that grows horizontally above the ground and forms a new plant. (15)

running bond. A pattern in which bricks are laid end to end and staggered between rows. (27)

S

saddle grafting. A plant joining method in which the *scion* is cut in the shape of a saddle and sits on top of the rootstock. (16)

safety data sheet (SDS). A document that contains information on the potential health impacts of a chemical or other dangerous substance. Also referred to as a *material safety data sheet (MSDS)*. (6)

safety hazard. Anything on a job site that can cause injury, illness, or death. (6)

samara. A dry *indehiscent* fruit in which part of the fruit wall is extended to form a wing. (8)

sandy soil. Soil that is dominated by a significant proportion of sand and has large-sized particles (2.0 mm to 0.05 mm) and pore spaces. (11)

sanitation. The process of keeping places or items free from dirt, *disease*, infection, and other substances that can cause illness. (22)

saturation. A situation in which water fills pore spaces in soil or *soilless media* to the point that there is no room for air in the pores. (11)

scaffold branches. Primary limbs that extend from the main trunk and form the tree canopy. (23)

scale. The size relationship between two objects, such as between a floral design and its surroundings, or among the various elements within a design. (25)

scarification. A process of physically removing part of the seed coat to allow *imbibition*. (13)

schizocarp. A dry *indehiscent* fruit that splits at maturity into two or more seeded parts. It is found in the carrot family. (8)

school-to-career plan. A document that lists one or more career goals and the steps or activities and resources needed to achieve those goals. (5)

scientific name. A two-word name that includes a *genus* and specific epithet for a plant *species*. (7)

scion. A young shoot or twig of a plant, often used for the upper portion of a graft. (16)

sclerenchyma. Plant *cells* that have thick, lignified walls that provide physical strength and support to stems, especially in woody plants. (8)

sclerite. A small, rigid plate that is part of the exoskeleton of an insect. (30)

scouting. The regular checking of a field, garden, orchard, lawn, greenhouse, or other area that identifies potential pest problems early while they can still be managed and thereby reduce crop loss and control costs. (29)

sedge. A type of weed that is characterized by parallel venation and a triangular, solid stem. (32)

seedbed. A specially prepared space for seed germination. (13)

seed blend. A mixture of various *cultivars* of grass seed. (28)

seedhead. The cluster of seeds within a plant's *inflorescence*. (28)

seedlot. Seeds of a particular crop gathered at one time, with similar germination rates and other characteristics. (13)

seed spreader. A tool that places seeds in an area. (28)

selective herbicide. A chemical that can kill some types of plants, but not others. (32)

selective market coverage. A distribution strategy in which a company sells its products using a limited number of locations. (5)

self-compatible weed. A weed that does not need to cross-pollinate in order to set seed; a self-pollinating weed. (32)

senescence. The ripening or biological aging of harvested crops. (4)

sepal. A modified leaf that surrounds and protects a flower bud. (8)

separation. A propagation method of *bulbs* or *corms* that multiply and can be propagated by gently pulling apart the new *bulbs* or *corms*. (15)

serpentine layering. A variation of compound layering in which shoots are laid horizontally to the ground and some buds are exposed while others are covered with soil in an alternating pattern. (15)

service learning. A strategy that integrates community service with instruction and reflection to enrich the learning experience, teach civic responsibility, and strengthen communities. (2)

setae. The hairs that cover some insects; used to send signals to the insect to react in some way. (30)

shutter. An area of the greenhouse that can be opened or closed to allow or prevent air exchange between the greenhouse and the outside *environment*. (18)

side-dressing. A fertilizer application method in which fertilizer is applied between rows of young plants. (12)

side-veneer grafting. A plant joining method in which a flap of bark is removed from the side of a plant stem and a *scion* is attached to the cut area. Once the graft has healed, the top portion of the rootstock is cut off. (16)

sign. An indication of a *disease* given by the observable presence of a pathogen or its parts. (31)

signal word. Language, such as *caution*, *warning*, and *danger* on a *pesticide* label used to call attention to potential threats to human health. (33)

silique. A dry *dehiscent* fruit with a seedpod that splits along two sides while the seed inside remains attached to part of the fruit structure. (8)

silviculture. The science or practice of planting, managing, repairing, and caring for forests; the cultivation of trees. Also referred to as *forestry*. (4)

simplicity. The quality of being plain, natural, or uncomplicated. (24)

skim. To read selected parts of a text looking for the main ideas. (3)

slit seeder. A device that creates a slit in the *thatch* and root zone while also planting seed. (28)

slope orientation. The direction a slope of land faces (north, south, east, west). (10)

slow-release fertilizer. Small, pellet-sized material of various compositions that is added to soil to increase fertility. (11)

small business. A company that is independently owned and operated, is organized for profit, and is not dominant in its field. (5)

SMART goals. Objectives that are specific, measurable, attainable, realistic, and timely. (1)

socket pot. A container used in *pot-in-pot* production that holds another container with plant material and is located in the ground. (21)

sod. A collection of grass plants that form a ground cover, often used for sports areas, homes, and industrial sites. (4)

softscape. The plant materials in a landscape. (27)

softwood. Soft, succulent, new growth on woody plants. (14)

soil auger. A tool used to pull soil samples from as deep as ten feet. (11)

soilless media. A sterile mix of natural ingredients used to raise plants in greenhouses, cutting beds, and containers; lighter than soil with more pore spaces; most commonly referred to as potting soil. (11)

soil pH. The acidity or alkalinity of a soil. (11)

soil pore space. A gap or open area between solid soil components that may be filled with air or water. (11)

soil probe. A tool used to pull a soil core from the earth. (11)

soil structure. The binding together of sand, silt, and clay particles into *aggregates* called *peds*. (11)

soil survey. A comprehensive study of the soil of an area. (11)

soil texture. The proportion of different sizes of mineral particles present in a soil. (11)

solarization. A weed control technique that uses the radiant heat from the sun to kill weed populations. (32)

soluble. Able to be dissolved in water. (12)

somaclonal variation. Changes seen in plants that have been produced by tissue culture or those that have been subcultured. (17)

sorption. A term used to describe both adsorption of *ions* to soil particles and absorption of nutrient *ions* into a plant. A *sorption zone* describes both of these processes that occur simultaneously. (12)

species. The lowest and least inclusive ranking of plant classification. (7)

specifications. Guidelines that include details for a project, such as plant selection, site preparation, and soil preparation. (27)

specific epithet. The second half of a *scientific name* for a plant *species*. It is usually descriptive of a plant feature or in honor of someone's name or a place. (7)

spermatheca. A special reproductive structure found in female insects that stores sperm. (30)

splice grafting. A method of joining plants in which a simple diagonal cut of the same length and angle is made on the rootstock and *scion*. They are placed on top of each other with the cambial layers aligned. (16)

sport. A type of tissue *mutation* that occurs in somatic (*vegetative*) *cells*. (9)

sports turf. Grass that is used for recreation and competitive sport purposes. (28)

spot seeding. Reseeding bare spots larger than 6″ (15.2 cm). (28)

spur. A cane that has been pruned to have only a few nodes. (23)

square foot gardening. A system of gardening in which every plant is allotted a square foot of space in a garden. (20)

stack bond. A simple design in which bricks are placed side by side without staggering. (27)

stamen. The male portion of the flower that contains the *anther* and the *filament*. (8)

standard. A benchmark or level for what is acceptable. (5)

Star Award. Prizes and recognitions given to those students who have exemplary SAE projects and are earning an FFA degree. (2)

static pressure. The force exerted by a still liquid or gas. (18)
stem cutting. A portion of the shoot. It may include the tip or just a section of stem. (14)
stem stripper. A tool used to remove prickles (thorn-like epidermal tissue) from stems of plants such as roses. (25)
stock plant. Plant material kept specifically for the purpose of propagation. (14)
stolon. Horizontal stems that grow aboveground and produce plants or *tubers*. (15)
stoloniferous. A plant that produces *stolons*, or aboveground lateral stems. (28)
stomata. Pores in epidermal plant *cells* that open to allow for gas exchange and control water loss. (8)
strain. Long-term impact of a pressure or tension. (6)
strategic business plan. A document that states the mission of the business, examines its current condition, sets goals, and outlines strategies for achieving the goals. (5)
stratification. A moist, chilling treatment used to break *dormancy* in seeds of certain plant *species*. (10)
stress. Short-term impact of a pressure or tension. (6)
stretcher. A brick that is laid in a design to display its long side. (27)
stroma. The aqueous space outside the stacks of *thylakoids*. (8, 9)
student interest survey. A questionnaire that helps a student identify his or her interests in agricultural education and what SAE project would be best suited for that student. (2)
student resources inventory. A questionnaire that helps students identify tools, supplies, and other resources they have access to for an SAE project. (2)
stylet. A modified mouthpart that allows insects to pierce the epidermal layer of a plant or animal and withdraw liquids. (30)
subculturing. Establishing a new culture by moving some *cells* from an existing culture. (17)

suberin. A waxy substance that seals wounds on plants. (14)
suberize. To form a waxy substance (called *suberin*) that protects plants from *desiccation* and pathogens. (15)
substrate. A growing medium or the material in which plant roots grow. (21)
succulent. A type of plant that has thick, fleshy leaves and is usually native to arid regions of the world. (26)
sucker. A shoot that grows from an *adventitious* bud on a root or any shoot that grows near the base of a plant. (15)
suction lysometer. A device used to pull a solution out of a given material. (21)
sump pump. A motor that pulls water from one location to another. (18)
sump tank. A holding container or reservoir for water. (18)
sunscald. A condition in which plants are damaged due to heat, cold, humidity, or intense sunlight. (10)
superphosphate. Rock phosphate that has been treated with phosphoric acid to increase the amount of plant available phosphorus. (12)
supervised agricultural experience (SAE). A project that involves hands-on learning in agriculture and natural resources and that is developed by the student. (1, 2)
supplemental SAE. A hands-on learning project that enhances agricultural skills and knowledge and takes less than eight hours. (2)
surface horizon. The fertile, upper, outermost layer of soil. Also referred to as the *A horizon* or *topsoil*. (11)
susceptible host. A host that is prone to *disease*. (31)
sustainable agriculture. Farming or producing plant and animal products in ways that promote the health of people, animals, and the *environment*. (4)
sustainable nursery. A *nursery* that incorporates growing practices that will maintain or better the *environment* for today and the future. (21)

symmetrical design. A design in which a nearly identical or mirrored arrangement of plant materials is used on both sides of the central vertical axis (CVA). (25)
symptom. A plant's reaction, such as wilt or blight, that is an indication of a *disease*. (31)
systemic herbicide. A chemical that can translocate, or move, within the plant from the point of entry to kill the plant. (32)
systemic insecticide. A *pesticide* that translocates throughout the plant, killing any insects that feed on the plant. (26)
systemic pesticide. A chemical that is translocated through a plant's vascular system; targets insects with piercing and sucking mouthparts. (33)

T

tact. The ability to deal with others in a sensitive manner. (1)
tannin. A naturally occurring chemical compound found in wine grapes (and other plants) that imparts a bitterness, astringency, and complexity to wine. (23)
taxonomy. The science of naming and classifying organisms. (7)
T-design. A floral arrangement created by combining a horizontal and vertical design that resembles an inverted T. (25)
tegmina. Parchment-like forewings found in grasshoppers and cockroaches. (30)
terrarium. A closed type of planting structure that is self-contained and provides a good *environment* for tender, hard-to-grow plants. (26)
texture. The quality of how a surface feels or appears, such as coarse, rough, smooth, or fine. (24)
thatch. A layer of dead grass and *organic matter* that forms a mat beneath leaf blades and above the root zone. (28)
thermodormancy. A secondary *dormancy* that inhibits germination. (13)

thermoperiod. The relationship between day and night temperatures and plant growth. (10)

thermostat. A device that measures temperature. (18)

thesis. A statement or theory that is proposed and then discussed to prove or disprove it. (3)

thinning cut. A cut used to prevent crowding in fruit trees by removing an entire branch back to its base. (23)

thylakoid. A disc-shaped sac surrounded by membranes on which the light reactions of *photosynthesis* take place. (8, 9)

tiller. A stem that develops from the *crown* of the parent plant and grows upward within the enclosing leaf sheath. (28)

tillering. The process of new stems or *tillers* being created by a mother plant. (32)

tilth. A general term to describe good soil quality for crop growth, including texture, structure, and pore space. (12)

tissue culture. A collection of techniques used to grow or maintain plants in a nutrient medium under sterile conditions. (17)

title block. A component of a landscape plan that gives information about the client, landscape firm, date of preparation, and possibly additional elements, such as the drawing scale. (27)

tone. The quality of a message that reflects the writer's attitude or mood. (3)

tonoplast. A permeable membrane that surrounds the *vacuole* and regulates the entrance and exit of cell sap, waste products, pigments, and other liquids. (8)

topiary. An ornamental form in which shrubs are sheared into interesting shapes and imaginative characters, such as animals. (27)

topography. The elevations and slope of a specific land area. (18)

topsoil. The fertile, upper, outermost layer of soil. Also referred to as the *surface horizon* or *A horizon*. (11)

topworking. The process of *grafting* new *scion* material onto a mature tree. (16)

totipotency. The ability of a *cell* to *differentiate* into any type of other cell. (17)

toxicity. A level or degree to which something is poisonous. (31)

training agreement. A signed contract that helps the student, teacher, parents or guardians, and the employers understand the objectives and goals of the SAE placement. Also called a *training plan*. (2)

training plan. A signed contract that helps the student, teacher, parents or guardians, and the employers understand the objectives and goals of the SAE placement. Also called a *training agreement*. (2)

transgenic. An organism into which genetic material from an unrelated organism has been introduced. (13)

transition zone. A growing area that can grow both warm- and cool-season grasses. (28)

translocation. The movement of sugars within the plant. (9)

transpiration. The process in which water is pulled upward from the roots through *xylem cells* and released into the air in a gaseous state. (9)

tree plot. A carefully measured area of trees. (4)

tuber. A swollen, underground stem that is a storage organ for the plant. (15)

tubercle. A small, aerial *tuber* produced in the axil of a leaf. (15)

tuberize. The process of developing new *tubers*, which are swollen, underground stems that are storage organs for the plant. (15)

tunicate. A type of *bulb* that has a dry, papery covering or tunic. (15)

turf. Grass and the surface layer of soil held together by the grass roots. (28)

turfgrass. A collection of grass plants that form a ground cover, often used for sports areas, homes, and industrial sites. (4)

turfgrass breeding. Selecting plants with desirable traits and combining them to develop improved *cultivars*. (28)

turfgrass industry. Businesses that grow seeds, *sods*, and grasses for recreational areas, home, and commercial, utility, and sports uses. (28)

turgor. The pressure within a plant *cell* that helps provide rigidity and support to plant structures. (8, 9)

U

unity. The relationship among elements in a work of art that makes them appear to belong together or function as a whole. (25)

utility turf. Grass that is used for spaces such as roads, airports, and parks. (28)

V

vacuole. A large cavity found within plant *cells* that stores cell sap, waste products, pigments, or other liquids. (8)

value. A principle or standard. (5)

variety. A form or subclassification of a *species* that is slightly different, but not different enough to warrant a new species. (7)

vector. An organism that transmits a *disease* or parasite from one plant to another. (29)

vegetative. The stems, leaves, and roots of a plant harvested for flavorings, foods, perfumes, or medicines; nonreproductive parts of the plant (not flowers, fruits, or seeds). (4)

vegetative propagation. Horticultural method of starting new plants from existing plant material. (14)

vendor. A company that sells services, goods, or supplies. (5)

vent. A part of the greenhouse that can be opened to release built-up heat to the outside *environment*, reduce condensation, and improve airflow. (18)

vermicompost. A type of *compost* in which worms as well as microbes and *bacteria* are used to turn *organic matter* into fertilizer. (2)

vermiculite. A lightweight material made of small pieces of mica that readily absorbs water. (11)

vernalization. Exposure of plants to low temperatures in order to stimulate flowering. (10)
vernation. The arrangement of leaves within the bud. (28)
vertical gardening. A method of growing plants on a vertical surface, such as a wall or trellis. (20)
verticutting. A method of *thatch* control done by a machine that slices into *thatch* layers to create space for water penetration and air exchange. (28)
viable. Alive; a living seed. (13)
viniculture. The cultivation of grapes to be eaten fresh and to be used for making juices, raisins, jams, jellies, and wines. Also referred to as *viticulture*. (4)
virus. A microscopic organism that infects and lives off other organisms. (31)
vision. A description of an organization's goals for the long term. (5)
viticulture. The cultivation of grapes to be eaten fresh and to be used for making juices, raisins, jams, jellies, and wines. Also referred to as *viniculture*. (4)
viticulturist. A professional who specializes in the cultivation of grapes to be eaten fresh and to be used for making juices, raisins, jams, jellies, and wines. (4)

vivipary. A phenomenon in which seeds germinate inside their fruit without maturation drying. Term used to describe organisms that give birth to live offspring that have developed in the mother's body. (13, 30)
voice. The style of expression or degree of formality used in a message. (3)
volatilization. A process in which fertilizer components change from a solid to a vapor and are released into the atmosphere. (28)

W

water culture system. A method of *hydroponics* in which plants in trays float with their roots submerged in a nutrient solution. Also referred to as floating hydroponics. (20)
water sprout. Plant stems that grow vertically from other stems or branches. (21)
weathering. The physical, chemical, and biological process that creates soil. (11)
wedge grafting. A plant joining method in which the *scion* is cut into a V-shaped wedge and the rootstock is cut to form a V-shape to hold the *scion*. (16)

whip-and-tongue grafting. A method of joining plants in which the rootstock is cut with two notches and the *scion* material is cut to match and create a tight fit. (16)
whorl. A spiral pattern of limbs. (23)
winterization. Protecting plants from cold and harsh winter conditions by covering with material, packing tightly, or placing in a sheltered place. (21)
works cited. A summary list of all research and sources that are used to create a text or presentation. (3)
wound-induced root. Root that develops after a cutting is made in direct response to a wounding from the severing of the stem piece or leaf from the parent plant. (14)

X

xylem. Plant tissue that conducts water and nutrients from the roots throughout the plant and can provide structural support to the plant. (8)

Z

zygote. In sexual reproduction, a *diploid cell* formed by the fusion of male and female *haploid* nuclei. (9)

Index

4-H, 9–10
abiotic, 838
aboveground system for PNP, 549
abscised, 370
abscisic acid, 345
absolute scale (design), 642
accident prevention, 164–166
acclimatization, 438
achene, 216
actin filaments, 203
action thresholds (IPM), 791
active ingredients (pesticide), 882, 885
acute toxicity, 890
adenosine triphosphate (ATP), 201, 228
   production, 230–231
adhesion, 287
adsorbed, 290
adventitious, 206
   roots, 207, 366
   shoot and bud, 367–368
advertising, 129–134
aeration, 283, 767
aeroponics, 510, 512
aerosol sprays, 882
AET (Agricultural Experience Tracker), 24
aggregate
   fruits, 214
   forms, 514
   hydroponics, 511
   peds, 285
aggregation pheromones, 815
agricultural beneficials, 816–820
agricultural business manager, 53
agricultural communications, 423
agricultural education teacher, 24–25
Agricultural Experience Tracker (AET), 24
agricultural inspector, 53–54
agricultural lawyer, 895
agricultural leadership, 2–25
   organizations for youth, 9–11
agricultural literacy, 40
agricultural pests, 816–820, 878
Agricultural Proficiency Award, 50–51
agricultural promotion, 19
agricultural research
   entomologist, 823

Agricultural Worker Protection Standard, 886
agriscience fair categories, 40
agriscience internship, 38
agronomist, 93
agronomy, 92, 93
agVets, 516
A horizon, 282
AHS (American Horticulture Society), 112
air drainage, 264
air layering, 389–390
air roots, 207
alarm pheromones, 815
algae, 657
   agar, 431, 434
   water gardens, 657
algaecides, 881
alleles, 239
allomones, 813, 815–816
AmericanHort, 54, 113
American Horticulture Society (AHS), 112
American Society of Horticulture Science (ASHS), 112
American Society of Landscape Architects (ASLA), 632
ametablous metamorphosis, 812
ammonification, 322
analogous (color), 641
angiosperms, 183, 184, 212
   monocots and dicots, 217
Animal and Plant Health Inspection Service (APHIS), 845
animal hazards, 162
animal wastes, 326–327
   compost safety, 299
anions, 290
annotate, 72
annotated bibliography, 74
antennae, 805
anther, 212
anthracnose, 846
anti-transpirant, 793
APHIS (Animal and Plant Health Inspection Service), 845
apical meristem, 203, 436
apiculturists, 38
apomixes, 341
approach grafting, 418–419

aquaculture, 509
aquaponics, 514–518
   history, 515–516
   system manager, 530
aquascaping, 652
aquatic creatures, 517
aquatic plants, 654–657
   grower, 658–659
arboretum director, 379
arm (tree branch), 620
aseptic, 432
asexual propagation, 366
   cellular division, 236
   layering, 386
   micropropagation, 430
ASHS (American Society of Horticulture Science), 112
aspirated thermostat, 466
assimilation (plant nutrition), 323
assistant vineyard manager, 423
asymmetrical designs, 667
athletic fields (turfgrass), 103
ATP (adenosine triphosphate), 201, 228
   production, 230–231
attractants (IPM), 793
audience (communication), 64–65
auricle, 753
automated irrigation sensors, 488
autotrophs, 198
auxin, 377–378
bacillus thuringiensis (Bt), 107
   integrated pest management, 787
   microbial control, 795
   plant incorporated protectant, 881
   transgenic cultivars, 355
   use with mosquitoes, 657
bacteria, 783, 834, 835
   aquaponics, 515, 518
   as microbrial control measure, 795
   bacillus thuringiensis (Bt), 107
   biological hazard, 161
   establishment, 840
   growth, 840
   microorganisms in soil, 288–289
   nitrogen cycle, 322–324
   reproduction, 840
   rhizobia, 289
   transmission, 818
bacterial blight, 847
bacterial diseases, 709–710

bacterial secretions, 841
bait (insect collection), 821
balance (design)
  floral design, 666
  interiorscapes, 697
  landscape design, 642
balled and burlapped (B&B), 728
banding, 330
bare root
  landscape materials, 728
  planting method, 100
  plants, 495
bark (periderm), 205, 296
  uses, 204
bark grafting, 416
bark slipping, 413
base plan, 634
Beagle Brigade, 845
bedding plants, 494, 499
beekeeper, 38
bench cut, 612
benches (greenhouse), 463–464
bench grafting, 409
beneficials, 793
  allomones, 815–816
  insects, 818–820
  organisms, 844
berry, 214
B horizon, 282
bibliography, 76
biennial, 262
billing, 636
binomial nomenclature, 181
biochar, 301
biochemical pesticides, 881
biodiesel, 35, 36
biofilm, 518
biological controls, 793–795, 869
biological hazards, 159, 161–163
biomass, burning for fuel, 462
biopesticides, 880–881
biopharming, 514
bioplastics, 303
bio-stimulant, 489
biotechnology, 93
biotic, 838
blackberries, 605
blade, 749
blanching, 252
blights, 846–847
blueberries, 606
bodies of water, 265
  gardens, 653
bond (paper), 648
bonsai, 696
boron (B), 318
botanical classification system, 181–188

botanist, 92, 218
botany, 92–93
  disciplines, 199
  study, 198–199
bouquets, 682
  increasing shelf life, 684
boutonniere, 668, 682
bows, 678
  tying, 679
brambles, 96, 605
brand, 124
brick patterns, 723
bridge grafting, 417
broadcasting, 330
broadleaf weed, 865–866
brown waste, 299
browning, 842. See necrosis.
bud arrangements, 208
budding, 404–423
  aftercare, 423
  grafting, 420–423
buds, 208
budwood, 420
bulblets, 394
bulbs, 394–395, 496
bulk density, 286
bunch-type weeds, 866
burrowing organisms, 290
business structures, 124–125
butterfly scales, 809
C4 plants, 229
CAD (computer-aided design), 637
calcium (Ca), 316
callus, 436
  tissue, 366
CAM (crassulacean acid metabolism), 229
cane, 620
cankers, 847–848
capillary mat, 466, 555
capillary water, 287
capsule, 216
carbon (C), 313
carbon footprint (of food), 105
career development
  agricultural communications, 44
  creed speaking, 23
  events, 22–23
  floriculture, 678
  marketing plan event, 146
career documents, 136–139
career exploration, 41, 146
careers, 894–895
  agricultural education teacher, 24–25
  agricultural leadership, 24–25
  agricultural management, 52–54
  aquaponic systems, 530–531

  disease management, 852–853
  environmental horticulture, 271–272
  floral design, 684–685
  grafting and budding, 423
  greenhouse production, 500–501
  greenhouse structure, 471–472
  horticultural communication, 82–84
  horticultural safety, 172–173
  horticulture, 112–113
  horticulture business, 146–148
  insects, related to, 823
  integrated pest management, 796–797
  interior plantscaping, 710–711
  landscape design, 658–659
  landscape installation and maintenance, 738
  layering and division, 398
  micropropagation, 440–442
  nursery production, 561–563
  olericulture, 593
  plant biology, 218
  plant growth and development, 241–242
  plant nutrition, 332
  plant taxonomy, 190–191
  pomology, 624
  seed propagation, 358
  soil science, 302–304
  stem and leaf propagation, 378–379
  turfgrass management, 772–773
  weed management, 870–871
carnivorous plants, 314
carotenoid, 227
Carver, George Washington, 201
caryopsis, 216
casparian strip, 234
cation, 290
cation exchange capacity (CEC), 290
CDC (Centers for Disease Control and Prevention), 157, 158, 167, 173
cedar-apple rust, 851
cell, 199
  expansion, 340
  membrane, 199
  nucleus, 200
cellular division, 236
cellulose, 199
Centers for Disease Control and Prevention (CDC), 157, 158, 167, 173
central leader system, 613
certified organic growers, 586
chapter banquet, 20–21
chapter recruitment, 19

character development, 18–19
Chavez, Cesar, 157
chelate, 319
chemical applicators, 871
chemical control, 796, 869
chemical dormancy, 345
chemical hazards, 159–161
chemical scarification, 345
chemical signals, 813–816
chilling injury, 261
chimeras, 241
chip budding, 421
chlorine (Cl), 318, 702
chlorophyll, 200, 227
chloroplast, 200
chlorosis, 292, 842
C horizon, 282
chroma (intensity), 671
chromatin, 237
chromoplasts, 201
chromosomes, 237
chronic toxicity, 890
Cincinnati Zoo's CREW CyroBioBank®, 437
circling pot bound, 542
circular (design), 673–674
circulation fans, 458
citizenship, 19
class, 184–185
classification, 180
clayey soils, 283
clean room, 432
cleft grafting, 412–413
climate, 281
    fruit and nut production, 609–610
    greenhouse operation, 452
    microclimates, 264–265
    topography, 617
clones, 430
club root, 848
CO_2 levels, 229
coarse (texture), 640
coconut coir, 295, 514
    vs. peat moss, 297
cohesion, 231, 287
cold frame, 456
collar type, 753
collenchyma, 206
color, 641, 670–671
    complementary, 641
    harmonies, 671
    intensity, 671
    schemes, 671
    theory, 641
commercial pollination, 819
common name, 181
communication, 60–84
    with audiences, 79–80

community development, 19
community engagement, 42
community supported agriculture (CSA), 108
compass, 646
complete fertilizer, 328, 483–484
complete metamorphosis, 812
compost, 286, 326
    safety, 163
composting operator, 332
compound layering, 388
computer-aided design (CAD), 637
concept plan, 634
concrete, 463
cones, 623
confidence, 4
consumption rates, 269
contact herbicide, 870
contact pesticide, 878
container gardening, 519
container grown, 727–728
container plants, 496–497
container water gardens, 653–654
container-grown production, 541–544
containers, 542
    capacity, 489
    drainage capability, 301–302
    durability, 300–301
    engineering, 543
    floral design, 676–681
    greenhouses, 491–493
    recycling programs, 491
    size, 300
    soils, 300–302
    types, 542
contracting, 635
control measures, 785–787
cool cells, 460
cool temperature, 343
cooling pad systems, 470
cool-season grasses, 756–757
cool-season vegetables, 578–579
cooperative extension service agents, 25
copper (CU), 318–319
cordon, 620
core ideology, 127
cormels, 396
corms, 396
corsage, 677, 682
cortex, 206
cotyledons, 185
cover crop, 286, 299, 868
cranberries, 607
crassulacean acid metabolism (CAM), 229
creed speaking, 23
critical night interval (CNI), 481

critical thinking, 69–73
crops, 514, 539
    advisor, 593
    competition, 868
    consultants, 871
    inputs, 489–493
    rotation, 584–585, 868
    selection, 545, 549
    to be cultivated, 452
crown, 546
    division, 393
crown gall, 849
cryopreservation scientist, 441–442
CSA (community supported agriculture), 108
culling, 524
culls (fruit), 608
cultivar, 188
    selection, 580–581
cut flowers, 498–499
cutting (propagation), 366
cyclic photoperiodic lighting, 482
cytochromes, 319
cytokinesis, 236
cytokinins, 377–378
cytoplasm, 200
cytoskeleton, 203
cytospora canker, 848
daily light integral (DLI), 252
damping off, 848
day-neutral plants, 253, 481
deciduous, 369
DED (Dutch elm disease), 832
dedifferentiates, 366
deep water culture (DWC), 510, 512
defensive allomones, 815–816
deficiency, 313
defoliation, 842
degree days, 263
dehiscent, 216
delegate, 8
denitrification, 323
deoxyribonucleic acid (DNA), 200, 835
Department of Labor (DOL), 167
Department of Transportation (DOT), 173
desiccation, 341
design
    elements, 666–673, 695–697
    planning, 648
    planting, 727–730
    principles, 666–673, 695–699
    problem and objective, 633
    process, 632–637, 699
    sequence, 633–636
desorbed, 290
diapause, 813

dichotomous key, 188
dicots. *See* dicotyledoneae.
dicotyledoneae (dicots), 184, 217
DIF treatment, 263, 486
differentiate, 366
dioecious, 213
diploid, 237
direct sales, 132
disease, 770, 832, 846–852
    anthracnose, 846
    bacterial blight, 847
    cedar-apple rust 851
    club root, 848
    crown gall, 849
    cycle, 839–841
    cytospora canker, 848
    damping off, 848
    development, 833–835
    early blight, 846
    entrances, 840
    establishment, 840
    fire blight, 846
    fusarium wilt, 852
    galls, 849
    growth, 840
    identification, 926–927
    index, 846–852
    late blight, 847
    leaf blister, 849
    leaf curl, 849
    leaf spot, 849
    management, 830–846
    mildew, 849
    mold, 850
    nectria canker, 848
    reproduction, 840
    root-knot nematode, 850
    rot, 850
    rust, 850
    scab, 851
    Setward's wilt, 852
    signs and symptoms, 841–843
    smut, 851
    testing, 837
    tobacco mosaic virus (TMV), 851
    triangle, 833–835
    types, 837–839
    verticuillium wilt, 852
    virus, 851–852
dish garden, 698
dissemination, 840–841
distal, 375
DNA (deoxyribonucleic acid), 200, 835
documentation, 75–76
DOL (Department of Labor), 167
domain, 182–183
dominant (trait), 239
dormancy, 261

drafting triangles, 647
drawing board, 636, 645
    for computer-aided design, 636–637
drawing instruments, 644–647
drawing scale, 719–729
drip irrigation, 270–271
    greenhouse, 465
    landscape design, 650
    nursery production, 554–555
    thinking green, 127
    vegetable production, 575
drip system, 510
dripline, 574
    hydroponic drip system, 511
    root zone, 269
drupe, 214
dry creek bed installation, 655
dry fruits, 216
Dutch elm disease (DED), 832
dwarfing, 843
EAB (emerald ash borer), 782
early American labor, 156–157
early blight, 846
eave (greenhouse), 455
ebb and flood system, 466, 510
E. coli, 785
ecologist, 191
economic development, 19
economic impacts, 111
economic injury level (EIL), 791
edging (landscape), 734–735
edible horticulture, 93–105
education, continuing, 8, 112
education entomologist, 823
effluent, 514
EIL (economy injury level), 791
electrical conductivity (EC), 558
    meter, 290
elytra, 809
e-mails, 143
embryogenesis, 811
emerald ash borer (EAB), 782
emitter, 465
empathy, 5
emphasis (design), 643, 699
emulsifiable concentrates, 883
endogenous contamination, 434
endoplasmic reticulum (ER), 202
enologists, 96
entomologist, 833
entomology, 804
entrepreneur, 123
entrepreneurship SAE, 34–35
    awards, 50
environmental hazards, 885–886
Environmental Protection Agency (EPA), 173, 797, 880, 884
envisioned future, 127

EPA (Environmental Protection Agency), 173, 797, 880, 884
EPA registration number, 884
epidermis, 204
equipment, 348
    checking and maintaining, 168–170, 469–471
    identification, 169, 552, 913–923
equipment salesperson, 562
ergonomic hazards, 159, 163
erosion control, 749
    specialist, 304
espalier, 615
ethylene, 593
etiolation, 252
evaporative cooling, 460–461
evapotranspiration, 110, 232
even span greenhouse, 455
evergreen, 500
exhaust fans, 458
exoskeleton, 804
experiential learning, 30–54. *See also* SAE.
explant, 433
exploratory SAE, 40–41
    awards, 50
extensive green roof system, 519
Fair Labor Standards Act (FLSA), 157
family, 186
fans, 458, 470
farm advisor, 853
farm coordinator, 271
farmers markets, 107–108
fertigation, 315, 331
fertilization
    plant reproduction, 237–239
    turfgrass, 758, 764–765, 772
fertilizers, 577
    application methods, 330–332, 732
    calculations, 328–329
    formulations and programs, 765
    grade, 328
    injectors, 468
    organic and inorganic, 485
    post-plant, 557
    pre-plant, 557
    rates, 731–732
    sales representative, 332
    timing, 732
FFA. *See* National FFA Organization.
fibrous roots, 207
field capacity, 287
field nurseries, 348–349
    seedbeds, 349
field seeding, 346–348
field-grown method, 541
field-grown production, 544–547
filament, 212

filler flower, 672
fire blight, 846
first aid, 892
   instructions, 886
flagging, 488
flaming (weed control), 562, 868
Flemish bond brick pattern, 723
fleshy fruits, 214
FLO (fungal-like organism), 835
floater plants, 655
flood floor, 465
floral design, 664–685
   texture, 672–673
   types, 673–675
floral designer, 685
floral foam, 680
floral knives, 676–677
floral merchandiser, 685
floricane-fruiting, 608
floriculture, 97–98
flower arrangements, 682–683
flowers, 212–213
FLSA (Fair Labor Standards Act), 157
focal point, 668–669
fogger, 467
foliage plants, 497–498
foliar application, 332
follicle, 216
food products and processing systems (FPP), 40
food safety manager, 624
forced hot air heating, 462
forest stand, 93
foresters, 93
forestry, 92–93
form, 639–640, 672
   geometric, 639
   naturalistic, 639
   plant forms, 639–640
form flower, 672
found objects, 525
4-H, 9–10
fragrance, 673
freezing injury, 261
fruit, 214–216
   harvest and storage, 616, 623–624
   health benefits, 600–601
   maintenance, 607–608
   markets, US, 601–602
   nutrients, 600–601
   production, 598–624
   thinning, 614
fruit and nut growers, 96
fruit nursery propagator, 398
fruiting bodies, 841
functional diagram, 634
fungal diseases, 709
fungal-like organism (FLO), 835

fungi, 783, 835
   treating with baking soda, 837
fungicides, 845, 879–880
fungus, 832
   beneficial, 317
fusarium wilt, 852
Future Farmers of America. See National FFA Organization.
gable, 455
gall-producing insects, 818
galls, 842–843, 849
garbage can garden, 391
garden blogger, 84
garden club, 525, 588
garden writer and speaker, 83–84
gardens
   lighting, 727
   maintenance, 651–652
   native soil, 294–295
   water features, 726
GDP (gross domestic product), 126
general safety hazards, 159, 163
genes, 237
genetic diversity, 433
genetically modified organism (GMO), 355
genetically resistant plants, 844
genome, 237
genotype, 240
genus, 186–187
geocarpy, 95
geometric designs, 673–674
geophytes, 393
   division and separation, 393–398
geotextiles, 297
germination
   environmental conditions, 342–344
   rates, 342
   temperature, 259
germplasm, 354
girdling, 386
glass, 457
glazing, 457
gluing, 823
glycolysis, 230
GMO (genetically modified organism), 355
goals, 6, 48, 123
golf courses, 103, 749
golgi bodies, 202–203
good agricultural practice (GAP), 586–587
gothic arch, 455
grafting, 404–423
   aftercare, 423
   approach grafting, 418–419
   bark grafting, 416
   benefits, 406–408

   bridge grafting, 417
   budding, 404–423
   cleft grafting, 412–413
   hole insertion grafting (HIG), 415
   inarch grafting, 418
   saddle grafting, 414
   side-veneer grafting, 419
   splice grafting, 410
   timing, 409
   wedge grafting, 413
   whip grafting, 410
   whip-n-tongue grafting, 410
grana, 227
granular pellets, 883
grapes, 618–620, 622–623
   cultivation, 97
   terminology, 620
   training systems, 621
graphics, 635, 637
grassy weeds, 866
gravel, 462
gravitational water, 287
green industry, 98, 109–112
green manures, 325
green roof systems, 519
green walls, 526–527
green waste, 299
greenhouse
   air, 482–483
   components, 457–469
   construction worker, 471
   controls, 466
   covering materials, 457–458, 469
   covers, 256–257
   crops, 496–500
   curtains, 468
   customer service careers, 501
   design, 256
   drainage and soil quality, 451
   engineer, 471
   environmental requirements, 480–489
   equipment, 466–469
   floors, 462–463
   growing structures, 456
   heating equipment, 470
   heating systems, 461–462
   humidity, 835
   maintenance, 470–471
   manager, 271
   medium guidelines, 490
   operation and maintenance, 448–472
   orientation, 453
   planning, 450–454
   production, 349–353, 478–501
   range, 455
   structures, 454–457, 469

greening pins, 681
grid paper, 647
grooming (plant), 706
gross domestic product (GDP), 126
ground level ozone, 521
grower, 624
growing practices, 96
growing site, 573–574
growing space calculations, 492
growing structures, 559–560
growth
    environmental conditions, 248–272
    habit, 755
    media, 430, 434, 523
guard cells, 204
gymnosperms, 183–184
habitat modification, 785
hand watering, 554
handling vegetables, 590
haploid, 237
harassment and discrimination, 172
hardening off, 262, 376
hardscapes, 101
    brick patios and pathways, 722–723
    installation, 721–727
    lines, 638
hardwood, 369
    cuttings, 369–372
harmony, 670
harvest, 544, 550, 608–609
harvesting, 107, 547
haustorium, 836
hazards
    biological, 159, 161–163
    chemical, 159–161
    ergonomic, 159, 163
    general, 159
    physical, 159–160
    safety, 159–164
    work organization, 164
headers, 723
heading cut, 612
heat and cold stress, 486
heat island effect, 521
heat stress, 262
hedgegrow design, 674
heirlooms, 588–589
hemelytra, 809
hemimetabolous metamorphosis, 812
herbaceous, 369
    cuttings, 374
    grafts, 415
herbaria, 188–190
herbarium, 188
herbarium director/curator, 190
herbicide-resistant weeds, 786
herbicides, 768, 869, 879

herbs, 94
herringbone brick pattern, 723
heterotrophs, 198
high tunnels, 456
    organic gardening, 457
    season extension, 265–267
    vegetable production, 583–584
high-density orchard, 96
high-quality seed, 347
hilling, 390
histodifferentiation, 340
hole insertion grafting (HIG), 415
holiday centerpieces, 682
holometabolous metamorphosis, 812–813
home lawns, 102
homemade floral preservative, 675
homozygous, 239
honey locust, 369
hop plants, 618–620, 622–623
horizons, 281–282
horizontal air fan (HAF), 458
horticultural business management, 120–148
horticultural extension agent, 82–83
horticultural oils, 708
horticultural risk consultant, 173
horticultural safety manager, 173
horticultural sales representative, 148
horticultural technician, 711
horticulture, 92
    business consultant, 147–148
    illustrator, 378–379
    impacts, 110
    industry, 90–113
    in the twenty-first century, 506–531
    materials, 522–523
    organizations, 112
horticulturist, 92
hot water heating, 461
hotbed, 456
hue, 670
human resources, 19
humidity, 700–701
humidstats, 467
hybrids, 355, 587–588
hydathodes, 840
hydrogen (H), 312
hydrometer, 283
hydromulch, 772
hydroponics, 508–514
    history, 508–510
    on Wake Island, 510
hydroponic systems, 510–512
    components, 512–514
hyphae, 841
iced storage, 591
imbibition, 341

immune (disease resistant), 834
imperfect flowers, 212–213
improvement SAE, 41–42
in vitro, 430
inarch grafting, 418
incomplete dominance, 240
incomplete fertilizer, 483–484
incomplete metamorphosis, 812
indehiscent, 216
independent learning, 82
Indian paintbrush, 836
indoor plants
    diseases, 708–710
    lighting types, 704
    watering, 701–703
induced dormancy, 344
induced mutations, 241
industrial and municipal wastes, 327
industrial nitrogen fixation, 323
industrial revolution, 156–157
industry information, 125
infiltration (water), 286
inflorescence, 213, 754
inflorescent types, 213
information literacy, 62, 80–83
    horticulture, 60–84
infrared heating systems, 461
initiative fostering, 7
injury, 839
inoculation, 839
inoculum, 839
inorganic fertilizers, 327–328
inorganic mulches, 297–298
inorganic nutrients, 234
insecticidal soaps, 708
insecticides, 878
insects, 802–823
    anatomy, 804–810
    collecting, 820–823
    eggs and embryos, 811
    growth and development, 811–813
    integrated pest management, 781–784
    internal systems, 811
    pests, 707–708
    pinning, 822–823
    preserving, 821–823
    removal, 792
    scouting, 787
    soil-dwelling, 290
    traps, 788–789, 793
    turfgrass, 770
inserts (cell pack), 492
insoluble fertilizers, 484–485
insoluble fiber, 571
integrated pest management (IPM), 106, 778–797, 843–844
    action thresholds, 791

careers, 796–797
control measures, 785–787
corrective actions, 792–796
creating a program, 780
for lawns, 762
for turf, 768
fruit and nut production, 615
greenhouse pest control, 489
insects, 781
inspection and monitoring, 787–791
landscape design, 652
managing plant diseases, 843–845
nursery production, 561
plant pathogens, 783–784
vegetable production, 586
weeds, 781, 867–870
intensive gardening, 529
intensive green roof system, 519
intensive market coverage, 133
interconnection, 643
interior plantscaping, 690–711
    balance, 697–698
    color, 696
    designer, 710
    environmental requirements, 700–706
    form, 695–696
    interiorscapers, 102
    interiorscapes, 101, 692
    interiorscaping, 101–102
    locations, 693–695
    texture, 696
intermediate filaments, 203
International Code of Botanical Nomenclature, 181
International Society of Horticulture Science (ISHS), 113
internodes, 208
internship SAE, 36–38
interplanting, 584
interspecific, 415
interstock, 408
interveinal areas, 317
interview. See job interview.
intuition, 5
invertebrate pests, 784
IPM. See integrated pest management.
Irish Potato Famine, 848
iron (Fe), 319
irrigation, 269–271
    application timing and amounts, 650
    drainage, 523
    drip, 127, 270–271, 575
    efficiency, 649–650
    for frost protection, 267
    fruit and nut production, 605, 623
    greenhouses, 464–466

growth, 269–271
landscape design, 651
nursery production, 553–556
overhead, 464–465
rooftop garden, 523
seed propagation, 352
specialist, 658
sprinkler, 270, 575–576
subirrigation, 465–466
surface, 269
timing, 763–764
turfgrass, 760–761, 763–764, 771
ISHS (International Society of Horticulture Science), 113
Jefferson, Thomas, 725
job interview, 139–143
    follow up, 143
    mistakes, 143
    practice questions, 142
    preparing, 140–141
    taking part, 141–143
Julian, Percy, 482
juvenile stage, 253
Ketchup 'n' Fries™ plant, 409
killing jars, 821
kingdom, 183
kiwifruit, 619, 622, 624
labeling insect specimens, 823
labels
    fertilizer, 484
    plant, 491–493
labor laws, 170–172
lag phase, 341
laminar flow hood, 432
land judging, 282
landraces, 353–354
landscape architect, 632
landscape company owner, 173
landscape contractor, 635, 738
landscape design, 630–659
    base plan, 634
    bed lines, 638
    designer, 633
    elements and principles, 637–644
    plans, 718–720
    rendering, 720
    steps, 632–636
    tools, 644–648
landscape maintenance, 730–736
    established trees and shrubs, 731
    fertilizing, 731–732
    installation, 716–738
    plan, 636
landscape manager/groundskeeper, 738
landscapers, 540
landscaping, 101
    business, 736–738

licensing and shipping regulations, 540
lasagna composting, 529
late blight, 847
lawns, 746, 749
    establishment, 757–762
    mowers, 761, 766–767
lawyer, agricultural chemical company, 895
layering, 386
    propagation, 386–387
    techniques, 388–391
layering and division, 384–398
LC_{50}, 891
LD_{50}, 890
leachate, 557
leaching, 575
leadership
    characteristics, 4–5
    development, 8–9
    events, 21–22
    FFA, 14–17
    path and skills development, 5–9
leaf-bud cutting, 369, 375
lean-to greenhouse, 455
leaves, 210–212
    adaptation, 212
    anatomy, 210–211
    and stem propagation, 366–368
    blade, 752
    blister, 849
    blower, safety, 160
    curl, 849
    cuttings, 369, 374–375
    mesophyll, 206
    spot, 849
    venation, 210
legend (drawing), 719–720
legume, 216
lenticels, 840
lethal concentration (LC_{50}), 891
lethal dose (LD_{50}), 890
letters of application, 136–137
leucoplasts, 201
level site, 451
light
    forced flowering, 254
    greenhouse, 467, 480–482
    growth, 250–255
    indoor plants, 703–705
    moles, 481
    optimizing, 255–258
    pollution, 254
    quality, 250–251
    quantity, 252
    seed propagation, 344
    traps, 821
light-dependent reaction, 227–228

light-independent reactions, 228–229
ligule, 753
limited liability company (LLC), 125
limited liability partnership (LLP), 125
line design, 675
line flower, 672
line-mass continuum designs, 674–675
line-mass (floral) designs, 675
liners, 492, 495
liquid concentrates, 883
living roots, 111
living wall designer, 530
loamy soils, 283
local food production, 107–108
localized heating, 268
locavores, 107
long-day plants, 253, 481
lons, 290
loose media, 527
louvers, 459–461
Love in a Puff, 342
low tunnels
    constructing, 583
    season extension, 265–267
    vegetable production, 582–583
loyalty, 5
lysosomes, 203
machine safety devices, 165
macronutrients, 312
macropores, 287
magnesium (Mg), 316–317
mail-order companies, 539
maintenance procedures, 471
mandibles, 806
manganese (Mn), 319–320
manometer, 459
market niche, 538–541
marketing, 129–134
    four Ps, 130–134
    opportunities, 454
    process, 130
market outlook, 540–541
mass design, 675
mass flower, 672
material safety data sheet (MSDS), 167
mat media, 527
maturation drying, 341
measurement tools, 644
measuring wheels, 644
mechanical control, 867
mechanical scarification, 345
mechanical seeders, 351
mechanical transplanters, 545
mechanics (floral), 666, 676–681
media, 351, 489
media manufacturing careers, 303
medium composition, 437

medium (texture), 641
meiosis, 237–238
meristem, 203
meristematic, 203
metamorphosis, 812–813
microbial control, 795
microbial pesticides, 880
microclimates, 264–265
microfilaments, 203
microgreens, 35, 498
micronutrients, 312, 317–321
microorganisms, 288–289
micropores, 287
micropropagation, 430
    acclimatization, 438–439
    advantages and disadvantages, 431–433
    environmental conditions, 438
    environmental requirements, 433–434
    history, 430–431
    initiation or establishment, 435–436
    labor and equipment costs, 432
    lab technician, 440
    multiplication, 436–438
    rooting, 438
    selection and cultivation of stock plants, 435
    stages, 434–439
    sustainable products, 431
micropyle, 811
microtubules, 203
mildew, 849
mineral nutrient uptake, 321–324
mission statement, 127–128
mistletoe, 836, 865
miticides, 879
mitochondria, 201
mitosis, 236
mixed fertilizers, 328
MLA Handbook for Writers of Research Papers, 76
modified central leader, 613–614
moisture loss, 592
moisture stress, 763
mold, 850
moles of light, 481
molluscicides, 880
Monticello (Thomas Jefferson), 725
molybdenum (Mo), 320
monochromatic (color), 641
monocots. See monocotyledoneae.
monocotyledoneae (monocots), 184, 217
monoecious, 213
monophagus, 817
morphological dormancy, 345
morphology, 181

mosaic, 842
mosquitoes, 657
mottling, 842
mound (stool) layering, 390
mower, 761, 766–767
mowing, 761, 766–767, 772, 868
    and fertilizing, 651
muck, 603
mulch, 296–300, 651, 705
    calculating, 651
    organic and inorganic, 296
    soil nutrition, 298
    vegetable production, 581–582
mulching, 287, 735–736, 868
mulchmat, 300
mutations, 240
mycelium, 841
mycorrhiza, 289
NADPH (nicotinamide adenine dinucleotide phosphate), 228
NALP (National Association of Landscape Professionals), 113
nanometer (nm), 250
National Association of Landscape Professionals (NALP), 113
National FFA Organization, 11
    activity levels, 20–23
    agriscience fair, 377
    degrees, 16–17
    emblems and motto, 12–13
    leadership development, 14–17
    national organization, 11–23
    national organization history, 11–23
    officers, 14–16
    program of activities, 18–19
    structure, 12–15
    timeline, 12
National FFA Week, 20
National Institute of Occupational Safety and Health (NIOSH), 157–158, 167, 173
National Junior Horticulture Association (NJHA), 10–11
natural layering, 391–393
natural predators, 819
necrosis, 252, 842–843
nectria canker, 848
nematicides, 880
nematodes, 604, 782, 836, 841
newly planted trees and shrubs, 730–731
nickel (Ni), 320
nicotinamide adenine dinucleotide phosphate (NADPH), 228
night interruption (NI), 482

nitrification, 323
nitrogen (N), 313–314
    application, 314
    aquaponics, 514–515
    cycle, 322–323
    loss, 323
    sources, 328
NJHA (National Junior Horticulture Association), 10–11
nodes, 208
nonselective herbicides, 870
nontunicate, 394
nontunicate bulbs, 394–395
nonvascular plants, 183–184
north arrow, 719
no-till, 867
nucleolus, 200
nursery, 99–100
    environmental management, 559–561
    grower, 100, 501
    inspector, 379
    installation, 548–549
    liners, 377
    maintenance, 544, 546
    manager, 563
    owner, 540
    production methods, 551
    propagator, 358
    types, 539–540
nut, 216
    maintenance, 607–608
    production, 598–624
nutrient film technique (NFT), 510–511
nutrient management, 556–558
    fruit and nut production, 614, 622
    vegetable production, 576–577
nutrients, 483–485
    analysis, 557
    cycle, 321–322
    management, 556–558
    mobility, 323–324
    monitoring, 557–558
    sources, 324–328
obstruction, 451–452
Occupational Safety and Health Administration (OSHA), 157–158, 167, 173
ocelli, 806
offsets, 392
oil bodies, 202
olericulture, 94–95, 573
oligophagus, 817
Olmsted, Frederick Law, 633
ommatidia, 806
oomycetes, 780, 784
open center system, 612

operating costs, 453
orchard manager, 398
order (taxonomy), 185
organic chemical herbicides, 70
organic edibles, 105–106
organic foods, 105
organic materials, 325–327
organic matter, 280
organic mulches, 298–300
organic pesticide, 879–880
organic production, 105–107
organic weed control, 762
organisms, 281
    detrimental to plants, 836
    disease-causing, 835–837
organization skills, 7–8
ornamental horticulture, 93–105
ornamental plants, 104
oscillating sprinkler, 554
overhead (costs), 131
overhead irrigation, 267, 554
overhead watering, 464–465
overseeding (turfgrass), 771
ovipositor, 810
oxygen (O), 313, 343
Palace of Versailles, 639
panicle, 754
papaya ringspot virus, 844
paper point, 823
papers (drawing), 647–648
paraphrase, 75–76
parasitic drain, 39, 49
parasitic plant damage, 817
parasitic plants, 836
parasitic predators, 820
parasitic wasps, 106
parasitic weeds, 864
parenchyma, 206
parent material, 281
parliamentary procedures, 14
parthenogenesis, 811
patch budding, 420
pathogen, 834
    environment, 835
    transmission, 818
pathologist, 833
peanuts, 95
pedologists, 285
pedology, 302
peds, 285
pelleting, 347
pencils, 645
perched water table, 576
perennial, 99, 500
perforated convection tubes, 462
performance standards, 129
perlite, 295
perseverance, 5

personal protective equipment (PPE), 161, 845, 885
    selecting, 887
perspective (balance), 642
pesticide chemist, 894
pesticides, 781
    active ingredients, 882, 885
    amount to use, 887–888
    application, 796, 886–890
    applicator certification, 886–887
    container reuse, 894
    disposal, 886, 893–894
    eye exposure, 892
    formulations, 881–883
    inhalation, 892
    labels, 883–886
    management, 887–888
    mixing, 888
    poisoning symptoms, 891
    safety, 167, 876–895
    sales (associate), 797
    spills, 892
    storage, 886, 893–894
    types, 878–881
pests, 780–784
    agricultural, 816–820, 878
    barriers, 792–793
    control, 489
    feeding behaviors and plant damage, 817–820
    identification, 790, 814, 924–925
    inspection and monitoring, 787–791
    management, 106, 615, 623
petal, 212
PGRs (plant growth regulators), 377–378, 486
    application and benefits, 490
pH
    changing, 577
    optimal ranges, 291–292
    paper, 291
    soil, 291
    testing, 293
phase change, 460
phenols, 345
phenotype, 240
pheromones, 788
    chemical signals, 813–816
    traps and attractants, 793
phloem, 205
phosphorus (P), 315, 328–329
photoblastic, 251
    seed germination, 251
photocell, 467
photodormancy, 345
photons, 252
photoperiod, 250, 253–255
photoperiodic lighting, 481–482

photoperiodism, 253, 481
photosynthesis, 198, 226–229
phototropism, 254–255
phylum, 183–184
physical hazards, 159–160
physiological dormancy, 345–346
phytohormones, 377
phytophagus, 816
phytoremediation, 317
phytosanitary certification, 540–541
phytotoxicity, 332
PIP (plant incorporated protectant), 880–881
pistil, 212
pistillate flowers, 212–213
pitfall traps, 821
pith, 206
placement SAE, 36–38
    awards, 50
plagiarism, 74–75
    documentation, 74–76
plant biology, 196–218
    cells, 199–203
    parts, 206–218
    tissues, 203–206
plant breeding, 93
    breeder, 242
    induced mutations, 241
    principles, 239–241
    process, 240
plant buyer, 562
plant development, 224–242
plant disorders
    abiotic, 838
    biotic, 838
    disease symptoms, 839
    identification, 814
    types, 837–838
plant growth, 224–242
    optimal pH ranges, 291
plant growth regulators (PGRs), 377–378, 486
    application and benefits, 490
plant hardiness zones, 260
plant identification (glossary), 900–912
plant incorporated protectant (PIP), 880–881
plant keys, 188
plant lines, 638
plant management (interiorscape), 706–710
plant material
    greenhouse production, 493–496
    integrated pest management, 786–787
    landscape installation, 727–728
    securing in vases, 680
    stem and leaf cuttings, 368–376
    vegetable production, 587–589
plant mutations, 240–241
plant nutrition, 310–332, 706
    essentials, 312–321
plant pathogens, 780, 783–784
plant physiologist, 241–242, 852
plant residues, 325
plant responses to temperature, 260–262
plant science, 92–93
plant selection
    landscape design, 648–649
    light availability, 257–258
plant spacing, 560
    orientation, 255
plant stakes, 681
plant tag technician, 500–501
plant taxonomist, 191
plant taxonomy, 178–191
Plant Variety Protection Act (PVPA), 356–357
planters, 525–526
planting
    date, 868
    depth, 347–348
    hole preparation, 729
    maintenance, 549–550
    methods, 759–760
    nursery production, 543, 545–546
    site, 580
    small fruits, 605–607
    timing, 347, 349, 610–611, 728
    tree fruits and nuts, 610–611
    vine fruits, 618–619
plantlets, 367
plants
    biological hazards, 162–163
    changing names, 186
    for aquaponics, 517
    for rooftop gardens, 523
    stress, 261
plasma membrane, 199
plasmalemma, 234
plastic, 463
plasticulture, 297
plastid, 200
plug production, 250–251
plug trays, 492–493
plugging, 760, 771
plugs, 343, 495
plumule, 342
PNP (pot-in-pot), 100, 541
    production, 547–550
pockets (vertical gardening), 524
poinsettia production, 467
polarity, 375
pollination, 608
polycarbonate, 257, 469
    fiberglass, 458
polyethylene, 257
    film, 457
    plastic film, 469
polyphagus, 817
pome, 214
pomologists, 95
pomology, 95–96, 602
Poplar Forest (Thomas Jefferson), 725
porosity, 286
porous concrete, 462
post-emergent herbicide, 870
postharvest, 589–593
pot hangers, 525
potassium (K), 314–315, 329
    sources, 328
pot-in-pot (PNP), 100, 541
    production, 547–550
potting media, 705
potting medium pH, 705
PPE (personal protective equipment), 161, 845, 885
    selecting, 888
precautionary statements, 885
pre-emergent herbicide, 870
preformed roots, 367
preliminary design, 635
presentation, 65, 81
    methods, 76–80
prevention (weed), 867
price, 130–131
primary dormancy, 344
primary macronutrients, 313–315
priming, 343
primocane-fruiting, 608
principal strategy, 128–129
priority, 7
proboscis, 807
procrastination, 7
product, 130–131
production methods
    good agricultural practice (GAP), 586–587
    nursery, 541–551
    vegetable, 584–589
professional certifications, 135
professionalism, 134–136
profit margins, 129
program of activities (POA), 18
project plan, 67–68
promotion, 130, 133–134
proofreading, 68–69
propagate, 100
propagation, 395
proportion, 642, 667–668
    interior plantscaping, 698
protoplast, 440

proximal, 375
pruning, 255, 620
    bonsai, 696
    cuts, 611–612, 733–734
    espalier, 96, 615
    fruit and nut production, 611–612
    grape vines, 620
    indoor plants, 706–707
    landscape maintenance, 732–734
    maintenance, 546
    nursery production, 560–561
    renewal, 734
    root, 546
    small fruits, 607–608
    summer versus dormant, 614
    timing, 732–733
    training, 607–608, 611–614
pseudobulb, 398
public relations, 19
public speaking, 79
pumps and lighting, 656
purpose (in writing), 63–64
pyrethins, 879
pyrolysis, 301
pyruvate, 230
Q10 temperature coefficient, 258–260
quarantines, 845
quick dip method, 378
quiescent, 341
Quonset greenhouse, 455
raceme, 754
radicle, 341
radicle protrusion, 341
rain gardens, 652–653
rain or moisture sensors, 650
rainwater capture, 650
rainwater catchment, 487
raised bed gardening, 528–530
raised bed media, 528–529
raspberries, 605
ratio, 325
Raulston, J.C., 373
Recessive (trait), 239
recordkeeping
    and evaluation, 790–791
    SAE, 39, 49–50
recultured (explants), 434
recycling, 49
    paper, 63
    programs, 53
reel mower, 766
REI (restricted entry interval), 889–890
relative humidity (RH), 232–233
relative scale, 642
relief (rock and soil), 281
renewal pruning, 734
repetition (design), 643
    interior plantscaping, 698–699

reproduction, 235–239
    rate, 818
research, 69–73
    process, 70–73
research and experimentation SAE, 38–40
    awards, 50
research scientist, 797
reseller sales, 132
resistant hosts, 834
respiration, 230–231
respire, 294
restricted entry interval (REI), 889–890
résumé, 136–139
    myths, 138, 140–141
    tips, 138, 140–141
retailers, 539
retaining walls, 723–725
retractable roof greenhouse, 460
reviewing, 68–69
revising, 68–69
re-wholesalers, 539
RH (relative humidity), 232–233
rhizobia, 289
rhizomatous, 755
rhizomes, 397–398, 866
    stolons, 754
rhythm, 669
ribbon, 677–678
ribonucleic acid (RNA), 200, 835
ribosome, 202
ridge and furrow greenhouse, 455
ridge greenhouse, 455
Riverbend nursery, 520
RNA (ribonucleic acid), 200, 835
road access, 452
Robert's Rules of Order, 14
rodenticides, 881
rooftop gardening, 519–524
    access, 521
    cost, 522
    maintenance, 524
    planning, 520–524
    structural capability, 520–521
    techniques, 519
rooting medium, 376–377
root-knot nematode, 850
roots, 206–207
    cuttings, 375–376
    pruning, 546
    zone, 269
rose picker's disease, 834
rosette, 862
rot (disease management), 850
rotary mower, 766
row covers, 582
runners, 391–392
running bond, 723

runoff management, 555
rust, 850
saddle grafting, 414
SAE (supervised agricultural experience), 11, 23–24, 30–54
    agricultural education, 32–34
    awards and recognitions, 50–52
    coordination, 24, 48–49
    exploratory SAE, 40–41
    history, 32–33
    improvement SAE, 41–42
    investigation, 24, 44
    placement SAE, 36–38
    plan, 46–48
    planning, 24
    program improvement, 24
    program process, 42–50
    purpose, 33–34
    recordkeeping, 24, 39, 49–50
    research and experimentation SAE, 38–40
    types, 34–42
SAFE working conditions, 167–168
safety
    aerosol cans, 882
    agricultural worker protection standard, 886
    animals, 162
    call 811 before digging, 169
    carbon monoxide, 461
    compost and manures, 163
    compost material, 299
    food safety, 105
    hand washing, 435
    hazards, 159–164
    leaf blower, 160
    machine safety devices, 165
    material safety data sheet (MSDS), 167
    micropropagation, 434
    organic pesticides, 106
    pesticide containers, 894
    pesticide emergencies, 893
    pesticide labels, 167
    pesticides, 69
    PPE, 887
    practicing, 167–168
    presentation safety, 78
    programs, 887
    safety data sheet (SDS), 166–167
    tools, 154–173
    toxic weeds, 862
    vapors and gases, 161
    walk-behind mower, 103
    wind and pesticides, 889
    workers, 154–173
safety and health agencies, 157–158
safety data sheet (SDS), 166–167

saguaro cactus, 640
samara, 216
sandy soils, 283
sanitation, 301, 592, 785
sawtooth greenhouse, 456
scab, 851
scaffold branches, 612
scale, 645–646, 667–668
scarification, 344
schizocarp, 216
school-to-career plan, 144–146
science teacher, 218
scientific name, 187
scion, 406
scissors, 676
sclerenchyma, 206
sclerites, 804
scouting, 787
season extension, 265–267
    vegetables, 579–584
secondary dormancy, 344
secondary macronutrients, 316–321
sedge, 865–866
seedbed, 346
    preparation, 347
seedheads, 751, 754
seeding, 759, 770–771
seedless vascular plants, 184
seedlot, 342
seeds, 217–218, 493–494
    blend, 759
    coat dormancy, 344–345
    dormancy, 344–346
    germination, 341–344
    main parts, 217
    morphology, 340–341
    production, 355–358, 751
    propagation, 338–358
    sales representative, 358
    saving, 357
    selection, 353–355
    spreader, 759
    stages of development, 340–341
    storage, 358
    techniques, 346–353
    treatments, 347
    viability experiment, 863
selective herbicides, 870
selective market coverage, 133
self-compatible weeds, 861
self-improvement, 6–7
semihardwood cuttings, 372–373
senior project designer, 711
sepal, 212
separation, 394
serpentine layering, 389
service learning, 42
setae, 804

Setward's wilt, 852
sex pheromones, 815
sexual reproduction, 236–239
shade, 580, 670
shade cloth, 468
shears, 676
sheath, 752
shoot, 620
shoreline plants, 656
short-day plants, 253, 481
shutters, 459–461
sick plants versus injured plants, 838–839
side-dressing, 330
side-veneer grafting, 419
signal words, 885
silique, 216
silviculture, 93
simple layering, 388
simplicity, 644
site preparation
    for planting, 617
    landscape installation, 721
    lawn establishment, 757–759
    small fruits, 604–605
    tree fruits and nuts, 609–610
    vine fruits, 617
site selection
    growth conditions, 264–265
    nursery production, 538
    PNP production, 548
    small fruits, 602–603
    trees and fruits, 609–610
skim, 71
skull and crossbones, 885
slit seeder, 771
slope, 603
slope orientation, 265
slow-release fertilizer, 296, 331
small business, 122–123
small fruits, 602–609
SMART goals, 6
smut, 851
social responsibility, 82
sod, 102
    establishment, 611
    growers, 103–104
    production, 750–751
    sales associate, 772
    sodding, 760
softscape, 722
softwood, 369
    cuttings, 373–374
soil, 576, 603, 610, 617
    amendments, 648
    analysis, 324–325
    and soilless media, 293–296
    biological properties, 288–290

    bulk density, 286–287
    chemical properties, 290–293
    color, 288
    defined, 280–282
    fertility, 604
    injection, 331
    media, 278–304
    nutrition and conservation, 106–107
    pH, 291–292, 324
    physical properties, 282–288
    pore spaces, 280
    preparation, 758
    probe, 292
    report, 293
    sampling, 789
    science, 93
    scientists, 302
    structure, 285–287
    surveys, 285
    temperature, 263–264
    testing, 292–293
    texture, 282–285
    water, 287
soil auger, 293
soil-dwelling insects, 290
soil-forming factors, 280–281
soilless media, 293, 295–296
solarization, 868
soluble, 314
soluble fertilizers, 484–485
soluble fiber, 571
solutes, movement, 234–235
somaclonal variation, 437
sorption, 324
spacing (plants), 348, 584
specialized stem structures, 208–209
species, 187
specific epithet, 187
specifications (landscape design), 720
spermatheca, 811
spike, 754
splice grafting, 410
spores, 841
    host plant tissue replacing, 843
sports, 241
sports turf, 747, 750
    manager, 773
spot seeding, 771
sprigging, 760, 771
sprinkler irrigation, 270, 575–576
spur, 620
square (design), 674
square foot gardening, 529
squirting cucumbers, 349
stack bond, 723
stakes, 681
stamen, 212
staminate flowers, 212–213

standard concrete, 462
standards, 129
Star awards, 51–52
statement of practical treatment, 886
static pressure, 459
steam heating, 461
stem and leaf cuttings
    plant material used, 368–376
    propagation, 364–379
stems, 207–209
    cutting, 369
stem stripper, 677
stock plant, 368, 420
stoloniferous, 755
stolons, 392, 866
    rhizomes, 754
stomata, 204
    disease, 840
storage organ development, 254
straight (line), 638
strategic business plans, 123–128
    internal resources, 126–127
strategic environment, 124–126
stratification, 261, 346
straw bale gardening, 529–530
strawberries, 605–606
    growing in winter, 266
stress, 159
stretchers, 723
stroma, 200, 227
structural media, 527
student development, 18
student interest survey, 44–46
student resources inventory, 46–47
stylets, 807
subculturing, 437
suberin, 367
suberize, 396
subirrigation, 465–466
substrate, 543
    management, 558–559
subsurface drip, 764
succulent, 698
suckers, 392–393
suction lysometer, 557
sulfur (S), 317
sump pump, 460
sump tank, 460
sunscald, 261
superphosphate, 328
supervised agricultural experience. *See* SAE.
supplemental lighting, 257, 480
supplemental SAE, 42
supply and demand, 432
support groups, 19
supporting industries, 751
surface horizon, 282

surface irrigation, 269–270
susceptible host, 834
sustainability, 48
sustainable agriculture, 106–107
sustainable agriculture extension agent, 272
sustainable nursery production, 552–561
sustainable production, 105–107
sweep nets, 787, 821
symmetrical (balance), 642
symmetrical designs, 667
symmetry, 667
systemic herbicide, 870
systemic insecticide, 708
systemic pesticide, 878
tact, 5
tags (plant), 491–493
tannins, 623
tape and glue, 681
taproots, 206–207
taxonomy, 180–181, 816
T-budding, 421–422
T-design, 674
Tegmina, 809
temperature, 233, 603, 700
    greenhouse, 485–486
    growth, 258–264
    managing, 264–268
    seed propagation, 342–343
    vegetables, 577–579, 591–592
templates, 647
terrarium, 701
    planting, 702
territorial pheromones, 815
textural classes, 283
texture, 640–641, 672–673
thatch, 767
thatch control, 767–768
thermoperiod, 263
thermostat, 458, 466
thesis, 67
thinning cut, 611
thylakoid, 200, 227
tillage, 867
tiller, 754
tillering, 866
time, 281
tint, 670
tip layers, 391
tissue analysis, 324–325
tissue culture, 430
    micropropagation, 428–442
title block, 719
tobacco mosaic virus (TMV), 851–852
tone, 65, 671
tonoplast, 202

tools, 676–681
    landscape design, 644–648
    maintaining, 168–170
topography, climate, 617
topiary, 734
topic, 66–67
topography, 451
topsoil, 282
topworking, 406
totipotency, 430
toxicity, 890–893
    defined, 841
    types, 890
toxicity categories, chemicals, 891
T-pins, 681
tracing paper, 647
training, 619–622
    indoor plants, 706–707
training agreement, 37
training plans, 37
transgenic, 355
transgenic cultivars, 355
transition zone, 755
transitional turf, 756–757
transit level, 644
translocation, 235
transpiration, 229, 231–234
    affecting factors, 232–234
    carbon dioxide entry, 232
    physical structures, 233–234
    water uptake and nutrient access, 232
transplanting, 352–353
transplants, 580, 589
traps, 793
trays, 491–493, 524
tree circus, 408
tree fruits and nuts, 609–616
Tree Gator®, 110
tree plot, 93
trellising, 619–622
trench layering, 391
triangular (design), 674
trichomes, 233
trough system, 466
trunk, 620
T-square, 646
tubercles, 396
tuberize, 396
tuberous roots, 397
tubers, 392, 396–397, 496
tunicate, 394
tunicate bulbs, 394
turf, 747
    applications, 749–751
    failure, 769
    industry, 747–748
    maintenance, 763–768

morphology, 752–754
painting, 649
pests, 769–770
renovation, 768
selection, 757
sustainability, 748
timing, 757, 769
types, 754–757
use, 649
turfgrass, 102
  after care, 771
  benefits, 748
  breeder, 773
  industry, 102–104, 747–751
  management, 744–773
  morphology, 751–757
  nitrogen, 764–765
  phosphorus, 765
  potassium, 765
  types, 751–757
turgor, 202, 232
underwater plants, 655
uneven span greenhouse, 455
United States Department of Agriculture (USDA), 105, 260, 797, 845
United States Department of Labor (DOL), 157–158
unity (design), 643–644, 669
  interior plantscaping, 699
unity of three, 643
unprofessional traits and behaviors, 136
unrooted cuttings, 494
urban heat island effect, 521
US gross domestic product (GDP), 126
USDA Economic Research Service (ERS), 601
USDA Natural Resources Conservation Service (NRCS), 302
USDA Agriculture Research Service (ARS) National Center for Genetic Resources Preservation (NCGRP), 441
USDA (United States Department of Agriculture), 105, 260, 797, 845
utility turf, 747, 750
vacuoles, 202
vapors and gases, safety, 161
variety, 187–188
vectors, 818
  disease, 840
vegetables, 94–95
  and herbs, 498
  environmental factors, 573–584
  grower, 95, 593
  health benefits, 570–571
  production, 568–593
  storage, 591–593
  US market, 571–573

vegetative, 94, 366
vendors, 129
vents, 459–461
vermicompost, 35
vermiculite, 295
vernalization, 262
vernation, 752–753
Versailles, Palace of, 639
vertebrate pests, 784
vertical gardening, 524–527
verticuillium wilt, 852
verticutting, 767
viable, 341
vine fruits, 616–624
viniculture, 96
virtual herbarium, 188
virus, 834, 835
  disease index, 851–852
  integrated pest management, 783–784
  microbial control, 795
  papaya ringspot, 844
vision, 127–128
  creating, 6
visual observation, 789–790
viticulture, 96–97
viticulturists, 96–97
viviparous aphids, 811
vivipary, 341, 811
voice, 65
volatilization, 765
Wake Island, 510
warm temperature, 343
warm-season grasses, 755–757
warm-season vegetables, 577–578
warning, signal word, 885
water, 233, 343
  greenhouses, 486–488
  indoor plants, 701–703
  management, 553–556
  plant growth, 268–271
  quality, 487
  treatment, 555–556
  vegetables, 574–576
water culture, 510, 512
water features, 654
  pumps, 657
water gardens, 652–657
  fish and snails, 654
  maintenance, 656–657
  overwintering, 654
watering, 348, 352, 730
  decisions, 488
  efficient, 731
water sprouts, 546
water-wise landscape design, 648–652
weathering, 280
weather stations, 468–469
wedge grafting, 413

weeds, 781, 858–871
  benefits, 863
  biology, 864–865
  characteristics, 861–863
  cost, 861
  counts, 789
  cultural control, 868
  environmental conditions, 862
  fruit and nut production, 604–605
  herbicide resistant, 786
  identification, 865–866, 928–930
  impact, 860–861
  management, 657, 867–870
  organic control, 762
  removal, 562, 792
  science, 93
  scouting, 787
  seeds, 861–862
  turf renovation, 769–770
  types, 864
wettable powders, 883
whip grafting, 410
whip-n-tongue grafting, 410
wholesale distributor, 684–685
wholesalers, 539
whorl, 613
wick system, 510
wild populations, 354–355
wilting, 841
wilts, 852
wind, 233
windbreaks, 581
wind machines, 268
winterization, 544, 560
wire cutters, 677
wire (floral design), 681
wood greenhouse benches, 463
wood picks, 681
woody ornamentals, 371
work organization hazards, 164
worker and tool safety, 154–173
workplace safety documents, 166–167
works cited, 76
wound-induced roots, 367
wrapping and staking, 729–730
wreaths, 682
written communication, 62–69
xeriscapes, 103
xeriscaping, 648
xylem, 205
yellowing, 842. See chlorosis.
young workers, 171
  job duties, 171
  responsibilities, 172
  work hours, 170–171
yucca moth, 820
zinc (Zn), 320–321
zones of growth, 755
zygote, 237

by the human body, so consume vegetables in many different colors, **Figure 22-2**. Nearly all vegetables are naturally low in fat and calories, and cholesterol free. Nutrients found in vegetables include protein, potassium, iron, calcium, dietary fiber, folate (folic acid), vitamin A, vitamin C, vitamin E, vitamin K, and beta-carotene.

Potassium may help to maintain healthy blood pressure, critical to heart function, and muscle contraction. Vegetable sources of potassium include spinach, Swiss chard, sweet potatoes, kale, white potatoes, white beans, beet greens, soybeans, lima beans, lentils, kidney beans, Brussel sprouts, zucchini, asparagus, and green beans.

Dasha Petrenko/Shutterstock.com

Figure 22-2. Each type of vegetable contains specific nutrients so try to eat a variety of vegetables every day.

Many vegetables contain dietary fiber, the indigestible portion of plant-based food. Dietary fiber can be soluble or insoluble. *Soluble fiber* dissolves in water, changing form in the digestive tract. *Insoluble fiber* does not change form in the digestive tract. Fiber is important for proper bowel function. It can also help reduce blood cholesterol levels and may lower the risk of heart disease. Fiber-rich vegetables can also provide a feeling of fullness with fewer calories. Vegetables high in soluble fiber include legumes, beans, dried peas, and lentils. Vegetables containing insoluble fiber include carrots, cucumbers, zucchini, celery, and tomatoes.

Folate, or folic acid, aids in red blood cell production. Folate minimizes the risk of birth defects during fetal development. Vegetables rich in folate include lentils, spinach, black beans, sunflower seeds, turnip greens, broccoli, and peanuts.

Vitamin A promotes good eyesight and night vision. It keeps skin and mucous membranes healthy and helps to protect against infections. Dietary vitamin A, or beta-carotene, may lower the risk for cancer.

Vitamin C helps growth and repair of tissues to heal cuts and wounds. It helps maintain teeth and gum health and aids in iron absorption. Vegetables that are rich in vitamin C include broccoli, bell peppers, yellow snap beans, cabbage, Brussels sprouts, cauliflower, collard greens, okra, onions, potatoes, sweet potatoes, tomatoes, spinach, radishes, and rutabagas.

"Go vegetable heavy. Reverse the psychology of your plate by making meat the side dish and vegetables the main course."
—Bobby Flay

Vegetable Markets in the United States

The USDA measures economic value and the amount of vegetable production by gathering data from hundreds of independent markets within the food marketing system. Vegetables and pulses (seeds from legumes) are considered specialty crops, and recent data from US farm cash receipts from the sale of vegetables and pulses (including potatoes) averaged $17.4 billion.

Corner Question

What vegetable has more protein than steak?

This is 14% of US crop cash receipts according to the National Agricultural Statistic Service. Vegetables are grown on nearly 2.8 million acres across the country. The states with the largest production of vegetables for fresh and processed food markets include California, Florida, Arizona, Georgia, Washington, Michigan, Wisconsin, Minnesota, Idaho, Oregon, and New York, **Figure 22-3**.

The vegetable industry is classified into two major end uses: fresh market foods and processed food. Processed foods can be further divided into foods that are canned, frozen, or dehydrated. In vegetable production, management practices, cultivar selection, and other factors differ greatly between growing for a fresh food or processed food market. For example, cultivars grown for processed foods are better adapted to mechanical harvesting, but they often lack the characteristics needed for fresh market sale, such as flavor and texture. Vegetables grown for processing are grown based on contractual arrangements between growers and processors. About half of all vegetable production is used in the processed food market.

According to the USDA Census of Agriculture, most US vegetable farms are individually owned and relatively small. About 75% of vegetable farms use fewer than 15 acres. Only a few farms, however, account for the majority of commercial vegetable sales. About 9% of vegetable farm operations are responsible for 90% of the value of vegetables sold. More than half of all vegetable production acreage is irrigated.

Mirec/Shutterstock.com

Figure 22-3. This map shows the states that have the highest production of vegetables for both fresh and processed markets.

Domestic vegetable production occurs seasonally due to climate, with the largest harvests happening in the summer and fall. During the winter, vegetables are imported from other countries to supplement domestic supplies. This results in increased choices for consumers but competition for US growers. For example, during the winter and spring, Florida produces most domestic warm-season vegetables, such as fresh tomatoes. Fresh tomatoes are also imported at this time. Most imports come from Mexico and Canada and are primarily greenhouse grown. These imports compete directly with winter and early spring products from Florida.

Statistics indicate that local food markets are a small but growing portion of US agricultural production. Smaller farms find that directly marketing crops to consumers accounts for a higher percentage of their sales than it does for larger farms. The local food movement involves more than just local vegetable production. It includes a joint effort to build more local, self-reliant food economies. Sustainable food production, processing, distribution, and consumption are integrated to enhance an area's economic, environmental, and social health.

Vegetable production is expected to increase because the current emphasis on health and nutrition will result in growing consumer demand. A number of national campaigns are focused on increasing awareness and consumption of fresh fruits and vegetables. These campaigns are aimed at consumers and address the benefits of a healthy diet that contains vegetables.

Environmental Factors

Several environmental factors impact vegetable production. These factors include the growing site, water, soil, nutrient management, temperature, and methods for extending the growing season.

Growing Site

Vegetable production, or *olericulture*, is the practice of producing edible vegetable crops. It starts with selecting a good growing site. A good location is essential for a productive, high-yielding vegetable plot. Selecting a location for vegetable production varies between home gardeners and commercial growers. However, a number of elements remain the same. Vegetables need at least six to ten hours of direct sunlight every day, **Figure 22-4**. Southern exposures warm faster in cooler spring temperatures. In flat fields, some growers will mound their rows and plant on the southerly side of the mound. Data suggests there is an average 30% heat gain in total heat from using this method.

Figure 22-4. Vegetables need a minimum of six hours of sunlight a day and preferably eight to ten hours.

Figure 22-5. Windbreaks, such as this fence, can prevent water loss and plant injury and create warmer temperatures.

The sun changes location in the sky throughout the year. Selecting a garden site that is free of trees and other sources of shade will impact the amount of sun that reaches the vegetables. Trees also compete for space, water, nutrients, and oxygen with a garden. The garden should be beyond the dripline of the foliage. The *dripline* is the outermost circumference of a tree's branches where water drips from the leaves onto the ground. Gardeners with limited space sometimes plant crops in the landscape, rotating crops among different spaces each season to minimize pests.

Air movement among plants is important to manage diseases and avoid frost injury from cold pockets. Situating a site to provide proper air drainage, or the movement of cold air through a field, minimizes potential cold damage. In windy sites, plants can suffer wind damage or dry out. Windy sites are also prone to soil erosion, making them undesirable locations for growing vegetables. Finding a space where natural windbreaks occur (or can be planted or constructed) will also minimize injury from the wind, **Figure 22-5**. A windbreak also allows warm temperatures to accumulate, reduces transpiration, and creates a microclimate conducive for growing tender vegetables.

Water

Vegetables are 80%–95% water. For optimal yield and quality, many commercial and home gardeners irrigate their plants. Most vegetables are rather shallow-rooted. Up to 1.5″ (4 cm) of water is required each week during warm temperatures to maintain most vegetable crops. Watering needs decrease to 3/4″ (2 cm) per week during cooler seasons. This amount is often supplied by rainfall.

For small gardens, locating the vegetable plot near a hose or faucet allows for simple hand watering. For larger gardens and commercial growers, however, irrigation systems provide a reliable and efficient water delivery method. There are a number of different irrigation systems, including drip irrigation and sprinkler irrigation. Choosing a commercial sprinkler system depends on a number of factors:

- Growing site conditions—soil type, drainage, erosion potential, availability of power, topography, distance from water source.
- Water considerations—availability, quantity, quality, costs to develop a water supply, annual crop water requirements.
- Crop requirements—yield potential, frost protection, and cultural practices related to planting, pest management, and harvesting.
- Infrastructure concerns—labor requirements, labor availability, initial irrigation investment, annual operating costs.

> "An onion can make people cry, but there's never been a vegetable that can make people laugh."
> —Will Rogers

Drip Irrigation

Drip irrigation distributes water directly to the plant roots, **Figure 22-6**. It is used exclusively in plasticulture (the practice of using plastic mulch) and requires less water than other methods. Drip irrigation systems generally include a main irrigation line coming from the water source and multiple driplines that extend down the rows. Tiny holes in the dripline release water directly to the roots of the plant.

Vadym Zaitsev/Shutterstock.com

Figure 22-6. Drip irrigation is an efficient method for delivering water in commercial vegetable fields.

A drip irrigation system has several benefits. Water is used efficiently. Because not as much water is required, growers are able to pull from low-volume water sources. Fertigation works well with drip irrigation, reducing soil erosion and nutrient *leaching* (the movement of nutrients out of the soil into the groundwater). The risk of disease is minimized because foliage remains dry. The use of low-pressure pumps reduces energy costs and requirements. Due to automation in the system, minimal labor is required. Drip irrigation systems effectively control weeds because water is not applied between rows. Finally, harvesting can continue during irrigation because areas between rows remain dry.

Some disadvantages of drip irrigation include annual installation costs, clogged lines, water filtration needed to minimize clogging, no frost protection, and more maintenance than some other systems.

Sprinkler Irrigation

Sprinkler irrigation systems water crops from overhead, applying water to both the plant foliage and the soil. Sprinkler irrigation systems range from simple sprinkler attachments for garden hoses to complex systems with underground piping and pop-up spigots. Depending on crop value, many fields have permanent piping for irrigation. A pump pulls water from a reservoir and sends it to the main line. Lateral lines run off the main line to reach the crops. Sprinkler irrigation in commercial vegetable production includes using hand-propelled sprinklers, solid set sprinklers, **Figure 22-7**, and hand-propelled or travelling big gun sprinklers. All systems deliver overhead water; the method used depends on the grower's needs and preferences.

dnaveh/Shutterstock.com

Figure 22-7. Sprinklers are commonly used in irrigation and may sometimes be used to provide frost protection for vegetable crops.

Sprinkler systems can cost less than drip irrigation systems and provide frost protection for some crops. In some cases, the labor requirements for setup and operation are lower than for other systems. The system may be adapted to any shape, size, and contour of a field. Many sprinkler systems can provide fertigation as well. Overhead watering from sprinkler irrigation can, however, increase the chances for disease due to wet foliage and splashing up of soil particles that may have plant pathogens attached.

Soil

An ideal soil for producing vegetables is a sandy loam. Almost any soil, however, can be used to grow vegetables when an effort is made to improve the soil. Clay soils tend to hold too much moisture and crust. However, they do hold nutrients well. Managing the soil structure by adding aggregate particles can improve water drainage, **Figure 22-8**. Sandy soil has limited water-holding capacity, is nutrient poor, and tends to be dry, but it also has large pore spaces for air and warms earlier in the spring. Cultural practices, such as adding organic matter, play a significant role in improving soil structure, which in turn increases infiltration, water- and nutrient-holding capacity, and drainage. Organic matter levels can be raised by incorporating thick layers of well-rotted leaves, compost, old horse manure, and peat moss in the spring before preparing the soil and again in the fall after harvest. Green manure crops, such as annual rye, ryegrass, and wheat, can be planted as winter cover crops and turned under in the spring.

Ruud Morijn Photographer/Shutterstock.com

Figure 22-8. Clay soil can be suitable for vegetable production if organic matter is added.

Avoid wet soils that have a high water table, a seasonally high water table, or a perched water table. A *perched water table* is groundwater that is temporarily located above unsaturated soil due to compaction or some other soil formation factor. Soils that lie only a few inches above bedrock should also be avoided. Avoid working in the garden or field when the soil is too wet. Often in late winter and early spring, snow melt, rain, and cool temperatures keep the soil moisture high. Soil that is worked or tilled when it is too wet forms large, hard clods, which are difficult to break up and unsuitable for creating a smooth seedbed. Once the soil is ready, work it by turning it over with a shovel, tilling, or simply loosening to a depth of at least 6″–7″ (15 cm–18 cm). Then rake it smooth before sowing or planting.

Nutrient Management

Vegetables, like any other plants, have specific nutrient and pH requirements. A soil test can determine the nutrient and pH levels of the soil. The results will determine whether to add nutrients and alter the pH.

Most vegetables grow best in a pH range of 6–6.8. Nutrients vary by crop, but generally a complete fertilizer (one that contains nitrogen, phosphorus, and potassium) will be sufficient for growing needs.

Changing pH

A soil's pH can be amended to be more acidic or alkaline and to make more nutrients available for plant uptake. Liming is the most common way to raise pH levels. Liming involves adding ground lime (calcium carbonate and other materials) to the soil using some sort of spreader. Soil test reports will recommend the amount of lime to add to the tested soil. In alkaline soils, acidic soil amendments are added. These amendments can include pine bark, peat moss, or elemental sulfur. Apply sulfur with caution since applying too much can injure plants.

Fertilizer

Based on the results of soil tests and recommended rates of application, fertilizers are typically applied before or at planting time. Fertilizers are broadcast and then tilled into the soil at a depth of 3″–4″ (7.5–10 cm). Broadcast one-half to two-thirds of the fertilizer over the garden and incorporate it into the soil. Band the remaining fertilizer in furrowed rows 3″ (7.5 cm) from the seeds or transplants. Generally, fertilizer applications are spread throughout the growing season in multiple applications. In a home garden, fertilizer should be applied through side dressing every four to six weeks. Side dressing places the fertilizer on both sides of the vegetable row about 4″ to 6″ (10 cm to 15 cm) from the plants. In irrigated commercial operations, fertigation may deliver nutrients in small amounts daily.

Fertilizer can come from both organic materials and mineral sources. Examples of organic material (not specifically approved for organic production) include manure, cover crops, bonemeal, blood meal, soybean meal, alfalfa meal, and seaweed. Mineral sources used in both conventional and organic production include different mineral forms of nitrogen, potassium, and phosphorus along with other macro and micronutrients. Always follow directions on fertilizer labels regarding application and safety.

Temperature

Temperature plays a key role in vegetable production by determining what can be grown in a given season. In some cases temperature can be manipulated to extend the growing season. There are two categories of vegetables: warm-season vegetables and cool-season vegetables.

Warm-Season Vegetables

As the name suggests, warm-season vegetables thrive in warmer temperatures and risk chilling or freezing injury if exposed to cool temperatures. Injuries happen most often when temperatures drop below 45°F–50°F (7°C–10°C) and include rapid respiration, molds, and rot on plants. The injuries can also cause bitter flavors. Cucumbers, tomatoes, and tropical fruits and foliage are sensitive to chilling injuries.

Many warm-season vegetables need a lengthy growing season of 80–100 days to produce fruit. Plant warm-season vegetables as soon as soil temperatures warm to avoid extreme summer temperatures that can affect fruit set and development. Soil temperatures generally warm after danger of frost in the spring. Within warm-season vegetables the range of minimum soil temperatures for germination is between 50°F and 60°F (10°C and 15.5°C), **Figure 22-9**. Seed tends to rot rather than germinate if soils are cool and wet. Beans and corn are particularly susceptible to these conditions. Okra, peppers, and eggplant can germinate at 60°F (15.5°C), but at an optimal temperature of 85°F (30°C), they will germinate quickly.

Warm-season vegetables include beans (bush, pole, and lima), melons, sweet corn, cucumber, eggplant, okra, field peas, peppers, pumpkins, squash (summer and winter), sweet potatoes, tomatoes, and watermelon.

Cool-Season Vegetables

Cool-season vegetables grow during colder months. In mild climates, this creates two growing seasons: fall and spring. In northern climates, spring is generally when cool-season vegetables are planted unless season extension systems are used. Cool-season vegetables prefer growing temperatures between 60°F and 80°F (15.5°C and 27°C). Optimum ranges vary among crops.

Planting Guide for Warm-Season Vegetables

Crop	Minimum Germination Temperature	Optimum Germination Temperature	Maximum Germination Temperature	Plant Spacing	Planting Depth	Days to Germination	Typical Days to Harvest	Age of Transplants (weeks)
Beans	50°	80°	90°	6" or 4" × 12"	1"–1 1/2"	6–14	60	
Cantaloupe	60°	90°	100°	36"–48"	1"–1 1/2"	3–12	85	2–3
Corn	50°	80°	100°	12" × 30" or 9" × 36"	1"–1 1/2"	5–10	60–90	
Cucumbers	60°	90°	100°	6" trellised 24"–36" not trellised	1"	6–10	55	2–3
Eggplant	60°	80°	90°	18"–24"	1/4"	7–14	60T	6–9
Pepper	60°	80°	90°	15"–18"	1/4"	10–20	70T	6–8
Tomato	50°	80°	100°	24" between trellised plants	1/4"	6–14	65T	5–7
Summer Squash	60°	90°	100°	36"–48"	1"–1 1/2"	3–12	50	2–3
Winter Squash	60°	90°	100°	36"–48"	1"–1 1/2"	6–10	100	2–3
Watermelons	60°	90°	110°	36"–48"	1"–1 1/2"	3–12	85	2–3

Goodheart-Willcox Publisher

Figure 22-9. Warm-season vegetables have minimum and optimum temperatures for germination.

When temperatures start to rise, many species tend to bolt, or develop a flowering stalk that can produce off-flavors and bitterness, **Figure 22-10**.

Cool-season vegetables can germinate when soil temperatures are at 35°F–40°F (1.5°C–4.5°C) with optimal temperatures being higher, **Figure 22-11**. Many cool-season vegetables can reach a harvestable age fairly rapidly and seeds can be planted multiple times. For example, lettuce or other leafy greens can be sown or transplanted every few weeks up to a certain point for multiple harvests.

Cool-season vegetables include beets, broccoli, cabbage, carrots, cauliflower, collards, kohlrabi, leeks, lettuce, onions, parsnips, peas, potatoes, radishes, rutabagas, spinach, Swiss chard, and turnips.

Season Extension

Season extension techniques can be as simple as selecting early maturing varieties of a vegetable. More complex systems include using high tunnels, row covers, or other methods.

AN NGUYEN/Shutterstock.com

Figure 22-10. When temperatures increase, lettuce bolts, becoming bitter and inedible.

Planting Guide for Cool-Season Vegetables

Crop	Minimum Germination Temperature	Optimum Germination Temperature	Maximum Germination Temperature	Plant Spacing	Planting Depth	Days to Germination	Typical Days to Harvest	Age of Transplants (weeks)
Beets	40°	80°	90°	4"–6"	3/4"–1"	7–10	60	
Broccoli	40°	80°	90°	18"	1/2"	3–10	65	5–7
Cabbage	40°	80°	90°	18"	1/2"	3–10	85	5–7
Carrots	40°	80°	90°	2"–3"	1/4"	10–17	70	
Cauliflower	40°	80°	90°	18"	1/2"	3–10	65	5–7
Kohlrabi	40°	80°	90°	7"–9"	1/2"	3–10	50	
Leeks	40°	80°	90°	4"–6"	1/4"	7–12	120	
Lettuce (leaf types)	35°	70°	70°	7"–9"	1/4"	4–10	60	
Onion (green)	35°	80°	90°	2"–3"	1/4"	7–12	60	
Onions, dry (seed)	35°	80°	90°	4"–6"	1/4"	7–12	110	
Parsnips	35°	70°	90°	5"–6"	1/2"	15–25	70	
Peas	40°	70°	80°	4"–6"	1"	6–15	65	
Potatoes	45°			12"–15"	4"–6"		125	
Radish	40°	80°	90°	2"–3"	1/2"	3–10	30	
Spinach	40°	70°	70°	4"–6"	1/2"	6–14	40	
Swiss chard	40°	85°	95°	7"–9"	1"	7–10	60	
Turnips	40°	80°	100°	4"–6"	1/2"	3–10	50	

Goodheart-Willcox Publisher

Figure 22-11. Cool-season vegetables have a minimum and optimum temperature range for germination.

The purpose of season extension methods is to lengthen the growing season by harvesting crops earlier in the spring and continuing to produce and harvest later into the fall and early winter. Season extension techniques moderate temperatures to protect crops from damage from extreme heat or cold. For commercial growers, season extension allows potential year-round income and employment, high yields, and customer retention and attraction. Season extension can involve several factors including the planting site, shade, transplants, cultivar selection, windbreaks, mulches, row covers, low tunnels, and high tunnels.

Planting Site

The location of the planting site can greatly influence temperature, creating microclimates or allowing management of warmer or cooler air. As mentioned earlier, a south-facing slope will stay warmer in the late fall and warm up more quickly in the early spring. Wind exposure at the top of a hill and cold pockets at a bottom of a hill can be a challenge and should be avoided. The soil provides a large reservoir of heat. Dark soils will warm up sooner in the spring than light soils. A similar effect can be achieved by mulching soils. In certain areas, such as the fruit-growing regions around the Great Lakes in Michigan and New York, large bodies of water provide a strong buffer, lessening extreme temperatures.

Shade

Natural or artificial shade can be created to moderate temperatures during summer heat. Using shade allows the production of heat-sensitive crops, such as lettuce, spinach, and other leafy greens, further into the growing season. Shade can be created in a variety of ways including lathe houses, shade cloth, or even planting underneath solar panels.

Transplants

Using transplants is a key season extension technique, providing as much as a three- to four-week head start on the season compared to direct seeding. Transplants may be able to establish themselves more quickly and outcompete weeds. However, transplants are more expensive to buy or grow than seeds.

Cultivar Selection

Cultivar selection is another way to extend the season and may involve an additional cost. Choosing early-maturing cultivars can result in earlier harvests, offering growers an edge in competitive markets, **Figure 22-12**. Selecting cultivars with different maturity dates allows harvesting to be staggered.

Kingarion/Shutterstock.com

Figure 22-12. This radish cultivar can be harvested in as few as 22–25 days.

Staggering the planting dates can also lengthen the season. Harvest dates can be calculated using the days to maturity information on a seed packet and the planting date. Selecting cultivars that are heat tolerant and others that are more cold tolerant will lengthen the seasons into summer or into the fall and early winter.

Windbreaks

Windbreaks are structures, plants, or features of the landscape that serve as a physical barrier against wind. Windbreaks can protect tender crops from cooling winds and injury. Fences, brush piles, brambles, trees, fence rows, cropping strips, shrubs, stone walls, and snow fences make effective windbreaks. Rye planted in strips has been found to be an economical windbreak in intensive vegetable plots. In some cases, windbreaks can provide a desirable crop, such as blueberries or figs. The decorative branches of willows grown as a windbreak can also be used in floral designs, **Figure 22-13**. Place windbreaks perpendicular to prevailing winds.

Hallgerd/Shutterstock.com

Figure 22-13. Willows can be planted as a windbreak and then harvested for use in floral design.

Mulches

Plasticulture is the practice of using plastic mulch in vegetable and small fruit production, **Figure 22-14**. Benefits include increased soil warming in the spring, weed suppression, enhanced insect control, and high-quality crops. Research has shown that crops grown with black plastic mulch can be harvested 7 to 21 days earlier than those produced on bare ground. Black plastic is the most common color used, but clear, white, and red plastics are also used. Each has its own benefits and drawbacks. For example, clear plastic can quickly raise soil temperatures but it also encourages weed growth. All plastic mulches require irrigation, usually in the form of drip irrigation. Plastic mulches are not biodegradable. They are generally used for only one season and then discarded. New plastic must be purchased the following season. Specialized equipment is used to lay plastic mulch in commercial gardens.

chungking/Shutterstock.com

Figure 22-14. Plasticulture is the practice of laying thin plastic over a row. It can warm the soil early in the spring.

Organic mulches, such as newspaper, decomposed leaves, straw, and wood chips, are also used. In a home garden, these materials may be a good fit. However, depending on the type of mulch, they may delay soil warming.

Row Covers

A *row cover* is a piece of lightweight fabric (usually spun polyester) that is laid on top of plants to provide protection from frost or insects, **Figure 22-15**. Row covers allow rain, sunlight, and air to penetrate. The covers can be secured against the wind by burying the edges in soil or weighing them down with heavy materials or pins. Different material thicknesses offer varying degrees of frost protection. Row covers are used in the early spring and removed as the plants mature. They may be used again in late fall until harvest. Row covers can be used in conjunction with other season extension approaches.

Row covers can offer effective insect control, creating a physical barrier between plants and insect pests. Aphids, leafhoppers, cabbage worm, and cabbage looper can be well controlled with row covers. However, pests that spend part of their life cycle in soil can use the row cover as a screen that protects them from natural predators. Row covers may increase weeds as well.

Olaf Speier/Shutterstock.com

Figure 22-15. Row covers trap radiant heat from the soil and keep plants from freezing in the early spring or late fall.

Low Tunnels

Low tunnels consist of PVC pipes or wire formed into hoops and covered with clear plastic or fabric row covers that run the length of the vegetable row, **Figure 22-16**. Low tunnels provide another method for moderating temperatures, excluding insect pests, and providing wind protection. Row covers or plastic mulch can be used inside low tunnels. In colder climates, low tunnels may be used within a high tunnel to provide additional protection from chilling or freezing injury during the winter. Low tunnels need to be monitored because the heat they trap can cause daytime temperatures to rise to injurious levels. Opening the sides of low tunnels will provide ventilation and may help cool the air. Crops that require pollination, such as tomatoes, eggplants, squash, and peppers, may only be in the low tunnel for three to four weeks before the tunnel is removed for the rest of the growing season.

sch_o_n/Shutterstock.com

Figure 22-16. A low tunnel is a season extension method that provides protection from chilling temperatures.

Corner Question

Are sweet potatoes and yams the same thing?

Hands-On Horticulture

Constructing Low Tunnels

Many low tunnel hoops are constructed by using wire or PVC pipe. No. 9 wire is cut in 65"–72" (165 cm–185 cm) lengths using bolt cutters. The ends of the hoops are inserted 6"–8" (15 cm–20 cm) into the soil and spaced about 48" (120 cm) apart. The cover is laid across the hoops with the edges buried in the soil. PVC pipes can be used in a similar manner to form hoops. Metal rebar stakes are pounded into the ground and serve as holding pegs for the PVC pipes.

Akiyoko/Shutterstock.com

High Tunnels

A high tunnel (also called a hoop house) is similar to a greenhouse; it is typically covered in a polyethylene plastic and enables nearly year-round vegetable production, **Figure 22-17**. Common high tunnel styles are Quonset or Gothic. They are usually single, stand-alone structures, but multiple bay tunnels are also used. Some high tunnels are made to be movable, whereas others are stationary.

High tunnel frames can be made with metal pipe, wood, or PVC. They are covered with one or two layers of polyethylene. When there are two layers, the space between the layers is filled with air to provide greater insulation. Like low tunnels, daytime temperatures can rise significantly in high tunnels. The tunnels need to be ventilated, either by rolling up the sides or lowering drop-down walls. Most high tunnels do not have permanent heat sources, but some may use portable heat. Some growers even use compost to provide a source of heat. Plants growing in the high tunnels are typically planted directly into the soil. The soil is tilled using a small rototiller or, in some cases, a small tractor. Plants in high tunnels can be covered with row covers or low tunnels to maximize the capture of radiant heat.

paul prescott/Shutterstock.com

Figure 22-17. High tunnels give growers more opportunity to sell their vegetables.

High tunnels offer several advantages to growers. Growers have opportunities for earlier and later market dates, can sometimes produce twice the amount of produce as for field-grown crops, and can have high-quality vegetables. High tunnels can also be a strategy for pest control, providing a physical barrier against many insects and plant pathogens.

High tunnels also have some disadvantages. They can limit the ability to move crops to different areas, they can have insect and disease problems, and they require an initial investment to build.

Production Methods

Growing vegetables requires a number of different techniques and practices to maintain soil quality, reduce risk of pests, encourage high yields, and optimize food safety. Growers and gardeners have a range of production choices available in which factors such as economics, environmental implications, and personal philosophy all play a role.

Spacing

Proper spacing between plants is important for proper plant growth, ease of cultivation, weed suppression, disease minimization, and optimal yields. Spacing requires a balance between efficiently using space while allowing for airflow and reception of light for photosynthesis. Each vegetable has a recommended spacing between plants within a row and between rows. *Interplanting* is an intensive production method in which early maturing crops are planted between rows of later-, or long-season, crops. For example, peas or radishes, both early maturing crops, may be planted between rows where tomatoes, peppers, cabbage, or corn will be grown. Vertical gardening is another approach often used by home gardeners to maximize planting space. Vertical gardening uses poles, trellises, nets, strings, or cages to train and support vines and other sprawling plants. Vegetables suited to this method include peas, pole beans, cucumbers, tomatoes, and melons.

Crop Rotation

Crop rotation is a production practice in which vegetables of the same plant family are planted in different locations every growing season. Crop rotation helps minimize pests and plant pathogens and manage soil fertility. Many insect pests and diseases can overwinter in the soil. Their populations can be significantly diminished if their preferred host plant is not immediately available to them. Crop rotation can also increase soil nutrients. Plantings of nitrogen-fixing legumes, such as snap beans, lima beans, and field peas, can be followed in the next season by high-feeding crops, such as corn or tomatoes. Roots crops, such as carrots, beets, parsnips, and rutabagas, can improve soil structure. They should be followed by shallow-rooted crops, such as salad greens, that benefit from soil improvement.

Large growers will have more room to manage crop rotation, but even home gardeners can move vegetables around each year. Make a planting map and record which plant families are placed where each season to create a system for future plantings. **Figure 22-18** categorizes various plant families.

Family Name	Crops	Ornamentals	Weeds
Apiaceae	caraway, carrots, celery, chervil, cilantro, dill, fennel, parsley, parsnips	Trachymene, Bupleurum	poison hemlock, wild carrot
Asteraceae	artichoke, chamomile, chicory, dandelion, Echinacea, endive, escarole, Jerusalem artichoke, lettuce, radicchio, safflower, sunflowers, tarragon	aster, cosmos, marigold, mums, Rudbeckia, yarrow, zinnia	chicory, cocklebur, dandelion, goldenrod, thistles
Brassicaceae, Cruciferae (cabbage or mustard family)	bok choy, broccoli, Brussels sprouts, cabbage, cauliflower, Chinese cabbage, collards, cress, horseradish, kale, kohlrabi, mustard, pak choi, radish, rapeseed, rutabaga, turnips, watercress	alyssum, kale, candytuft, stock, yellowcress, garden yellow rocket	bittercress, mustards, pepperweed, shepherd's purse, swinecress
Chenopodiaceae	beets, chard, spinach, sugar beets		Kochia, lamb's-quarter
Cucurbitaceae	cucumber, gourds, melons, pumpkin, summer squash, watermelon, winter squash		
Ericaceae	blueberries, cranberries	azalea, heather, rhododendron	
Fabaceae	alfalfa, beans, birdsfoot trefoil, black medic, clovers, cowpea, edamame, fava beans, garbanzo bean, hairy vetch, lentils, peanut, peas, soybean, vetches		black medic, common vetch
Lamiaceae	basil, catnip, lavender, marjoram, mints, oregano, rosemary, sage, thyme	bells-of-Ireland, salvia	catnip, henbit, mints
Liliaceae	asparagus, chives, garlic, leeks, onions, shallot	daffodils, daylily, hosta, hyacinth, tulip	
Poaceae	barley, corn, fescue, millet, oats, rice, rye, ryegrass, sorghum, timothy	ornamental grasses	brome, barnyard grass, crabgrass, fall panicum, foxtail, Johnson grass, quackgrass
Polygonaceae	buckwheat, rhubarb		knotweed, smartweed
Rosaceae	apples, apricots, blackberries, cherries, nectarines, peaches, pears, plums, raspberries, strawberries	multiflora rose	
Solanaceae	eggplant, peppers (bell and chile), potatoes, tobacco, tomatillo, tomatoes	Million Bells®, petunia	buffalobur, groundcherry, henbane, horsenettle, jimsonweed, nightshade

Goodheart-Willcox Publisher

Figure 22-18. This table lists the family name for most commonly grown vegetables. Use it when considering crop rotation.

Thinking Green

Certified Organic Growers

Organic farmers, ranchers, and food processors conform to a defined set of standards for organic food and fiber production. Following the standards for organic vegetable production, farms that have passed an audit may use the USDA organic seal. The seal certifies the following were not used: irradiation, sewage sludge, synthetic fertilizers, prohibited pesticides, and genetically modified organisms. These standards cover the product from start to finish. They include soil and water quality, pest control, livestock practices, and food additive rules. All farmers, whether certified organic, pesticide free, sustainable, or conventional, tend to use overlapping production methods that protect and enhance the environment and provide profits for the growers. Many organic growers work toward a triple bottom line, one that positively impacts the environment, economics, and social good. Organic certification exposes growers to markets that can demand higher prices from consumers, making organic certification both a philosophical and economical decision.

Lance Cheung/UDSA

Integrated Pest Management

Integrated pest management (IPM) is an environmentally sensitive approach to managing pests that uses common sense, economical practices and results in the least possible hazard to people, property, and the environment. IPM practices include cultural, physical, biological, chemical, and genetic tactics to control or minimize damage by pest organisms. A number of pest management practices for vegetable crops exists, including row covers, crop rotation, spacing, and cultivar selection. Tactics for managing pests are specific to each crop and each pest and should be researched to find the best control methods.

Good Agricultural Practices

A *good agricultural practice (GAP)* is a farm-level production method used to ensure that fresh produce is safe for human consumption. GAP production and postharvest guidelines are designed to minimize the risk of foodborne disease contamination on fresh produce. Harmful microbes such as E. coli, salmonella, and listeria (among many others) can contaminate fresh produce at any stage of production: growing, harvesting, processing, packaging, or marketing. The major source of microbial contamination on fresh produce is associated with human or animal feces.

Good agricultural practices consist of voluntary procedures that address potential sources of contamination in the soil and water and on hands and surfaces. Many buyers of produce, from public schools to grocery stores, require that growers have a GAP certification.

Certification comes from a third-party audit and approves the documented grower practices related to these factors:

- Clean hands. Hand washing is essential for preventing foodborne illness. Hand washing with soap and clean running water is the best practice for minimizing food safety risk. Also consider installing hand-washing stations for customers at U-pick operations.
- Clean soil. Follow specific steps to reduce the risk of introducing microbial contaminants to the soil. Animal manure can carry human pathogens that can be eliminated through proper composting, storage, and application timing. Wild and domestic animals also carry the risk of introducing harmful microbes and should be physically barred from planting areas.
- Clean water. Practices should be in place for all water used in irrigation to minimize microbial hazards and meet minimum quality standards. All water used in washing, cooling, and processing of fresh produce needs to be potable.
- Clean surfaces. Surfaces used in packing, processing, storing, and transporting food should all be washed and sanitized regularly. Farms that have both produce and livestock should take specific steps to avoid any contamination.

Plant Material

Selecting vegetables and cultivars to grow is an important step in planning production gardens or fields, **Figure 22-19.** Different cultivars or varieties have unique characteristics, such as disease resistance, early maturity, superior flavor, or long shelf life. Gardeners and growers will often plant high-performing standbys (plants proven to do well) and try one or two new cultivars for a few years to see how their production compares. Growing a cultivar for multiple seasons allows the gardener to evaluate the plant over time and discount the environmental differences that may affect growth in one year.

photogal/Shutterstock.com

Figure 22-19. A garden plan will help you determine what you want to grow and how much space you will need.

Hybrids

Many vegetable crops grown commercially and in home gardens are considered hybrid cultivars. Hybrid cultivars have parent plants with desirable traits. These plants are cross-pollinated to produce seeds that are called hybrids. The hybrids contain a combination of traits that may include increased yield, vigor, nutritional content, ornamental value, stress tolerance, and growth habit.

Thinking Green

Warren County High School Garden Club

A group of ninth graders in Warren County, North Carolina, trekked outside with the rest of their class on an August morning to tour their school garden. They discovered that what their teacher called a garden was really an overgrown mess, with a lot of potential. They wanted to fix the space but knew it would take a lot of work. By the time they had returned to their classroom, they had formed a garden club. The club planned to meet after school to refurbish the garden, giving students a space to be proud of.

The following year, the Warren County High School Garden Club began selling their produce, mostly to teachers and family members. As they learned more and more about growing food, they also learned that they could sell more produce (and make more money) if they had access to a larger market. Why not sell to their school cafeteria? The students decided to work toward a Good Agricultural Practices (GAP) certification. Following these simple food safety measures allowed them to sell produce to their school cafeteria, greatly expanding their market.

First they created a farm safety manual, outlining what food safety precautions they would take to manage risk in their school garden. They then put those practices into use. They built hand-washing stations in their greenhouse, devised a three-bin system for washing harvesting tools, and helped build a 10′ electric fence to deter deer and other wildlife. In order to become officially GAP certified, the students invited a GAP auditor to the garden from the North Carolina Department of Agriculture. He inspected their food safety procedures in action, watched them harvest their produce, and observed their diligent recordkeeping. In the end, the students' garden passed the inspection, allowing them to sell their produce directly to their school! Students say the best part is eating their own delicious, fresh produce for lunch!

Hybrid seed production is time-consuming and expensive. It can take years of breeding to develop the desired traits. The plant breeder holds exclusive rights and proprietary knowledge in the production of the hybrid. Growers and gardeners have to purchase new hybrid seed every year because collected seed from the hybrid will not be true to type. Seed, however, is often one of the least expensive production costs.

Browsing seed catalogs can be both fun and overwhelming. With so many choices, it can be hard to determine what cultivar is the best to grow. All America Selection (AAS) is a nonprofit organization that evaluates and selects cultivars that show superior performance. They have conducted trials with hundreds of cultivars across the United States and Canada since the 1930s. AAS plants have been judged to be reliable and have significantly improved qualities.

Heirlooms

Before plant breeding became a big business, many home gardeners and farmers would save seed from year to year. Through a process of evaluation, they would identify plants that showed exceptional traits. As with cultivars, some of these traits would include good flavor and texture, tolerance to local conditions or disease, or a unique appearance, **Figure 22-20**.

Tom Grundy/Shutterstock.com

Figure 22-20. Heirloom vegetables are prized for their good flavor and texture, particularly in tomatoes.

These plants from saved seeds are considered heirloom plants. Historically, heirloom plants were passed from one generation of growers to the next. Today, there is a renewed interest in heirlooms.

Heirlooms come from plants that are open-pollinated, meaning that plants freely cross-pollinate and are part of a fairly stable genetic population. Offspring of the open-pollinated plants will not be exactly like the parent plants, but they will be very similar. There are a number of well-known heirlooms, including 'Kentucky Wonder' pole bean, 'Nantes' carrot, 'Black Beauty' eggplant, 'Black Seeded Simpson' lettuce, 'California Wonder' pepper, and 'Brandywine' and 'Roma' tomatoes. Many heirlooms are prized for their intense flavor and fragrance. Many heirlooms, however, do not always have a long shelf life, as in tomatoes, for example. The fresh market for heirlooms is often limited to farmers markets or any other market where they do not stay on the shelf for more than a day.

Transplants

Transplants give growers and gardeners a head start in reaching a harvestable crop. Transplants should be healthy, vigorous, and free of any noticeable nutrient deficiency or pest problem. Vegetable transplants have an ideal age and size at which they should be transplanted. When transplanted at that ideal juncture, they can begin active growth immediately and are fairly tolerant of environmental stresses, **Figure 22-21**. The best age to transplant tomatoes, for example, is six to eight weeks. Plants less than six weeks are more susceptible to damage from wind or low temperatures (below 45°F or 7°C) and water stress. On the other hand, plants older than 10 weeks have a relatively large aboveground mass that has already started producing flowers and may be heading into the reproductive phase of growth. These plants will still produce tomatoes but at a much lower rate than their full potential.

Postharvest Handling and Storage

Postharvest is a process that involves the handling processes of a crop immediately following harvest. Postharvest processes include cooling, cleaning, sorting, packing, storing, and shipping. The goal of postharvest practices is to maintain the highest level of quality, including overall appearance, uniformity, absence of damage or blemishes (discoloration, harvest injury, insects), good taste and optimal flavor, and proper texture. For each type of vegetable, there is a definition of quality. For example, quality strawberries

Age of Transplants for Ideal Growth

Plant	Age
Broccoli	6–7 weeks
Cabbage	6–7 weeks
Cauliflower	6–8 weeks
Celery	9–12 weeks
Cucumber	2–3 weeks
Eggplant	8–10 weeks
Endive	5–7 weeks
Muskmelon	2–3 weeks
Onion	9–12 weeks
Pepper	8–10 weeks
Squash	2–3 weeks
Tomato	6–8 weeks

Goodheart-Willcox Publisher

Figure 22-21. Planning is required to determine the proper time to plant seeds indoors to reach a transplantable size for the garden.

and peaches must be sweet, snap beans and sweet corn must be tender, and carrots and snap peas must be crisp. Several factors can reduce postharvest quality including:

- Harvesting at the incorrect stage of maturity.
- Careless handling during harvest, packing, and shipping.
- Poor sanitation.
- Delayed or suboptimal cooling.
- Shipping or storing above or below optimal temperature.
- Lack of proper relative humidity.
- Exposure to ethylene gas (for some crops).

As soon as a crop is harvested, the deterioration process begins. Physiological processes are still occurring in the plant, including respiration and other chemical changes.

Figure 22-22. This unripe pumpkin has not developed a hardened rind. It will have a diminished shelf life if it is harvested at this point.

Postharvest quality begins in the field with management practices. These practices produce the healthiest crops with the greatest opportunity for a longer shelf life. Scheduling harvest at the appropriate time is also important. If a crop is stressed from too much or too little water or is physically damaged, the crops may be more susceptible to postharvest plant pathogens that cause decay and rot. Harvesting vegetables that are too ripe or not ripe enough will also diminish shelf life, **Figure 22-22**. Many resources are available that detail maturity indicators by crop. Crops should be harvested according to these indicators. By following best growing practices, growers and home gardeners may be able to increase the length of time that vegetables remain fresh.

Handling

Cooler temperatures slow the rate of respiration. Harvest generally occurs during the coolest time of the day, which is in the early morning or the evening. After harvest, produce should be handled gently to minimize bruising. Damage to the vegetables, such as skin breaks, bruises, spots, and cuts, provides an entry for plant pathogens that cause decay. Moisture loss is also increased when vegetables are damaged. Damage can be reduced by proper handling, harvesting at optimal maturity for storage, harvesting when crops are dry, and minimizing the handling of the crops. For example, to avoid overhandling, growers might place harvested vegetables in their final storage container in the field. This is known as *field packing*. Packaging containers are designed to minimize physical damage to the crop.

STEM Connection

Iced Storage

Knowing which produce will suffer when stored with ice will reduce premature spoilage and financial losses.

Produce that can be iced: artichokes, asparagus, beets, broccoli, cantaloupes, carrots, cauliflower, endive, green onions, leafy greens, radishes, spinach, sweet corn, watermelon.

Produce that is damaged by direct contact with ice: blueberries, cucumbers, garlic, green beans, herbs, okra, onions, raspberries, romaine lettuce, squash, strawberries, tomatoes.

Storage

Several factors are important in postharvest storage. Temperature, moisture loss, sanitation, and the presence of ethylene can affect the quality and shelf life of stored vegetables.

Temperature

Temperature is the single most important factor in maintaining quality after harvest. Vegetables should be brought to their proper storage temperature as soon as possible after harvest. Cooler temperatures slow respiration rates and slow the loss of sugars, fats, and proteins that affect flavor, salable weight, and shelf life. Most postharvest systems include refrigerated storage, which delays crop deterioration. Factors that refrigeration slows includes:

- Aging, which leads to softening, textural changes, and color changes.
- Undesirable respiration, which produces heat that can speed deterioration.
- Moisture loss with subsequent wilting.
- Spoilage caused by pathogenic bacteria, fungi, and yeasts.
- Unwanted growth, such as sprouting in onions.

Cooling facilities on the farm allow growers to cool and store produce, eliminating the need to sell crops immediately and provides growers with more market flexibility. Cooling facilities can be costly. Harvested produce should be precooled to remove any field heat. *Field heat* is the residual heat that crops hold from air temperatures and from the sun. Precooling is most important for crops with high respiration rates, such as Brussels sprouts, green onions, snap beans, asparagus, broccoli, mushrooms, peas, and sweet corn.

Precooling can be done in a number of ways. In room cooling, produce is placed in an insulated room with refrigeration units. Cooling this way takes time, but this can be the same room where the produce is ultimately stored. The packaging should be ventilated to allow cold air to move through it. Forced air cooling adds fans to refrigerated rooms. The fans pull cool air through the packaged produce. This method is 75%–90% faster than room cooling only.

Hydrocooling is a process in which harvested vegetables are dumped into a cold water tank or cold water is run over produce. This method efficiently removes heat and cleans produce. Crops such as berries, potatoes, sweet potatoes, onions, and garlic cannot tolerate being wet and so should not be hydrocooled. Hydrocooling also increases the risk for disease.

Top or liquid ice cooling is used on some vegetables, such as asparagus, broccoli, cauliflower, green onions, leafy greens, and sweet corn. Crushed ice is placed on top of the packaging that covers the produce. In liquid icing, a slurry of water and ice is injected into produce packages. Icing is especially effective on perishable items that cannot be readily cooled by other methods.

Vacuum cooling creates a vacuum in the chamber that holds the produce. Water within the plant evaporates and removes the heat. Produce can be rapidly cooled this way, but it can be an expensive method. It is used on only certain crops, such as lettuce and other leafy greens.

Many vegetables have an optimal storage temperature at just above freezing. Some vegetables can be injured by low temperatures, preferring slightly warmer temperatures. Basil, cucumbers, eggplants, pumpkins, summer squash, okra, and sweet potatoes are very sensitive to chilling injury, **Figure 22-23**. Moderately sensitive crops include snap beans, muskmelons, peppers, winter squash, tomatoes, and watermelons. Both time and temperature play a role in chilling injury. Many people store tomatoes in the refrigerator, and they will be fine when first removed. However, after a few days in warmer temperatures, chilling symptoms will emerge, which can include skin blemishes, loss of texture, and pitting.

Lissandra Melo/Shutterstock.com

Figure 22-23. Sweet potatoes develop odd flavors and undesirable texture when exposed to chilling temperatures.

Moisture Loss

Relative humidity in the postharvest environment directly influences water loss in the harvested crop. Water loss causes shriveling, wilting, a loss of texture (limpness), diminished crispness and juiciness, and softening. Many vegetables are sold by weight, and water loss will reduce profits. For most vegetables, postharvest quality is optimized with a high relative humidity (80%–95%). However, high humidity also encourages disease growth. Cool storage temperatures reduce disease growth. Combine high humidity and cool storage temperatures with sanitation and other methods to reduce risks. Refrigeration removes moisture from the air, often requiring the addition of moisture to storage facilities.

Sanitation

Sanitation is the process of keeping places or items free from dirt, disease, infection, and other substances that can cause illness. Proper sanitation practices reduce risks of postharvest diseases and lessen food safety risks from foodborne illnesses. Using a disinfectant in wash water and precooling water can help to prevent both postharvest diseases and foodborne illnesses.

Ethylene

Ethylene is a naturally occurring chemical emitted in varying quantities by fruits and vegetables and by decaying plant materials. Ethylene is given off by some fruits as they ripen. If these commodities are stored near crops that are sensitive to ethylene, they can promote unwanted or faster ripening of the sensitive crop. This can result in a loss of quality, reduced shelf life, and specific symptoms of injury. For example, carrots and parsnips exposed to ethylene will turn bitter, **Figure 22-24**. Cucumbers and squash will soften more quickly when exposed to ethylene. Tomatoes are ethylene producers.

Perutskyi Petro/Shutterstock.com

Figure 22-24. Carrots will turn bitter if exposed to ethylene.

Careers in Olericulture

Olericulture offers many different career possibilities. Two careers related to olericulture are vegetable grower and crop advisor.

Vegetable Grower

Growing vegetables for the fresh market or for processing remains one of the most important jobs in agriculture. The responsibilities of a vegetable grower include making decisions related to production, marketing, financing, and human resources management. On the production end, growers and their employees manage the soil and plant, cultivate, and harvest vegetables. Growers may use a broad range of tools from specialized equipment, such as tractors and implements, to shovels, trowels, hoes, and other hand tools. A grower's duties can include tilling the soil; applying fertilizers and pesticides; transplanting, weeding, thinning, irrigating, or pruning crops; and cleaning, packing, and loading harvested products.

Crop Advisor

A crop advisor or consultant advises growers, farm managers, and farm operators regarding their agronomic production practices. Many crop advisors work for large agricultural companies. By supporting growers these companies are able to meet sales goals and objectives by increasing the yields for a contracted crop. A crop advisor may market products to growers by making calls, providing advising or consulting services, and monitoring individual grower programs with respect to pesticides, fertilizers, irrigation, tillage, seed, and related areas. An ideal crop advisor has a bachelor's degree from an accredited four-year college or university program, plus three to five years of experience.

CHAPTER 22 Review and Assessment

Chapter Summary

- Vegetables contain important vitamins and minerals and are an essential part of a healthy diet. Vegetables can reduce the risk of heart disease, heart attacks, strokes, and certain cancers. They can lower blood pressure, decrease bone loss, and potentially lower calorie intake.
- Vegetables average $17.4 billion in annual sales and are grown on nearly 2.8 million acres across the country. Vegetables are destined for either fresh or processed food markets.
- A number of factors influence vegetable production including location, water, soil, nutrient management, temperature, and growing season extension methods.
- Most vegetables are irrigated. Common irrigation practices include drip, or trickle, irrigation and different kinds of sprinkler irrigation.
- A good soil for producing vegetables is a sandy loam. Almost any soil, however, can be used to grow vegetables with significant effort toward soil improvement.
- For optimal yields, a nutrient management plan is critical. Growers should change the pH of the soil or add nutrients as recommended by a soil test report.
- Temperature plays a large role in vegetable production by determining what can be grown in a given season. Vegetables are either warm-season or cool-season and should be planted accordingly.
- Season extension techniques will lengthen the growing season and provide earlier or later harvest. Season extension can be achieved using shade, transplants, cultivar selection, windbreaks, mulches, row covers, low tunnels, and high tunnels.
- Crop production methods, such as using proper spacing, rotating plant families, managing pests, and choosing the right plant material, optimize space and reduce risk of infection by plant pathogens. Using good agricultural practices that address potential sources of contamination is also important.
- Postharvest storage and handling may include cooling, cleaning, sorting, packing, storing, and shipping the harvested crop. A number of factors reduce postharvest quality. Through management of temperature and humidity, the deterioration can be slowed.

Words to Know

Match the key terms from the chapter to the correct definition.

A. crop rotation
B. dripline
C. good agricultural practice (GAP)
D. interplanting
E. leaching
F. low tunnel
G. olericulture
H. perched water table
I. row cover
J. sanitation

1. A lightweight fabric that is laid on top of vegetables to provide protection from frost or insects.
2. The movement of nutrients out of the soil into the groundwater.
3. The process of keeping places or items free from dirt, disease, infection, and other substances that can cause illness.
4. The outermost circumference of a tree's branches where water drips from the leaves onto the ground.
5. A method of planting early maturing crops next to later maturing crops to maximize space efficiency.
6. A production practice in which vegetables of the same plant family are planted in different locations every growing season.
7. Groundwater that is temporarily located above unsaturated soil due to compaction or some other soil formation factor.
8. The practice of producing edible vegetable crops.
9. A farm-level production method used to ensure that fresh produce is safe for human consumption.
10. A season extension technique in which small hoops are placed over rows and covered in fabric or plastic.

Know and Understand

Answer the following questions using the information provided in this chapter.

1. Why is eating vegetables an essential part of a healthy diet?
2. What role does dietary fiber play in the body and in what vegetables can it be found?
3. Which US states are top producers of vegetables?
4. How are the products in the vegetable industry classified for end uses?
5. Describe the mix of small and large farms that produce vegetables in the United States.
6. What is involved in the local food movement?
7. What are some environmental factors that impact vegetable production?
8. What are some considerations for selecting a location for a vegetable garden?
9. What are some benefits of using drip irrigation for growing vegetables?
10. What is the ideal soil for growing vegetables? How can other soils be managed to grow vegetables?

11. What effect can cold temperatures have on warm-season vegetables?
12. What is the purpose of season extension methods and what are some factors that relate to season extension?
13. How can cultivar selection for a vegetable garden extend the growing season?
14. List some advantages and disadvantages of using high tunnels.
15. Why is crop rotation an important production method?
16. What is the purpose of good agricultural practices (GAP) guidelines and what are four factors related to GAP?
17. What is a hybrid plant and why would a grower select hybrid plants when planning a vegetable garden?
18. List factors that can reduce postharvest quality of vegetables.
19. What role does temperature play in postharvest storage and handling?
20. Describe some of the duties and activities of a vegetable grower.

Thinking Critically

1. You are an employee who sells vegetables at a farmers market. A customer comes in with a complaint about how his produce rotted shortly after bringing it home. Determine and write the most appropriate response to this customer. Follow these guidelines:
   - Make sure you obtain a clear understanding of the complaint.
   - Restate the complaint using less negative terms.
   - Change the complaint into a question.
   - Explore alternative solutions.
   - Solve the problem.
   - Use proper grammar and spelling in your writing.
2. Your vegetable farm has been contracted to grow enough produce for a local foods promotion held once a week within the school system. The weather has not been cooperative, and your vegetables are behind schedule. You will not have enough harvestable material to meet your agreement. What will you do? List three possibilities to solve this problem.

STEM and Academic Activities

1. **Science.** Choose four different vegetable plants that interest you. Find the common name and the scientific name of each, using binomial nomenclature. Record this information in a table with three columns: Common Name, Scientific Name, and Example. In the Example column, either draw a picture of each plant or attach a photograph of the plant.
2. **Science.** Conduct research to learn more about why hybridizing tends to breed the flavor out of plants. Write a report on your findings.
3. **Technology.** Research environmental issues related to olericulture and learn how technology has played a role in the advancement of olericulture. Choose two specific topics and write a report explaining how technology has helped (or hurt) efforts in these specific areas to become more environmentally friendly.

4. **Math.** Your parents have permitted you to till a 10' × 30' (3m × 9m) garden plot to grow vegetables for the family. Determine the vegetables you would like to grow and create a plan. How many plants of each different vegetable can you fit into your garden?
5. **Social Studies.** Different societies and cultures have different food cultures. Conduct research to learn how food culture in the United States is different from that in other countries. What factors may have caused these differences?

Communicating about Horticulture

1. **Reading and Writing.** Working in groups of two or three, read the "Growing Site" section. Work together to summarize and write down the main points on study cards. Quiz each other on each concept.
2. **Listening and Speaking.** Make a collage. Using pictures from magazines or free online resources, create a collage that helps you remember ways that vegetables provide significant health benefits. Show and discuss your collage in a group of four to five classmates. Are the other members of your group able to determine the benefits that you tried to represent?
3. **Reading and Speaking.** Using your textbook, library resources, and the Internet, research a nontraditional vegetable. Focus on its geography and the way in which it is grown. Where is it grown? How is it is prepared? Which culture uses the vegetable? If possible, create a dish that incorporates the vegetable. If plant materials are not available, create a poster to illustrate your findings and dish. Present your findings to your classmates.

SAE Opportunities

1. **Exploratory.** Job shadow a vegetable grower for a small, mid-sized, or large farm.
2. **Experimental.** Grow vegetables in two small plots. In one plot, space your vegetables as suggested by the manufacturer of the seed or plant. In the other plot, practice interplanting. What were your yields in each plot? Which plot had a greater yield?
3. **Experimental.** Research which vegetables will grow well in a high tunnel or low tunnel in your region. Try growing a few of the different types and determine which ones yield the best results.
4. **Exploratory.** Create a list of vegetable transplants that could be sold in your school's plant sale. Create a schedule for sowing these plants from seeds to be ready in time for a spring market. Give cultural information, including planting depth and time for transplant production. Determine how many seeds of each plant you will need to purchase and what the cost will be to fill your greenhouse.
5. **Entrepreneurship.** Grow vegetables in your school garden and sell them to customers. Where is your market? How will you advertise?

photogal/Shutterstock.com

CHAPTER 23
Fruit and Nut Production

Chapter Outcomes

After studying this chapter, you will be able to:
- List the reasons fruits and nuts are important for health.
- Describe the market opportunities for fruit and nut production.
- Discuss environmental requirements, production, and harvesting for small fruits.
- Discuss environmental requirements, production, and harvesting for tree fruits and nuts.
- Discuss environmental requirements, production, and harvesting for vine fruits.
- List careers related to the pomology industry.

Words to Know

bench cut	floricane-fruiting	primocane-fruiting
bine	heading cut	scaffold branches
central leader system	modified central leader system	spur
cone		tannin
cordon	muck	thinning cut
cull	nematode	whorl
espalier	open center system	

Before You Read

Before you read the chapter, interview someone in horticulture or pomology. Ask the person why it is important to know about the chapter topic and how this topic affects the workplace. Take notes during the interview. As you read the chapter, highlight the items from your notes that are discussed in the chapter.

While studying this chapter, look for the activity icon to:

- **Practice** vocabulary terms with Words to Know activities.
- **Expand** learning with identification activities.
- **Reinforce** what you learn by completing Know and Understand questions.

www.g-wlearning.com/agriculture

Summer-ripened, aromatic peaches; tart and tangy blueberries; and crispy, crunchy apples—fruits are among the favorite foods of both kids and adults. Fruits and nuts consist of a wide variety of plants. Examples include small strawberry plants, various bushes, vines such as grape and kiwi, peaches and apples that grow on small trees, and pecans that grow on very large trees. Some fruit and nut trees have a relatively short life span of only a few years. Others may survive and produce for more than a hundred years. Determining what can be grown depends significantly on factors such as climate, soils, and available water. Commercial growers must also consider market opportunities when deciding what to grow.

Fruits for Health

Fruits are an important part of an overall healthy diet. Consuming at least five servings of fruits and vegetables a day, choosing whole grains, and eating low-fat dairy foods and varied protein sources are a recipe for a healthy lifestyle. Being physically active is important as well.

Health Benefits

Research has shown that people who eat more fruits and vegetables as part of a healthy diet may diminish their risk for some chronic diseases. Fruits provide minerals and nutrients important for the proper functioning of your body, **Figure 23-1**.

S-F/Shutterstock.com

Figure 23-1. Fruits are an important part of an overall healthy diet for everyone.

- Eating fruits as part of a healthy diet may reduce risks for heart disease (including heart attack and stroke), obesity, type 2 diabetes, and certain types of cancers.
- Eating potassium-rich fruits as part of a healthy diet may lower blood pressure. Eating fruits may also reduce the risk of developing kidney stones and help to decrease bone loss.
- Eating fruits may be useful in lowering calorie intake.

Nutrients

Fruits contain many nutrients that are needed for normal growth and development. Eating fruits can also reduce the risk of diet-related diseases. Fruits and nuts are a key to good health for these reasons:

- Most fruits are naturally low in fat, sodium, and calories.
- Fruits are cholesterol free.
- Fruits contain many essential nutrients including potassium, dietary fiber, vitamin C, and folate (folic acid).

- Diets rich in potassium may help maintain healthy blood pressure. Sources of potassium in fruit include dried apricots, cherries, and kiwi.
- Whole or cut-up fruits contain dietary fiber that along with a healthy diet may lower blood cholesterol levels and may lessen the risk of heart disease. Fiber is important for proper bowel function and provides a "full" feeling. Good sources of fiber include apples, blackberries, pears, raspberries, bananas, blueberries, figs, dates, guavas, kiwi, oranges, plums, and most nuts.
- Vitamin C is important for tissue growth and repair. It helps heal cuts and wounds and keeps teeth and gums healthy. Fruits that contain high levels vitamin C include apricots, blackberries, gooseberries, grapefruit, guavas, kiwis, lemons, limes, oranges, papayas, pineapples, pummelos, raspberries, strawberries, and tangerines.
- Folate (folic acid) helps the body form red blood cells. Pregnant women often take folic acid supplements to reduce the risk of birth defects. A good fruit-based source of folate is strawberries.
- Unsaturated fats are the "good" fats found in nuts. Both monounsaturated and polyunsaturated fats may lower bad cholesterol levels.
- Omega-3 fatty acids are found in many nuts. Omega-3 fatty acids are a healthy form of fatty acids that may help prevent dangerous heart rhythms that can lead to heart attacks, **Figure 23-2**.

Did You Know?
Luxury fruits bring in thousands of dollars in revenue in Japan. One Japanese Yubari cantaloupe sold at auction for more than $23,000.

Igor Palamarchuk/Shutterstock.com

Figure 23-2. English walnuts are a rich source of omega-3 fatty acids, which are good for heart health.

Fruit Markets in the United States

According to the USDA Economic Research Service (ERS), the typical American eats about 270 pounds of fruit and tree nuts each year. Comprised of a wide variety of commodities, the US fruit and tree nuts industry is an important part of the agriculture sector with an annual average of $18 billion in cash receipts. Production occurs on less than 2% of total harvested cropland. Small family or individual farm operations continue to dominate US fruit and tree nut production.

The largest fruit-producing states (excluding tree nuts) include California, Florida, and Washington. California has about 50% of the harvested fruit acreage. California's climate is conducive to growing fruit crops. Fruits produced in California include grapes, strawberries, peaches, nectarines, and kiwifruits, as well as apples, pears, plums, and sweet cherries.

Figure 23-3. Florida produces the most citrus in the United States.

California is second in citrus production. Florida grows nearly 25% of the harvested fruit acreage with its primary acreage in citrus, **Figure 23-3**. Washington State has 10% of the nation's fruit production and is the largest apple producer. This state also grows grapes and pears. Michigan, New York, Oregon, Pennsylvania, and Texas are also important fruit-producing states. Collectively, they have more than 10% of all US fruit acreage.

Fruit crops are sold in both fresh food and processed food markets. Fruits commonly found in fresh markets include avocados, bananas, nectarines, kiwifruits, strawberries, tangerines, sweet cherries, apples, pears, and lemons. While still sold fresh, the following fruits tend to be produced for the processed foods market: oranges, grapefruits, grapes, apricots, figs, prunes, peaches, tart cherries, and most berries (including blueberries and cranberries). Processed fruit products include canned fruit, frozen fruit, fruit juice, dried fruit, and wine. In both fresh and processed form, oranges, grapes, apples, bananas, and pineapples are the top five fruits consumed in the United States.

Nut tree production in the United States has increased significantly over the past 30 years, with annual production greater than two billion pounds. In the United States, a person eats, on average, three pounds of shelled tree nuts per year. Tree nut production in the United States generates nearly $4 billion annually, with highest sales for almonds, walnuts, pistachios, and pecans. California produces nearly 90% of all US tree nuts. Georgia, New Mexico, and Texas produce nearly 75% of the US pecan crop. Oregon produces significant numbers of hazelnuts, and macadamia nuts are grown in Hawaii. The top three nuts consumed in the United States are almonds, walnuts, and pecans.

Did You Know?
Temperatures in subtropical areas are not cold enough to break down the chlorophyll in the skin of citrus fruits. This means oranges grown in subtropical areas do not turn orange, even when ripe.

Small Fruits

Small fruits consist primarily of berries. Types of berries include blueberries, strawberries, blackberries, raspberries, and cranberries. Currants, gooseberries, and other small specialty crops, such as lingonberries and elderberries, are also small fruits. Berry production can be profitable for commercial growers and very rewarding for the home gardener. However, growing berries requires intense management and a considerable investment of time and resources. Some small fruit plantings, such as strawberries, produce fruit for as little as one or two years, while crops such as blueberries produce fruit for as many as 40 or 50 years.

Site Selection

Pomology is the science of fruit production. Successfully growing small fruits starts with finding the optimal location. Consider factors such as temperature, soil, slope, and water availability when selecting a site.

Temperature

Berry crops can be sensitive to both high and low temperatures. Many berries require a chilling period to break dormancy; requirements depend on the species and cultivar. Some examples of general requirements for chilling include:

- Strawberries—200–300 hours.
- Blueberries—650–850 hours.
- Blackberries—700 hours.
- Raspberries—800–1700 hours.
- Currants and gooseberries—800–1500 hours.
- Cranberries—2000 hours.

> **Corner Question**
>
> How are oranges from subtropical areas treated so they will turn orange?

In northern climates, chilling requirements are easily met. In the southern states, however, lack of chilling time may constrain production. Conversely, some small fruits, such as blackberries, some types of blueberries, and fall raspberries, may only grow well in slightly warmer winter temperatures. Temperatures below –20°F (–29°C) will kill blueberry shoots and flower buds. Blackberries cannot tolerate temperatures lower than –5°F (–20°C). Good air drainage is necessary to avoid frost injuries in the spring, particularly strawberries.

Slope

Avoid sloping land with a grade greater than 5%, as there is a risk of erosion and cultivation and irrigation are difficult. Moderate slopes (3%–5%) provide good air drainage, lessening the risk of chilling or freezing injury. Southern slopes can be problematic because earlier, warmer temperatures in the spring may encourage earlier blooming that can be damaged by frost. Western slopes often have winds that can desiccate small fruits. Windbreaks can diminish damaging winds, but the site should still allow for good air circulation.

Soil

Most small fruits prefer well-drained soil. Avoid heavy, poorly drained soils with high water tables. Subsoil drainage or raised beds may provide drainage on wetter sites. Strawberries, raspberries, blackberries, gooseberries, currants, and elderberries can tolerate a wide range of soil types if sufficient nutrients are available or added. They prefer soils that are high in organic matter (2%–4%) and have a pH of 6.0–6.8. Sandy loam or loamy soils are optimal for small fruit production. Blueberries are native to North America and tend to like poor, acidic soils, **Figure 23-4**. Blueberries' ideal pH is 4.2–4.8. Soils often require management to lower the soil pH for good production. Blueberries like loamy sand with high organic matter (greater than 4%) and can be grown in muck soils. *Muck* is a soil composed primarily of decomposed organic matter. Cranberries also prefer acidic soil with a pH between 4.0 and 5.5, and organic (muck) or sandy soil with a water table near the soil surface. Cranberry marshes may need to be developed by creating a perched water table.

Maria Dryfhout/Shutterstock.com

Figure 23-4. Blueberries are one of the few fruits native to the United States.

Site Preparation

Most small fruits are perennial crops. Strawberries are the exception. They are planted as an annual crop in the southeast. Thoroughly preparing a site can ensure years of successful cultivation.

Soil and Fertility

Test soils to determine nutrient levels and pH using proper soil testing methods. Work the soil with a rototiller, disk, or harrow several times in the year before planting to destroy weeds and loosen the soil. Just prior to planting, plow the soil again. For most small fruits, applications of potassium, phosphorus, magnesium, and calcium prior to planting (as given in the soil test recommendations) will help establish a productive plot. Insoluble phosphorus moves slowly in the soil and may take years to reach the root zone. However, it can be mixed into the root zone before planting. Organic materials used for nutrients, such as animal manures and legumes, can offer a source of slow-release nitrogen. This material should be incorporated into the soil prior to planting. Many animal manures also contain sufficient levels of potassium, phosphorus, and calcium. It can be difficult to make these nutrients available to plants after the plants are established. The soil pH should also be considered. It takes at least a year for lime to raise soil pH and a year for sulfur to acidify the soil.

Weeds and Other Pests

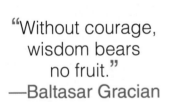

"Without courage, wisdom bears no fruit."
—Baltasar Gracian

Perennial weeds cause the greatest economic losses in berry crops and so should be eliminated in preparation for planting. Weeds can be hard to control after planting and can encourage other pest populations. Few herbicide options are labeled for use in small fruits. A number of strategies can be used to control weed populations, beginning one to three years before planting. These strategies include:

- Crop rotation, particularly with cover crops.
- Broad spectrum post-emergent herbicide application the season before planting.
- Repeated cultivation (tilling).
- Covering the soil with black plastic for several months.
- Soil fumigation, which fills an area with pesticide gas. This practice is currently allowed, but will likely be restricted in the future due to environmental concerns.

Many small fruits are susceptible to plant parasitic **nematodes** (roundworms that live in the soil), which are considered significant pests. Nematodes may build up in association with crops such as legumes, clover, alfalfa, vetch, and other weeds. Growers should determine if there are nematodes in the soil and take steps to minimize them if damaging levels are present. The soil-borne fungi *Verticillium* (which causes wilt) and *Phytophthora* (which causes rotting) are sometimes found in fields that have previously been planted with certain crops, **Figure 23-5**.

Miyuki Satake/Shutterstock.com

Figure 23-5. Much of the damage caused by nematodes occurs in plant roots.

These crops include tomato, potato, tobacco, eggplant, pepper, cucurbits, strawberry, black raspberry, and blackberry cultivars.

Irrigation

Water is the most critical factor for optimal fruit growth and development in small fruits. Prior to planting, thoroughly moisten the transplants or bare roots of bramble crops. Water plants immediately after planting. For both commercial and home fruit plantings, provide irrigation throughout the growing season, especially near harvest time. A shortage of water near harvest can limit fruit size and impact the health of the plants for production the next year. Most small fruits are shallow rooted so drip or trickle irrigation will efficiently deliver water, **Figure 23-6**.

Max Lindenthaler/Shutterstock.com

Figure 23-6. Drip irrigation for blueberries and other small fruits is an efficient and effective way to meet water requirements for these crops.

Planting

Planting methods vary for different types of small fruits. Proper planting time, soil nutrients, and soil pH are important considerations for successful crops.

Brambles: Blackberries and Raspberries

Most brambles are planted in the early spring. Commercial growers generally use bare root, dormant nursery stock. Bare rootstock should be healthy and free of disease and insects. Rootstock is planted within a day or two of purchase. Dry bare roots can be soaked for several hours to rehydrate the plant before planting. Holes for planting should be large enough to allow the root system to spread. Rootstock is covered to a depth of 2″–3″ (5 cm–7.5 cm), and the soil around the base of the plant is pressed firmly into place. The stem can be cut off, leaving 3″–4″ (7.5 cm–10 cm) of stem, from which new growth will emerge. Water the plants thoroughly after planting.

Some growers use tissue culture plugs because they are supposed to be completely disease free, including free of viruses. The plugs may need to be hardened off prior to planting. Once a plug is ready for planting, set it in the hole so the top of the plug is even with the soil. Place a thin layer of soil on top around the plug. Water plants thoroughly after planting.

Strawberries

Strawberry planting varies depending on climate. In the southeastern United States, strawberries are grown as an annual crop commercially. Plants are set in the ground from September to early November and bear fruit in April, May, and June. In northern climates, strawberries are grown as perennial crops. Planting occurs as early in the spring as possible, after the ground will no longer freeze. All plants (plugs or bare roots) should be well watered and planted within a day or two of arrival. Make a hole that is large enough for the roots to spread out. Plant the strawberries so that the crowns are even with the soil surface. Press the soil down firmly but gently after planting.

> "You've got to go out on a limb sometimes because that's where the fruit is."
> —Will Rogers

Figure 23-7. Tighter spacing of strawberries using the double-row hill system increases yields.

Strawberries that are vigorous should be spaced 24″–30″ (60 cm–75 cm) apart. Cultivars with moderate to low vigor are set 18″–24″ (45 cm–60 cm) apart in the row. The rows are generally placed 36″–48″ (90 cm–120 cm) apart. Most of the runners from mother plants are permitted to grow during the first season. In a double-row hill system, the plants are placed 12″ (30 cm) apart in double rows, with plants alternating, **Figure 23-7**. Rows are set with 30″–42″ (75 cm–100 cm) of space between rows. Runners are cut as they form to create large individual hills that have high yield and good quality fruit.

The process of growing strawberries using plasticulture begins with the soil bed being prepared, drip irrigation laid, and the plastic pulled over the soil's surface. Holes are then cut or punched into the plastic at the preferred spacing. Plants, roots, or plugs are planted in the openings in the plastic.

Blueberries

Prior to planting blueberries, the soil will likely need to be amended. Using the soil test recommendations, bring the soil pH to an appropriate level. To transplant successfully, choose nursery plants that are two or three years old. Keep the plant roots moist at all times. Dig a hole large enough for the roots to be able to spread. Plant the root ball or bare root plant so it is level with the soil surface.

Five main types of blueberries are grown in the United States: northern highbush, southern highbush, lowbush, rabbiteye, and half-high (a cross between highbush and lowbush), **Figure 23-8**. Lowbush blueberry (*Vaccinium angustifolium*) cultivars are grown in Maine and Canada. Space plants 12″–24″ (30 cm–60 cm) apart in rows. Northern highbush (*Vaccinium corymbosum*) cultivars are grown primarily in the Midwest, mid-Atlantic, Northeast, and Northwest. They are spaced 48″–60″ (120 cm–150 cm) apart in a row, with 8′–10′ (2.4 m–3 m) between rows. Rabbiteye blueberries (*Vaccinium ashei*) are native to the southeastern United States and are planted 72″ (150 cm) apart in a row with 10′–12′ (3 m–3.7 m) between rows. Most blueberries are planted in late winter or early spring depending on the geographic region. Top growth of the plants should be pruned, leaving one to three of the most vigorous upright shoots. In the first year, remove flower buds (plump rounded buds) to drive energy toward root and shoot development rather than berry production.

Figure 23-8. Lowbush blueberries are native to the northern United States and are an important commercial crop in Maine and Canada.

Cranberries

Commercial cranberry bogs require special site preparation to create soil conditions that allow for 24″ (60 cm) of flooding. In backyard gardens, use a plastic liner to hold water. Cranberries are planted from rooted cuttings or nursery containers, generally in the fall in northern climates. Space one-year-old plants about 18″–30″ (45 cm–75 cm) apart. Space cuttings 9″ (23 cm) apart throughout the planting bed.

Maintenance

All small fruits require management as they grow and develop through the seasons. All crops need proper nutrition, irrigation, pest management, and pruning. Each crop also has specific growing needs that should be researched. In general, plants require nutrients throughout the growing season. Using a synthetic fertilizer or other means, applications and amounts of nutrients should be appropriately timed for a crop. Water is usually supplemented through irrigation since small fruits are particularly sensitive to drought stress. All small fruits may have weed problems that need to be addressed. Monitoring and management of insect pests and other plant pathogens, such as fungi, bacteria, nematodes, and viruses, is important. More detailed information can be found in cooperative extension service publications for each crop.

Pruning and Training

Bramble crops need the most pruning and training. Other small fruits, such as blueberries and gooseberries, may need to be pruned to remove weak, dense, and dead growth to allow for rejuvenation and optimal growth. Blackberries and raspberries are trained and pruned for use in commercial and home gardens. Trellising minimizes labor and maximizes yields. A trellis can be as simple as a fence or other supportive structure. Commercial growers tend to choose an I-trellis, a V-trellis, or a shift trellis, **Figure 23-9**.

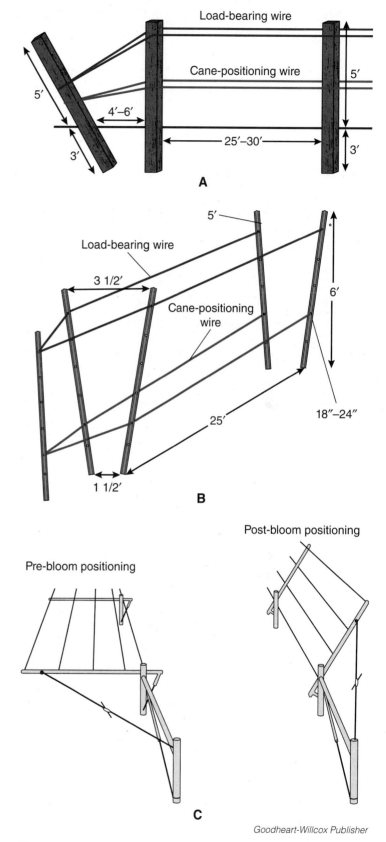

Goodheart-Willcox Publisher

Figure 23-9. Types of trellises. A—An I-trellis, or hedgerow, with an end-post configuration. B—The most commonly used trellis is the V-trellis. C—The shift trellis allows for easier harvesting.

Generally, the trellis has wires stretched between posts, and the fruiting canes are tied to these wires in the spring. The erect varieties are tied where the canes cross the wires. Trailing types have canes tied horizontally along the wires. Each trellis has advantages and disadvantages and can be modified. *Floricane-fruiting* cultivars (brambles that produce fruit on second year canes only) need pruning several times a year. *Primocane-fruiting* cultivars (brambles that produce fruit at the top of first-year canes in late summer), such as raspberries, need to be pruned only once a year during the winter or early spring.

Pollination

In home garden production, most small fruit pollination needs are met by wild or local hives of honeybees or native pollinators, or the crop is self-fertile. For commercial producers, berries can be improved by cross-pollination, **Figure 23-10**. Pollination needs for several berries are listed below:

- Most blackberries are self-fertile, but a few require cross-pollination.
- Fruit set and crop yield for blueberries are improved by cross-pollination.
- Currants are self-fertile.
- Most varieties of gooseberries, black raspberries, and strawberries are self-fertile.
- Most red raspberries are self-fertile, but their size is improved by cross-pollination.

V. J. Matthew/Shutterstock.com

Figure 23-10. Commercial beehives brought into a field for cross-pollination can help increase fruit set and size.

Harvest

Small fruits are highly perishable. Harvesting for strawberries and brambles is done frequently when fruit is ripe (every day or every two or three days). Most harvesting is done in the morning when the berries are still firm and temperatures are cooler. To optimize postharvest quality, fruit is picked gently and placed into picking containers. *Culls*, undesirable, low-quality fruit that is composted or thrown away, go into a different container. Blackberries and raspberries bruise easily. They are packaged only two or three berries deep. Field heat is removed through refrigerated cooling as soon as the fruit is harvested.

Blueberries should be picked every five to seven days for the best quality. Berries are placed in small buckets or shallow trays not more than 4"–5" (10 cm–13 cm) deep to prevent crushing. For the processing market, blueberries may be mechanically harvested. Ripe currants and gooseberries can remain on plants for as much as a week, allowing all fruit to reach maturity before harvesting.

Harvesting cranberries for the fresh market differs from harvesting cranberries destined for processing. Fresh fruits are harvested with a picking machine that has tines. The tines comb through the vines and catch the fruit. The fruit is sent along a conveyor to be packaged. The wet cranberries are dried in boxes with ventilation and refrigerated. Grading is done as they are packaged for retail sale. Cranberries that will be processed for juice, dried fruit, or other products are wet harvested. In wet harvesting, beds are flooded with water.

A machine with a beater or a slipper is driven through the bed to remove the berries. The berries float to the surface and are moved or pumped into a truck.

Tree Fruits and Nuts

Tree fruits and nuts are investments that need time to produce a harvestable crop. Most tree fruit and nut orchards are productive for decades. Tree fruits are classified by the type of fruit produced. Apples and pears are pome fruits. Cherries, peaches, plums, nectarines, and apricots are stone fruits, or fruits that have a hard pit in their center. Citrus tree fruits include oranges, grapefruits, tangerines, lemons, and limes. Other tree fruits include persimmons, pomegranates, figs, quince, and avocados, **Figure 23-11**. Tree nuts include walnuts, almonds, cashews, pistachios, pecans, hickory nuts, filberts, and hazelnuts.

lauraslens/Shutterstock.com

Figure 23-11. California produces about 95% of avocados grown in the United States.

Home gardeners might consider planting fruit and nut trees as specimen trees in the landscape, instead of ornamental evergreens or deciduous plants. Using fruit or nut trees as shade trees provide an edible harvest and a cool, refreshing space to sit and relax. Pecans, walnuts, standard apples, and persimmons are beautiful trees that create shade. A fruiting hedge of figs or filberts provides a fence or barrier and a crop to eat. This can also serve as a windbreak for other plants in the yard.

Site Selection and Preparation

Growing tree fruits and nuts successfully starts with finding an optimal location. Considerations for selecting a site include climate and soil.

Climate

Climate is a very important factor when considering what to plant in a backyard or commercial orchard. Many tree fruits have chilling requirements, while others cannot tolerate freezing temperatures. Some locations may not have a long enough growing season. For these reasons, it is essential to determine what can be grown in a given area to ensure successful planting. For example, cherry trees generally do not thrive where summers are long and hot or where winter temperatures are high for short periods, such as much of the southeastern United States. Most sweet and tart cherry varieties have chilling requirements of about 1000 hours. This means they need about 1000 hours of temperatures between 35°F–55°F (1.5°C–13°C) during the winter. The cumulative hours of chilling are required to break dormancy so buds can develop in response to warm spring temperatures. Areas such as southern California only reach 100 to 400 chilling hours. Predominant cherry-producing regions in Washington state and Michigan have more than 1200 chilling hours on average every year. Each tree fruit and nut crop has different chilling requirements to break dormancy.

Did You Know?

Black walnuts are not only delicious but have a number of interesting uses. The shells can be used as an abrasive cleaning agent for jet engines, a filter in smokestacks, and a filler for dynamite.

> **Did You Know?**
> Some cherry growers use fans and even helicopters to dry fruit after a rainfall. Ripe fruit will swell and split as it absorbs additional moisture.

Climate also plays a role in the ability of pests and diseases to develop in tree fruits and nuts. For example, many of the tree fruit and nut production centers are in parts of California that have low humidity. Low humidity reduces the risk of diseases that prefer moist, humid environments.

Soils

Fruit and nut trees should be planted in well-drained, medium-textured, and fairly fertile soil. Avoid poorly drained soils as they inhibit oxygen from reaching tree roots, encouraging the growth of root rot pathogens. It is nearly impossible to establish and maintain a good orchard on a site with poorly drained soil. Adding organic matter in the form of compost, manure, straw, and sawdust can amend soils to increase their capacity to hold water and nutrients. Woody materials may create nitrogen deficiencies in the soil and may also bring weed seeds and other pathogens.

Test soils that will be used for planting new fruit trees. If recommended, soils should be adequately limed and fertilized as much as two years prior to planting. Orchard subsoils should be loosened to a depth of 24″–36″ (60 cm–90 cm) to break up any hardpans and to encourage the movement of lime and nutrients to the roots. Recommended soil pH is generally around 6.5 for most crops. Fruit and nuts trees are generally not fertilized at planting or in the first year of growth.

Planting

When choosing a cultivar for a fruit or nut tree, consider hardiness, disease resistance, yield, flavor, fruit size and appearance, and shelf life. Depending on the type of fruit or nut and the cultivar, pollinizer plants may also be needed to aid in pollination and fruit set. Home gardeners should consider the space requirements of any fruit or nut trees they wish to plant.

Almost all fruit and nut trees are grafted or budded in the nursery to a cultivar that will bear fruit or nuts. For example, all apple cultivars (for example, 'Honeycrisp' or 'Gala') are grafted onto a rootstock. Apple rootstocks are classified as dwarf, semi-dwarf, or standard. A backyard gardener might select an apple cultivar that is grafted onto a dwarf rootstock for a small tree that will be easy to pick, prune, and protect against pests. Rootstock material also provides additional characteristics, such as disease resistance, soil adaptability, and stress tolerance, and should also be chosen with this in mind.

Planting Timing

Climate determines the best time to plant tree fruits and nuts. In mild climates, trees may be planted in the late fall or early winter. In northern climates, bare root trees are planted in early to late spring. This timing gives roots enough time to establish themselves and allows plants to develop into stronger, larger trees. Young fruit trees are commonly shipped bare root, with the exposed roots wrapped in moist sawdust. The trees should be planted as soon as possible after purchase. Since potted trees can be thoroughly watered before shipping, nurseries ship potted trees to limit plant stress during transport and transplanting.

Dig a hole that is twice the size of the root system, loosening the soil on the sides of the hole. Prune any roots that are too long to avoid girdling.

If the tree is grafted, plant the tree with the graft union above the soil surface, usually about 2″–6″ (5 cm–15 cm). After the tree is in place, fill the hole with soil that has been mixed with well-aged compost or other organic matter. Water the tree thoroughly. Provide about three to five gallons of water per week until the tree is established. Many growers will prune the top of tree after planting to promote vigorous growth.

Sod Establishment

Many fruit tree orchards plant permanent sod covers between rows to prevent soil compaction, decrease surface water runoff, increase soil-water infiltration rates, minimize wind and water erosion, maintain or increase soil organic matter content, conserve plant nutrients, and to make it easier to move sprayers and other equipment during wet periods. Fescue, or a beneficial insect attraction mix, is the most commonly planted grass. Establish the cover using seed or sod at least six months prior to planting trees. It should also be managed for weed control.

Pruning and Training

Proper pruning and training of fruit trees is critical in the development of a strong tree structure that can support fruit production. Correctly shaped trees will yield high-quality fruit sooner and will last considerably longer than incorrectly shaped trees. Regular pruning and training will also optimize light infiltration to the developing flower buds and fruit, maintain tree vigor and size, and allow air movement through the tree to minimize pest problems. *Training* is a practice that permits tree growth to be guided into a specific form. *Pruning* removes parts of a tree to correct or preserve tree structure.

Types of Pruning Cuts

Three main types of cuts are used during training and pruning:
- ***Thinning cut***—removes an entire branch back to its base. Thinning cuts are used to eliminate crowding among branches, **Figure 23-12**.

Goodheart-Willcox Publisher

Figure 23-12. Thinning cuts remove crowding within a fruit or nut tree.

Figure 23-13. Heading cuts promote invigorated growth and the emergence of strong lateral branches.

- *Heading cut*—removes only the terminal portion of a branch to just above a bud or side branch. This cut invigorates growth of lower buds and creates a much stronger branch. Lateral secondary branches emerge, **Figure 23-13**.
- *Bench cut*—removes vigorous, upright limbs back to side branches. These cuts can open the center of the tree, allowing branches to spread outward. This is an aggressive cut and should be used only when necessary.

Open Center System

The *open center system* is a fruit tree training method that keeps the center of the tree free of large branches, allowing sunlight to penetrate the lower fruiting wood. This commonly used method results in a tree that is vase-shaped, **Figure 23-14**. This method of pruning is generally used on almond, apricot, cherry, fig, nectarine, peach, plum, and prune trees. In the first growing season with newly planted trees, three to five shoots are selected to train into the *scaffold branches* (primary limbs that extend from the main trunk and form the tree canopy). All other shoots are cut back to 4″–6″ (10 cm–15 cm). Scaffold branches are spaced several centimeters apart vertically and spread evenly around the trunk. Avoid upright limbs with narrow angles because they can break easily. A 45° angle is the most desirable. The lowest branch should be about 24″–32″ (60 cm–80 cm) above the ground.

Heading cuts are made to scaffold branches during the dormant season of the first two years to promote continued lateral branching and to strengthen the scaffold branches. Remove all other strong branches to minimize competition and keep the open structure. During the dormant season of the third year, the primary scaffold branches can be pruned so that secondary scaffold branches arise. After secondary branches grow to 24″–30″ (60 cm–75 cm), they can be headed to develop two to three tertiary branches. Manage mature trees with thinning cuts to avoid crowding and to remove any undesirable upright growth.

Figure 23-14. An open center system in fruit trees allows for increased airflow and light infiltration for fruit development.

Central Leader System

A *central leader system* is a training method for fruit trees that is characterized by one main, upright trunk called the leader. This method allows maximum light penetration and is often used for apple, pear, plum, and pecan trees, **Figure 23-15**. Pruning the tree to a height of 30″–36″ (75 cm–90 cm) will balance the top and roots. It will also encourage buds just below the cut to grow and form scaffold branches. In newly planted trees, select the most upright shoot as the central leader. Of the remaining shoots, select three to four branches spaced uniformly around the trunk to make up the scaffold *whorl* (a spiral pattern of limbs). Above each scaffold branch, leave an open area of 18″–24″ (45 cm–60 cm) to allow light to penetrate to all lower leaves and fruit. Scaffold branches should alternate up the leader, creating these light slots.

The scaffold branches or lateral branches may need to be spread to a wider crotch angle to provide stronger, supportive branches for bearing fruit loads in future years. Toothpicks or clothespins can be used to angle the young branches open to 50°–60°. Maintaining the central leader includes dormant pruning with a heading cut to approximately 24″–30″ (60 cm–75 cm) above the highest whorl of scaffold branches. This promotes continued branching and scaffold whorl development. Dormant pruning will also eliminate weak, damaged, dead, or diseased wood, as well as branches growing downward or upright. Unbranched lateral branches should be cut back.

In trees that have not been trained properly, very upright branch angles can still be spread out to minimize the risk of limb breakage. Branches can be tied down with weights. A board with notches cut at each end can be used to hold the spread branches in place.

Goodheart-Willcox Publisher

Figure 23-15. A tree with a central leader system has a strong main trunk that provides optimal light penetration.

Modified Central Leader

A *modified central leader system* is a variation of the central leader form. In this system, the central leader is removed after five to nine main lateral branches develop, **Figure 23-16**. This method is used on walnuts and persimmons. It can help manage fire blight in apples and pears. If one branch has fire blight, it can be removed without significant loss to the tree.

As with central leader training, the most upright and vigorous shoot is developed into the leader. The other branches are cut back except for two or three lateral branches that are well spaced around the tree. Several branches can be developed and should be offset from each other. The branches should be well established before the central leader is removed. In young pear trees, prune back to about 28″ (70 cm) at the time of planting. Four to six wide-angled branches (greater than 45°) should be selected as the primary scaffolds and should be located at least 4″–12″ (10 cm–30 cm) apart. The lowest branch should be 20″ (50 cm) from the ground, and the branches should whorl upward around the trunk. Some branches may be selected the second year. This is done for walnuts and persimmons, with the spacing being slightly different for larger trees.

Goodheart-Willcox Publisher

Figure 23-16. A tree with the modified central leader system has multiple main trunks.

Using spreaders to ensure a proper angle will result in trees that are stronger, stiffer, and have a better structure. To encourage secondary scaffold branches, follow the same process as for the central leader. Cut primary scaffold branches and select secondary branches. Be sure to make proper thinning cuts and remove any vigorous growth.

Summer versus Dormant Pruning

The time for pruning varies with the type of plant and its fruiting habits, as well as the geographic location. Pruning in spring and summer allows growers to train young trees and can shorten the time to full fruit production. Shoots of young trees can be bent and staked to make them grow at the proper angles and in the proper directions. On mature trees, summer pruning mainly involves removing vigorous upright shoots that are not needed as permanent branches and heading or thinning shoots to control tree height and develop branches. It also slows the growth of the tree.

Pruning during dormancy in the winter months will encourage greater growth and branching during the growing season, **Figure 23-17**. To foster the development of scaffold branches, prune the central leader above the highest scaffold whorl during the dormant season. Dormant pruning should be done after the risk of a severe freeze, generally in late winter or early spring. Also make any thinning cuts to remove dead, diseased branches at this time.

pryzmat/Shutterstock.com

Figure 23-17. Dormant pruning encourages vigorous growth in the spring.

Did You Know?
Apples float in water because they are twenty-five percent air.

Fruit Thinning

Fruit thinning is done on many fruit and nut trees to produce a desirable fruit size and to ensure annual bloom. Trees that benefit from thinning include apple and pear, stone fruit (plums, peaches, apricots, nectarines, and cherries), some heavy-bearing citrus (mandarin orange), and some nut trees such as pecans. Some trees naturally drop fruit, thinning themselves. Many trees produce a crop only every other year unless they are thinned. Alternate bearing years may also bring unmarketable nuts, small fruit size, or physiological issues such as increased susceptibility to cold damage. Trees with brittle wood may also risk breakage if a fruit crop is too heavy.

Thinning can be done by hand or using a pole to reach the fruit. Thin apples and pears to one fruit per cluster and space clusters 6″–8″ (15 cm–20 cm) apart. The space between peaches should be about the size of a hand, 6″–10″ (15 cm–25 cm). Pecan trees are shaken to remove up to half the crop load, although this varies by cultivar. The timing for thinning depends on the crop and weather conditions, but it generally happens soon after the fruit has set.

Nutrient Management

Orchards generally use less fertilizer than other crops. High levels of nutrients can stimulate excessive growth and decrease fruit quality. An annual soil analysis keeps growers and gardeners informed about the levels of nutrients present in the soil and the soil acidity. Fruit and nut trees require nutrient applications throughout their lifetime using recommendations from soil tests.

STEM Connection

Espalier

Espalier is a training method in which a plant grows flat against a wall, lattice, or other support. Espalier is a beautiful and functional horticultural technique that trains plants into a single plane and optimizes space. The fruit tree itself can be a fence if trained on wire between posts and can block unsightly views or provide an attractive screen. Most fruit trees respond to espalier training, but apples, pears and plums are the most commonly used. Landscape plants can be used as well. The entire process can take as long as 10 years with yearly maintenance.

In espalier training, tree branches are typically trained along the wires of a trellis. The trellis supports may be steel posts or wooden fence posts and generally have wire running between them. The tree is planted 6"–10" (15 cm–25 cm) from the wall to provide enough room for the roots to grow and to allow for air circulation and pest control.

To use the espalier method successfully, the tree must be in its first year or two of growth. Older trees are more difficult to train. Select an uppermost new shoot, which is usually growing in an upright position, and tie it into position as a central leader. Then select two shoots that will emerge below the central leader and tie them loosely with plastic ties to the bottom wire, with one going in each direction. Cut back any other shoots to short stubs. In tying a shoot to a wire, do not bend it so far downward that the tip is lower than the attachment point on the trunk. If possible, arrange the shoot so the tip is a few centimeters higher than where it meets the trunk. The uppermost shoot should grow well past the second wire by the end of the first growing season.

Steve Heap/Shutterstock.com

During dormancy in the following season, cut back the central stem just below the second wire. As lateral shoots emerge just below the cut, again tie these to the second wire, with one going in each direction. The branches that are trained to the first wire won't need much pruning the second year, except to maintain terminal growth and to prevent vigorous upright shoot growth. Repeat this process during the following years of training until the basic espalier framework is in place. When the central leader reaches the top wire, bend it in one direction and tie it to the top wire or cut it just below the top wire.

Pest Management

Unless properly managed, insects and diseases can seriously damage fruit trees. Practicing proper integrated pest management, such as scouting and monitoring, allows growers and gardeners to properly identify pests so that proper control measures can be used. Pests can be managed with multiple methods, including cultural, biological, chemical, and physical or mechanical control. Fruit trees are susceptible to a host of pest challenges. Pests must be managed before severe crop loss and tree damage can occur. Contact a local cooperative extension service office to learn about the most common pests for each crop and the best management practices to control them.

Corner Question

What is the only apple native to North America?

Harvesting and Storage

Fruit must be harvested at the proper stage of maturity in order to maintain its nutrients and postharvest quality. Factors that maintain shelf life and quality include timely harvesting, cooling, proper handling, and storage. Each commodity has different maturity indicators, which are far too numerous to list here. However, general indicators can include color changes, size, starch and sugar content, firmness, taste, and natural falling (nuts). Ripening times may vary from year to year depending on environmental conditions.

Fruits and nuts can be harvested by hand or with mechanical harvesters. Fruit intended for the fresh market are most often picked by hand, which is an intensive and expensive process. Hand-picked fruits include apple, peach, pear, apricot, mango, avocado, litchi, kiwi, olive, and sweet cherry. Many tree nuts are harvested with mechanical shakers. A large machine wraps a pliers-like mechanism around a tree trunk and vibrates the tree, causing the nuts to drop, **Figure 23-18**. Mechanical pickers are then driven across the ground to collect the nuts (similar to a golf ball picker on a driving range). Citrus fruits and cherries may be mechanically harvested, especially if intended for processing as juice or dried fruit.

Vine Fruits

Plants that grow as vines require some type of structural support and training to produce a harvestable crop. Grapes are the most commercially important vine crop. Kiwifruit, hops, and passion fruit are minor specialty crops that have somewhat similar growing needs. Grapes are an ancient crop. They are grown for fresh table grapes or processed into jams, juices, and wine. Many native and nonnative grapes can be easily grown by home gardeners. While there are a few hardy kiwifruit vines for gardeners in warmer climates, most commercial kiwi production occurs in California.

Michael Parker, North Carolina State University

Figure 23-18. Mechanical harvesting is done on many nut trees, including pecans.

Hops, used primarily in the production of beer, are produced in the Pacific Northwest and Idaho, with smaller production pockets elsewhere. Passion fruit is a specialty vine crop grown for both the fresh and processed market. Most production is done outside of the United States, **Figure 23-19**.

Site Preparation and Planting

Vineyard (both grape and kiwifruit) and hopyard establishment involves careful planning, thorough site preparation, design, planting, and trellis construction. In establishing a perennial crop, preparing the site well will minimize costs and labor for maintenance and harvest later on.

bouybin/Shutterstock.com

Figure 23-19. Passion fruit is produced primarily outside of the United States, but makes an interesting vine crop for home gardeners in warm climates.

Climate and Topography

Climate is the long-term prevailing weather of a region or site. The climate of a vineyard or hopyard is influenced by temperature, precipitation, winds, and other conditions. Large bodies of water or mountains can also affect the climate. For example, many grape-growing regions are close to large bodies of water such as the Great Lakes or the Finger Lakes. Grapevines require at least 165 frost-free days to develop cold hardiness to protect them from killing frost that damages plant tissues. Low temperatures are a consideration for many grapes and will determine what type of grape and cultivars can be grown in a given area.

Sloping land has advantages for trellised vine crops. Slopes allow for adequate air drainage. Slopes can also reduce the risk of spring frost injury dramatically. However, slopes greater than 15% are at risk of erosion and nutrient runoff, and can be dangerous when using machinery.

Soil

Establish vine crops in well-drained, loamy soils. Many grapes can manage on a wide range of soils, and rootstocks can increase the adaptability to more marginal soils. Avoid wet soils for all plants. Hops grow best in soil with a pH of 6–7. Depending on the grape, the pH range extends from 5 to 7. *Vitis labrusca* (American) grapes can tolerate pH of 5.5–6.0. Hybrid and *Vitis vinifera* (European) grapes prefer 6.0–6.5 pH. *Vitis rotundifolia* (muscadine) grapes require 5.5–6.5 pH. Organic matter builds soil structure, improves the soil's ability to hold water and nutrients, and can be added prior to planting. Using the results from a soil test, planting sites should be prepared at least a year before planting so that pH and fertility can be adjusted. Many growers will till the subsoil to break up any hardpans and to incorporate lime.

Planting

Planting times and methods vary for grapes, hop plants, and kiwi plants.

Grapes

Many different types of grapes are grown in the United States, with each one having a characteristic suitable for the fresh food or processed food market. Two species native to North America are the Concord grape, *Vitis labrusca* ('Concord' and 'Niagara' cultivars) and the summer grape, *Vitis aestivalis* ('Norton' and 'Delaware' cultivars). Both are used for table grapes, juice, jam, and jellies. A grape native to the southeastern United States is the muscadine grape, *Vitis rotundifolia*, **Figure 23-20**. Several seedless grape varieties have been developed from these native species. Wine grapes from the European grape, *Vitis vinifera*, or from French-American hybrids are grown in many parts of the country, most notably in California and the Pacific Northwest. Many of the hybrids have been developed to adapt to cooler climates in areas such as the Midwest and Northeast. Some are more tolerant to southern climates. Most *vinifera* are grafted onto American rootstocks for disease resistance and soil adaptability.

Melinda Fawver/Shutterstock.com

Figure 23-20. The muscadine grape is native to North America and is well-suited for production in southeastern states.

Grapevines are generally planted in late winter to early spring as bare root plants or container-grown plants. Dig holes for vines one day prior to planting. Holes should be about 12"–18" (30 cm–45 cm) deep. Set the grapevines in the hole with the graft union remaining several centimeters above the soil surface. Vines that are not grafted, such as muscadines, can be placed in the hole so the crown is even with the soil line. Spacing of vine rows varies depending on the level of vigor, but generally plants are placed 6'–10' (1.8 m–3 m) apart, more for muscadines. Water each vine thoroughly after planting.

Hop Plants

Hop plants (*Humulus lupulus*) are perennial vines that have male and female flowers growing on separate plants, **Figure 23-21**. Hop plants die back in the winter and produce annual *bines* (long, flexible climbing stems) from a rhizome. They are most often propagated vegetatively from rhizomes or softwood cuttings. Rhizomes are cut into 6"–8" (15 cm–20 cm) lengths and are transplanted directly into the soil or stored in a cool place. Softwood cuttings are taken from stem lengths with one to two nodes and two leaves.

sivivolk/Shutterstock.com

Figure 23-21. Hop plants are grown mostly in the Pacific Northwest, but there are emerging pockets in the upper Midwest and Northeast.

Hop plants should be planted into recently tilled rows in early spring. The most common plant spacing is 7′ (2.1 m) apart in a row with 7′ (2.1 m) between rows. An alternate spacing is 3.5′ (1 m) apart in a row with 8′–15′ (2.4 m–4.5 m) between rows, for an average of 800 to 1000 plants per acre. Hop rhizomes are planted horizontally with the bud side up and covered with 1″–2″ (2.5 cm–5 cm) of soil. Once the bines grow to 18″–24″ (45 cm–60 cm), select four bines from each rhizome to keep and thin the remainder. Two bines are then trained up each of the support strings in a clockwise direction.

Kiwifruit

Kiwifruit (*Actinidia deliciosa*) is a perennial vine that bears large fruit on female flowering vines. The fruit has a characteristic fuzzy brown skin. Fruit texture is similar to that of a strawberry and the flavor resembles a blend of strawberry and pineapple. Hardy kiwifruit (*Actinidia arguta, Actinidia kolomikta*) are the size of a grape. Unlike the skin of *Actinidia deliciosa*, hardy kiwifruit has smooth, edible skin. Kiwifruit are planted from potted cuttings and trained on a trellis system. Some kiwis are also grafted, but limited research has been done in this area.

Plant kiwifruit vines 10′ (3 m) apart in the spring after the danger of frost has past. Plant one male plant for every eight to nine female plants. Create a hole large enough to accommodate the roots. Minor trimming of the roots is acceptable if needed. Plant vines just deep enough to cover the top roots. Never mound soil around the plant, even when the plant is established. Vines should be thoroughly watered after transplanting.

Training and Trellising

Vine crops are grown on various trellis systems. These systems provide vertical support for the growing plant, **Figure 23-22**.

Grapes

Vineyard trellises are an investment that should last a minimum of 20 years and be strong enough to support large plants and heavy fruit crops. A typical grape trellis is a canopy training system. It includes two to seven wires strung between 5′–6′ (1.5 m–1.8 m) wooden posts, with end posts for anchoring. Trellis designs should:

- Be strong and long-lasting.
- Support the trunk, cordons, arms, canes, and foliage.
- Provide the maximum sunlight penetration to leaves and buds.
- Be economical to build.
- Be permanent, easy to repair, and require limited annual maintenance.
- Be adaptable to modern mechanical pruning and harvesting machines.

mrfotos/Shutterstock.com

Figure 23-22. Trellis types vary by orchard, but all are designed to guide growth and provide support.

Grapes are trained to maximize exposure of leaf area to sunlight, create a desirable environment within the canopy, and promote uniform breaking of buds. Grape training is also important for effective and efficient vineyard operations regarding equipment traffic, fruit harvesting, pesticide application, and dormant pruning. Training grape vines takes as long as three years. Start by establishing the *cordons* (major branches that come from the trunk) and encouraging the development of shoots and fruiting *spurs* (canes that have has been pruned to have only a few nodes). **Figure 23-23** details some of the more common types of training systems that have been developed and for which situations they may be useful. There are numerous training systems that can be used, depending on the situation.

Grape Terminology

- **Cane**—mature, woody shoot after leaves fall.
- **Cordon** or **arm**—major branches off the trunk from which canes or spurs emerge. Usually situated horizontally along a wire. Fully developed cordons can support spurs and canes. Cordons can extend in one or two directions from the trunk. They are kept for several years.
- **Pruning**—cuts that remove parts of a plant to manage crop size, fruit quality, and vegetative growth. Cane renewal or cane replacement involves removing canes that bore fruit the previous year and selecting canes that grew the previous year for fruiting in the current year. Spur pruning removes short canes that fruited the previous year and replaces them with spurs on the cordon as a site for fruiting and shoot growth the current year.
- **Shoot**—growth from a bud on a cane, spur, cordon, or trunk. A shoot has leaves and tendrils and may also bear fruit. As a shoot matures and drops its leaves in the fall, it is then called a cane.
- **Spur**—a cane pruned to five or fewer nodes. A renewal spur has one or two buds for cane production, and a fruiting spur is chosen to produce fruiting shoots.
- **Training**—development of the vine structure on the trellis. Training will maximize sunlight penetration to fruit and foliage.
- **Trunk**—semipermanent, aboveground, vertically oriented stem of a vine plant.

Hop Plants

Hop plants have bines that wrap shoots clockwise around a trellis or support wire. Many have downward-pointing bristles called trichomes to aid in their ascent. Vertically growing hop bines produce more flowers than those growing horizontally. This means that a taller trellis system will result in greater yields. The overhead trellis system is most often used for commercial production with 16′–20′ (4.9 m–5.5 m) systems. Poles made from cedar, fir, or pine are set 3′ (0.9 m) in the ground and are connected by overhead wire cables that run down and across the rows. Edge poles are placed along the end of the rows and installed with a 60°–65° outward angle. Guy wires are fastened to anchors in the ground to brace the system. On the lower 24″–36″ (60 cm–90 cm), leaves and lateral branches are generally stripped from the bine to promote airflow and reduce spread of diseases. Stripping can be done manually, chemically, or even using livestock.

Common Training Systems for Grapes

Training System	Advantage	Disadvantage	Recommended
High Wire Cordon (Hudson River Umbrella) 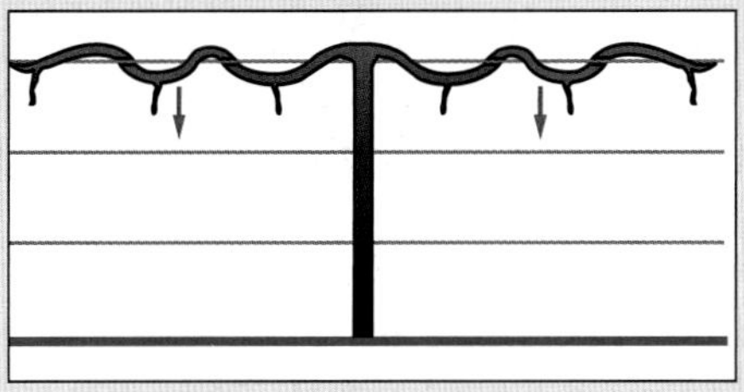	• Simple trellis installation • Large trunk and cordon provide reservoir for carbohydrate storage • Annual vine tying is minimized • Fruit high for sun exposure • Allows for mechanical pruning	• Large trunk and cordon can be difficult to replace if winter injury occurs • Old cordons hard to remove and may be a source of disease	• Hybrid or native cultivars where inexpensive production is desired
VSP (Vertical Shoot Positioning) 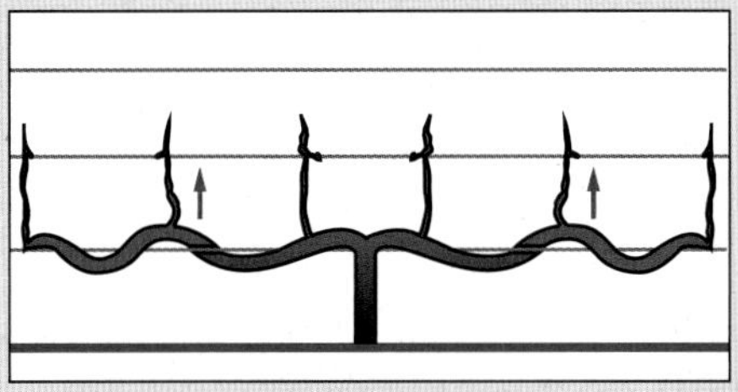	• Simple vine concept makes for easier replacement after winter injury and annual management • Little tying • Adaptable to mechanical pruning	• Fruiting spurs may be lower quality • Fruiting zone can be diminished compared to other systems	• Good for vines with small size or moderate to lower vigor *vinifera* where fewer nodes are required
Umbrella Kniffin 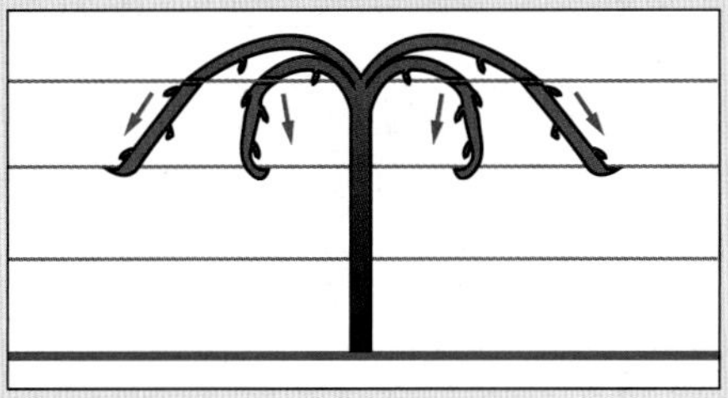	• All canes originate from upper arms, thereby optimizing sunlight • Apical dominance minimized • Adaptable to mechanical harvest	• Cane tying required • No mechanized pruning	• Hybrid and native cultivars susceptible to cold damage, or white *vinifera* intended for bulk production
Geneva Double Curtain 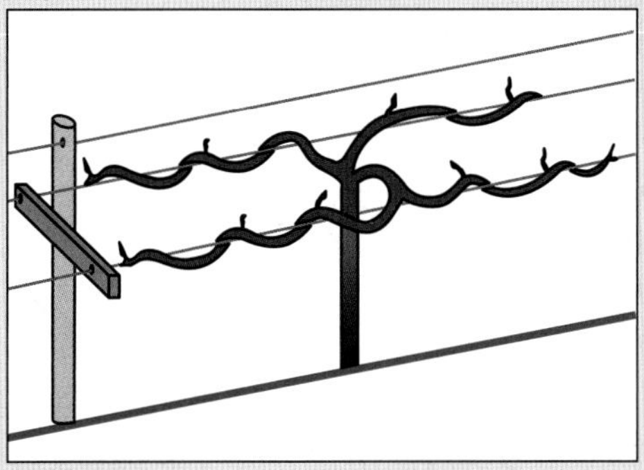	• Good for large canopies • Increases light penetration to horizontally divided canopy • Increases yield per acre, reducing production cost per ton • Can be mechanically pruned and harvested	• More costly to install and manage • Requires wider row spacing	• Highly vigorous hybrid or native cultivars

Goodheart-Willcox Publisher

Figure 23-23. Common training systems for grapes.

Kiwifruit

Kiwifruit pruning and vine training are done to reduce shade and vine growth to encourage better fruit bud development. Most vines are trained to a single trunk. Once the trunk reaches the top of the trellis, two canes are selected to be the cordons (main arms) to grow in opposite directions. Fruiting canes will originate from these cordons and produce four to six fruits at their base, **Figure 23-24** Male vines are trained to promote maximum flowering. Once blooms are finished, canes are cut back to promote the growth of short, new canes on the trellis.

Mehmet Cetin/Shutterstock.com

Figure 23-24. Properly trained kiwifruit will set four to six fruits per fruiting cane.

Nutrient Management

The best way to determine crop nutrient needs is to take annual soil tests. Nutrient needs vary depending on soil quality, cultivars, the growing region, and other environmental factors. In general, the following recommendations provide a baseline understanding of vine crop nutrient needs. Supplement this information with specific soil test results and additional research into what is suitable for a particular area.

Hop Plants

The nitrogen requirement for hop plants is approximately 100–150 lb per acre. Hop plants have low phosphorus needs. They only remove 20–30 lb of phosphorus per acre, which should be replaced. Potassium is required at a rate of approximately 150 lb per acre.

Grapes

With established grapes, annual applications of nitrogen may be the greatest nutrient need. Apply ammonium nitrate at the rate of 0.3 lb (or the equivalent in nitrogen from another source) per vine in an area extending from 4′ to 6′ (2.2 m to 1.8 m) from the trunk. Nitrogen application should be made in late winter to early spring before growth resumes. Magnesium deficiency may occur and may be a function of low pH. This can be corrected with liming. If pH is not the issue, magnesium sulfate can be sprayed on the vines.

Kiwifruit

Fertilizer application is important for kiwifruit and consists mostly of nitrogen fertilizer applied two or more times per year. For fruiting vines, use 1 lb of nitrogen per plant or 150 lb per acre with 50%–60% of nitrogen applied in March to April and the balance in May, June, and July.

Irrigation

Most grapes, hop plants, and kiwi are irrigated. Many use drip or trickle irrigation. Others may use a sprinkler system that can both irrigate and provide frost protection. Similar to systems described above, irrigation comes from a mainline and separates into different lines along each row. Irrigation optimizes fruit quality and minimizes stress on the plants to keep them productive for years.

Pest Management

All vine crops have pests, such as weeds, insects, and diseases, that need to be monitored and managed throughout the growing season. There are many cultural, biological, physical, and chemical practices that are used by growers and are specific for each crop. Detailed crop information is available through cooperative extension service offices.

Harvest and Storage

Harvesting and storage methods also vary for grapes, hop plants, and kiwi plants.

Grapes

Grapes are not picked until they are fully ripe because their quality does not improve after harvesting. The best indication for table grape maturity is flavor. In wine grapes, characteristics of aroma, flavor, tannins, sugars, and acids are specific to each variety and should be monitored. *Tannins* are naturally occurring chemical compounds found in wine grapes (and other plants) that impart a bitterness, astringency, and complexity to the final product. Many grapes are mechanically harvested, sorted, crushed, and processed for juice or wine. Grapes may also be harvested by hand. Each cluster is cut with a knife and carefully packed to avoid crushing, **Figure 23-25**. Table grapes will keep for several weeks if they are picked carefully and stored in a cool, well-ventilated place.

CoolR/Shutterstock.com

Figure 23-25. Hand harvesting fruit crops such as grapes maintains optimal postharvest quality.

Hop Plants

In commercial hop production, bines are mechanically harvested to collect the female *cones* (fruit). In home gardens, hand harvesting is common. Vines can either be cut down completely or cones can be picked as they reach maturity. Mature hop cones feel light and dry, and spring back when squeezed. The lupulin glands at the base of the cone petals are golden-yellow, have a sticky residue, and are fragrant when squeezed. Hop cones must be dried after harvesting. They can be forced air dried between 130°F and 150°F (55°C and 65°C), then stored in a cool, dry, dark freezer or refrigerator.

Kiwifruit

Commercial kiwifruit (*Actinidia deliciosa*) is hand-picked when sugar levels are at 7% and the fruit is at a hard stage (14 to 20 lb of pressure). Fruit are cooled and stored at 32°F (0°C). Refrigerated cold storage can keep fruit firm for three to six months. Avoid exposure to ethylene which causes kiwifruit to soften and have a shortened shelf life.

Careers in Pomology

Pomology offers several interesting career options. Two careers related to pomology are grower and food safety manager.

Grower

Growing fruit for the fresh food market or processed food market is an important agriculture career. From vineyards to orchards or integrated farms, growers make decisions regarding production, marketing, financing, and human resources management. On the production end, growers manage the soil, plant, cultivate, prune and train plants, and harvest fruit. Growers use a broad range of machinery and tools, from tractors and irrigation equipment to trellising equipment and pruning shears, **Figure 23-26**. Duties of a grower may include managing the soil and applying fertilizers; transplanting, weeding, irrigating, or pruning crops; applying pesticides; and cleaning, packing, and loading harvested products.

Phillip Minnis/Shutterstock.com

Figure 23-26. Growers may purchase or lease machinery during planting or harvest.

Food Safety Manager

A food safety manager works for an agricultural food company to ensure that all suppliers follow food safety requirements. The food safety manager will often act as a liaison between departments in the company, suppliers, customers, and regulatory agencies. Responsibilities of a food safety manager may include supervising and training food safety specialists and food safety samplers who work in the field. This person ensures that all agricultural food company suppliers are approved and have third-party audits for good agricultural practices (GAPs). If there is a food recall, the manager coordinates documentation retrieval and works as a liaison with the US Food and Drug Administration (FDA). Food safety managers must be informed about food safety and related federal and state requirements to ensure they are implemented and explained to all responsible parties. This position usually requires an advanced degree in agriculture, biology, or microbiology and requires at least five years of related experience.

CHAPTER 23 Review and Assessment

Chapter Summary

- Fruits and nuts contain important vitamins and minerals and are an essential part of a healthy diet. Eating fruits can also reduce the risk of diet-related diseases.
- Tree fruits and nuts average $18 billion in annual sales. Fruit crops are sold into both fresh food and processed food markets.
- Small fruits consist primarily of blueberries, strawberries, blackberries, raspberries, and cranberries. Considerations for selecting a growing site include factors such as temperature, soil, and slope.
- Most small fruits are perennial crops. Soil fertility and pH, weeds and other pests, and irrigation are factors that affect production of small fruits.
- Planting methods and timing vary for different types of small fruits. All small fruits require proper nutrition, irrigation, pest management, and pruning.
- Small fruits are highly perishable. Strawberries and brambles are harvested repeatedly as fruit ripens. Cranberries may be harvested with a picking machine or by flooding the growing beds.
- Tree fruits include apples, pears, peaches, cherries, apricots, plums, citrus fruits; tree nuts include cashews, pistachios, pecans, hazelnuts, and macadamia nuts. Tree fruits have specific chilling requirements and prefer well-drained soils and full sun.
- The timing of planting for tree fruits and nuts depends on the climate. Many fruit tree orchards establish permanent sod covers between rows.
- Proper pruning and training of fruit trees is a critical factor in the development of a strong tree structure that can support fruit production.
- Three types of pruning cuts are thinning cuts, heading cuts, and bench cuts.
- Open center, central leader, and modified central leader are the most common training methods.
- Fruit thinning is done to produce a desirable fruit size and to ensure annual bloom of many fruit and nut trees. Fruits and nuts can be harvested by hand or with mechanical harvesters.
- Grapes are the most commercially important vine crop. Kiwifruit, hops, and passion fruit are minor specialty crops. Climate, slope, and soil are important considerations for the growing site.
- Vine crops are planted in late winter to early spring. They are grown on various trellis systems. Nutrient needs vary depending on soil quality and cultivars. Most grapes, hop plants, and kiwi are irrigated, often with drip irrigation.
- Grapes should not be picked until they are fully ripe because they do not improve in quality after harvested. In commercial hop production, bines are mechanically harvested to collect the fruit. Commercial kiwifruit is hand-picked.
- Pomology offers several interesting career options including grower and food safety manager.

Words to Know

Match the key terms from the chapter to the correct definition.

A. bench cut
B. bine
C. central leader system
D. cone
E. cordon
F. cull
G. espalier
H. floricane-fruiting
I. heading cut
J. modified central leader system
K. muck
L. nematode
M. open center system
N. primocane-fruiting
O. scaffold branch
P. spur
Q. tannin
R. thinning cut
S. whorl

1. The major branches that come from the trunk of a vine.
2. A training method in which a plant grows flat against a wall, lattice, or other support.
3. A cut that removes upright limbs back to side branches.
4. A soil composed of primarily decomposed organic matter.
5. Primary limbs that extend from the main trunk and form the tree canopy.
6. A roundworm that lives in the soil, some of which are plant parasitic and considered significant pests.
7. Naturally occurring chemical compounds found in wine grapes (and other plants) that impart a bitterness, astringency, and complexity to wine.
8. A fruit tree training method that keeps the center of the tree free of large branches, allowing sunlight to penetrate the lower fruiting wood.
9. A spiral pattern of limbs.
10. A training method for fruit trees that is characterized by one main, upright trunk called the leader.
11. An undesirable, low-quality fruit that is composted or thrown away.
12. A long, flexible climbing stem.
13. A variation of the central leader form in which the central leader is removed after five to nine main lateral branches develop.
14. Brambles that produce fruit at the top of first-year canes in late summer.
15. A cut used to prevent crowding in fruit trees by removing an entire branch back to its base.
16. A cane that has been pruned to have only a few nodes.
17. Brambles that produce fruit on second-year canes only.
18. The fruit of a hop vine.
19. A cut that removes the terminal portion of a branch to just above a bud or side branch, promoting growth and lateral branch emergence.

Know and Understand

Answer the following questions using the information provided in this chapter.

1. What are some health benefits of eating fruits?
2. What role does vitamin C play in the body and in what fruits can it be found?
3. What are the largest fruit-producing states in the United States?
4. Name some common fruits produced for the fresh food market and for the processed food market.
5. What are some examples of chilling requirements for small fruits?
6. What type of surface slope is good for growing small fruits and what slopes should be avoided?
7. What type of soil is good for growing small fruits?
8. What nutrients should be applied (based on soil test recommendations) when preparing plots for small fruits?
9. Describe techniques used to minimize potential weed problems prior to establishing small fruit beds.
10. Discuss the importance of water for growing small fruits.
11. Why do growers use tissue culture plugs in bramble production?
12. Describe how the process of growing strawberries using plasticulture begins.
13. At what time of year are blueberries planted and how should they be pruned?
14. Explain how cranberries are harvested for the processed food market.
15. What type of soil is best for growing fruit and nut trees and how should soil for new plantings be prepared?
16. Describe the three main types of pruning cuts used for fruit trees.
17. Why would a grower use a modified central leader training system for a fruit or nut tree?
18. Why might a grower prune fruit and nut trees in the spring or summer?
19. What is the purpose of fruit thinning and what are some fruits and nuts that benefit from this practice?
20. What is the most commercially important vine crop and what are some specialty vine crops?
21. How should hop plants be planted?
22. What are characteristics to consider when designing a trellis for vine fruits?
23. How are grapes harvested and how should table grapes be stored?

Thinking Critically

1. A local architectural firm has just finished restoring a home that was originally built in the 1800s and will now be used to house a local history museum. The city and the architectural firm are jointly holding an open house to showcase the restoration. The architectural firm has hired you to plant heirloom fruits and nuts on the grounds. What type or types of fruits and nuts are appropriate? What might you need to take into consideration when filling this contract?
2. Fruits and nuts are susceptible to a number of different diseases. Imagine that production in your region is threatened with a new disease. What options are available? How would you begin to determine what to do?

STEM and Academic Activities

1. **Science.** Find out more about flower pollination. Why do some fruits need neighboring plants in order to produce blooms or fruits? How does this impact fruit production? Write a report summarizing your findings.
2. **Technology.** Investigate technologies that allow for mechanical harvesting of some fruits and nuts. Prepare a three-paragraph summary of your findings.
3. **Engineering.** Vine crops require trellising to support growth and optimal yield. Research the different trellis and training methods and construct a trellis system that will provide maximum light infiltration and airflow among the vines.
4. **Math.** The cafeteria manager at your school would like to purchase local apples for student lunches. They are required by federal law to provide a one-fourth cup serving per student. There are 1000 students in your high school. How many servings could typically be prepared from one apple? How many apples would your cafeteria manager need to order to provide a serving to every student?
5. **Language Arts.** Complete an oral history by interviewing a fruit grower. If you are unable to interview someone, read one or more case studies about fruit farmers or pomologists from reliable Internet or library resources. How does the information you learned from the interview or reading supplement what you have learned in this chapter? Write a detailed summary of your interview or reading, describing how being involved in fruit production influenced the person's life.

Communicating about Horticulture

1. **Listening and Speaking.** Working in small groups, create a poster by drawing a tree fruit training system with labels illustrating the basic components. Toward the edges of the poster, draw the types of pruning cuts that are made. Use arrows and labels to indicate where these cuts are usually made. Present your poster to your peers.
2. **Reading and Speaking.** In small groups, discuss the photographs in the chapter. Identify, in your own words, what is being shown in each photograph. Discuss the effectiveness of the illustrations compared to the text description.

3. **Listening and Speaking.** Working with a partner, compare and contrast the different types of fruit and nut crops. Consider the perspective of manufacturers, growers, merchandisers, and consumers. In what situations would one commodity be preferable to the other? Record the key points of your discussion. Hold a class discussion. Compare your responses to those of your classmates.

SAE Opportunities

1. **Exploratory.** Job shadow a fruit grower for a small, mid-sized, or large orchard. What are the daily responsibilities for this job?
2. **Experimental.** Take three young fruit trees and train each one using a different training method. What impacts did the training have on fruit set the following year? Which method yielded more?
3. **Experimental.** Research which fruits and nuts will grow well in your region. Try growing a few of the different types in your school garden and grounds and determine which fruits thrive.
4. **Exploratory.** Create a list of processed fruit products that could be sold in your school's plant sale. What food safety guidelines do you need to follow? What regulations do you need to comply with? Create a guide for processing fruits for your agriculture and food classes. Include information relevant to this project for future students.
5. **Entrepreneurship.** Grow small fruits in your school garden and sell them to customers. Where is your market? How will you advertise? Can you provide enough produce to make this feasible? What other ways could you supplement your fruit sales?

Nejron Photo

CHAPTER 24
Landscape Design

Chapter Outcomes
After studying this chapter, you will be able to:
- Describe the landscape design process.
- Identify the elements and principles of landscape design.
- List tools used by landscape designers and architects.
- Discuss water-wise gardening.
- Describe methods of water gardening.
- List careers related to landscape design.

Words to Know

absolute scale	functional diagram	preliminary design
aquascaping	graphic	rain garden
base plan	interconnection	relative scale
color theory	landscape architect	repetition
concept plan	landscape contractor	simplicity
container water garden	landscape designer	texture
emphasis	perspective	

Before You Read
The summary at the end of the chapter highlights the most important concepts. Read the chapter and write a summary of it in your own words. Then, compare your summary to the summary in the text.

While studying this chapter, look for the activity icon to:

- **Practice** vocabulary terms with Words to Know activities.
- **Expand** learning with identification activities.
- **Reinforce** what you learn by completing Know and Understand questions.

www.g-wlearning.com/agriculture

*I*f you were asked to describe one famous American landscape design or garden, you might consider Central Park in New York City. This lush landmass nestled below the New York City skyline is a green retreat for millions of residents and tourists, alike. The original design and concept of this green space was created by a man many experts refer to as the father of American Landscape architecture. Frederick Law Olmsted designed Central Park and also dozens of college and university campuses; private homes, such as the Biltmore in Asheville, North Carolina; and many designs housed at US and local parks, such as the National Zoo in Washington, DC.

Olmsted helped found the American Society of Landscape Architects (ASLA) and forged conservation movements across the country. A *landscape architect* is a professionally trained and educated individual with knowledge of plant material and engineering. In 1898, Olmsted and several other men founded the ASLA to act as a voice for landscape architects throughout the United States. This organization sponsored education efforts and provided leadership promoting the incorporation of principles of landscape design into American life.

Jim Lopes/Shutterstock.com

Figure 24-1. Frederick Law Olmsted fought to save Niagara Falls (now a national landmark and park) from being industrialized.

Frederick Law Olmsted's thorough understanding of conserving American natural resources and native landscapes helped to form many parks that are still in use today. Many industrialists set sites on Niagara Falls for manufacturing, but Olmsted worked to preserve this landmark, **Figure 24-1**. He continued to advocate for conservation throughout his career, leading to the formation of more than 600 landscape projects under his direction throughout his lifetime.

The Design Process

The design process begins with objectives for a landscape. The designer must understand the objectives for the site and its conditions. With this knowledge, the designer creates a landscape plan to include plants, hardscape elements, and other features. The designer incorporates elements and principles of design to create a functional and aesthetically appealing landscape that can be maintained and enjoyed by the client or public.

Landscape Design Steps

Whether a customer is coming to a professional designer, or a homeowner is attempting to make an area more functional, there are certain steps to follow to create a suitable landscape plan. These steps include establishing the problem and objective, succeeding in the design sequence, accepting the final plan, contracting the final plan, performing the billing services, and setting up a maintenance plan.

Problem and Objective

The landscape design must first start with a need. There is no reason to install a landscape if there is no problem to solve or goal to achieve for the site. Whether it is for aesthetics; to help detour water runoff or erosion; or to create a space that will function for recreation, learning, or edible production, there should be a goal for the landscape.

A *landscape designer* is a person with horticultural expertise, formally or informally educated, who often works on small landscape projects. During this first stage of the landscape design process, the designer may be working with a customer. This stage can also be referred to as the client interview. Designers must identify the limitations of the site and determine the problem to be solved or goal to be achieved. The designer will then create an objective for the design.

The objective for a landscape plan creates a goal for the site. This goal should state a solution to the need or problem. The client can determine the objective for the site, the designer can assist the client, or the designer can act alone to form an objective suitable for the needs of the future landscape.

Design Sequence

The design sequence step includes a great deal of communication and exploration between the designer and the client. This process permits idea exchange between the designer and the client until a final plan is reached.

Corner Question
Who is Martha Schwartz?

History Connection

Frederick Law Olmsted

Frederick Law Olmsted had to interview his client, George Vanderbilt, to determine the objectives for the 125,000-acre Biltmore estate. Olmsted was not impressed with the surrounding land and suggested a solution to the site's problem: create a park, manage the timber, grow plants in their own nursery, and reestablish the deforested site that would become the Biltmore Estate. Today, this land is now part of the Pisgah National Forest.

Tomf688/Wikipedia/GFDL

Corner Question
What historical pieces or artifacts should a landscape designer or architect look for on an older piece of land?

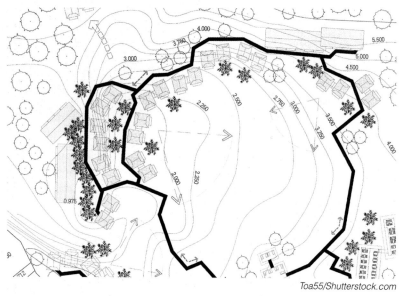

Figure 24-2. A site plan provides the basic map or foundation plan for a landscape design.

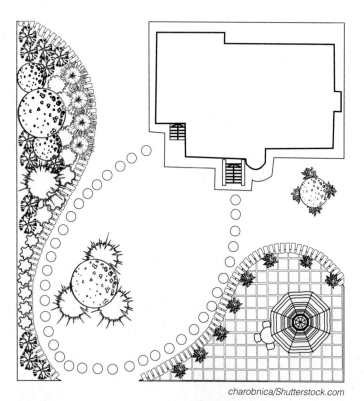

Figure 24-3. A functional diagram begins to create spaces, in the shapes of bubbles.

The steps of the design sequence include site analysis, functional diagrams, concept plans, preliminary designs, and the final landscape plan.

During the site analysis phase of the design sequence, a designer creates a site survey and analysis. He or she measures the site and takes note of all existing plants that will be saved and standing structures. The designer also investigates soil quality and creates a thorough base plan. A *base plan* is a map showing a site with the details of the landscape and its structures, **Figure 24-2**. This plan is the foundation for all further planning. Special attention should be paid to sun direction, surrounding roads, existing water and utility lines, historical features or markers, unique habitats or ecosystems, and prevailing winds. A plot map should be acquired that shows the actual survey of the land. These items together will make up what is known as the base plan or base map. This is a bird's-eye view of the site and is known as the plan view.

A *functional diagram* is a series of loosely drawn circles that divide spaces into areas that will be decided on by the designer, **Figure 24-3**. These circles represent the size of the area, but the scope can be further developed into outdoor rooms. Functional diagrams are also known as bubble diagrams.

A *concept plan* is a map or design that shows the general layout and possibilities for a landscape. This plan is more detailed than functional diagrams. This stage may also include the creation of outdoor rooms. Outdoor rooms move space from inside of the building or home to the outdoors. This method creates a continuous flow that can include spaces such as private, public, recreational, or utility areas. A private area can include a secluded porch or kitchen space. A public room will be visible from various vantage points. A recreational space can include a pool or playground, and the utility space may function for storage or garbage disposal.

The *preliminary design* is a map that breaks down the concept plan and includes more thorough details. This stage, also called a draft design, depends on the size and complexity of the site. Here, elements and principles of design lead to selection of materials. Several layers and approaches are made that include changes, deletions, and substitutions. The focus on meeting the objectives of the design and exchange with the design client continue throughout this part of the design program.

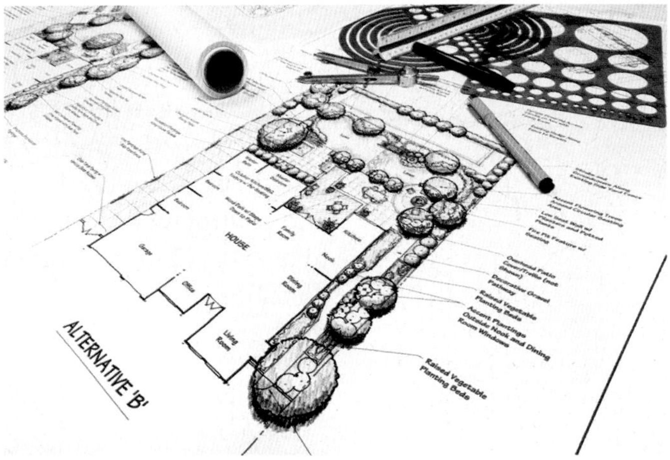

The final plan is a map that reflects reactions from the client fused with the design objectives, **Figure 24-4**. This final plan includes all details and is a road map for the installation of the design at every level. It also includes plant selections, specified features, and *graphics* (pictorial representations) meant to impress the client. Often, these final designs are colorful interpretations made by the designer to represent the future landscape.

Figure 24-4. A final design plan represents a map that can be followed by those installing the landscape.

Acceptance

When a client accepts the final plan, the following phase of the landscape process focuses more on business practices and transactions. At this point, a landscape designer's job is over. Some landscape designers and architects work alone and offer no further services, but many landscape design businesses also provide design installation and maintenance services for design customers.

"All gardening is a landscape painting."
—William Kent

Contracting

The final plan is presented to a landscape contractor for a bid. A *landscape contractor* is an individual who installs and maintains landscape elements. This bid is a quote of how much the installation will cost the customer. Projects range from thousands to many millions of dollars, depending on the scope and size. Usually, various contractors will bid and provide a plan for the installation. The customer will then choose the company to implement the design and settle on a price.

Landscape contractors vary in size and specialty. A landscape contractor may only install plant material and not garden features. The landscape contractor can subcontract or hire other businesses to finish work that is out of the company's range. For example, a landscape design usually includes irrigation and hardscapes. A subcontractor who specializes in each of these areas can be hired to install each of these components to make the overall installation and construction a success, **Figure 24-5**.

Figure 24-5. A subcontractor installs elements of the landscape design. Subcontractors are often hired by the general contractor for the project.

Corner Question

Where is the largest pool in the world?

Billing

An essential part of landscape design and installation work rests in the details. These details usually include billing and finances. A landscape crew cannot be paid if money is not collected. An understanding of business principles is essential to every landscape company. Landscape designers and contractors must employ individuals who offer business expertise to oversee financial transactions. Money must be collected by clients and distributed to all of the employees and the companies who worked together on the project.

Maintenance Plan

What happens after the landscape is installed? Who takes care of this new beautiful site? Landscape maintenance workers can continue to cultivate gardens and keep them in their best shape.

Paying particular attention to the business or financial management of a landscape maintenance plan benefits the future of a green industry business. A contract should be made to outline what services will be provided, when they will take place, and when payment will be collected. All of these elements ensure that both the customers and the providers of the landscape services are content.

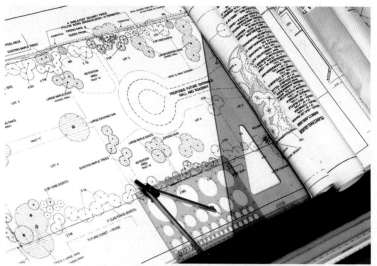

Steve Bower/Shutterstock.com

Figure 24-6. A landscape design may be hand-drawn using paper, drafting tools, and a drawing board.

Drawing Board or Computer-Aided Design

Like all artists, a landscape designer or architect must find a method to express landscape plans. Designers may choose to draw their designs by hand using a drawing board or to create designs using a computer, **Figure 24-6**. Computer-aided designs are created using a number of software systems or programs on the computer. This method of design became popular in the late twentieth century and is continuously reinventing itself in the twenty-first century through the progression of technologies.

Drawing Board

A drawing board is a flattened space that lends itself to drawing. Designers employ various forms of paper or canvases on top of their preferred drawing board. This could be a desk, a wooden sketch board, or a table. Many designers prefer to work on an elevated platform where they can stand or have a raised seat. Several of these design workspaces (which include drawing boards) comprise a studio.

Computer-Aided Design

Computer-aided design software allows users to create landscape designs for many types of projects, **Figure 24-7**. Graphic images of completed projects appear so real that some untrained eyes cannot determine whether the product is an actual photograph or a designer's interpretation created using computer-aided design. Many designs are drawn in three dimensions and allow the viewer to see all angles of the design.

Graphics

The graphics (symbols) on a landscape plan act as two-dimensional representations of real-life plant material or features. The landscape plan is a map that contains symbols. The symbols are the graphics. The landscape plan also has a key that tells what the graphics represent. For example, a circle with serrated edges may represent an evergreen perennial shrub or an ornamental grass. A key created by the landscape designer or architect tells what the symbols represent. Actual plant names or other identification points often accompany a graphic on a landscape plan.

Each landscape designer or architect has his or her own interpretation of commonly used graphics. Much like handwriting, the designer will add something to the graphic that makes it unique to that artist. Like all art, a hand-drawn landscape plan can be traced back to the designer or artist based on the manner in which the graphics appear.

Crystal Home/Shutterstock.com

Figure 24-7. Computer-aided design software can be used by designers to create a landscape plan.

Elements and Principles of Landscape Design

Creating a pleasing and functional landscape does not happen by chance. It is planned and executed using the elements and principles of design.

A talented and knowledgeable landscape designer or architect forms landscape plans by employing elements and principles of design. The principles guide the designer or architect in solving a problem or achieving an objective for the site. The result of a well-executed landscape plan should be an enjoyable, functional, and sustainable landscape, **Figure 24-8**.

The features of a landscape can be physically described by the visual qualities and are referred to as the elements of design. These include line, form, texture, and color.

Chaloemphan/Shutterstock.com

Figure 24-8. A functional landscape plan should create a beautiful outdoor space with purpose.

Corner Question

How many colleges and universities in the United States offer landscape architecture programs?

The principles of design include the fundamental concepts of composition, including balance, proportion, repetition, emphasis, and unity. These principles serve as guidelines to arrange or organize the features to create an aesthetically pleasing or beautiful landscape.

Line

A line defines the borders of a space. In a garden, the lines help to create rooms and help individuals relate to the landscape. The properties of a line govern how people will respond emotionally and physically to the landscape. Lines can be straight, curved, vertical, or horizontal.

Adisa/Shutterstock.com

Straight

Straight lines serve to direct the eye to a focal point. Lines appear formal and produce rigid structure to a space. Symmetrical designs employ straight lines. Lines are often found in features (hardscape materials), **Figure 24-9A**.

Curved

A curved line produces a more diverse impact than a formal straight line in a space. A curved line helps a person to meander through a space. A large curved line creates a much different effect from small curved lines. Curves align themselves with nature and partner well with asymmetrical designs, which are informal and relaxed, **Figure 24-9B**.

Elena Elisseeva/Shutterstock.com

Vertical

Vertical lines guide the eye upward. This action persuades one to think that the space is larger than it may really be. Tall trees and structures, such as poles meant to trellis vines, are examples of vertical lines.

Horizontal

A horizontal line guides the eye laterally. These lines can divide or unite spaces. Examples include paths, hedges, or structural walls. When looking at a landscape design in plan view (bird's-eye view), horizontal lines define garden beds and hardscapes, **Figure 24-9C**. There are three types of horizontal lines:

- Plant lines—the boundaries of plant material in relation to one another and the garden bed or hardscape materials.
- Bed lines—edges where the plant bed encounters another material, such as hardscape materials.
- Hardscape lines—examples include a wall, fence, or edge of the material.

rodho/Shutterstock.com

Figure 24-9. A—The straight lines created by these trees and shrubs are geometric and formal. B—These curved lines are informal. They generate a relaxed mood. C—Pavers, sidewalks, and border plantings can be constructed to produce horizontal lines.

Form

An outline that encloses a space produces a shape. A three-dimensional mass of that shape is referred to as the form. A person encounters form throughout the landscape. Plants, features, and water provide forms that help to establish a space visually. A negative space (void) between plants can help to define a plant's form.

Corner Question

What is the longest straight line you can sail in the world?

Geometric

Geometric (formal) forms include circles, squares, and polygons. Circles can vary infinitely. Full, half, semi-circles, ovals, ellipses, arcs, and segments are all circular forms. Circles entice the eye into the center, perhaps leading to a focal point.

The strong lines and formality of a square provide a great deal of opportunity for complexity in a landscape. Features, such as stepping stones, pavers, and bricks, are easily incorporated into landscapes.

Polygons are many-sided and can include triangles or hexagons, **Figure 24-10**. These forms can be complex and should be used in limited quantities because overuse can be distracting in a landscape.

Serg Zastavkin/Shutterstock.com

Figure 24-10. Various shapes, such as these polygons that have many sides, are used in a variety of garden designs.

Naturalistic

Forms that imitate nature can be used in a landscape.
- Meandering lines—imitate the lines of rivers and streams.
- Organic edges—imitate the lines of plants, rocks, and other natural materials.
- Fragmented edges—imitate edges of pavers or bricks. Often, edges will gradually disappear into the landscape.

Plant Forms

The form of a plant is its distinguishable aspect. A mass of plants (several plants clustered together) can create a new form altogether. A unique plant form serves as a possible focal point in the landscape.

History Connection Palace of Versailles

One of the most famous gardens in the world, at the Palace of Versailles, was designed for King Louis XIV by André Le Nôtre. The gardens began design in 1661, and the gardens had developed to the point of inauguration in 1664. The gardens were not finished, however, at this time. Construction continued for over a century. The gardens of Versailles are known for their formality and immense water features.

Brian Kinney/Shutterstock.com

Corner Question

How popular are variegated plants compared to green plants?

Plant forms will vary depending on the perspective of the viewer. A form of a tree perceived while sitting under its shade vastly differs from the form perceived when standing at a distance. Forms of plants can include trees, shrubs, and ground covers.

- Trees. Forms of trees include oval, columnar, vase, round, weeping, and pyramidal. Tree forms not only create an aesthetic, they also can be functional. A weeping tree can provide a unique space for cover, and a series of columnar trees can create a screen.
- Shrubs. Forms of shrubs involve upright, arching, rounded, mounding, vase, spikey, cascading, and irregular shapes. Particular attention to the form must be paid, especially if the designer wants to form a mass or grouping. A rounded or spikey plant may generate an excellent mass, but a vase-shaped shrub form may produce a better focal point.
- Ground covers. Technique where producing a mass of ground covers achieves a more functional and attractive landscape component. Matting, spreading, clumping, sprawling, and short spikes comprise ground cover forms.

Texture

Texture adds interest and contrast to a landscape, **Figure 24-11**. The *texture* of a plant or a feature refers to how coarse, rough, smooth, or fine a surface feels or appears.

Coarse

Plants and hardscape materials exhibiting coarse texture are bold, rough, or interesting and draw attention. Plants with large leaves, variegation (difference in colored zones of the plant), irregular shapes, and thorns lend themselves to coarse texture. Hardscape materials that are roughly cut, large, aged, or appear unfinished contribute to coarse texture.

SDeming/Shutterstock.com

Figure 24-11. Horticulturists and garden enthusiasts create texture in garden spaces through use of varied leaf shapes, branching, and variegation.

Fine

Skinny leaves, delicate flowers, long stems, and vines are examples of fine textures. These are the wispy or light and airy feeling plants. This texture helps to overstate distance and make the area appear more spacious. Fine-textured plants include bamboo, grasses, Japanese maples, and ferns. Smooth hardscapes, reflecting pools, and water features with a fine mist also appear to have a fine texture.

STEM Connection

The Saguaro Cactus

The saguaro cactus (*Carnegiea gigantea*) is known as the cowboy cactus of the southwestern United States desert landscape. These extremely drought-tolerant plants can shrink and swell by 25% in girth during periods of rain and drought. Their deep root systems help them to survive dry and windy periods of time.

Ingrid Curry/Shutterstock.com

Medium

Plants and hardscapes that can be considered neither coarse nor fine textures fall into the grey zone known as medium texture. A plant that is rounded with average spacing of leaves and insignificant color or attributes is an example of medium texture. Hardscape materials, such as flagstone, finished wood of a patio or pergola, and concrete surrounding a pool, are examples of medium texture. The medium texture plants and hardscapes help to unify their coarse and fine texture counterparts of the landscape.

Color

Corner Question

What is the color scheme of a garden that has all of the colors within the color wheel?

Many people value color in a landscape. Color is the most visible characteristic of the landscape, but it is the least permanent. *Color theory* is a body of practical guidance for using and mixing colors. Color theory produces an arrangement of colors known as a color scheme. Three basic color schemes are complementary, analogous, and monochromatic.

Complementary

Greatly contrasting colors, those directly across from one another on the color wheel, make up a complementary color scheme. Common complementary color schemes are red and green or blue and orange. Complementary color schemes occur naturally in plant material, **Figure 24-12A**.

Analogous

Three to five colors that are adjacent to one another on the color wheel are called analogous. This color scheme sometimes is referred to as harmonious, **Figure 24-12B**. An example of this color scheme would be blue, blue-violet, and violet.

Did You Know?

No "true black" flower has been found in existence. There are deep shades of red and purple that are nearly black, but there are no real black flowers.

Monochromatic

A monochromatic color scheme involves only one color but has different shades or hues within that color scheme, **Figure 24-12C**. Green is always present in landscapes, but inclusion of an orange color and different light or dark versions of this color creates a monochromatic effect.

A *Leena Robinson/Shutterstock.com*

B *happykamill/Shutterstock.com*

C *ying/Shutterstock.com*

Figure 24-12. A—Complementary colors can be found in various plantings. This bird of paradise naturally has complementary colors, blue and orange, within its unique flower. B—This bed of zinnias is an analogous array of color. Oranges, pinks, reds, and yellows are adjacent on the color wheel. C—A monochromatic color scheme employs only one color, such as this white garden.

Figure 24-13. This garden's geometric pattern is symmetrical and very formal.

Figure 24-14. This asymmetrical space has reflections on both sides of the creek. The sides are balanced but they are not the exact same.

"Happiness is not a matter of intensity but of balance, order, rhythm, and intensity."
—Thomas Merton

Balance

Balance is the distribution of components to create aesthetic on both sides of the design. Balance can produce the symmetry or lack of symmetry on the left and right side of the landscape.

Symmetrical

A symmetrical balance crafts a feeling of formality. The landscape acts as a mirror or reflection of itself in relation to a central point, **Figure 24-13**. This stately landscape demands particular care and attention from horticulturists. The sides must remain similar to maintain the appearance of symmetry.

Asymmetrical

An asymmetrical balance differs from one side of an axis to the other. The effect of this technique produces varying sides, but the weight of each side balances and keeps equilibrium. The overall effect of asymmetry yields interest, mystery, and curiosity, **Figure 24-14**.

Perspective

Perspective refers to the size and placement of objects in a landscape to create depth and distance in a design. The balance of the foreground, midground, and background in a landscape contribute to perspective. The designer can create focal points or attention at any of these points. Designers often choose to compose a design with either the foreground (closest to the viewer) or the background dominant.

Proportion

The size of an object in relation to other objects constructs proportion. The human size determines the proportion for landscape because the sizes of other objects are relative to the human frame.

Absolute Scale

Absolute scale is the true size of an object. It relates the proportional value of the landscape features to a fixed structure. For example, a small tree next to a house will cause the house to appear large. On the other hand, large trees around the same house will cause the house to appear small. The house is the same; but in relation to the trees, it can appear small or large.

Relative Scale

The relative size of objects within a landscape is the *relative scale*. For example, the size of a dwarfed apple tree may seem fairly small for a person that is over six feet tall; however, this same tree may seem large to a toddler.

Repetition

Repeatedly using elements or features in a landscape invents a pattern or sequence that pleases the human eye. Too little or too much *repetition* (repeated elements within the landscape) can create a poorly designed landscape. Some repetition produces a harmonious feel through replicating form, texture, color, or lines in a landscape. Plant materials, hardscapes, or other features can be easily echoed through countless efforts that can include replicating patterns. However, to add depth and interest, slight variances through subtle changes of color, texture, or size can be used, **Figure 24-15**.

Jamelle Lugge/Shutterstock.com

Figure 24-15. Repetition, an element of design, can be noted with these repeated arbors bracing bright yellow flowering vines.

Emphasis

Emphasis refers to the prominence given to something, such as the focal point in a design. Emphasis is achieved through the use of varying colors, textures, sizes, shapes, groupings, and materials. A bold sculpture can create an area of emphasis. A plant that comes into bloom draws attention. Emphasis in a garden should change throughout the year as the seasons change. A plant that was dominant one week while in bloom may remain subordinate the remaining 51 weeks. A landscape constantly changes; and the emphasis does as well.

Did You Know?
Plants that draw attention are referred to as *specimen plants*. The plants are often rich in one or more of the following: color, shape, and texture.

Unity

Unity is often referred to as harmony. Achieving this principle of design occurs only by linking all elements and principles of design together in a consistent composition. Much like a puzzle, all of the elements must fit together to form one cohesive, harmonious design. Manipulating interconnection, unity of three, and simplicity can pave the path for a landscape rich with unity.

Interconnection

A seamless linkage between diverse features to create blending is known as *interconnection*. For example, plants and other elements can be related to one another in a landscape. Continuing lines or linking spaces together promotes interconnection.

Unity of Three

Organisms in nature seek balance. Grouping plants or other garden elements in threes or other odd-number groupings gives a strong sense of balance and unity. Odd numbers permit greater variances in heights and are seen as one group that cannot be visually divided with ease. These odd-number principles contribute to an overall feeling of unity, **Figure 24-16**.

kezza/Shutterstock.com

Figure 24-16. The "unity of three" principle works better in the landscape than using even numbers of plantings.

> **Corner Question**
>
> What style of gardening or landscaping is known for its simplistic style?

Simplicity

Simplicity is the quality of being plain, natural, or uncomplicated. Keeping a landscape simple stimulates clarity and provides a revelation for the purpose of a design. Careful and intentional deletions of elements within a landscape design help to promote simplicity. Too many variables in any design can create a feeling of anxiousness. Simplicity promotes a feeling of being grounded and relaxed. Designers must keep this in mind when creating garden spaces.

Tools of Landscape Design

Whether you plan to create a landscape plan by hand or use a computer, you will begin with a foundation of measurement. Precise measurement, whether metric or with an American rule, must be followed by the designer. Designers use various tools to create a two-dimensional site plan reflecting what exists in the landscape.

Measurement Tools

Measuring a site requires precision. Close attention to detail, keeping measurements in the same unit, and a keen understanding of geometry work together to produce a precise representation of the landscape. A skilled landscape designer will use various tools to aid in accurate compilation of measurements on the site.

Measuring Wheels

A measuring wheel, or surveyor's wheel, is a measuring device that measures a distance from one space to another. The operator walks beside the measuring wheel and records the measurements, **Figure 24-17**.

PRILL/Shutterstock.com

Figure 24-17. A measuring wheel helps landscape professionals accurately measure areas that will then be transferred to design plans.

Transit Levels

Surveyors use a transit level to measure land elevations, **Figure 24-18**. A surveyor looks through this telescope-like device and reads elevation measurements.

Drawing Instruments

After the measurements have been made, a site plan showing specific measurements of the area must be formed. A designer or architect employs various tools and instruments to create the landscape plan. These tools may include drawing boards, pencils, erasers, scales, T-squares, compasses, templates, and drafting triangles.

Dmitry Kalinovsky/Shutterstock.com

Figure 24-18. A transit level is a measuring tool that helps landscape professionals measure elevations and distances.

Drawing Boards

A flat surface, such as a desktop or table, can act as a drawing board. Usually, architects and designers prefer to work on a large drawing board space that does not inhibit the ability to reach the design from all sides.

Pencils

Architects and designers use two types of pencils: mechanical and drawing pencils.

- Mechanical pencils—range in thickness of lead (ranging from 0.2 mm to 0.9 mm). These pencils are advantageous in that there is no need to sharpen the lead.
- Drawing pencils—lead encased in wood that must be repeatedly sharpened.

Whether it is a mechanical or drawing pencil, the pencil lead (made of graphite) has varying degrees of hardness. A soft lead creates dark, heavy marks, while a hard lead forms light, finer marks. A European scale illustrates the varying degrees of hardness (H) and blackness (B).

- The H scale—the higher the Hs (1H to 10H), the harder the lead and the lighter and finer the mark.
- The B scale—the higher the Bs (1B to 10B), the softer and darker the mark.

The American #2 pencil is an HB pencil. Most designers and architects like an HB to a 2H. These pencils are less likely to smear on paper than a 10B. See **Figure 24-19**.

Erasers

Erasers can potentially be used to fix a mistake. They also can make a big mistake bigger. A poor-quality eraser can tear the paper or cause increased smudging. Erasers should only be used during the final stages of design. In functional diagrams or preliminary designs, there is no need to erase. Just write over your mistakes and move on. In the final plan, a mistake can be erased with a vinyl eraser for the best effect.

Scales

An architect's scale or engineer's scale is a specialized, triangular ruler. The scale is ordinarily 12" long (ranges from 4" to 36" long) and includes multiple units of length to facilitate drafting. An architect's scale is a fractional scheme differing from the engineer's scale, which uses a decimal scaling system (more precise).

Corner Question

What did people use before rubber erasers to remove graphite marks?

Did You Know?

Graphite is not only used for pencils but also for insulation, barriers for brakes, rockets, and fire doors.

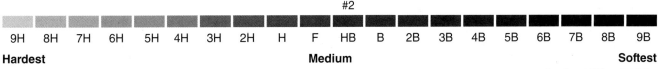

Goodheart-Willcox Publisher

Figure 24-19. The varying hardnesses of pencils help landscape professionals create their perfect plan. What is the most common pencil used by landscape architects and designers?

Andrey Popov/Shutterstock.com

Figure 24-20. This designer is using an engineer's scale.

Common measurements of the architect's scale are 16 (full), 3, 1-1/2, 1, 3/4, 1/2, 3/8, 1/4, 1/16, 1/8, and 3/32. For example, a 1/4" means that for every 1/4", it represents an amount determined by the architect or designer. If the architect or designer says that 1/4" = 1', then 1" on the landscape plan is equal to 4' in the landscape. Another example is that 1/8" = 10', meaning 1" on the landscape plan equals 80' in the landscape. The most common scale used by architects and designers is 1/8.

The engineer's scale is used by some architects and designers and differs a little. The scale will include 1/20, 1/30, 1/40, 1/50, and 1/60. The designer will use the scale 1/10, where 1" equals 10'. This scale (1/10) is also the most commonly used scale for landscape designers and architects who use an engineer's scale. See **Figure 24-20**.

T-Squares

A T-square is a device that is used for drawing vertical and horizontal lines, **Figure 24-21A**. This technical drawing device resembles a *T* shape. A component of the device, called the *stock* or *head*, rests on the edge of the drawing board. The perpendicular bar, called the *blade* or *shaft*, guides the designer in drawing lines. T-squares range in size from 18" to 42" (46 cm to 107 cm).

Compasses

A compass is an adjustable drawing device used to draw circles or arcs, **Figure 24-21B**. This technical device is comprised of two parts connected by a hinge. The hinge can be adjusted to increase or decrease the radius of the circle. A spiked point rests stationary on the paper, and the compass spins via the hinge. Meanwhile, a pencil or pen attached to the other arm of the compass outlines the circle or the arc.

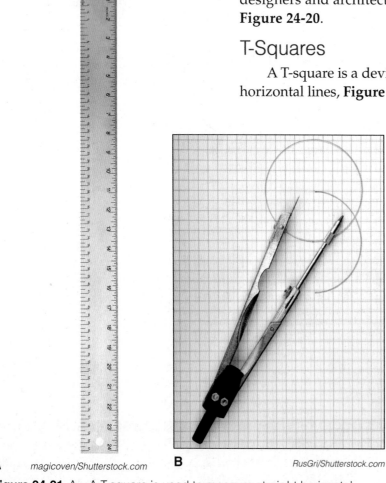

A *magicoven/Shutterstock.com*   B *RusGri/Shutterstock.com*

Figure 24-21. A—A T-square is used to measure straight horizontal, vertical, and 90° angles. B—A compass can be used to create circles of varying sizes by landscape designers and architects.

Templates

Designers and architects use templates to insert circle or other graphic representations on the landscape plan quickly, **Figure 24-22A**. A circle template, French curve (template with many different curves), or a landscape template with varying graphics can be used.

Drafting Triangles

A drafting triangle, also known as a set square, provides a straight edge or a right angle for a baseline on a landscape plan. These triangles usually are made of plastic or metal and have an additional triangle cut out in the center, **Figure 24-22B**.

Paper

The paper that will be used by an architect or designer depends on the objective for the final design. If the goal is to sketch or take notes, then a grid paper or tracing paper may be used. As the design becomes closer to final, it can be transferred to vellum. The final design may be on a bond paper (used in printers).

Grid Paper

Designers who are just starting out may feel more comfortable using a grid paper to lay out the base plan. A grid with blue markings can be photocopied, and none of the lines will show. A designer can also use the grid to help create a scale. Each square could represent a measurement of the land. For example, one block could equal 1', or one block could equal 10'. The grid paper may make calculations easier, as well, for a beginning designer.

Tracing Paper

Tracing paper, often referred to as trash paper in the trade, can be an efficient and fairly inexpensive way to add semitransparent layers to a base plan. With each layer of tracing paper, details increase until a somewhat finalized plan is reached. The tracing paper allows for mistakes to be drawn over easily, using a new layer; there is no need for erasing. Tracing paper can be found in white or manila.

Vellum

Designers may also use vellum for their design sequence. Vellum paper, manufactured from cotton, wood pulp, or other synthetic fibers, ranges in thickness and transparency. Vellum is sturdier than tracing paper. Unlike tracing paper, vellum holds up well to repeated erasing and works well with pencils, inks, markers, colored pencils, and watercolors.

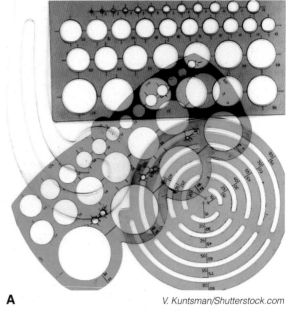

A

V. Kuntsman/Shutterstock.com

B

Sergio Stakhnyk/Shutterstock.com

Figure 24-22. A—A template with various shapes and sizes can quickly aid in landscape design. B—A triangle is used to create right angles.

Corner Question

Where was the first form of paper found?

Corner Question

How much trash that goes into landfills can be recycled?

Bond

Once a designer is ready for the final copy, he or she will transfer the entire final design plan to bond paper. The bond paper is similar to copy or printer paper, but is available in large sheets or rolls. Once the final design is placed on the bond paper, the architect or designer may also choose to render the design with color to enhance the illusions. The final design may now be offered to the client for presentation.

Water-Wise Landscape Design

When planning a landscape, the designer must ensure that water is conserved. Water-wise gardening, also called *xeriscaping*, incorporates seven basic principles that help to conserve water, **Figure 24-23**. These principles include planning and design, soil analysis and amendments, plant selection, turf use, efficient irrigation, mulch, and maintenance.

Bruce C. Murray/Shutterstock.com

Figure 24-23. Water-wise gardening uses plants requiring little regular water. These succulent and drought-tolerant plants can still create a beautiful landscape design.

Planning and Design

Planning and creating an effective design are the first steps to water-wise gardening. Designing a water-efficient garden is much easier today than decades ago. Drought-tolerant plants are readily available in wholesale and retail garden centers and nurseries around the world. In addition, many technical advances in horticulture, such as cisterns, drip irrigation devices, and computer software, have helped to reduce water consumption. Attention must still be given to the elements and principles of design with careful attention to water conservation.

> "The nation that destroys its soil destroys itself."
> —Franklin D. Roosevelt

Soil Analysis and Amendments

Soil analysis should coincide with the planning and design of a water-wise garden. Obtain a soil test to determine the soil pH and identify the nutrients available to plants. Amend the soil, and be sure to add organic matter to help promote water retention. Adding 4″–6″ (10 cm–15 cm) of organic matter provides pore spaces for the water to be stored for root absorption. Annual amendments will only further promote soil health and increase water retention.

Plant Selection

Succulents, cacti, and other drought-tolerant plant choices flooded the horticultural market within the past decade. Consumers find these plants attractive choices for their aesthetics and because of their low maintenance.

Choosing native plants (plants originating locally) and exotic plants that thrive in your region will help to conserve water, **Figure 24-24**. These plants are most likely well-suited for your growing zone and soil type, are hardier to local pests, and usually require less fertilizer application.

Turf Use

Most turfgrasses require more irrigation than many other water-wise plant choices. Turf breeding efforts continually create newer, drought-tolerant varieties. Turf selection must be carefully researched and planned to ensure the best turf for the site. Depending on the growing climate, warm- or cool-season grasses will thrive. Research the most drought-tolerant variety for the zone.

Consider the area that will be covered in turf in relation to irrigation design. Create irrigation patterns that will not allow watering impervious sites, such as concrete or pavement. Every drop of water should reach the turf and nothing else.

Arina P. Habich/Shutterstock.com

Figure 24-24. Today's availability and assortment of drought-tolerant plants surpasses any horticulturist's wildest dreams.

Efficient Irrigation

Irrigation is the application of water to plants. Irrigation methods can vary, but the method used must ensure that water reaches the plants roots. Generally, many plants require an inch of rainwater per week throughout the peak growing seasons. Water-loving plants, such as sedges, require greater quantities of water. Succulent plants, such as hens and chicks, require much less water.

Thinking Green

Turf Painting

Forcing cool-season turf into dormancy during the heat of summer can help with conserving water. The look of dormant turf, however, may not be that appealing to a homeowner or businessperson. Why not paint it? Painting turfgrass green can help the lawn look nice until invigorating fall rains.

For drought-stricken regions of the United States, this practice could take place at any time of the year. The average homeowner will spend only $300 to paint the grass and create several months of curb appeal.

YardGreen, Inc.

Corner Question

How much money will a rain sensor save in irrigation costs?

Rain or Moisture Sensors

Technological advances in horticultural irrigation continue to further efforts to conserve water. Rain and moisture sensors send signals to irrigation devices to indicate when a plant needs water. Rain sensors relay signals to automated irrigation heads to remain off. Moisture sensors measure the amount of water in soil and can dictate when to irrigate. Both of these devices can help to prevent overwatering or unnecessary irrigation.

Application Timing and Amounts

Water-wise irrigation takes place only upon necessity. If a plant needs to be irrigated, the water is applied in a conscientious manner. Watering efficiently uses only enough water, preventing overwatering, runoff, and erosion. This can ultimately save water. Follow these guidelines for watering:

- Apply water in the morning or at dusk to avoid evaporation and water loss.
- Group like plants together (plants needing the same amount of water).
- Install drip irrigation or use soaker hoses.
- Use watering bags.
- Apply water directly to roots.

Drip Irrigation

Compared to sprinkler irrigation, drip irrigation uses much less water. Water is released slowly from emitters, bubblers, or hoses and is released directly on the soil. Water does not evaporate or run off. Plants are healthier, and fewer nutrients are leached from the soil than with overhead irrigation.

Rainwater Capture

Rainfall recharges groundwater. When the rain hits roofs and impervious surfaces (pavement or concrete), the water runs off and is often collected through a storm water system. In this process, water particles pick up pollutants and soil particles. Rainwater can be captured instead of lost.

Water catchment systems can be small or large in scale, **Figure 24-25**. Small-scale systems (homes or buildings) can have rain barrels attached to gutters. Cisterns (large water-holding vessels) can store great quantities of water. Retaining ponds can be constructed as well. Additionally, rainwater gardens can be planted to catch rainwater and also filter pollutants from the water.

Pack-Shot/Shutterstock.com

Figure 24-25. A rain barrel or cistern can help to reduce the amount of fresh water that is needed to irrigate plants. Simply catching water off the roof of a building and then using this water can significantly lessen the burden of using fresh water for plants.

Mulch

As you read in Chapter 11 (*Soils and Media*), mulching not only improves the physical appearance of the landscape, but it also contributes to numerous environmental benefits. Mulches composed of organic or inorganic materials can be utilized in water-wise gardening. Organic mulches, such as wood, leaf, bark, and straw, will break down over time and add nutrition and structure to underlying soils. Inorganic mulch, such as gravel, is commonly used in water-wise gardening. Based on the objective for the mulch, selections will vary, but all types will help to conserve water. Be sure to purchase mulch from reputable sellers.

Mulch should be applied at various depths depending on the plants grown. Spread mulch 2″ (5 cm) deep for annuals and perennials. Spread mulch 3″–4″ (7 cm–10 cm) for woody trees and shrubs.

Regardless of the type of plant, be sure to keep mulch away from the crown of the plant. This will help prevent pest problems.

Corner Question

What does *xeric* mean?

Maintenance

Water-wise gardening requires much less maintenance than many other garden forms. A well-planned xeriscape can reduce maintenance by more than 50% through less fertilizing, mowing, and irrigation.

Mowing and Fertilizing

Mowing plants at the appropriate height can help to insulate the soil and maintain water. Fewer times mowing also helps the environment through less fuel consumption and less atmospheric pollutants. Adding grass clippings to the soil will also add nutrients and help retain soil moisture.

Applying the proper nutrients to the soil helps to make healthier plants. Using organic fertilizers will also help to amend the soil's structure and allow a soil to hold more water for plant use.

Did You Know?

You can add compost to your soil by adding small piles to areas throughout your lawn and raking it into the grass.

Irrigation

Overwatering drought-tolerant plants has more potential to cause problems than underwatering. Carefully monitor irrigation systems to ensure they are running efficiently so water is not being lost. Excessive irrigation also leads to leaching (loss) of nutrients in the soil environment.

STEM Connection: Calculating Mulch

Mulch is sold by the cubic yard (yd^3). This is a measurement of volume. Remember that one yard is equal to 3′. When calculating cubic yards of mulch, begin by calculating the volume in cubic feet (ft^3). To determine how much mulch you need for an area, measure the area (width × length), and multiply by the depth, which must be expressed in feet.

Now, take that number and divide it by 27 (a cubic yard is 3′ × 3′ × 3′ = 27 ft^3). This value will tell you how many cubic yards of mulch you will need.

Example: 20′ long × 4′ wide × 0.3′ deep = 24 ft^3

$$\frac{24 \text{ ft}^3}{27 \text{ ft}^3/\text{yd}^3} = 0.89 \text{ yd}^3 \text{ of mulch}$$

Corner Question

What is *agricultural nonpoint source pollution*?

Did You Know?

In the seventeenth century, tobacco, arsenic, and mixtures of herbs became the most widely used pesticide concoctions. Today, some of these same ingredients are used in commercially available products to combat pests.

Integrated Pest Management (IPM)

Integrated pest management (IPM) is an approach to managing pests that uses common sense, economical practices and results in the least possible hazard to people, property, and the environment. IPM contributes to plant health and water conservation. Horticulturists who use IPM have healthier plants. When a plant is hardier and has vigorous growth, it can withstand periods of drought. A plant that is already stressed by pests will suffer greatly from drought stress.

Water Garden Landscape Design

The tranquil sound of running water or the added interest of goldfish in a small body of water is often associated with established gardens, **Figure 24-26**. Sales of aquatic plant materials, fish, and water garden materials increased as more home gardeners choose to include water features in their gardens. Water gardens also remain a staple feature in public gardens, arboretums, and estate gardens.

Attractive and functional water gardening can also take place on ponds and wetlands, known as aquascapes. *Aquascaping* is the craft of arranging aquatic plants and hardscape materials in a pleasing manner. Aquascaping adds beauty to the landscape through the inclusion of ornamental plants into the blank canvases of bodies of water.

Whether creating a garden with a water feature or adding plant life to a pond, the possibilities for water gardening are endless. Not only does this form of gardening add zest to spaces, but it can also increase the value of the property because most individuals realize the value of a water garden on a property. Water gardening can take many forms in the landscape, including rain gardens, ponds, lakes, streams, container water gardens, waterfalls, and water fountains.

focal point/Shutterstock.com

Figure 24-26. A waterscape full of life and movement provides garden enthusiasts with hours of entertainment.

Rain Gardens

A *rain garden* is a landscape that collects runoff water and pollutants, filters out pollutants, and slowly releases the water. Rain gardens deliver a gardening solution to excessive runoff during and after rainfall related to impervious surfaces (concrete, pavement, and rooftops). The rain garden is placed in a depression that collects the runoff and allows the water to penetrate the underlying soil slowly. This gardening system helps to reduce pollution and erosion and allows water to be purified before reaching other bodies of water.

Rain gardens flourish with native plants that are well-suited for wet conditions, **Figure 24-27**. The root systems of the plants should vary in depth and allow the plants to survive both wet and dry conditions associated with a rain garden that can have saturated or dry soils. Rain gardens are attractive solutions to ecological problems associated with urban rooftops, parking lots, driveways, and compacted lawns.

Bodies of Water

When thinking of water elements in gardens, you may immediately picture a pond or stream. Ponds have several purposes in the landscape that may be functional or aesthetic in nature. A pond is a small body of water, either artificial or natural, that can be used to retain water for irrigation purposes, hold fish for recreation or cultivation, or provide habitat for aquatic organisms. A pond can also be used for ornamental or landscape purposes. When plants are cultivated purposefully in the pond landscape, this is known as aquascaping. Many plants can be grown in a pond aquascape, including lily, iris, hyacinth, lotus, and many more species.

Peter Turner Photography/Shutterstock.com

Figure 24-27. A rain or bog garden contains plants that can tolerate consistently wet conditions. These plants also help to purify the runoff and polluted rainwater from surrounding areas, such as a parking lot.

Other bodies of water can provide habitats for aquascapes. Streams, rivers, and creeks can provide movement in a landscape that creates interest and lines within the landscape design. These moving bodies of water can be somewhat easier to manage than standing bodies of water that are more likely to grow algae.

Container Water Gardens

Smaller aquatic gardens have gained recent popularity in home gardens. A *container water garden* is a water element that is housed in a vessel and can contain plants and aquatic animals, **Figure 24-28**. To install a container garden, follow these simple guidelines:
- Choose containers varying in size (between 15 and 25 gallons) and makeup (metal, plastic, or ceramic) to provide manageable canvases for container water gardens.
- Use containers with dark interiors to discourage algae growth.
- Select aquatic plant material. Many aquatic plants require full sun or at least six hours of direct sunlight. Placement in locations with less than this amount of light will result in fewer blooms occurring less often.
- Plant aquatic plants in a heavy clay medium and top with gravel in a separate container. Submerge the crown of the plant so that it is placed at that specific plant's preferred depth. Approximately 50%–60% of the water surface should be covered with plant material.

Jillwt/Shutterstock.com

Figure 24-28. A container water garden includes plants or animals in a vessel holding water.

Corner Question

What is the difference between a pond and a lake?

Corner Question

How do snails and slugs differ?

Fish and Snails

Fish and snails can help regulate the ecosystem of the container water garden. These organisms help balance this miniature ecosystem by eating algae, decaying plant waste, and mosquito larvae. A container of at least 10 gallons can handle one mature goldfish. Fish must be properly fed and monitored for health.

Overwintering

Many of the organisms used in aquascaping will need to be overwintered properly. If your area has winter temperatures that are below freezing, the plants must be taken out of the water garden, and the remaining water from the container should be removed. Any fish will need to be moved to an aquarium. Plants can be moved to a tub filled with water and placed in a garage or basement. These plants will enter a phase of dormancy and can be moved back to the container garden when temperatures rise in the spring. Some plants may be better treated as annuals, with new plants purchased the following spring.

Tatiana Ganapolskaya/Shutterstock.com

Figure 24-29. Fountains help to increase sound and interest within a garden. This new element, however, does not come without a price tag and additional maintenance issues.

Water Features

Water features are elements of the garden that provide actively flowing water on various scales, **Figure 24-29**. These features can include fountains or waterfalls.

A water fountain is an artificial water element that must include a pump to pull water from a basin and force the water to be distributed in a pattern. Fountains can vary in size, material, and effect. Every fountain has a unique purpose in the garden. It might be a focal point or simply add texture through a fine water mist.

A waterfall forces water to cascade over an impervious material, such as gravel or rocks. Again, water is pumped from a basin and taken to a higher point where it is released and flows over the rocks. This creates a unique sound effect and feeling that accompanies the landscape.

Water fountains and waterfalls must recycle water to be sustainable. In areas where a drought is occurring or in places that have desert-like conditions, the inclusion of a water feature that is a fountain or waterfall should be carefully considered. Water must be continually refilled into basins, as some is lost to evaporation and splashing. In areas where water is in extremely high demand, a water fountain or waterfall is not a sustainable choice.

Aquatic Plants

Aquatic plants thrive in conditions where the roots are submerged in water. These plants have adapted to these unique living conditions. Aquatic plants grow at different points in the ecosystem and include underwater, floater, and shoreline plants.

Hands-On Horticulture

Installing a Dry Creek Bed

A dry creek bed offers a smooth and soothing appearance in a garden without the application of water. This approach does not actually use water in the design but gives the same aesthetic and effect. To install a dry creek bed:

1. Determine the size and lines of the creek bed. Excavate 12″–15″ (30 cm–38 cm) deep. Use a backhoe or shovel. (Remove the soil, but save it for another project or add it to an existing garden.)
2. Tamp the ground and create a firm surface for the creek bed. Place a landscape fabric or geotextile over the area to prevent weed growth.
3. Cover the sides and the base with 1/2″ (1.3 cm) of crushed pea gravel.
4. Add river rock along the sides of the base to create a natural look. Incorporate varying sizes of rocks with smooth edges.
5. Include a few large rocks or boulders on each side of the dry creek bed.
6. Create a bridge over the dry creek bed with large, flat pieces of stone, or build a wooden bridge.

D Matarazzo/Shutterstock.com

Corner Question

What world-renowned artist captured water lilies in his paintings?

Underwater Plants

Plants growing entirely underwater are also referred to as oxygenators. They do supply some oxygen to the water; however, they consume oxygen 24 hours a day (like all plants) through respiration. They also can benefit the aquatic environment by removing nutrients that would otherwise promote the growth of algae. Examples are eelgrass (*Vallisneria* sp.), fanwort (*Cabomba canadensis*), and parrot feather (*Myriophyllum aquaticum*), **Figure 24-30A**.

Floater Plants

The root systems of floater plants hang down into the water. Like other aquatic plants, they filter out nutrients that would otherwise support algae growth. The plants also provide shade that is needed for aquatic organisms, such as fish and frogs. The shade provided by their leaves also prevents light that algae needs for growth from entering the water system.

Examples of floater plants are water hyacinth (*Eichhornia crassipes*) and water lettuce (*Pistia stratiotes*), **Figure 24-30B**. These beautiful plants root themselves in the subsurface mud, but their leaves and flowers float on the water surface. Hardy water lilies and lotuses will die back to their root zone in winter when freezing occurs. Tropical water lilies and lotuses cannot survive freezing temperatures. All lilies and lotuses thrive in full sun, which promotes heavy flowering. Some dwarf varieties are available for container water gardens.

A
Reece with a C/Shutterstock.com

B
Phanumassu Sang-ngam/Shutterstock.com

Figure 24-30. A—Parrot feather is a plant that is used in aquascapes for above and below water plantings. B—Floater plants, such as this water lettuce, float on the top of the water. Be careful: some of these plants can overtake an area and may be too invasive to be used in your planting.

Shoreline Plants

- Arrowhead—*Sagittaria* sp.
- Blue flag iris—*Iris versicolor*
- Canna—*Canna X generalis*
- Cattail—*Typha* spp.
- Chameleon plant—*Houttuynia cordata*
- Corkscrew rush—*Juncus* sp.
- Creeping jenny—*Lysimachia nummularia*
- Hibiscus—*Hibiscus moscheutos*
- Horsetail—*Equisetum* sp.
- Lobelia—*Lobelia cardinalis*
- Paper reed—*Cyperus papyrus*
- Pitcher plant—*Sarracenia* sp.
- Rain lily—*Zephyranthes* sp.
- Swamp lily—*Crinum americanum*
- Sweet flag—*Acorus calamus*
- Taro—*Colocasia esculenta*

Goodheart-Willcox Publisher

Figure 24-31. A listing of various plants can help the landscape designer create a beautiful and sustainable aquatic landscape.

Shoreline Plants

Some shoreline plants do best in areas where the crown of the plant is submerged under several inches of water. Others prefer to have their roots covered in water or in saturated soils. If using these plants in a container water garden, be sure to plant the crown at the appropriate depth of water. Refer to **Figure 24-31** for a list of shoreline plants.

Water Garden Maintenance

A simple water garden is fairly low-maintenance. Refilling containers with water, overwintering plants and aquatic organisms, and using IPM are the primary maintenance tasks associated with water gardens. When water features (fountains and waterfalls) are included, the maintenance element of a water garden increases.

Follow these simple tips to keep a water garden thriving:

- Add water when needed.
- Feed the fish.
- Skim dead plant material from the surface and bottom to help prevent excessive decay. Otherwise, water can become brown or tea colored.
- Overwinter plant materials (as you choose or for species that are needed) and fish.

Pumps and Lighting

Be sure that water is always freely flowing from a water feature. Add additional water if needed. If water still does not flow freely, check to see if the pump is operating. Unplug the pump and remove anything that may be covering the filter. Attach the pump to the power source, and see if the water begins to flow. If not, unplug it again and see if there is any blockage in the tubing that forces the water from the pump to the point of release. Again, attach it to a power source, and see if this changes the situation. Keep the pump free of any debris and ensure that there is adequate water always flowing to the pump.

Lighting contributes greatly to the elements of a water garden. At night, water that is illuminated adds extra energy and focus to a landscape design, **Figure 24-32**.

Alexey Stiop/Shutterstock.com

Figure 24-32. Landscape lighting adds more elements to the landscape and creates a new feeling for evening garden enjoyment.

Algae and Weeds

Algae and weeds need water, light, and nutrients to grow. Removing one or all of these elements will prevent the growth of these organisms. Water cannot be removed from the scenario, but the amount of nutrients and the light source available can be limited. Additional nutrients can be absorbed by cultivated aquatic plants. These same plants can also shade the water environment to prevent other algae or plant growth. Very few algaecides or herbicides can be used in an aquatic environment. The best management practice for algae and weed growth is to suppress or prevent the growth of these invasive species, **Figure 24-33**.

Mosquitoes

Mosquito larvae are a wonderful food source for many fish and frogs. Adding these organisms to a standing body of water can prevent increased populations of mosquitoes. Not all aquatic sites provide a habitat suitable for fish and frogs. How do you prevent or treat mosquito larvae if you cannot have fish or frogs? Mosquitoes can be controlled through a number of organic means. Simply overfilling a water feature, such as a container garden, will cause the immature mosquitoes to run off and fall into areas without standing water that will lead to their death. Using *Bacillus thuringiensis* (a naturally occurring bacteria) kills the larvae without injuring plants, other aquatic fish or organisms, or pets that may drink from the water. Also, vegetable oil droplets placed on the surface of the water will prevent mosquitoes to some extent. This method does not work where fountains or running water is in place.

Candace Hinton/Shutterstock.com

Figure 24-33. Preventing the growth of algae, which is found in green water, is much easier than treating surfaces. Reducing the amount of sun and excessive nutrients (fertilizers) as well as lowering temperatures can reduce the amount of algae.

Safety Note

Water Feature Pumps

Water and electricity do not mix. Be sure that whatever pumps or lighting system you use are specifically intended for water gardening. Read all operational manuals thoroughly. Disconnect the device from its power source while installing or servicing it.

Igor Kovalchuk/Shutterstock.com

Corner Question

For what human diseases are mosquitoes a carrier?

Careers in Landscape Design

The landscaping industry includes a variety of horticultural careers. Some occupations involve only office work or management of employees while others require a great deal of manual labor. For example, a landscape architect designs landscape plans, and an irrigation specialist (subcontractor) installs elements in the landscape that an architect or designer has planned. In addition, growers must cultivate aquatic plants for the various water garden applications that a landscape architect may incorporate into his or her design.

Candus Camera/Shutterstock.com

Figure 24-34. An irrigation specialist designs, installs, and maintains irrigation equipment in a number of different settings.

Irrigation Specialist

Irrigation specialists help design, install, and maintain irrigation designs of residential and commercial properties, **Figure 24-34**. These individuals, sometimes referred to as irrigation technicians, may work on new or older construction. These individuals are also skilled at repairing or retrofitting outdated equipment. This job requires little office time and much contact time with job sites and crews. Irrigation specialists may often move from one job site to another to lend their expertise.

Most irrigation specialists need a two-year degree or a certificate to fulfill educational requirements demanded by employers. Specialists should have a thorough understanding of engineering, soil science, and plant science. There is projected growth in this field in the future. As more individuals and businesses realize the concerns associated with water availability and conservation, irrigation specialists will continue to be demanded in the workforce.

Aquatic Plant Grower

Growers can specialize in just one species of plant or an entire genre, such as aquatic plants. A company with a specialization grows plant materials that have similar cultivation requirements. Aquatic plants all need water. Growing plants that have that one same need is much easier than growing succulents, bedding plants, perennials, roses, and everything in between.

A specialist grower, such as those who cultivate only aquatics, is usually inspired to grow those plants because of a sincere interest in the plants. Some growers may grow to love that type of plant, or some may have the facilities that are well-suited for the cultivation of that species.

Growing aquatic plants usually requires cultivation in beds or benches that retain water. As long as the plants are sitting in the water on a Friday and the containers can hold the water for a couple of days, an aquatic grower may have more opportunity for weekends off and vacations. With the increased interest in installing water gardens, there is a continued demand for aquatic plants. Growers should take this opportunity to fill this unique market niche.

Career Connection

Alex Ramirez
Design Workshop

Alex Ramirez

Alex Ramirez always had an artistic ability, a knack for creativity, and a drive to work. He was a college baseball player and was also studying horticulture. Injury derailed his future in baseball but not in horticultural pursuits. He focused on landscape architecture at Louisiana State University (LSU). He graduated from LSU and accepted a position as a landscape architect at the innovative Design Workshop's Austin office.

As a landscape architect, Alex works on projects of various scales. One project Alex collaborated on was a sustainable landscape design plan for Houston's Bagby Street Renovation in Midtown. This site often flooded, and solutions were needed for this ongoing problem. Here, he and his firm created rain gardens that caught 33% of all water shed at the site.

Ramirez suggests that students interested in landscape architecture travel to broaden their minds and their skill set. Seeing other places helps to provide more tools be used when tackling a landscape project. Above all, Alex urges students to challenge themselves and others. "Be competitive," he says.

Ramirez continues to grow as a designer. Future landscape projects of all scopes and scales will require his unique perspective, artistic ability, and leadership at Design Workshop.

CHAPTER 24
Review and Assessment

Chapter Summary

- The landscape design process begins with stating the problem to be solved or the objective to be achieved for the landscape.
- The steps of the design sequence include site analysis, functional diagrams, concept plans, preliminary designs, and the final landscape plan.
- When a client accepts the final plan, it is presented to a landscape contractor for a bid. A landscape contractor may install all plants and garden features, or the contractor may subcontract some jobs to other companies.
- Designers may choose to draw their designs by hand using a drawing board or to create designs using a computer. Designers may use graphics on a landscape plan to represent plants or features.
- A landscape plan is a fusion of art and science created using design elements and principles. The elements of design include line, form, texture, and color. The principles of design include balance, proportion, repetition, emphasis, and unity.
- Many tools are employed by landscape designers and architects. Measurement tools may include measuring wheels and transit levels. Drawing tools may include drawing boards, pencils, erasers, scales, T-squares, compasses, templates, and drafting triangles. Various types of paper are used at different stages of design creation.
- Water-wise gardening incorporates seven basic principles that help to conserve water. These principles include planning and design, soil analysis and amendments, plant selection, turf use, efficient irrigation, mulch, and maintenance.
- Water gardening can take many forms in the landscape, including rain gardens, ponds, lakes, streams, container water gardens, waterfalls, and water fountains.
- Aquatic plants grow at different points in the ecosystem and include underwater, floater, and shoreline plants.
- Refilling containers with water, overwintering plants and aquatic organisms, and using IPM are the primary maintenance tasks associated with water gardens. When water features are included, the maintenance element of a water garden increases.
- The landscaping industry includes a variety of careers. Landscape architect, irrigation specialist, and aquatic plant grower are some of the occupations to consider when investigating landscape careers.

Words to Know

Match the key terms from the chapter to the correct definition.

A. absolute scale
B. aquascaping
C. base plan
D. color theory
E. concept plan
F. container water garden
G. emphasis
H. functional diagram
I. graphic
J. interconnection
K. landscape architect
L. landscape contractor
M. landscape designer
N. perspective
O. preliminary design
P. rain garden
Q. relative scale
R. repetition
S. simplicity
T. texture

1. A map showing a site with the details of the landscape and its structures.
2. A map or design that shows the general layout and possibilities for a landscape.
3. A series of loosely drawn circles that divide spaces into areas that will be decided on by the designer.
4. A landscape map that breaks down the concept plan and includes more thorough details.
5. The size of an element of the landscape in relation to another object.
6. The quality of how a surface feels or appears, such as coarse, rough, smooth, or fine.
7. A landscape that collects runoff water and pollutants, filters out pollutants, and slowly releases the water.
8. An individual who installs and maintains landscape elements.
9. A seamless linkage between diverse features to create blending, such as the way plants and other elements are related to one another in a landscape.
10. A body of practical guidance for using and mixing colors.
11. The true size of an object.
12. A water element that is housed in a vessel and can contain plants and aquatic animals.
13. The craft of arranging aquatic plants and hardscape materials in a pleasing manner.
14. A professionally trained and educated individual with knowledge of plant material and engineering.
15. The prominence given to something, such as a focal point in a design.
16. The size and placement of objects to create depth and distance in a design, such as the foreground, midground, and background in a landscape.
17. A pictorial representation of an item in a landscape plan.
18. A person with horticultural expertise, formally or informally educated, who often works on small landscape projects.
19. The recurrence of an event or thing, such as repeated elements within the landscape.
20. The quality of being plain, natural, or uncomplicated.

Know and Understand

Answer the following questions using the information provided in this chapter.

1. What are the steps of the design sequence?
2. Why do design contract companies often use subcontractors for landscape installation?
3. What are two methods designers may use to create landscape plans?
4. What are the elements and the principles of landscape design?
5. What are the four types of lines in landscape design?
6. What are the three types of plant forms?
7. List and describe three color schemes often used in landscape design.
8. What is the purpose of a transit level?
9. What are some tools a designer may use to create a landscape design drawing?
10. What is a good choice of hardness and darkness for a pencil that would be used by a landscape designer or architect?
11. How does an architect's scale differ from an engineer's scale?
12. What are the seven principles of water-wise gardening?
13. What are two devices that can help to prevent overwatering or unnecessary irrigation?
14. How is mowing involved in landscape maintenance?
15. What is integrated pest management and what benefits does it provide for a landscape?
16. What are some forms that water gardening can take in a landscape?
17. How can rain gardens combat pollution?
18. What are three maintenance tasks associated with water gardens?
19. What are some safety steps you can take related to water feature pumps?
20. What are some job duties of an irrigation specialist and what training or education is needed for this job?

Thinking Critically

1. In today's urban society, there is sometimes very little room to grow plants. What could a landscape designer do for someone who has only a small patio in the back or the front of a row-type home in which to create a garden space?
2. You have decided to collect water in a cistern or rain barrel. You attach a hose and realize the water pressure that comes out of the hose is very low. It takes a great deal of time for you to water your plants. What could you do to increase the pressure of the water flow that is being released from your water catchment system?

STEM and Academic Activities

1. **Science.** A site plan must be designed for a future garden at your school. The area is oddly shaped and represents two egg-shaped areas connected together. What is the best approach for measuring this garden space and transferring the measurements to a base map?

2. **Science.** When designing a garden for children, what elements and principles of design must be carefully observed? How would this garden differ from another garden when designing the areas related to measurement?
3. **Math.** A water garden that is circular has a diameter of 10′ (3 m). The depth of the pond is 4′ (1.2 m). If you were to fill this garden from empty to full, how many gallons of water would be required? If your hose flows at a rate of 5 gallons per minute, how long will it take to fill up your pond?
4. **Social Science.** Contact your state's chapter of the American Society of Landscape Architects (ASLA). Ask them to identify local ASLA members in your area. Contact a member and ask her or him how ASLA impacts her or his work.
5. **Language Arts.** Choose a famous garden and research the design elements and principles of this landscape. Write a statement that you think would best reflect the design of this garden. Include the objectives for this space and what you think the intentions of the designer or architect were.

Communicating about Horticulture

1. **Writing and Speaking.** Visit a local garden or landscape that is well known in your community, region, or state. Take photos and create a three- to five-minute presentation to educate your class or FFA chapter about your recent visit.
2. **Writing and Speaking.** Create a multimedia infographic about aquatic landscapes.

SAE Opportunities

1. **Exploratory.** Job shadow a person who installs hardscapes.
2. **Experimental.** Compare how light duration impacts flowering of water lilies.
3. **Exploratory.** Interview landscape designers and landscape architects. Create a list of colleges and universities that offer these programs of studies in your state. Develop a virtual poster that explains how these careers differ and the educational opportunities available to students in your state. Link this to your school's website or your college counseling website.
4. **Entrepreneurship.** Develop a water gardening installation service.
5. **Placement.** Get a job working for a company that sells hardscape materials.

Toa55/Shutterstock.com

CHAPTER 25
Floral Design

Chapter Outcomes
After studying this chapter, you will be able to:
- Describe the principles and elements of floral design.
- Identify types of floral designs.
- Describe containers, tools, and materials used in floral design.
- Explain how flower arrangements are constructed.
- List careers related to floral design.

Words to Know ↗

asymmetrical design	harmony	proportion
boutonniere	hedgerow design	rhythm
color harmony	line	scale
color scheme	line design	stem stripper
corsage	line flower	symmetrical design
filler flower	line-mass design	T-design
focal point	mass design	unity
form	mass flower	
form flower	mechanics	

Before You Read
Before reading this chapter, review the objectives. Based on this information, write down two or three items that you think are important to note while you are reading.

While studying this chapter, look for the activity icon to:

- **Practice** vocabulary terms with Words to Know activities.
- **Expand** learning with identification activities.
- **Reinforce** what you learn by completing Know and Understand questions.

www.g-wlearning.com/agriculture

Human beings have been using flowers and plant materials for adornment for thousands of years. Today, floral design and floriculture is a multimillion dollar industry. The floriculture industry includes the cultivation of cut flowers and foliage, bedding plants, houseplants, and potted or container plants. The floriculture industry is comprised of growers, wholesalers, retailers, distributors, importers, and designers, as well as the support businesses that supply equipment and accessories. Although the growing techniques, plant nutrition, and growing conditions discussed in this text apply to plants grown for the floriculture industry, there are many processing and storage techniques specific to the cultivation of these plants—far more than can be covered in one chapter. Therefore, this chapter instead focuses on the basics of floral design.

Principles and Elements of Design

Did You Know?
The principles of design are also referred to as the "laws of beauty" and apply to all forms of art, including landscape architecture and design.

Floral designs are often used to decorate spaces for formal occasions, such as parties or weddings. Many people send flower arrangements to friends or family to express congratulations or caring, and to make holidays and special occasions, such as birthdays and anniversaries, more special. Floral designers use principles and elements of design to create a design that is appropriate for the occasion, space, and audience it is intended to please.

Principles of Design

The principles of design establish guidelines for creating floral arrangements that are pleasing to the eye. While many people have a natural talent for design, principles of design can also be learned with time and practice. Principles of floral design include balance, symmetry, proportion, scale, focal point, rhythm, harmony, and unity.

Balance

Balance creates a sense of stability. Arrangements must be physically and visually balanced. A physically balanced design is stabilized with mechanics. *Mechanics* are the devices and techniques used to keep floral placements secure in a design. Mechanics are also used to provide support, anchorage, and footings to the arrangement as a whole. Poor mechanics contribute to poor balance. To ensure a design does not fall over, careful attention must be paid to the engineering and security of the footings. A designer must also consider factors that may contribute to physical failures, including weather and human contact, and construct measures to prevent the fall or collapse of the floral design, **Figure 25-1**.

In floral designs with visual balance, the elements are arranged so that no one part of the design overpowers other parts. Visual balance must address the color, shapes, sizes, and patterns used in the design.

KMW Photography/Shutterstock.com

Figure 25-1. The thin stem of this tall table arrangement allows guests to see and speak easily to those across the table. However, the designer must ensure the arrangement is well-balanced to prevent it from easily falling or being knocked over.

Symmetry

Floral designs are divided into visually balanced parts related to a central vertical axis (CVA). Balance can be expressed as symmetrical or asymmetrical.

In *symmetrical designs*, a nearly identical or mirrored arrangement of plant materials is used on both sides of the CVA. Symmetrical designs invoke a feeling of formality. Repetitions of flowers, foliage, and accents create the symmetry. Triangular, round, or fan-shaped arrangements work well with symmetrical designs, **Figure 25-2A**. A symmetrical arrangement can often be found on opposing sides of an entrance or an altar.

A   Andrew Mayovskyy/Shutterstock.com   B   Chris Bankhead/Shutterstock.com

Figure 25-2. A—This flower arrangement is a symmetrical, equilateral triangle design. B—This right triangle design is not symmetrical, but the weight on either side of the CVA is proportional.

In *asymmetrical designs*, the weight of the design is distributed equally on the both sides of the CVA, but the sides are not mirror images. The CVA does not have to divide the arrangement into equal sides. An asymmetrical design lacks formality and often parallels what occurs in nature. Asymmetrical designs create balance by repeating color, sizes, shapes, and patterns. If one side of the axis uses a particular flower of one color, the other side may have another flower of another color. Common asymmetrical floral designs include crescents, Hogarth curves (S-shaped lines), and right triangles, **Figure 25-2B**. Bridal bouquets, tabletop arrangements, and many other designs may use asymmetrical balance.

Did You Know?
Butterfly wings are examples of natural symmetrical design.

Proportion

The relative sizes and amounts of elements in a design denote the *proportion*. Various elements, such as the flowers, foliage, accents, and container, combine and must be proportionate to one another for a visually pleasing result.

Proportion is important within a design. The relative proportion of a design to the space in which it is placed is also important, **Figure 25-3**. For example, a Valentine's Day teddy bear that dwarfs the overall arrangement does not have pleasing proportion. An arrangement with small button mums coupled with very large foliage leaves also lacks pleasing proportion.

Konstantin Goldenberg/Shutterstock.com

Figure 25-3. A floral design should be proportional or to scale with its environment. Following the rule of thirds, this arrangement is dominant, taking over two-thirds of the space within view. *How would this same arrangement fit on a small café table?*

STEM Connection

Proportion or Scale?

Proportion and scale are often confused. Proportion is the relativity of the materials to one another. For example, a flower arrangement that has 15 flowers that are proportional to a container can be increased in size to meet the requirements of a larger space. If the place where the arrangement will be placed needs to be three times as large, then 45 flowers and a different container can be used. A plant may be proportional to the container (2–3 times as large as the container) but much too small, or not to scale, for the space it is to occupy. For example, a single daisy in a vase would suit a table meant for two people, but would look lost on a banquet table set for eight.

Africa Studio/Shutterstock.com

Did You Know?

In competitive flower shows, a miniature flower arrangement cannot exceed 5″ (12.7 cm) in height. Therefore, to follow the element of proportion, a container should be no larger than 2″ (5.08 cm) tall.

Floral designers follow simple guidelines to achieve proportion in arrangements:

- Plant materials (flowers and foliage) should be 1 1/2 to 2 times the size (height, length, or width) of the container used.
- Applying the rule of thirds to get a rough estimate for arrangement sizes. The space is first divided into thirds. The arrangement can be dominant and fill two-thirds of the space, or subordinate and fill one-third of the space. A dominate arrangement is the point of emphasis in a space whereas a subordinate arrangement rests in the background.

Scale

A *boutonniere* is a small floral arrangement that is worn on the lapel area of clothing. A boutonniere may be small compared to the size of a grown man, but that same boutonniere pinned to the lapel of a child will appear much larger, **Figure 25-4**. *Scale* is the size relationship between two objects, such as between a floral design and its surroundings, or among the various elements within a design. A flower arrangement meant for a ballroom or banquet hall must be large so that it will not seem too small for the space. That same arrangement would be much too large for a dining room table in a home.

Focal Point

What catches your eye in a design? Is it a certain color? Is it a specific flower? The *focal point* of a floral design is the area of dominance or emphasis to which the eyes naturally travel.

oliveromg/Shutterstock.com

Figure 25-4. The ring bearer's boutonniere is the same size as the one on the groom's lapel. The flower appears larger on the boy due to scale.

The focal point will capture the viewer's attention and highlight a particular part of the design. Various flowers, foliage, or materials can be used to create a focal point, **Figure 25-5**. A bright or contrasting color, a unique form, or even a piece of art or a figurine can be a focal point. Materials that are unique or add interesting details to a design are called accents. Accents can stress or emphasize the focal point of the design.

Rhythm

Rhythm is a pattern of repeating or alternating elements in a design that suggests a sense of movement. Flowers and materials used in a design can create a pattern or guide the eye by the way they are spaced. Colors, shapes, textures, and lines can be placed in intervals. Flowers that are tightly spaced in a design create a slow rhythm. As the flowers are spaced further apart, the rate of movement or rhythm increases. The rhythm of the design should lead the eye to a focal point.

Olinchuk/Shutterstock.com

Figure 25-5. The gerbera daisy is the focal point of this arrangement. The viewer's eye is automatically drawn to the large orange flower.

Unity

Unity is the relationship among elements in a work of art that makes them appear to belong together or function as a whole. When a floral design evokes a feeling of unity, all of its elements work together. To create unity, designers use principles of design and also work to include:

- Repetition—reusing flowers, foliage, accessories, colors, textures, and other elements. Items can be reused similarly or exactly, **Figure 25-6A**.
- Proximity—the way elements of the design are placed in relation to one another's space. The designer can group or sort similar pieces together in a way that is visually pleasing.
- Transition—the gradual changes from one point unto another. Gradients of color or size can help transition components of a design. An example is having a container, such as a woven basket, flow into the flower arrangement, **Figure 25-6B**.

"The essence of the beautiful is unity in variety."
—Somerset Maugham

A infinity21/Shutterstock.com   B   natrot/Shutterstock.com

Figure 25-6. A—Repetition contributes to a sense of unity and harmony. Here the vases and materials are repeated. B—These flowers transition from one type to another.

Harmony

Harmony is a pleasing interaction or blending among the elements of a design. Viewers recognize harmony by the uniqueness of different elements that contribute to the overall effect. The color, texture, shapes, and materials blend together and create a whole design. The principles that contribute to the creation of harmony in a design include simplicity, theme, and similarity.

- Simplicity. A design with very few elements is simple. To incorporate simplicity in an arrangement, designers may omit elaborate accessories and focus on the floral materials, **Figure 25-7A**.
- Theme. The subject or focus of a piece is the theme. For a fairy tale wedding theme, the designer might use a glass slipper as the container or adorn the space with Cinderella figurines. Using a theme in the floral design contributes to the theme of the overall event or occasion.
- Similarity. Items that are similar resemble one another but may not be identical. Using similar items in a design contributes to the harmony of the piece while also providing variety, **Figure 25-7B**. For example, for a brunch a designer might use varying tea cups filled with the same combination of white flowers. The teacups provide some differences but the decorations are united by the similarity of the type of container and the flowers used.

pullia/Shutterstock.com

Alex Andrei/Shutterstock.com

Figure 25-7. A—An overall feeling of simplicity and harmony is achieved in this design. B—Although the designs are different, the similarity of plant materials, shapes, and sizes helps to contribute to a sense of harmony.

Elements of Design

The elements of design are all of the things that go into creating an arrangement. In floral design, the elements are used to create an artistic expression using floral materials. The elements of floral design include color, line, form, and texture.

Color

Color is the element that impacts most viewers immediately. Color likes or dislikes help determine whether or not someone likes an item and heavily impacts purchases. A pure *hue* (a color on the color spectrum) has not been lightened, darkened, or grayed. Adding black, gray, or white to the pure hue will change the purity or value (making it darker or lighter). A *tint* is a hue that has been lightened, and a *shade* is one that has been darkened. For example, red is a pure hue. A tint of red is pink, and a shade of red is maroon.

Corner Question

How many Crayola® crayon colors are there?

When gray is added to a hue, it weakens the color and results in a *tone*. A viewer recognizes the intensity (also called chroma) in the brightness or dullness of a hue, which is created by mixing the hue with other colors.

Color harmonies (also called *color schemes*) create a guideline for combining colors in a design. A color wheel may be used to choose or explain color harmonies, **Figure 25-8**. Designers may use several types of color schemes:

- In a monochromatic color scheme, one color that varies in value and intensity is used throughout the design. For example, bright yellow roses and both light and dark yellow carnations could be used as the base of a monochromatic arrangement. Monochromatic arrangements generate feelings of peace and unity.
- In an analogous color scheme, three or more neighboring colors on the color wheel are used throughout the design. Analogous arrangements feed resemblance, or similarities of color, without being monotonous. Fall weddings and other fall events can easily use analogous color schemes, **Figure 25-9**.
- In a complementary color scheme, two hues opposite each other on the color wheel are used to create dramatic contrast. A commonly used example is red and green. Nature often replicates this color scheme. For example, red holly berries contrast with the green luster of foliage.
- A polychromatic color scheme uses five or more hues. Flowers of many colors and shades are included in this type of color scheme.

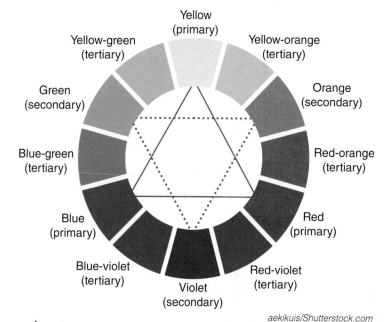

aekikuis/Shutterstock.com

Figure 25-8. The color wheel is a useful tool for establishing a color scheme for designers and customers.

> "Color is my day-long obsession, joy, and torment."
> —Claude Monet

Line

Line represents a path that constructs the foundation, shape, and form of a floral arrangement. Lines can be straight or curved, horizontal or vertical, or any combination thereof. The actual lines of flower stems can aid in constructing the overall design. Different types of lines contribute to different moods. A horizontal line creates a restful feeling and is often used for sympathy or funeral pieces. Vertical lines evoke a feeling of energy. They force the viewer's eye upward or downward toward a focal point.

Maria Sbytova/Shutterstock.com

Figure 25-9. An analogous color scheme uses three or four colors that are next to one another on the color wheel.

Form

The overall shape of an arrangement and the shapes of the flowers or foliage are referred to as the *form*. The shape of the arrangement will be discussed later in this chapter with geometric designs. Flower and foliage forms are classified into four main categories:

- *Line flower*—a long stem or spike. Examples are bells of Ireland, snapdragon, gladiolus, horse's tail grass, and delphinium, **Figure 25-10A**. This is the first plant form put into the design to establish the height and width or general framework of the design.
- *Mass flower*—a closed-form, single flower that has a dense, round shape. Carnations, roses, peonies, mums, and galax or pittosporum for foliage are mass flowers, **Figure 25-10B**. Mass flowers fill in spaces around the line materials and make up the majority of the arrangement.
- *Form flower*—a flower that is unique in some manner (color, shape, texture, or size). These are usually exotic and expensive flowers. Examples include orchid, bird of paradise, anthurium, and gloriosa lily, **Figure 25-10C**. Unique leaves may also be used.
- *Filler flower*—fine or airy materials used to fill in voids and often made of very small clusters of blooms or branched inflorescence. Statice, asters, wax flowers and baby's breath are common filler flowers, **Figure 25-10D**. Ferns may be used as filler foliage.

Texture

The visual and physical qualities of a material compose its texture. Therefore, texture is experienced by both sight and touch. For example, a thistle's texture appears coarse and is prickly to the touch.

A. Line Flower
Tatiana Volgutova/Shutterstock.com

B. Mass Flower
LiuSol/Shutterstock.com

C. Form Flower
max777/Shutterstock.com

D. Filler Flower
Susii/Shutterstock.com

Figure 25-10. A—Delphinium have a strong vertical form and are available in shades of blue, pink, white, and purple. B—Mums are available in many colors and are one of the standard mass flowers. C—The gloriosa lily is a popular form flower because of its unusual shape and bold colors. D—Baby's breath is a one of the most commonly used filler flowers.

STEM Connection | Fragrance

Flowers need to attract pollinators to help them create offspring. Flowers evolved to produce fragrances as lures for specific pollinators. Some flowers exude a sweet aroma while others give off the smell of rotting meat (such as the carrion flower illustrated below). Regardless of how sweet or putrid the fragrance, there is a natural "partner" who will be lured by it and help the plant reproduce.

A floral designer must consider their customers' sense of smell when choosing flowers for a specific arrangement. Some people enjoy fragrant flowers whereas others do not. Flowers such as stocks may be too sweet or overpowering for some people, but a gardenia may be a welcome fragrance. Potted spring bulbs, such as paperwhites, have a unique smell that may be offensive to some but enjoyable to others. Designers should talk to their customers about fragrances to ensure that both understand what fragrances should be avoided or included.

Graeme Knox/Shutterstock.com

An arrangement's visual impact can be manipulated by adding various textures to the piece. As each component of an arrangement (plant materials, accessories, and containers) brings its own texture to a design, each piece must be carefully considered. It is recommended that at least three varying textures should be used. Some ways to add texture include:

- Using more than one type of foliage.
- Selecting containers and decorations that offer texture.
- Varying flower texture when including more than one flower with the same form.
- Using flowers at different growth stages (bud, slightly opened, and fully open).

Types of Floral Design

Floral designs are often categorized by style. Some design styles are associated with countries or regions, time periods, or design shapes. Some designs have been popular during more than one period of history and many traditional designs are used as the basis for new, modern designs. However, most designs can be reduced to their basic geometric shape and their position on the line-mass continuum.

Geometric Designs

When classifying arrangements, designers often categorize their work based on the overall shape. Floral designs are configured based on shapes such as circles, triangles, and squares.

Circular

The circle is the most common geometric shape used by florists, **Figure 25-11**. Variations of this shape can include crescents, Hogarth curves, ovals, fans, and semicircles.

Palo_ok /Shutterstock.com

Figure 25-11. Handheld bouquets typically use a circular shape. When viewed from the top, the bouquet appears circular. When viewed from the side, the rounded shape is visible.

Corner Question

What type of triangles make up the Egyptian pyramids?

The Hogarth curve (an S-shaped circular design) may be constructed in a taller container, often with a platform or pedestal.

Triangular

A triangular design has three sides and can vary greatly in shape. A triangle can be isosceles (two sides of equal length), scalene (no equal sides), or equilateral (equal lengths on all sides), **Figure 25-12**. Triangles may also be a right triangle, where one of the angles is 90°. Triangular designs can be symmetrical or asymmetrical. Stacking or inverting two triangles together creates a diamond-like shape and is known as a double-ended triangle design. This double-ended triangle may be assembled vertically or horizontally. Triangular designs are often configured for viewing from the front and sides, but not from the back.

Square

Linear designs, such as vertical and horizontal arrangements, are based on the square design. A less common design created by combining a horizontal and vertical design is known as the *T-design*. This shape resembles an inverted T. Another design constructs a box-like effect and is referred to as the *hedgerow design*. The flowers and foliage of the hedgerow design are massed to create a cubed effect by placing plant material at the same level. This design mimics a row of hedges in a garden.

Line-Mass Continuum Designs

Classifying floral arrangements based on the line and mass within a design creates three basic design classes: line designs, mass design, and line-mass designs. Every floral design falls somewhere along the line-mass continuum, **Figure 25-13**.

A — Everything/Shutterstock.com   B — Rob Hainer/Shutterstock.com   C — Mayovskyy Andrew/Shutterstock.com

Figure 25-12. A—An isosceles triangle has two sides of equal length. B—A right angle triangle has one 90° angle. C—An equilateral triangle has equal lengths on all sides.

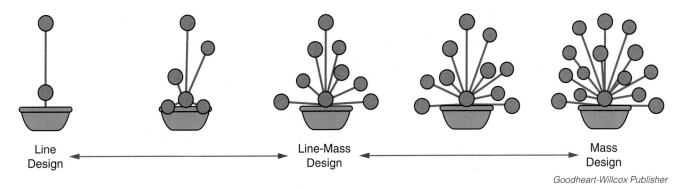

Figure 25-13. The line-mass continuum.

Line

A *line design* focuses on the lines rather than the flowers or foliage. Simplicity is key in this design format. Great expanses of negative space (voids in the design) coupled with precise placement of flowers and foliage, characterize line designs. Line direction can vary from lines parallel to one another to a curvilinear design. Asian styles, such as ikebana, use line designs that manage negative space.

Line-Mass Designs

A mixture of line and mass designs is the *line-mass design*. These designs depend on both line and mass characters. This design can often be seen in an equilateral triangle.

Mass

A *mass design* is a floral arrangement in which form and mass are more important than the individual elements or lines. This style relies on the weight or form of plant material. Massing flowers or placing heavy quantities of flowers in close proximity to one another creates a mass design. These arrangements are designed to be viewed from any angle.

Hands-On Horticulture

Homemade Floral Preservative

A floral preservative extends the life of cut flowers. Commercially purchased floral preservatives can be found in small packages that accompany freshly arranged flowers or bouquets. Making your own floral preservative is easy and inexpensive. Follow one of these simple recipes.

Recipe 1—Mix one part of lemon-lime soda (not diet, the flowers need the sugar) with three parts water. Add 1/4 teaspoon household bleach to every quart (four cups) of solution you made. Use this final solution instead of plain water.

Recipe 2—Add two tablespoons of lemon juice to four cups (one quart) of warm water. Add one tablespoon of sugar and 1/4 teaspoon bleach. Add 1/4 teaspoon bleach to every quart of solution every four days.

Containers, Tools, and Mechanics

A good floral arrangement should be visually pleasing and should endure for its intended duration of time. To extend the life of a floral arrangement, a container must hold water for the plant materials. The container must also support the plant materials securely. A proper design should not easily fall over but rather be secure and stable. To create a stable design, the appropriate container must be selected and the right tools used to create the arrangement.

Containers

The main purpose of the container is to hold water (treated with floral preservative) and support plant material. The container may add to the design or, if improperly selected, will take away from it. Consider the color and texture of the container and include this in the overall design.

Although the container cost is usually only a fraction of the overall cost of the arrangement, it is still an important factor in its selection. Using simple, inexpensive glass containers help florists maintain a higher profit margin and reduce the cost for the consumer, **Figure 25-14**. Basic glass containers are versatile, durable, and easily cleaned. Since the stems and water of the arrangement are usually visible in a glass vessel, it is important to keep the stems neat and the water clear.

Almost any container that can be stabilized and modified to hold water can be used to hold an arrangement. Look for unusual and attractive containers made of ceramic, plastic, wicker, and metal to complement your design.

Africa Studio/Shutterstock.com

Figure 25-14. Although glass vases can easily be acquired in nearly every shape and size, most florists use simple, versatile vases for most of their arrangements.

Tools

In floral design, as with many other tasks, using the right tool for the job can make the task much easier. Florists must cut plant material, foam, and tape. To cut materials properly, a florist must use extremely sharp tools. These tools include floral knives, scissors, shears, wire cutters, and stem strippers.

Floral Knives

A floral knife should be used to make exact cuts to plant materials. These knives can have folding or stationary blades. Plant material that is herbaceous (not woody) should be cut with a floral knife. A floral knife makes a clean, smooth cut and will not damage the stem. Floral knives should be used only on plant material. Use other knives or scissors for cutting items such as tape or wood supports. Knives can be sharpened for unlimited use.

Scissors and Shears

Designers use scissors or shears for a number of floral design tasks. Ribbon, fabric, netting, and other materials must be cut with sharp shears. Multipurpose utility scissors can cut very fine gauges of wire. Pruning shears are heavy-duty scissors with a spring action that are built to cut woody stems of plants. By-pass (scissors-like) or anvil (pinch type) shears can be used, depending on the preference of the designer.

Safety Note

Floral Knives

Use caution and proper techniques when working with floral knives. Follow your teacher's instructions and keep the following in mind when working with a floral knife:
- Stay focused on what you are cutting, and always cut in a direction away from yourself.
- Keep the knife in a safe accessible position when working, and properly store the knife when you are finished.
- Keep blades sharpened. You are more likely to cut yourself with a dull blade because you have to use more force to get the blade to cut.
- Keep knives clean to prevent the transmission of bacteria.
- If you cut yourself, immediately seek or apply first aid. Clean your wound and dress it properly. If working in class, always tell your instructor if you have cut yourself, regardless of the size of the cut.

Corner Question

What rock band misidentified the dermal appendages of a rose in the song *Every Rose Has Its Thorn*?

Pruning shears can be sharpened and must be properly cared for to ensure a long life. Only plant materials, not wire or fabric, should be cut with pruning shears.

Wire Cutters

Wire cutters cut metal by pinching. Artificial flowers and wires can be cut using a good pair of wire cutters, **Figure 25-15**. Every florist must ensure they have a pair nearby.

Stem Strippers

The stems of roses have prickles (thorn-like epidermal tissue), which can injure a person's skin. Some varieties of roses have many more prickles than others. A *stem stripper* is a tool that can be used to remove prickles or thorns from the stems of plants. Stem strippers can also be used to remove leaves that would fall below the water line in an arrangement. The skin of the stem (epidermis) should not be damaged when removing prickles or leaves. Damaged skin will affect the ability for a plant to take up water.

Ribbon

Not all floral designs incorporate a bow or a ribbon; however, many arrangements include or may even focus on a ribbon. A ribbon may be methodically looped to create a bow. Bows are often used to adorn wreaths or potted flowers, such as Easter lilies or Christmas cacti. A bow may also be part of a *corsage* (an arrangement of flowers pinned to clothing or worn on the wrist), **Figure 25-16**. The ribbon adds texture, color, and pattern.

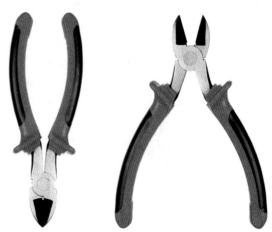

Andrei Kuzmik/Shutterstock.com

Figure 25-15. Wire cutters are an essential tool for florists.

Fleming Photography/Shutterstock.com

Figure 25-16. A bow can be a substantial part of a corsage design.

AgEd Connection — Floriculture CDE

At a local, regional, or state level, the artistry and engineering of manufacturing the perfect floral design bow can be a competition. A participant is given the materials to create a bow. The participant has only five or six minutes to construct an arrangement with good mechanics and artistry. Check with your advisor and see if a competition exists in your state for making a florist's bow.

Types of Ribbon

Designers commonly use nylon and acetate satin ribbon. Nylon ribbons often have wired edges, which helps them hold their shape in bows or arrangements. The wire may also help keep the fabric from absorbing moisture. Satin ribbons can be single-faced (one side is shiny and the other is dull) or double-faced (both sides are shiny). Velvet ribbon is often used at Christmas. Ribbon made with burlap may be a festive choice for fall.

Sizes of Ribbon

Ribbon is sold in bolts of 25, 40, or 100 yards. The size of ribbon is based on the width of the material. The higher the number of the ribbon, the wider the ribbon. Florist ribbons can include sizes #1, #2, #3, #5, #9, #40, and #100. The #1 size ribbon is 6/16″ (.95 cm) wide. The #100 size ribbon is 4″ (10.16 cm) wide. Various arrangements will include unique sizes of ribbons. For instance, a #1 ribbon may be used to make a corsage, a #9 or #40 for a bow on a wreath, and a #100 for a casket arrangement at a funeral.

Bows

Over the course of a designer's career, he or she will most likely make thousands of bows. Designers make bows using many different materials and techniques. Types of bows can include graduated loop bows, layered bows, puffy bows, tailored bows, and pinch bows. The pinch bow is commonly considered the standard technique for making floral design bows.

Mechanics

The materials that are used to secure items in place and the methods used are known as the *mechanics* of a floral design. Mechanics are concealed by precisely placing plant materials and ornaments. A viewer should not be able to easily see how the arrangement is secured. Materials used to secure the components of a floral arrangement include floral foam, tape, glue, wire, pins, picks, and stakes.

Hands-On Horticulture

Tying a Bow

Images: Jodie Riedel/Goodheart-Willcox Publisher

The materials needed for this bow are floral stem wire (22 gauge), green floral tape, ribbon (bolt of #9), and scissors. The measurements given may be modified to suit different size ribbons. Prepare the floral stem wire before you begin tying your bow by cutting it to size and wrapping it with the floral tape. Set the wire aside.

1. Measure 1″–2″ from the end of the ribbon. Hold the ribbon at this point between the thumb and forefinger of your "holding" hand. Pinch the ribbon and turn 180°. Continue pinching with the shiny side of the ribbon below the pinch and the dull side above the pinch/turn.

2. Measure the next 2″–3″ and pinch/twist 180° with the thumb and forefinger of your opposite hand. Bring your hands together to form a loop with the ribbon. Grip the loop between the thumb and forefinger of your holding hand. Only the shiny sides of the ribbon should be exposed.

3. Measure the next 6″ and pinch/twist with the thumb and forefinger of your free hand. Fold the ribbon over and pinch/twist with your holding hand. After making the first two loops, you should see a figure eight.

4. Measure another 6″ and fold the ribbon over to form another loop. Pinch/twist and stack directly underneath the previous loop. Continue forming new loops in the same manner until you have at least four loops on each side.

5. Once the four loops are on each side, create a tail that will be placed directly under your last loop. Cut a piece of ribbon approximately 12″–14″ long. With the shiny side up, stake this underneath the last pinch/twist. Insert the floral stem wire through the central loop on top of the bow and bend it to form a U shape.

6. Bend the loose ends of the wire around the bottom of the bow and through the U part of the wire to form a knot. Pull to tighten. Separate the loose ends of the wire and pull opposite of each other to tighten. Hold the wire and twist in a clockwise motion several times to secure. This can then be secured around a stake and placed in a potted plant or on an appropriate wreath or floral arrangement.

7. Cut the tails of the ribbon as desired.

8. Fluff the loops of your bow to create a concentric circle, with a full appearance.

Floral Foam

Floral foam is used to hold plant materials in place in arrangements using containers other than glass vases. Hydrophilic foam (material that can absorb water) also provides a means of holding water for fresh flower arrangements. Hydrophobic foam, such as those made of urethane and polystyrene, resists water and is used for artificial or dried-flower arrangements.

Wet floral foam is a hydrophilic material that absorbs water and can be used to help keep cut flowers and other plant materials in an arrangement fresh. This type of foam is easily dented (compressed) and should be handled with care. Once the foam is cut to size, it must be soaked in clean water that is deeper than the height of the block. When placed in the water, the block will float until its pores fill with water. Once the block is saturated, it will sink below the water surface. This process takes only about a minute. The floral foam can be cut to fit securely in the container and to make the top of the foam level with the container, **Figure 25-17A**.

Wet floral foam may also be secured to the container using waterproof tape. The outside of the container must be dry for the tape to adhere securely, **Figure 25-17B**. If the foam is to be glued to the container, it must be glued before being soaked in water.

A
Agnes Kantaruk/Shutterstock.com

B
Goodheart-Willcox Publisher

Figure 25-17. A—Floral foam may be cut to fit securely in a container. B—Tape may also be used to secure floral foam in place.

Securing Plant Material in Vases

Designers use a matrix or grid to secure the plant material stems in vase arrangements. The most commonly used grid is a system of woven or interlocking pieces of adhesive tape, metal, or wire. The grid can be attached to the clean and dry rim of a glass container, **Figure 25-18**. You may also interlace plant material stems to secure flowers and foliage in place.

Decorative materials, such as marbles, beads, fruit, and rocks, may also be used to hold stems in place and add to the design. These materials hold the stems in place but can also hasten the development of bacteria and fungi in the water. These materials should be used for floral arrangements that are expected to live only a short time.

Goodheart-Willcox Publisher

Figure 25-18. Grids can be constructed in vases by using tape or wire.

Did You Know?
Wet floral foam should not be forced under water. Forcing the foam under water may leave the inner part dry and ultimately result in wilted plant materials.

Safety Note
Materials in Water

Using a variety of materials in water can create a dramatic effect but can also add weight to the container. Glass containers with thin walls can easily be broken. They may make the container too heavy for safe transport. To prevent breakage and ensure safe transport, use glass vessels with thick walls when using other materials within the container.

Tape and Glue

Several types of tape, glue, and adhesives are used in floral design.

- *Waterproof tape* is an adhesive tape that is manufactured in 1/4″ (.635 cm) and 1/2″ (1.27 cm) widths. This can be found in various colors.
- *Floral tape* (also referred to as *stem wrap* or *bowl tape*) is made of paraffin-coated crepe paper. It will adhere to itself when stretched and wrapped. This tape is often used to cover metal wire or stems of plant material.
- *Hot glue* is plastic glue that is heated by a glue gun or skillet that can be used to secure floral materials or mechanics in a design. Hot glue is used only for embellishments, foam, or dried flowers. It cannot be used for live plant materials.
- *Adhesives* are substances that adhere plant materials and mechanics. Some adhesives may only be used with dry materials while others may be used to secure wet plant materials. Do not get adhesive on the ends of stems or they will be unable to take up water.

Safety Note

Hot Glue

Hot glue can be dangerous. Use low-temperature glues until you are more experienced using a glue gun or glue skillet.

"I was reading a book… "the history of glue." I couldn't put it down."
—Tom Vine

Wire

Wire ranges in size and length. Florist wire is sold in 12″ (30.48 cm) and 18″ (45.72 cm) lengths, as well as on spools or paddles. The thickness of the wire (also called *gauge*) is made to meet the needs of various floral mechanic applications. The higher the number of the florist wire, the thinner the wire. Gauges run from 16 to 32. Florist wire is painted dark green to blend in with the design and to prevent rusting and breakage.

Pins, Picks, and Stakes

Pins are used to secure materials or hold small arrangements in place.

- *Greening pins* are made of wire and bent into a u-shape. They are commonly used to secure items into foam.
- A *corsage pin* or *boutonniere pin* is a straight pin with a decorative pearl on the end. This pin is used to secure the corsage or boutonniere to the lapel of a jacket or the front of a blouse.
- *T-pins* are pins shaped like a T and can be used to place fabric or ornaments into a design.
- *Wood picks* are small sticks that are pointed on one end and used to secure floral items in an arrangement. These wooden picks can have pieces of wire attached so that flowers or ornaments can be affixed to the pick and then secured into foam or the grid of the arrangement. Bows and other ornaments can be attached to picks and inserted into wreaths, potted plants, centerpieces, and other arrangements, **Figure 25-19**.
- *Plant stakes* are small stick-like supports used to hold up plant material in a floral arrangement or potted plant. They can also be secured to the floral design with wire, glue, or an adhesive.

Jodie Riedel/Goodheart-Willcox Publisher

Figure 25-19. Bows may be attached to wood picks with tape or wire.

Did You Know?
The ancient Greeks wore corsages at weddings to ward off evil spirits.

Flower Arrangements

Some floral arrangements, such as corsages and boutonnieres, are worn by a person. Others can be carried, such as a bridal bouquet or pomander. The majority of flower arrangements are stationary and are used to beautify a room, send a friendly message, or express sympathy.

Corsages and Boutonnieres

Corsages and boutonnieres are created for special occasions, such as formal dances and weddings. The design can be very simple to extremely elegant and unique, **Figure 25-20**. Although corsages and boutonnieres are not usually intended for more than a day's use, they should be well constructed and last throughout the event. The materials should be lightweight so they can be pinned to clothing and not pull at the fabric. The piece should not restrict the wearer's ability to move about freely at the event.

Falcona/Shutterstock.com

Figure 25-20. A boutonniere may have one or several flowers, as well as accent pieces.

Bouquets

Bouquets are also designed for special occasions such as weddings and pageants and must be constructed to last the duration of the event. The basic geometric shape of a bouquet is round, but there are many variations of this shape. Waterfall, cascade, and crescent shapes are simple and subtle changes from the typical round design, **Figure 25-21**.

Did You Know?
Centuries ago, people would tear pieces from the bride's gown and flowers to reap the benefits of her good fortune. The bride would try to escape from the crowd by tossing her bouquet as she ran away. Today, the bouquet is tossed with the belief that the woman who catches it will be the next to marry.

A   *titov dmitriy/Shutterstock.com*

B   *Ekaterina Pokrovskaya/Shutterstock.com*

Figure 25-21. A—This hand-tied bouquet has a traditional round shape. B—The flowers in this cascade bouquet flow from the round, main part of the bouquet.

Holiday Centerpieces

Many people decorate their homes for holiday seasons. Christmas, Hanukah, Thanksgiving, Halloween, Fourth of July, Easter, St. Patrick's Day, and Valentine's Day are commonly celebrated in homes throughout the United States. To enhance the holiday spirit, many people place a holiday centerpiece on the dining table. The holiday centerpiece contains plant materials that accompany the season and novelties or ornaments that help to create additional holiday sentiment, **Figure 25-22**.

Wreaths

Many florists construct wreaths for various holidays. Some nurseries and greenhouses offer wreaths to supplement off-season sales as well. Wreaths are constructed using a variety of plant materials (which usually vary by season) and different types of frames.

- Boxwoods, pines, firs, arborvitae, cedar, magnolias, and many other evergreen plant materials can be used to create a long-lasting wreath, **Figure 25-23A**.
- Fruits and nuts, such as lemons, apples, crabapples, pears, pecans, walnuts, and pine cones, are used in fall and winter wreaths, **Figure 25-23B**.
- Wreaths usually include a bow and ornaments to complete their festive look. Wreaths can be constructed with a box wreath frame that is hand-wired or with a wreath frame system.

Goodheart-Willcox Publisher

Figure 25-22. Holiday flowers may include traditional and/or exotic flowers along with seasonal accents, such as the gold-painted pine cones.

A *EQRoy/Shutterstock.com*   **B** *AGfoto/Shutterstock.com*

Figure 25-23. A—Various types of evergreens make long-lasting holiday wreaths. B—A fruit and nut wreath is a unique addition to the holiday season.

Hands-On Horticulture

Increasing the Shelf Life of Bouquets

Proper care and handling of a bouquet of fresh flowers and foliage can make the bouquet remain attractive longer.

If the bouquet is already in water:
1. Place the bouquet in a draft-free area (away from heating and cooling devices). Keep it away from ripening fruit.
2. Change the water every two days, and clean the container with hot water and soap. Add 1/4 teaspoon bleach or commercial floral preservative to every quart of water in the container.
3. Quickly remove any leaves or petals that fall into the water. Water, along with bacteria and fungi, speeds decomposition of organic matter.

If the bouquet is not in water:
1. Keep the bouquet upright (if possible) until it is placed in its container.
2. Cut the ends of stems with clean pruning tools at 45° angles to fit the size of the container. The container should be one-third of the entire height of the arrangement.
3. Remove any leaves that will come in contact with the water inside the container.
4. Place the bouquet in a cool place, away from heating and cooling vents. Keep it away from ripening fruit.
5. Change the water every other day and add floral preservative or 1/4 teaspoon bleach for every quart of water.

Careers

The floriculture industry offers many interesting career options. Opportunities range from the growing and production end to the marketing and distribution of the finished product. Wholesale distributors and floral merchandisers are just two of many career opportunities available in floriculture.

Wholesale Distributor

A wholesale distributor purchases plant materials from around the globe and sells the materials to retailers, florists, and other businesses. A wholesaler also sells containers, floral design supplies, ornaments, ribbons, and other accessories as well as equipment such as buckets and cutting tools.

A wholesale distributor usually employs a number of people in full-time and part-time positions. These positions include those responsible for receiving, processing, and storing plant materials; designers; buyers; and managers. In smaller companies, one person may be responsible for multiple tasks. This is especially true of managerial positions. The manager of a wholesale distributor company may be responsible for purchasing, receiving, and distributing products. The manager may need to keep financial records, hire and train employees, and establish work schedules.

A wholesale distribution manager must be an excellent communicator with supervisors, employees, and customers. The ability to manage a budget, use computer software, and demonstrate math and analytical skills is important for a person in this position. This person must also be able to do some laborious tasks, such as lifting plant materials.

Wholesale florist distribution managers can be found in major cities around the world, **Figure 25-24**. Hours for this career may include day or night shifts.

Floral Merchandiser

Wholesale floral companies create bouquets at a warehouse and then distribute the bouquets and other plant materials to retailers around the country. The floral merchandise is shipped to retail stores and placed in display cases. The floral merchandiser travels to retail stores and ensures that the stock is fresh and displayed in an effective and attractive manner. The store's supply of ready-made bouquets must be replenished four to five times per week. Other duties for this position include cleaning the display unit, replenishing supplies, taking a weekly inventory, and developing good relationships with store personnel. Some horticultural knowledge is helpful. Job training for this position usually occurs within the company.

Jordan Tan/Shutterstock.com

Figure 25-24. Wholesale buyers may need to travel to international markets, such as this one in Amsterdam, to purchase plant materials.

Career Connection — Anna Passarelli

Floral Designer

Anna Passarelli became interested in floriculture while attending an urban high school in Raleigh, North Carolina. Anna had always enjoyed her mother's garden and was further inspired by her horticulture lessons. She earned an Agricultural Business Management and Horticulture degree from the Agriculture Institute at North Carolina State University and worked in the landscaping industry for several years before she found her true calling: designing wedding flowers.

Anna started Simply Elegant, a wedding and event design company. Originally, Anna was focused on wedding flowers for brides on smaller budgets. Today, however, her company has grown and now offers dozens of wedding options to fit a wide range of budgets for the bride's big day. Anna communicates with her customers and audience through fresh images of the day's flowers posted to various social media sites.

Anna suggests students interested in floral design should begin by studying and learning to identify the many types of flowers. She suggests,

Lauren Jolly Photography

Anna Passarelli, a former FFA member, owns her own floral design business called Simply Elegant.

"Students need to learn that a rose is not just a rose. There are thousands of them out there to explore, and some roses don't even look like roses." She also suggests students should begin working with small arrangements such as corsages and boutonnieres.

CHAPTER 25 Review and Assessment

Chapter Summary

- Floral designs are often used to decorate spaces for formal occasions, to express congratulations or caring to friends or family, and to decorate homes for holidays and special occasions.
- Floral designers construct floral arrangements based on design principles and elements. Design principles include balance, symmetry, proportion, scale, focal point, rhythm, unity, and harmony.
- Mechanics are the devices and techniques used to keep floral placements secure and stable in a design. Poor mechanics, support, anchorage, or footings contribute to poor balance.
- Floral designs can be symmetrical or asymmetrical. Flowers, foliage, or other materials can be used to create a focal point in a design.
- Design elements include color, line, form, and texture. The element that impacts most viewers immediately is color.
- Floral designs are often categorized by style. Categories of floral design include geometric designs, line-mass continuum designs, traditional designs, and contemporary designs.
- Designers use various containers, tools, and materials to construct floral designs. Types of containers for floral arrangements include glass, ceramic, plastic, wicker, and metal. Floral knives, scissors, shears, wire cutters, and stem strippers are tools used by designers.
- Some materials that can be used to secure the mechanics and other components of a floral arrangement include floral foam, tape, glue, wire, pins, picks, and stakes.
- Common floral arrangements are those that are worn or used for special occasions and holidays. Corsages, boutonnieres, and bouquets can be worn or carried. Floral centerpieces and wreaths are used for events, special occasions, and holidays.
- The floriculture industry offers many interesting career options including wholesale distributors and floral merchandisers.

Words to Know

Match the key terms from the chapter to the correct definition.

A. asymmetrical design
B. boutonniere
C. color harmony
D. corsage
E. filler flower
F. focal point
G. form
H. form flower
I. harmony
J. hedgerow design
K. line
L. line design
M. line flower
N. line-mass design
O. mass design
P. mass flower
Q. mechanics
R. proportion
S. rhythm
T. scale
U. stem stripper
V. symmetrical design
W. T-design
X. unity

1. A pleasing interaction or blending among the elements of a design.
2. A design in which a nearly identical or mirrored arrangement of plant materials is used on both sides of the central vertical axis (CVA).
3. The relationship among elements in a work of art that makes them appear to belong together or function as a whole.
4. A floral arrangement in which form and mass are more important than the individual elements or lines.
5. A floral design that is patterned after a garden hedgerow.
6. A floral arrangement created by combining a horizontal and vertical design that resembles an inverted T.
7. A flower that is unique in some manner (color, shape, texture, or size).
8. A pattern of repeating or alternating elements in a design that suggests a sense of movement.
9. The size relationship between two objects, such as between a floral design and its surroundings, or among the various elements within a design.
10. A small floral arrangement that is worn on the lapel area of clothing.
11. A tool used to remove prickles from stems of plants.
12. An arrangement of flowers pinned to clothing or worn on the wrist.
13. Fine or airy materials used to fill in voids and often made of very small clusters of blooms or branched inflorescence.
14. A closed-form, single flower that has a dense, round shape.
15. The visual path that creates the foundation, shape, and form of a floral arrangement.
16. A floral arrangement that emphasizes lines rather than the plant material.
17. Guidelines for combining colors and color values in a design.
18. The area of dominance or emphasis, as within a floral design, to which the eyes naturally travel.
19. The devices and techniques used to keep floral placements secure and stable in a design.
20. A design in which the visual weight is equal on both sides of the central vertical axis (CVA), but the sides are not mirror images.

21. A floral arrangement in which a mixture of line and mass designs is used.
22. A flower that has a long stem, a spike, or linear form.
23. The overall shape of a floral arrangement or the shapes of the individual materials used in an arrangement.
24. The relative sizes and amounts of elements in a design.

Know and Understand

Answer the following questions using the information provided in this chapter.

1. What are some reasons or occasions for which floral arrangements are used?
2. What are the principles of design and how are they used in floral design?
3. How are symmetrical and asymmetrical floral arrangements different?
4. Generally, how much larger should plant material be than the container to maintain pleasing proportion?
5. In a floral arrangement, what types of materials can act as the focal point?
6. Describe briefly three underlying principles that contribute to the effect of harmony in a floral design.
7. Describe four color schemes that may be used in floral design.
8. Describe four forms of flowers and foliage.
9. What is the rule of thumb for incorporating texture into floral arrangements?
10. What are the three geometric shapes used for floral designs?
11. What are the characteristics of a floral arrangement that uses line design?
12. What are the characteristics of a floral arrangement that uses mass design?
13. What are some types of containers that can be used for floral arrangements?
14. What are some tools that floral designers use in creating floral arrangements?
15. What are some safety procedures to follow when using floral knives?
16. Describe a career development event that involves making bows.
17. What are some materials that can be used to secure the components of a floral arrangement?
18. What can designers do to improve safety when using glass containers for floral arrangements?
19. What are some job duties of a floral merchandiser?

Thinking Critically

1. You recently hurt someone's feelings and want to give the person flowers as a way of saying that you are sorry. What flower or flowers would be best suited for this type of expression?
2. Your boyfriend or girlfriend has invited you to the prom. There are several expenses involved, such as buying clothes, dinner, tickets, and pictures. What could you suggest to help alleviate the costs associated with the corsage or boutonniere besides just omitting this floral arrangement from the night altogether?

STEM and Academic Activities

1. **Math.** Determine how many flowers must be purchased for a dining hall used for a wedding if 40 tables will have 30 flowers each on the centerpieces. In addition, the cake will need 20 flowers, the bride and groom's table will need 100 flowers, and the two bathrooms will each have 10 flowers. What is the total needed for the wedding reception?
2. **Engineering.** Design a flower arrangement for a table that is 30' long that will be in a banquet room feeding 22 guests. The room's colors are neutral and the event is celebratory.
3. **Math.** A dozen roses arranged at Valentine's Day can cost up to $75 for delivery to a customer from a florist. A dozen roses that are arranged at a wholesaler or box store will cost less than half that price. What causes the difference in price between a florist and a retail store?
4. **Social Science.** Contact a local florist and ask them to donate leftover floral magazines or publications for use at your school.
5. **Language Arts.** Compose a list of essential flowers and arrangements for a successful prom season. Justify your choices of flowers.

Communicating about Horticulture

1. **Reading.** Create an idea board for a specific holiday or event. Be sure to include ideas for flower arrangements, colors, and what should be included to make this event a floral design success.
2. **Reading and Speaking.** Recently, there has been a surge in the production of locally grown cut flowers. Research a farm in your area or state that sells cut flowers. Contact the grower and see what flowers are being grown right now at this farm.

SAE Opportunities

1. **Exploratory.** Job shadow a floral designer.
2. **Experimental.** Use various types of glues or adhesives for a corsage or boutonniere. Determine which type of adhesive holds up to the most wear and damage.
3. **Exploratory.** Create a survey to determine your peer's attitudes toward fresh flowers. Construct questions that use scales to better analyze the data. Use your data to create a report with findings and suggestions for the floral industry in relation to your generation.
4. **Entrepreneurship.** Develop a floral design service.
5. **Placement.** Get a job working at a floral shop.

Jodi Riedel/Goodheart-Willcox Publisher

CHAPTER 26
Interior Plantscaping

Chapter Outcomes

After studying this chapter, you will be able to:
- List the benefits of and different settings for interior plantscapes.
- Summarize the elements and principles of interior plantscape design.
- Describe the environmental requirements for indoor plants.
- Describe the methods for indoor plant management.
- Discuss the business opportunities and careers available in interior plantscaping.

Words to Know

- bonsai
- chlorine
- horticultural oil
- insecticidal soap
- interior plantscaping
- interiorscape
- succulent
- systemic insecticide
- terrarium

Before You Read

Review the chapter headings and use them to create an outline for taking notes during reading and class discussion. Under each heading, list any term highlighted in bold italic type. Write two questions that you expect the chapter to answer.

While studying this chapter, look for the activity icon to:
- **Practice** vocabulary terms with Words to Know activities.
- **Expand** learning with identification activities.
- **Reinforce** what you learn by completing Know and Understand questions.

www.g-wlearning.com/agriculture

*I*magine a classroom filled with colorful foliage, interesting flowers, and hanging, draping plants. Compare this image to a typical classroom that is full of desks and books but devoid of plant life. Which space would you prefer? Indoor environments rich in plant material provide a professional, warm interior that invites visitors to feel welcome and comfortable.

Interior Plantscaping

From elaborate planters inside a luxury hotel to simple houseplants at the front desk of an office building, plants can transform a space. *Interior plantscaping* is the practice of designing, installing, and maintaining indoor spaces with plants. It provides a variety of benefits. Interior plantscaping can:

- Enhance a person's well-being.
- Create a positive, professional tone and image.
- Offer a sense of calm and stability.
- Build a cozy, intimate setting.
- Raise employee productivity.
- Lower workplace stress.
- Purify indoor air.
- Increase attentiveness.

An interior plantscape can be created in a wide range of designs. Plantings can be placed in a number of different sites. An installation of plants, or landscape, designed for and installed inside a building is called an *interiorscape*. Settings where an interiorscape might be found include health care facilities, hotels, shopping malls, offices, restaurants, homes, zoos, and botanical gardens, **Figure 26-1**.

Did You Know?

The Mall of America in Minneapolis, Minnesota, has more than 30,000 live plants and 400 live trees. Some of these plants climb as high as 35′ (10.6 m) tall.

Karen Grigoryan/Shutterstock.com

Figure 26-1. This botanical garden conservatory is a beautiful example of a commercial interiorscape installation.

Health Care Facilities

Hospitals, waiting rooms, and other health care facilities can be stressful places. Professional plantings in these areas can provide a sense of well-being, **Figure 26-2**. Studies have shown that indoor plants can speed surgical recovery; lower heart rate and blood pressure; lessen fatigue and anxiety; and decrease colds, headaches, coughs, sore throats, and flu-like symptoms.

Elizabeth Driscoll/Goodheart-Willcox Publisher

Figure 26-2. Exposure to indoor plants helps hospital patients recuperate more quickly.

Hotels

Innovative indoor plant designs in hotels offer a sense of luxury and hospitality to guests. Beautiful planters often flank hotel entrances to welcome visitors. Hotels may use a well-designed interiorscape to accent or define casual seating areas or to soften large structural pillars or columns. Many of these designs are rotated throughout the seasons, offering interiorscape companies opportunities for increased business. Large resorts create elaborate displays to create a luxurious tone.

Shopping Malls

Shopping malls provide opportunities to create exciting and impressive interiorscape designs. Attractive displays can encourage consumers to linger longer in the mall, increasing revenue for mall tenants, **Figure 26-3**. Plants can highlight a store or vendor, increasing customer traffic. Store leases may increase based on the value that plants provide, thus improving the value of mall real estate.

Photograph by Yuko Frazier, Senior Project Manager at Ambius

Figure 26-3. Customers stay longer and have more positive experiences at shopping malls with beautifully landscaped interiors.

Offices

A company's corporate image is built on client perceptions and interactions. A thoughtfully designed and maintained interiorscape helps present a professional appearance and may help build confidence in the company. Planters filled with beautiful plants communicate a sense of prosperity and stability. Having artificial plants, or no plants at all, can send a message that a company is in poor financial condition and trying to skimp on resources.

Restaurants

As consumers dine out more often, restaurants compete for their business. Interiorscapes can provide a unique atmosphere and experience that may be absent in competing restaurants. Placing plants between booths or tables creates a living screen that offers a sense of privacy and reduces noise levels. Plants used in restaurants must be clean and free of grease, dust, and yellowing or dead leaves.

Homes

With an increase in busy lifestyles and an emphasis on entertaining, more homeowners are hiring skilled interior plantscape designers to create interiorscapes for their homes. As you flip through the pages of design magazines, you will notice that stylish houses all contain lush displays of flowers and plants. Plants may be a focal point or they may be accent plants placed throughout common rooms in the house, **Figure 26-4**. For the homeowners interested in horticulture, designing an indoor plant environment and tending to it can be a rewarding pursuit.

cycreation/Shutterstock.com

Figure 26-4. Busy homeowners appreciate the value of interiorscapes that accent their interior design style.

Botanical Gardens and Zoos

Botanical gardens and arboretums are known for extensive, thoughtful outdoor plantings. Many gardens have conservatories that feature an interesting and unusual collection of indoor plants, **Figure 26-5**. Conservatories are sources of inspiration for interiorscaping, often pushing boundaries and exhibiting rare plants. Visitor centers also showcase unique specimens representative of the garden.

bouybin/Shutterstock.com

Figure 26-5. Conservatories serve as educational and inspiration resources for interiorscape professionals.

The focus of a zoo is its animals, but it can also include exhibits with plants. Zoo exhibits filled with plants give visitors the sense they are seeing animals in their native habitats. In bird conservatories and in other animal exhibits, native plants may be grown as a source of food and habitat for the animals.

Design Principles

Interior plantscaping requires an understanding of basic design principles to create a beautiful planting that is functional and aesthetically appealing. Applying the fundamentals of design to an indoor environment allows designers to showcase plants in an effective way. Designers must have a thorough knowledge of the types of plants suitable for indoor locations. They must also understand light requirements, the level of care and maintenance required, and ways to handle pest problems. Blending design and knowledge of plant biology permits designers to decorate and enhance an interior space.

Elements of Design

The elements of design are physical qualities that can be readily seen. The elements of design include line, form, texture, and color.

Line

A line is a mark (or an implied mark) that spans the distance between two points. The lines used in a plantscape help direct the way people will respond emotionally and physically to the plantscape. In interiorscaping, the structure of a container is defined by a line. Square containers have straight lines that may evoke a modern aesthetic or formal feeling, **Figure 26-6**. A container that is rounded or curved may create a sense of casualness or intimacy. Lines can also be created within the container through plant placement. Vertical and horizontal lines can extend the design. A tall vertical wall, such as a planted tapestry that extends to a ceiling, can make a bold statement, **Figure 26-7**. A horizontal line guides the eye laterally and can divide spaces or work to unite spaces.

Elizabeth Driscoll/Goodheart-Willcox Publisher

Figure 26-6. Square or rectangular containers that have straight lines evoke a sense of formality.

Form

Form is the shape of a plant, accessory (sculpture, for example), or water feature. Some common plant forms include columnar, oval, vase, weeping, pyramid, round, mounding, upright, spikey, cascading, irregular, matted, clumping, or sprawling.

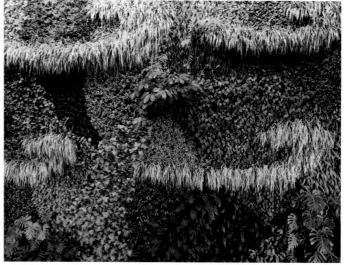

chris kolaczan/Shutterstock.com

Figure 26-7. Green, living walls are exciting features that amplify any organizational space.

STEM Connection

Bonsai

Bonsai is an ancient Japanese art form of pruning and manipulating a plant into miniature form. This art form has been in existence for as far back as 1000 years. Japanese plant enthusiasts view this practice as a form of meditation, both in the act of bonsai and in the viewing. Bonsai takes patience, creativity, and long-term cultivation and shaping to reach the intended form. Bonsai plants are masterpieces of art and horticulture and are dramatic features in the right interiorscapes. They can be a high-end feature in the appropriate interiorscape setting, or they can be used as a seasonal display.

The United States National Arboretum in Washington, D.C., has an exceptional display of bonsai. In 1976, to commemorate the US Bicentennial, more than 53 bonsai were donated to the arboretum from the Nippon Bonsai Association in Japan. In the collection was a Japanese white pine (*Pinus parviflora*) that was approximately 400 years old. Both the Japanese art of bonsai and the Chinese equivalent of penjing continue to amaze and inspire visitors each year.

The technique of bonsai consists of keeping a plant bound to its shallow pot by pruning the roots and pinching the shoots in a strategic and

Joel_420/Shutterstock.com

consistent manner. Time and experience help gardeners to determine how much pruning is appropriate.

The plant material for bonsai depends on the indoor or outdoor environment. For the interior landscape, common plants include weeping fig, parasol plant, bougainvillea, acacia, hibiscus, gardenia, *Ixora*, pomegranate, jade plant, pistachio, and olive.

Bonsai styles include formal upright, informal upright, twin trunk, slanting, windswept, weeping, cascade, semicascade, and root over rock. Branches are pruned and wired into place to train to a particular style. Bonsai is a fascinating endeavor and many useful books fully illustrate the concepts and techniques used to create these masterpieces.

Pekka Nikonen/Shutterstock.com

Figure 26-8. This snake plant (*Sanseveria*) has a naturally formal structure with its upright, stiff leaves.

Each form intentionally nudges a viewer's eye or draws a certain emotion from the observer. For example, a plant with an upright form, such as the mother-in-law's tongue (*Sansevieria pearsonii*), suggests a formality or a stiffness that can fit well with certain designs, **Figure 26-8**. A cascading form, such as a *Philodendron cordatum*, softens an edge and draws the eyes downward. Thoughtful selection of plants for their form is essential to build an exciting design.

Texture

Texture refers to the surface quality of the plant or feature. Texture adds visual interest and contrast to a design. For example, a plant's texture may be coarse, fine, glossy, rough, or smooth. Texture can be viewed from a distance and through close observation of unique characteristics. For example, you might be eating lunch on a bench next to a planter and notice the soft, fine, hairy texture of a *Tradescantia* ('Baby Bunny Bellies'). This plant gives visual cues of friendliness and a sense of invitation.

Color

Color is a powerful and noticeable element that can create strong moods and feelings. Color theory explains using the color wheel as a guide to color use and the way colors should be used in relation to one another. The colors in interiorscaping primarily come from the foliage of a plant. Most plants suited to an indoor environment do not produce an abundance of flowers. They do, however, have an array of greens, yellow-greens, blues, pinks, and other colors that will fit any design, **Figure 26-9**. Many indoor plants produce beautiful flowers, but these plants are typically grown for their foliage. Flowers are often used in a seasonal or temporary display. Orchids and begonias are good examples of long-lasting, beautiful flowers that may be in a design for only one to three months.

Innovations in plant breeding have led to a diverse selection of foliage plants. Crotons now come in a mosaic of colors. Rex begonias have everything from darkened leaf margins, to polka-dots on leaves, to bright pink and purple foliage, **Figure 26-10**. Other plants, such as a golden *Philodendron xanadu*, have a gorgeous saturation of color.

Principles of Design

The principles of design involve using the elements of design (line, form, texture, and color) and applying them in unique and creative ways to form an interiorscape. The principles of design used to create interiorscapes include balance, proportion, repetition, emphasis, and unity.

Balance

Balance refers to the concept of symmetry, with the sides of a design being similar in size and shape. In a design with formal balance, the left and right sides are exactly the same, **Figure 26-11**. This gives a sense of good form. In a design with informal balance, the left and right sides are not the same. This gives a plantscape movement and creates a sense of curiosity in the observer. Asymmetrical designs are intentionally unbalanced.

Photograph by Yuko Frazier, Project Manager at Ambius

Figure 26-9. Indoor plants have a myriad of colorful foliage that can fit into any design.

PeterVrabel/Shutterstock.com

Figure 26-10. Rex begonias have lively, flamboyant leaves and do not require much water or light.

Photograph by Yuko Frazier, Project Manager at Ambius

Figure 26-11. Formal designs will often be the same on both sides, creating an organized and thoughtful space.

Photograph by Yuko Frazier, Project Manager at Ambius

Figure 26-12. Correct proportions should enhance a design. These tall trees enhance the tall, open atrium.

They give a sense of discord, pushing possibilities and asking an observer to look carefully. Perspective is also created with balance by placing plants in the foreground, midground, and background of a design.

Proportion

The size of an object in relation to other objects is the concept of proportion. In interiorscaping, designers relate the size of plants to the space they occupy, **Figure 26-12**. Plants should fill a space in a way that is relative and appropriate to the surroundings. For example, if a building has a window wall that is 15′ (4.5 m) tall and 15′ (4.5 m) wide, using plants that may reach only 2′ (0.5 m) is inadequate. A much taller and wider planting is needed to complement this large space.

Repetition

Repetition is the design practice of using elements or features in a recurring way. This builds patterns or sequences that people find pleasing. Repetition creates a rhythm that provides interest and a sense of movement.

Hands-On Horticulture

How to Make a Dish Garden

Some spaces are not large enough for an elaborate and large interiorscape. In these cases, consider creating a dish garden. You can apply the elements and principles of design on a small scale in a dish garden. Succulents are ideal plants for a dish garden. A **succulent** is a type of plant that has thick, fleshy leaves and is usually native to arid regions of the world. Succulents have unique colors, textures, and forms, and can be arranged in artful ways. Succulents also require very little water and are easily maintained for many years.

Good drainage is essential for a dish garden. Use a soilless potting medium mixed with compost that will drain well. Use a shallow dish that has holes for drainage. Tall, tree-like jade plants can anchor the background. Small, bushy plants, such as a zebra plant

Parinya Hirunthitima/Shutterstock.com

(*Haworthia attenuata*), may be a good choice for the midground. Finish with a creeping plant, such as sedum, in the foreground.

Some repetition harmonizes a design, pulling together repeats of form, texture, color, or lines throughout the plantscape, **Figure 26-13**. Many designers use groupings of three, five, or seven plants that can be echoed within a planting.

Emphasis

Emphasis is the dominance or subordination of elements. For example, a brightly colored *Dracaena*, **Figure 26-14**, can be a dominant feature. A more subdued, monochromatic *Aglaonema* can be a subordinate feature in the same design. Emphasis occurs through the strategic use of varying colors, textures, sizes, shapes, groupings, and materials. Often a unique and intriguing sculpture is the dominant focal point, and the plants play a supporting role. In other instances, a group of vivid bromeliads may bloom for a few weeks and become a seasonal emphasis. The remainder of the year, these plants would have a subordinate emphasis.

Unity

The principle of unity involves bringing all design elements together in a harmonious fashion. In a design with unity, separate features and ideas are used together in a way that strengthens the larger, whole design. A good designer will practice simplicity and clarity, striking the right balance between interest and the ability for an observer to process and understand the composition.

Perspectives - Jeff Smith/Shutterstock.com

Figure 26-13. Repetitive plants or features create rhythm and movement in a design. What repetitive groupings do you see in this planting?

marylooo/Shutterstock.com

Figure 26-14. This brightly colored *Dracena* provides a dominant focal point in a vibrant planting.

The Process of Design

A good designer begins by listening to the client. Through an interview or conversation with the client, the designer must learn the objectives for the site and the site conditions. The designer must also know the project budget. The designer then uses the elements and principles of design to compose a plantscape that artfully and functionally includes plants and other features, such as water, stones, and sculptures, to meet the objectives for the site. In response to the client's comments and suggestions, the designer will develop a final rendered plan with details, illustrations, and graphics meant to impress the client and sell the design.

"Design must reflect the practical and aesthetic in business, but above all...good design must primarily serve people."
—Thomas J. Watson

Environmental Requirements

The plants available for use by an interiorscaper are much different than those used by a home gardener or commercial grower. Most often, houseplants are used for interior plantscaping. In general, the climates from which most houseplants come are tropical and have consistently warm conditions with low light. Many houseplants are understory plants, growing in dappled light conditions. This makes them adaptable to an interior environment, **Figure 26-15**. Examples of these types of plants include philodendrons, pothos, figs, and dieffenbachias. Some houseplants, such as *Sansevieria*, aloes, *Peperomia*, and cacti, come from regions that are arid and have high light levels. In addition to plants, the elements for creating and maintaining an interiorscape include temperature, humidity, water, light, potting media, mulch, and plant nutrition. Structures and containers for the plants are also needed.

AustralianCamera/Shutterstock.com

Figure 26-15. Most houseplants used in interiorscapes were understory trees in their native tropical habitats and do not require high light intensities.

Temperature

Nearly all indoor plants will thrive in a temperature range between 55°F and 75°F (13°C and 24°C). Most home and work environments provide an optimal environment for the coexistence of people and houseplants in the range of 65°F–75°F (18°C–24°C). While few houseplants can thrive at temperatures greater than 75°F (24°C), some can manage with the addition of significant humidity. Most plants are tolerant of slight temperature changes outside their optimal range and will withstand a drop of a few degrees. A sudden shift in temperatures can be damaging or fatal. During extended periods of building vacancy, such as in a school over winter break, the thermostat may be set at a lower temperature. Plants selected for such a setting should be tolerant of temperature extreme. Check optimal growing temperatures carefully when choosing a plant.

Humidity

Many houseplants come from tropical locales that have humidity levels of 70%–90%. Colder air holds less water and requires less moisture to raise humidity levels. Warmer temperatures, however, can hold greater amounts of water (water vapor). Commercial buildings or homes kept at temperatures higher than 65°F (18°C) will have low levels of relative humidity. Many foliage plants and flowering plants will suffer if humidity levels are not maintained at a proper level. Plants prefer humidity levels between 40% and 60%. Plants show symptoms of leaf tip browning, shriveling, aborted buds and flowers, yellow leaf margins, and leaf fall if there is not enough humidity.

To raise the humidity levels, plants can be misted, placed in rooms with high humidity conditions (such as a bathroom or kitchen), or placed in a terrarium.

A *terrarium* is a closed type of planting structure that is self-contained and provides a good environment for tender, hard-to-grow plants. Plants may also be grouped together, forming a small microclimate with the desired humidity conditions for the plants. Moisture coming from damp potting medium or a tray filled with pebbles and water will raise humidity concentrations as well.

If there is too much humidity, plants (especially succulents) may mold or rot. Houseplants that originate in desert conditions require humidity levels of only 10%–30%, and ambient humidity levels are adequate. In considering design, group plants with similar environmental needs together. For example, place plants (such as succulents) that require high light, low humidity, and low water in one container and tropical plants in another. This will simplify plant maintenance.

Water

Watering indoor plants is a skill that requires experience and practice, **Figure 26-16**. All plants need water, but their needs vary and can change throughout the seasons. Many houseplants die from overwatering. The soil becomes saturated, and there is no oxygen for root respiration. The result is yellowing, curling, wilted leaves that turn brown at the tips and die. Leaves may become soft and rotten or show poor growth. Roots will turn brown and mushy under too much water.

Understanding the watering needs of plants is the first step in cultivating long-term plant health. Plants can be grouped into four watering categories:

Iakov Filimonov/Shutterstock.com

Figure 26-16. Understanding the different watering needs of plants will ensure a healthy, long-lived interiorscape.

- Dry in winter plants. During the cool temperatures of winter, cacti and succulents prefer very dry soil. Their growth slows during this time and they require significantly less water than during the spring and summer months.
- Moist/dry plants. Most foliage plants, cacti, and succulents fall into this category. As these plants actively grow during high light and warmer temperatures of spring, summer, and early fall, they should be watered thoroughly, allowed to dry, and watered again. During the cool temperatures of winter, cacti and succulents prefer very dry soil. Their growth slows during this time and they require significantly less water than during the spring and summer months. Foliage plants also benefit from reduced watering (although not to the extent of cacti and succulents.
- Moist plants. Most flowering plants belong to this group and prefer soil that is moist (not wet) all the time. Water thoroughly each time, and as the surface dries, water again. The growing medium should not be saturated, but it should be moist. If you press down on the soil and water surfaces, the medium is too wet.
- Wet plants. Plants in this group tend to originally come from wet environments. Very few plants can tolerate wet soil. A few that can include *Acorus*, azalea, and *Cyperus*. Water these plants thoroughly and frequently.

Hands-On Horticulture

Planting a Terrarium

A terrarium is a tightly closed, glass or plastic container filled with small plants suitable for growing indoors. Terrariums may also be a transparent container for growing and displaying plants. Terrariums were initially used as a way to nurture tender, hard-to-grow plants. The concept has been expanded to include fun, artful plantings that showcase horticultural specimens. In a closed terrarium, high levels of humidity are perfect for certain plants. Plants need minimal watering as elements cycle through the air to the plant, then to the soil, and then around again.

aon168/Shutterstock.com

To create a terrarium, choose an open or closed (depending on intended plantings) container and fill it first with drainage material that will allow excess water to collect. Drainage material may be activated charcoal, rocks, pebbles, or marbles. Follow the drainage layer with a layer of moistened sphagnum moss. This will prevent the potting medium from sifting into the drainage material. Add potting medium that is sterile, well drained, and high in organic matter. A mix typically contains a blend of peat moss, vermiculite, and perlite.

Select plants for a terrarium that have similar environmental requirements and are fairly small in stature. Larger plants may be used, but they will require regular trimming. Many houseplants are suitable for a terrarium and can thrive in the warm, moist conditions. A few examples include:

- African violet (*Saintpaulia ionantha*)
- Arrowhead vine (*Syngonium podophyllum*)
- Bird's nest fern (*Asplenium nidus*)
- Bird's nest sansevieria (*Sansevieria trifasciata*)
- Bromeliads (*Cryptanthus, Billbergia, Aechmea* species)
- Buddhist pine (*Podocarpus macrophyllus*)
- Creeping fig (*Ficus pumila*)
- Croton (*Codiaeum variegatum*)
- Dracena (*Dracena godseffiana*)
- Fern (*Pteris* species)
- Ivy (*Hedera helix*)
- Miniature peperomia (*Pilea depressa* and *Peperomia* species)
- Mother fern or parsley fern (*Asplenium bulbiferum*)
- Nerve plant (*Fittonia verschaffeltii*)
- Parlor palm (*Chamaedorea elegans*)
- Peacock plant (*Calathea makoyana*)
- Philodendron (*Philodendron* species)
- Pothos (*Scindapsus aureus*)
- Prayer plant (*Maranta* species)
- Selaginella (*Selaginella* species)
- Strawberry begonia (*Saxifraga sarmentosa*)

Once planted, add water to the container until you see the water begin to seep through the sphagnum moss. Do not add too much water. If too much water is added, allow the terrarium to dry out for several days. Water only when the soil surface is dry. Terrariums function like a mini-ecosystem when closed. The water condenses on the sides of the glass and slides into the soil where it is taken in by the plants. Plants transpire and release water vapor back into the terrarium. Place terrariums in a location with plenty of natural light but out of the direct sun.

Tap water is an acceptable source for watering most plants. Chlorine, often found in tap water, can impact sensitive plants. **Chlorine** is a harsh chemical that can be used as a bleach or disinfectant for water purification. Leave tap water out overnight to lessen the concentration of the chlorine. Rainwater or distilled water (that does not contain chlorine) can also be used in place of tap water. If hard water (alkaline water) is consistently used, a white crust may appear on the soil surface and look unsightly.

Indoor plants may be watered in a number of ways. For smaller installations, using a watering can or a portable watering system allows maintenance crews to easily reach plants. A portable watering system allows control of the type of water used. This system does not infringe on clients to use their facilities and interrupt their daily routines. It has a pressurized tank and delivers water through a hose. Users may have the ability to exchange nozzle sizes. Many large plantscapes (such as those in malls or other busy retail outlets) have irrigation systems available and require less labor. If a container has a tray beneath it to capture any drained water, the water should be emptied after a half hour. The immersion method is appropriate for individual potted plants or small containers and for plants that are sensitive to water on their leaves, such as African violets, gloxinia, and cyclamen. A pot is placed into a container of water and allowed to soak until the soil is saturated. Then the pot is permitted to drain before it is returned to the display.

One common risk for container plants is that the potting medium dries out and shrinks away from the pot edges. If this happens, submerge the plant in water and allow the water to fully permeate the medium. Keep the potting medium moist and do not allow it to significantly dry out. In some cases, the potting medium will also form a crust on the surface. Scratch or poke holes to allow water infiltration, and place the pot in a bucket of water to saturate the soil.

Another watering risk is the buildup of fertilizer and chlorine salts. Salts can burn plants and reduce their growth. If possible, technicians managing the irrigation needs of an installed plantscape should allow containers to leach to reduce the accumulation of solids and salts. Leaching is a watering practice in which excessive amounts of water are applied to a plant's container to carry excess salts through the drainage holes. In potted plants, the saucers may be emptied. In plantscape revitalization, all potting medium may be removed and replaced.

> **Corner Question**
>
> How do cacti survive in the desert without water?

Light

Most plants prefer lighting conditions similar to those in their native environments. The indoor setting should provide light intensity and light duration to meet the plant's needs. Most interiorscape plants require 12–16 hours of natural light or supplemental light each day for normal growth and development. Light intensity needs vary among species, and different levels of light are available in different locations inside a building. A sunny, south-facing window in the summer has a much different light intensity than a north-facing window in the winter, **Figure 26-17**. Light in the middle of a room will also be much different than light in the far corner of a room. The light intensity in a spot only a few feet away from a window may be as much as 95% less than light intensity at the window. The challenge for designers of interiorscapes is to select the proper plant for the given light conditions.

Rob Byron/Shutterstock.com

Figure 26-17. Plants that need high light intensities should be placed in a window that faces south.

Plants can be grouped into categories according to their lighting needs. Types of lighting for indoor plants include:

- Full sun—maximum amount of light in an indoor setting. Plants should be no more than 24″ (60 cm) from a south-facing window. These conditions can scald sensitive plants, especially during the midday in summer. Cacti, succulents, and pelargoniums can withstand this light intensity. If moderate shade is provided for peak intensities, other plants can tolerate this light. These plants include cacti, succulents, citrus, and oxalis.
- Some direct sun—brightly illuminated areas with some sunlight reaching the plants during the day. This may be an east- or west-facing window that receives morning or afternoon light. Spots that are situated 24″ (60 cm) away from a south-facing window have similar light intensities. Many flowering plants and some foliage plants thrive in this type of location. Examples of plants that enjoy this lighting include *Cordyline*, *Hoya*, *Sansevieria*, and *Tradescantia*.
- Bright but sunless—good light but no direct sun. Plants grow well when placed about 5′ (1.5 m) from a south-facing window or a large window that receives little to no direct sunlight. Examples include *Anthurium*, asparagus, rex begonia, bromeliads, spider plants, and *Monstera*, **Figure 26-18**.
- Semishade—moderate light, usually 5′–8′ (1.5 m–2.5 m) from a sunny, south-facing window or near a sunless window. Many foliage plants will prosper in these conditions, although flowering plants rarely receive enough light to initiate flower buds. Examples include *Dracaena fragrans*, *Dracaena marginata*, *Fatshedera*, and *Ficus*.
- Shade—poorly lit by natural sunlight or have only artificial light (not intentional supplemental light). A few plants can grow well in these conditions. Some semishade plants can be used as temporary exhibits and rotated out after a few months, **Figure 26-19**. Examples include *Aglaonema*, *Aspidistra*, *Fittonia*, and *Philodendron scandens*.

Many interiorscape installations in offices or other commercial settings rely exclusively on artificial light. In some cases, clients will need plants in places that have no windows and no natural light. Plants are amazingly adaptable and can tolerate such conditions and even thrive for months or years in this kind of environment. Considerations should be given to duration and light intensity. Will the lights be turned off at night or over long holidays? Should supplemental lighting be installed, and in what ways will this fit with the overall design?

PRILL/Shutterstock.com

Figure 26-18. The lush, large leaves of *Monstera* grow well in bright areas with indirect sunlight.

JAROON MAGNUCH/Shutterstock.com

Figure 26-19. Even rooms with no windows can still be brightened by a plant such as pothos.

Plants in these situations may need to be switched out periodically if they are not flourishing. Using a flowering plant for a short period of time can be another way to brighten these spaces. Orchids, African violets, cyclamens, and other flowering plants would be good in these locations for a temporary display, **Figure 26-20**.

Potting Media

Commonly used potting medium is a general mixture containing materials such as peat moss, perlite, vermiculite, sand, bark, coir, and compost. A blend of these materials creates well-drained soil with large pore spaces for good aeration. Some indoor plants may prefer different mixes. For example, cacti and succulents thrive in soil mixes that have a high percentage of sand to allow for excellent drainage. An orchid prefers a mix that drains well but also holds water, such as a mix that contains peat and bark.

Viacheslav Nikolaenko/Shutterstock.com

Figure 26-20. Orchids make a splashy and vivid seasonal display to brighten a dark location.

Since most interiorscapes are permanent or semipermanent displays, soilless media can decompose and compact over time. Compaction is detrimental to plant health because gas-exchange is reduced and root growth will be poor. Plants should be repotted or containers amended to maintain optimal growing conditions.

Potting Medium pH

For most indoor plants, a potting medium pH range of 5.5–6.5 maximizes growth and development. PH changes can be caused by fluoridated water use, fertilization applications, and decomposition of soilless mixtures. Conduct annual soil tests to ensure the pH is at an appropriate level.

Mulch

Mulches may be added to a display to decrease the loss of moisture in the potting medium and improve the look of the display. In large plantscapes, mulches that may be normally used outside (such as a shredded bark mixture) can also be used indoors, but the mulch must be sterilized to minimize pathogens. Rocks or pebbles (depending on scale) may also be used. In smaller plantings, using "moss" provides visual appeal. Spanish moss, which is actually not a moss, but from the genus *Tillandsia*, is commonly used, **Figure 26-21**. Other mosses used include sheet moss, reindeer mosses, and bun moss.

Rob Hainer/Shutterstock.com

Figure 26-21. Spanish moss is harvested from trees and sold commercially for use in the floral and interiorscape industry as a container covering.

Plant Nutrition

All plants have nutritional needs that must be met for growth and development. Indoor plants typically require a complete fertilizer such as a 10-10-10 to supply nutrient needs. Some specialty or flowering plants have very specific nutrient needs. Some initial nutrient can be met by compost in the potting medium, but the nutrients will eventually be depleted and must be supplemented through another source. Many interiorscape technicians add fertilizer to their water and deliver it each time they irrigate. Other methods of adding nutrients include using a pelleted slow-release fertilizer or granulated fertilizer that is directly added to the soil. Slow-release fertilizer gradually sends out nutrients over time. A granulated fertilizer will need to be reapplied in a shorter period of time, but it may reach the roots sooner.

Plant Management

Commercial interiorscapes should be consistently monitored for plants that are becoming unsightly or infested with pests. Interiorscapes are intended to enhance a setting. If yellowed or dead leaves litter the soil line or insects start crawling along the plants, it detracts from the objective of the design. Proper management of a plantscape requires horticultural technicians to monitor and manage any problems that may arise.

Hrynevich Yury/Shutterstock.com

Figure 26-22. Keep interiorscapes looking attractive by removing yellowing or dead plant material and dusting and polishing leaves.

Grooming

Plants in permanent installations can collect dust, drop leaves, and develop an unkempt appearance. Dust detracts from the design and reduces photosynthesis and gas exchange in the leaves. For these reasons, plants should be groomed regularly to remove dust and yellowing or dead leaves or flowers, **Figure 26-22**. Remove dust regularly using clean water and a damp cloth. Support each leaf as you wipe it. Leaves can also be polished using commercial products or a simple application of olive oil to add shine to the leaves.

Training and Pruning

In some cases, plants in an interiorscape may need to be supported or trained. An artful and complementary trellis with hidden clips or ties can support climbing plants.

In an interiorscape, pruning enhances growth in a desired way and removes unwanted material. Pruning includes pinching out growing points to encourage bushier growth.

As coleus plants begin to flower, the inflorescence stalk should be pinched to foster foliage growth and prevent flowering. Vigorous plant material can be cut back in a way that preserves the aesthetics of the plant. If pruning is overdone, remove the plant and replace it with another specimen. The overly pruned plant can recuperate until it is fit for another installation. Trimming dead leaves and damaged plant material is also essential to maintain a professional look.

Insect Pests

Managing insect pests requires multiple strategies to effectively control populations. The first step is to set up a routine monitoring of all commercial sites. Closely inspect and observe plants while watering or during separate visits. If an insect pest is found, properly identify the pest to determine appropriate methods for management. The threshold of tolerance for insects in a commercial planting may vary by site. A high-end luxury hotel may have a zero tolerance for any insect pest. In a botanical garden, however, more insects may be tolerated and even used as part of an education program, **Figure 26-23**.

popular business/Shutterstock.com

Figure 26-23. Mealybugs are a common indoor plant pest that can be controlled using systemic insecticides.

Pest management may include cultural, mechanical, biological, and chemical methods. The choice for a method depends in part on the site. Would store managers in a retail mall want to see biological control envelopes hanging discreetly in plantings? Selecting plants that have few pest problems and are well-suited to the site provides a healthy start for resisting insects. In some cases, hand removal may be appropriate and could be done after public hours to minimize disturbances. In severe cases, plants may be removed for treatment and returned once they are healthy.

Controlling insect pests using pesticides can diminish populations and allow plants to continue to grow. Employees of interiorscape companies who apply pesticides must successfully pass a state pesticide applicator license examination. Pesticide applicators should follow safe-use practices. This means following label recommendations precisely. Chemicals should be used only in the way they are labeled, specifically for indoor use, and on the plants that the pesticide is labeled for. Applicators should wear the recommended personal protection equipment and mix the pesticides according to the label specifications. Erroneous application and rate can harm plant material and people. Consult with the state cooperative extension service for up-to-date pesticide information.

Did You Know?

The same plant that grows coffee beans also makes a great houseplant. The coffee plant is great for experienced or novice gardeners as it is easy and interesting to grow. With rich, glossy leaves, it is a perfect large houseplant for a sunny window. It may even grow enough berries for a cup of coffee.

Chemical controls commonly used in interiorscapes to manage pests include insecticidal soaps, horticultural oils, and systemic insecticides.

- *Insecticidal soaps* are synthetic soap solutions used to kill insect pests. They are effective, safe, and popular. They are, however, still pesticides and should be applied according to the label. They are specially formulated to disrupt the cellular membranes of insects. Insects must be thoroughly covered with the solution, and applications must be repeated over time.
- *Horticultural oils* are pesticide treatments consisting of different types of oils used to kill insect pests. Horticultural oils are usually highly refined petroleum oils that block an insect's spiracles, or breathing organs. Neem oil is another horticultural oil that comes from the neem plant (*Azadirachta indica*) and disrupts feeding and reproduction. Some vegetable oils, including cottonseed and soybean oils, have also shown promise in controlling mites and other insects.
- *Systemic insecticides* are pesticides that translocate throughout the plant, killing any insects that feed on the plant. As piercing and sucking insects try to feed on the plant, their nervous systems are attacked, causing them to die. Some systemic insecticides have a reentry period. This means that after they are applied, a certain amount of time must pass before it is safe to be in contact with plants (unless wearing proper safety gear). Pesticide application should be done on the weekend, when a building is closed, or at other times when there is no risk of harming others.

Diseases

In addition to monitoring for insects, plants should also be monitored for diseases. Plant diseases occur when all three elements of the disease triangle are present: pathogens, the right environmental conditions, and a host plant. Environmental conditions that influence disease development depend on the pathogen. These conditions include saturated soil from overwatering, temperatures that are too warm or cold, or high levels of humidity or air movement.

A commonsense approach to disease management is to avoid the conditions that foster disease. Carefully water the soil to minimize splashing water on the foliage. Pathogens can dwell in the soil and be carried by water. They are also attracted to standing water. Provide ample space between plants to optimize airflow. Stagnant air may allow disease to flourish and spread. Disease management is a balance between providing an optimal environment for plants to thrive and limiting opportunities for pathogens to take hold.

Sanitation is an important practice for reducing incidences of disease. Leaves and other plant parts that show symptoms of disease should be swiftly removed to reduce the spread of disease. Equipment such as pruners, rakes, and cloths should be sterilized between working with diseased and clean plants, **Figure 26-24**. In some cases, a plant may be so infested with disease that it must be removed and discarded.

Alena Brozova/Shutterstock.com

Figure 26-24. Sterilize tools after trimming diseased plant material.

Fungal Diseases

Many plant pathogens that impact indoor plants come from fungi. Fungi produce airborne or waterborne spores that can spread easily in moist, humid conditions. Fungi are heterotrophs that feed off plants, causing deterioration of tissues, and result in rots, molds, and mildews. Common fungal diseases include anthracnose (*Colletotrichum* and *Gloeosporium*), powdery mildew (*Oidium*), and root and stem rots (caused by *Rhizoctonia*, *Pythium*, *Botrytis*, *Phytophthora*, *Alternaria*, and *Sclerotinia*). Symptoms of these diseases vary widely and may include yellowing and browning leaf tips, leaf spots and lesions, wilting, rotting stems and roots, or white powdery growth, **Figure 26-25**.

Figure 26-25. A common symptom of the fungal disease powdery mildew is a white, powdery growth on leaves.

Disease control begins with monitoring, prevention, and sanitation (cultural control). If a disease spreads rapidly, a fungicide treatment may be necessary to eliminate the disease. Only a certified pesticide applicator may use a fungicide. Instructions on the product label must be followed exactly. The fungicide should be registered for indoor use and specifically list the type of plant requiring treatment. Due to the low thresholds of many interiorscapes, removing and replacing the diseased plants is often the best solution.

Bacterial Diseases

Bacterial diseases occur most frequently when plant material is stressed and surface moisture provides a point of entry into the plant. Wounds or damaged plant parts provide easy access for bacteria as well. Warm and humid environments, such as indoor pools and spas, have higher rates of a bacterial infection than cooler, less humid interiors.

Bacterial leaf spot is the most common bacterial disease affecting houseplants. Symptoms include water-soaked spots. The spots sometimes have a yellow halo and a sticky ooze, **Figure 26-26**. The spots enlarge and merge together as the disease continues to spread under high moisture condition.

Figure 26-26. Bacterial leaf spot leaves a water-soaked spot on the leaf.

Under drier conditions, the spots do not enlarge but dry out and turn reddish brown, creating a speckled appearance. Remove all diseased plant material. Avoid plant overcrowding, water splashing on leaves, and low temperatures. Some pesticides that control bacterial diseases may be registered for indoor plants. Consult with the local cooperative extension service for more information.

Interior Plantscaping Business and Careers

Interior plantscaping is a niche business within horticulture industry. It can be rewarding and lucrative with thoughtful business planning and determination. Growing an interiorscape company requires strong relationship-building skills to find new clients and retain current customers. An interiorscape business owner must give compelling reasons and show through experience how interiorscapes enhance the professional appearance and tone of a company. Successful owners have self-initiative. Interaction between clients and company employees should be positive, warm, and friendly, letting customers know that they are valued. Excellent communications between the interiorscape company and the client creates trust. When there is a problem, such as a diseased plant, the client will trust the interiorscape company to fix the problem.

Interiorscape firms should focus on providing quality plants and superior service. Installations that brim with healthy, attractive plants generate potential customers and satisfy the client who paid for the plantscape. Consistent and discreet plant care should be maintained with a professionalism that reflects well on the company. To be successful, new interiorscape companies require experience and knowledge of the industry, sources for high-quality plant material, design skills, and an understanding of plant care. The ability to communicate the value of indoor plants to potential and existing clients is also important. Two interesting careers in the interiorscaping industry include interiorscape design and horticultural technology.

Interiorscape Designer

Interiorscape designers enjoy creative positions that allow them to take basic design principles and create plans that bring together color, form, rhythm, and balance. They consider the objectives for a site and the project budget to create a plan for an interiorscape that a client will approve of and value. Designers often study landscape design in college or university. This training gives them skills needed to express design plans using multiple types of media, from paper and ink to computer- assisted design software. Designers will complement or highlight architecture, interior design styles, and client ideas in their design. Depending on the size of the company, a designer may also be the person who sells the design to the client.

Designers should be able to communicate their ideas, provide information and quotes in a timely manner, and follow through with installation and maintenance once a contract is set.

Horticultural Technician

Horticultural technicians care for the plants in an interiorscape. They conduct routine services, such as watering, trimming, and grooming. They evaluate plants for diseases, insects, and other issues that may detract from the planting. They also monitor plant nutrient needs, apply fertilizers as needed, replace unsightly plants, remove leaf litter and other trash, and dust plant containers. If certified, they may apply pesticides and replace or revitalize mulches. Many technicians have a high school degree or an associate degree with related work experience.

Career Connection: Yuko Frazier

Senior Project Designer, Ambius

Growing up in urban Japan, Yuko Frazier, always felt the tug of nature. She came to the United States to attend college, first to study sculpture and then to study horticulture. Yuko always wanted to do something to bring people and nature closer. Combining her fine art skills with a degree in landscape design enabled her to enter the unique field of interior landscaping.

Yuko is a project designer for Ambius and travels the world consulting with commercial and retail companies to design and manage interior landscape installations. She works with a wide range of professionals including property developers, general contractors, architects, and interior designers. Yuko creates interior landscape designs and displays and then prepares design proposals, reviews and processes contracts, orders products, and schedules production and installation crews. Her interiorscapes can be found in a variety of locations, from shopping malls to cruise ships.

Yuko Frazier, Project Manager at Ambius

Yuko remains motivated by the sense of accomplishment she feels when a project is done that creates a beautiful plant-filled space and connects people to a nature experience. For young people considering this field, Yuko describes interior landscaping as a niche industry compared to outdoor landscaping and, for this reason, there are many opportunities for creativity.

CHAPTER 26
Review and Assessment

Chapter Summary

- Interior plantscaping is the practice of designing, installing, and maintaining indoor spaces with plants. Interiorscapes can provide benefits such as an enhanced sense of well-being, a professional tone and image, and lower workplace stress.
- Interiorscapes may be found in health care facilities, hotels, shopping malls, offices, restaurants, homes, botanical gardens, and zoos.
- Design elements and principles guide the crafting of an interiorscape. The elements of design include line, form, texture, and color. The principles of design used to create interiorscapes include balance, proportion, repetition, emphasis, and unity.
- Indoor plants should be suited to the available temperature and humidity. Nearly all indoor plants will thrive in a temperature range of 55°F–75°F (13°C–24°C). Plants prefer humidity levels between 40% and 60%.
- All plants need water. Their needs vary from plant to plant and each plant's needs vary based on the season.
- Indoor plants may be watered using a watering can or a portable watering system. Many large plantscapes have irrigation systems.
- The indoor setting should provide light intensity and light duration to meet the plant's needs. Most interiorscape plants require a light duration of 12–16 hours per day.
- For indoor plants, the potting medium is a general mixture containing materials such as peat moss, perlite, and compost. The medium should be well drained.
- Indoor plants typically require a complete fertilizer such as a 10-10-10. Nutrients may be supplied by adding fertilizer to the water or using slow-release or granulated fertilizer.
- Managing plantscapes includes grooming, pruning, training, and monitoring and controlling for insect pests and diseases. Management of pests may include cultural, mechanical, biological, and chemical methods.
- Disease control includes monitoring, prevention, and sanitation (cultural control). Plant diseases may be caused by fungi or bacteria.
- Interiorscape firms should focus on providing quality plants and superior service. Installations with attractive plants generate potential new customers.
- Many careers are available in interiorscaping. Examples of these careers include owner of an interiorscape business, sales associate for interiorscapes, interiorscape designer, and horticultural technician.

Words to Know

Match the key terms from the chapter to the correct definition.

A. bonsai
B. chlorine
C. horticultural oil
D. insecticidal soap
E. interior plantscaping
F. interiorscape
G. succulent
H. systemic insecticide
I. terrarium

1. A type of plant that has thick, fleshy leaves and is usually native to arid regions of the world.
2. A pesticide treatment consisting of different types of oils used to kill insect pests.
3. The practice of designing, installing, and maintaining indoor spaces with plants.
4. A closed type of planting structure that is self-contained and provides a good environment for tender, hard-to-grow plants.
5. The Japanese art of pruning and manipulating plants into miniature form.
6. A pesticide that translocates throughout the plant, killing any insects that feed on the plant.
7. An installation of plants inside a building.
8. A harsh chemical that can be used as a bleach or disinfectant for water purification.
9. A synthetic soap solution used to kill insect pests.

Know and Understand

Answer the following questions using the information provided in this chapter.

1. What are some benefits interior plantscapes can provide?
2. What are some settings where an interiorscape may be found?
3. What are four elements of design used in creating plantscapes?
4. Describe the design element *texture* and give examples of different textures a plant may have.
5. What are the principles of design used to create interiorscapes?
6. Describe the use of repetition as a design element in interiorscapes.
7. What are ways to create emphasis in an interiorscape?
8. What elements are involved in creating and maintaining an interiorscape?
9. What is the optimal temperature range for nearly all indoor plants and what may happen to plants if the temperature is outside this range?
10. What can be done to raise the humidity level for plants grown indoors?
11. What types of plants should be selected for growing in a terrarium? Name two examples of such plants.
12. Describe the effects excess water may have on indoor plants.
13. Briefly describe five types of lighting conditions that may be used for indoor plants.
14. What type of fertilizer is typically required for indoor plants?

15. Why should plants in an interiorscape be groomed regularly?
16. What is the purpose of pruning plants in an interiorscape?
17. Name three chemical control options for managing insect pests in an interiorscape.
18. What are some fungal diseases that affect indoor plants and what are some symptoms of these diseases?
19. What are some symptoms of bacterial leaf spot in houseplants?
20. What do new interiorscape companies need to be successful?
21. What duties do horticultural technicians perform as related to interiorscapes?

Thinking Critically

1. You find a population of insects in an interiorscape at the mall. It is Friday morning, and the weekend is the busiest time at the mall. What are your options for pest control? What will you choose to do and why?
2. Consider the list of indoor plants that are available for an interiorscape. Draw a plan for a space that is 30 ft² (9 m²). What plants will you use? What decisions led you to select the plants you did? What were the factors that you considered for optimal design and maintenance of the space?

STEM and Academic Activities

1. **Science.** Research recent croton cultivars. Choose three cultivars and write a two-page report comparing the differences and similarities of each new variety.
2. **Science.** A client has asked you to create an entire wall of plants as a dramatic and bold design statement. Determine what materials would be available to support the weight of numerous plants and potting material. How will this installation be watered? Draw a design that incorporates these considerations and annotate with suggested supplies including plants, growing medium, and building materials.
3. **Math.** You have been asked to design a large planter for a hotel lobby. The planter is 5′ × 3′ (1.5 m × 0.9 m). Research the size of different indoor plants. Create a list of plants you want to include and how many you will need of each type to fill the planter.
4. **Social Science.** Conduct research about the use of terrariums during the Victorian era. If you were a greenhouse owner in Victorian England, what types of terrariums would you make? What plants might you use most?
5. **Language Arts.** Write an essay explaining how the selection of foliage and flowering plant materials affects the design process.

Communicating about Horticulture

1. **Reading and Speaking.** In small groups, discuss the captions in the chapter. Describe, in your own words, what each caption is describing. Discuss the effectiveness of the captions compared to the chapter text.

2. **Listening and Speaking.** Interview an interiorscape firm owner. Ask the person to describe a typical day at work. Here are some questions you might ask: How does a person open and run an interiorscape business? Is it difficult to get financing? What is the work environment like? What are the job duties? What kinds of delivery or storage problems are dealt with? What other types of professionals do you work with? Report your findings to the class, giving reasons why you would or would not want to pursue a career similar to that of the person you interviewed.
3. **Reading and Speaking.** As an interiorscape designer's assistant, create an informational pamphlet on the types of potted flowering plants that are seasonally available through the designer's business. Research the history of the types of potted flowers for some background information. Include images in your pamphlet.

SAE Opportunities

1. **Exploratory.** Job shadow an interiorscape designer. What are the daily responsibilities? What do you like or not like about this position? What educational degrees and experiences are required to have this position?

Liz Driscoll/Goodheart-Willcox Publisher

2. **Experimental.** Your classroom has no windows. Determine which plants grow well in an indoor environment with little light. Purchase three different kinds of plants that you think will be able to do well in your classroom. Make weekly observations of how the plants perform. At the end of two to three months, determine if any are suitable for your classroom. What made the plants suitable or unsuitable?
3. **Experimental.** Research the biological controls that could be used in the large atrium of a conference center. Are they easily available? How effective are they? What are the benefits and costs of using these methods to control insect pests on your interiorscape?
4. **Exploratory.** Visit a local greenhouse that grows houseplants. Ask the owner or grower to share information about the production practices used at the greenhouse. What methods are used to cultivate a high-quality crop? Why would this be important for an interiorscape company?
5. **Exploratory.** Investigate propagation methods for 10 different houseplants. How much would it cost to propagate and grow houseplants for a hypothetical interiorscape business compared to purchasing the plants wholesale?

CHAPTER 27
Landscape Installation and Maintenance

Chapter Outcomes
After studying this chapter, you will be able to:
- Interpret the elements of a landscape design plan.
- Describe the process of hardscape installations.
- Illustrate how to properly install plant material.
- Explain the fundamentals for maintaining a healthy landscape.
- Discuss the process of starting a landscape business.
- List careers related to landscaping.

Words to Know 🔗

basket weave	legend	stack bond
drawing scale	north arrow	stretcher
Flemish bond	renewal pruning	title block
guys	running bond	topiary
header	softscape	
herringbone	specifications	

Before You Read
Skim the Know and Understand questions at the end of the chapter first before you read the chapter. Use them to help you focus on the most important concepts as you read the chapter.

While studying this chapter, look for the activity icon to:

- **Practice** vocabulary terms with Words to Know activities.
- **Expand** learning with identification activities.
- **Reinforce** what you learn by completing Know and Understand questions.

www.g-wlearning.com/agriculture

Looking at a beautifully rendered landscape design plan brings to mind vivid images of plants growing across a yard, **Figure 27-1**. Most clients relate to these artistic expressions of their landscape objectives. However, landscape companies require more detailed plans to create the landscape. Landscape designers will create multiple plans that contain information from planting lists, site evaluations, and other guidelines. These plans will enable a landscape contractor to build and plant the landscape.

Landscape Design Plans

The installation of a landscape occurs once a client and a designer have agreed on a final plan. The plan may be sent to landscaping companies that are asked to submit bids (prices and time lines) for completing the project. Some companies provide both design and construction services. The company that installs the landscape needs to understand how to interpret the design and complete the construction or installation activities. These activities can include properly planting the trees, shrubs, and flowers as well as constructing other features, such as walkways.

Landscape plans are used to turn ideas into visual concepts. They contain detailed plans and drawings that allow landscaping companies to understand the overall intentions of the landscape designer. The landscape

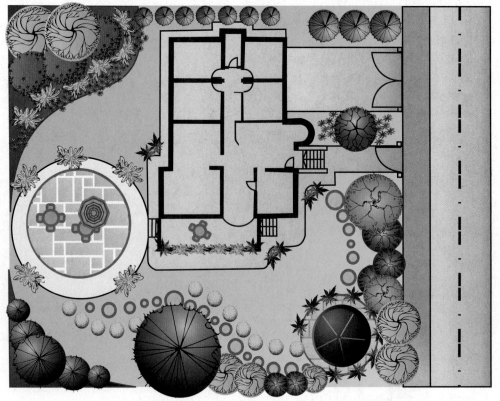

charobnica/Shutterstock.com

Figure 27-1. A landscape design illustrates the rich potential of plants to transform a space.

plan will illustrate the types and locations of materials (such as plants, features, and irrigation) that are needed and provide the basis for estimating the quantity of materials needed to complete the design.

Reading a landscape plan is like reading a map. It shows the placement of individual plants as well as representing where paths, patios, and water features will go. Plants are generally illustrated as symbols, **Figure 27-2**, and grouped together in mass plantings or as single specimens. The plants may be labeled directly on the plan or abbreviated and defined in the legend. A *legend* is an element on the design plan that describes any symbols or abbreviations used in the plan. A hardscape is an element built into the landscape, such as a patio or pathway. Hardscapes are clearly drawn to an appropriate scale on the design plans.

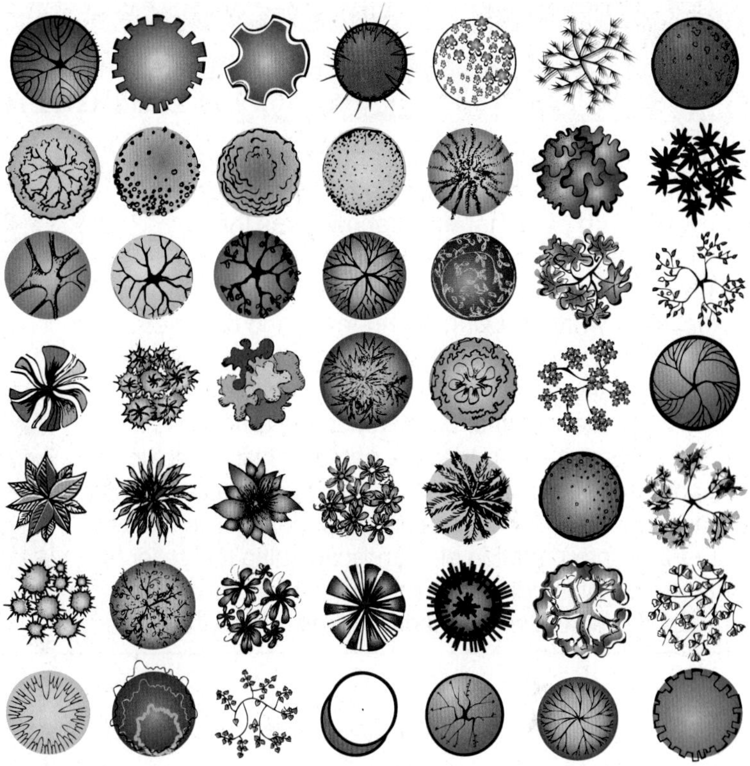

Few design plans will show all the necessary information, so often there are multiple plans. These may include planting, grading, drainage, irrigation, and construction plans. The design plan generally includes a number of different elements, such as these:

- Title block.
- North arrow.
- Legend.
- Specifications.
- Drawing scale.

Figure 27-2. On the landscape design, plants are represented symbolically to show the quantity and spacing.

Title Block

The *title block* is a natural starting point for identifying the key information about the project. It is usually located at the bottom or the side of the plan and may include the project name, project location, client's name, designer's name, date of preparation, sheet title, sheet number, and drawing scale. The *drawing scale* specifies the relationship between the distances on the plan to actual distances on the site.

North Arrow

The *north arrow* is a simple graphic or symbol, usually in the form of an arrow or another artful representation, that points north, **Figure 27-3**. It is used for orienting the plan at the site of construction. Many landscape designers will draw this symbol in creative and clever ways.

Figure 27-3. The north arrow gives orientation to the design and ensures proper placing of all landscape materials.

STEM Connection — Rendering a Landscape Plan

Many clients still prefer hand-drawn landscape plans and rendering. A rendered drawing creates an advanced plan that really showcases the designer's artistry. A rendering adds color and artistic interpretation to what would otherwise be a black-and-white drawing. Hand rendering can also impact computer-aided designs by softening the look or helping the drawing to look more personal.

Computer-aided or digital rendering is now a standard practice in landscape design and architecture. Various software programs help designers create realistic renderings. The objective of digital rendering is to create an artistic interpretation of a landscape plan that reflects the finesse of the designer and still communicates the overall feeling of what the landscape would be if it existed.

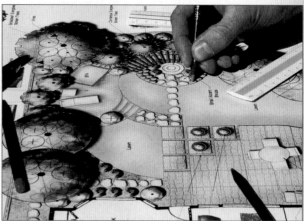
Scott E. Feuer/Shutterstock.com

Legend

The legend indicates what the symbols or abbreviations on the plan represent. Rather than listing the entire name of a plant directly on the design, many planting plans use an abbreviation for the name of the plant. For example, if a plan includes a willow oak tree, the abbreviation could be *WO*. Some designers may use a scientific name. Willow oak, which is *Quercus phellos*, may be abbreviated *QUph*. The legend may also indicate the size of the plant container. For example, if the plan specifies planting seven abelia, the legend may include using plants in containers that are 3 gallons in size.

Specifications

The design plan will include instructions and requirements that must be followed when creating the landscape design. The *specifications* are guidelines for some or all of the following: plant selection, site preparation, excavation, maintenance requirements, planting techniques, mulching, and soil preparation. This amount of detail is intended to give clear directions to installers to ensure a high level of quality. In some cases, the specifications may be written in a separate manual, such as a plant manual.

Drawing Scale

The design plan is naturally drawn at a much smaller scale than the actual area to be planted. As stated earlier, the drawing scale shows the relationship between the distances on the plan to actual distances on the site. A scale bar is typically a line marked like a ruler in units proportional to the map's scale. For example, a map may use inches to equal feet. Scales are often shown as ratios, such as 1:50, which means each map unit equals 50 actual units. Units may be inches, feet, meters, or yards.

Site Preparation

If a site needs to be prepared prior to hardscaping and planting, these details may be included as a grading plan or irrigation plan. Demolition of unwanted structures or removal of undesirable plant material also needs to occur. If a site has hills, bumps, valleys, or slopes, a certain amount of grading to level the surface may be necessary. A grading plan shows the contours and grade elevations for existing and proposed ground surface elevations at the site. This enables contractors to set up survey equipment to mark the changes and then to use a grader with a large blade that essentially scrapes the soil into the adjusted elevation, **Figure 27-4**.

Cathy Kovarik/Shutterstock.com

Figure 27-4. Grading may be necessary to level the ground in some settings.

Many sites have drainage issues that need to be addressed. Without the proper installation of a good drainage system, plants, features, or structures may be damaged. Patios can lift, wooden structures can rot, plants may drown in saturated soils, or drainage water may overwhelm building foundations. Drain systems collect water close to buildings and other structures, such as retaining walls, and send the water to catch basins set in the lawns. Water from roof gutters and downspouts is often directed into the drainage system as well. Drainage water is routed to the street storm water drains or an approved natural runoff location. However, many places across the country may not need a drainage system. The landscape designer determines this information.

Preparations for irrigation can occur at the same time as work for the drainage system. The installation of mainline pressure pipes, valve wiring, and some of the lateral irrigation pipes can be laid into the same deep trenches as the drainage components. Once all the pipes have been installed, the trench is backfilled with soil and lightly compacted.

> "Life is like a landscape. You live in the midst of it but can describe it only from the vantage point of distance."
> —Charles Lindbergh

Hardscape Installation

Most residential and commercial landscapes feature a combination of hardscapes and ornamental plants. Hardscapes are any built element placed in the landscape. Hardscape features can significantly impact the tone and use of a space. These components can enhance the landscape by providing opportunities for entertainment and recreation, providing a focal point in the landscape, minimizing erosion, resolving privacy or security challenges, and making areas more accessible and easier to maintain. Hardscape structures may include retaining walls, patios, fences, arbors, gazebos, water fountains, gazing pools, sidewalks, or pathways, **Figure 27-5**.

Elena Elisseeva/Shutterstock.com

Figure 27-5. Retaining walls are beautiful and functional hardscape features that complement the landscape.

722 Horticulture Today

> "Logic will get you from A to B, but imagination will take you everywhere."
> —Albert Einstein

There are many different hardscape materials, such as gravel, pebbles, sand, bricks, wood, rocks, stones, pavers, and cement. Other elements, such as a birdbath, swing set, gazing ball, pool, spa, or sculpture, are also part of the hardscape. Hardscaping follows basic design principles and elements, blending colors, textures, forms and lines with balance, proportion, repetition, emphasis, and unity.

The installation of a hardscape usually precedes the planting of trees and shrubs in the landscape. Hardscaping most typically involves the construction of pathways, patios, retaining walls, and water features. There are numerous other types of hardscape elements. The three projects described in the following sections will provide a foundation for understanding the concept of hardscape installation.

Brick Patios and Pathways

Bricks provide a classic hardscape medium for patios, pathways, and garden borders, **Figure 27-6**. Brick offers a natural complement to the *softscape* (herbaceous and woody plants in the landscape). Many patios and walkways are constructed with brick, although pavers, poured concrete, or crushed gravel or stone may be used.

To build a brick patio or pathway, use a brick designed for paving. Using the design plan, determine how many bricks are needed for the entire job, and order them from a single batch. Bricks vary significantly, and there may not be similar colors or sizes available later to match an initial purchase. Common brick sizes range from 8" to 8 1/4" (20.3 cm to 21 cm) long, 3 3/4" to 3 7/8" wide (9.5 cm to 9.8 cm), and 2 1/4" to 2 1/2" (5.7 cm to 6.4 cm) deep.

Did You Know?
More than 80% of brick sold in the United States is for residential construction. About 16% is for nonresidential purposes, and only 3% is used for paving applications.

A jeep2499/Shutterstock.com

B pics721/Shutterstock.com

Figure 27-6. A—Hardscaped pathways offer greater accessibility to the landscape and lend movement and interest. B—Brick patios are enviable features of an integrated landscape design.

Bricks can be placed in numerous patterns that provide interest and give a sense of movement. Typical brick patterns include:

- *Running bond*. The most common brick pattern, the running bond is used in multiple settings. Bricks are laid end to end and staggered between rows. This pattern is economical as there is little waste and minimal cutting, **Figure 27-7A**.
- *Stack bond*. The stack bond is the simplest of designs. Bricks are placed side by side with no staggering. This design is most commonly used for pathways, **Figure 27-7B**.
- *Flemish bond*. The Flemish bond pattern has been in existence for hundreds of years. This pattern is laid in layers consisting of stretchers and headers in each layer. *Stretchers* are bricks laid along the wall with their sides showing. *Headers* are bricks laid across the wall with their ends showing. The pattern alternates in each layer with the header brick being centered between the stretchers above and below, **Figure 27-7C**.
- *Basket weave*. The basket weave is also a very old pattern that has been used for centuries. Two bricks are laid horizontally next to two vertically laid bricks to look as though they are weaving in and out of each other, **Figure 27-7D**.
- *Herringbone*. The herringbone is the strongest pattern. The herringbone is an interlocking pattern of bricks set at 45° or 90° angles continuously across the area to be paved. The bricks are tightly interlocked, **Figure 27-7E**.

Prepare the area to be paved by digging out existing soil and grading as appropriate. There should be a depth of at least 6″ (15.2 cm) to fill with gravel, sand, and brick. With a level soil bed, pour a 3″–4″ (7.6 cm–10 cm) layer of gravel or crushed stone and firmly tamp it down. On top of the gravel, spread a 1/2″ (1.27 cm) layer of sand and level it carefully. Being cautious not to step on the sand, place the bricks on top in the desired pattern. Work from the center toward the sides. Using a tamper or wooden hammer, firmly tamp the bricks as they are set.

Use as many whole bricks as possible and only cut bricks to fit as needed. Most patios are not cemented together with mortar; rather they are placed so tightly within a border that they can be laid with closed joints. With closed joints, the bricks are pushed tightly together and covered with sand. The sand is then swept into the cracks. Patios can also be laid with mortar. Mortar is applied to an existing or new concrete slab, and the bricks are tapped into place. After the bricks have been leveled, mortar is packed into the joints. The excess mortar is scraped off, and the mortar is allowed to set. Dried mortar can be washed off brick surfaces as needed.

Retaining Walls

Retaining walls are used in both commercial and residential settings as a way to hold soil on a sloped site or simply as an aesthetic way to define a garden bed or frame a patio. For example, in a hillside setting, a retaining wall can make the land more usable. An area can be leveled and a retaining

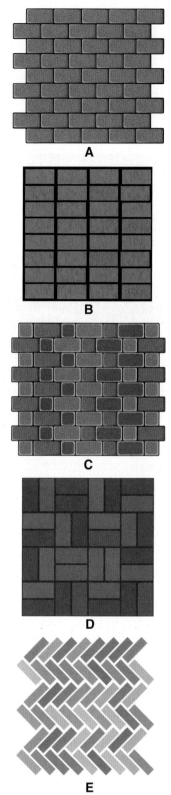

Liubou Yasiukovich/Shutterstock.com

Figure 27-7.
A—Running bond.
B—Stack bond. C—Flemish bond. D—Basket weave.
E—Herringbone.

Corner Question

How many stones do you think were used to create the Great Pyramid of Giza?

wall used to hold back the remaining soil from the slope. Retaining walls also manage storm water runoff and limit soil erosion. Retaining walls may be just a freestanding wall that delivers seating in an entertainment area. They can also be used as a design feature to physically divide spaces in the landscape. Retaining walls can be constructed using a variety of materials. Common materials include:

- Cement block. Solid and semi-solid blocks are strong and versatile. Many systems are designed to fit together with clips or pins, and they can be split easily to fit the wall. Many cement blocks have a rough face that provides a nice look. Smooth concrete blocks can be the foundation material and be faced with stone or brick.

- Rocks and stones. For centuries, stone walls on farms have complemented the rural landscape of the northeast. Farmers would use stone from their fields and construct walls to define their properties. There are a wide variety of choices that will fit just about any style of yard and garden, **Figure 27-8**. Stone is usually the most expensive material. Skilled workers are needed to construct stone walls.

Jorge Salcedo/Shutterstock.com

Figure 27-8. Rocks and stones may be expensive, requiring skilled craftsmen to construct a classic look.

- Wood. Using standard timbers, a wood wall can easily incorporate angular walls and steps. Timbers are readily available, inexpensive, and easy to build with. Timbers are placed on a gravel base to provide good drainage behind the wall. They are set with long spikes or screws. Timbers should be treated to be suitable for ground contact and sealed to reduce rotting.

Most retaining walls are professionally constructed to account for the engineering considerations of holding significant volumes of soil and managing steep sloping land. How best to work with heavy materials and the height of the wall must also be considered. Simpler, smaller projects around a school or home can be fun projects that are easily doable. To begin crafting the wall, mark the site according to the landscape design using stakes to mark the borders. String is wound through the stakes and pulled tightly.

Begin the excavation of the soil to prepare the base on which the block will be laid. As with a brick patio or path, the foundation is essential to the integrity of the wall structure. The depth of the trench depends on the height of the wall, but generally for every 8″ (20 cm) of wall, 1″ (2.5 cm) of soil should be removed. For example, a wall that is 4′ (1.2 m) high should be set in a base that is 6″ (15.2 cm) deep. Make the trench level, and compact the soil with a tamping tool.

Retaining walls are set on a base material of angular, sharp rocks, such as crushed gravel. The pointed pieces will lock together when compacted and ensures the stability of the wall. Lay a 4″–6″ (10 cm–15.2 cm) layer of the base gravel, level it, and compact it firmly so there is no shifting of material. Finish the base by spreading a 1″ (2.5 cm) layer of sand on top of the gravel.

Lay the first row of blocks or stones on the base material. The blocks should be placed with the edges touching and be level on all sides (front to back and side to side). A mallet can be used to tap each block into place. If the blocks are not level, lift the block and slip more sand under the low side. Reset the block and check for level placement. The first row sets up the remaining rows to be level. The second and subsequent rows are staggered with each block straddling the joint line below. A masonry chisel or diamond saw can split a block in two to keep the correct spacing on the wall.

Backfill material, such as crushed gravel, is used to strengthen the wall after each row is laid. Backfill also helps with water drainage, allowing the water to move out and away from the wall. The backfill should be laid after the completion of each row and be thoroughly compacted before starting a new row.

Most retaining walls have some sort of cap, **Figure 27-9**. For some material, such as stone, a cap may not be needed. For concrete block walls, caps can be used to fit on top and protect the wall from the elements of weather. Cap materials are glued into place onto dry and clean blocks with masonry adhesive.

Corner Question

Which gardens does the BBC consider to be the best in the United States?

Elena Elisseeva/Shutterstock.com

Figure 27-9. Caps finish off a wall in an elegant and functional way.

History Connection

Thomas Jefferson's Monticello and Poplar Forest

The third President of the United States was a skilled architect and horticulturist. He designed Monticello and Poplar Forest, his two residences in the Blue Ridge Mountains. Both homes were renowned for their geometry and form, and many of his gardens followed a natural form.

Jefferson's horticultural mentor, Bernard McMahon, wrote a book titled *American Gardener's Calendar* (1806). Jefferson considered this book his guide for horticulture. Within the book was an essay titled "Ornamental Designs and Plantings." McMahon wrote the book to illustrate principles of design where naturalistic and informal styles are followed to mimic a natural design style. This

Jill Lang/Shutterstock.com

method included designing plantings in "clumps" or clusters of irregular or curved beds filled tightly with shrubs, trees, and flowers.

Thomas Jefferson followed McMahon's written works to create gardens at Monticello and Poplar Forest that mimicked natural designs.

Water Features

Water in the landscape evokes feelings of tranquility and reflection. Water features can soften undesirable noises and provide water for wildlife or habitats for animals. Ponds, fountains, small streams, and even swimming pools and spas are water features that can be designed for the landscape, **Figure 27-10**. A swimming pool may not be appropriate for a commercial landscape, such as one for an office building. However, a small pond or fountain can enhance the workplace and provide a soothing focal point where employees can gather for lunch or an informal meeting.

Antonina Potapenko/Shutterstock.com

Figure 27-10. Small, constructed streams generate soothing sights and sounds for any space.

Most water features require the professional services of a landscape contractor for installation. Most features have common elements and steps needed for construction. Site preparation, lining material, installation of a water circulating pump, and access to electricity for the pump to run are typically involved. For a pond (of any size) or a pool, the soil should be excavated and the space lined with an appropriate liner or preformed insert. Ponds can be installed with flexible waterproof liners that come in varying thicknesses or with premade plastic shells. Pools have similar construction. They can be a premade fiberglass structure or lined with concrete. A fountain should be placed on a stable base that has been created in a similar fashion as a patio or pathway foundation.

Once the liner, insert, and/or fountain has been installed in the landscape, a circulating pump should be connected. The pump will continuously move water and provide oxygen in the aquatic environment. For a fountain, this pump moves water throughout the fountain and constantly recirculates the water. In ponds and pools, the addition of a filter will allow for contaminants large and small to be caught. The filter should be frequently cleaned.

Ponds and small streams can be stocked with fish and aquatic or water-loving plants, **Figure 27-11**. Many landscape plans will detail which plants should be included and how they should be arranged. Pond plants are generally potted in a container and placed within the pond. Some species that will thrive in this environment include cannas, sedges, taro plants, cattails, calla lilies, Siberian and Japanese irises, horsetails, umbrella grass, and water poppies. Many plants can float on the surface of the water, such as American lotuses, water lettuces, water ferns, water hyacinths, and water lilies.

Elena Elisseeva/Shutterstock.com

Figure 27-11. Aquatic plants enhance any water feature.

Thinking Green

Lighting in the Garden

Garden lighting can provide a yard with style for entertaining, relaxing, beauty, security, and safety. Many professionally installed landscapes include lighting features that are wired to a power source. They may be discreet lights on stairs for navigation or more visible focal points, such as an outdoor chandelier over an eating area. Lights serve to highlight special spaces to be viewed in the dusk and dark hours. Advances in technology have created affordable solar lights. Using small solar panels, these lights charge during the daylight hours. In the evening, they exude a soft glow that enhances the landscape.

Grisha Bruev/Shutterstock.com

Planting the Design

After the installation of the hardscape, the site needs to be prepared prior to installing plants. Amending soil with organic materials, such as compost and manure, and tilling it in will provide a medium that permits plants to flourish. If irrigation systems are to be installed, this step is completed as well before any plant material goes into the ground. The irrigation pipe that was previously laid is now extended throughout the garden with driplines and/or sprinkler heads being set as listed in the design plan. Once the irrigation system has been established, it is finally the time to plant.

Plant Material

The landscape plan will detail the quantity and size of the plant material to be ordered. Woody ornamental trees and shrubs are available in several different ways, including in containers, balled and burlapped (B&B), and bare root.

Container Grown

Almost 80% of all nursery crops grown are planted in containers, **Figure 27-12**. Container plants are cultivated with soilless media in a variety of container types. The plants are then grown in greenhouses, tunnels, or on the ground in nursery operations. Trees, shrubs, and herbaceous ornamentals grown as container plants experience little transplant shock.

Winning7799/Shutterstock.com

Figure 27-12. Most landscape plants are grown at nurseries in containers.

Container plants can easily be shipped long distances and are available all year for most climates. Container plants may become pot bound and will need to be pruned. Pruning the roots will help prevent circling.

Balled and Burlapped

Balled and burlapped (B&B) plants are grown in a nursery field and then dug with soil still attached around the roots, **Figure 27-13**. This ball of soil is wrapped with burlap that is secured with twine or a wire basket. B&B stock generally has little transplant shock because the roots are not disturbed and are protected by the soil. The weight of the soil makes B&B trees and shrubs very heavy and expensive to ship. They are often locally grown and sold. Using plants from the local area can be an advantage as the plant material is well conditioned to the local climate. Due to their large size and heavy nature, B&B materials are usually installed with special landscaping equipment.

J. Bicking/Shutterstock.com

Figure 27-13. The balled and burlapped transplant method is excellent for reducing transplant shock in large specimens.

Bare Root

Bare root trees are generally grown in loose, sandy soil. Then they are dug and stored without any soil around their roots. Trees and shrubs can be purchased as bare root specimens and then planted directly into the ground. Bare root trees can have up to 200% more roots than B&B or container trees, depending on the soil and production methods. Bare root trees and shrubs may cost less because fewer labor hours and materials are needed and shipping charges are reduced, as there is not heavy soil to add weight.

Since bare root materials do not have any soil to protect them from drying out, they should be planted very soon after arrival. There is a narrower window for planting, with the range falling in the spring and late fall. Not all specimens are available as bare root nor are they available in large sizes.

"The best time to plant a tree was 20 years ago. The second best time is now."
—Chinese Proverb

Timing of Planting

When to plant the landscape depends on the region's climate. In many of the milder areas across the country, container plants and B&B plants with well-developed root systems can be planted throughout the year. In the southern states, a fall planting may be most preferable to allow a plant to establish itself before the heat of the summer growing season. In the colder climates, planting primarily occurs in the spring season, although there are many exceptions of shrubs and trees that can successfully be planted in the fall. The spring season is best for planting bare root plants and broadleaf evergreens, such as hollies.

Preparing the Hole and Planting

Preparing the hole properly prior to planting is very important. The hole should be two to three times wider than the size of the root system of the shrub or tree, **Figure 27-14**. A hole larger than the plant allows for good root growth, which leads to a well-established plant. As the shovel slides down into the soil, the sides of the hole can become smooth with no pore spaces. Rough the sides of the hole to create open pore spaces to permit root growth. The hole should be dug to a depth that is the same as the root ball of the plant. The plant should not be set deeper to avoid excess moisture and low oxygen levels.

Zigzag Mountain Art/Shutterstock.com

Figure 27-14. To properly plant a tree or shrub, make sure the hole is two to three times larger than and of equal depth to the root ball.

In poorly drained soils, drainage considerations should already have been made in site preparation. If this was not done, drainage tiles can be installed. Rocks or gravel should not be laid in the bottom of the hole to improve drainage unless they are tied to the drain tile. In sandy soils with low water-holding capacity, the addition of organic matter, such as compost, can increase moisture levels for transplanted material. There is little need for amendments in good soils, but the use of compost can be beneficial in poor soils. The planting holes are generally filled with the same soil material that was dug out. Adding too many soil conditioners can create uneven moisture gradients and limit root growth to just the planting hole.

> "A doctor can bury his mistakes but an architect can only advise his clients to plant vines."
> —Frank Lloyd Wright

Wrapping and Staking

Once a tree or shrub is installed, it may need extra support as the roots grow and it becomes established in the landscape. After planting, wrapping and staking permit the plants to weather windy conditions, temperature extremes, and other potential hazards to growth.

Wrapping sensitive tree species can protect them from injury by sunscald or frost cracking. This may also prevent mechanical injury from mowers or trimmers and damage from rodents. Tree wrap is readily available at nursery centers and can be removed once the tree has grown for two to three years, **Figure 27-15**. Some orchard growers use a diluted white paint to cover the southwest side of the tree, but this is much less acceptable in the aesthetics of a landscape. Good design can place trees in locations that minimize southwestern exposure in the winter.

A *Photo by Tim McCabe, USDA Natural Resources Conservation Service*

B *Photo by Paul Fusco, USDA Natural Resources Conservation Service*

Figure 27-15. A—Wrapping may prevent tender woody specimens from sunscald, mechanical injury, and animals. B—Tubing may also be used to protect young saplings.

Newly planted trees may need additional support through staking. Staking can provide a young tree support, anchorage, and protection on windy sites. Unstaked trees will grow faster and stronger than staked trees. Stakes should remain for only one to two years until the tree is strong enough to stand alone. The slings that are wrapped around the tree trunks and branches can girdle the tree if not eventually removed. Large transplanted trees, the kind that a tree-spade moves, may require guys. *Guys* are heavy wires or ropes attached to the tree and anchored to the ground to add stability. This is especially true of shallow-rooted and top-heavy trees, such as magnolias.

Landscape Maintenance

The care of a newly installed landscape is critical to long-term growth and development of the plant material and the integrity of the hardscape, **Figure 27-16**. Landscapes require an investment in time and money. They may be maintained by a homeowner or through the contracted services of a professional landscape company. Often, landscape businesses, known as design–build firms, offer everything from design to maintenance options. Some landscape businesses offer only maintenance. The horticulture industry is not a "one size fits all" system of operations. There are countless green industry businesses that can meet the objectives of the customer. Unique maintenance plans can be tailored to sustain landscapes for each customer.

Vadim Ratnikov/Shutterstock.com

Figure 27-16. A great deal of time, labor, and money go into new landscaping. Properly installed hardscapes will endure for long periods of time if they are maintained regularly.

For plants to properly mature and thrive, a number of essential maintenance tasks should be frequently considered, such as watering, fertilizing, pruning, edging, and mulching.

Watering

As newly planted trees and shrubs are becoming established, growers must make sure the plant's needs for water are met. Watering is critical to long-term plant health as plants mature, especially during hot, dry conditions.

Newly Planted Trees and Shrubs

Prior to planting, trees and shrubs should be thoroughly moistened. If a bare root, B&B, or container plant arrives dry, it should be watered and only planted once rehydrated. Once planted, water thoroughly and repeat the following day. This watering will remove large air pockets and allow the soil to settle. To promote optimal growth, new trees and shrubs should be watered frequently, at least once to twice a week (if there is no natural rainfall) for the first month.

Thinking Green

Efficient Watering

Standing around with a hose all day watering trees and shrubs can be a chore. Homemade and commercial watering solutions make it easy to deliver proper care to your plants and still allow you to enjoy other gardening tasks that may await. Most nurseries carry bags that have been designed to hold 3 to 5 gallons of water and release it slowly over time. Alternatively, a home gardener may take a bucket and drill holes (1/8" [3.2 mm] diameter) on the side close to the bottom. Fill the bucket with water, and it will slowly trickle out onto the plant material.

bikeriderlondon/Shutterstock.com

Following the first month, reduce watering to a weekly soaking of about 10 gallons of water for trees and large shrubs. Larger trees will need more water; small shrubs will need less. In general, the amount of water should be enough to thoroughly saturate the root ball.

Established Trees and Shrubs

To maintain optimal growth of established trees and shrubs, the soil should be soaked to a depth of 12" (30.4 cm) underneath the canopy and beyond the dripline. This should occur every three to four weeks if there is no significant precipitation. In many landscape beds, irrigation lines may be installed and programmed to deliver water when needed. For a home gardener, soaker hoses can be laid and left on for a couple of hours until the soil is sufficiently moistened.

Fertilizing

Fertilization is an important component of nurturing healthy and vibrant plant growth and development. Fertilization supplements naturally occurring essential mineral elements in the soil to maintain an optimum supply for plant growth. A soil test provides critical information to determine most effectively what the nutrient needs are for the plants. Growers should look for visual symptoms, such as chlorotic leaves, reduced leaf size and retention, early fall color, and leaf drop as well as diminished plant growth and vigor, that indicate nutrient deficiencies, **Figure 27-17**.

Fertilizer Rates

Nutrient needs vary by plant species, time of year, soil type, and soil pH. Fertilizer use rates vary with the age of plants. Younger trees and shrubs generally require higher rates of nitrogen than mature plants. For best results, research should be conducted into specific fertilizer requirements for different kinds of plants in the landscape.

Leonora Enking

Figure 27-17. Nutrient-deficient trees and shrubs may show chlorosis or yellowing of leaves.

In general, 1–6 lb of actual nitrogen per 1000ft^2 per year is needed to maintain the health of woody plants in most landscape settings. Evergreen shrubs and trees need less (1–3 lb). Deciduous trees and shrubs usually require more (3–6 lb). To reduce the risk of fertilizer injury, this total amount should be divided into two or more portions and used in two or more applications throughout the growing season. Soil test recommendations will have the best suggestions. Generally, woody plants respond to fertilizers with ratios of 4:1:2, 3:1:2, 4:1:1, or 3:1:1.

Fertilizer Timing

Applications twice a year are suggested for optimal landscape maintenance. Woody plants can uptake nutrients as long as the temperature is above 40°F. The optimal time to fertilize is in the fall and the spring. The spring application should be prior to active growth, and the fall application should be a month prior to a killing frost. Late summer applications can stimulate undesired growth and delay acclimation to winter conditions.

Methods of Application

The application of fertilizer should be spread evenly over the entire root zone, which extends two to three times the span of the canopy. The fertilizer should be scattered on top of the soil or mulched and then watered lightly. Most granular fertilizers will move through the mulch and soil, and there is little need to work it in. Many commercial fertilizers come in a container that allows for this method of application.

Not all fertilizer is spread in granular form. In mature trees, holds are made (with special tools called a drill or punch bar) into the soil around the tree, and fertilizer is placed in the hole. Most of a tree's feeder roots (where active nutrient uptake occurs) are near the surface. Fertilizer can also be applied with a liquid soil injection, fertilizer stakes, tree injection, or less commonly a foliar spray.

> **Safety Note**
>
> **Fertilizer Safety**
>
> Fertilizers should be handled carefully using gloves and some sort of mechanism to shake it evenly. Sweep up any fertilizer that lands on driveways or walks to minimize contact by others. Avoid washing fertilizer into ditches or storm drains.

Pruning

It is necessary to properly prune trees and shrubs. Without pruning, trees and shrubs can become overgrown, grow weak, and lose vigor. Pruning provides a host of benefits for trees and shrubs, including:

- Maintaining or reducing plant size.
- Removing unwanted growth.
- Removing weakened, diseased, dead, or broken branches.
- Motivating growth of flowers and fruit.
- Revitalizing older plants.
- Preventing damage to property from falling branches.
- Forming plants into shapes or desired growth.

Timing for Pruning

The plant's flowering, fruiting, or growth habits determine the time to prune. For pruning purposes, flowering woody trees and shrubs are in one of three categories.

The first category includes trees and shrubs that bear flowers in the spring and should be pruned immediately after they flower. The flower buds on these species develop from the previous season's growth. If pruning were to occur prior to flowering, the flower buds would be removed, and there would be no flowers. Some examples include dogwood, redbud, flowering quince, lilac, viburnum, and magnolia, **Figure 27-18**.

Some trees and shrubs flower during the summer months and will form flower buds the following spring. Prune these plants before new growth starts, in winter or early spring. A few of these plants are beautyberry, hydrangea, stewartia, butterfly bush, and sumac.

Marie C Fields/Shutterstock.com

Figure 27-18. Redbuds should be pruned after their spring flowering.

Other plants frequently form flower buds and may be lightly pruned both before and after flowering. This can expand flower and fruit production, sometimes even yielding a second bloom during the year. Mahonia, cotoneaster, and weigela are members of this category.

Types of Pruning Cuts

Three types of cuts are used in the maintenance of woody ornamentals. The primary cuts used are pinching, thinning, and heading back. Pinching is usually done to remove the growing tip of a shrub to manage plant size. Pinching may also be done to pine candles to thicken growth. Thinning removes branches back to a main trunk and allows for greater light penetration, **Figure 27-19**.

Goodheart-Willcox Publisher

Figure 27-19. Thinning cuts open the interior canopy of a tree or shrub and allow more light penetration.

Copyright Goodheart-Willcox Co., Inc.

Barbara Fair, NCSU Landscape Extension Specialist

Figure 27-20. The three-cut method should be used on larger diameter branches to avoid breakage and irreparable wounds.

Thinning cuts remove overcrowding and enhance the plant's structure. Heading back is a cut that shortens branches to a good bud or lateral branch. For example, the flush of growth that yew shrubs exhibit can be headed back to keep the prominent form of the established specimen.

A three-cut method can be used on large trees that have branch diameters of more than an inch. The first cut starts about 6″–12″ from the branch union or collar (depending on the branch diameter), with a cut going underneath one-third to halfway through the branch. This is followed by a second cut from on top of the branch until it meets the undercut, **Figure 27-20**. A final, third cut is made at the branch union. This method prevents a large branch from breaking off the trunk and creating a significant injury that may not heal.

Renewal pruning is the process of removing one-third of a plant's old mature stems per season from the ground level. Renewal pruning can help invigorate old shrubs with new growth. It can also be used on some shrubs that may show symptoms of dieback from freezing winter temperatures. Young vigorous branches are selected, including water sprouts, to develop into strong new branches. This occurs over the course of three to five years until a new satisfactory form emerges.

Some shrubs and a few trees are suited to be trained into hedges or topiaries. Hedges should be pruned to have a wide base and narrower top. This shape allows for optimal sunlight penetration and avoids straggly, poor growth. *Topiary* is an ornamental form in which shrubs are sheared into interesting shapes, such as animals, **Figure 27-21**. Topiary is most typically done on boxwood, juniper, yew, and privet. A topiary starts by crafting a basic form through strategic pruning. Early pruning is followed by shearing of leaves to eventually develop a recognizable living sculpture.

Edging

The edge of a landscape defines the space between two areas. Most commonly, an edge may be between a landscape or garden bed and a lawn. Edging materials, such as plastic, metal, bricks, or some other artful material, can create this border. In large landscape areas, a small motorized edging machine is used. It carves into turf and the ground. As the operator walks the line detailed in the design, the edges of the bed are made.

Corner Question

How have hedges been used in entertainment?

Konstantin Kuznetsov/Shutterstock.com

Figure 27-21. Topiaries can be delightful and whimsical additions to any landscape and require significant attention to proper pruning techniques.

The process of edging the line between the spaces must be done fairly frequent during the active growing season. Edging can be done by hand by using a shovel and digging into the soil in a line that follows the edge. The unwanted plant growth, such as turfgrass or weeds, is removed to create a clean edge. A hand tool (also called an edger) can be used as well. It has a rounded blade with spikes that sinks into the soil and pulls up a soil clod. The clods are crumbled back into the earth. Many landscape companies and gardeners will use a trimmer to maintain a crisp and clean edge.

Mulching

Mulching is important for healthy landscape plants. Mulch provides numerous benefits to the landscape including:
- Retains moisture.
- Improves the soil structure (with organic mulches).
- Prevents soil crusting.
- Lessens soil compaction.
- Increases aesthetic appeal.
- Reduces weed growth.
- Moderates soil temperatures to be cooler in the summer and warmer in the winter.
- Minimizes soil splashing to reduce incidences of soilborne disease.
- Limits soil erosion.

As discussed in previous chapters, there are two basic types of mulches: organic and inorganic. Organic mulches are made from living and once-living materials, such as decomposed leaves, shredded bark, pine bark, wood chips, pine needles, grass clippings, newspapers, and straw.

Organic mulches will decompose over time. They may need to be replaced after one growing season or after a few years, depending on the material. Inorganic mulches include materials such as gravel, pebbles, black plastic, and landscape fabrics. While plastic and fabrics do not decompose, they do break down over time and need to be replaced.

Landscaping Business

The landscaping industry is a service industry. A landscaping business needs clients who value the services offered by the company. Like any successful business, a landscaping firm needs to assess the opportunities and challenges prior to starting. A landscaper may be excellent at horticulture, but she or he also needs to understand how to operate a business.

Establishing a Business

Starting a business begins with determining whether there is a need for the products and services the business desires to sell. If several similar businesses operate in the area, another business selling what others are already offering may not find enough customers to be successful. A business needs to identify the products and services it will provide, who the customers will be, and what competition exists. A landscape firm may decide to specialize only in hardscaping after determining that other area businesses do not offer these services, **Figure 27-22**. Other companies may offer an entire range of design, construction, installation, and maintenance services because there are few other companies in their region.

JPL Designs/Shutterstock.com

Figure 27-22. A company may specialize in just one aspect of landscaping, such as hardscaping.

Product and Services

Landscape companies can offer customers a wide array of services and products. Determining the right products to offer is an important consideration for a company. Some companies may find success by offering landscape design and installation of the design, including hardscaping, irrigation, managing subcontractors, and maintenance. Maintenance may even include snow removal and lawn care services. Another business may choose to be solely a landscape contracting firm, bidding on design projects, and managing the installation, **Figure 27-23**. Each business has to weigh factors of competition, available customers, cost of products and services, and desired growth of the business over time.

Delpixel/Shutterstock.com

Figure 27-23. Landscape contracting companies may focus on just installation of designs.

Identifying Customers

A landscape company needs to identify potential clients. Homeowners, places of worship, educational institutions, commercial property owners, other businesses, city parks, and recreational areas may all be possible customers. Satisfied homeowners may be a good source for word-of-mouth advertising. Homeowner or condominium associations may contract for design and build services for common areas and individual units. Nearly all commercial businesses that are not in the business of landscape will hire companies to install a landscape and to maintain the landscape and the lawn. Customer service should be a priority for any business and especially for a service-oriented company. Communicating clearly and meeting customer needs helps attract new customers and retain current ones.

Assessing Competition

An important part of creating a strategic business plan for a new company is assessing the competition. A highly competitive environment may leave little room for a new business to be established. An alternative may be to purchase an existing company and restructure it to fit the needs of the new owners. Businesses may intentionally offer products and services that will set them apart from the competition. A focus on green and sustainable landscape design and maintenance could be a key factor for a customer selecting one firm over another. Another design and construction company may feature landscapes that incorporate native plant material or edible plants. Finding and highlighting a competitive advantage can be essential for long-term stability and success.

> "Whatever the mind of man can conceive and believe, it can achieve."
> —Napoleon Hill

Evaluating Risks

Every business field has inherent risks that must be considered. Business risks include uncertainty in profits and the events or circumstances that could cause a business to fail or lose money. These risks may be internal to the organization or external. Examples of an internal risk may be poor organizational structure or management. Many landscaping companies hire Spanish-speaking employees. If the company has no manager who can communicate job needs to employees effectively, it could result in a loss of time, money, and additional contracts. Additionally, a landscape company needs contracts with clients to stay in business. The company also needs employees to fulfill the contractual obligations. Small businesses pose a risk of not having enough jobs to support the employees that have been hired or not having the needed employees to complete jobs.

External risks may come from economic factors, political factors, and even environmental factors. Landscaping is a service industry. It may not be perceived as critical to everyday living, especially during an economic downturn. Homeowners may be concerned about being able to make a mortgage payment, which leaves little room to consider a landscape design installation. Government regulations may pose a risk. In times of drought, political leaders may find little value in allowing irrigation to water lawns when water is needed to produce food crops or drive industry. Weighing both internal and external risks should influence a thoughtful business plan that considers and addresses these factors to ensure future success.

Careers in Landscape Installation and Maintenance

Many rewarding careers are associated with the landscape industry. Two of these careers are landscape contractor and landscape manager or groundskeeper.

Landscape Contractor

A landscape contractor typically is a supervisory position. A person in this position organizes and directs the activities of crew members engaged in landscaping activities. These activities may include planting and maintaining ornamental trees, shrubs, flowers, and lawn, and applying fertilizers and pesticides. A landscape contractor may also coordinate activities of employees installing hardscape features. These activities may include installing retaining walls, pathways, and patios, and performing other activities in translating the landscape design. Responsibilities may involve communication with clients, finding and sharing price ranges of materials, and preparing estimates based on labor, material, and machine costs. Landscape contractors should have a two- or four-year degree in a field related to horticulture, preferably landscape construction, or equivalent experience. The ability to speak multiple languages is very desirable.

Landscape Manager/Groundskeeper

After a landscape is installed, it must be maintained to provide the highest degree of satisfaction for clients. The management of landscapes is a rapidly expanding part of the green industry and offers several job opportunities. Increasing focus on water conservation and environmentally sustainable landscapes is currently leading to expanding opportunities. Landscape managers typically perform a variety of tasks. These tasks may include laying sod, mowing, trimming, planting, watering, fertilizing, digging, raking, installing sprinklers, and installing and maintaining masonry projects. This position is available to high school graduates with experience as well as graduates with an associate's degree in landscape management or contracting.

CHAPTER 27 Review and Assessment

Chapter Summary

- Landscape design plans show the placement of individual plants, groupings of plants, and hardscape features. The plan contains a title block, north arrow, legend, specifications, and drawing scale.
- Prior to planting, the landscape site must be prepared by removing unwanted material, grading if needed, laying drainage lines, and preparing for irrigation.
- Hardscape installation precedes planting. Each hardscape element needs a foundation to ensure durability over time. Hardscape elements may include patios, walkways, retaining walls, gazebos, and water features.
- A number of hardscape materials can be used that add to the design and allow ease of construction. Examples include brick, cement block, poured cement, stone, rock, wood, gravel, and pebbles.
- Plant material may come in containers, balled and burlapped, or as bare root material. Landscape plants are typically planted in the fall or spring and should be placed in a hole that is at least two to three times wider than the root ball.
- After planting, wrapping and staking permit the plants to weather windy conditions, temperature extremes, and other potential hazards to growth.
- Landscape maintenance is critical for a healthy landscape that lasts for decades. Key maintenance tasks include watering, fertilizing, pruning, edging, and mulching.
- Watering is critical to long-term plant health, especially during hot, dry conditions. Fertilization supplements naturally occurring essential mineral elements in the soil to maintain an optimum supply for plant growth.
- Without pruning, trees and shrubs can become overgrown, grow weak, and lose vigor. The timing of pruning is generally based on the flowering, fruiting, or growth habits of a plant.
- Every landscaping company should begin with a strategic business plan that includes a mission statement and a vision for the company. Selecting products and services to offer, identifying customers and markets, understanding the competition, and evaluating risks are essential tasks for the business.
- Many rewarding careers are associated with the landscape industry. Two of these careers are landscape contractor and landscape manager or groundskeeper.

Words to Know

Match the key terms from the chapter to the correct definition.

A. basket weave
B. drawing scale
C. Flemish bond
D. guys
E. header
F. herringbone
G. legend
H. north arrow
I. renewal pruning
J. running bond
K. softscape
L. specifications
M. stack bond
N. stretcher
O. title block
P. topiary

1. A brick design with bricks laid along the wall showing their sides and those laid across the wall showing their ends.
2. An interlocking pattern of bricks set at 45° or 90° angles.
3. A pattern in which bricks are laid end to end and staggered between rows.
4. A simple design in which bricks are placed side by side without staggering.
5. An ornamental form in which shrubs are sheared into interesting shapes and imaginative characters, such as animals.
6. A design pattern in which two bricks are laid horizontally next to two vertically laid bricks to look as though they are weaving in and out of each other.
7. Heavy wires or ropes that are attached to an object (such as a tree) and anchored in the ground to provide stability.
8. An element that describes symbols or abbreviations used in a document, such as a design plan.
9. A component of a landscape plan that gives information about the client, landscape firm, date of preparation, and possibly additional elements, such as the drawing scale.
10. The plant materials in a landscape.
11. An orientation point on a landscape design that shows the direction north.
12. The relationship between the distances on a plan to actual distances on a site.
13. The process of removing one-third of a plant's old mature stems per season from the ground level in order to stimulate new growth.
14. Guidelines that include details for a project, such as plant selection, site preparation, and soil preparation.
15. A brick that is laid in a design to display its long side.
16. A brick that is laid in a design with its end or short side exposed on the design.

Know and Understand

Answer the following questions using the information provided in this chapter.

1. What five elements does a landscape plan generally contain?
2. What is the objective of a digital rendering of a landscape plan?
3. What are seven types of specifications that may be included in a drawing plan?
4. What is the purpose of a drawing scale on a design plan?
5. What steps may be needed to prepare the site prior to hardscaping and planting?
6. How do hardscapes enhance the landscape?
7. What are some structures that may be part of a hardscape?
8. What are some patterns that are typically used for brick patios and which pattern is the strongest?
9. Describe the materials used and the method for creating a base for a retaining wall.
10. List the three most common ways landscape plants can be purchased.
11. When should container plants and balled and burlapped (B&B) plants be planted in milder areas of the United States?
12. Describe the appropriate size and other considerations for a hole for planting a shrub.
13. What is a design-build landscape firm?
14. What are some maintenance tasks that are essential for a landscape?
15. How often do newly planted trees and shrubs need to be watered?
16. What are some visual symptoms in plants growers should look for that may indicate nutrient deficiencies?
17. In general, how much nitrogen should be applied to maintain growth for woody plants, evergreen shrubs, and deciduous trees and shrubs?
18. Describe the safe handling of fertilizers.
19. What are some benefits that come from pruning landscape plants?
20. When should summer-flowering trees and shrubs be pruned?
21. Describe three types of pruning cuts used for maintenance of woody ornamentals.
22. What benefits do mulches provide to the landscape?
23. What are the typical job duties of a landscape manager/groundskeeper and what training or education is needed for this job?

Thinking Critically

1. A local businessperson would like to increase entertaining at her home for prospective customers. She tells you that her home is "modernistic" with stark white walls, chrome-and-glass furniture, and abstract paintings. She would like you to install a patio that complements her house. What materials would you use? What patterns might be suitable? Plan a design that may be appropriate for her style.

2. A client would like you to install a landscape design at his house. After testing the soil, you realize that it is poorly drained and will require a significant drainage system. This is something that the designer overlooked. How will you approach your client and relay this information? How could you present it in a way that still allows you to keep your contract for the job?

STEM and Academic Activities

1. **Engineering.** Retaining walls require significant engineering to hold sloped land with volumes of soil and to maintain aesthetic appeal. Obtain a set of specifications for a retaining wall and review the guidelines for its construction. What materials and equipment would you need to implement this plan?
2. **Science.** Investigate the science behind mulching. Write a three-page report explaining the benefits of using mulch. Describe materials that make good mulches and explain why.
3. **Science.** Examine the physiology behind pruning. How do plants respond to different pruning cuts? Find a shrub on your school grounds, backyard, or community and try out a few of these pruning cuts (with supervision). What plant growth responses do you observe?
4. **Math.** A client has asked you to create a small patio that is 10′ × 10′ (3 m × 3 m) using 4″ × 8″ bricks laid out in a basket weave pattern. Draw the design on graph paper and determine how many bricks you would need to complete the job.
5. **Math.** Acquire a landscape design plan that details the plants to be installed. Determine how many plants of each species are needed and the sizes needed for each. Research online and find prices according to your list. How much would the plants cost?

Communicating about Horticulture

1. **Speaking.** With a peer, role-play the following situation: A client cannot have the brickway pattern that she wants due to restrictions by the historical society. The designer must explain why she cannot have the pattern and present her with an alternative plan. One student plays the role of the designer and the other acts as the client. Discuss reasons why the historical society would place restrictions on walkway designs. As the designer explains the restrictions and alternatives, the client should ask questions if the explanation is unclear. Switch roles and repeat the activity.
2. **Reading and Speaking.** Some organizations or associations provide mentoring services for small businesses as a membership benefit. Form a small group with two or three of your peers and collect informational materials from landscape contracting and design associations that provide these services. Analyze the data in these materials based on the knowledge gained from this chapter. Make inferences about the services available and recommend the best ones to the class.

3. **Reading and Writing.** The ability to read and interpret information is an important workplace skill. You work for a designer who is considering pitching a proposal to the local city government to design and install a park. The designer wants you to evaluate and interpret some research on past designers and the contractors that were used. Locate three reliable resources for the most current information on designs for a city park. Read and interpret the information. Write a report summarizing your findings in an organized manner.

SAE Opportunities

1. **Exploratory.** Job shadow a landscape contractor. What are the daily responsibilities for this job? What do you like or not like about this position? What education and experience are required to have this position?
2. **Exploratory.** Inventory the landscape ornamentals on your school campus. What specimens need pruning? What could you do to facilitate their care?
3. **Experimental.** Research the ways that a pond could be installed in the landscape. What materials are needed? Are they easily available? How effective are they? What are the benefits and costs of using these materials? Secure permission and funding and create a small pond at your school or home.
4. **Exploratory.** Visit a local landscape design and building firm. Ask the owners to share their business management practices. Go with an employee to meet a client and observe their interactions. Write a short report to share your findings.
5. **Exploratory.** Create specifications for installing a large balled and burlapped tree. Include diagrams and useful information for anyone who may be doing the planting.

Iakov Filimonov/Shutterstock.com

CHAPTER 28: Turfgrass Management

Chapter Outcomes
After studying this chapter, you will be able to:
- Describe the turfgrass industry and turf applications.
- Describe turf morphology and types of turf.
- Discuss methods of lawn establishment.
- Explain how turf is maintained.
- List methods of turfgrass renovation.
- List careers related to turfgrass management.

Words to Know

auricle	rhizomatous	tiller
blade	rotary mower	transition zone
collar	seed blend	turf
herbicide	seed spreader	turfgrass breeding
lawn	seedhead	turfgrass industry
ligule	slit seeder	utility turf
moisture stress	sports turf	vernation
overseeding	spot seeding	verticutting
plugging device	stoloniferous	volatilization
reel mower	thatch	

Before You Read
Arrange a study session to read the chapter aloud with a classmate. Take turns reading each section. Stop at the end of each section to discuss what you think its main points are. Take notes of your study session to share with the class.

While studying this chapter, look for the activity icon to:

- **Practice** vocabulary terms with Words to Know activities.
- **Expand** learning with identification activities.
- **Reinforce** what you learn by completing Know and Understand questions.

www.g-wlearning.com/agriculture

FloridaStock/Shutterstock.com; Benoit Daoust/Shutterstock.com; topseller/Shutterstock.com; Matt Browne/Shutterstock.com; Svetlana Turchenick/Shutterstock.com

Owners of English estates began to establish *lawns* (an expanse of turfgrass used for recreation or beauty) during the seventeenth and eighteenth century. Only the very wealthy could afford to install and maintain lawns, **Figure 28-1**. American settlers' lawns, in contrast, consisted of dirt, cottage gardens, or some crops growing outside their doorways. Wealthy travelers to Europe saw the beautiful lawns in those countries and wanted to replicate them in the United States.

Growing turfgrass in the United States at the turn of the twentieth century was difficult. Grass seed was not commercially available as it is today. At that time, lawns consisted of native grasses that were untidy and weed-like. These were the only grasses that could thrive in the soil and environmental conditions present in the United States.

> "There is not a sprig of grass that shoots uninteresting to me."
> —Thomas Jefferson

In 1915, the United States Department of Agriculture (USDA) and the United States Golf Association combined efforts to find a grass variety that would thrive here. Researchers tested bermuda grass from Africa, bluegrass from Europe, and a mixture of fescues and bentgrass. They were looking for an attractive, durable, and easy-to-grow grass suited for the United States. Almost two decades later, the USDA found suitable grasses for lawn cultivation. Many other lawn-growing challenges emerged, including pest problems, irrigation, fertilization, and mowing.

The home lawn revolution began in the middle of the twentieth century when many homeowners gained easy access to lawn care equipment. Garden clubs pushed the home lawn movement by promoting lawn care as a form of gardening. Clubs sponsored competitions to convince homeowners that a green, weed-free lawn was the ideal environment to surround a home or business.

pisaphotography/Shutterstock.com

Figure 28-1. London's Hyde Park opened to the public in 1637. It is well known for its beautiful lawns, trees, and plantings.

History Connection

Sheep on the White House Lawn

Before the lawn mower, livestock or scythes were used to cut grass. A flock of sheep grazed on the White House lawn during World War I to show Americans a way to trim their lawns, feed their flock, and provide wool (which was scarce) while soldiers fought overseas.

Everett Historical/Shutterstock.com

The lawn, once a symbol of wealth, is now more easily attained by average Americans. Multimillion-dollar turfgrass companies manage lawns across the country. The competitive nature of some homeowners makes the home lawn a wonderful canvas for contests of many forms. Families use this space for countless activities from throwing a ball to roasting hot dogs.

Turfgrass Industry

Turf is grass and the surface layer of soil held together by the grass roots. This agricultural crop is cultivated the same as vegetables and flowers. The total area of the United States covered in turf is more than 50,000 square miles, making this the largest crop grown. Turf is beneficial to the environment and to people, **Figure 28-2**.

Turf Industry

The *turfgrass industry* consists of businesses that grow seeds, sods, and grasses for recreational areas, homes, and commercial, utility, and sports uses. The turfgrass industry contributes nearly $40 billion annually to the US economy. Approximately 65% of total turf areas in the United States are lawn. The other 35% of turf area is utility turf, golf courses, and athletic fields. *Utility turf* is grass that is used for spaces such as roads, airports, and parks. About 50 million American homes include a lawn in their landscape. Nearly 50 million more lawns are located at schools, churches, parks, cemeteries, businesses, and hospitals.

A huge component of the turfgrass industry is turf for athletic uses. *Sports turf* is grass that is used for recreation and competitive sport purposes. Golf course turf is one type of sports turf. More than 50% of the total golf courses in the world are in the United States. An estimated 20,000 sports fields in the United States are grown with turf.

Fotokostic/Shutterstock.com

Figure 28-2. Turfgrass is enjoyed all over the world, especially for sports-related purposes.

Corner Question

Which early US presidents had home lawns?

Corner Question

What size lawn is needed to release enough oxygen to sustain a family of four people?

The turfgrass industry has nearly 600,000 full- and part-time employees. Nearly 100 universities and colleges offer turfgrass programs that provide formal training in science and management. Numerous industry professional organizations, publications, and other resources on a national and international level deliver research information to a diverse audience of turf enthusiasts and specialists.

The turfgrass industry is a top agricultural business in many states. Continued attention to sustainable practices and future research ensures that there is no slowing of this industry's growth.

Benefits

Turf provides many benefits to people and the environment. Turf does the following:

- Reduces noise pollution.
- Decreases allergens related to pollen.
- Stabilizes dust particles and reduces mud.
- Provides safety in vehicle operations and contact sports.
- Lowers fire hazards.
- Moderates temperature fluctuations and reduces the urban heat island effect via *evapotranspiration*.
- Functions in *carbon sequestration* (a process in which carbon dioxide is removed from the atmosphere and held in solid or liquid form).
- Improves soil through healthy root system and microorganism activity.
- Prevents erosion and runoff while aiding in flood control.
- Boosts ground water recharge, helping to absorb up to six times more rainwater than other crops.

The environmental impacts of turfgrass management can be destructive, however, if growers use poor growing practices. Turf sustainability is discussed in the next section.

Turf Sustainability

Some environmental enthusiasts believe that turf cultivation should be avoided because of issues related to irrigation, pesticides, and fertilizer. Conservationists may think that this crop is the cause of too many environmental problems. However, poor management practices result in negative outcomes for the environment.

Turf growers should use sustainable practices just as horticultural growers do. Excessive use of water, fertilizer, and pesticides is harmful to the environment regardless of the crop being produced. Because there is so much turf that blankets the United States and so many people who are trying to manage this crop, using sustainable practices is especially important. Business owners, landscapers, and homeowners must consider sustainable practices to ensure that the environmental benefits of turf outweigh any negative impacts associated with mismanagement.

Did You Know?

Combining the front lawns of eight average-sized homes provides the same cooling effect as about 70 tons of air conditioning.

Turf Applications

The four main applications or uses of turf are lawns, golf courses, utility turf, and sports turf. In addition, sod and seed production are part of this industry. Several industries that support or are associated with the turf industry include pesticide, fertilizer, irrigation, equipment, and information technology companies.

Corner Question

Which mowing pattern is better: spiral or rows?

Lawns

Many Americans value a beautiful lawn whether it is in a park, business, garden, or yard. An individual *blade* (the flat, leaf part) of grass on a lawn lives an average of 40 days. Lawns provide opportunities to connect with nature and breathe in fresh air. They also provide a clean space for a picnic. Can you imagine your favorite park without green grass and only pavement or concrete?

Golf Courses

The type of golf played today can be traced back to Scotland in the fifteenth century. Today, this sport is popular around the globe and is played by many types of people. Golf can be played on a 9-hole or 18-hole course with varying levels of difficulty. Golf courses often change the locations of the holes to provide variety for players. A golf course is composed of many different components and each is managed to impact the game in some manner.

Career Connection

Andy Smith
Erosion Control, Eco Turf

Andy Smith, a former high-school FFA member and tobacco farmer, did not know that he would end up owning an erosion control company in Raleigh, North Carolina. After college, Smith worked for a hotel and then sold real estate. After Hurricane Fran, Smith went to work on a golf course repairing storm damage. He found that he loved working on the golf course. He quickly merged his agricultural roots, understanding of land development, and new love for turf management into a job with a company that installed golf courses around the country.

In 2007, golf course installation nearly stopped. Smith began working for an erosion control company and used his knowledge, experience, and network contacts to sell erosion control materials to golf courses around the country. He is now the owner of Eco Turf, a company that sells erosion control materials.

Jodi Riedel/Goodheart-Willcox Publisher

Smith's tenacity for hard work, his ability to learn quickly, and his enthusiasm for working with other people helped him get where he is today. He urges students interested in the turfgrass industry to further their education in a two- or four-year program and to participate in internships in the industry. "You have to love what you do," says Andy. "So many hours of the day are spent working; it is important to enjoy your job." Andy Smith did not know where his career path would lead him, but he has enjoyed the journey so far.

Corner Question

Which professional football stadiums have real turfgrass?

Sports Turf

Sports turf provides a recreational space for physical activity while protecting the soil. Sports turf provides space to play soccer, lacrosse, football, and baseball, or just to run or walk and enjoy being outdoors. These activities can help increase physical and mental health. Turfgrass areas help decrease sports injuries (compared to playing on hard surfaces) and increase community participation and pride.

Utility Turf

Utility turf serves a purpose in the landscape. Utility turf is found in places where soil must be conserved or stay in place. Utility turf prevents erosion and helps to increase water penetration in an area. Places where utility turf is used include airport runways, closed landfills, and alongside roads, **Figure 28-3**.

ziher/Shutterstock.com

Figure 28-3. This airstrip is covered in utility turf. Can you explain why the soil on this airstrip needs to stay in place?

Sod Production

Sod producers grow a solid platform of turfgrass. The harvested grass along with roots and a thin layer of topsoil is known as sod.

Career Connection

Todd Lawrence
Golf Course Superintendent

Todd Lawrence works in the turfgrass industry as the golf course superintendent for a tournament course for professional golfers. Lawrence and his staff use best management practices combined with a National Audubon certification. The Audubon golf course certification program helps golf courses protect the environment while preserving the natural heritage of the game.

Lawrence did not begin his career track with the goal of golf course management in mind. While managing test fields on a research farm as an intern, a researcher encouraged Lawrence to consider a career in the turfgrass industry. Lawrence decided to take a turfgrass course while studying agronomy at North Carolina State University. He began working on a golf course while in college and realized he could not only

Jodi Riedel/Goodheart-Willcox Publisher

grow turfgrass, but also enjoy a unique work environment that focused on recreation. Lawrence saw his future in managing golf courses.

Lawrence's educational background, hard work, and ability to work well with others earned him jobs with increased responsibility. He eventually was hired as the superintendent of the newly built tournament course. Lawrence is extremely proud of the course and appreciates his crew's work and their commitment to excellence for both golfers and the surrounding community.

Many sod producers also transport and install the sod. Once harvested, sod is placed onto prepared soil and immediately provides the look of an established lawn, **Figure 28-4**.

Seed Production

Since the Dust Bowl of the 1930s, people in the United States have realized the importance of protecting soil. The most economical and common method of establishing turf is to use turfgrass seed. Growers, mostly located in the Pacific Northwest, grow various cool-season turf varieties through flowering and fruiting. The turf crop is harvested and the seed is then cleaned, analyzed, bagged, and distributed.

Sutichak Yachiangkham/Shutterstock.com

Figure 28-4. Sod installation creates a complete lawn in a short time.

Supporting Industries

Various industries help the turfgrass industry to function more efficiently. Turfgrass growers use integrated pest management (IPM) and sustainable growing practices to cultivate the highest quality turf crop. Growers use pesticides, fertilizers, irrigation, equipment, and information technology to promote plant growth. These industries include key businesses that produce the following:

- Chemicals—fertilizer, insecticide, fungicide, rodenticide, and molluscicide.
- Irrigation—hoses, valves, timers, couplings, heads, and rain sensors.
- Equipment—tractors, mowers, rollers, golf carts, painting devices, aerators, sod cutters, utility vehicles, spray applicators, dethatchers, and seeders.
- Information technology—computer software, global positioning systems (GPS), and bar coding.

Thinking Green

Freshkills Park

Freshkills Park, a 2200-acre site in New York City, was once a landfill. It is now scheduled to become a park. Areas inside the park will be covered in utility turf. The park will be three times the size of Central Park in New York City.

Turfgrass Morphology and Types

Turfgrass belongs to the grass family known as Poaceae. Turfgrass has unique characteristics that distinguish it from other plants. These characteristics include:

- Grass leaves have linear-shaped leaves that are elongated. Leaves have parallel veins.
- Roots are fibrous systems.
- Flowers and *seedheads* (a cluster of seeds within a plant's inflorescence) are not easily visible when the turf is mowed.
- Grass stems are located at the base of the leaves called the crown.

As with all plants, some identifying characteristics help to differentiate one turf from another.

Turf Morphology

Morphology is a branch of biology dealing with the study of the form and structure of organisms and their specific structural features. Grasses have unique morphological structures that include:

- Leaf blade.
- Sheath.
- Vernation.
- Ligule.
- Auricle.
- Collar type.
- Rhizome.
- Stolon.
- Tiller.
- Inflorescence and seedhead.

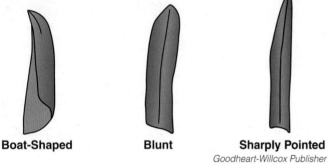

Figure 28-5. Leaf blade tip shapes.

Leaf Blade

The turf's leaf blade is also known as the broad portion of the leaf. Blades can be distinguished based upon shape, size, and color. All leaf blades have parallel veins. However, the width and lengths of mature blades vary. Blade tips can be boat shaped, blunt, or sharply pointed, **Figure 28-5**. Boat-shaped blade tips are cupped and curved upward. Blunt blade tips are rounded and end abruptly. Sharply pointed blade tips are long with a sharp point.

Sheath

The basal portion of the grass that surrounds the stem is known as the sheath. Sheath types can be round and flattened. A round sheath is cylinder-shaped, circular in cross-section, and rolls easily between your fingers. A flattened sheath is compacted, elliptical in cross-section, and tends to flip from side to side rather than roll between your fingers.

Sheath margins (edges) are one of three types. In *closed margins*, an overlapping portion of the sheath cannot be seen. The *open sheath margin* forms a small V-shape. In the *split with overlapping margins*, grass blades are not closed (fused) together, but are overlapping, **Figure 28-6**.

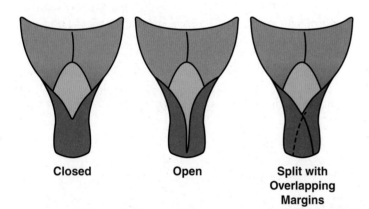

Figure 28-6. Turfgrass leaf sheath margins.

Vernation

The arrangement of the leaves within the bud (youngest leaf) and the surrounding sheath is known as the *vernation*. Vernation usually is described as rolled or folded, but some plants have both features. To determine the type of vernation, place the leaves, surrounded by the sheath, between your thumb and index finger, and roll the plant back and forth. If it moves as a pencil would, and has a bumpy path, it is folded. *Folded vernation* leaves fold lengthwise in a V-shape, with the margins meeting but not overlapping.

If it moves like a straw it is rolled. *Rolled vernation* margins overlap due to curled leaves, **Figure 28-7**.

Collar Type

The *collar* is a band at the backside of a leaf where the leaf blade and sheath meet. Collars are usually a lighter color than the rest of the grass blade. Collars are either divided by the midrib or continuous.

Auricles

A small, claw-like appendage that projects from a grass blade is known as an *auricle*, **Figure 28-8A**. Auricles can be absent, claw-like, or rudimentary. On claw-like auricles pointy appendages edge the grass stem. Rudimentary auricles have partial appendages that do not enclose the grass stem.

Ligules

A thin structure that clasps the top of the leaf sheath (where the blade and sheath join) is the *ligule*. The ligule structure starts from the auricle. It can be seen when the blade is bent backwards, **Figure 28-8B**. Ligules can be absent, hairy, or membranous. Membranous ligules are transparent, flexible, and can vary in height.

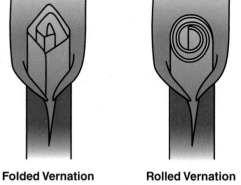

Folded Vernation   **Rolled Vernation**
Goodheart-Willcox Publisher

Figure 28-7. The vernation of turf varies and can be rolled or folded.

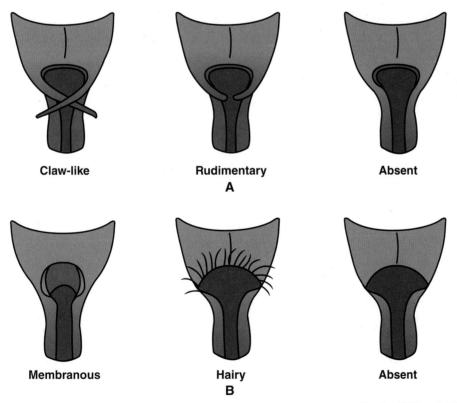

Goodheart-Willcox Publisher

Figure 28-8. A—Auricles are appendages that project from each side of the collar. B—Ligules are located near the back of the collar.

Did You Know?

Some rhizomes are edible. Ginger and turmeric are both rhizomes that make flavorful additions to many dishes.

Rhizomes and Stolons

A rhizome is a modified stem structure that grows horizontally at or near the soil's surface. It produces roots on the bottom and shoots on the top. These structures can begin at the main stem. Rhizomes allow grass plants to spread horizontally.

Stolons are lateral stems at or just below the surface of the soil that produce new plants from buds, or nodes. Stolons allow a grass plant to spread laterally. Stolons often root at the nodes, producing new plants. A turfgrass plant can have both rhizomes and stolons.

Tillers

A *tiller* is a stem that develops from the crown of the parent plant and grows upward within the enclosing leaf sheath of the parent plant, **Figure 28-9**. Multiple tillers can grow from an initial seedling, forming dense tufts of grass.

Inflorescence and Seedheads

The flowering and reproductive parts of a grass plant are known as the inflorescence. The fruiting structure is known as the seedhead. This morphological structure makes identification of turf much easier. However, when turf is mowed, seedheads may be lost for identification purposes. Inflorescence and seedhead shapes include panicle, raceme, and spike, **Figure 28-10**.

- Panicle—flowers are less densely clustered than in raceme or spike types of seedheads. Panicles often have a triangular shape. The seedhead is branched with multiple seeding branches clustered around the stem.
- Raceme—several slender spikes are attached along the end of a stem. Individual spikes can be attached at one point or along the top of the stem in an alternating fashion.
- Spike—an unbranched seedhead is located at the end of a stem.

©iStock.com/design56

Figure 28-9. Tillers are smaller plants that form as an offshoot from the central plant.

Types of Turf

The types of turf that a grower selects for an athletic field, lawn, or golf course depend on how that space will be used. Growth habit, zone of growth, and specific characteristics of the turf all contribute to the final selection of the grass.

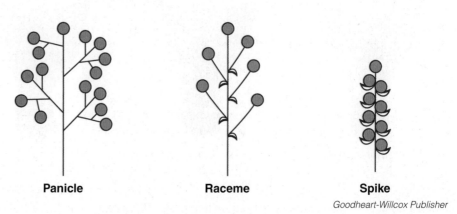

Goodheart-Willcox Publisher

Figure 28-10. Inflorescence and seedhead shapes help to identify types of turfgrass.

Growth Habit

Depending upon the growing conditions, a turf with a specific growth habit can be useful in a specific site. The growth habit of a grass can also help identify turf. Types of turfgrass growth habits are rhizomatous, stoloniferous, or bunch type.

- *Rhizomatous* plants produce rhizomes, or underground lateral stems. Rhizomatous grass quickly recovers from heavy traffic.
- *Stoloniferous* plants produce stolons (aboveground lateral stems).
- Bunch-type grasses produce only tillers and take more time to recover from injury. They are not well suited for athletic fields or heavily trafficked areas.

Corner Question

What is indicated by a blue tag on a seed container?

Zones of Growth

The zone of growth refers to a site's location and the climatic conditions of the location. Temperature, humidity, and available moisture all factor into zones of growth in the United States, **Figure 28-11**. The United States can be divided into three zones: warm season, cool season, and *transition zone* (a growing area that can grow both warm- and cool-season grasses).

Warm-Season Grasses

Warm-season grasses are best suited for the warmer southern regions of the United States. These plants are propagated using the vegetative parts of the plant and are planted in late spring or early summer.

Mirec/Shutterstock.com

Figure 28-11. The United States is divided into three zones: warm season, cool season, and transition.

Corner Question

How many seeds of bermuda grass are in one pound?

The turf goes dormant when the air temperature is below freezing during the winter season, usually after the first frost. These turf types can tolerate very high temperatures. However, they can be damaged by cold temperatures, leading to winterkill. Generally, these grasses have low shade tolerance but good drought tolerance. Examples of warm-season grasses include:

- Bahia grass (*Paspalum notatum*)—well adapted to mild coastal climates, coarse textured, best for low-quality and maintenance turf.
- Bermuda grass (*Cynodon* spp.)—popular warm-season turfgrass, used for all quality levels.
- Centipede grass (*Eremochloa ophiuroides*)—light green, medium-textured, low maintenance, tolerates shade, lacks tolerance for high traffic.
- St. Augustine grass (*Stenotaphrum secundatum*)—coarse-textured grass, tolerant of traffic, shade, and salt, rarely available as seed.
- Zoysia grass (*Zoysia japonica*)—hardy turf, tolerates high heat and humidity, withstands cold temperatures, long dormant period, slow growing, low maintenance.

"A warm smile is the universal language of kindness."
—William Arthur Ward

Cool-Season Grasses

Cool-season turfgrasses thrive in temperatures between 60°F and 80°F (15.6°C and 26.6°C). These grasses are well suited for the northern United States. These grasses do not show signs of dormancy during the winter season but grow very gradually. Cool-season turfgrasses are propagated by seed and planted in the fall. They show signs of stress in heat and humidity. The United States produces the largest quantity of cool-season grass seed and many other areas of the world rely on this production. Examples of cool-season grasses include:

- Kentucky bluegrass (*Poa pratensis*)—rhizomatous, medium- to fine-textured grass, deep green color, goes dormant during summer months, recovers well from most droughts.
- Tall fescue (*Festuca arundiacea*)—bunch-type grass, coarse texture, breeding efforts have improved drought tolerance and pest resistance.
- Perennial ryegrass (*Lolium perenne*)—bunch-type grass, medium texture, tolerates partial shade, excellent wear tolerance.
- Fine fescues (*Festuca* spp.)—shade and drought tolerant, fine texture, often mixed with other cool-season grass seeds.

Transitional Turf

A transition zone is an area where warm- and cool-season grasses can be adapted to survive the climate. Fewer cool- and warm-season grasses thrive in transition zones. Winterkill and summer heat and drought stress are disadvantages associated with this zone.

Cold-tolerant warm-season turfgrasses and heat-tolerant cool-season grasses are well suited for lawns in the transition zone. Zoysia grass and bermuda grass are the two most popular warm-season lawn turfs for the transition zone. Tall fescue is the most popular cool-season lawn turf for the transition zone.

Management strategies to help grasses thrive in this zone include changing fertilization programs, increasing mowing heights, and avoiding shady conditions for warm-season grasses. Cool-season grasses are adapted by reducing nitrogen use, watering deeply and infrequently during mornings, increasing mowing height, and following integrated pest management (IPM) strategies during summer heat. The best turfgrasses for the transition zone include Kentucky bluegrass, bermuda grass, tall fescue, and fine fescue.

Lawn Establishment

To achieve a lush lawn, several steps must be followed. The appropriate turf must be selected, the site prepared, the soil conditioned, the turf planted, irrigation operated, fertilizer applied, and pest management employed.

Turf Selection and Timing

When choosing a site for turfgrass, consider the appropriate type of turf and the time of year to cultivate.

Warm-Season Grasses

Growers have better success establishing warm-season grasses during the warmer months of the year. Most warm-season grass seeds germinate and thrive from spring to midsummer. Warm-season sod will not produce roots unless soil temperatures exceed 55°F (12.78°C) for several weeks. Sprigs or plugs, however, do well in the same conditions.

Cool-Season Grasses

Cool-season grasses do best when installed during the cooler months of the year. Seeding is usually done from August to September in most of the United States, **Figure 28-12**. Later seeding usually fails due to winter injury or winterkill. Sowing seeds in late winter or spring does not give the turf enough time to grow the deep roots needed to help plants survive the heat (and possible drought) of summers. Cool-season sods do best when the growing season is cool but the ground is not frozen.

Site Preparation

Site preparation is important in order to establish quality turf. To properly prepare a site, pay particular attention to details. Follow these steps to prepare a site for turf planting:

1. Remove any trash, debris, or rocks that are larger than a quarter.
2. Grade the site, if possible, to no more than a 15%–20% grade. Avoid planting turf on steep grades or heavily shaded areas. If grading is required, remove and store the topsoil. Replace the topsoil when grading is completed. A 2%–3% slope around buildings will ensure that water drains properly away from structures.

> **Did You Know?**
> Turfgrasses can be cultivated in all US states and territories.

pryzmat/Shutterstock.com

Figure 28-12. Planting seeds is an economical way to establish turfgrass.

3. Install a drainage system for sites with poor drainage.
4. Remove weeds.
5. Water the area and look for areas of uneven settling. Note any areas with standing water. Reshape where needed.
6. Mix one to two cubic yards of compost per 1000 ft² per 1000 ft^2 into the top 8″ (20.3 cm) of soil.
7. Follow the soil preparation steps.

Soil Preparation

Proper soil preparation ensures healthier turfgrass and is worth the investment of time, and resources. Without the results of a soil test, there is no guide to adjusting pH or adding nutrients. Growers can only add nutrients based on a guess. This is not a logical way to establish turf, and is not a good practice for the environment. Growers should always test the soil and amend it according to test results.

Take soil samples from the area where turf will be established. If there is more than one area that will be planted, take soil samples from those areas. Follow proper soil sampling protocol. Based on the results of the soil test:

1. Adjust the pH of the soil using lime or sulfur. Add fertilizer according to the suggestions. Incorporate into the top 8″ (20.3 cm) of topsoil using a disc or rototiller.
2. Rake the site. Soil particles should be no larger than a quarter; pea-size is best. Hand raking is the best method to break up soil pieces and level the soil.
3. Irrigate the soil to allow settling. A well-timed rain shower will also work as irrigation.
4. Roll the soil before using your chosen planting method.

Hands-On Horticulture

Turf Soil Testing Protocol

Contact your local cooperative extension service agent for specifics regarding soil testing in your state, the materials needed to submit an application, and any costs associated with this service. In general, follow these procedures in taking a soil sample for analysis:

1. Gather a clean trowel, a ruler, and a clean plastic bucket.
2. Sample your site in 15 to 20 spots. Avoid any spots that look as if they do not represent the sample as a whole (where there is standing water or no vegetation, for example).
3. Take 1 pint of soil from a depth of about 4″ for lawns (6″ for crops, trees, and shrubs).
4. Combine all the samples together and mix thoroughly.
5. Send approximately one pint of the soil mixture in for soil testing. Often, your cooperative extension service agent will supply a specific bag or box for the sample, along with a sample information sheet.

sharon kingston/Shutterstock.com

Fertilization

A soil test will help determine exactly how much fertilizer to add to a site for turf establishment. General recommendations for fertilizer applications before planting can be used, but they are not nearly as accurate and can lead to cultural problems in turf establishment. If no soil test was done, follow these guidelines:

- Add limestone or sulfur according to the suggested rate per 1000 ft^2.
- Apply a starter fertilizer (higher in phosphorus). The second number of the fertilizer NPK (nitrogen, phosphorus, potassium) analysis denotes the percentage of phosphorus.

Nitrogen promotes green growth and color, phosphorus promotes rooting, and potassium enhances vigor and overall hardiness.

Examples of common starter fertilizer percentages are 5-10-10, 10-20-20, or 18-24-6.

Planting Methods

Determining which method to use to establish a lawn or turf site requires thought and planning. Once you have chosen a turf that is well suited for your site, you can then select a planting method based on the availability of that turf species.

Seeding

Purchase top-quality seed to avoid future problems with turf establishment. Select a seed or *seed blend* (a mixture of various cultivars of grass seed) based on the information on the seed bag. Be sure the grass seed is free of noxious weeds, contains low percentages of other grass seed, and is a certified seed (indicated with a blue tag). The blue certified tag indicates that the seeds meet established industry standards and have been analyzed to ensure seed purity.

To successfully sow grass in a seed bed (a site prepared for starting turf), follow these steps:

1. Determine the amount of seed that is required for the area.
2. Use a seed spreader to apply one-half of the grass seed, working in one direction. A *seed spreader* is a tool that distributes seeds on a prepared area, **Figure 28-13**. The spreader uniformly distributes seeds or other granular materials.
3. Apply the remaining grass seed perpendicular to the first pass.
4. Lightly cover the seed by gently raking or dragging the soil with a mat or chain-link fence panel.
5. Roll the soil to ensure contact between soil particles and the seed.
6. Mulch the seed with straw. Use one bale per 1000 ft^2 for warm-season grasses and one bale per 2000 ft^2 for cool-season grasses. This helps to retain moisture, insulate the seeds, and prevent erosion.
7. Roll the mulch.
8. Irrigate gently. Ensure that the soil remains moist at all times through germination and until maintenance begins.

pryzmat/Shutterstock.com

Figure 28-13. Seeds can be planted using a seed spreader.

> **Corner Question**
>
> What is straw?

Sodding

Sodding is using strips of turf grown at another site and transplanting that grass to the current site. This produces an instant lawn. Purchase sod that is certified and grown by reputable grower. Install sod on the site as soon as possible after harvesting. Follow these steps for sod installation:

1. Thoroughly moisten the area where the sod will be laid.
2. If the sod cannot be installed within 24 hours of receiving it, it should be unstacked, unrolled, placed in the shade, and watered.
3. Start sodding along a straight edge (sidewalk or driveway) and stagger the strips in a brick-like pattern, **Figure 28-14**.
4. Use a knife to trim pieces to fit. Lay sod lengthwise across slopes.
5. Roll the lawn after all pieces are transplanted to ensure soil contact.
6. Irrigate gently and keep evenly moist until the sod is well established and the sod cannot be pulled up.

Vadim Nikolaevitch/Shutterstock.com

Figure 28-14. Sod is installed by placing strips in a brick-like pattern.

Plugging and Sprigging

Some species of warm-season grasses do not produce viable (live) seed or are sterile. In such cases, plants must be used to establish turf. Laying sod is expensive. Less costly alternatives are plugging and sprigging. In plugging, mature pieces, or plugs, of sod are planted in the soil. Pieces are normally about 2″ (50.8 cm) or larger and planted on 6″–12″ (15.2 cm–30.4 cm) centers.

Sprigging involves broadcasting pieces of grass (cut stems) over an entire area. The sprigs are then pressed into the soil 1″ (2.5 cm) deep by hand or roller. Three to five bushels of sprigs are required for every 1000 ft^2 for minimal coverage, or 5–10 bushels per 1000 ft^2 for rapid establishment.

Irrigation

One of the most important factors when establishing plant material of any type is irrigation. The top 2″ (50.8 cm) of soil must remain constantly moist when establishing turf. To ensure this, lightly water the area two to three times a day for the first month.

As the turf establishes itself and plants develop roots, they require less frequent, but deeper watering. After the third mowing, water to a depth of 8″ (20.3 cm) about once a week or when the turf shows signs of water stress (change in color). Most turf needs an average of 1″ (2.5 cm) of rain a week or supplemental irrigation to equal that amount, **Figure 28-15**.

topseller/Shutterstock.com

Figure 28-15. Most turf requires at least 1″ (2.5 cm) of rain or irrigation per week to maintain active growth.

This general rule depends on the water-holding capacity of the soil and surrounding environmental conditions. Determine when, how much, and whether to water depending on the established turf's needs.

Mowing

Once grass height is 50% higher than the preferred height, begin mowing. For example, when Kentucky bluegrass reaches 3″ (7.6 cm) in height, it is mowed to a height of 2″ (5 cm). The amount of mowing needed depends on the quantity of turf growth. Growth depends on temperature, moisture, fertility, the rate of growth for a species, and time. In general, however, turf should be mowed when the leaf surface is 50% higher than the suggested mowing height.

Mowing safety is a serious concern. Follow a few simple rules to maintain your safety.

- Use a well-maintained mower and check all safety features before operating.
- Use a lawn mower with a sharp, balanced blade.
- Mow when grass is dry to prevent slipping or spreading turf disease.
- If clippings are heavy enough to shade grass, bag the clippings or rake and remove them from the lawn. Otherwise, leave grass clippings on the turf to decompose and release vital nutrients needed for soil and plant health, **Figure 28-16**.

Corner Question

How much water does one leaky hose spigot waste each year?

danyimages/Shutterstock.com

Figure 28-16. Mow turf safely by following safety protocols.

Thinking Green

Lawn Mowers

Every sunny weekend that turf is growing, approximately 54 million people in the United States mow their lawns. The EPA estimates that a gas-powered lawn mower emits as much pollution as eleven new cars being driven for one hour each. What can be done to keep grass manicured without polluting the air?

Electric lawn mowers use electricity as their fuel source, creating less pollution during operation than gas mowers. Electric mowers are quiet, and do not need oil changes or tune-ups. Remember, however, that the electricity needed is generated by burning fossil fuels or natural resources, which creates air pollution.

Ints Vikmanis/Shutterstock.com

A reasonable environmental solution for mowing is a manual lawn mower. This mower allows you to be the engine and the power behind the machine, unlike gas or electric mowers. Only your energy is expended, and no air pollution is created.

Integrated Pest Management for Lawns

Establishing new turf will require some pest management. Before planting turf, weeds must be removed and any pests that live in the soil, such as voles and moles, must be removed. Prevention is always easier than treatment. Proper pest identification is essential to management. Follow these strategies for integrated pest management (IPM):

- Scout for weeds and pests. Look in the soil and surrounding area for perennial and annual weeds. Determine if there are animals dwelling in the soil that might cause problems. Check for grubs in the soil that will later emerge as beetles.
- Use cultural mechanisms to prevent problems. Use weed-free seed or vegetative sources. Clean the soil. Remove any areas where pests can breed or live.
- Create mechanisms to trap pests. Use traps for insects, animals, and other pests that can impact developing turfgrass.
- Use natural barriers. Find or establish trees that could screen your site from others that could contaminate the location.
- Use biological controls to combat pests such as voles and moles, **Figure 28-17**.
- Use chemicals only when necessary. Whether you are considering a fungicide, herbicide, insecticide, or other chemical, use it only as a last resort. Try to use organic chemicals first and then apply other chemicals when all other methods of IPM have been exhausted. Follow all chemical labels when applying any pesticide.

sauletas/Shutterstock.com

Figure 28-17. Dirt mounds indicate a mole infestation in this lawn.

Thinking Green

Organic Weed Control

Weeds are best controlled through prevention before they grow. Pre-emergent herbicides contain chemicals that prevent weed seeds (or any seed) from germinating. One organic product that has been used with some success on smaller lawns is corn gluten meal, an organic herbicide. This material not only inhibits weed growth, it also acts as a soil amendment. Application timing is critical. Corn gluten must be applied just before seedlings emerge. While it is a pre-emergent, it works to actually prevent the seedling (just after it emerges from the seed coat) from being able to grow by dehydrating the roots. To achieve good results, this product must be applied at a rate of 20 pounds per 1000 ft^2.

Nika Novak/Shutterstock.com

That is a lot of corn gluten meal for an acre (43,560 ft^2). How much corn gluten meal would be required for one football field?

AgEd Connection
Equipment and Supplies Identification

As part of a career development event, students must identify materials and equipment used in the industry. The tools and materials may be presented physically or as photographs. Use the illustrated glossary (page 913) and the e-flashcards on your textbook's student companion website at www.g-wlearning.com/agriculture to help you study.

Turf Maintenance

A lawn enters the maintenance life cycle when it has been mowed at least three times. At this point, the lawn is considered established, and it enters a new phase of its life. During the maintenance phase, irrigation, fertilization, mowing, aeration, thatch control, and pest management are important for healthy turf.

Irrigation

Many growers need turf that is drought tolerant. Ordinances and legislation often do not permit unlimited use of water. Growers of turfgrass must be water wise and conserve this resource. Turfgrass should not be watered haphazardly. Indiscriminate watering leads to problems with fungus, fertilization, and other cultivation issues.

When to Irrigate

Turfgrass should not be irrigated on a set schedule. A stand of grass should only be watered when a significant portion of the lawn exhibits signs of moisture stress, **Figure 28-18**. *Moisture stress* is a condition caused by lack of water. Signs of the condition include changing leaf blade color; wilted, folded, or curled leaf tips; or depression marks that remain in the turf from foot traffic.

To properly irrigate turfgrass, follow these guidelines:
- Water in the early morning. This will reduce the amount of time that water will stand on the grass and can reduce disease risk. Water will not be wasted to immediate evaporation (as it would be during a hot afternoon).
- Use overhead irrigation with oscillating or rotating sprinkler heads for best delivery.
- Monitor the site to ensure that sprinkler heads are only irrigating turf and not any other surfaces. Be sure that heads are not unevenly watering the site.

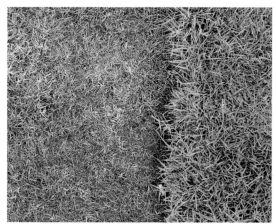

komkrit Preechachanwate/Shutterstock.com

Figure 28-18. Moisture stress can lead to permanent injury to turf.

How Much to Irrigate

Turf requires about an inch of water per week. It is best not to apply that amount all at once, however. Follow these guidelines for irrigation:

- Measure the amount of water being delivered to an area. Deliver 0.5" (1.27 cm) of water every four days in most cases. Use a rain gauge to measure water, **Figure 28-19**. Most soils cannot absorb more than 0.5" (1.27 cm) of water per hour.
- Water turf to a depth of 8" to encourage deep root growth.
- If watering cool-season grasses in summer, slowly reduce water to coax the turf into dormancy. Allow drought symptoms to appear between infrequent irrigation cycles. During dormancy, reduce fertilization and mowing as well.
- Water a dormant cool-season lawn about 0.25" (0.6 cm) every three weeks to keep it alive during summer months. When cool temperatures return in the fall, irrigate or allow rain to irrigate the site to help promote new growth.

Darryl Brooks/Shutterstock.com

Figure 28-19. Place a rain gauge near a sprinkler head to measure the amount of water being applied.

Fertilization

Following soil test recommendations prevents unnecessary fertilizer application. Responsible turfgrass growers realize the importance of using fertilizers in a sustainable and sensible manner. No single fertilizer application program is suited for all forms of turfgrass cultivation.

Conduct soil tests at least every other year to determine the amount of nitrogen, phosphorus, potassium, lime, or sulfur to add. If a soil test is not done, use a complete fertilizer with ratios such as 4:1:2, but this is a weak alternative. An example of a complete fertilizer is one with a nutrient analysis of 16-4-8.

Nitrogen in Turfgrass

Nitrogen comes in several forms. It is important to recognize how quickly the nitrogen is released into the soil and under what conditions. Improper application of a nitrogen fertilizer can be detrimental to turf and the environment. In some cases, improper applications can be irreparable, requiring installation of new turf as a replacement.

Thinking Green

Subsurface Drip

Subsurface irrigation conserves water and produces healthy turf. A newly introduced system contains a drip pipe with patented copper material inside that prevents roots from growing into the driplines. The drip pipe is installed 6"–12" (15.2 cm–30.4 cm) apart in the lines. The driplines are then fed into a feeder line that supplies the water. The driplines water the roots below the surface. This reduces runoff, evaporation, and water consumption. This irrigation is initially expensive and labor intensive, but it produces healthy turf that requires much less water.

Quick-release sources of nitrogen are also called fast acting, soluble, or quickly or readily available. These fertilizers concentrations of nitrogen range from 11% to 46%. They are less expensive than slow-release fertilizers, react readily with water, and produce rapid turf greening. Frequent applications at low concentrations will prevent the grass from burning. Water the fertilizer in immediately after application to prevent volatilization of nitrogen into ammonia. *Volatilization* is a process in which fertilizer components change from a solid to a vapor and are released into the atmosphere. The forms of this fertilizer include urea, nitrate, and ammonium.

Slow-release sources of nitrogen are called controlled release, slowly available, or insoluble. They provide a longer, slower release of nitrogen and are less likely to burn grass. They are also less likely to leach into groundwater. Slow-release nutrient sources are expensive and less effective than other types. They include organic sources, polymer, and sulfur-coated fertilizer.

> "Failure is the fertilizer for success."
> —Dennis Waitley

Phosphorus in Turfgrass

Phosphorus helps promote a healthy root system and seed production. Recently, there has been an effort to limit phosphorus because there is concern that this nutrient impairs water health. Avoid applying this nutrient near water sources or where runoff can occur.

When a plant is phosphorus deficient, leaf blades usually appear purple. Phosphorus is very immobile in soils and can take months to move just a couple of inches. Phosphorus sources include superphosphate (treating rock phosphate with acids) and bone meal (harvested from animal bones).

Did You Know?
Nitrogen is the fifth most abundant element in the universe.

Potassium in Turfgrass

Potassium is responsible for several physiological processes in plants, which contributes to plant hardiness and vigor. Potassium deficient plants are sensitive to drought, cold temperatures, and disease.

Introduce potassium to turf using inorganic or natural organic fertilizers. The potassium in fertilizer comes mostly from inorganic sources including muriate of potash (potassium chloride) and sulfate of potash (potassium sulfate). These fertilizers are water-soluble. Unlike phosphorus, potassium is not considered to be a major contributor to water pollution.

Fertilizer Formulations and Programs

No single fertilizer formulation or program is ideal for all settings. Lawns, athletic fields, and golf courses all have their own specific objectives for turf growth, **Figure 28-20**. In addition, fertilizer needs and applications depend on the type of grass that is grown, soil conditions, time of year fertilizer is applied, and location in the United States. Turfgrass species differ in the amount of fertilizer, especially nitrogen fertilizer, that is required for optimum performance. This makes creating a general plan for a fertilizer program nearly impossible. Check with your cooperative extension agency or state land-grant university for specific information for the types of grass and soils in your region.

Bochkarev Photography/Shutterstock.com

Figure 28-20. Fertilizer formulations and programs vary. Check with your local cooperative extension service to find specific information regarding fertilizer application for your region.

Mowing

Most mowing is done using rotary (centrifugal) or reel (cylinder) mowers. A *rotary mower* is a piece of equipment that uses a blade that cuts grass by sucking and tearing the grass blade. Think of a helicopter blade. That is the same action used to cut turfgrass with a rotary mower. A *reel mower* is a piece of equipment that uses scissor-like action to cut grass blades, **Figure 28-21**. These mowers are best suited for grasses that will be cut to less than 1" (2.5 cm).

Follow these simple mowing operation tips for a better cutting experience and healthy turf:

- Never mow wet grass.
- Mow only when needed. Mow when the grass is 50% taller than the desired height (usually weekly or biweekly during peak season).
- If the lawn is too high, meaning you will remove more than 50% of the leaf blades, cut off one-fourth to one-half of the present growth. Mow again to the proper height in one or two days.
- Mow across slopes, never up and down slopes.
- Do not mow slopes greater than 15% using a push mower or 20% using a riding lawn mower.
- Mow in one direction and then mow perpendicular to that direction the next time you mow. This prevents compaction and makes the turf appear more even.

Kitch Bain/Shutterstock.com

Figure 28-21. A reel mower uses scissor-like action to mow grass.

Corner Question

What is the largest mower on the planet?

Safety Note

Mower Safety and Maintenance

Mower safety and maintenance work hand in hand. Paying close attention to both will ensure easier mower operation.

- Read the owner's manual before operating.
- Always shut off the lawn mower when not actively mowing.
- Wear the proper personal protective equipment (PPE): goggles, sun protection, insect repellent, tightly laced (close-toed) shoes, and appropriate, fitted clothing.
- Watch for obstacles.
- Disconnect spark plug before servicing the lawn mower.
- Keep mower blades sharp.
- Balance the blades. Dull and unbalanced blades will reduce the quality of the cut and can lead to jagged ends. Damaged grass blades create sites for disease to enter the turf.
- Mow away from the cord when using an electric lawn mower.
- Do not operate blades while going over gravel or pavement.
- Do not spray cold water on a hot engine.
- Never leave a mower unattended when the engine is running.

- Leave short grass clippings on the lawn to decompose. These clippings place valuable nutrients back into the soil, **Figure 28-22**.
- If grass clippings are long and will shade the grass beneath, remove the clippings.

Collected grass clippings can be placed in gardens, around trees, or added to compost piles. Yard waste accounts for 20% of the waste in the landfills. Composting your yard waste rather than putting it out for trash collection will keep it out of landfills. Contact your cooperative extension service agent for other ideas about how to use yard waste.

Tony Campbell/Shutterstock.com

Figure 28-22. Leaving grass clippings on turf puts nitrogen back into the soil, increases microbial activity, and amends the soil structure.

Aeration

Compaction-prone soils and high-traffic turf areas should be aerated. The process of aeration (also referred to as plugging or coring) is done using a machine that pulls out cores or plugs of soil. The holes left in the soil allow air to penetrate into the root zone of the turf.

Aeration is done during active growth when turf can quickly recover from the process. Aerate cool-season grasses in fall or early spring. Aerate warm-season grasses in late spring or early summer. Rent an aerating machine to do the work yourself or hire a landscaping or turf maintenance company to do the job for you.

Thatch Control

Thatch is a layer of dead grass and organic material that forms a mat beneath leaf blades and above the root zone. Some thatch is beneficial; it can help retain moisture and add nutrients to the soil. Too much thatch can prevent water penetration and proper air exchange with roots. Excess thatch can also be a breeding site for pests.

Verticutting is a method of thatch control done by a machine that slices into thatch layers to create space for water penetration and air exchange. This process is also called power raking. Some grasses produce a great deal of thatch, which is desirable on athletic fields. Thatch can help prevent injury when athletes fall to the ground.

Thinking Green

Keep Mower Blades Sharp

Sharp mower blades are beneficial for the mowing equipment, the operator, and the environment. Sharpened blades reduce vibration and noise and lengthen the mower's life. This makes operation easier for the person mowing the grass. In addition, keeping blades sharp can reduce fuel consumption by as much as 22%.

Corner Question

How many times per year is a putting green aerated on newly installed sod?

Figure 28-23. A verticutter slices into the thatch and helps to prevent thatch buildup.

Jodi Riedel/Goodheart-Willcox Publisher

Thatch that exceeds 0.75″ (1.9 cm) should be power raked or cored. Verticut warm-season grasses during the spring. Specialized equipment can be rented for this job, **Figure 28-23**, or you can hire a landscaping company to do the work.

Other methods to help prevent and control thatch include:

- Reduce fertilizer applications.
- Monitor and adjust the pH with lime or sulfur applications.
- Mow only when needed.
- Make sure clippings are decomposing. If thatch is building up, collect clippings and compost elsewhere (other than on the turf site).

Integrated Pest Management for Turf

Practices used during turf establishment for integrated pest management (IPM) should be continued once the turf is established. The IPM program should include these activities:

- Scout for pests, such as insects, diseases (fungus, bacteria, virus, and parasites), weeds, and living organisms (moles, voles, woodchucks, gophers, and slugs).
- Correctly identify pests.
- Determine your threshold for the pest.
- Choose a method of control.
- Evaluate the results.

Prevention is always better than treating a pest in the IPM process. To help prevent pest problems, properly cultivate the site, select turf appropriate for the site, apply but do not overuse fertilizers, apply appropriate pH adjustments, prevent excessive thatch buildup, use best mowing techniques, and use precision irrigation.

Thinking Green

Herbicides

Weeds can become an issue in any garden or turfgrass site. Trying to prevent weeds is a good practice but weed control will most likely become necessary. An *herbicide* is a chemical that will kill a plant. A post-emergent herbicide is applied after weeds have emerged. Nonselective herbicides kill all plant material. Selective herbicides target and kill specific plants.

One selective herbicide is the chemical 2, 4-D. This chemical kills broadleaf plants (those with intersecting veins) but will not harm grasses. The herbicide (2, 4-D) can be used on fescue, zoysia, or bluegrass turf. Plants such as dandelions will be killed, but the grass will not be harmed.

Carefully research, select, and apply herbicides according to label directions. The label of a pesticide contains important warnings and use instructions that should not be overlooked. The person who applies a pesticide is responsible for any harmful effects caused by the application.

Turf Renovation

There are times when turf renovation is necessary. A newly planted lawn may not work out as planned, **Figure 28-24**. Maybe the school football field has become uneven and dangerous for players. These situations will require turfgrass renovation.

Turf Failure

There are several common mistakes or problems that create turf problems. Before the turf can be renovated, the underlying cause of the failure must be identified. Possible causes for turf failure include:

- The wrong species or variety of turf was selected for the site. For example, a nondurable species was planted in a high-traffic area.
- Poor establishment procedures were used.
- Improper mowing, fertilization, irrigation, or other management techniques were used.
- Excessive thatch buildup occurred.
- The turf was infested with pests, such as weeds, insects, or diseases.

After identifying the underlying cause of the turf failure, create and enact a plan to deal with the issues.

komkrit Preechachanwate/Shutterstock.com

Figure 28-24. Conduct a thorough inspection and analysis of failed turf to determine the cause before spending additional time or resources on renovation.

"I have not failed. I have just found 10,000 ways that will not work."
—Thomas Edison

When to Renovate

Renovation timing depends on the type of turfgrass that is grown. Cool-season grasses should be renovated in the late summer to early fall. Warm-season grasses should be renovated between late spring and early summer. Trying to renovate at times that are not suited for the turf can make repair difficult.

Pests and Turf Renovation

Pests are a leading causes of turf failure. Pests must be controlled before a lawn can be renovated. Common pests include weeds, insects, or diseases.

Weeds

Weeds (unwanted plants growing in a space) steal nutrients, space, water, and light from turfgrass. Methods for controlling weeds include:

- Hand pulling.
- Using pre-emergent herbicides (applied before the weeds emerge) to prevent new weeds.
- Using selective or nonselective post-emergent herbicides to kill existing weeds.
- Using organic methods such as flaming, solar sterilization (using clear plastic to heat the site), or the use of goats or sheep for grazing.

Most often, a site that will be renovated will be seeded. In this case, pre-emergent herbicides cannot be used as this will prevent the turf seed from germinating.

Corner Question

What percentage of all plants is considered weeds?

Post-emergent herbicides can control annual and perennial weeds. Most herbicides should be applied four to six weeks before using additional renovation techniques. Some sites are so weed infested that the entire area should be sprayed with a nonselective herbicide to kill everything on the site, including turf. More than one application of herbicide may be needed to kill stubborn weeds. Wait at least seven days or as directed by the herbicide label to observe results before respraying. Young weeds are always easier to control than mature weeds (with flowers or seeds). Carefully read all label information before applying herbicides for the best results. Always follow labeled safety directions to comply with legal use of the product.

Insects

Insects can completely destroy turfgrass. White grubs attack roots, and caterpillars eat the leaf blades, **Figure 28-25**. Depending on the insect, IPM can be used to control the pest. Biological controls, such as a bacteria known as *Bacillus thuringiensis*, work well against caterpillars and grubs. Organic and synthetic chemicals can be applied to targeted pests with good results when used correctly. Growers should apply chemicals when pests are most susceptible, select the least dangerous chemical, and follow the chemical label.

Stephen Farhall/Shutterstock.com

Figure 28-25. Grubs are larvae of various beetles. They live below the turf and eat the roots.

Diseases

Diseases, often fungi, plague many turf sites, **Figure 28-26**. Brown patch, dollar spot, and pythium can stress turf and lead to death. Proper cultivation techniques, such as reducing nitrogen applications, can help prevent disease infestations. Fungicides applications may also help. Apply them according to chemical label instructions. Pay attention to environmental conditions and spray rates before and during application.

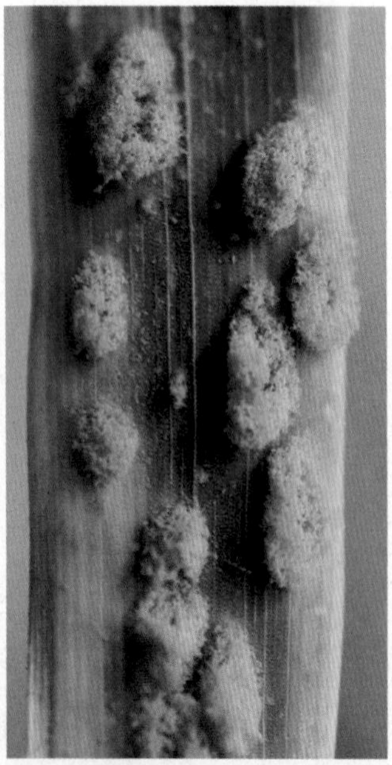

Damian Herde/Shutterstock.com

Figure 28-26. Various fungi can grow on turf and cause significant growth problems.

Seeding

Prepare the site properly before planting turf seed. Follow these steps:
1. Remove weeds in and around the turf site. An easy method for removal is to mow on the lowest setting and collect all clippings.
2. Remove thatch so that the soil is exposed. Use rakes, hoes, or power equipment such as a dethatcher. This reduces competition and allows good light penetration to emerging seeds.
3. Prepare the soil, including proper lime, sulfur, and fertilizer application. Apply nutrients and lime or sulfur based on soil test results. Apply evenly over turf site.
4. Loosen 4″–6″ (10.1 cm to 15.2 cm) of soil with a rake, hoe, or cultivator. Fill in low areas with existing soil or a topsoil blend. Smooth out soil and ensure that the soil aggregates are no larger than a marble. On large sites, use an aerator to bring cores to the surface. Allow the cores to dry, pulverize them with a mower, and then rake flat.

STEM Connection: Overseeding

In some regions, such as the warm-season grass zones, turf growers may choose to overseed during the dormant season. **Overseeding** is sowing grass seed over the thatch of dormant grass. The overseeded grass grows up through the brown, dormant grass, providing color in the cooler months. When the temperatures begin to climb in the spring, the overseeded grass begins to die back and the dormant, warm-season grass replaces it.

Small bare spots less than 6″ (15.2 cm) in diameter do not need to be replanted. Instead, repair these spots using proper cultivation techniques, including fertilizer application, irrigation, and pest management.

Reseed bare spots larger than 6″ (15.2 cm). This is known as *spot seeding*. To renovate large barren areas, follow the same procedures used for establishing turf from seed.

If you do not want to do the project by hand, use a power rake or vertical slicer. These devices slice into the ground and allow air and water to penetrate the soil. A slit seeder can also be used to ensure soil to seed contact. A *slit seeder* is a device that creates a slit in the thatch and root zone while also planting seed.

Plugging

A turf with rhizomatous or stoloniferous growth habits can be renovated using plugs that contain turf and some soil with roots. Place the plugs on 6″ or 12″ (15.2 or 30.4 cm) centers. Collect plugs from healthy areas and place in bare areas using a shovel or a *plugging device* (equipment that pulls pieces of leaf, stem, and root material from the turf).

Sprigging

Areas greater than 15,000 ft² that already have rhizomatous or stoloniferous grass in place can be sprigged to fill in bare areas. Spread sprigs (containing the rhizome or stolon for the selected turf) over the surface of a well-prepared site. The sprig should be inserted into the soil at least 0.5″–1″ (1.2 cm–2.5 cm) deep. Roll the site to ensure good contact between the rhizome or stolon and the soil.

After Care

After turf has been renovated, the principles of good plant management must take place. These management activities involve irrigation, fertilization, and mowing.

Irrigation

Lightly water renovated turf several times a day. As the turf grows, reduce irrigation frequency but increase irrigation time. The goal is to create deep and vigorous root systems. Keep soil constantly moist through the third mowing. After the third mowing, irrigate two to three times a week until the soil is saturated to at least 6″ (15.2 cm) in depth.

"The grass may look greener on the other side of the fence, but when you get there it is usually artificial turf."
—Author Unknown

Did You Know?
Turf used for the forty-ninth Super Bowl® in Phoenix, Arizona, was grown in Alabama and then shipped to the University of Phoenix Stadium. The turf was shipped on 33 refrigerated trucks for three days, and then crews from 30 stadiums joined efforts to install the turf.

Thinking Green

Hydromulch

The USDA's Agricultural Research Service (ARS) constantly works to improve techniques and practices to improve soil conservation. One soil conservation method is growing grass in bare areas. Hydromulch is a material that is a mixture of seed, mulch, water, binding ingredients, and other materials. This slurry mixture is sprayed over the bare soil, and grass begins to grow. The ARS tests materials that can help make this method less expensive and more successful.

USDA Agricultural Research Service (ARS)

Fertilization

Warm-season and cool-season grasses require fertilizer of different quantities at different times of the year. The type of grass also determines how much fertilizer is applied. Contact your local cooperative extension service agent or land-grant university for guidelines on fertilizer applications. These organizations have done extensive research about proper and sustainable fertilizer applications for optimum plant growth.

Mowing

Follow the safety operation and maintenance guidelines given earlier in the chapter. Continue to mow existing turf on the renovated site. For newly planted seed or vegetative structures, mow when leaf blade height is 50% more than ideal height.

Careers

Many careers in the turfgrass industry include turf maintenance or equipment operation, but other careers are available in the turfgrass industry. Some of these careers include sod sales associate, sports turf manager, and turfgrass breeder.

Sod Sales Associate

Sod sales associates work for sod growers. They help customers choose products. A sales associate for a sod company must be well informed about the product and all its characteristics. Customers depend on the sales associate to know about the product, from where it can be grown to price and installation information. Sod sales associates must be able to work with both expert and novice horticulturists. In addition, the associate must be friendly and organized.

Sod sales associates need a high school diploma or associate degree, but a four-year degree in an agriculture-related field is preferred. An emphasis in turfgrass education also is beneficial.

Sports Turf Manager

Turf is the centerpiece for many sports. It must be installed, established, and maintained according to specifications. Sports turf can be found everywhere in the United States.

Sports turf managers work with their crews to ensure that turf is irrigated, fertilized, aerated, and mowed. Paint is also sometimes applied to fields based on rules and for team logos and branding requirements, **Figure 28-27**. Sports turf managers must have experience in athletic field management. They must pay attention to detail, work well with others, and be willing to work at various days and times of the year.

Sports turf management positions require a college degree in an agriculture-related field, typically turfgrass management. In addition to the educational requirements, many sports turf managers have work experience gained through internships. Internship opportunities provide job seekers with a competitive edge over other applicants.

Toa55/Shutterstock.com

Figure 28-27. A sports turf manager's duties can include painting the turf before an athletic event.

Career Connection

Dr. Melodee Fraser
Turfgrass Breeder

Dr. Melodee Fraser was raised in a family of turf lovers. Her father was a superintendent of a golf course in Indiana, and all of his children are currently employed in the turfgrass industry. Dr. Fraser went to Mississippi State University (MSU) and was the first woman to graduate with a turfgrass management degree. Originally, Dr. Fraser wanted to work as a golf course superintendent like her father, but her passion for science and analysis changed her focus to scientific inquiry. She continued her studies at Rutgers University where her research focused on turfgrass breeding. **Turfgrass breeding** involves selecting plants with desirable traits and combining them to develop improved cultivars.

Dr. Fraser was hired by Pure Seed Testing of Oregon to open a southeastern research station in Rolesville, North Carolina. She breeds grasses to resist brown patch disease. Once she and her team select improved varieties, the company licenses the new varieties to other seed companies that then grow, harvest, and sell the licensed seeds.

Jodi Riedel/Goodheart-Willcox Publisher

Her focus is to breed turfgrass that requires the least maintenance. The ideal turfgrass requires fewer fertilizer, irrigation, and pesticide applications, making it more sustainable. Water scarcity, water salinity (of irrigation or at coastal sites), and legislative limitations on chemical availability provide challenges for turf breeders at Pure Seed Testing.

Dr. Fraser feels rewarded by her work breeding turfgrass. Technological advances, such as computer software that analyses digital images of turf, increases the objectivity of trial evaluations and keeps her work fresh. She enjoys new and ongoing experiments involving turfgrass. She also has the opportunity to travel the world, working with a wonderful group of turfgrass professionals.

CHAPTER 28 Review and Assessment

Chapter Summary

- The turfgrass industry consists of businesses that grow seeds, sods, and grasses for homes, commercial spaces, utility uses, and sports uses. The use of turfgrass provides benefits that are helpful to people and the environment, such as reduced noise pollution and soil erosion.
- Turf growers should use sustainable practices. Excessive use of water, fertilizer, and pesticides is harmful to the environment.
- There are four main applications or uses for turf: lawns, golf courses, utility turf, and sports turf. Sod and seed production are also part of this industry. Companies that provide pesticide, fertilizer, irrigation, equipment, and information technology are associated with the turfgrass industry.
- Turfgrass has unique, identifying characteristics. These characteristics include specific traits regarding the leaf, roots, stem, flower, seedhead, and growth habit.
- Warm-season grasses and cool-season grasses are best suited for specific growing zones.
- To establish a lawn, the appropriate turf must be selected, the site must be prepared, the soil conditioned, the turf planted, irrigation operated, fertilizer applied, and pest management employed. Proper mowing is also important.
- The method for establishing turf depends on the grass selected and the site. A lawn can be established using seeding, sodding, or plugging. Follow soil test results to determine the amount of fertilizer to add to a site for turf establishment.
- Turfgrass must be maintained. During the maintenance phase, irrigation, fertilization, mowing, aeration, thatch control, and pest management are important for healthy turf.
- Turfgrass managers can choose a number of methods to renovate a lawn. These methods include clearing all plant material from the site or reestablishing turf in bare patches. The first step is to identify the reason for the turf's distress and then act accordingly to replant or replace turf to the site.
- The turfgrass industry offers careers in installation, maintenance, sales, and planning. Three careers related to the turfgrass industry are turfgrass breeder, sod sales associate, and sports turf manager.

Words to Know

Match the key terms from the chapter to the correct definition.

A. auricle
B. blade
C. collar
D. lawn
E. ligule
F. moisture stress
G. overseeding
H. plugging device
I. reel mower
J. rotary mower
K. seed blend
L. seed spreader
M. seedhead
N. slit seeder
O. sports turf
P. spot seeding
Q. thatch
R. tiller
S. transition zone
T. turfgrass
U. turf breeding
V. turfgrass industry
W. utility turf
X. vernation
Y. verticutting
Z. volatilization

1. A method of thatch control done by a machine that slices into thatch layers to create space for water penetration and air exchange.
2. Selecting plants with desirable traits and combining them to develop improved cultivars.
3. Businesses that grow seeds, sods, and grasses for recreational areas, homes, and commercial, utility, and sports uses.
4. A layer of dead grass and organic matter that forms a mat beneath leaf blades and above the root zone.
5. A condition in turfgrass caused by lack of water.
6. Sowing grass seed over the thatch of dormant grass.
7. A small, claw-like appendage that projects from the collar of a grass blade.
8. The flat, leaf part of the grass.
9. A piece of equipment that uses a blade that cuts grass by sucking and tearing the grass blades.
10. A device that creates a slit in the thatch and root zone while also planting seed.
11. A stem that develops from the crown of the parent plant and grows upward within the enclosing leaf sheath.
12. A process in which fertilizer components change from a solid to a vapor and are released into the atmosphere.
13. Grass that is used for spaces such as roads, airports, and parks.
14. The arrangement of leaves within the bud.
15. A piece of equipment that uses scissor-like action to cut grass blades.
16. Equipment that pulls pieces of leaf, stem, and root material from the turf.
17. An expanse of turfgrass that is used for recreation or beauty.
18. The band at the backside of a leaf where the leaf blade and sheath meet.
19. A tool that places seeds in an area.
20. A mixture of various cultivars of grass seed.
21. A thin structure that clasps the top of the leaf sheath.
22. Grass that is used for recreation and competitive sports purposes.
23. The cluster of seeds within a plant's inflorescence.
24. Reseeding bare spots larger than 6″ (15.2 cm).
25. Grass and the surface layer of soil held together by the grass roots.
26. A growing area that can grow both warm-season and cool-season grasses.

Know and Understand

Answer the following questions using the information provided in this chapter.

1. What do businesses in the turfgrass industry produce and how much does this industry contribute to the US economy?
2. What percent of the turf area in the United States is for lawns and what are the other uses of turf?
3. What are some benefits of turf for people and for the environment?
4. What are some unique characteristics of turfgrass that separate it from other plants?
5. List the unique morphological structures of grasses that can be used to identify them.
6. Describe the three shapes that a grass leaf blade may have.
7. Describe the three growth habits of turfgrasses.
8. How are warm-season grasses affected if they are grown in an area where temperatures are too cold?
9. Name four grasses suitable for the transition zone.
10. When is the best time of year to establish a lawn with cool-season grasses and what are two methods used?
11. Why should a grower use certified seed for establishing a lawn?
12. Describe how to establish a lawn using sod.
13. What steps must be taken to prepare a site for turf planting?
14. Why should growers do a soil test before installing turf?
15. What procedures should be followed in taking a soil sample for analysis?
16. What is the general rule for watering established turf?
17. What are some integrated pest management strategies used to benefit turf growth?
18. What activities are important for turfgrass sites during the maintenance phase?
19. Why is there no single fertilizer formulation program that is suitable for all settings?
20. What personal protective equipment should be worn when operating a lawn mower?
21. What are some common pests that may be a problem in managing turfgrass sites?
22. What are some possible causes of turf failure that may make renovation necessary?

Thinking Critically

1. You are driving through your neighborhood and see someone applying a chemical to a lawn while it is raining. What would you do?
2. It is not a good idea to leave gas in small-engine equipment, such as mowers, for more than 30 days. What should you do at the end of a season if your mower still has gas in the tank?

STEM and Academic Activities

1. **Math.** Your school wants to apply a pre-emergent herbicide to prevent weeds from growing, and they have heard that corn gluten meal is an organic option. Corn gluten meal requires an application of 20 pounds per 1000 ft^2. How much will need to be applied to a football field that is nearly one acre with the end zones? How much will this cost if corn gluten meal is $12 for 5 pounds?

2. **Math.** You recently received the soil test results from your cooperative extension service agent. The results indicate that your lawn is lacking nitrogen. The suggested application rate is 1 lb of nitrogen per acre. Your yard is one-half acre. How much nitrogen must you apply? If you have a 50-pound bag of 12-4-8 fertilizer, how many pounds of this fertilizer will you need to apply to your entire lawn?
3. **Engineering.** Design an irrigation system that could be used on a football field.
4. **Social Science.** Contact a local golf course or sports turf site and schedule a visit. Create a video of your visit. Ask employees at the site how they practice sustainability and integrated pest management. Create your own list of questions to ask as well.
5. **Language Arts.** Write a one-page response to someone who believes that turfgrass is responsible for your area's water pollution. Be sure to support your response with facts.

Communicating about Horticulture

1. **Speaking and Listening.** Divide into groups of four or five students. Each group should choose one of the following topics: seeding, sodding, plugging and sprigging, fertilizer, or irrigation. Using your textbook as a starting point, research your topic and prepare a report on how it impacts turfgrass management. As a group, deliver your reports to the rest of the class. Take notes while other students give their reports. Ask questions about any details that you would like clarified.
2. **Writing.** Identify insect pests that may be a problem in your area or state and prepare one-page information sheets about those pests. What do the pests look like? Which turfgrass species do they attack? At what time of year are these insect pests most likely to be seen? What can be done to control or eliminate the pest? To make your information sheets engaging and informative, use images from the Internet. Ask permission to post these information sheets in your school, in public libraries, or on your school's website.

SAE Opportunities

1. **Exploratory.** Job shadow a turfgrass employee.
2. **Experimental.** Grow a plot of turf and apply different fertilizer concentrations to specific parts of the plot. Analyze the differences in growth. Record your results.
3. **Exploratory.** Interview people in your community regarding their perception of the turfgrass industry. After completing the interviews, create an educational brochure that will help participants understand the many environmental and social benefits of turfgrass.
4. **Entrepreneurship.** Develop a lawn mowing service.
5. **Placement.** Get a job working at a sports field or golf course.

Ana Ado/Shutterstock.com

CHAPTER 29
Integrated Pest Management

Chapter Outcomes
After studying this chapter, you will be able to:
- Explain the concept of integrated pest management.
- Describe the different types of pests.
- Summarize the steps for pest prevention.
- Describe pest inspection and monitoring.
- Define action thresholds.
- Discuss different methods for corrective actions.
- Describe careers related to integrated pest management.

Words to Know
- action threshold
- anti-transpirant
- beneficial
- economic injury level (EIL)
- integrated pest management (IPM)
- oomycete
- pest
- pesticide
- pheromone
- plant pathogen
- scouting
- vector

Before You Read
As you read the chapter, put sticky notes next to the sections where you have questions. Write your questions on the sticky notes. Discuss the questions with your classmates or teacher.

While studying this chapter, look for the activity icon 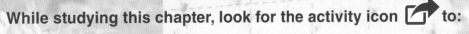 to:

- **Practice** vocabulary terms with Words to Know activities.
- **Expand** learning with identification activities.
- **Reinforce** what you learn by completing Know and Understand questions.

www.g-wlearning.com/agriculture

hans engbers/Shutterstock.com; Kosobu/Shutterstock.com; Svetlana Zhukova/Shutterstock.com; Vadym Zaitsev/Shutterstock.com; Sumet Baosin/Shutterstock.com

From leaf-sucking aphids and root-eating nematodes to plant-smothering weeds, garden pests abound. A plant *pest* is any organism that damages or disrupts plant growth. Opportunities constantly arise for pests to threaten and disrupt the productivity of farms, gardens, and urban landscapes. *Integrated pest management (IPM)* is an environmentally sensitive approach to controlling pests. The goal of IPM is to effectively control pests with the least possible hazards to people, property, and the environment.

Using a well-planned IPM provides growers with a variety of strategies. Growers can use a combination of cultural, physical, biological, and chemical controls to help manage pests. These controls include the use of natural enemies, disease- and pest-resistant cultivars, and crop rotation. The main benefits of IPM are the minimization of hazards to people and the environment as well as effective pest management.

Creating an IPM

Growers and gardeners use a variety of strategies to manage a given pest situation most effectively. Once the pests have been properly identified, growers can examine the environmental factors that are allowing the pests to thrive, and then shift conditions to make them unfavorable for the pests, **Figure 29-1**. Pest management decisions are based on a variety of factors, including:

- Size of the affected geographic area.
- Estimated number of pests.
- Life cycle stage of the majority of pests.
- Type of plants or crop.
- Whether the affected area is urban or rural.
- Grade or slope of the affected area.
- Time of year.
- Cost relative to the effectiveness of the strategy.

Videowokart/Shutterstock.com

Figure 29-1. Planting disease-resistant plants, such as the grape tomato cultivar 'Juliet', is one way to control pests.

In IPM, all components are important. They should work together seamlessly and be used appropriately to prevent pests from causing too much damage to plants and crops. By understanding how to use best practices to manage pests, growers can protect natural resources and produce abundant food, fiber, and fuel crops.

Pests

As stated earlier, agricultural/horticultural pests are organisms that damage or disrupt the development of plants or crops. These same pests may damage homes or other structures. Pests may also include organisms that impact human or animal health. Pests can include weeds, birds, rodents, mammals, insects, snails, nematodes, and *plant pathogens* (organisms that cause disease in plants, including bacteria, fungi, viruses, and oomycetes).

Corner Question

Can you name a few pests that impact human health?

Oomycetes are filamentous protists, including water molds and downy mildews that can cause diseases, such as blights and rots. Additionally, some pests are *vectors*, or organisms that transmit diseases or parasites from one plant to another.

Weeds

Weeds are undesirable plants in gardens, farms, orchards, lawns and landscapes, **Figure 29-2**. Weeds become a problem when they reduce crop yields, lower crop quality, or take over areas of a landscape. Weeds compete with crops and other desired plants for basic needs, including moisture, light, and nutrients. Their competitive success depends on the specific crop involved and how that crop is managed. Depending on the species, weeds can also have characteristics such as:

- Fast growth rate.
- Vigorous nutrient uptake.
- Tolerances to stresses, such as drought.
- Dormancy mechanisms that allow staggered and long-term germination.

If weeds with fast growth rate exist in a crop that has a slow growth rate, conditions are right for weeds to become a problem. Some weeds attract detrimental insects, reproduce rapidly, and persist in fields for years. Weeds ultimately reduce crop yield and quality, negatively impacting profits. Farms and gardens often have many types of weeds in one area. For this reason, growers use not just one control practice, but rather an integrated approach to management.

©iStock.com/PFMphotostock
Figure 29-2. English ivy may be desirable in some situations, such as in this container garden. However, it can be a prolific weed, taking over natural areas.

Insects

Insects are a class of invertebrates that can be pervasive and damaging pests to agricultural crops and ornamental plants. They have a small body size that requires minimal resources for survival, and many insects have the ability to fly to avoid predation. Most insects have the ability to produce large numbers of offspring as well as multiple generations within a growing season. These factors, as well as the development of pesticide resistance, make it challenging to manage pest populations, **Figure 29-3**. (A *pesticide* is a chemical that is used to control insects, weeds, fungi, rodents, and other pests.) Insects have a variety of feeding mechanisms, such as chewing, piercing, and sucking, that damage the plant by removing material or draining fluids. Many insects use plants as a place to lay their eggs, which hatch into larvae that use the plant for food. Some insects inject pathogens into the plant while feeding, causing diseases to spread quickly among and between fields of crops.

Randimal/Shutterstock.com

Vespa/Shutterstock.com
Figure 29-3. Aphids are a plant pest that can rapidly produce multiple generations in a single growing season. In addition to the physical damage they cause, aphids may also spread viruses.

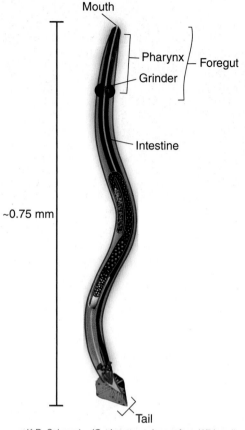

Figure 29-4. Plant parasitic nematodes punctures hole in plant cells, withdrawing nutrients and diminishing plant growth.

K.D. Schroeder (C. elegans male.svg from Wikimedia Commons; License: CC-BY-SA 3.0)

Nematodes

Nematodes are tiny invertebrate roundworms that live in soil and water habitats. Free-living nematodes feed on bacteria, fungi, and other nematodes. Many nematodes are parasites of animals and plants. Plant parasitic nematodes are small (less than one millimeter in length) and eel-shaped with unsegmented bodies that are generally translucent and microscopic in size, **Figure 29-4.** Most plant parasitic nematodes have a hollow stylet or spear used to puncture holes in plant cells and withdraw nutrients. These holes allow the injection of damaging proteins and metabolites into the plant cells. As many nematodes live in the soil, much of their damage occurs on the roots of the plant. The nematode feeding process results in dead or distorted roots, smaller root mass, lesion formation, tissue breakdown, swellings, and galls.

Plant diseases caused by nematodes may be mistaken for other issues or conditions. For example, a root-feeding nematode species that limits the ability of plant roots to uptake water and nutrients will cause nutrient and water deficiency symptoms aboveground that are similar to those caused by other pests, diseases, or environmental conditions. When treatments for other possible causes have not worked, and midday wilting occurs, consider nematodes. Nematodes destroy vascular tissue in plants, so the plant cannot take up enough water on a hot day to keep from wilting; in the mornings and evenings, when heat is not as much of an issue, the affected plant appears fine (not wilted). The damage done by nematodes can also provide a pathway for other plant pathogens.

> "Ants make up two-thirds of the biomass of all the insects. There are millions of species of organisms, and we know almost nothing about them."
> —E. O. Wilson

STEM Connection: Emerald Ash Borer

The emerald ash borer (EAB) is a wood-boring beetle that was accidentally introduced to the United States in 2002 from Asia. Since its arrival, this invasive pest has killed millions of ash trees (*Fraxinus* spp.). In southeast Michigan, where it was first introduced, more than 99% of ash trees with stems greater than 1" (2.5 cm) in diameter have been killed.

The EAB lays eggs in bark cracks and crevices. The newly hatched larvae bore through the outer bark and begin feeding in galleries in the phloem and cambium. The feeding disrupts the ability of trees to transport nutrients and water, eventually girdling branches and the trunk. Scientists increasingly believe that EAB could cause local extinction of ash species across North America. Ash trees are the most widely distributed tree genera, and the loss of any ash species brings devastating economic and ecological impacts. Ecologically, ash trees are a sole food source for a number of species, making coextinction a real possibility.

Pennsylvania Department of Conservation and Natural Resources - Forestry, Bugwood.org

Plant Pathogens

Plant pathogens are organisms (including bacteria, fungi, viruses, and oomycetes) that cause disease. Note that many of these types of organisms are not plant pathogens and can be beneficial in the agricultural ecosystem. Each requires different management methods that are more fully discussed in Chapter 31, *Disease Management*. This chapter will offer a brief summary of each pathogen.

Bacteria

Bacteria are single-celled microorganisms that range from beneficial decomposers to pathogens of plants and animals. Almost all plant pathogenic bacteria develop mostly in the host plant as a parasite on the plant surface tissue or on dead or decaying organic matter in the soil. They can cause galls, cankers, wilts, leaf spots, blights, scabs and various other symptoms, **Figure 29-5**. Bacteria thrive in conditions such as high humidity, crowding of plants, and poor air circulation around plants. Bacterial diseases can be difficult to control.

phomphan/Shutterstock.com

Figure 29-5. Citrus canker is caused by the bacteria *Xanthomonas axonopodis*, which causes lesions on leaves and fruit.

Fungi

More than half of all plant diseases are caused by plant pathogenic fungi, which creates significant annual losses in horticultural and agricultural production. Diseases, such as mildews and rusts, have a specialized biology. They grow only on living plants, **Figure 29-6**. Other pathogenic fungi, such as those that cause root rots and wilts, can grow on both living and dead material. These usually are soil inhabitants and cannot be reached easily for control.

Scot Nelson

Figure 29-6. Powdery mildew is caused by fungi and is easily identifiable by its white powdery spots.

Viruses

Viruses are obligate parasites, which means they must have a suitable host to complete their life cycle. In their simplest form, they include a nucleic acid surrounded by a protein coat. Different viruses have different hosts. Some can tolerate a wide range of species as hosts, whereas others are very specific. For example, the cucumber mosaic virus has a broad range of hosts and can infect more than 85 plant families, **Figure 29-7**.

A *Scot Nelson* B *Scot Nelson*

Figure 29-7. Cucumbers are susceptible to cucumber mosaic virus; however, so are more than 1000 plant species within 85 different plant families. A—Passion fruit leaves exhibiting symptoms of the virus. B—Zinnia leaves exhibiting symptoms of the virus.

Did You Know?

Not all viruses are detrimental to plants. Some create different colors in flower petals or affect the shape of leaves, creating highly desirable ornamental characteristics.

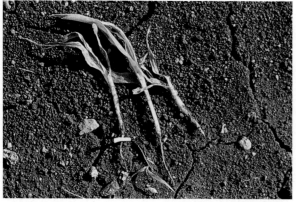

Figure 29-8. Young, tender seedlings (such as these corn seedlings) are often susceptible to a disease called damping off, caused by the water mold *Pythium*. Avoiding overwatering can reduce risk.

Figure 29-9. The leaves of these hostas have been damaged by slug feeding. Although the plant will survive, the ornamental value is diminished.

Figure 29-10. Moles can be significant pests, creating dirt mounds and surface tunnels, and killing plants by scraping away soil from roots and eating roots and bulbs.

Depending on the combination of virus, host, and environmental conditions (the disease triangle), a plant's response to infection may range from a symptomless condition to severe disease and plant death.

Oomycetes

Oomycetes, also known as water molds, can be both terrestrial and aquatic, with the terrestrial species primarily consisting of parasites of plants. Oomycetes, such as *Phytopthora* and *Pythium*, tend to cause plants to rot, **Figure 29-8**. Avoiding saturated soils and extremely wet conditions can minimize incidences of oomycetes.

Other Invertebrate Pests

Insects are not the only organisms that are horticultural pests. Mites, snails, and sowbugs can be damaging to plant material as well. Mites are tiny arachnids (related to spiders) that can damage a wide variety of crops, including fruit trees, vines, berries, vegetables, and ornamental plants. Spider mites cause stippling, discoloration, and leaf drop. Other pest mites cause distorted growth of leaves, shoots, buds, or fruit. Snails and slugs are mollusks that emerge at night and cause damage by chewing the leaves and flowers of many garden plants and fruit, **Figure 29-9**. Slugs and snails have similar structures and biology (slugs lack a shell). Similar management methods (eliminating moisture and hiding spots, trapping, setting up barriers, and handpicking) are used for both. Sowbugs and pillbugs are important decomposers, feeding on dead and decaying plant matter. They can, however, sometimes feed on tender young seedlings, new roots, and lower leaves.

Vertebrates

Vertebrates, such as birds, rodents, and other mammals, can be pests. Wildlife can be destructive in gardens and farms and can be a source of risk for foodborne illnesses. Vertebrate pests include deer, raccoons, rabbits, opossums, and rodents (such as squirrels, mice, rats, voles, and moles, **Figure 29-10**). Birds can also be considered pests, feeding on fruits just as they begin to ripen. Most control methods involve keeping pests away from an area, such as using fencing, netting, devices that frighten pests away, or traps.

STEM Connection: E. Coli Outbreak Linked to Deer

Pest management of deer is about preventing damage to plants and about preventing the transmission of disease to humans. In Oregon, an outbreak of E. coli (*Escherichia coli*) that sickened 15 people (one fatally) was linked to samples of deer feces found in a commercial strawberry field. Growers can keep deer from entering crop fields with effective fencing. Although deer normally will not jump a 6′ (1.8 m) fence, they can clear an 8′ (2.4 m) fence on level ground. Effective deer fences may range from 7′ to 11′ (2.1 m to 3.6 m). The type of fencing depends on cost, terrain, and other needs.

Fences can be constructed using high-tensile wire or woven mesh and are more effective than electric fences. Deer can sometimes crawl under or through a fence. The fence should be secure near the ground with no gaps. In order for the wire to remain tight, the vertical posts should be no more than 6′ to 8′ (1.8 m to 2.4 m) apart. Consulting with an agricultural fencing company is beneficial when planning fencing.

Control Measures

A number of best practices, widely considered as cultural control measures, create an environment that does not permit some pests to thrive.

Sanitation

Sanitation techniques reduce pests, their habitat, and their alternate hosts. All infected plant residues (disease and pests) should be removed from fields and gardens. This practice helps reduce the chance of a recurring infestation. Weeds can be mowed or destroyed before they go to seed. Removing volunteer plants that are prone to pests or places that pests can overwinter is also important. Equipment and tools should be frequently cleaned to minimize the transport of any pest organisms. Pests can be excluded indoors by modifying points of entry (doors, vents, and other building openings) using screening or other barriers.

Habitat Modification

Changing conditions or environments that are suitable for pests can significantly reduce populations. For example, proper irrigation timing and delivery can reduce diseases that thrive in wet or damp environments, such as those where plants are consistently overwatered. Irrigation water should also be free of any pathogenic organisms. Removing their sources of food, water, and shelter will lessen the pests' ability to survive. For example, not planting gooseberries or currants (which serve as an alternate host for white pine blister rust) can reduce the risk of white pines becoming infected, **Figure 29-11**. Crop rotation can prevent buildup of certain pests. Scheduling the timing of planting and harvest may help to miss critical windows when pests are present.

A ©iStock.com/TKphotography64

B ©iStock.com/elzeva

C Marek Argent

Figure 29-11. Gooseberry (A) and currant (B) plantings can serve as an alternate host for white pine blister rust (C), which can be devastating for native stands of white pines.

Plant Material

Growers can provide plants a stronger start for growing healthy by beginning with plant materials that are free of viruses, bacteria, fungi, or weed seeds. Growers may also use plant material that has been bred with more resistance to certain diseases. For example, a grower may choose tomato cultivars that offer resistance to minimize the incidence of late blight caused by the plant pathogen *Phytopthora*, **Figure 29-12**.

Some agronomic and horticultural crops have been genetically modified for increased resistance to certain insect pests or diseases. A genetically modified organism, or GMO, is a plant with a genome that has been transformed by genetic engineering. Crops such as corn, cotton, and soybeans have been injected with a gene that is resistant to an herbicide called glyphosate.

Vadym Zaitsev/Shutterstock.com

Figure 29-12. Late blight caused by *Phytopthora*. Plant breeders have developed hybrids that have resistance to some diseases, such as late blight.

STEM Connection

Herbicide-Resistant Weeds
Palmer Amaranth

Native to the southwestern United States, Palmer amaranth, a type of pigweed, has become a calamitous weed problem in the south and has recently spread to the upper Midwest. Palmer amaranth is the most competitive and aggressive pigweed species and can grow as much as 2.5″ (6.4 cm) a day in southern states. Competition from Palmer amaranth can reduce soybean yields by as much as 80%.

Prolific seed production has perpetuated the establishment and spread of Palmer amaranth. A single female Palmer amaranth in the southeastern United States can produce nearly 500,000 seeds per plant, and even as much as one million seeds in optimal growing conditions. The number of seeds produced varies depending on the amount of competition the plant is experiencing and the region in which it is growing. Palmer amaranth has male and female flowers on different plants (dioecious), increasing the genetic diversity of this species by forcing cross-pollination. This also facilitates the spread of herbicide resistance in seeds. This type of pigweed emerges throughout the growing season, requiring constant monitoring. Since the late 1980s, Palmer amaranth has evolved resistance to five different types of herbicides. As the selection pressure from other herbicides increases, multiple resistant populations will evolve.

IPM is very important in managing Palmer amaranth. Herbicides need to be used in combination with other methods, such as deep tilling of the soil to bury the seeds, which are very short-lived. Rotating crops, hand weeding, planting a cover crop of rye, monitoring fields and ditches, and harvesting infested fields last can also help stop the spread of this plant.

University of Delaware, Carvel

When the herbicide glyphosate is sprayed on a field, the weeds die and the resistant crops remain. Another example of genetic modification is plants that have been injected with the soil-dwelling bacterium *Bacillus thuringiensis* (Bt). Bt acts as a pesticide toward any predation by insects. In some plants, the genes are "turned on" or "silenced" to promote resistance to a particular pest. In recent years, a challenge to some of these crops has been the development of resistance by the pest organisms.

Inspection and Monitoring

After doing as much as possible to prevent pests, sites should be monitored through regular observation for the appearance of insects, weeds, diseases, and other pests. This will allow growers and gardeners to locate, identify, and rank the severity of pest outbreaks. Pest populations vary between locations, and from year to year. Consistent monitoring will allow early detection of pests and provide a better chance to prevent populations from increasing.

Scouting

Scouting is the regular checking of a field, garden, orchard, lawn, greenhouse, or other area that identifies potential pest problems early while they can still be managed. This practice reduces crop loss and control costs. Scouting methods vary by situation and by crop. Scouting should occur throughout the season. Notes should be taken on environmental conditions, beneficial insects, pest insects, diseases, weeds, crop growth stage, and crop health. This information can be collected over years, creating a field history that tracks common pest problems. Scouting methods include the use of sweep nets, insect traps, soil sampling, weed counts, and visual observations.

Sweep Nets

Sampling with a sweep net is a common practice used to inventory fields for various agricultural and horticultural pests, **Figure 29-13**. To use a sweep net, swing it in an arc so the net rim sweeps the top 6" to 15" (15 cm to 40 cm) of plant growth, depending on the type of crop. After a few sweeps, the net is inspected for pests.

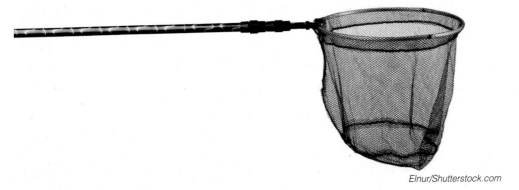

Elnur/Shutterstock.com

Figure 29-13. Sweep nets are tools that allow fields to be surveyed for both pest and beneficial insects.

Insect Traps

Several types of insect traps can be used to monitor and control pests. The data gathered with these traps is valuable for making decisions about pest management. Some examples include sticky traps, pheromone traps, pitfall traps, and light traps.

- Sticky traps use a material coated with glue to trap and hold insects. They are used in floriculture and nursery industries to catch and monitor insects and other pests. Sticky traps provide a relative measure of insect populations and can allow early detection of pests before damage is observed in crops, **Figure 29-14**.

Martchan/Shutterstock.com   *Melpomene/Shutterstock.com*

Figure 29-14. Sticky traps are commonly used in greenhouses and nurseries as a monitoring tool for pest presence and populations.

- Pheromone traps are baited with sex *pheromones*, the chemicals released by female insects to attract mates, **Figure 29-15**. (Pheromones may also be synthetically produced for use in traps.) Pheromone traps are valuable tools for monitoring pest populations. They allow early pest detection, identify areas of pest infestations, and help growers record the population size and increase. Pheromone traps can be used in a variety of situations, including orchards, crop fields, and forests.

Meryll/Shutterstock.com   *Juergen Faelchle/Shutterstock.com*

Figure 29-15. Pheromone traps are another method for luring pest insects, observing population levels, and gathering data used to make decisions regarding pest management.

- Pitfall traps help growers detect early activity. They are used to monitor walking and crawling soil and litter arthropods, particularly those that are active at night. A container is sunk into the ground so that its rim is even with the soil surface. Insects and other organisms are captured when they fall into the trap. To prevent escape, the traps usually contain a killing/preserving agent, such as soapy water or ethyl alcohol.
- Light traps use a funnel with a light source. The light is attractive to some insects, such as moths. The insects become trapped, allowing for inventory. Light traps are less commonly used than other methods.

Soil Sampling

Soil cores are pulled from fields, usually to a depth of 12" (30 cm), and surveyed for populations of plant parasitic nematodes, plant pathogens, and other soil-dwelling insect pests. Management decisions can be made after proper identification and population counts of pests. Depending on the pest being sampled, soils may be collected at different times of the year.

Weed Counts

Scouting for weeds allows growers to accurately determine weed diversity and populations. Some weeds are not competitive with specific crops, and their presence does not need to be managed. Others weeds are highly competitive and may be best controlled at a certain stage in their life cycle. Growers use scouting methods established for specific crops to monitor weed populations.

Generally, scouting occurs throughout the season at regular intervals. Approximately 10 stops may be made in each field. At each stop, weeds are counted in a section 12" (30 cm) wide. The number of stops can vary depending on the size of the field, **Figure 29-16**. It is important to obtain enough representative samples to understand the situation. Additionally, fields that have different environmental conditions, such as wet, dry, and shady areas, should be sampled. The number and size of broadleaf weeds and grassy weeds are recorded separately. If weeds are present in only a concentrated area of the field, the locations should be marked to be managed separately.

Visual Observation

As fields are scouted for weeds or insects, disease should also be monitored. Stunting, discoloration, poor stands, and girdling are examples of visual symptoms of diseases. Experienced scouts will understand the

Goodheart-Willcox Publisher

Figure 29-16. Scout fields in regular intervals throughout the season to get an accurate understanding of pest levels.

growing conditions and management and should be able to rule out nutrient deficiencies or other environmental stressors. If growers suspect that plants are infected with a disease, samples can be collected for research or testing. Many cooperative extension service offices and their affiliated land-grant universities have plant diagnostic clinics that may be able to identify the disease.

Pest Identification

For pest management to be effective, pests must be accurately identified before corrective actions can be taken. A mistake in identification can lead to improper control tactics that cost time and money, and may present unnecessary risks to people or the environment. Different types of pests can cause similar damage, so unless the pest is correctly identified, the control program may target the wrong pest. As different pests are more susceptible to controls at certain life-cycle stages, proper identification will ensure the most effective control method and timing can be selected.

Although there are many online resources for pest identification, visual comparisons with online images are inadequate for accurate identification. Contact an expert from your state's cooperative extension service, land-grant university, or pest management association for assistance in identifying pests.

Recordkeeping and Evaluation

Recordkeeping is a vital component of an IPM program. Maintaining good records will help solve pest problems and offer a historical perspective of pests. Data collected through scouting can be mapped and archived to allow growers to predict seasonal pest problems and prepare for the IPM methods that will give the best control. Detailed records should also be kept on pest management methods used, including cultural, physical, biological, and chemical controls.

AgEd Connection

Pests and Disorders Identification

As part of a career development event, students must identify certain common pests and/or the symptoms and damage they cause. The pest may be presented as an intact specimen, with a photograph, or as a preserved specimen (insect mount). Some of the common greenhouse and nursery pests that may be included on the identification list include aphids, bagworms, borers, leafhoppers, leaf miners, scale, spider mites, snails, slugs, whiteflies, and white grubs. Use the illustrated glossary (page 924) and the e-flashcards on your textbook's student companion website (www.g-wlearning.com/agriculture) to help you study and identify these common pests.